Improve Your Grade!

Access included with any new book.

Get help with exam prep! REGISTER NOW!

Registration will let you:

- Prepare for exams by taking online quizzes.
- Visit *Hazard City* and explore the role an environmental geologist plays in the field.
- Expand your understanding of important issues related to your environment.

www.mygeoscienceplace.com

TO REGISTER

1. Go to www.mygeoscienceplace.com
2. Click "Register."
3. Follow the on-screen instructions to create your login name and password.

Your Access Code is:

Note: If there is no silver foil covering the access code, it may already have been redeemed, and therefore may no longer be valid. In that case, you can purchase access online using a major credit card. To do so, go to www.mygeoscienceplace.com, click on "Buy Access," and follow the on-screen instructions.

TO LOG IN

1. Go to www.mygeoscienceplace.com
2. Click "Log In."
3. Pick your book cover.
4. Enter your login name and password.

Hint:
Remember to bookmark the site after you log in.

Technical Support:
http://247pearsoned.custhelp.com

Stop!

Did you register? Don't miss out!

Turn back one page to register and get help preparing for exams... whenever and wherever you need it.

www.mygeoscienceplace.com

LIVING WITH EARTH

AN INTRODUCTION TO ENVIRONMENTAL GEOLOGY

LIVING WITH EARTH

AN INTRODUCTION TO ENVIRONMENTAL GEOLOGY

Travis Hudson

American Geological Institute
National Association of Geoscience Teachers

PRENTICE HALL
BOSTON COLUMBUS INDIANAPOLIS NEW YORK SAN FRANCISCO
UPPER SADDLE RIVER AMSTERDAM CAPE TOWN DUBAI LONDON
MADRID MILAN MUNICH PARIS MONTRÉAL TORONTO DELHI
MEXICO CITY SÃO PAULO SYDNEY HONG KONG SEOUL
SINGAPORE TAIPEI TOKYO

Acquisitions Editor: *Andrew Dunaway*
Editor in Chief, Geosciences and Chemistry: *Nicole Folchetti*
VP/Executive Director, Development: *Carol Trueheart*
Marketing Manager: *Maureen McLaughlin*
Development Editor: *Daniel Schiller*
Project Editor: *Crissy Dudonis*
Editorial Assistant: *Kristen Sanchez*
Marketing Assistant: *Nicola Houston*
Managing Editor, Geosciences and Chemistry: *Gina M. Cheselka*
Project Manager: *Edward Thomas*
Proofreader: *Donna Young*
Full Service/Composition: *Prepare, Inc.*
Production Editor, Full Service: *Francesca Monaco/Prepare*
Art Director: *Maureen Eide*
Cover and Interior Design: *Jill Little*
Senior Technical Art Specialist: *Connie Long*
Art Advisor: *Jay McElroy*
Art Studio: *Spatial Graphics*
Photo Research Manager: *Elaine Soares*
Photo Researcher: *Kristin Piljay*
Senior Manufacturing and Operations Manager: *Nick Sklitsis*
Operations Supervisor: *Amanda A. Smith*
Senior Media Producer: *Angela Bernhardt*
Media Producer: *Ziki Dekel*
Senior Media Production Supervisor: *Liz Winer*
Associate Media Project Manager: *David Chavez*

Cover, title, and half-title page photo: *Diamond Head, Hawaii. (Photo by Mick Roessler / Age Fotostock, Inc.)*

Pearson Prentice Hall
Pearson Education, Inc.
Upper Saddle River, New Jersey 07458

Library of Congress Cataloging-in-Publication Data
Hudson, Travis,
Living with Earth : an introduction to environmental geology / Travis Hudson.
p. cm.
ISBN 978-0-13-142447-0
1. Environmental geology—Textbooks. I. Title.

QE38.H83 2011
550—dc22 2009039701

Printed in the United States of America
10 9 8 7 6 5 4 3 2 1

ISBN-10: 0-13-142447-5 / ISBN-13: 978-0-13-142447-0 (Student Edition)
ISBN-10: 0-32-169636-0 / ISBN-13: 978-0-32-169636-6 (Books à la Carte)

Prentice Hall
is an imprint of

www.pearsonhighered.com

AUTHOR TEAM

About the Author

Travis Hudson is an applied and research geologist with 40 years of diverse experience studying Earth and its relation to people. While completing graduate school at Stanford University, he began his career as a research geologist with the U.S. Geological Survey in Alaska. His regional understanding of Alaska was put to good use when he became a mineral explorationist for a private company. He subsequently served as a research director for an oil company, studying regional tectonics and basin evolution, and as an exploration manager on Alaska's North Slope, where he helped discover several oil fields. As his company had inherited significant environmental problems from its mining division, environmental remediation technology became his next focus. While managing environmental cleanups at mining-related sites, he studied environmental laws and standards, worked with regulators, and took on many community education responsibilities.

Since 1996 Travis has been a consulting research geologist for the U.S. Geological Survey, studying crustal character in Alaska; a field geologist exploring for mineral deposits in Alaska; and the Director of Environmental Affairs for the American Geological Institute (AGI). At AGI he coordinated the development and publication of the Environmental Awareness Series, richly illustrated 64-page books designed to educate citizens and policy-makers about the insights that Earth Science can contribute to our understanding of environmental issues. He is the author of *Metal Mining and the Environment* in that series, as well as many scientific contributions.

Travis lives with his wife Patti in Sitka, Alaska. When he is not writing or working in the field he is often fishing on the Kenai, staking out his claim to a niche at the top of the food chain alongside the local bears.

About the American Geological Institute

The AGI is a nonprofit federation of 46 geoscientific and professional associations that represents more than 120,000 geologists, geophysicists, and other Earth scientists. Founded in 1948, AGI provides information services to geoscientists, serves as a voice for shared interests in the profession, plays a major role in the strengthening of Earth Science education, and strives to increase public awareness of the vital role the geosciences play in society's use of resources and interaction with the environment. The AGI, through its broad connections with the professional Earth Science community, assembled a panel of experts to advise on the technical content of *Living with Earth* and review its development. The expert panel members are active scientists with long research careers in the subjects covered by the textbook. Their participation has helped ensure that the scientific content of *Living with Earth* is complete, current, and accurate.

About the National Association of Geoscience Teachers

The NAGT was established in 1938 to foster improvement in the teaching of Earth Sciences at all levels of formal and informal instruction, to emphasize the cultural significance of the Earth Sciences, and to distribute knowledge of this field to the general public. Members of NAGT who serve on the advisory board of *Living with Earth* are Earth Science and environmental geology teachers. These teachers helped to define the scope of the book and establish pedagogic guidelines for its development, provided consultation during its creation, and reviewed text materials. The many lessons from their classroom experiences have enriched *Living with Earth* in a variety of ways, helping it to be a more engaging and thought-provoking book.

BRIEF CONTENTS

CONTENTS

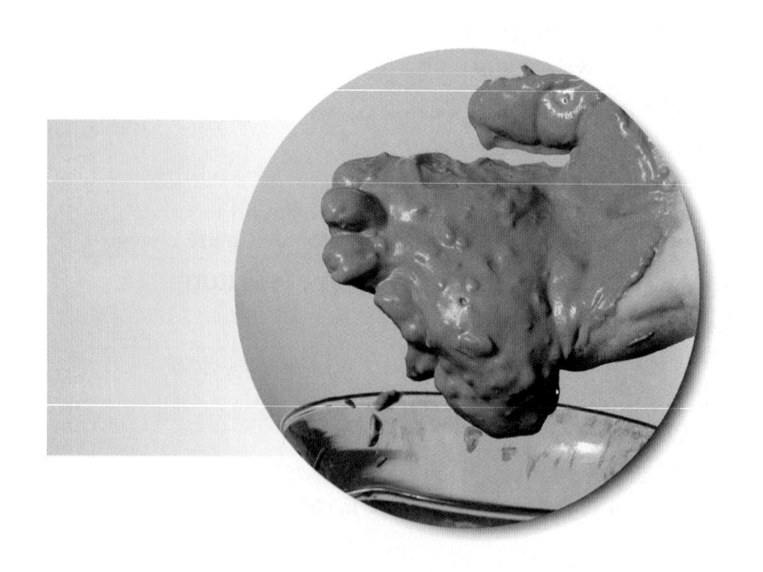

CHAPTER 5
Earthquakes 113

CHAPTER 6
Volcanoes 151

CHAPTER 7
Rivers and Flooding 189

CHAPTER 8
Unstable Land 223

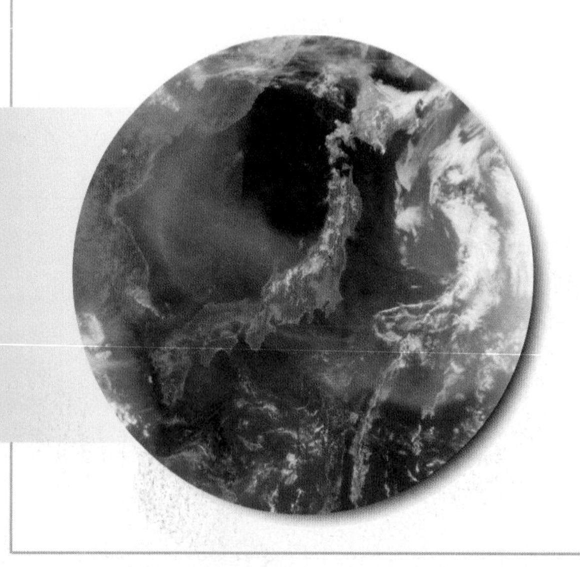

EXXON
Self
Regular
Unleaded
429 9/10
Plus
Unleaded
444 9/10
Supreme
Unleaded
458 9/10

CHAPTER 14

CHAPTER 15

PREFACE

Why This Text?

In the Indian Ocean region, a massive earthquake triggers a tsunami that kills more than 250,000 people. . . . On the Gulf Coast of the United States, a hurricane ravages New Orleans. . . . Proposed oil exploration in Alaska's Arctic National Wildlife Refuge is debated in Congress as gasoline prices rise. . . . In Southern California, a landslide devastates the community of La Conchita. . . .

These are just a few examples of how people's interactions with the Earth are in the news—and why students should be taking an environmental geology class. Environmental geology provides special opportunities to learn how Earth works and how people are connected to environmental outcomes—many that become the news.

Even people with little or no interest in science can find themselves captivated by the power and fascination of the changing Earth around them. But these changes need not be dramatic and front-page news to have significance in people's lives. Nor do they have to involve destruction or hazards; some of our most important transactions with Earth involve our use of the planet's resources—its soil, its water, its mineral wealth. It is on these resources—and their wise use—that our lifestyles, and indeed our lives, depend. This book's purpose is to help people who do not have a background in science—who may find environmental geology to be one of the few science classes they ever take—to better understand how they interact with Earth and how their actions play a role in determining Earth's environmental health.

My perspectives are probably similar to those of many other Earth scientists. I am fascinated by Earth systems processes and history, I am confident of the positive role Earth Science can play in people's lives, and I am excited to help others know Earth better. I want as many people as possible to understand how our planet changes and how they can live more sustainably with it.

My public education experiences—directly in homes and communities, and through development of the American Geological Institute's *Environmental Awareness Series*—has led me to a consistent approach in presenting environmental issues in this book. Understanding how environmental problems originate is always the foundation for explaining how we can avoid, mitigate, or remediate them.

You will probably find that there is more environment and less geology in *Living with Earth* than in similar texts. This is deliberate. *Living with Earth* emphasizes people's connections to environmental issues. It provides the scientific foundations needed to understand Earth's history, structure, and processes, as well as the way in which Earth systems interact. It illustrates in case after case how science can help us by providing not just insights but also tools for wisely utilizing Earth's resources and ways of mitigating its hazards. In the end, though, the take-home messages are about people's options, choices, and roles in environmental outcomes. Everyone has a part in determining Earth's environmental health. It is individual or collective actions that both cause environmental problems and determine how effectively we deal with them.

A Dynamic Approach

Living with Earth is organized around two unifying, dynamic perspectives that emphasize *interaction:*

- *How the various Earth systems interact with one another*
- *How Earth interacts with people—in particular,*
    - How Earth affects people, providing resources to meet our needs, but also creating a variety of hazards.
    - How people affect Earth, both deliberately and inadvertently, in positive and negative ways.
    - The concept of stewardship: How we can live with Earth in a responsible manner to solve today's problems while assuring a sustainable future for generations to come.

Although the basic organization and topic coverage of the chapters in *Living with Earth* will be familiar to instructors, the text is conceived and written within an Earth systems framework. Reinforced by numerous examples, this perspective helps students fully understand the causes and effects of environmental change—and why seemingly unrelated and apparently inconsequential actions can have far-reaching results. They will consistently encounter environmental consequences within a framework that explains how:

- Earth is dynamic and changing. Natural environmental change is normal, ubiquitous, and inevitable.
- Earth systems interactions cause environmental changes. These interactions are the reason that small or local changes can aggregate into broadly significant ones.
- People interact with Earth systems. These interactions have both positive and negative consequences and make the concepts of stewardship and sustainability especially relevant.

A Student-Centered Approach

I believe that *Living with Earth,* more than any other text, addresses key concerns from the student's point of view. There are several ways this book proactively connects students to their role as stakeholders in Earth's future. Throughout the book, I have tried to maintain a strong focus on why this book and this course matter to students. Even more important, there is a sustained emphasis on what students can do—to find out more, to make informed decisions, to get involved in issues affecting their lives, their communities, their planet, and their futures.

- **EXAMPLES OF PEOPLE'S CONNECTIONS TO ENVIRONMENTAL OUTCOMES** are woven into the *Living with Earth* provides many examples and case studies in narrative form, telling the stories of how communities grapple with decisions concerning natural hazards, natural resources, and pollution. These stories highlight the complexities, uncertainties, and value choices people confront as they work out ways to "live with Earth." Students learn by example that decisions have costs and consequences. These are lessons they can take with them and apply in their own lives.
- **FEATURES THAT ENGAGE STUDENTS IN ACTIVE LEARNING**—exploring information resources they can use, independently investigating topics, and contemplating very real dilemmas—are central to this text. Their inclusion is based on my strong conviction that reading an environmental geology textbook should not be a passive chore. Students should be engaged by what they read. They should be encouraged to reach conclusions independently about environmental issues that affect them. This active approach requires them to frame issues clearly, gather information, analyze it, weigh multiple solutions, and make informed choices. Special features of the text of *Living with Earth*—*You Make the Call, In the News,* and *What You Can Do*—help students to engage in this form of active, inquiry-based learning. These features are integral parts of the text—they are not relegated to sidebar boxes that are easily overlooked or ignored.
- **WEB RESOURCES** that augment the content, provide guidance, or help maintain topic currency are embedded in the text directly where they are relevant. These take the form of *search terms*—titles that will direct an Internet search to a specific Web page, not a list of potentially out-of-date URLs for hard-to-use home pages. The search titles enable students to immediately access Internet resources in the way they are used to using the Web.
- **INSIGHT QUESTIONS** dealing with *Earth Systems Interactions* and *Sustainability* are distributed throughout the text, prompting students to keep these key perspectives in mind and apply them to what they are reading. Answers are provided at the end of each chapter.

I am fundamentally an optimist and have faith that understanding and awareness enable people to recognize problems and effectively deal with them. Environmental concerns covered in *Living with Earth* are inherently complex technical, social, and economic issues and simple solutions are conspicuously uncommon if not absent. On the other hand, the resilience of people and Earth systems provides confidence that a better future is possible. *Living with Earth* is written and designed to help accomplish this.

Acknowledgments

Patrick Lynch and Dan Kaveney were the acquisitions editors who recognized its potential and fostered its initial development—a process that continued under their successors, Chris Rapp, Dru Peters, and Andy Dunaway. Daniel Pendick, a professional science writer, provided initial drafts of parts of Chapters 3, 5, 6, 7, 8, and 9. Ann Heath managed the project through several difficult transitions and her encouragement and support. Crissy Dudonis and Kristen Sanchez have been—and continue to be—enormously helpful in many and varied ways.

Once the project approached the production phase, Pearson marshaled what seemed an army of expertise to support its completion. The development editor, Daniel Schiller, was outstanding not only in guiding me through the process but also in polishing the manuscript, identifying flaws, and creating many needed refinements. Dan is the person most responsible for the coherent, well-designed book that evolved from my initial drafts. The illustrations were especially critical to our pedagogical and aesthetic aims. Dan, Jay McElroy, and Connie Long at Pearson, together with Kevin Lear and his associates at Spatial Graphics, were in making the drawings and maps a major strength of this book. The extraordinary photographic program that complements the art owes an immense debt to the creativity and perseverance of our dauntless photo researcher, Kristin Piljay. Maureen Eide deserves credit for the design of the text (including the beautiful cover). I am deeply grateful to our production team: Managing Editor Gina Cheselka, Project Manager Ed Thomas, Production Editor Francesca Monaco, Copy Editor Angela Pica, the eagle-eyed Donna Young, and all the people at Preparé who worked so hard to make this book a physical reality. I would also like to acknowledge the contribution of marketing manager Maureen McLaughlin, who, though a relative newcomer to the project, truly hit the ground running. Her efforts on behalf of this text are greatly appreciated. And last, but emphatically not least, I must express my gratitude to Nicole Folchetti, Editor in Chief for Geosciences and Chemistry, who not only supported this project through many vicissitudes but played an active role in bringing it to fruition. To all of these people, heartfelt thanks.

Another large group that helped this book is the reviewers. Their contributions actually started before the writing did. A team of NAGT members convened an initial workshop that laid out the book's organization and emphasized the need for the inquiry-based features that are now an integral part of the text. Many teachers subsequently reviewed chapter drafts and helped firm up many scope and organizational choices. Active researchers served as technical reviewers of many chapter drafts as well. All the reviewers, whose names are listed in the following section, provided checks on accuracy and completeness and many of the examples in the book came from reviewer guidance. A book like this cannot be developed without thorough and conscientious review and I appreciate them very much.

Living with Earth is now completed with this first edition publication but I think of it as an ongoing project. I am sure that continuing development, based on the experience of students and teachers, can make this a better book. Your feedback on any aspect of *Living with Earth* is always welcome.

Living with Earth NAGT Advisory Board

Jill Whitman, chair
Pacific Lutheran University
Scott Burns
Portland State University
Jennifer Rivers Coombs
Northeastern University
P. Geoffrey Feiss
College of William and Mary
Judi Kusnick
California State University, Sacramento
Michael T. May
Western Kentucky University
Jeffrey W. Niemitz
Dickinson College
Mary Poulton
University of Arizona
Michael C. Roberts
Simon Fraser University

Living with Earth Expert Panel

Dale H. Easley (Water resources)
University of New Orleans
Thomas L. Holzer (Unstable land)
U.S. Geological Survey
Robert Kleinmann (Minerals)
U.S. Department of Energy
Murray H. Milford (Soils)
Texas A and M University (ret.)
Thomas P. Miller (Volcanoes)
U.S. Geological Survey
William J. Neal (Coastal processes)
Grand Valley State University
Orrin H. Pilkey (Coastal processes)
Duke University
George Plafker (Earthquakes, megatsunamis)
U.S. Geological Survey
Geoff Plumlee (Minerals and health)
U.S. Geological Survey
John Renton (Geochemistry)
West Virginia University
Russell Slayback (Water resources)
Leggette, Brashears, & Graham
Steven M. Stanley (Earth system history)
University of Hawaii

General and Specialist Reviewers

Christine Aide
Southeast Missouri State University
Michael Aide
Southeast Missouri State University
Joseph Allen
Concord College
Carlos Aramburú
Universidad de Oviedo
Gail Ashley
Rutgers University
Charly Bank
University of Toronto
David Bazard
College of the Redwoods
Robert Benson
Adams State College
Prajukti Bhattacharyya
University of Wisconsin, Whitewater
Patrick Burkhart
Slippery Rock University of Pennsylvania
Tom Bush
Pierce College
Don Byerly
University of Tennessee, Knoxville
Elizabeth Catlos
University of Texas at Austin
Michael Caudill
Hocking College
Sean W. Chamberlin
Fullerton College
Chun-Yen Chang
Norwegian University of Science and Technology
Jeff Chaumba
University of Georgia
Paul Cheney
University of West Florida
Robert Cicerone
Bridgewater State College
Mitch Colgan
College of Charleston
Cathy Connor
University of Alaska, Southeast
Ellen Cowan
Appalachian State University
Anna Cruse
Oklahoma State University
John Dembosky
Methodist University
Chuck DeMets
University of Wisconsin, Madison
Bruce Douglas
Indiana University, Bloomington
Carolyn Dowling
Arkansas State University
Laurie Duncan
Austin Community College
Steven Dunn
Mount Holyoke College
Karen Duston
San Jacinto College
Norlene Emerson
University of Wisconsin, Richland
Yoram Epstein
Kent State University
Kevin Evans
Missouri State University
Larry Fegel
Grand Valley State University
Anthony Feig
Central Michigan University
Bill Fenner
University of Saint Thomas
Tom Fitz
Northland College
Kenneth Galli
Boston College
Christoph Geiss
Trinity College
Michael Gibson
University of Tennessee, Martin
Alan Goldin
Westminster College
Michael Grossman
Southern Illinois University
Maureen Haberfield
Tallahassee Community College
Mark Hafen
University of South Florida
Robyn Hannigan
Arkansas State University
Michael Harrison
Tennessee Technological University
Frederick Heck
Ferris State University
Martin Helmke
West Chester University of Pennsylvania
Eric Henry
University of North Carolina, Wilmington
Judith Hepburn
Boston College
Alice Hoersch
La Salle University
Steven A. Hovan
Indiana University of Pennsylvania
Michael Kelley
Georgia Southern University
Jeffrey Knott
California State University, Fullerton
Gary Kocurek
University of Texas, Austin
Peter Kolesar
Utah State University
Lawrence Krissek
Ohio State University
J. Richard Kyle
University of Texas, Austin
Venkat Lakshmi
University of South Carolina
Gene Lene
Saint Mary's College
Steven K. Lower
Ohio State University
José Martinez
University of Buenos Aires
Paul McCarthy
University of Alaska, Fairbanks
Michael McKinney
University of Tennessee, Knoxville
Kirsten Menking
Vassar College
Katherine Milla
Florida A & M University
David Miller
Clark State Community College
Sadredin Moosavi
Tulane University
Peter Nabelek
University of Missouri, Columbia
Jennifer Nelson
Indiana University–Purdue University, Indianapolis
Mark Noll
State University of New York, Brockport
Edward Nuhfer
California State University, Channel Islands
Mark Ouimette
Hardin Simmons University
Evan Paleologos
University of South Carolina
Terry Panhorst
University of Mississippi
Daniel Pardieck
Lander University
Marta Patino Douce
University of Georgia
Darryll Pederson
University of Nebraska, Lincoln
Mauri Pelto
Nichols College
Alyson Ponomarenko
San Diego City College
Alan Price
University of Wisconsin
Eric Riggs
San Diego State University
Michael Roden
University of Georgia
Lawrence P. Rudd
Nevada State College
Barbara Savage
San Jacinto College
Keegan Schmidt
Lewis-Clark State College
Rich Schultz
Elmhurst College
Frank Schwartz
Ohio State University
Gina Marie Seegers Szablewski
University of Wisconsin, Milwaukee
Laura Serpa
University of Texas, El Paso
Kurt Shoemaker
Shawnee State University
Robert Shuster
University of Nebraska-Omaha
Laura Smart
Grand Valley State University
Joseph Smoak
University of South Florida
Peter Thompson
University of New Hampshire
Slawek Tulaczyk
University of California, Santa Cruz
Mark Turski
Plymouth State University
Sarah Ulerick
Lane Community College
Bill Ussler
Monterey Bay Aquarium Research Institute
Cynthia Venn
Bloomsburg University
Natasha Vidic
University of Wisconsin, Oshkosh
Jack Vitek
Oklahoma State University
Laura Wetzel
Eckerd College
Carol Wicks
University of Missouri, Columbia
Clark R. Wilson
University of Texas, Austin
Michael C. Wizevich
Central Connecticut State University
Joe Yelderman
Baylor University

FEATURES OF THIS TEXT

Living with Earth was conceived with the student in mind. All of its main features have in common the goal of fostering active involvement with the material presented in the text. Their aim is not merely to assert but to demonstrate vividly why this material is compelling, important, and relevant to the lives of those who will use this book.

CHAPTER-OPENING VIGNETTES, many of them dramatic and all of them highly engaging, capture student interest immediately while also introducing important chapter themes.

CHAPTER

6

VOLCANOES

On December 14, 1989, a KLM 747-400 jumbo jet high over the Talkeetna Mountains of Alaska starts its final approach to Anchorage International Airport after an uneventful over-the-pole journey from Amsterdam. With 231 passengers and 14 crew members on board, KLM-867 is a routine flight. The pilot at the controls radios Anchorage Air Traffic Control (ATC):

"KLM-867 heavy is reaching level 250 heading 140."

Heavy indeed. The aircraft, just recently off the assembly line, has a maximum takeoff weight of 438 tons. It is a wide-body jumbo jet, a "heavy" in the jargon of commercial aviation. It is one of the most advanced and safest flying machines ever built. But KLM-867 is flying into a problem. Mount Redoubt, a volcano 176 kilometers (109 mi) southwest of Anchorage, is erupting (Figure 6-1).

ATC: "Do you have good sight of the ash plume?"

Mount Redoubt is one of the 40 historically active volcanoes strung along the Aleutian Islands chain. It has blasted a plume of fine pulverized lava and hot gases more than 8 kilometers (5 mi) into the atmosphere. The prevailing winds have blown this mass of ash into the path of KLM-867.

Previously, the crew had requested a flight path they assumed would steer them clear of the ash plume, a hazard to aircraft that many other crews before them had learned to respect. In 1982, for example, a British Airways 747 cruising at 11,300 meters (37,000 ft) with 241 passengers flew into an ash cloud from Galunggung volcano in Indonesia. All four engines flamed out and the aircraft glided for 16 minutes without power, falling to an altitude of 3800 meters (12,500 ft) before the crew was able to restart the engines. A month later, another 747 lost power after encountering ash from the same volcano. It landed with only one engine still operating.

Madrid County has increased from about 400 to more than 20,000 people. There are now some 12 million people living in the affected zone, including densely populated cities like St. Louis and Memphis. The lower Mississippi is also a key shipping lane, so earthquake damage here could have national repercussions. Catastrophic earthquakes like those of 1811–1812 may recur every 500 years or more, but there is a much higher probability of moderate shocks. | The Mississippi Valley-"Whole Lotta Shakin' Goin' On"

Intraplate earthquakes, such as those of the New Madrid area, can be large and cause widespread damage. In continental interiors, the crust is old, thick, and strong. When it breaks, it can unleash large amounts of seismic energy and transmit it long distances.

r the biosphere, times ne extinct. These are -23 shows, there have been at least five major ones during the last 540 million years. The most recent and most famous mass extinction is the one during which the dinosaurs died out, some 65 million years ago. | Debating the dinosaur extinction Scientists think this mass extinction resulted from the impact of a large asteroid on what is now the Yucatan Peninsula of Mexico (Figure 2-24). Earth was not a closed system 65 million years ago! The impact of this large asteroid darkened the skies with dust and caused severe global climate changes that were just too much for the

SEARCH TERMS, incorporated right into the text, give students easy in-context access to relevant online resources, allowing them to learn more and become actively engaged with the subject. Highlighted by their distinctive design, these terms take the form of titles that will direct an Internet search to specific Web pages (rather than unreliable URLs).

lished cleanup levels for Butte community soils. The soil in yards that contain lead concentrations above the lead action level of 1200 ppm must be removed and placed in an appropriate disposal area.

What are some ways in which Earth systems are related to environmental standards?

ANSWERS TO IN-CHAPTER INSIGHT QUESTIONS

P. 67 *What are some of the Earth systems interactions that occur at a hydrothermal vent on the ocean floor?*

- Parts of the hydrosphere are heated by the geosphere (energy transfer).
- Hot water dissolves material from rocks in the geosphere (material transfer)
- Vent water precipitates material on the seafloor (material transfer)
- Biosphere eats precipitated material (energy transfer).

P. 76 *What are some of the Earth systems interactions that can occur when an explosive volcano erupts?*

- Atmosphere composition can be changed (material transfer); the volcanic eruption can affect global climate.
- Groundwater in the hydrosphere can be heated by shallow magma (energy transfer).
- Biosphere habitat can be destroyed or changed.
- Hydrosphere reservoirs can be contaminated (material transfer).

IN-CHAPTER INSIGHT QUESTIONS dealing with **Earth Systems Interactions** and **Sustainability** are distributed throughout the text, prompting students to keep these key perspectives in mind and apply them to what they are reading. (Answers are provided at the end of each chapter.)

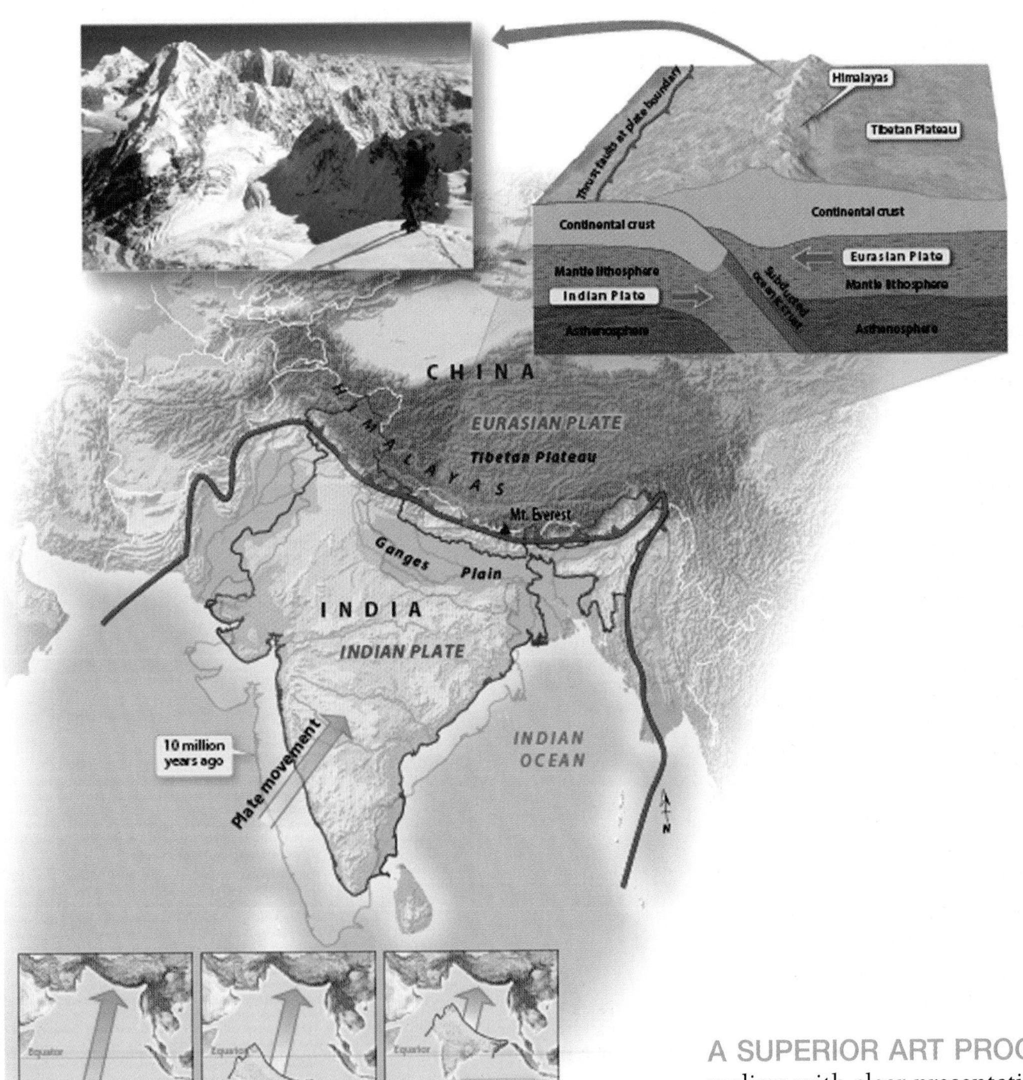

A SUPERIOR ART PROGRAM combines unprecedented realism with clear presentation of concepts, aided by annotations and labels designed to make the art as self-explanatory as possible. Spectacular photographs set a new standard in their combination of pedagogical utility with striking beauty.

THREE TYPES OF ENGAGING, INQUIRY BASED DISCUSSIONS—*What You Can Do, In the News,* and *You Make the Call*—relate text material directly to the concerns of students and foster their personal involvement. These features encourage students to find out more, make informed decisions, and get involved in issues affecting their lives, their communities, and their futures.

- Discussions are integrated into the flow of the exposition rather than segregated into "boxes."
- Virtually all include search terms that guide students to Web resources.
- Most are illustrated with screen grabs of relevant websites to pique students' curiosity and whet their appetite for further exploration.

What You Can Do

Map Earthquake Intensity

You can actually help make an earthquake intensity map. If you experience an earthquake, go to the USGS Earthquake Hazards Program website and submit information about the earthquake. "Did you feel it?" The USGS uses such information to generate a Community Internet Intensity Map that is updated every few minutes following a major earthquake. In this way, you can be involved in real-time earthquake studies. The intensity map you help make will evolve before your eyes as your neighbors and others submit information in the aftermath of an earthquake.

USGS science for a changing world
USGS Home
Contact USGS
Search USGS
Earthquake Hazards Program
Home Earthquake Center Regional Information About Earthquakes Research & Monitoring Other Resources
You are here: Home » Earthquake Center » Did You Feel It?
Latest Earthquakes
USA
World
EQ Notification Service
Feeds & Data
Animations
Recent Earthquakes Last 8-30 Days
Earthquake Archives
Lists & Maps
Search EQ Database
EQ Summary Posters
Scientific Data
About EQ Maps
Did You Feel It?
Archives
About the Maps
Website Feedback
Disclaimer
FAQ
Fast Moment Tensors
Media Info
PAGER
Seismogram Displays
ShakeMaps

Did You Feel It?
Report Unknown Event
Jump to World
Jump to US
Jump To Alaska
Jump to Hawaii
Events - Last 24 Hours
View Archives

MMI	Mag	Location	Event Time	Event ID	Response
I	4.8	NEAR THE NORTH COAST OF PAPUA, INDONESIA	2009-05-19 12:32:03 UTC	US2009GVAU	0
II	2.1	NORTHERN CALIFORNIA	2009-05-19 09:44:39 UTC	NC51221841	4
II	1.9	GREATER LOS ANGELES AREA, CALIFORNIA	2009-05-19 07:04:34 UTC	CI10411017	2
I	5.1	MID-INDIAN RIDGE	2009-05-19 07:04:16 UTC	US2009GVAI	0
I	4.9	WESTERN SAUDI ARABIA	2009-05-19 06:38:33 UTC	US2009GVAJ	0
III	2.0	SOUTHERN CALIFORNIA	2009-05-19 04:51:51 UTC	CI10410945	7
III	3.4	BAJA CALIFORNIA, MEXICO	2009-05-18 19:24:06 UTC	CI10410705	3

What You Can Do discussions suggest ways in which students can both explore text topics more deeply and apply what they are learning in their daily lives.

In the News

Protecting the Spotted Owl

Actions to protect the spotted owl are still a work in progress. Litigation settlements require federal agencies to develop forest management plans that protect spotted owl habitat. However, recent independent reviews of the proposed plans have concluded that they will not adequately protect this species. Spotted owl plan fails peer review What happens next is not clear. It will take many years to resolve this issue.

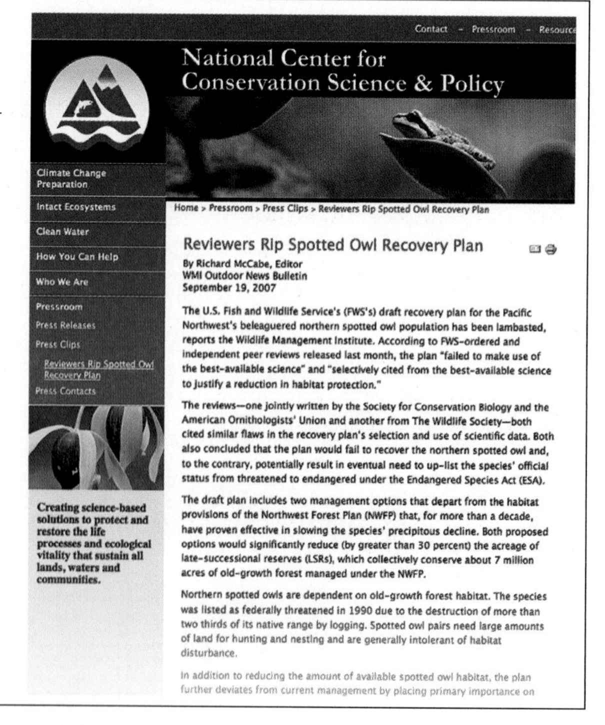

Contact - Pressroom - Resources
National Center for Conservation Science & Policy
Climate Change Preparation
Intact Ecosystems
Clean Water
How You Can Help
Who We Are
Pressroom
Press Releases
Press Clips
Reviewers Rip Spotted Owl Recovery Plan
Press Contacts
Creating science-based solutions to protect and restore the life processes and ecological vitality that sustain all lands, waters and communities.

Home > Pressroom > Press Clips > Reviewers Rip Spotted Owl Recovery Plan

Reviewers Rip Spotted Owl Recovery Plan

By Richard McCabe, Editor
WMI Outdoor News Bulletin
September 19, 2007

The U.S. Fish and Wildlife Service's (FWS's) draft recovery plan for the Pacific Northwest's beleaguered northern spotted owl population has been lambasted, reports the Wildlife Management Institute. According to FWS-ordered and independent peer reviews released last month, the plan "failed to make use of the best-available science" and "selectively cited from the best-available science to justify a reduction in habitat protection."

The reviews—one jointly written by the Society for Conservation Biology and the American Ornithologists' Union and another from The Wildlife Society—both cited similar flaws in the recovery plan's selection and use of scientific data. Both also concluded that the plan would fail to recover the northern spotted owl and, to the contrary, potentially result in eventual need to up-list the species' official status from threatened to endangered under the Endangered Species Act (ESA).

The draft plan includes two management options that depart from the habitat provisions of the Northwest Forest Plan (NWFP) that, for more than a decade, have proven effective in slowing the species' precipitous decline. Both proposed options would significantly reduce (by greater than 30 percent) the acreage of late-successional reserves (LSRs), which collectively conserve about 7 million acres of old-growth forest managed under the NWFP.

Northern spotted owls are dependent on old-growth forest habitat. The species was listed as federally threatened in 1990 due to the destruction of more than two thirds of its native range by logging. Spotted owl pairs need large amounts of land for hunting and nesting and are generally intolerant of habitat disturbance.

In addition to reducing the amount of available spotted owl habitat, the plan further deviates from current management by placing primary importance on

In the News discussions highlight contemporary environmental issues covered in print or broadcast media—particularly issues with ongoing story lines that present dilemmas for people and involve problems that could remain debated for many years.

You Make the Call

Is Ethanol a Sustainable Energy Resource?

The increasing price of gasoline and our country's dependence on foreign sources of petroleum have made alternative transportation fuels, such as ethanol, attractive to many people. The fact that ethanol is a renewable energy resource makes it appear even more promising. One result of ethanol's appeal was the 2005 Energy Bill, which mandated that use of biofuels (mostly ethanol) reach 7.5 billion gallons by 2012. In order to meet this goal, the gasoline you use in your car will have to contain 2% to 10% ethanol, and in some cases more.

Ethanol is an alcohol distilled from plants, mostly corn in the United States. (It is the alcohol in alcoholic beverages.) This is why it's called a renewable energy resource. All people need to do to produce more fuel is grow a new crop. But is growing corn, especially at the scale needed to supply significant amounts of transportation fuel, really sustainable?

Several aspects of large-scale corn farming for ethanol production raise concerns about sustainability. Large amounts of water are needed to grow and process corn, and more farmland would need to be dedicated to corn production. Moreover, extensive use of pesticides and herbicides accompany corn farming (corn is grown on about one-fourth of U.S. croplands, but requires two-thirds of the total herbicide use). Corn is a row crop, and soil erosion rates are higher for corn fields than for several other types of crops. Growing corn in the United States has also required large amounts of fertilizer to maintain crop yields.

From the standpoint of sustainability, the effect of corn farming on soil nutrient levels and quality are significant concerns.

- What do you think is happening to nutrient levels and biodiversity in corn fields?
- Do you think soil quality is being sustained in these fields?
- Over the long term, do you think ethanol is a sustainable energy resource?

You Make the Call discussions challenge students to confront environmental issues that may not have a "right" answer. These exercises in open-ended inquiry encourage students to consider the risks and benefits, costs and consequences, and varied solutions that actual individuals, communities, and nations have had to consider.

MEDIA RESOURCES

For the Instructor

Instructor Resource Center on DVD

Included are three PowerPoint® presentations for each chapter: (1) **Lecture Outline**—lecture presentations based on the outline of the text to get you up and running as quickly as possible; (2) **Art only**—every illustration and most of the photos in the text, in order, pre-loaded onto PowerPoint slides; (3) **Animations**—high quality animations of key geologic processes.

Instructor Manual

Provides chapter summaries, chapter outlines, objectives as well as answers to the end-of-chapter questions in the text. Also includes weblinks for further study.

Test Bank

Includes multiple choice, true/false, and short-answer test questions based on the text.

For the Student

PEARSON mygeoscience™ place

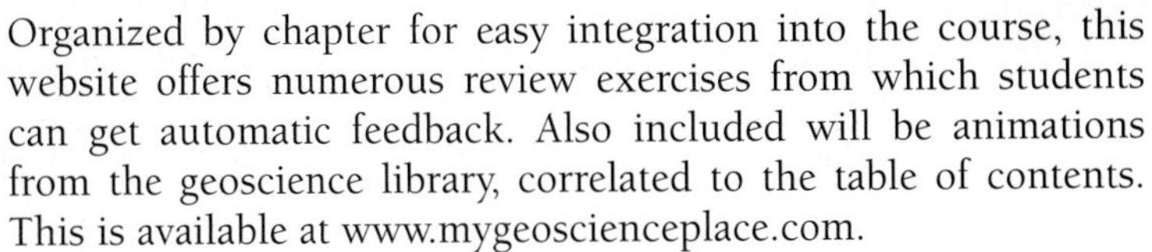

Companion Website

Organized by chapter for easy integration into the course, this website offers numerous review exercises from which students can get automatic feedback. Also included will be animations from the geoscience library, correlated to the table of contents. This is available at www.mygeoscienceplace.com.

Hazard City

Provides 11 meaningful, easy-to-assign, easy-to-grade, class-tested assignments. The idealized town of Hazard City provides the setting in which students take on the role of a practicing geologist gathering and analyzing data, evaluating risk, and making assessments and recommendations. This is available at www.mygeoscienceplace.com

About our Sustainability Initiatives

This book is carefully crafted to minimize environmental impact. The materials used to manufacture this book originated from sources committed to responsible forestry practices. The paper is FSC certified. The binding, cover, and paper come from facilities that minimize waste, energy consumption, and the use of harmful chemicals.

Pearson closes the loop by recycling every out-of-date text returned to our warehouse. We pulp the books, and the pulp is used to produce items such as paper coffee cups and shopping bags. In addition, Pearson aims to become the first carbon neutral educational publishing company.

Pearson is also supporting student sustainability efforts, through our Sustainable Solutions Awards, our Student Sustainability Summits, and our Student Activity Fund.

The future holds great promise for reducing our impact on Earth's environment, and Pearson is proud to be leading the way. We strive to publish the best books with the most up-to-date and accurate content, and to do so in ways that minimize our impact on Earth.

Prentice Hall
is an imprint of
PEARSON

This book is dedicated to MARCUS E. MILLING—Executive Director of the American Geological Institute from 1992 to 2006, a strong supporter of education at all levels (including outreach to the general public), and the man whose vision established the project that led to *Living with Earth*.

FIGURE 1-1 ▶ Earth as Seen from Space
This photograph of Earth, known as "the blue marble," was taken from Apollo 17 on December 7, 1972. What can you tell about our home from 55,000 kilometers away?

CHAPTER

1

WHAT DOES "LIVING WITH EARTH" MEAN?

In the early morning hours of December 7, 1972, the three-man crew of Apollo 17—Eugene Cernan, Ron Evans, and Harrison "Jack" Schmidt—blasted off to land a human on the Moon. At first they zoomed around Earth at 29,000 kilometers per hour (18,000 mph) in 90-minute orbits and the scenes below passed by very quickly. Then they left Earth's orbit and headed for the Moon. About five hours after initial liftoff, and 55,000 kilometers (34,000 mi) into space, they took what is now the most widely distributed photograph in the world—it is shown in Figure 1-1.

NASA credits the entire Apollo 17 crew for this photograph, but Jack Schmidt probably snapped the picture. Schmidt, a geologist, was especially focused on observing Earth as they departed. He even gave weather descriptions as the crew sped away. But it was fellow astronaut Eugene Cernan who memorably radioed back to mission control, ". . . we'd like to confirm, from the crew of Apollo 17, that the world is round."

At a distance of 55,000 kilometers, Earth is clearly a sphere. What else could you observe about Earth from that distance? If you were like Jack Schmidt, you could readily identify the continents, oceans, and regional features like the Red Sea. But what if you were a traveler from another galaxy, and the Apollo 17 vantage point of 55,000 kilometers away were as close as you would ever come to Earth? What would you observe then?

Look closely at the photo of Earth. You can probably distinguish the oceans from the continents. If you were completely unfamiliar with Earth, your first observations might identify:

1. Areas with swirling white patterns,
2. Large deep-blue and smooth expanses,
3. Brown or reddish-brown rough-looking areas, and
4. Green-tinted areas.

These very first observations identify the four basic components of Earth—(1) its *atmosphere*, with constantly moving and changing white clouds, (2) blue areas where waters of the *hydrosphere* cover the surface, (3) brownish landmasses of the *geosphere*, and (4) green areas where the living organisms that comprise the *biosphere* are abundant. Atmosphere, hydrosphere, geosphere, and biosphere are Earth's four most basic components, or what we call *Earth systems* in this book, and they are readily apparent even from 55,000 kilometers away.

Perhaps more important, the "blue marble" in the photo shows that Earth is an isolated place. With a few exceptions, such as journeys like that of Apollo 17, people are stuck on the space-capsule Earth, and we are all doing a lot more than just going along for the ride.

IN THIS CHAPTER YOU WILL LEARN:

- How environmental geology is a part of your life
- How human population, resource consumption, and technology are factors that influence people's impact on the environment
- The costs of Earth's impacts, such as earthquakes and floods, on people
- How Earth systems are defined and how they interact
- How science works and how it will play a part in your future
- The meaning of sustainability and why it is important for people to achieve

1.1 Environmental Geology and You

Look again at the photo of Earth. What factors do you think significantly influence global environmental conditions, things like climate, air quality, and the availability of fresh water? Perhaps surprisingly, one of the most important turns out to be the part of the biosphere that includes you—people! People are so numerous and cause so many changes in Earth's systems that they now influence environmental conditions on a global scale. And in turn, Earth systems affect people. Thousands of people die each year in earthquakes, floods, and other natural disasters. Environmental studies relate to you because Earth and people interact, and those interactions have consequences.

Environmental geology is the part of Earth Science that studies the interactions between people and Earth. It is rooted in understanding the natural processes that shape and change Earth, processes that cause earthquakes, landslides, or floods, for example (Figure 1-2a). Environmental geology also investigates how people affect Earth, especially through their use of natural resources (such as soil, fuel, minerals, air, and water) and by the manner in which they dispose of their wastes. By studying environmental geology, we learn to become better stewards of our planet and its life-sustaining natural resources.

The "environment" in environmental geology is broadly defined. It includes all the physical and biological components that people commonly associate with the word **environment**, such as rivers, mountains, and forests. It also includes the changes people make to Earth, such as damming rivers to control floods, clearing forests to create farms, or constructing buildings, roads, bridges, and the rest of the infrastructure that supports our communities (Figure 1-2b). As we consider the parts of the environment that people create or change, we will also consider the personal, social, economic, cultural, and political factors that shape human actions. Thus, "environment" in the context of environmental geology encompasses all of the natural and human factors that have come to shape the world.

The goal of *Living with Earth* is to help you understand and, more important, use the information it contains. Studying environmental geology is not merely an academic exercise. People enter this field specifically to help other people and the planet. The lessons people have learned, sometimes in difficult ways, are valuable guidelines for you.

This book emphasizes four central themes to help you understand and apply what people have learned about living with Earth:

1. How people and Earth interact,
2. How Earth systems interact with each other,
3. How science helps people to understand and deal with issues related to Earth, and
4. How people can take steps to achieve a *sustainable* future.

The idea of *sustainability* is a critical concept in environmental studies and is one that we will use as a measure throughout *Living with Earth*. To say that something is **sustainable** means that it is capable of being continued with minimal long-term effect on the environment (Figure 1-3). In the context of people, it

(a)

(b)

FIGURE 1-2 ◀ Earth Affects People; People Change Earth
(a) Earth systems processes shape and change Earth, creating natural hazards such as severe floods. **(b)** But people also influence Earth—for example, by damming rivers to control floods, provide irrigation, and generate electric power.

FIGURE 1-3 ◀ Many Natural Resources Can Be Sustained
On farms like this experimental agricultural station in Peru, crops, fertilizers, irrigation methods, and tilling practices are carefully chosen to help maintain the soil's quality. Sustainable agriculture research and education

means that the needs of the present generation are met without compromising the ability of future generations to meet their needs. Thus, making good choices about how we live on and interact with Earth will help to determine the future of our planet.

1.2 How People and Earth Interact

Virtually everything people do has environmental consequences. How people provide food for themselves, how they construct buildings, how they move about, and where they choose to live all affect the environment in some way. Everyone on Earth is a player in this interaction, and the number of players is unfathomably large and growing.

HUMAN POPULATION

Do you know how many people you are going to have to share Earth with? Although predictions of future population require several assumptions, especially about the rates at which people are born and die, we can study general trends. As you can see in Figure 1-4, the human population grew faster than ever before during the last few hundred years. The highest average *growth rate* (birth rate minus death rate) of 2.1% was in the 1960s, and the greatest number of people added in a year, 86 million, was in the 1980s. At the end of 2008, the world population totaled 6.7 billion, and it was increasing by about 80 million people per year—a rate of 1.2%.

Concerns about the impact of an increasing number of people are having an effect on the number of births around the world. For example, many developed countries, such as Japan and most European countries, have essentially stable populations. Population growth has also been stemmed in Latin America, where more than 60% of the women of childbearing age (15–49) use some sort of modern contraception. National policies and economic changes in China have lowered that country's population growth rate from 2.6% in 1969 to 0.6% in 2004. The U.S. Census Bureau estimates that by 2050, the number of people on Earth will be almost 9.5 billion, or about 40% greater than it was in 2010. POPClocks[1] Because birth rates in general are going down, some researchers now estimate that the world population will level out at a little under 10 billion people by 2150 (see Figure 1-4).

Where will all these new people live, and how will the planet continue to provide for all of them? Between now and 2050, over 99% of the world's new people will be born in less-developed countries, which have lower standards of living and weaker economies than developed countries like the United States. In fact, by 2050, 9 out of 10 people will live in less-developed countries. The United States, with a growth rate of 0.6% but high anticipated levels of immigration, is the only developed country expected to experience significant population growth between now and 2050.

Another coming change is that more and more people will be living in cities. In 1960, there were only two cities in the world with more than 10 million people—New York and Tokyo. But by 2015 there are expected to be 26 of these mega-cities, most of them—like Lagos, Nigeria, shown in Figure 1-5—located in less-developed countries. As of 2006, about half the world's

[1]Many of the discussions in this book include highlighted titles for web searches. Enter one of these titles into a search engine such as Google and it will guide you to a specific website where you can find more information about the topic being discussed.

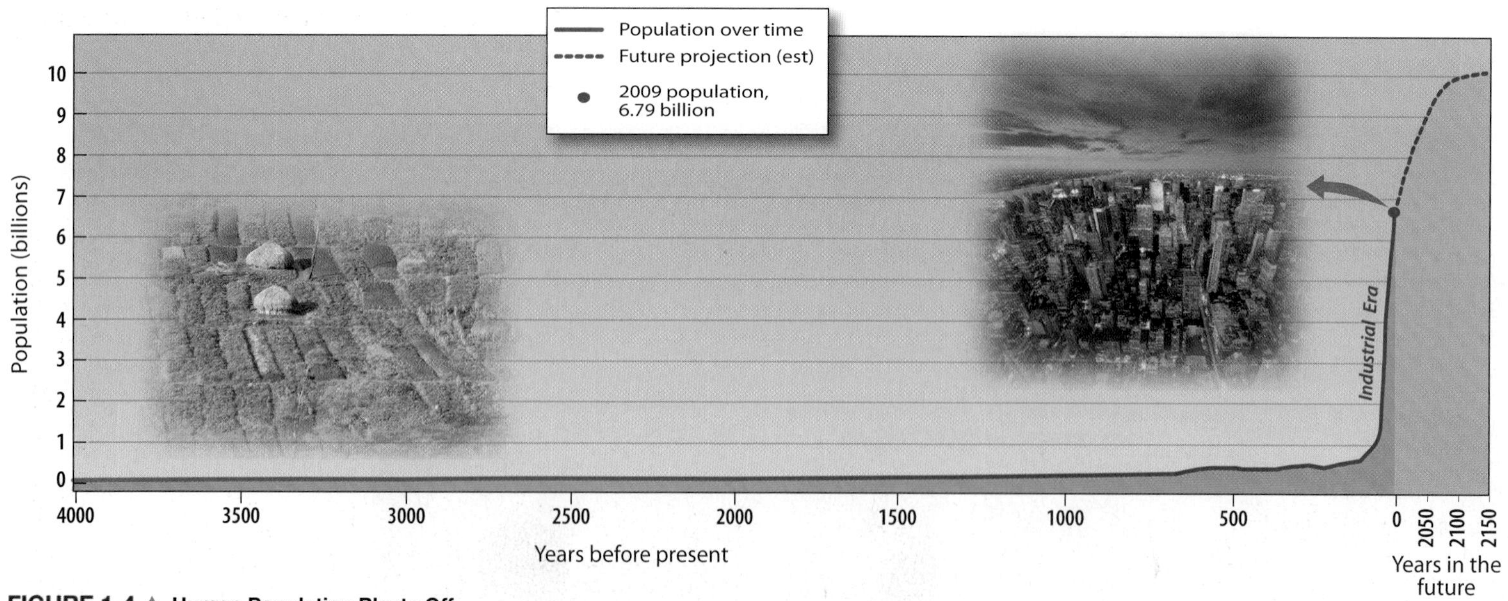

FIGURE 1-4 ▲ Human Population Blasts Off

After thousands of years of slow growth and gradual cultural and economic advance, the human population has exploded since the beginning of the industrial era. The human population may level off at about 10 billion people within the next century. In the meantime, this tremendous increase in population will stress Earth's resources and the environment. World population projections World population to 2300

FIGURE 1-5 ▲ More and More People Are Living in Cities
More than half the inhabitants of Earth now live in cities, and the proportion of urban dwellers is increasing. Twenty-six cities will soon have populations of over 10 million. Lagos, Nigeria, shown here, is home to well over 8 million people—perhaps as many as 15 million, by some estimates.

people lived in cities, but this fraction is expected to increase to 60% by 2030.

Environmentally, it could be helpful that so many of the world's people will live in cities. Well-managed cities promote more efficient energy use, advanced sanitation, and consolidated waste management, for example. Of course, achieving well-managed cities is an ongoing challenge, especially if they grow rapidly, but they provide many opportunities for dealing effectively with the environmental impacts of large numbers of people.

What is the significance of adding 3 billion or more people from an environmental perspective? In one way it's quite simple. The more people there are, the more changes they cause, the more wastes they generate, and the more of Earth's resources they use. However, as you probably suspect, it is not actually that simple. An increasing number of people is not the only issue because there is tremendous variation in the way that each individual impacts the environment. A person's *environmental impact* depends on how much of Earth's resources he or she consumes and how much waste is generated in the process.

RESOURCE CONSUMPTION

You are probably aware of how resource consumption varies among people. Perhaps you know a family with several cars and a relative or friend who doesn't drive at all. The consumption of meat in America averages over 91 kilograms (200 lbs) per year per person, but you may know someone who is a vegetarian. The average consumption of a commodity per year per person, called the *per capita consumption,* depends on several personal, social, cultural, and economic factors. However, if you look at the way that large numbers of people consume resources around the world, the key factor influencing consumption is economics. As people become wealthier (more affluent), they consume more (Figure 1-6). The United States is a prime example.

The United States is among the world's most affluent nations. Our *gross domestic product,* or *GDP* (the total market value of goods and services produced in a year), was 14.4 trillion dollars in 2008. The United States, with less than 5% of the world's human population, produced almost 24% of the world's

FIGURE 1-6 ▼ Wordly Goods
The possessions of a family in rural India **(a)** and suburban England **(b)**. As these photos show, families around the world vary tremendously in the amount of resources they use. This is why population alone is not a complete indicator of people's environmental impact.

(a)

(b)

GDP in 2008. To achieve this productivity, we consume tremendous amounts of natural resources, such as energy, as Figure 1-7 shows. We use about 25% of all the oil and natural gas that the world consumes each year. In fact, there are more cars per person (0.765) in the United States than anywhere else in the world. But times are changing, and other countries are catching up.

Affluence is increasing around the world. The economies of very populous but less-developed countries like India and China have expanded at 5 to 10 % annual rates in recent years. Globally, people's average annual income is increasing slightly faster than the world's population (1.2%). This increase in wealth, even small advances, translates into increased consumption. Examples of consumption outpacing global population growth include the number of cars (increasing at 2.8% per year), the number of TVs (increasing at 5.7% per year), and the number of cell phones (increasing at more than 10% per year). As the growth rate of the human population slows, consumption patterns will become an ever more significant influence on people's environmental impacts.

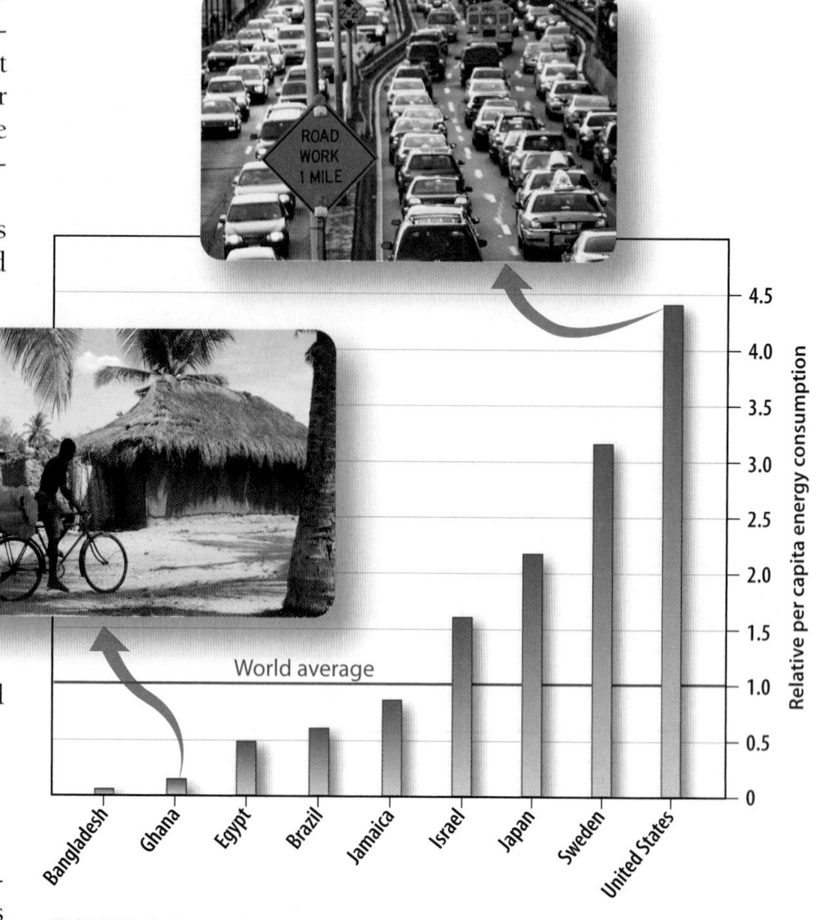

FIGURE 1-7 ▲ Relative Energy Consumption per Person for Selected Countries

The average citizen of the United States uses 20 times as much energy in a year as a citizen of Ghana, and 46 times as much energy as a citizen of Bangladesh. (Data from 2005.) Per capita energy consumption

THE TECHNOLOGY FACTOR

The combination of increasing population and increasing consumption means that people will use more and more of Earth's natural resources. The increasing commodity prices of recent years (copper prices have tripled, for example) are evidence that the global expansion of natural resource use is under way. How we exploit these resources and how we deal with the wastes that are generated in the process is another factor determining what people's environmental impacts will be. This is commonly called the technology factor.

To better understand how technology affects environmental outcomes, consider some options faced by growing cities:

- Municipal wastewater can be dumped directly into the environment (commonly in rivers and oceans) or can be treated and recycled.
- Transportation needs can be provided for with extensive street and highway systems for vehicle use or with mass transport systems using subways, buses, and trains.
- Solid waste can be dumped in landfills (or in the ocean in some cases), or it can be incinerated and recycled.

Technological options with different environmental consequences exist for all the principal ways that people acquire and use natural resources. For example, farming techniques determine how much soil erodes, how we use water determines how much is available, and the characteristics of our vehicles determine the amount of pollutants that are emitted to the atmosphere. Technological choices play a big role in determining people's impact on the environment (Figure 1-8).

The relationship between population, consumption, and technology in influencing people's environmental impacts was originally summarized by Paul Ehrlich and John Holdren in 1971 in the journal, *Science,* with the following equation:

$$I\ (\text{impact}) = P\ (\text{population}) \times A\ (\text{affluence}) \times T\ (\text{technology})$$

Over time we have learned that environmental impact is not necessarily directly proportional to these factors but the equation's key point, that multiple factors are at play, has had a powerful influence on how people evaluate and predict environmental impacts. The complexities of applying relationships between population, consumption, and technology to understanding present and future environmental impacts are discussed fully by the American Association for the Advancement of Science. AAAS atlas of population and environment—the theory of population-environment links Although understanding the key relationships and how they apply in specific cases is an ongoing challenge for researchers, the take-home message is that environmental impacts are not just a function of how many people there are. They also depend on other factors for which people make choices, especially what resources they consume and how they use them.

(a)

(b)

FIGURE 1-8 ▲ Technology Gives Us Options with Different Environmental Consequences
(a) When it comes to buying a car, people today have many choices—most of which can affect our environment. What factors do you consider when you make such a choice? **(b)** These 70,000 refrigerators are waiting for the crusher. What will be done with the steel and other materials they are made of? Is it better for the environment to buy a new, energy-efficient refrigerator or hold on to your old one for a while longer?

You Make the Call

How Much Is Enough?

Affluent people consume a lot. The great majority of Americans have homes, cars, and air-conditioning if they need it. They can buy food at local supermarkets every day, have the latest electronic gadgets, and seemingly travel where and when they want. As people consume more, they tend to need, or believe they need, more and more.

- Are there other things in life as important as what we consume?
- How much is enough, really?

There are no right and wrong answers to questions like these. Everyone has different needs, wants, and responsibilities. The point is that consumers, especially affluent consumers, make choices and these choices help determine the cumulative environmental impacts of society.

EARTH'S IMPACT ON PEOPLE

We have considered the most fundamental ways in which people impact Earth systems. In turn, Earth systems affect people. As you know, natural hazards, including hurricanes, earthquakes, landslides, volcanic eruptions, and floods, kill people and destroy property (Figure 1-9 on p. 9). Natural disasters result from the interaction of people with processes that change Earth—processes that have been changing Earth for millions, even hundreds of millions, of years. As a student of environmental geology, you will learn that people are getting better at predicting where, if not when, natural disasters will occur. However, natural disasters continue to affect people and their economic well-being.

In the News

The Increasing Costs of U.S. Natural Disasters

Natural disasters make the news because they affect people in a negative way. But you don't have to live near a volcano or be in the path of a hurricane to be affected by this type of disaster. Increasing amounts of the tax dollars you pay to the federal government are being spent to deal with the impacts of natural disasters.

The increasing costs of U.S. natural disasters

The federal government is the ultimate insurer for those affected by natural disasters. Through the Federal Emergency Management Agency (FEMA), money is made available to provide emergency services and rebuild communities. The amount of money used for these purposes is rapidly increasing—not just in the number of dollars but in their percentage of the gross domestic product (GDP) and of the federal budget, as you can see in the accompanying graph. In fact, the amount of money spent on disaster relief has more than tripled as a function of GDP during the last 50 years. And these costs don't include the expenses of dealing with Hurricane Katrina in 2005. With total costs adding up to about $200 billion, Katrina was the most expensive natural disaster in U.S. history.

The key cause of increased spending is the greater number of people affected by natural disasters. More and more people are choosing to live in vulnerable areas, such as hurricane-prone Florida and earthquake-prone California. And these people build expensive homes, communities, and other facilities.

In a way, federal disaster relief makes it easier for people to live in disaster-prone regions. It assists people who rebuild in hurricane paths or in places that suffer recurrent floods, for example. At the same time, science is helping people understand natural hazards and getting better at predicting where they will occur. Several chapters in this book will help you understand how people prepare for,

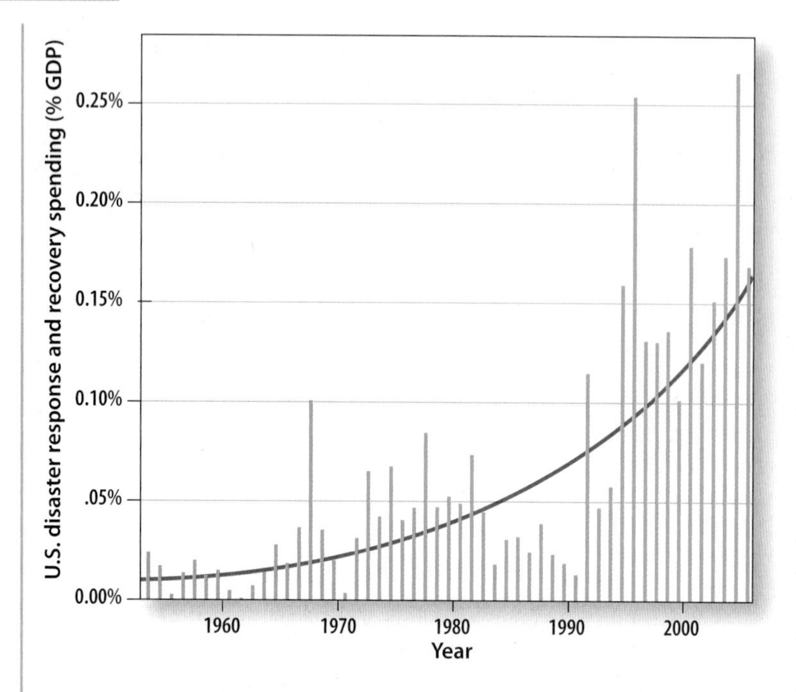

and mitigate damages from, natural disasters. But natural disasters will continue to be in the news, and as the figure illustrates, your tax dollars will still be used to deal with their effects.

Hurricane Katrina flooded large parts of New Orleans, causing more than 1800 deaths and property damage exceeding $200 billion.

1.3 How Earth Systems Interact

Earth has constantly changed since its beginning 4.5 billion years ago. Over that enormous period of time, it has evolved into a planet with a distinctive atmosphere and abundant water, populated by diverse life-forms that thrive just about everywhere on and near its surface. For Earth, natural environmental change is normal and inevitable. One goal of this book is to help you understand how and why Earth changes from a systems' perspective.

A **system** is a group of interacting, interrelated, or interdependent parts that together form a whole. Understanding each part of a system requires understanding its connections and interactions with all other parts of the system. The four systems interacting at the scale of the entire Earth are:

- The **geosphere** consists of the rock (solid and molten) and material derived from rock that make up Earth.
- The **atmosphere** consists of the gases that surround Earth.
- The **hydrosphere** consists of all the water (both liquid and frozen) on Earth's surface and underground.
- The **biosphere** consists of all the living organisms that inhabit Earth.

These **Earth systems** are constantly interacting with and affecting one another. You can examine particular aspects of systems to understand how they work. One key aspect is how—and the degree to which—energy and matter move among systems.

ENERGY AND SYSTEMS

All system changes and interactions involve energy. The main sources of energy that drive Earth's systems are the Sun's radiation, heat from Earth's interior, chemical reactions, and gravity. For example:

- Radiation from the Sun transfers energy to the oceans and atmosphere, causing severe weather such as hurricanes and tornados.
- Solar energy also drives chemical reactions that make possible the growth of green plants, the principal source of food in the biosphere.
- Heat from Earth's interior melts parts of the geosphere and helps produce volcanic eruptions.
- Gravity moves water in rivers to the ocean, gradually changing landscapes and inundating entire regions during floods.

Energy is stored in systems and transferred among systems, and it also drives movement of matter among systems. You will learn more about energy in Chapter 13.

(a)

(b)

(c)

FIGURE 1-9 ◀ **Natural Hazards** Earth influences people in a variety of ways. Some of them, such as earthquakes **(a)**, volcanic eruptions **(b)**, and landslides **(c)**, can be highly destructive.

MATTER AND SYSTEMS

Matter comprises all physical substances, the liquids, solids, and gases that things are made of. Like energy, matter is stored in systems, moved around in them, and transferred among them. Consider, for example, the element carbon. Carbon is involved in a set of transfers among systems, creating a vast, intricate cycle that takes place at local and global scales. The basic features of this cycle are outlined in Figure 1-10.

Carbon is stored in **reservoirs** in each of the four Earth systems. Because it is an integral component of the molecules that make life possible, it is stored in reservoirs throughout the biosphere—in plants, for example. In addition, dissolved carbon is stored in hydrosphere reservoirs such as rivers and oceans. Carbon is also stored in the atmosphere as a gas (mostly carbon dioxide), but the largest carbon reservoirs on Earth are in rocks, such as limestone (calcium carbonate), within the geosphere.

The geosphere contains 100 million billion tonnes of carbon. (A *tonne* is a metric ton, equal to 1000 kilograms, or 2205 lbs.) Compare this amount to that in two other important carbon reservoirs. All the carbon in land plants totals 560 billion tonnes, and all the carbon in the atmosphere totals about 760 billion tonnes (see Figure 1-10). These are large reservoirs, but rocks contain some 76,000 times more carbon than both of them together!

Within Earth systems, carbon can be quite mobile. When animals, from ants to zebras, eat food, they absorb and carry with them carbon-containing molecules that they use for energy and growth. Carbon in the atmosphere and hydrosphere is shifted about by winds and currents. Even carbon in rocks can move. Forces inside Earth can transfer rocks formed deep in the geosphere to Earth's surface.

As you can see in Figure 1-10, carbon is also transferred *among* systems. When plants obtain carbon dioxide from the atmosphere, they move it to the biosphere. Volcanoes, parts of the geosphere, release many gases, including carbon dioxide, into the atmosphere. Clams and other organisms in the biosphere use carbon in ocean water, part of the hydrosphere, to make shells and other structures. And carbon, deposited from ocean water onto the seafloor, can become rocks in the geosphere.

Flux is the rate of transfer of matter, such as carbon, among systems. Figure 1-10 shows that the rate at which land plants absorb carbon from the atmosphere, or the flux of carbon from the

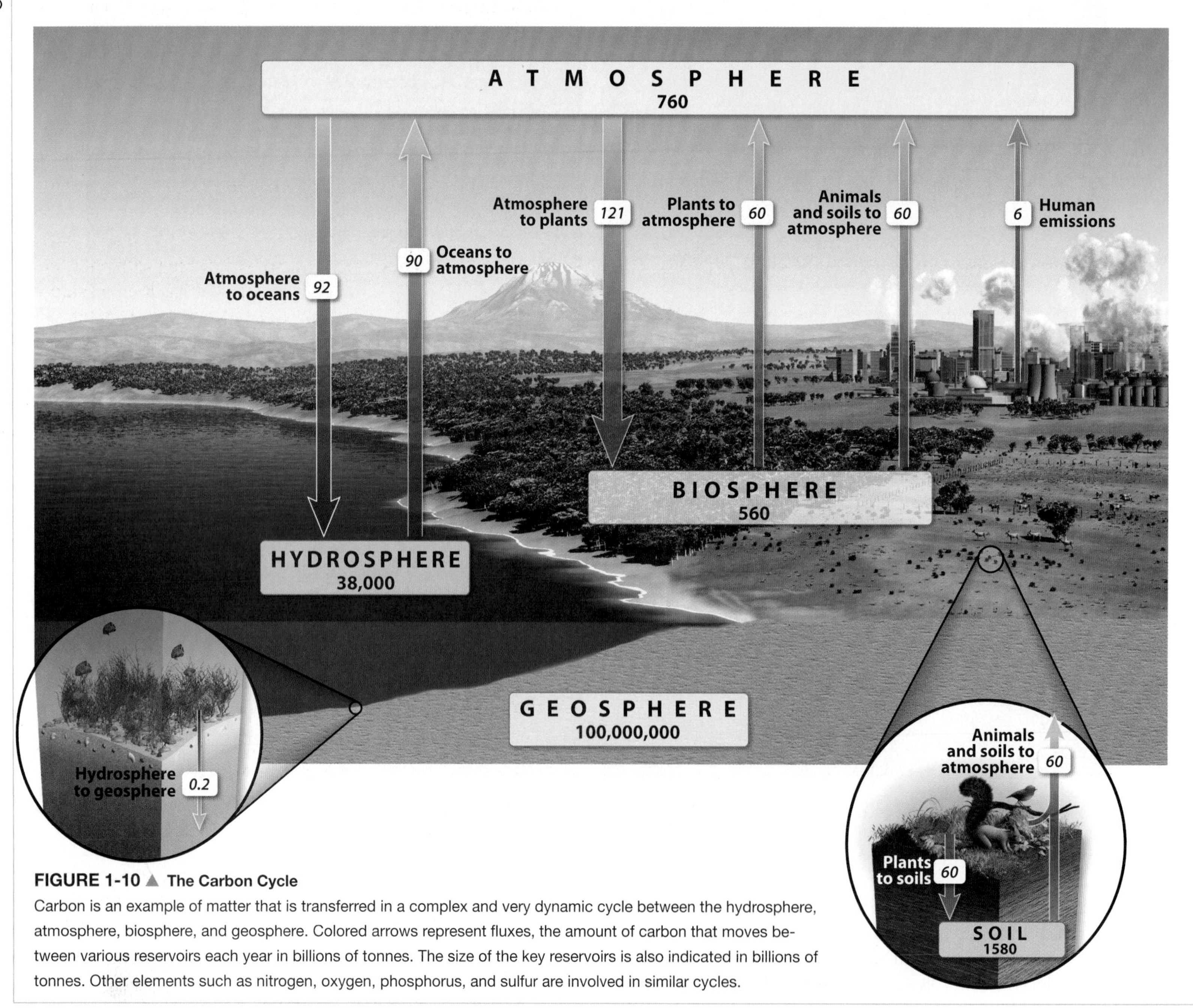

FIGURE 1-10 ▲ The Carbon Cycle
Carbon is an example of matter that is transferred in a complex and very dynamic cycle between the hydrosphere, atmosphere, biosphere, and geosphere. Colored arrows represent fluxes, the amount of carbon that moves between various reservoirs each year in billions of tonnes. The size of the key reservoirs is also indicated in billions of tonnes. Other elements such as nitrogen, oxygen, phosphorus, and sulfur are involved in similar cycles.

atmosphere to plants, is 121 billion tonnes per year. However, the 560 billion tonnes of carbon in land plants is only about 5 years' worth of the estimated flux to land plants from the atmosphere. If land plants obtain 121 billion tonnes of carbon each year from the atmosphere, why don't they contain more than five years worth of the carbon that is transferred to them in this way?

The reason is that carbon is transferred from land plants at the same time they are accumulating it. When they die, plants release about 60 billion tonnes of carbon per year to soils as they decompose. Moreover, plants need energy to grow and maintain their cellular functions. Plant respiration, the process of using the chemical energy that plants stored through photosynthesis, also transfers carbon dioxide to the atmosphere at a rate of about 60 billion tonnes per year. As you can see, then, plants get carbon from the atmosphere at about the same rate they give it back to soils and the atmosphere.

FIGURE 1-11 ▼ Reservoirs and Flux

(a) A closed system is one that exchanges only energy with its surroundings, whereas an open system can exchange matter as well. **(b)** If a reservoir is an open system, its size will be affected by the relative amounts of flux in and flux out. If these are equal, the size of the reservoir will remain constant. Such a reservoir is in a *steady state.* **(c)** If the flux into and out of a steady-state reservoir is small, the residence time for matter will be long. If the flux is large, the residence time will be short.

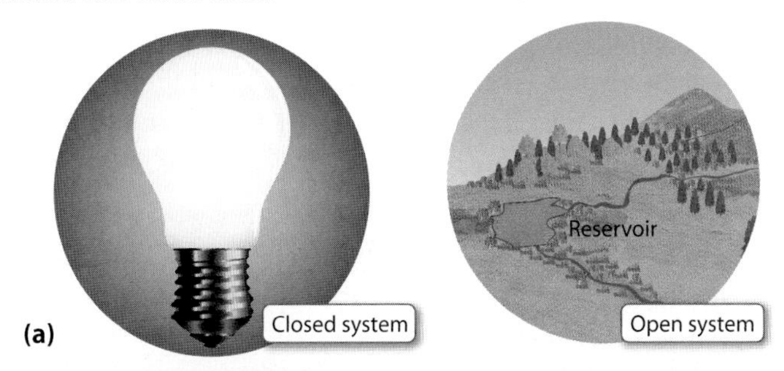

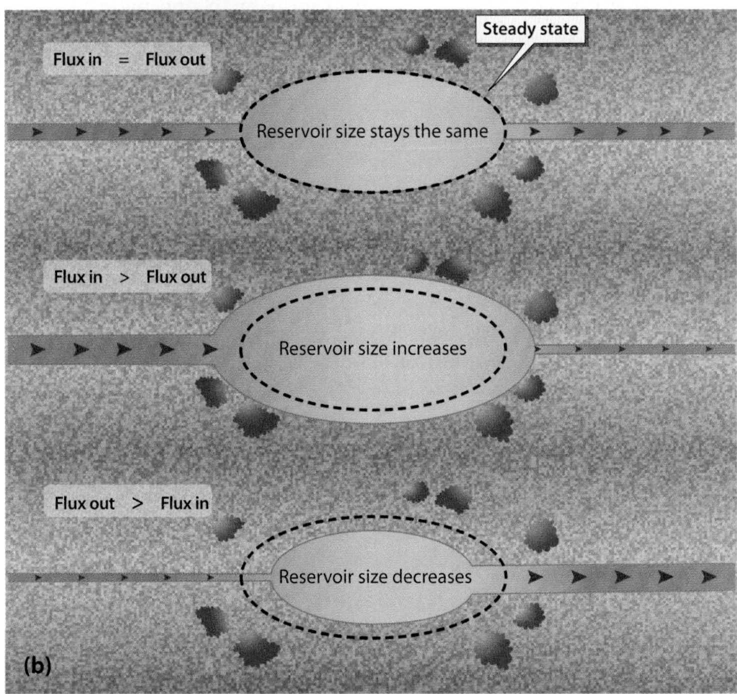

With all these transfers between reservoirs constantly taking place, the carbon atom's "home" seems to be a moving target. One measure of home for matter such as a carbon atom is the average amount of time it is contained in a specific reservoir. This is called its **residence time**. The equation for calculating residence time if the flux in is equal to the flux out is:

$$\text{Residence Time} = \frac{\text{Reservoir Size}}{\text{Flux In}} = \frac{\text{Reservoir Size}}{\text{Flux Out}}$$

The residence time of carbon in plants is about 4.7 years (560 billion tonnes/120 billion tonnes per year).

If a reservoir is very large and the flux of matter in or out is low, residence times can be very long—up to billions of years for some elements that become parts of rocks in the geosphere. If matter has a long residence time in a reservoir it is essentially trapped in a **sink**. The opposite situation develops if flux rates are high and the reservoir is relatively small. The residence time for a carbon atom in the atmosphere, for example, is only about three years.

Here are some examples of residence times for water in hydrosphere reservoirs:

- In the atmosphere—9 days
- In streams and rivers—17 days
- In groundwater—330 years
- In oceans—2,900 years

Residence times help us to understand why dealing with pollution can be particularly challenging. Pollution that has long residence times in large reservoirs is difficult and expensive to clean up.

Reservoirs, fluxes, and residence times help us understand interactions between systems of one type. However, there is another type of system that doesn't allow transfers of matter in or out.

TWO TYPES OF SYSTEMS

In *open systems*, matter and energy can move in or out. Examples of open systems include the human body, a lake, and an erupting volcano. *Closed systems* allow only energy (not matter) to be transferred in or out. A lightbulb is a simple example of a closed system (Figure 1-11a). Virtually all natural systems, including those discussed in this book, are open systems that readily interact and change, which makes them *dynamic systems.*

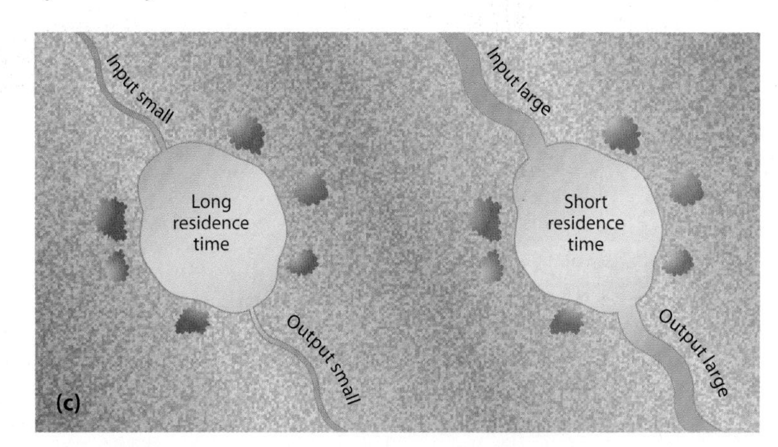

If transfers of energy and matter into and out of an open system are about the same, the system is said to be in a **steady state** (Figure 1-11b). For example, if the amounts of carbon entering and leaving the atmosphere are about the same, then the total amount of carbon in the atmosphere stays roughly constant—the atmosphere is in a steady state with respect to its carbon content. You may have heard, however, that carbon concentration is increasing in the atmosphere. This carbon, chiefly as carbon dioxide, has increased to levels that affect global climate, as you will learn in Chapters 2 and 14.

Changes in one part of an open system can affect other parts. For example, rain in distant mountains can cause devastating floods many kilometers (miles) away. Seemingly small or local changes can aggregate into broadly significant ones. Exhaust emissions from people's cars in southern California aggregate into air pollution throughout the American southwest. One of the great strengths of science is that it helps us understand how seemingly unrelated events or impacts are connected—how systems work.

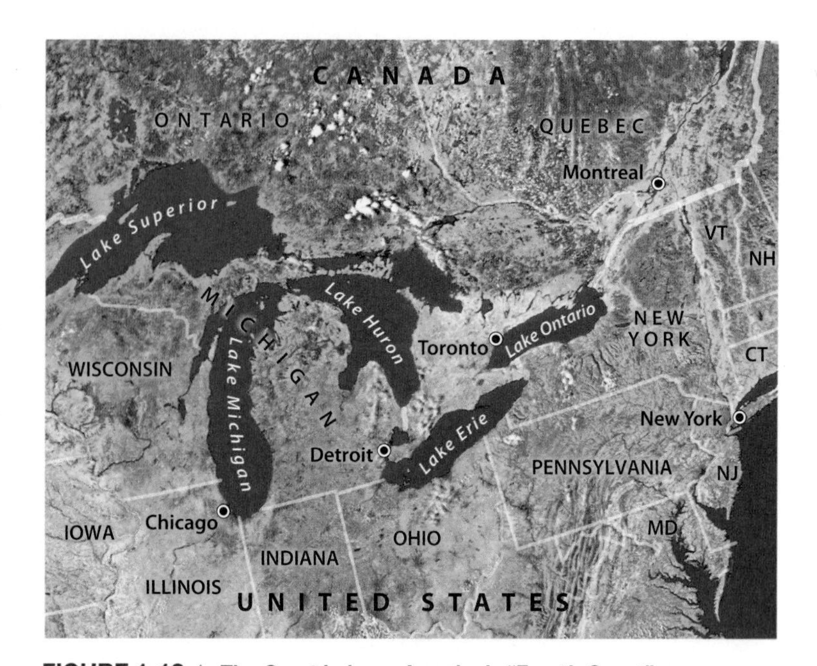

FIGURE 1-12 ▲ The Great Lakes—America's "Fourth Coast"
The Great Lakes, which are as much as 400 meters (1300 ft) deep in places, contain more than 20% of the world's freshwater resources. They are sometimes called America's fourth coast because they have so many miles of shoreline.

1.4 How Science Helps

Are science-related articles low on your list of interesting reading sources? Science may sometimes seem to be a remote intellectual activity with little relevance to you or your daily activities. This is far from the case when it comes to environmental geology. *Living with Earth* provides many examples of how science helps people to better understand and solve problems. Predicting volcanic eruptions (Chapter 6), showing where landslides are likely to occur (Chapter 8), and preventing or treating contaminated water (Chapter 10) are among the issues that will be discussed in the text. You can address challenges in your daily life the same way that scientists address challenges.

One goal of this book is to develop your understanding of how science can be part of the everyday decisions you make, particularly those that relate to interactions between people and the environment. *Living with Earth* will help you think critically, develop understanding, and reach knowledgeable decisions. These abilities characterize science and permeate the scientific method.

THE SCIENTIFIC METHOD

The **scientific method** is an approach to asking and answering questions that scientists use to explore and explain how the natural world operates. It is a process, or a way of solving problems, that everyone can use. This process has four general steps.

1. **Developing questions:** Examining observations and data about a subject commonly raises unanswered questions about it. For example, examining data on global water resources reveals that the Great Lakes (Figure 1-12) contain about 22% of the entire world's freshwater. The Great Lakes are a tremendous water resource. One question you might ask about these huge lakes is, how did they form?
2. **Developing hypotheses:** In science, answering questions begins with developing hypotheses. A **hypothesis** is a tentative explanation that is consistent with all we know about the situation. In fact, scientists often try to develop more than one hypothesis for each question or problem. They call these "multiple working hypotheses" and use them to help understand the range of answers that they need to consider.

   Lakes are inland depressions or *basins* filled with water. Hypotheses that tentatively explain how huge basins like the Great Lakes form include (1) deformation of the land surface, (2) erosion by rivers, (3) erosion by glaciers, (4) some combination of these processes, or (5) some other process.
3. **Testing hypotheses:** Hypotheses are used to make testable predictions about the situation. The predictions implied by the hypotheses can then be investigated to determine whether they are valid. In the case of the Great Lakes basins:
   - If land deformation formed them, then the land surface may have been displaced downward along breaks (faults) in the geosphere.
   - If river erosion formed them, then their water depths would not be lower than sea level.
   - If glacial erosion formed them, then deposits of glacial materials, variable water depths including some below sea level, and scoured land surfaces could be present.

Data and information are gathered to test and evaluate each hypothesis and its predictions. Data commonly come from making observations about the situation—determining whether the land surface around the Great Lakes has been displaced downward, measuring how deep the Great Lakes are, and studying the geological record to establish where and when glaciers were present in the Great Lakes region, for example.

4. **Evaluating results:** Finally, the hypothesis is reexamined to see if it is consistent with all the available data and information. In the case of the Great Lakes, studies show that many water depths are indeed below sea level, and that glaciers extensively covered the region just 20,000 years ago. The hypothesis that glacial processes formed the Great Lakes is supported by this data, and this hypothesis is then used to help develop a more detailed understanding of the lakes' history. If the data were not consistent with a glacial origin for the Great Lakes, then a new hypothesis would be developed and the process started over again.

The scientific method is an *iterative* process, which means that it repeats itself over and over again, as shown in Figure 1-13. Even if the initial tests support a hypothesis, scientists continue to test the hypothesis in more detail and further refine it. Was the ice in the glaciers thick and heavy enough to deeply scour into the land and even perhaps depress the land surface? As it turns out, the extensive ice accumulations of the last glaciation 20,000 years ago were up to 4 kilometers (2.5 mi) thick. Experiments and laboratory simulations (another way to test hypotheses) show that this amount of ice cover is very capable of scouring and even depressing the land. The iterative nature of the scientific method enables people to develop better and better answers to questions about Earth.

The scientific method as outlined here is an idealized example. In practice, scientists do not commonly start completely from scratch when asking questions. They use information and knowledge gained by others to constrain their questions and hypotheses. For example, scientists have long known that thick ice accumulations covered Canada and northern parts of America 20,000 years ago. They also learned how the ice melted and gradually receded northward to the present-day remnants that make up the Greenland ice sheet. A key part of being a good problem solver, whether you are a scientist or not, is understanding what others have already learned about the problem.

Ongoing observations of Earth give us the opportunity to continually refine hypotheses. In some cases, many extensively tested hypotheses become integrated in a well-accepted statement of relationships called a **theory**. Scientific laws, hypotheses, and theories To scientists, the word "theory" has a very different meaning from its more common usage. In common usage, a theory is like a hunch. For example, I have a theory that enrollments in environmental geology classes are increasing because of widespread concern about how people are impacting the environment. However, for scientists a theory explains a large set of observations and relationships that have been verified independently by many researchers. Scientific theories can be used to accurately predict natural relationships and are generally accepted as true. They are not a hunch but rather a powerful synthesis of diverse facts and relationships tested and then worded into a unifying explanation.

So, science helps us understand how Earth works. With this understanding we can begin to predict what the future will bring—where natural disasters are likely or what the effects of our waste disposal practices will be, for example. Believe it or not, science is in your future.

SCIENCE IN YOUR FUTURE

Every chapter in *Living with Earth* has examples of how science helps people better understand and deal with environmental issues. Here are just three examples that illustrate how important science will be to your future; (1) the availability of clean, fresh water, (2) the transition from oil to other sources of energy, and (3) the extent and effects of global climate change.

Availability of water

Everyone needs clean, fresh water. Earth's people need over 3 billion gallons a day just for drinking. In addition to

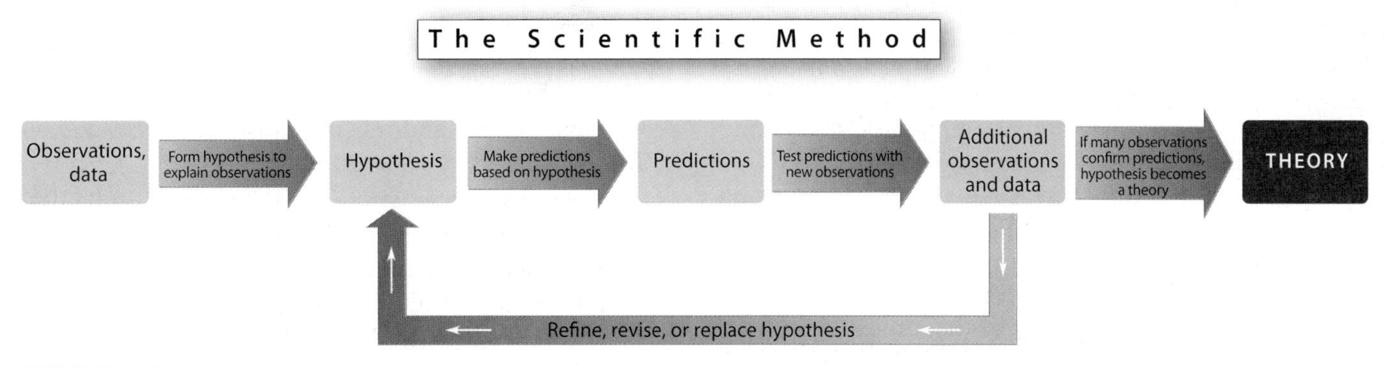

FIGURE 1-13 ▲ The Scientific Method

The scientific method is a way of investigating nature based on forming hypotheses and testing them against evidence. If a hypothesis is consistently supported by observations, it can be elevated to the status of a theory. While theories are not final—they can be modified if new evidence requires it—they represent the state of our knowledge and understanding at any given time.

FIGURE 1-14 ▲ A Thirsty World
Enormous and ever-increasing amounts of water are needed, not just for drinking and sanitation but also for industry and irrigation. In many places, water resources are already severely strained.

sustaining life, tremendous amounts of water are used for agricultural and industrial purposes (Figure 1-14). So much water is being used that freshwater resources are being stressed around the world. SAHRA—news watch Even in the United States, water resources are being depleted, some in ways that cannot be reversed (Chapter 10). Science will help us understand where our freshwater resources are located, how we can prevent water pollution, how we can clean polluted water, and how we can use our freshwater resources sustainably. The availability of clean, fresh water may not affect you now but, rest assured, this will be a critical issue in the future.

Transition from oil to other energy sources

Have you been affected by recent high prices for gasoline? Current high gasoline prices are one indication of the decreasing ability to provide the oil people need (Chapter 13). Expanding economies around the world are using more and more oil each day—and because the supply will not be replenished any time soon (it takes millions of years for natural processes to produce oil), we are using it up. Increasingly, oil will have to be replaced by other energy sources, and science will help us understand how to replace it in environmentally sustainable ways. For example, we will need to learn how to burn coal without emitting harmful pollutants to the atmosphere, how to dispose of radioactive wastes from nuclear power plants safely, and where to build wind farms that do not harm wildlife. In many cases, the fields of science and engineering will cooperate to solve these problems, and the innovations that help us transition from oil to other sources of energy will be rooted in science. This transition, as challenging as it is, is happening right now.

The effects of global climate change

Global climate change is also occurring today. It seems always to be in the news, especially with warnings about what global warming will bring. Most scientists are convinced that global warming is caused by increased concentrations of greenhouse gases (especially carbon dioxide) in the atmosphere. You will learn a lot more about greenhouse gases and climate change in Chapter 14, but the principal source of increased atmospheric carbon dioxide concentrations has been the burning of oil, natural gas, and coal. Tremendous scientific efforts are now under way to understand the causes of global warming, how global warming will evolve, and how people can mitigate and adapt to global warming. Scientists are hard at work trying to help everyone understand and deal with global warming.

Providing clean, fresh water, transitioning from oil to other energy sources, and understanding global warming are three major challenges that science will continue to help people deal with during your lifetime. Now that you have been introduced to Earth systems, it shouldn't be a surprise that there are connections between these three issues. Will global warming change precipitation patterns and influence the availability of fresh water? Will processes be developed that effectively limit the release of greenhouse gases from the burning of fuels, especially coal? Will we accurately predict the effects of global warming and learn how to adapt to them? Science is helping to answer these questions.

1.5 How to Achieve Sustainability in the Future

The enormous and growing human population is using land, depleting resources, and lowering environmental quality around the world. These changes in Earth systems follow directly from people's efforts to provide themselves and their families with shelter, food, water, clothing, and fuel. In Chapters 10 (Water), 11 (Soil), 12 (Minerals), 13 (Energy), and 14 (Atmosphere), you will learn specifically how people have come to need and rely on Earth's resources and, especially, how this affects the environment.

A **renewable resource** is one that will continue to be available because it is naturally replenished as fast or faster than it is being consumed. Solar power, for example, is a renewable resource as long as the Sun continues to shine. People today, however, are increasingly reliant on **nonrenewable resources**—ones that are not replenished as fast as they are being used. Fossil fuels for energy are the key example, but people have also used other resources, such as soil and water, in ways that have essentially made them nonrenewable, too. For example, it takes hundreds to thousands of years for soils to redevelop if they have been eroded or otherwise severely degraded. Some depleted groundwater resources may not be replenished for hundreds of years. Resources are not being used sustainably if they are consumed or otherwise depleted faster than natural processes can replenish them.

The concept of sustainability has been developed to help people better understand the role that their actions play in both their own and Earth's future condition. Sustainability is most commonly defined as providing for the present needs of people in ways that do not jeopardize meeting the needs of people in the future. This is easy to say, but it is not so easy to convert this concept into specific actions. Of course, the entire population will require basic needs like shelter, food, and clothing. What other specific needs must be met? At what level should all these requirements be met? Since placing the needs of future generations in balance with our current needs requires economic and social changes, achieving worldwide sustainability is a very challenging goal. Take the challenge of sustaining diversity within the biosphere, for example.

SUSTAINING BIODIVERSITY

Biodiversity, which refers to the full range of variability within the living world at all levels, including genomes, species, and ecosystems, is a measure of the health of the biosphere. Scientists believe that ecosystems with greater biodiversity are more resilient to stress because there are more alternative ways to capture energy, produce food, and recycle nutrients. Greater biodiversity is also valuable to people. For example, people eat more than 3,000 different plants. In addition, plants are the source of 25% of the prescription drugs sold in the United States alone. Environmental literacy council—value of biodiversity

One measure of biodiversity is the number of species. There are 1.4 million described or known species in the world, but estimates of the total number of species range from 10 million to as many as 30 million. Environmental literacy council—how many species are there? Species naturally become extinct. Most species that have evolved on Earth are now extinct, having lived, on average, for 2 to 10 million years. But some scientists think the current rate of extinction is high, and that habitat loss caused by people is playing a big role.

Ecology studies on islands have shown a key relation between habitat area and the number of species. As the habitat area decreases, the number of species also decreases. As Figure 1-15 shows, if one Caribbean island's area is 1/10 that of another island, it will have only half as many species. This is an important relationship: Other things being equal, species diversity is proportional to habitat area.

As people increasingly change the land and degrade, fragment, or destroy habitats, the number of species decreases. Of course, people can decrease biodiversity in other ways, such as by hunting, but the relationship seen in Figure 1-15 is very fundamental: The more habitat people change, the fewer species should be expected to survive.

What tradeoffs will we be able to accept to achieve sustainable biodiversity or at least try to mitigate our impact on biodiversity? Are we willing to pay more for food or even change our eating habits to protect habitats from farm development? Will we accept that some lands should be set aside to protect habitat rather than be used as places to farm, log, mine, or build homes? Many of the toughest challenges to sustaining biodiversity are rooted in the economic and social consequences of our choices. These are difficult choices in light of people's basic needs for food, energy, and shelter, and they will be even more difficult as the global human population grows.

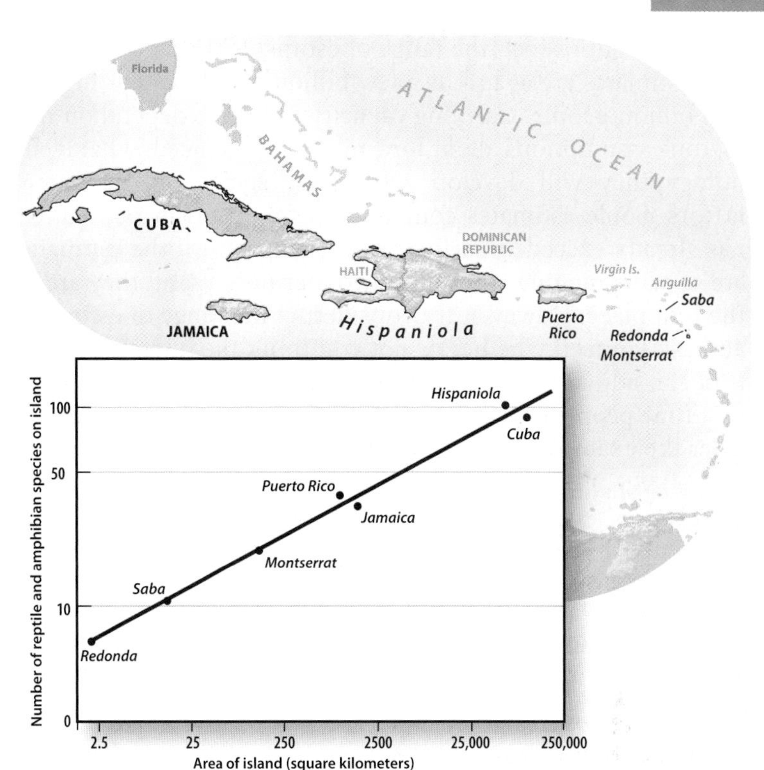

FIGURE 1-15 ▲ Biodiversity Depends on the Size of Habitat
Ecological studies on Caribbean islands show that biodiversity, as measured by the number of reptile and amphibian species, is related to island size—a reduction of 90% in area from one island to the next results in a 50% loss of species. Islands are excellent places for ecological studies because they are relatively isolated from outside influences.

CARRYING CAPACITY

Just how much land and other resources will people need to live sustainably? Or, said another way, how many people can Earth support without irreversible depletion of its resources? Earth's **carrying capacity** is the number of people that Earth can support sustainably at a defined level of economic and social well-being—in other words, at a specific standard of living. What should this standard of living be? The answer depends tremendously on a wide range of cultural and social perspectives, and there has never been consensus as to what it should be.

We can simply look at carrying capacity as a physical problem—calculating the quantity of resources Earth can provide at sustainable levels and comparing it to the number of people these resources could support. Many people have tried to estimate Earth's sustainable carrying capacity for humans.

As you might expect, the range of estimates is very wide. Current estimates are as low as a few billion people and as high as 100 billion people. Carrying capacity estimates depend on numerous assumptions about how much people need, what technology they will develop, and many social and economic factors. Some estimates conclude that the human population has already exceeded Earth's carrying capacity. If the estimates are so wide ranging, you might wonder how useful they are in the first place. However, try considering the range of estimates and ask yourself whether or not you think the human population can adjust to the planet's carrying capacity. This isn't the first time people have had to face carrying capacity limits. Consider the example of Easter Island.

EASTER ISLAND

Easter Island is a small and very remote island of volcanic rock in the South Pacific Ocean (Figure 1-16a). NOVA online secrets of Easter Island Pictures of huge stone statues, some of which are shown in Figure 1-16b, are the most common image we have of this now desolate place. These statues, or *moai*, are as tall as 9.8 meters (32 ft) and weigh as much as 74 tonnes (82 tons). They were erected to honor and show respect for ancestors and gods, thus helping to ensure an abundant food supply. By the time that residents stopped making them, 393 *moai* were mounted on platforms along the coast, 97 more were en route to their intended sites, and an additional 397 were being carved in a quarry several kilometers (miles) away.

The people of Easter Island were Polynesians who arrived there about 1100 years ago. The early inhabitants found a forest-covered island full of large palms and as many as 21 other tree species (Figure 1-16c)—all now extinct there. The trees were the foundation of the islanders' sustenance. They used them to make sea-going boats, as construction material for homes, and as sources of fiber for ropes needed to transport and erect *moai*. But when Easter Island was first visited by Europeans on April 5, 1722, the forests had disappeared.

The history of Easter Island is a sad but instructive one. As the island's population grew, it developed into a farming society, and people began to clear the forests. This worked for a while, as the population grew until about 1400 A.D. Then the big trees started to become scarcer. People cut them down to use as daily fuel, make canoes, provide more land for gardens, cremate the deceased, and make timber and ropes to transport and erect ever larger and more numerous *moai*.

FIGURE 1-16 ◀ Easter Island

(a) Easter Island is located in the southeastern Pacific Ocean, west of Chile. **(b)** The island is dotted with hundreds of *moai*, enormous stone sculptures weighing many tons, carved in honor of the inhabitants' gods. **(c)** When humans first arrived on Easter Island more than 1000 years ago, they found it covered with forests like these. Easter Island's society collapsed when all the great trees were gone. Intense efforts by the islanders to obtain divine help by carving more *moai* were unsuccessful.

Starting about 1300 A.D., agriculture expanded from the lowlands to just about everywhere possible. More and more trees were felled, with the inevitable consequences—by the 1600s, the forests were pretty much gone. The large trees, which could have been a sustainable resource if properly managed, became extinct instead.

Along with the forests went the stability of Easter Island society. The abundant resources that the inhabitants had initially discovered on and around the island were either gone or no longer accessible to them. The loss of the forests led to erosion and soil degradation that lowered agricultural productivity. And without trees to build the sturdy canoes needed for fishing, the people could no longer use the ocean as a source of offshore seafood such as dolphin and tuna. People starved, cannibalism developed, and civil war occurred. Loss of faith led to the toppling of moai, and by 1722 none were left standing. The culture that had created them had become extinct, along with the forests needed to sustain it.

The Easter Island people exceeded their island's sustainable carrying capacity and suffered greatly for it. As you read this book, you will encounter many examples of how people interact with Earth and its resources in ways that may or may not help to achieve sustainability, and in ways that challenge Earth's carrying capacity. There are a huge number of people on Earth, and the near future will only bring more. The good news appears to be that population growth will not go on indefinitely. But regardless of what Earth's carrying capacity actually is, it is clear that people will need to become better stewards of Earth's resources and its environment in order to achieve sustainability.

1.6 Understanding Your Role

Human interactions always involve choices—choices about what people do with their waste, where they live, what kinds of cars they drive, and many others. As you consider the environmental consequences of people's choices, you can relate them to issues of personal responsibility, involvement, and action. This book provides many examples of situations where informed individuals or groups of individuals have participated in discussions, helped build consensus around an issue, and shaped decisions within the larger community. You can be an informed citizen who participates regularly in this problem-solving process.

Examples in *Living with Earth* show how people have faced challenges and how their decisions have had both positive and negative consequences. Easter Island is a classic example of how shortsighted choices led to a society's demise. This book's examples are intended to stimulate your interest, develop your critical thinking skills, and allow you to apply the scientific method. As you study these examples, ask yourself the following kinds of questions:

- Why did the situation arise?
- How did it affect people?
- How did people affect the situation?
- How does the situation impact sustainability?
- What would I do in this situation?
- Am I contributing to a similar situation?
- What can I do as an individual or as part of a community of people to improve this situation?

Problems that arise from the interaction of people and Earth systems can also be opportunities, particularly for learning and preparing for the future. The choices illustrated in the book's examples range from personal actions to national policies. The kinds of choices that people have made in the past give you alternatives to consider as you make choices for the future.

Achieving sustainability is a tremendous challenge. Sustainability will not be accomplished unless people view Earth as the isolated space capsule that it is. As Jack Schmidt remarked from 55,000 kilometers away, "I'll tell you, if there ever was a fragile-appearing piece of blue in space, it's Earth right now." Earth is fragile in many ways. Sound stewards of Earth's valuable—and in some ways irreplaceable—resources and environments are needed. People's choices, including yours, will determine Earth's environmental future.

SUMMARY

Earth is like an isolated space capsule, and people have become the key factor influencing its environmental conditions. An Earth systems perspective helps people understand how their actions and choices, as well as natural processes, change Earth.

In This Chapter You Have Learned:

- Earth is a special place, a place that changes as its four global Earth systems—the atmosphere, hydrosphere, geosphere, and biosphere—interact.
- Environmental geology studies people's interactions with Earth. People affect Earth on a local and global scale as they use its atmosphere, hydrosphere, geosphere, and biosphere resources.
- Human population is increasing. Global population is estimated to increase 40% to 9.3 billion people by 2050. Most of these new people will live in cities. As human population increases, resource consumption will increase along with the generation of waste and pollution.

- People's consumption patterns and the technology they use to exploit natural resources and deal with the wastes they generate also have significant environmental consequences. Consumption patterns are likely to be more important than population growth in the future.
- Earth's open and dynamic systems interact by exchanging energy and matter among them. These interactions cause changes such as floods, volcanic eruptions, and species extinctions.
- The changing Earth affects people. Many people live where Earth systems interactions place them at risk. Flanks of volcanoes, river floodplains, and earthquake-prone regions are examples of places where Earth systems processes are hazardous to people.
- The human and economic costs of natural hazards are increasing and borne by everyone. Learning to live more safely with natural hazards is an ongoing challenge for people.
- Science can help solve problems, meet challenges, and increase our understanding of people's interactions with Earth. The scientific method iteratively questions, hypothesizes, predicts, acquires data and observations, and evaluates results to gain understanding. It's an approach that everyone can use to make better choices.
- Biodiversity is a valuable resource that people negatively impact by destroying, fragmenting, or degrading habitat.
- Earth's carrying capacity is the number of people that Earth can sustainably support at a specified standard of living. It is very challenging to estimate, but exceeding Earth's carrying capacity is likely to have significant negative consequences. The people of Easter Island irreversibly depleted valuable resources and exceeded their island's sustainable carrying capacity.
- People's choices will influence how significantly they affect Earth's environmental health. To achieve a sustainable future, one in which today's choices do not jeopardize the needs of people in the future, is a challenging goal. Choices that balance personal, social, economic, and environmental needs are complex and difficult.
- Every person plays a part in determining people's effects on the environment. People's choices, from personal actions to national policies, have had both positive and negative environmental consequences. Achieving a sustainable future, one with a human population approaching 10 billion people, depends on everyone understanding and contributing to the sound stewardship of Earth. Everyone can participate, and everyone has a role.

CHAPTER REVIEW

KEY TERMS

atmosphere (p. 8)
biodiversity (p. 15)
biosphere (p. 8)
carrying capacity (p. 15)
Earth systems (p. 8)
environment (p. 2)
flux (p. 11)
geosphere (p. 8)
hydrosphere (p. 8)
hypothesis (p. 12)
nonrenewable resource (p. 14)
renewable resource (p. 14)
reservoir (p. 10)
residence time (p. 11)
scientific method (p. 12)
sink (p. 11)
steady state (p. 12)
sustainable (p. 2)
system (p. 8)
theory (p. 13)

QUESTIONS FOR REVIEW

Check Your Understanding

1. The term "environment," as commonly used, usually refers to natural features such as rivers, mountains, and forests. How is this term used more broadly in Environmental Geology?
2. The concept of *sustainability* is central to Environmental Geology. What does this term mean in this context?
3. Human population and population growth has been identified as a major factor affecting the availability of resources needed to sustain human life. How does increasing human population influence the sustainability of these resources?
4. How do population, consumption, and technology factors influence people's impacts on the environment? How is the influence each of these factors likely to change in the next hundred years or so? Why?
5. What is the difference between renewable and nonrenewable resources, and how are these concepts related to sustainability?
6. *Systems* are a crucial concept in understanding Earth. What is a system?
7. Earth scientists have identified four distinct Earth systems that interact to create the environment in which we live. What are they? Describe each one.
8. The study of Earth systems is largely concerned with *interactions*. Briefly describe three interactions of Earth systems that are influenced by people.
9. In studying Earth systems, it is often necessary to identify *reservoirs* and *residence times* for materials. What are these, and why are they important to understand?

10. What is the difference between *open* and *closed* systems? Provide an example of each.
11. Earth is largely a closed system with respect to matter. What about with respect to energy? Provide examples with your answer.
12. This book is concerned with the study of how Earth's *dynamic systems* affect people. What does the word "dynamic" mean in this context? Why are Earth systems considered dynamic?
13. The scientific method depends critically on the development of *testable hypotheses*, or testable possible explanations, for observed phenomena. Why are testable explanations so crucial to science?
14. What is a scientific theory, and how are such theories developed?
15. This chapter discusses the issue of freshwater supplies for people on Earth. What are some factors that can make water availability a problem?
16. Biodiversity is a key measure of ecosystem health. What benefits does greater biodiversity typically deliver for any given ecosystem? What is the major threat to biodiversity?
17. Earth systems have an influence on human population through the concept of *carrying capacity*. Why do estimates of carrying capacity vary by a factor of almost 100?
18. The concept of *sustainability* seems simple but it is very challenging to apply, especially on an international scale. Suggest some reasons why this might be the case.
19. Human population is projected to level off over time. Why? What factors other than those mentioned in the text might be relevant to this growth scenario, and how might these projections be in error?
20. The densest populations on Earth reach over 20,000 people per square kilometer (52,000 people per mi^2). This is comparable to the density of people in a sold-out sports stadium, which covers about a square mile of area. Why are some challenges and potential advantages of such population densities?
21. Global human population is a major factor in the environmental sustainability of the planet. Give an example of the influence of large-scale population growth on each of the four Earth systems as discussed in the chapter.
22. How does the example of Easter Island illustrate the transition from sustainable to unsustainable living? Do you think that this example holds any relevance for modern societies? Why or why not?
23. Critics of Federal Disaster Relief often point out that our current policies for aiding devastated communities encourages people to rebuild in hazardous areas. Propose possible alternates to the current policy, and discuss their costs and benefits compared to the current approach.

FIGURE 2-1 ▶ Earth Systems In this photograph, you can see all four major systems of planet Earth: the solid *geosphere*; the *atmosphere* that surrounds us; the water of the *hydrosphere*; and the diverse life of the *biosphere*. Together, these make up a single larger system, much as the human body consists of a series of integrated systems, such as the skeletal, muscular, and cardiovascular systems (inset).

CHAPTER

2

EARTH SYSTEMS

You should be very familiar with at least one system—your body. Do you know what kind of system it is? Energy comes in when you eat food, and energy goes out when you pick up the trash, push the lawn mower, or shoot a basketball. Matter is transferred in and out, too—when you breathe in oxygen from the air, eat food, and flush your body's waste down the toilet. This makes the body an open system, as both matter and energy are transferred in and out. In fact, our bodies are both open and dynamic (changing) systems.

Scientists have found it helpful to subdivide the human body into parts that are also systems. On the scale of the entire human body, we can define a number of systems, including the muscular, respiratory, skeletal, and digestive systems. Systems in the human body interact. If the muscular system uses oxygen to perform a strenuous activity, then the respiratory system breathes harder. If the circulatory system gets dehydrated, then the nervous system tells us we are thirsty. We can't really understand how the human body works without understanding its systems and how they interact. It's the same thing with Earth and its systems.

Let's compare the human body and Earth systems, starting at the scale of the entire human body and the entire Earth. Both systems transfer energy in and out. We just saw that the human body is an open system in terms of energy. Earth receives energy from the Sun and loses energy as it cools very slowly. The body takes in and loses matter. Does Earth receive and lose matter, too? In a way, it does. Tiny amounts of the atmosphere are lost to space, and small particles from space (called *meteoroids* when they strike the atmosphere and *meteorites* if they reach the surface) bombard Earth every day. However, unless a large asteroid that is capable of making a big impact on the surface heads Earth's way, the amount of matter that Earth gains or loses is very small compared to its overall size. | Observation of meteoroid impacts by space-based sensors* Earth is best thought of as a closed system when it comes to matter.

Like the human body, Earth has parts that are systems, too. At the scale of the entire Earth, there are the four systems that were introduced in Chapter 1—the atmosphere, hydrosphere, geosphere, and biosphere (Figure 2-1). And like the systems in the human body, Earth's systems change and interact. We cannot understand Earth without understanding its major systems and their interactions.

*Many of the discussions in this book include highlighted titles for web searches. Enter one of these titles into a search engine such as Google and it will guide you to a specific website where you can find more information about the topic being discussed.

IN THIS CHAPTER, YOU WILL LEARN:

- The origin of the four Earth systems
- The structure and character of the four Earth systems
- The nature of the geologic time scale and the major periods of Earth history
- How the age of geosphere materials is determined
- The rates of some Earth systems processes

2.1 Earth's Geosphere

The geosphere is the solid Earth, together with some molten parts that lie deep within it. Although the solid Earth may seem to be just a large, stable mass, it moves and changes constantly. As we will see in Chapter 3, the geosphere's internal energy and dynamic processes have created mountains and have opened and closed ocean basins many times during Earth's long history. Minerals and rocks that make up the solid Earth are constantly recycled and reformed into new materials (Chapter 4). Movements within the geosphere cause natural hazards that affect people, especially earthquakes (Chapter 5) and volcanic eruptions (Chapter 6). The geosphere also provides the mineral and energy resources (Chapters 12 and 13) that people need. The geosphere is Earth's foundation.

THE GEOSPHERE'S ORIGIN

The origin of the geosphere is the story of the birth of our planet. The story begins in an immense cloud of gas and interstellar debris called a *nebula*, like the one shown in Figure 2-2a. About 4.5 billion years ago, a relatively dense region in the cloud began to collapse under its own gravity, as depicted in Figure 2-2b. The denser and hotter center became the Sun. The dust, rocks, and gases swirling around the Sun coalesced to form the planets of our *solar system*. The rocky planets—Mercury, Venus, Earth, and Mars— formed closer to the Sun, and the large gaseous planets—Jupiter, Saturn, Uranus, and Neptune—formed farther away. The third planet from the Sun is Earth (Figure 2-3).

The aggregation of nebular debris into a planet is a very violent process. Bodies ranging in size from dust to small planets collide to create larger bodies. As the initial Earth grew, it became hotter. Heat was generated by debris collisions, by gravitational compression as the aggregated debris compacted, and by decay of radioactive elements such as uranium, thorium, and potassium. The initial Earth was probably fairly uniform in composition and at least partly molten. The Moon formed after the early Earth had become a stable planet with a regular orbit around the sun.

Many scientists think that the origin of the Moon is tied to the impact of an especially large body, one perhaps about the size of Mars, with Earth. | PSR discoveries: hot idea: origin of the Earth and Moon This giant impact, when the Earth was about half its present size, would have ejected material into Earth's orbit that then collected together to form the Moon. Like Earth, the initial Moon was partly molten as collisions and aggregation continued to occur. But this process of aggregation was largely over within 50 to 100 million years.

Earth's very early surface probably looked like the present Moon's, extensively pockmarked with impact craters (Figure 2-4a on p. 24). A few such scars, produced by relatively recent (and now very rare) impacts, are visible on Earth today. (A famous example is shown in Figure 2-4b). None remain from the original heavy bombardment, however, as Earth's early surface was long ago destroyed by changes in the geosphere.

(a)

(b)

FIGURE 2-2 ▲ The Formation of Earth and Our Solar System
(a) Most stars, like the Sun, are born in clouds of gas and dust called nebulae. **(b)** The solar system was formed along with the Sun when part of a nebular cloud became dense enough to collapse.

FIGURE 2-3 ▼ Location of the Planets Relative to the Sun
Earth is the third planet from the Sun. The four small, rocky planets—Mercury, Venus, Earth, and Mars—are closer to the Sun than are the large gaseous ones—Jupiter, Saturn, Uranus, and Neptune. (Note that the distances are not drawn to scale. If they were, Earth would be almost 11 meters, or 36 ft, from the Sun, and Neptune about 330 meters, or 1080 ft.)

(a)

(b)

FIGURE 2-4 ▲ Hints of Earth's Early Appearance
(a) Impact craters on the far side of the Moon show what Earth's very early surface may have looked like. The Moon's far side, the oldest surface on the Moon, preserves very old impact structures possibly formed between 4 and 4.5 billion years ago. **(b)** Such scars are rare on today's Earth. Most extraterrestrial objects burn up from friction as they pass through the atmosphere, and the craters formed by those that do reach the surface are soon erased by natural processes such as erosion. This crater, 1.2 kilometers (0.75 mi) across, formed in the Arizona desert when an asteroid 20 meters (65 ft) in diameter struck Earth. The impact occurred a mere 50,000 years ago—recently enough for the evidence to still be visible.

THE COMPOSITIONAL STRUCTURE OF THE GEOSPHERE

The biggest change in Earth as aggregation neared its end was the separation of its interior into layers with different compositions. In its early, highly molten state, Earth had a fairly uniform makeup, but this didn't last long. Denser elements like iron and nickel sank to the center of the planet, and less dense elements like silicon, aluminum, oxygen, potassium, and sodium floated to the surface. When this process was complete, Earth had three compositional layers: the deep **core**, the **mantle**, and the thin **crust**, diagrammed in Figure 2-5.

The core

The core has two parts. The solid inner core, 1200 kilometers (745 mi) thick, accounts for 1.7% of Earth's mass and is composed of solid iron and some nickel (an iron-nickel alloy). The liquid outer core, 2250 kilometers (1400 mi) in thickness, makes up almost a third of Earth's mass (30.8%). Both parts of the core are mostly iron, some nickel, and up to 10% of lighter elements, possibly oxygen or sulfur. Geologists have inferred the composition of the core from studies of Earth's density, from the composition of iron-rich meteorites that are thought to be pieces of other planetary cores, and from the way in which *seismic (earthquake) waves* pass or don't pass through it. VisionLearning – Earth structure (You will learn more about seismic waves in Chapter 5.)

The mantle

Most of Earth's interior consists of its mantle (look again at Figure 2-5a). This layer, 2850 kilometers (1770 mi) thick, makes up approximately 67.1% of Earth's total mass. This huge part of the geosphere has some internal variations in composition, but it is mostly magnesium, silicon, oxygen, and iron (in order of decreasing abundance). Estimates of the mantle's composition, like that of the core, draw on studies of Earth's density, data from meteorites, and measurements of how seismic waves pass through it. Information is also provided by pieces of the upper mantle that have been brought to Earth's surface (in volcanic eruptions, for example). You can investigate one way to estimate the mantle's composition at Dynamic Earth – composition of the Earth – calculating mantle composition .

FIGURE 2-5 ◀ Compositional Structure of the Geosphere
(a) The geosphere's internal structure formed early in Earth's history. Its structure, based on composition, includes the inner and outer core, thick mantle, and thin crust. **(b)** The crust under continents is thicker and less dense than the crust under the oceans. Granite is an example of the kind of rock that is typically found in the continental crust. Oceanic crust consists primarily of volcanic rock called basalt.

The crust

The outer thin skin of Earth is its crust (Figure 2.5b). It is only 70 kilometers (43 mi) thick at its deepest point and is composed of rocks that can be examined at Earth's surface. There are two fundamentally different types of crust. The crust that makes up Earth's landmasses and their shallowly-submerged edges is called **continental crust**. In most places, it is about 30 to 40 kilometers (19 to 25 mi) thick, makes up approximately 0.37% of Earth's total mass, and is composed mainly of less dense elements including calcium, sodium, potassium, aluminum, silicon, and oxygen. A rock containing these elements that is typical of continental crust is **granite**. Geologists have determined that the oldest portions of continental crust formed more than 4 billion years ago.

The second type of crust underlies the oceans, and is therefore called **oceanic crust**. Oceanic crust is thin, iron-rich, and young compared to continental crust. It is commonly only 5 to 8 kilometers (3 to 5 mi) thick, and contains primarily iron, magnesium, calcium, silicon, and oxygen. It makes up only 0.01% of Earth's total mass, but it covers 61% of Earth's surface. The oldest known portions of oceanic crust are only about 180 million years old. Oceanic crust consists primarily of volcanic rock called **basalt**. Basalt is formed from partially melted mantle that migrates to Earth's surface at seafloor volcanoes. It is the most abundant volcanic rock on Earth, and it is denser, on average, than rocks of the continental crust because it contains more iron and magnesium.

Early Earth's crust was probably similar to oceanic crust in composition and, because the planet was still being bombarded and heated by aggregating nebular debris, highly molten. It wasn't until Earth had cooled enough for the initial crust to solidify that formation of the continental crust was possible. Chapter 3 explains how oceanic and continental crust formed, and Chapter 4 provides more details about their compositions.

THE PHYSICAL STRUCTURE OF THE GEOSPHERE

The compositional structure of the geosphere is one way to describe its internal character. Another way is to identify how physical properties, especially density, change within it. The *density* of a material, its mass per unit volume, can be measured with a balance scale or estimated from the velocity of seismic waves (earthquake vibrations; you will learn more about these in Chapter 5) that pass through it, since seismic waves travel faster through denser material. Changes in the velocity of seismic waves as they pass through Earth are used to define the internal physical structure of the geosphere. As shown in Figure 2-6, abrupt changes in seismic wave velocities mark the boundaries between the various regions of the core, mantle, and crust. In addition to the crust, these regions are the *inner core, outer core, lower mantle, mantle transition zone, upper mantle, asthenosphere*, and *lithosphere*.

The inner and outer core

The inner core is solid and the outer core is liquid. The very dense inner core is solid in spite of extremely high temperatures because of the very high pressures at Earth's center. The outer core is liquid because pressures are slightly lower in this region, but the temperatures are still high enough to be above the melting point of iron. We know the outer core is liquid because it drastically slows down fast seismic waves (see Figure 2-6) and other seismic waves cannot pass through it at all.

The lower mantle

The *lower mantle*, from 2900 kilometers (1800 mi) to 660 kilometers (410 mi) in depth, is solid rock that gradually increases in density (and seismic wave velocity) downward toward the boundary with the molten material of the outer core. Although upper and lower mantle rocks are similar in composition, the higher pressures in the lower mantle are thought to make the minerals there different from (denser than) those in the upper mantle. *Minerals*, naturally occurring solids with an orderly arrangement of atoms and a distinct chemical composition, are discussed in more detail in Chapter 4.

The mantle transition zone

The mantle *transition zone* is between 660 kilometers (410 mi) and 410 kilometers (250 mi) deep. Although it is sometimes considered part of the upper mantle, the transition zone is well defined by changes in seismic wave velocities at its boundaries, and so can be readily identified as a separate mantle layer. Within the transition zone, the density of many minerals gradually increases with depth as their atomic structures change to forms that are more stable at higher pressures. The transition zone is the part of the mantle where rocks like those in the upper mantle are being transformed into rocks like those in the lower mantle.

The upper mantle

The *upper mantle* lies above the transition zone and extends upward to the base of the crust. The boundary between the crust and upper mantle, as defined by an abrupt increase in seismic wave velocities as the waves pass downward through it, was one of the first internal geosphere boundaries to be recognized. Andrija Mohorovičić, a Croatian seismologist, first identified this boundary in 1909, and it has come to be known as the **Mohorovičić discontinuity** or just **Moho**, in his honor (see Figure 2-6).

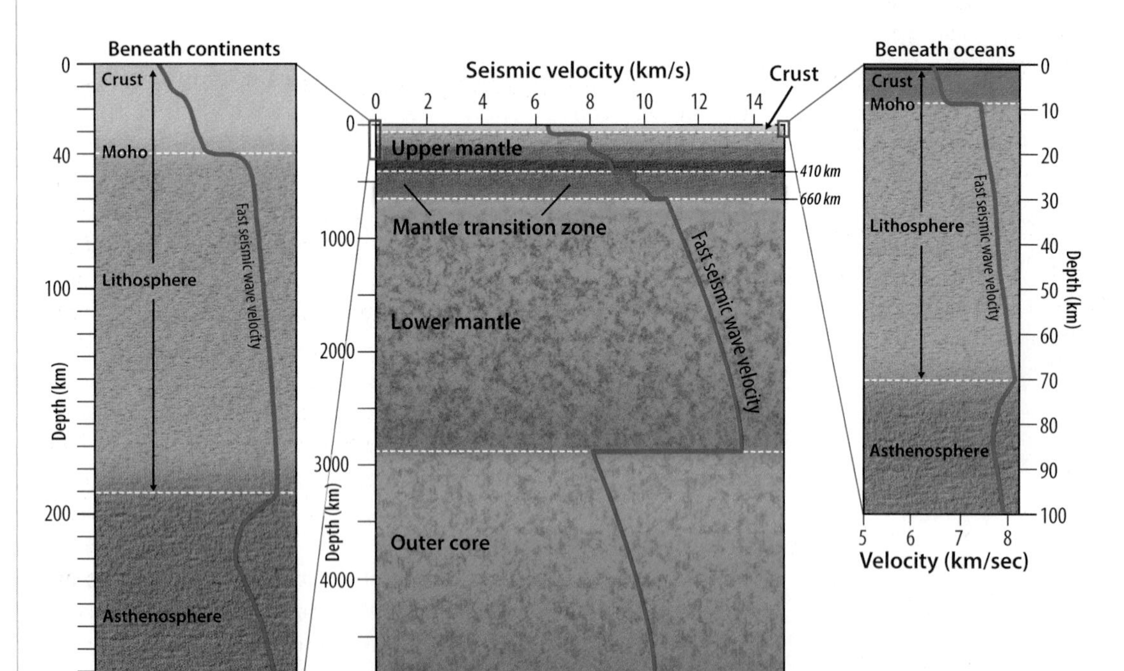

FIGURE 2-6 ◀ **The Physical Structure of the Geosphere**
Abrupt changes (discontinuities) in the velocity of seismic waves as they travel through the geosphere mark the boundaries between its physical subdivisions.

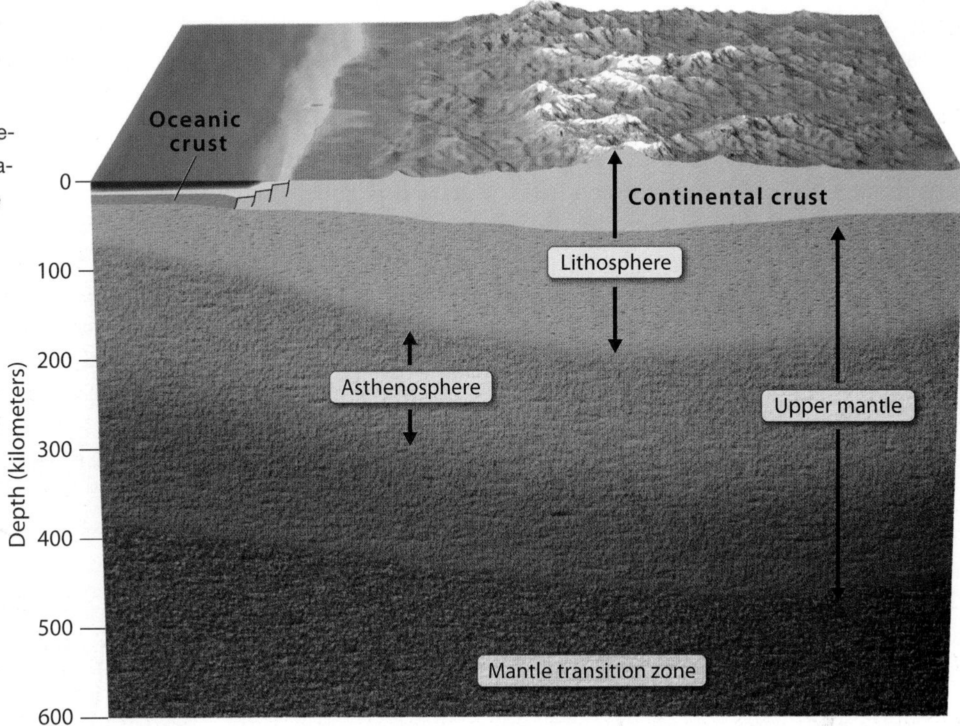

FIGURE 2-7 ▶ The Asthenosphere and Lithosphere
The asthenosphere is the part of the upper mantle where rocks are near their melting point and thus weak enough to flow under pressure. The top of the asthenosphere is well defined by seismic velocity data but its base appears to be gradational downward into rigid rocks of the upper mantle. The lithosphere consists of the upper part of the upper mantle and the crust, where rocks are strong and rigid.

The changes in seismic wave velocity at the top of the transition zone (410 kilometers deep) and at the Moho mark the upper mantle's boundaries. However, as you can see in Figure 2-6, there are other places where seismic wave velocities change significantly, even decrease, within the upper mantle. These other changes in seismic wave velocities suggest a different way of dividing the material of the upper mantle and crust: into layers called the **asthenosphere** and the **lithosphere**, diagrammed in Figure 2-7.

The asthenosphere

The asthenosphere is a distinctive part of the upper mantle directly below the lithosphere (see Figure 2-7). It is commonly present at depths between 250 and 72 kilometers (155 and 45 mi) but varies in thickness—its upper boundary can be as shallow as a few tens of kilometers under the oceans and as deep as 300 kilometers (190 mi) below the continents. The asthenosphere is mostly solid, but pliable (like putty) and can flow under pressure. This makes it weaker than the geosphere layers above and below it. The weakness develops because rocks are very close to their melting point within the asthenosphere. In fact, some parts of the asthenosphere may be molten. These physical conditions slow seismic waves. Therefore, the asthenosphere has also been called the upper mantle's *low velocity zone*. Because of its relative weakness, the asthenosphere is the part of the mantle where significant movements of material occur.

The lithosphere

The **lithosphere** is the shallowest physical layer in the geosphere. It is present from Earth's surface down to depths of a few tens of kilometers beneath the oceans and as much as 300 kilometers (190 mi) beneath the continents. The lithosphere is distinctive because it is made up of strong, rigid rocks that can break when they move. It includes rigid rocks of the oceanic crust, the continental crust, and the top portion of the underlying upper mantle (see Figure 2-7). Even though there are significant compositional differences between the three components of the lithosphere, they combine to define one physical layer.

The geosphere's internal physical structure is the key to understanding how the solid Earth moves and changes. The dynamic geosphere and the movements of material and energy within it influence all major surface features on Earth. These movements control earthquakes, volcanoes, and the distribution and character of oceans and continents. You will learn how the geosphere's physical layers, especially the asthenosphere and lithosphere, are involved in these dynamic processes in Chapter 3.

2.2 Earth's Atmosphere

The gases that surround Earth, what we commonly call air, comprise the atmosphere. The atmosphere extends from Earth's surface upward for many hundreds of kilometers. In a way, the atmosphere doesn't have a distinct top; it just gradually becomes less dense (has fewer gas molecules in a given volume) and merges with empty space. It also includes gases that fill voids in the near-surface soil and rock of the geosphere.

The atmosphere is arguably Earth's most dynamic system. Clouds carried along in the atmosphere move and change before our eyes. Movements of the atmosphere (winds) can be used to generate useful energy (Chapter 13)—but they can also be so fast, in hurricanes or tornados, that they uproot large trees and destroy strong buildings. The composition of the atmosphere is what keeps Earth's surface temperature within livable ranges for the biosphere. It supplies the oxygen we breathe and the carbon dioxide that plants need for survival.

THE ATMOSPHERE'S ORIGIN

Gases were part of the nebular cloud that condensed to form our solar system. Gases that are common in nebular clouds, and in the outer gaseous planets are hydrogen (H), helium (He), methane (CH_4), and ammonia (NH_3). Gas giants What some call Earth's *first atmosphere* probably included these gases. However, when early Earth was still very hot, perhaps before a liquid outer core (and the magnetic field it creates) was fully developed and before complete aggregation, gravity could not hold these gases closely around Earth. The first atmosphere is thought to have been lost to space during the final stages of Earth's formation and before cooling of the primordial crust.

The second atmosphere

Elements and compounds that vaporize easily, called *volatiles*, were also contained in the nebular material that formed the initial solid Earth. These components include hydrogen (H), carbon (C), and nitrogen (N) as well as compounds formed from these elements such as water (H_2O), carbon dioxide (CO_2), and ammonia (NH_3). These elements and compounds are characteristic of primitive meteorites called *carbonaceous chondrites* that are considered to be representative of the composition of the nebular cloud that condensed to form our solar system. (An interesting story about studying meteorites and their relation to Earth's origin is at Philip Bland – meteor man.) So, material in the early solid Earth contained volatile components that, if vaporized, could contribute to the formation of Earth's atmosphere.

Volcanoes were common on Earth's surface as it cooled and the initial crust solidified. Volcanism is a key way that matter, including volatile components, is transferred from within the geosphere to Earth's surface and the atmosphere. The volatiles included in molten rock that erupts from volcanoes are released into the atmosphere as the molten material solidifies. (Molten rock is called **magma** within the geosphere and **lava** on the surface.) The volcanic processes that transfer volatile components from the geosphere to the atmosphere are called *outgassing*.

We can estimate what early volcanism released to the atmosphere by studying the outgassing that is taking place at volcanoes like Italy's Mt. Etna, shown in Figure 2-8. Present-day volcanic eruptions release water vapor (H_2O), carbon dioxide (CO_2), nitrogen, and sulfur compounds (especially sulfur dioxide, SO_2) to the atmosphere. Gas compositions and tectonic setting For this reason, the atmosphere created by outgassing in Earth's early history is thought to have contained mostly water vapor and carbon dioxide along with nitrogen, sulfur compounds, some hydrogen, and compounds formed from the reactions of these gases, such as ammonia and methane. The atmosphere created by volcanic outgassing in the first billion years of our planet's history has been called Earth's *second atmosphere*.

As you will learn in Chapter 14, carbon dioxide, methane, and water vapor are gases that cause the atmosphere's temperature to rise. They are called **greenhouse gases**. The climate during the time of the second atmosphere was very warm. There were no polar ice accumulations, no oceans of liquid water, and oxygen, so critical to life as we know it today, was not present as free O_2 molecules.

FIGURE 2-8 ▲ Outgassing
Studying the release of gases by modern volcanoes, such as Mt. Etna in Sicily, shown here, helps us to understand how Earth's early atmosphere was formed.

The third atmosphere

Today's atmosphere (Table 2-1) is called Earth's *third atmosphere*. It took big changes to Earth's second atmosphere to make the third atmosphere. A lot of water and carbon dioxide had to be removed and a lot of nitrogen and oxygen added to evolve into the third atmosphere. The excess water went to the hydrosphere—mostly the oceans, but also as rivers, lakes, and streams. By 3.5 billion years ago, Earth's oceans had formed from water that precipitated from the second atmosphere as the geosphere began to cool.

Remember the carbon cycle shown in Figure 1-10? Once the world ocean existed, carbon dioxide from the atmosphere began to dissolve in its waters. Dissolved carbon dioxide then reacted with calcium to form solid *calcium carbonate*, which started to precipitate and accumulate on the seafloor. (You will

TABLE 2-1 COMPOSITION OF EARTH'S PRESENT ATMOSPHERE (ALSO KNOWN AS EARTH'S THIRD ATMOSPHERE)

Gas Name	Chemical Formula	Percent Volume
Nitrogen	N_2	78.08%
Oxygen	O_2	20.95%
Water*	H_2O	0 to 4%
Argon	Ar	0.93%
Carbon dioxide*	CO_2	0.0360%
Neon	Ne	0.0018%
Helium	He	0.0005%
Methane*	CH_4	0.00017%
Hydrogen	H_2	0.00005%
Nitrous oxide*	N_2O	0.00003%
Ozone*	O_3	0.000004%

*variable gases

learn more about calcium carbonate in Chapter 4.) Because many organisms incorporate calcium carbonate into their shells and skeletons (Figure 2-9a), which then accumulate on the seafloor (Figure 2-9b), the rate of precipitation increased once living plants and animals had evolved in the oceans. Most seafloor calcium carbonate accumulations gradually turn into rocks and become part of the geosphere (Figure 2-9c). (Limestone is the most common calcium carbonate rock.) The geosphere is the world's largest carbon reservoir and a global sink for carbon dioxide. The carbon dioxide concentration of the atmosphere slowly decreased as carbon dioxide was transferred through the oceans to the geosphere.

Nitrogen is the most abundant element in Earth's third atmosphere. Nitrogen is not a very reactive element, so as long as it doesn't escape into space, most of it just stays in the atmosphere as nitrogen gas, N_2. Over time, outgassing slowly

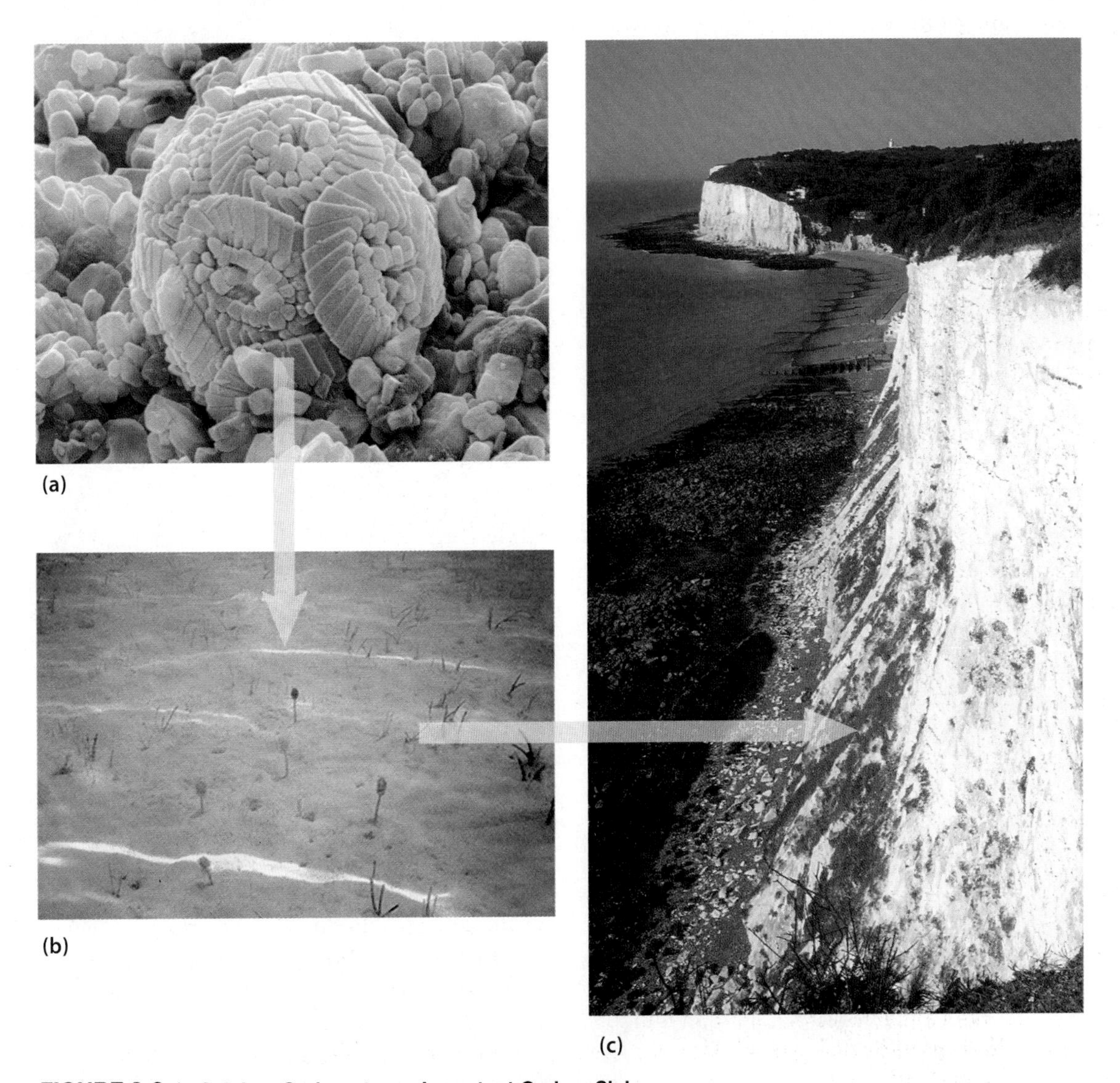

FIGURE 2-9 ▲ Calcium Carbonate, an Important Carbon Sink
Many organisms incorporate calcium carbonate into their shells and skeletons **(a)**, which then accumulate on the ocean floor **(b)**. During Earth's history, much of this material has been compressed into rock, like the famous white limestone cliffs that overlook the English Channel **(c)**, thus becoming a major geosphere sink for carbon.

increased the amount of nitrogen to its present high levels (see Table 2-1). But even though the concentration of nitrogen increased during Earth's early history, there was still no free oxygen. The origin of the atmosphere's free oxygen is tied to changes in the biosphere.

Among the earliest forms of life on Earth were microorganisms called *cyanobacteria* that are *photosynthetic*. Photosynthetic organisms use sunlight to convert carbon dioxide and water into food and oxygen, a process called **photosynthesis** (Figure 2-10a). Cyanobacteria started making oxygen about 3.5 billion years ago, but oxygen didn't increase in Earth's atmosphere for another billion years. What was going on? The answer is in the rocks. The newly formed oxygen reacted with abundant iron and sulfur that was on Earth's surface and dissolved into the early ocean. These chemical reactions formed solid minerals, especially iron oxides that accumulated in sediments to become rocks known as *banded iron formations*, shown in Figure 2-10b. Banded iron formations are sinks where Earth's early oxygen is stored today. The residence time for oxygen in banded iron formations is measured in billions of years.

Finally, about 2 billion years ago (about 1.5 billion years after cyanobacteria started generating oxygen), most of the material that readily reacted with the atmosphere's early oxygen was used up. Then the oxygen concentration of the atmosphere began to slowly increase. Photosynthetic organisms increased in abundance and by about 500 million years ago, when plants first appeared on the continents, the atmosphere's oxygen level had become approximately what it is today. Almost all animals must breathe in oxygen to survive, so these oxygen levels are thought to have caused a tremendous expansion of the biosphere's diversity at this time. | Origin of "breathable" atmosphere half a billion years ago | Since then, the atmosphere's composition has varied in small but important ways, especially in its amounts of greenhouse gases (Chapter 14).

What are some Earth systems interactions that contributed to making Earth's present atmosphere?*

*Possible answers to Earth Systems Interactions and Sustainability questions can be found at the end of the chapter.

THE COMPOSITIONAL STRUCTURE OF THE ATMOSPHERE

From its outer limits, where solar radiation first begins to affect the gases held by Earth's gravity, the atmosphere's overall thickness is typically 480 kilometers (300 mi) or more. However, because the atmosphere's gases are compressible, gravity pulls most of the atmosphere's mass (most of its gas molecules) down to low levels near Earth's surface. Ninety percent of the

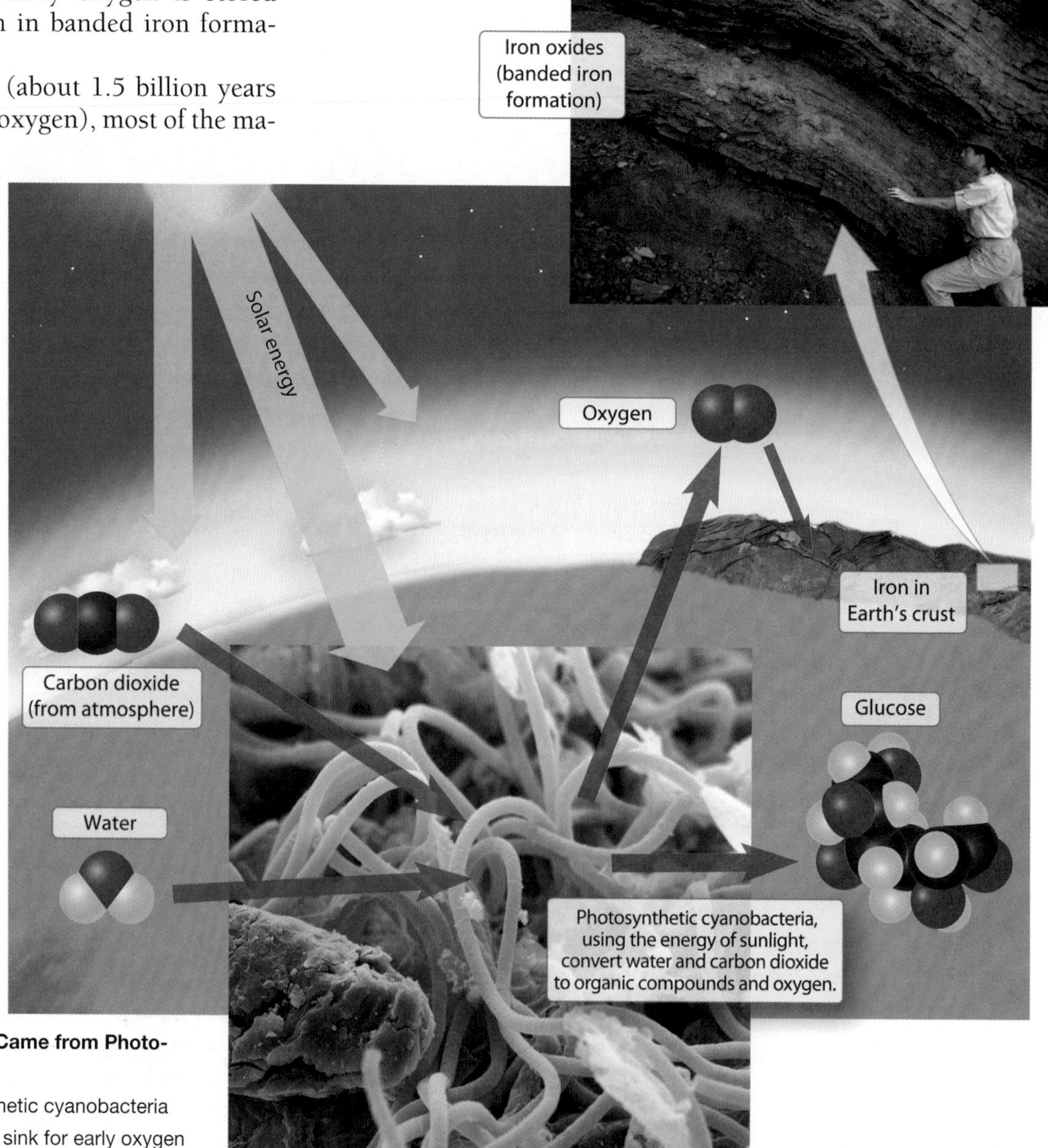

FIGURE 2-10 ▶ The Atmosphere's Oxygen Came from Photosynthetic Organisms
Earth's first free oxygen, produced by photosynthetic cyanobacteria **(a)**, was trapped in banded iron formations **(b)**, a sink for early oxygen in the geosphere.

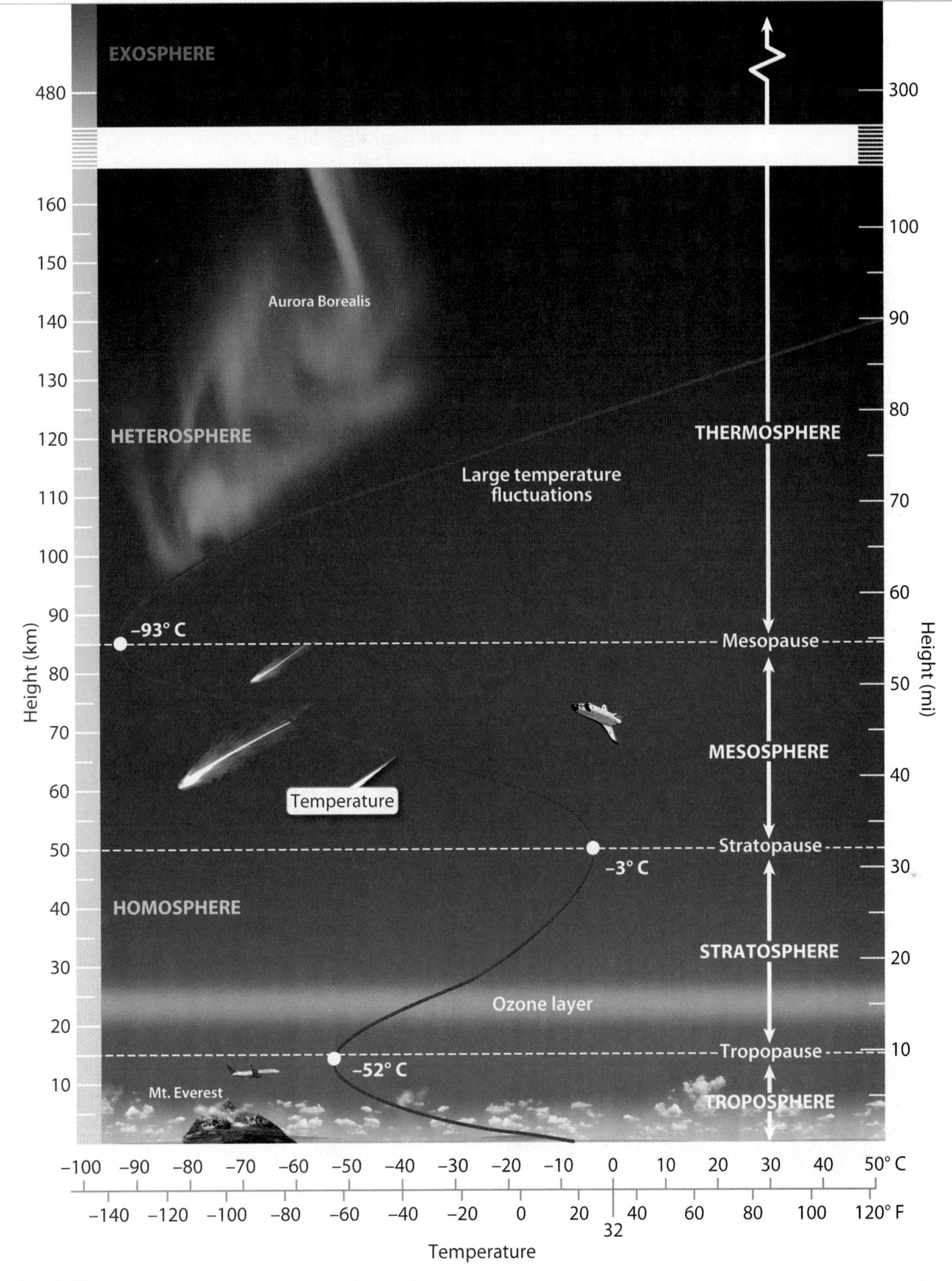

FIGURE 2-11 ▲ The Structure of the Atmosphere

Variations in composition and temperature define the atmosphere's structure. Note that the heterosphere, from 80 km to about 480 km, is not shown in full. The region above the heterosphere, where Earth's atmosphere merges with outer space, is sometimes called the exosphere.

atmosphere's mass is below 16 kilometers (10 mi), and all but 0.1% is below 50 kilometers (31 mi). Even so, the atmosphere has two layers that are defined by compositional variations.

The homosphere

Although the total amounts of the major gases (nitrogen, oxygen, and argon—see Table 2-1) decrease drastically from Earth's surface up to 85 kilometers (53 mi), these gases have the same proportions throughout this layer. This compositionally homogenous part of the atmosphere—the lower portion of Figure 2-11—is called the *homosphere*. Everywhere within the homosphere, the major gases are present in the proportions shown in Table 2-1. The concentrations of some minor but important constituents, such as water, carbon dioxide, and ozone, vary significantly within the homosphere, but its major constituents are well mixed.

The heterosphere

Above the homosphere, from about 85 kilometers (53 mi) to 480 kilometers (300 mi), the atmosphere actually contains very few gas molecules. Gravity causes those that are present to be segregated according to their mass. Gas molecules with the greater mass, like nitrogen (N_2), are concentrated lower, and atomic oxygen (O), helium (He), and hydrogen (H) are concentrated higher. The heterogeneous compositional character of this upper part of the atmosphere is why it is called the *heterosphere*.

THE TEMPERATURE STRUCTURE OF THE ATMOSPHERE

Interactions of solar radiation with Earth's surface and with the atmosphere produce temperature variations that define four atmospheric layers. From the surface upward, they are the *troposphere*, *stratosphere*, *mesosphere*, and *thermosphere*, shown in Figure 2-11.

The troposphere

The lowest atmospheric layer defined by temperature variations is the **troposphere**. Temperatures decrease from Earth's surface upward to the top of the troposphere, where temperatures become constant in a boundary zone called the *tropopause*. The troposphere's thickness varies from about 7 kilometers (4 mi) near the poles to as much as 17 kilometers (11 mi) near the equator. The temperature changes in the troposphere are related to the heating of Earth's surface by solar radiation. Sunlight that hits Earth's surface is radiated back to the atmosphere as heat energy that warms its lowest levels. This warm air is less dense than surrounding air and rises, causing currents and turbulence that mix the lower part of the homosphere. The rising air also expands and cools as it moves upward, causing the temperature to vary from warm at Earth's surface to cool at the tropopause.

The troposphere is where most of the action takes place within the atmosphere. It is where life lives, weather happens, and most of the clouds appear. About half of the mass of the entire atmosphere is in the lower 5 kilometers (3 mi) of the troposphere.

FIGURE 2-12 ▼ The View from the Stratosphere
This photo is taken from the cloudless stratosphere, looking down on the clouds in the troposphere below.

The stratosphere

The **stratosphere** is the layer above the troposphere. Temperatures increase upward through the stratosphere. The higher temperatures of this layer compared to those in the underlying troposphere prevent air from rising and crossing their boundary zone, the tropopause. The top of the stratosphere is marked by a temperature decrease. This boundary, called the *stratopause*, is at an altitude of about 50 kilometers (30 mi) above Earth's surface.

Temperature changes through the stratosphere are caused by the interaction of incoming solar radiation and oxygen. Common oxygen molecules (O_2) absorb short-wavelength ultraviolet radiation in the upper stratosphere and split apart into highly reactive oxygen atoms (O). These single oxygen atoms then bond with O_2 molecules to form **ozone** (O_3), oxygen molecules that contain three oxygen atoms. Ozone forms slowly in the stratosphere and can be destroyed by reactions with sunlight and other atmosphere components. (Ozone is discussed in more detail in Chapter 14.) Even though present in small amounts, ozone is concentrated in the lower stratosphere where it is not destroyed as rapidly. This part of the stratosphere is commonly called the **ozone layer**.

Ozone is a good absorber of longer-wavelength ultraviolet radiation from the Sun. (This helps people by preventing ultraviolet radiation, which can cause skin cancer, from being too intense on Earth's surface.) The formation of ozone and ozone's absorption of longer-wavelength ultraviolet radiation combine to control temperatures in the stratosphere.

The upward increase in temperature through the stratosphere prevents air from rising within it and causes it to be internally stratified or layered. The small amount of water vapor and stratified character of the stratosphere make it a generally cloudless part of the atmosphere where winds are usually parallel to Earth's surface. If you get a chance to fly across the country, look out the window. You may be close to the stratosphere, and the clouds far below will be in the troposphere (Figure 2-12).

The mesosphere

The layer above the stratosphere, from about 50 to 85 kilometers (30 to 53 mi) in altitude, is the **mesosphere**. Temperatures decrease upward through the mesosphere to a zone where the lowest temperatures in the atmosphere are present. This upper boundary zone of the mesosphere is called the *mesopause*.

The decreasing temperatures through the mesosphere reflect the low (and decreasing upward) concentration of gas molecules that absorb ultraviolet radiation. It is also related to the presence of very small amounts of carbon dioxide. This carbon dioxide absorbs and reradiates solar energy. However, because there are so few other gas molecules around to absorb this energy, it is lost to space. This cools the mesosphere, just the opposite of the effect carbon dioxide has in the troposphere. There, carbon dioxide acts as a greenhouse gas by trapping heat energy from Earth's surface and transferring it to other gas molecules in the atmosphere.

Although the concentration of gas molecules in the mesosphere is low, there are still enough of them to have some significant effects. Friction with gas molecules causes meteoroids to heat up as they pass through the mesosphere, creating the bright, fast-moving "shooting stars" (meteors) visible in clear night skies. Most of the meteoroids headed toward Earth burn up in the mesosphere. It is in this region, too, that the space shuttle starts flying upon reentry to the atmosphere.

The thermosphere

The **thermosphere** is the outer atmospheric layer, above the mesosphere. Temperatures increase and air molecules become fewer and fewer upward through the thermosphere as it gradually merges into space at altitudes of 480 kilometers (300 mi) or higher.

The thermosphere is essentially the same layer as the compositionally defined heterosphere. The temperature increase upward through the thermosphere is due to interaction of intense solar radiation with the increasingly sparse gas molecules and atoms (mostly nitrogen and oxygen). Because there are fewer molecules to absorb the radiation, there is more of it for each individual molecule to absorb. The gas atoms thus acquire, on average, more thermal energy, and their temperature rises.

The intensity of radiation emissions by the Sun can also vary and influence thermosphere temperatures. During times of strong solar radiation, temperatures as high as 2500° C (4532° F) can be reached in the thermosphere. Remember, though, that because there are so few molecules in the thermosphere, there is very little heat energy available to be transferred between objects. It wouldn't even feel warm up there.

Where solar radiation is intense, it can change gas molecules into charged particles called **ions**. Because a lot of the ionization that takes place in the atmosphere happens in the thermosphere, it is also called the **ionosphere**. Collisions of charged particles in the thermosphere make it home to the wonderful northern lights (aurora borealis) and southern lights (aurora australis)—the moving sheets and wisps of colored light visible at high latitudes on clear nights (Figure 2-13).

Sometimes it is helpful to recognize where the atmosphere merges with space as a separate layer, the *exosphere*. Any gas molecules here are as likely to head off into space (and exit the atmosphere) as not. Many satellites orbit Earth within the exosphere.

The atmosphere's structure, whether defined by composition or temperature, helps us understand how Earth receives energy from the Sun. Solar energy, interacting with Earth through the atmosphere, drives many Earth-system processes, some directly and some indirectly.

FIGURE 2-13 ▼ Northern Lights on a Clear Winter Night
Ions, formed by the collision of fast-moving charged particles ejected from the Sun with atoms of gas in the atmosphere, create the polar auroras.

2.3 Earth's Hydrosphere

Another area where there are direct connections between solar energy and Earth-system processes is located within Earth's hydrosphere. The hydrosphere consists of all the water in the oceans, on land in streams and lakes, in glaciers and other accumulations of ice, in the atmosphere and underground. Water covers 71% of Earth's surface, all the area that appears as blue on the "blue marble" that is Earth (Figure 1-1). Earth's distance from the Sun gives it a temperature range that allows water to exist here in three different phases: liquid, gas, and solid. And Earth is big enough that its gravity keeps water vapor from escaping the atmosphere and moving into space. Without the special planetary conditions that allow abundant water to be present in all its phases, life as we know it couldn't exist on Earth.

Movement of water is a key factor in Earth-system processes. As you will see, water is a major factor in wearing down and eroding the land (Chapter 4), soil development (Chapter 11), and the transfer of energy and matter among Earth systems. In fact, another often-used name for Earth is "the water planet."

ORIGIN OF EARTH'S WATER

Where did all of Earth's water come from? Water was a part of the original nebular debris that coalesced and combined to form Earth. Volcanic outgassing released volatile elements and compounds, including abundant water vapor, during formation of Earth's second atmosphere. Once the planet had cooled significantly, water vapor in the second atmosphere condensed and was able to remain liquid on Earth's surface. The oldest known rocks that formed from ocean sediments are 3.8 billion years old. These rocks confirm some compositional characteristics of the second atmosphere and tell us that oceans were present at this time. World's oldest sedimentary rocks show how Earth may have avoided becoming a giant snowball Once the oceans formed, the hydrosphere was in place.

Recently, scientists have begun to think that some of Earth's water may have arrived with comets and other water-rich space debris that collided with Earth early in its history. The role that comets may have played in developing the initial Earth is now the subject of much new research. One example is Deep Impact, an amazing space experiment and the focus of the *In the News* feature.

In The News

Deep Impact—Investigating Water's Origin

Jessica Sunshine seems like the perfect name for a scientist studying the solar system. She is a scientist on the team that hit a bull's-eye with one of the longest space shots in history. Solar system exploration: deep impact legacy site The Deep Impact space probe struck the comet Tempel 1 on July 4, 2005. NASA – deep impact Why would scientists want to drive a spacecraft into a comet? *Comets* are made up of debris left over from the birth of our solar system. Their location and composition can tell us about how materials were distributed and eventually aggregated to form the planets. They might even tell us where much of the water in Earth's hydrosphere came from.

Comets are icy masses that are abundant in the outer regions of the solar system. In addition to water in the form of ice, some comets in the very cold outer parts of the solar system also contain solid hydrocarbon compounds such as methane (CH_4), ethane (C_2H_6), and acetylene (C_2H_2)—potential building blocks of amino acids and other *organic* (carbon-bearing) compounds that make up living organisms. Understanding the composition of comets can help us determine how Earth's early atmosphere and hydrosphere originated and what they may have been like.

The Deep Impact team specifically wanted to know what the interior of a comet was like—not just its composition but its structure, too. The team collected data before and after the collision. The data included images of the comet that showed brightness variations on its surface, its temperature, and its surface features (topography). After the collision, the team determined the character of the material that was ejected from within the comet. Their initial results showed that there was water ice on the surface of the comet but not enough to be the source of all the water in the comet's surrounding cloud of gas and dust—its *coma*. The scientists were able to conclude that there must be much more water deep within Tempel 1.

The study of comets gets more and more interesting every day. Projects like Deep Impact, an amazing engineering and scientific feat, help us understand comets better. Scientists exploring the far reaches of the solar system with both land- and space-based telescopes find more and more comets as they continue to look. Stay tuned, more comet studies are likely to be in the news.

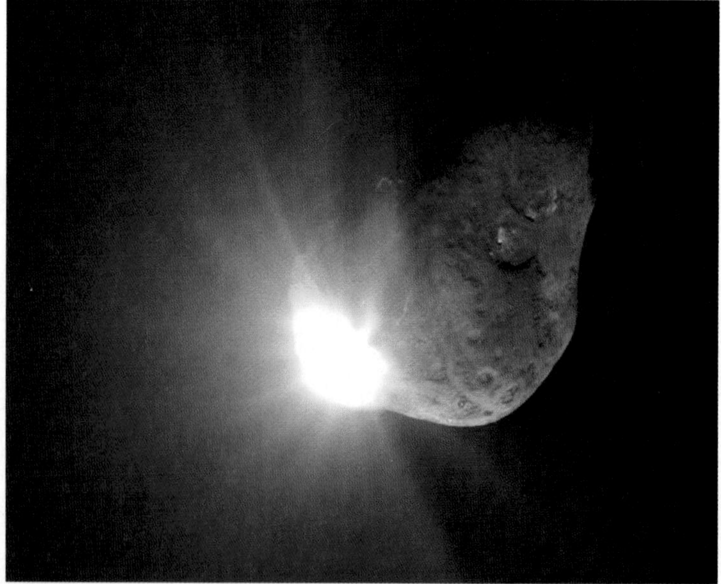

Because comets like Tempel 1 (upper photo) are debris left over from the birth of our solar system, understanding their composition can shed light on the origin of Earth's early atmosphere and hydrosphere. In 2005, the Deep Impact space probe struck Tempel 1, allowing scientists to determine the amount of ice on its surface and in its interior. The lower photo was taken shortly after impact.

Reservoirs in the hydrosphere

Overall, the hydrosphere doesn't have an internal structure like those of the geosphere or atmosphere, but it does have distinctive reservoirs, as shown in Figure 2-14. The world ocean is by far the largest reservoir in the hydrosphere. At any given moment, more than 97% of Earth's water molecules are in the ocean. Only 2.8% of Earth's water is fresh (non-salty), and only a small amount of that water is readily available for people to use. Freshwater is widely distributed in lakes, streams, rivers, underground, and in the atmosphere. People use lots of water from these reservoirs, but they are small compared to Earth's accumulations of solid freshwater in glaciers, ice caps, and ice sheets. Ice makes up 68.6% of Earth's freshwater. (Chapter 10 covers freshwater reservoirs and water use in more detail.)

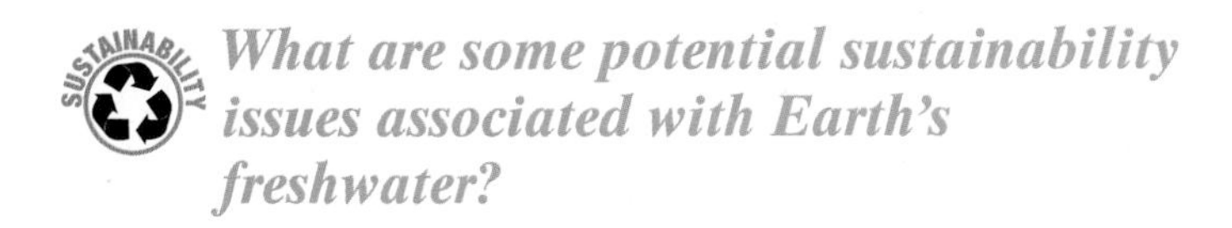

What are some potential sustainability issues associated with Earth's freshwater?

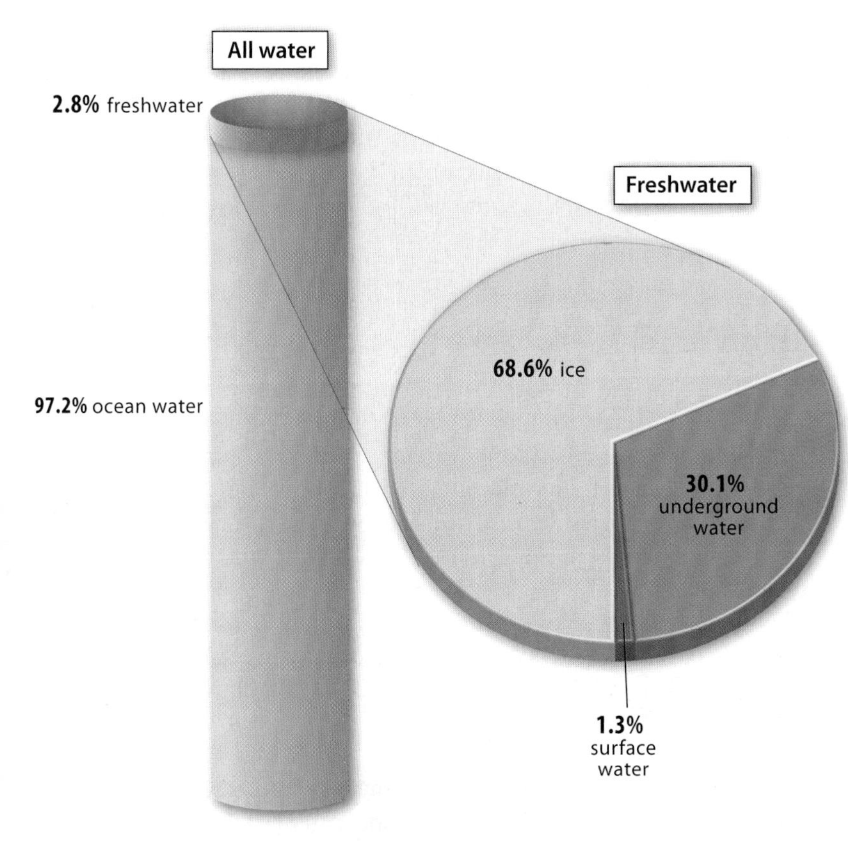

FIGURE 2-14 ▲ The Hydrosphere's Reservoirs

More than 97% of the hydrosphere consists of salt water in the oceans. Of the fresh (non-salty) water, almost 69% is frozen in glaciers, ice caps, and ice sheets, and 30% is underground. The rest—only 1.3%—can be found in many small reservoirs such as streams, rivers, and lakes. This fresh, liquid surface water thus makes up only about 0.036% of all the water on Earth.

The world ocean

The world ocean contains 97.2% of all the water in the hydrosphere. This huge reservoir is divided into different geographical parts (Pacific and Atlantic Oceans, for example), but it is a continuous body of water that covers 71% of Earth's surface. It is over 3 kilometers (2 mi) deep in over half this area. The world ocean is so large that it does have internal structure, definable by variations in characteristics like salinity and temperature. However, as one big reservoir, it essentially has two parts.

There is an upper layer, commonly about 200 meters (660 ft) thick, that is warmed by solar radiation and mixed by the waves and currents created by winds blowing across the ocean's surface. At depths below about 1000 meters (3000 ft), solar radiation has little effect, and water temperatures are low, commonly in the 0° to 4° C (32° to 39° F) range. The water can still be in motion at these depths because salinity and temperature differences change the water's density. Denser (colder and saltier) water sinks and slowly flows through the deep ocean and back to the surface in a global circulation pattern shown in Figure 2-15 on p. 36. Water molecules can reside in the deep ocean for a few thousand years.

The world ocean is a major influence on global climate. Liquid water's *specific heat capacity*, the amount of heat energy it takes to raise the temperature of one gram of water by one degree centigrade, is high compared to that of other substances. This is why it takes a lot of solar radiation to appreciably warm the world ocean, but once it's warm, it can retain its heat energy for a long time. It becomes a major energy reservoir. It is energy from this reservoir that drives important parts of the water cycle.

Glaciers, ice caps, and ice sheets

Wherever winter snow doesn't melt completely in the summer, **glaciers** can form. As snow accumulates in these places, it becomes thicker, compresses, and becomes ice in glaciers. The seasonal conditions that enable glaciers to form exist at high elevations in mountains and at high latitudes near the north and south poles. Mountain glaciers vary from small patches to large rivers of ice that slowly flow downslope. Ice movement carves deep valleys and creates many landforms that are distinctive of glacial regions, several of which can be seen in Figure 2-16 on p. 36.

Where glaciers coalesce and cover larger areas they become *ice caps* (less than 50,000 square kilometers, or 19,000 mi^2) and **ice sheets** (greater than square 50,000 square kilometers). Today, ice covers about 10% of Earth's land area. Most of this ice is contained within Earth's two giant ice sheets, one centered on Greenland and one on Antarctica.

The Greenland ice sheet is a remnant of a much larger ice sheet in the northern hemisphere that first developed about 3 million years ago. The most recent glacial advance covered very large areas with ice that gradually melted back to the present Greenland ice sheet between 20,000 and 6,000 years ago. The

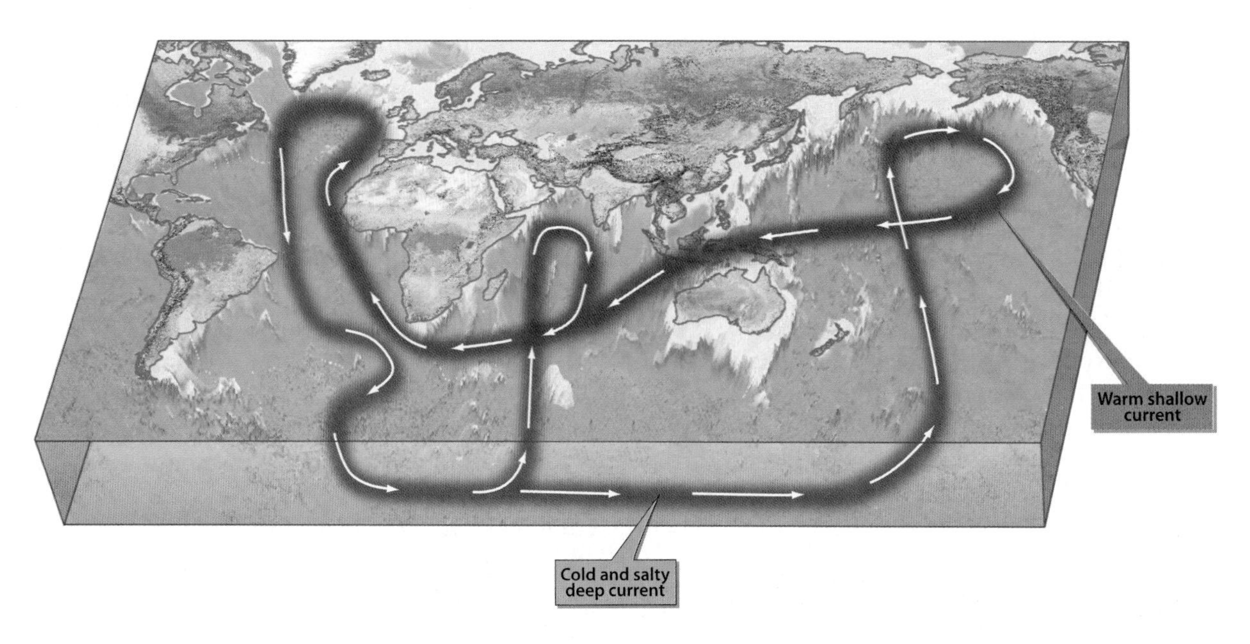

FIGURE 2-15 ▲ Movement of Water in the World Ocean
Currents, driven by winds and density differences, move water in the hydrosphere's largest reservoir, the oceans.

Greenland ice sheet covers 1.7 million square kilometers (0.66 million mi^2) and is over 3 kilometers (2 mi) thick in some places.

The Antarctic ice sheet is by far the largest accumulation of freshwater ice on Earth. Because it covers a large landmass (the continent of Antarctica) centered on the south pole, it has existed in one form or another for a very long time, at least 34 million years. | Geotimes – drilling back to the future | It covers 98% of Antarctica and has fringing *ice shelves* that float over the adjacent ocean along most of the continent's coast. The volume of the Antarctic ice sheet is 10 times greater than the volume of the Greenland ice sheet. How the size and character of the Antarctic ice sheet has varied over its long history is the subject of much ongoing research. | The Antarctic ice sheet – its initiation and evolution |

FIGURE 2-16 ▼ Glaciers in Southern Alaska
Glaciers carve deep valleys, move broken and ground-up rock along their base and margins, and coalesce into large masses of ice. These glaciers flow from an ice cap in the coastal mountains of southern Alaska.

THE WATER CYCLE

The hydrosphere interacts with other Earth systems in the water cycle, diagrammed in Figure 2-17. This cycle replenishes water resources that people need, and it is a major influence on weather and climate conditions. The water cycle transfers this valuable resource among its reservoirs in the oceans, the atmosphere, on land, and below land's surface, in groundwater.

The atmosphere over the oceans is a key part of the water cycle. It is here that the atmosphere obtains about 86% of its water vapor through evaporation. The Sun's radiation is greatest in tropical regions near the equator, where it warms large volumes of ocean water. This warm water in turn heats the overlying air and increases evaporation, transferring energy and water from the oceans to the atmosphere.

When air rises, it cools. This can cause water vapor to turn into tiny droplets and form clouds. Water vapor and clouds move around the atmosphere and carry water from the oceans to the land. Eventually, where the air temperature falls, as when air rises along mountains, the liquid water droplets coalesce and get large enough to fall to Earth's surface. The precipitation of rain, snow, or ice (hail) transfers water to the land.

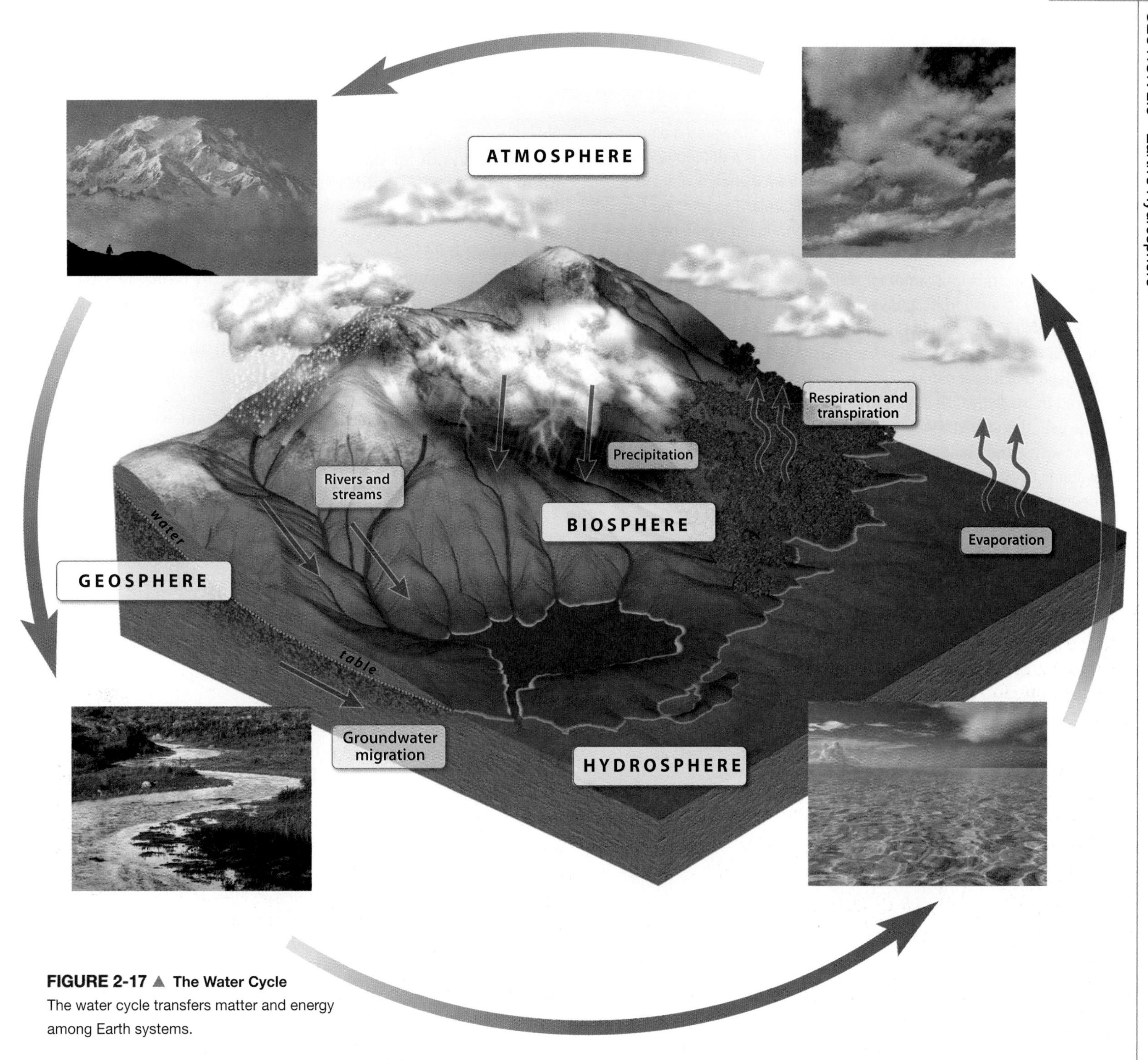

FIGURE 2-17 ▲ The Water Cycle
The water cycle transfers matter and energy among Earth systems.

Precipitation transfers water to three hydrosphere reservoirs on land—(1) ice accumulations in glaciers, ice caps, and ice sheets; (2) surface water in streams, rivers, and lakes; and (3) groundwater below the land surface (see Figure 2-17). Rivers and streams carry water back to the oceans; groundwater migrates back to the oceans in many coastal settings as well. Along its path back to the oceans, water is used by plants, animals, and people. However, living things in the biosphere only temporarily store water. They excrete it, release it when they die, or transfer it to the atmosphere through respiration (breathing). Transpiration of water from plant leaves and stems transfers water directly back to the atmosphere, too.

Consider the Earth-system processes and interactions in the water cycle. When water evaporates, energy and matter are transferred to the atmosphere. When water precipitates, energy and matter are transferred to the geosphere. As water migrates from land back to the ocean, it carries material with it and transfers matter to the oceans. Also, plants, animals, and people of the biosphere use water everywhere. In fact, in some places more water is needed than is available.

You Make the Call

Whose Water Is It?

People use surface water as it makes its way back to the oceans. The Colorado River illustrates this dramatically. The area drained by this river and its tributaries, shown in the map, is a 650,000-square-kilometer (250,000 mi^2) region stretching from its headwaters in southwestern Wyoming to the Gulf of California. The principal use of Colorado River water within its sparsely populated watershed is irrigation, both in its watershed and elsewhere in places like California (where it also provides water for millions of city dwellers). If you have ever enjoyed fresh fruits and vegetables from California, you may have consumed Colorado River water used to irrigate the crops.

The Colorado River, 2300 kilometers (1400 mi) long, is difficult to manage because it carries a relatively small amount of water compared to the large size of its watershed, and its flow varies a lot from one year to the next. People in seven states and Mexico use the river water, and competition for this limited resource is fierce. The water is used and reused so much that it becomes salty from repeated evaporation. The flow is almost nonexistent at the river's mouth on the Gulf of California.

The history of how people have managed Colorado River water is reviewed at | Sharing Colorado River water: history, public policy, and the Colorado River Compact .

- How well do you think people have managed Colorado River water?
- Should officials consider natural habitat needs in allocating Colorado River water? | The Lower Colorado River Basin: challenges of transboundary ecosystem management
- How can people consider sustainability when managing this water source?
- How would you manage Colorado River water?

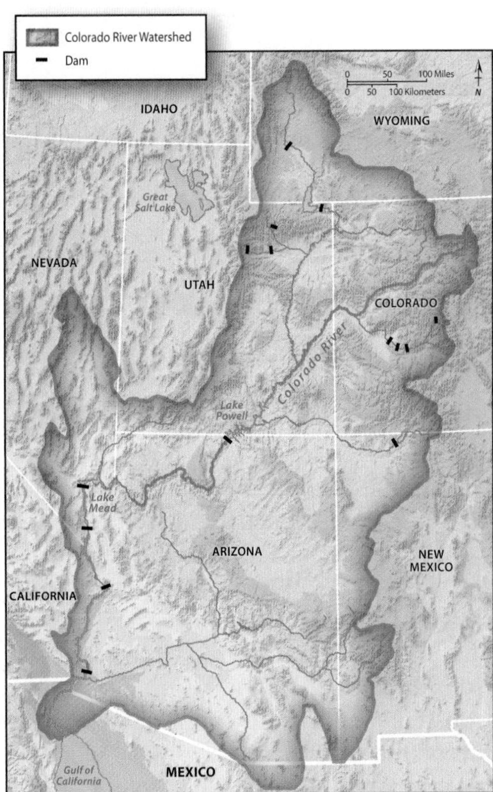

2.4 Earth's Biosphere

The biosphere consists of all life on Earth. Life inhabits every nook and cranny at or near Earth's surface. It exists underground, thrives in the deepest parts of the ocean, and flies or floats around in the atmosphere. Life comes in all sizes, from microscopic one-celled creatures to trees taller than the length of a football field.

Extensive study of the biosphere, from molecules within cells to vast, complex ecosystems, has determined how it is organized and how its many parts are related. | Evolution 101: the family tree Although we may understand the biosphere very well, we may not always appreciate its role in the Earth system as a whole and the many ways in which it helps people. Obviously the biosphere is the source of the food we consume in order to survive. But the biosphere also has key roles in the carbon cycle (Figure 1-10); as we will see, it helps soils form (Chapter 11), and it naturally cleans water in wetlands (Chapter 10). It captures energy from the Sun and makes it available to people—that's why we eat from it.

After billions of years of evolution, the biosphere has become very diverse and now includes large and growing numbers of people (see Chapter 1). How did this happen?

LIFE'S BEGINNINGS

Chemical and fossil evidence in ancient rocks provides insight into the origin of life on Earth. **Fossils** are remains and indications of former life that are preserved in rocks. | Fossils, geological time, and evolution Some typical examples are shown in Figure 2-18. Fossils can be actual remains, or just imprints of an organism or its hard parts, such as shells or dinosaur bones. Imprints of tracks or burrows left behind by an animal as it passed through an area are other examples. And fossils come in all sizes; many are microscopic. Organic remains, especially hard parts, that become buried in sediments are the most likely to become fossils.

Remnants of bacteria are the oldest fossils on Earth. Traces of thread-like, single-celled organisms have been preserved in 3.2-billion-year-old sedimentary rocks in Australia. *Stromatolites* are fossils found in even older rocks. These are structures built from layer after layer of muddy sediment trapped by mats of bacteria (Figure 2-19a). Stromatolites are found in sedimentary rocks as old as 3.5 billion years. Descendants of stromatolite-forming bacteria still inhabit some coastal environments (Figure 2-19b). Cyanobacteria, the first photosynthetic organisms, live in stromatolites today, but it is not clear if the bacteria that lived in very old stromatolites were photosynthetic.

So, fossils indicate that life existed as early as 3.5 billion years ago. The initial Earth was too hot and dynamic for life to exist until about 4.2 billion years ago. Therefore, sometime between 4.2 and 3.5 billion years ago, primitive life got started.

(a)

(b)

(c)

FIGURE 2-18 ◀ Fossilized Remains of Former Life
Fossils are remains or imprints of organisms, preserved in rocks. **(a)** Ammonites, members of a group of shelled animals related to modern squid and octopi, survived for some 350 million years. The one who left these remains floated in the ocean about 200 million years ago. **(b)** These fossilized bones belonged to a relative of the famous *T. rex*. Dinosaurs of this group lived from 170 to 65 million years ago. **(c)** Seed plants like the ones that left these impressions grew in Europe and North America some 280–300 million years ago.

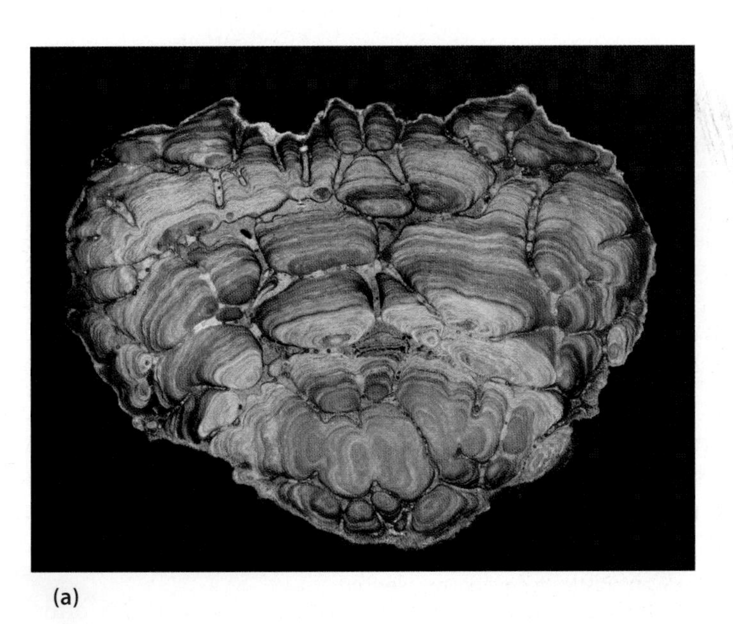
(a)

(b)

FIGURE 2-19 ▲ Stromatolites
Stromatolites—layers of muddy sediment trapped by films of microorganisms—are some of Earth's oldest fossils **(a)**, but they have descendants **(b)** that are alive today. The descendants grow in very salty coastal waters of Australia.

Scientists are still not sure how life began, but once it did, they know how it evolved.

EVOLUTION

Evolution is the change in organisms over successive generations that leads to new forms and functions, even new species. Our understanding of the theory of evolution has been developed from 300 years of scientific observations of the natural world. Many of the early breakthroughs were made by people who recognized connections between present and ancient life.

Establishing the foundation

In 1666, a Danish anatomist named Nicolas Steno (also known as Niels Stensen), who was working in Italy, was asked to examine a particularly large shark caught by local fisherman. Steno was a keen observer. As he dissected the shark, he realized that its teeth were strikingly similar to peculiar "tongue stones" that people collected from rocks. He concluded that tongue stones were in fact shark's teeth that had been incorporated in rocks when they first formed and subsequently turned into stone. Steno convincingly demonstrated that tongue stones were fossils that represented former life in former seas.

By the end of the eighteenth century, fossils themselves had become the focus of naturalist studies. The anatomy of vertebrate fossils was the specialty of Georges Cuvier, an early member of the French Museum of Natural History. By 1798, Cuvier had demonstrated that some vertebrate fossils represented species that no longer existed—for example, that mammoths differed from living elephants (Figure 2-20). Cuvier showed that some ancient life had become extinct.

Jean Baptiste Lamarck, a contemporary and colleague of Cuvier's at the museum, was a biologist assigned to "worms and insects" (invertebrates, a term that he originated)—a subject of investigation not then considered very important. However, his study of invertebrates combined with the advances in understanding of fossils at the time led Lamarck to greatly appreciate the complexity of ancient and present life. He also recognized that Earth was old and that life changed over time. Lamarck proposed that life changed from simple to more complex forms through interactions with the environment and by passing on acquired characteristics to offspring. For instance, successive generations of giraffes would develop longer and longer necks as they reached higher for leaves. Although Lamarck's ideas about *how* life changed were mainly philosophical and scientifically unsound in many cases, his proposals did constitute a theory of evolution.

By the 1800s, the stage was set. As is commonly the case in science, the new concepts and understanding provided by the contributions of Steno, Cuvier, and Lamarck were not the only foundations in place at this time. Others also recognized the significance of fossils, that Earth had an ancient history, and that natural processes might provide the best explanations for changes in the biosphere over time. These foundations were the ones that Charles Darwin built upon as he developed his own theory of evolution.

(a)

(b)

FIGURE 2-20 ▲ An Extinct Vertebrate and a Modern Relative

From the study of their fossil remains, Cuvier demonstrated that mammoths **(a)**, shown here in an artist's reconstruction, were an extinct species, different from modern elephants **(b)**.

Darwin and Wallace

Charles Darwin and a contemporary, Alfred Russel Wallace, both sought to explain how life changed over time by the process of **natural selection**. Wallace studied the distribution of species (biogeography) and independently developed his own ideas about natural selection. Darwin and Wallace communicated, shared specimens, and jointly presented their ideas on the subject at a scientific meeting in 1858. Darwin published *The Origin of Species* (originally titled *On the Origin of Species by Means of Natural Selection, or the Preservation of Favoured Races in the Struggle for Life*) in 1859, while Wallace continued his biogeography studies. It was Darwin's many years of investigations, careful syntheses, and clear explanations, brought together in *The Origin of Species*, that had an enormous influence on both scientific and popular thought.

In *The Origin of Species*, Darwin explained how species evolved. This contribution was the culmination of many years of investigations and diverse observations about biology, ecology, paleontology, and geology, but some of his best examples came from his early studies as a naturalist on a cruise of the H.M.S. *Beagle* (Figure 2-21). In 1835, the *Beagle* visited the Galápagos Islands at the end of a three-year exploration of South America. Because of their physical isolation, islands are particularly good places for ecological studies, and the Galápagos are exceptional in this regard. It was here that Darwin recognized how mockingbirds, finches, tortoises, and iguanas

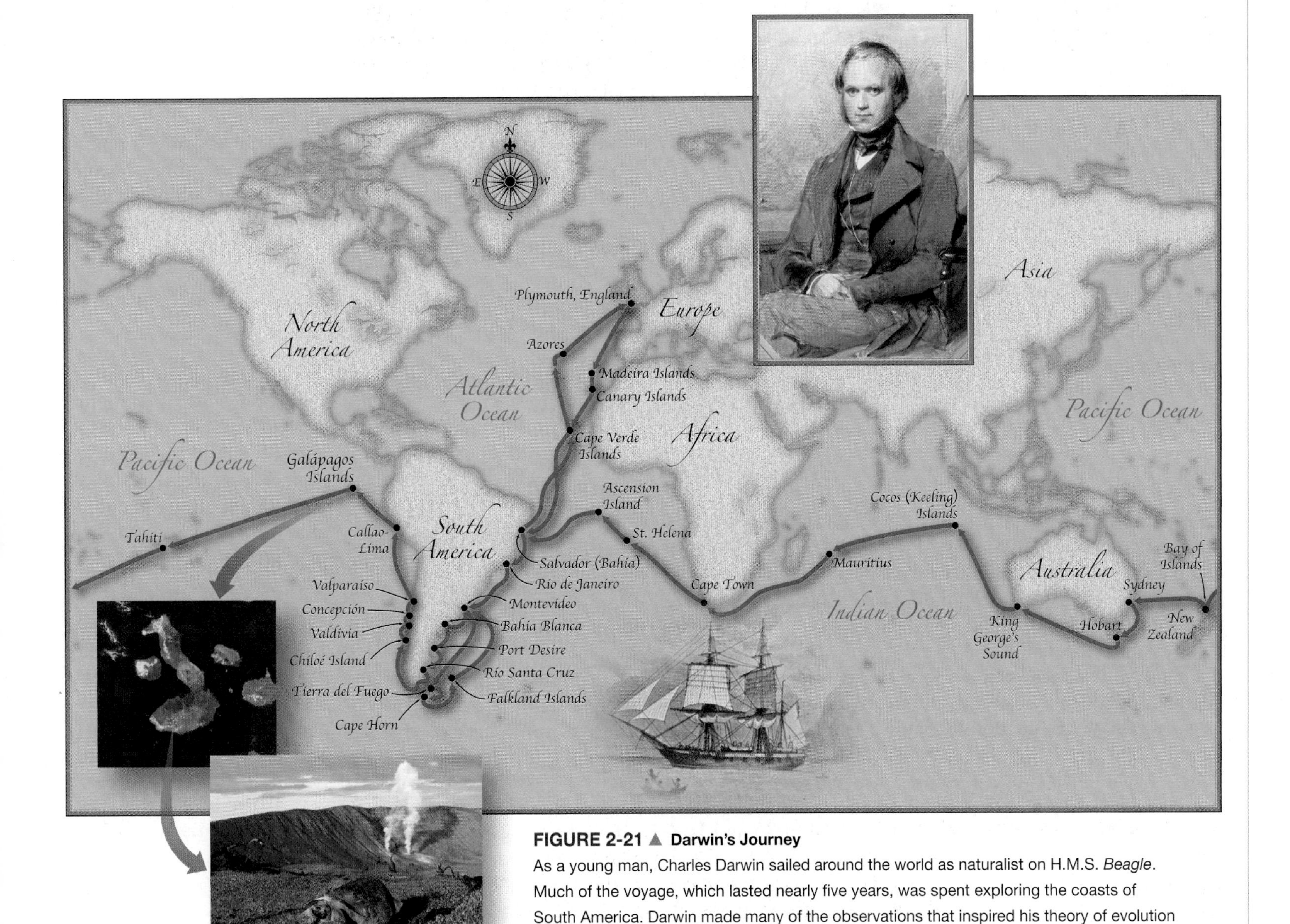

FIGURE 2-21 ▲ Darwin's Journey
As a young man, Charles Darwin sailed around the world as naturalist on H.M.S. *Beagle*. Much of the voyage, which lasted nearly five years, was spent exploring the coasts of South America. Darwin made many of the observations that inspired his theory of evolution by natural selection in the Galápagos Islands. In that isolated region, a unique and distinctive array of species has evolved as plants and animals have adapted to the varied environments on the different islands.

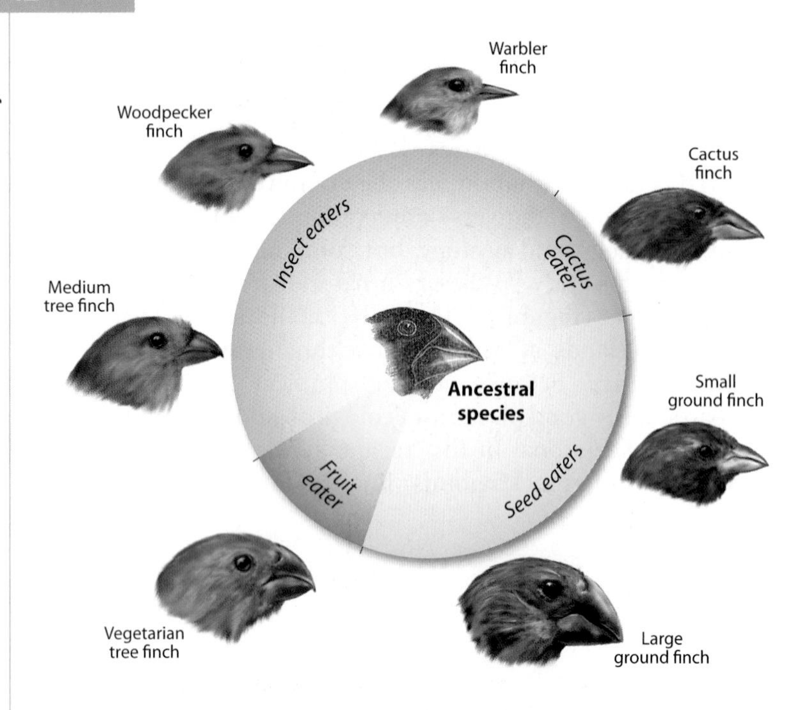

FIGURE 2-22 ▲ Seven of Darwin's Finches

Darwin discovered an amazing variety of closely related species on the Galápagos Islands in 1835. Thirteen species of finches had evolved different characteristics, especially of their beaks, to adapt to the food that was available on the different islands.

had developed distinct physical differences as they adapted to environmental conditions on different islands. For example, there are thirteen closely related but different species of finches on the Galápagos Islands (Figure 2-22). Darwin proposed that natural selection explained how these and other species evolved.

Natural selection

Natural selection is the principal mechanism behind evolutionary change in the biosphere. | Evolution and natural selection It is a natural process working within species populations that has three key parts: (1) variation in traits between individuals; (2) differential reproduction; and (3) the passing on, or descent, of traits through heredity.

Slight differences between individuals exist in all species populations. In some cases, a difference can be advantageous for survival or reproduction. Perhaps it helps an organism get more food (a stronger beak for birds that feed in places where there are many hard nuts or seeds), or escape predators better (camouflage coloring on an insect), or attract more mates. Individuals with these traits will tend to live longer and have more offspring than others in their population. As the advantageous trait is *selected for* in this way and passed on from generation to generation, individuals that have it will gradually come to make up a larger and larger proportion of the population. Many generations later, the population as a whole may have the advantageous trait; we say that the population overall has *evolved.*

Darwin saw that natural selection worked, but in his time the mechanism by which traits were inherited was not clear. Today we know how heredity works. It is through the transfer of replicable bits of genetic information, commonly called genes, that are encoded in molecules of DNA in our cells. These molecules are passed from parent to offspring as a part of reproduction. It is the natural variation of genetic information—through mutations in DNA molecules for example—that leads to new traits in individuals. Sometimes the new traits help survival, and sometimes they don't; in many cases, the new traits simply do not matter. But once in a while, the trait is helpful for survival, and natural selection increases the prevalence of this trait in the population. This can happen in populations of microscopic organisms (microbes) within days.

Microbes that cause disease evolve so rapidly that people can watch them change in the laboratory. Scientists grow populations of microbes in the lab and investigate their mutations over successive generations. These experimental evolution studies are aimed at determining how the microbes develop resistance to drugs. Scientists have found that microbes can evolve entirely new strains in just a few days or weeks.

The reason why disease-causing microbes evolve so quickly and become drug-resistant is because they are extremely abundant, reproduce rapidly, have high mutation rates, and face intense environmental pressure to adapt. Microbial evolution studies in the laboratory confirm how species evolve in general.

EXTINCTIONS

As well as some species are able to evolve and adapt to changing environmental conditions, this is not always enough to guarantee survival of the species. The number of identified species in the biosphere is now about 1.4 million. The actual total may be much higher, but species number has varied tremendously over life's history. Species are always dying out (becoming extinct), and new species are evolving from others.

Mass extinctions

There have been some very hard times for the biosphere, times when a large number of species became extinct. These are called **mass extinctions**, and as Figure 2-23 shows, there have been at least five major ones during the last 540 million years. The most recent and most famous mass extinction is the one during which the dinosaurs died out, some 65 million years ago. | Debating the dinosaur extinction Scientists think this mass extinction resulted from the impact of a large asteroid on what is now the Yucatan Peninsula of Mexico (Figure 2-24). Earth was not a closed system 65 million years ago! The impact of this large asteroid darkened the skies with dust and caused severe global climate changes that were just too much for the dinosaurs—and many other species—to handle.

The causes of the other mass extinctions are not entirely clear. However, all the mass extinctions show that the biosphere has not been entirely successful in facing some signifi-

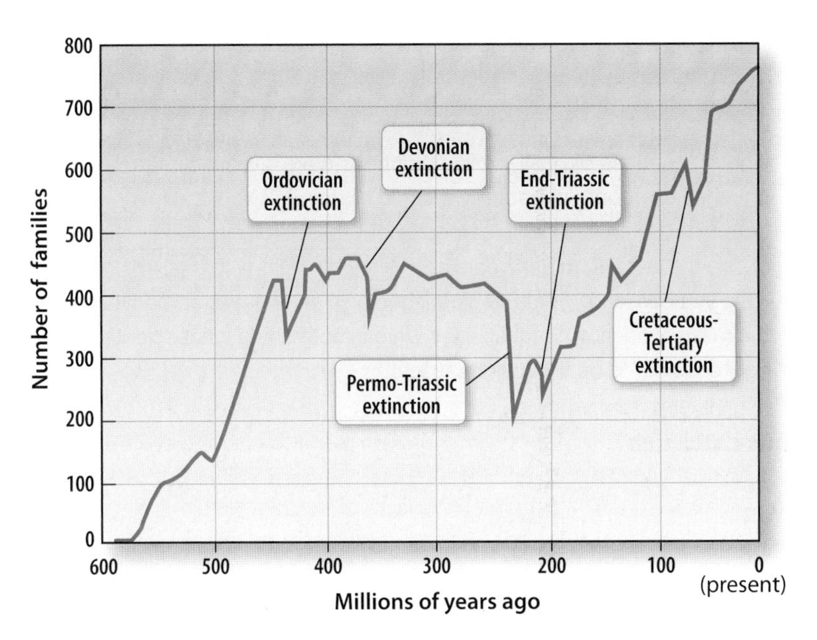

FIGURE 2-23 ▲ Mass Extinctions
Mass extinctions indicate very stressful times for the biosphere. There have been five major mass extinctions during the last 540 million years.

FIGURE 2-24 ▼ The End of the Dinosaurs
The dinosaurs died out in a mass extinction that occurred 65 million years ago. The cause of this event is thought to have been the impact of an asteroid, shown here in an artist's depiction **(a)**, in the Gulf of Mexico just off the coast of Mexico's Yucatan Peninsula **(b)**. The asteroid was probably at least 10 kilometers (6 mi) in diameter. Its impact released more energy than a million hydrogen bombs and blasted a crater greater than 160 kilometers (100 mi) in diameter, now buried under thick layers of sediment.

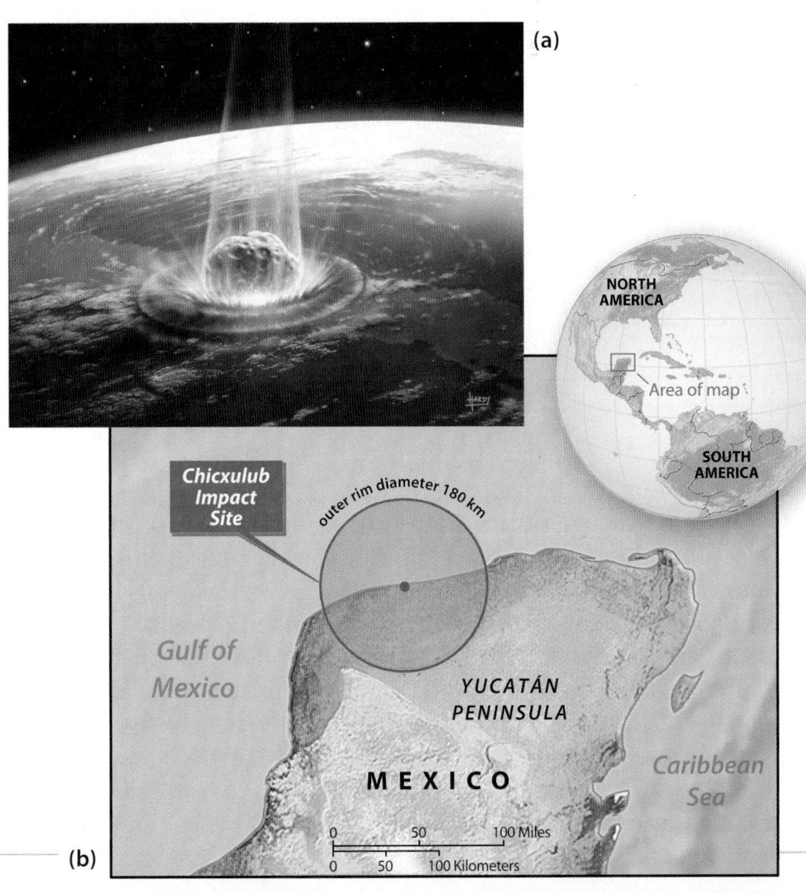

FIGURE 2-25 ▲ Martha: The Last Passenger Pigeon

cant, global-scale environmental changes in its past. Some scientists think that another mass extinction is under way today because estimates of extinction rates are as high as 70 to 700 species per year compared to the long-term average of 1 to 10 species per year. People are likely to be the primary cause of the current high extinction rates.

In Chapter 1, you learned how people decrease biodiversity by decreasing natural habitat. Around the world, people have also hunted species for food and other purposes to the point that they became extinct. Consider the case of the passenger pigeon.

The passenger pigeon

The passenger pigeon is probably the only species for which we know the exact date of its extinction. At 1:00 pm on September 1, 1914, Martha (Figure 2-25)—the last passenger pigeon that we know of—died in the Cincinnati Zoological Garden. But the fate of the passenger pigeon was determined years before. Passenger pigeons had a problem; they were easy to catch and good to eat.

Several billion passenger pigeons, 40 centimeters (16 in) tall and resembling mourning doves, lived in the forests of the Midwest when colonists first came to America. Passenger pigeons formed huge flocks that blackened the sky and took days to pass overhead during their migrations. At one time they may have been the most numerous birds on Earth. In the mid-1800s, passenger pigeons were increasingly being hunted for food, and people killed them by the tens of thousands for weeks on end. They were essentially gone before the end of the century. Loss of a species like the passenger pigeon is an irreversible consequence of people's actions.

What You Can Do

Investigate Mass Extinctions

Mass extinctions are global events that affect large parts of the biosphere. They are of particular interest to Earth scientists because they are such clearly identifiable episodes in Earth's history—in effect, global time markers. They are also of interest because of their possible connections to global change, especially major changes in global climate. By studying past times of global change, we may come to understand current changes better. But it is not yet clear what the causes of past mass extinctions were. Was it climate change? Meteorite impacts? Volcanic eruptions? You can investigate what scientists have learned about the major mass extinctions during the last 540 million years at Mass extinctions of the Phanerozoic. Here you can learn what the global geologic setting was at the time of the mass extinction, what species were affected, and what scientists now consider to be the major causes.

The Permian Mass Extinction

Fast Facts

- Permian Period (286-248 million years ago)
- Terrestrial faunal diversification occurred in the Permian
- 90-95% of marine species became extinct in the Permian

Geological Setting

With the formation of the super-continent Pangea in the Permian, continental area exceeded that of oceanic area for the first time in geological history. The result of this new global configuration was the extensive development and diversification of Permian terrestrial vertebrate fauna and accompanying reduction of Permian marine communities. Among terrestrial fauna affected included insects, amphibians, reptiles (which evolved during the Carboniferous), as well as the dominant terrestrial group, the therapsids (mammal-like reptiles). The terrestrial flora was predominantly composed of gymnosperms, including the conifers. Life in the seas was similar to that found in middle Devonian communities following the late Devonian crisis. Common groups included the brachiopods, ammonoids, gastropods, crinoids, bony fish, sharks, and fusulinid foraminifera. Corals and trilobites were also present, but were exceedingly rare.

Species Affected

The Permian mass extinction occurred about 248 million years ago and was the greatest mass extinction ever recorded in earth history; even larger than the previously discussed Ordovician and Devonian crises and the better known End Cretaceous extinction that felled the dinosaurs. Ninety to ninety-five percent of marine species were eliminated as a result of this Permian event. The primary marine and terrestrial victims included the fusulinid foraminifera, trilobites,rugose and tabulate corals, blastoids, acanthodians, placoderms, and pelycosaurs, which did not survive beyond the Permian boundary. Other groups that were substantially reduced included the bryozoans, brachiopods, ammonoids, sharks, bony fish, crinoids, eurypterids, ostracodes, and echinoderms.

- Causes of the Permian Extinction
- Mass Extinctions of the Phanerozoic Menu
- Main Menu

2.5 Understanding Geologic Time and Earth History

This book has so far presented ages of events and changes throughout Earth's history without much clarification of how we determine these ages. An ancient date, such as 3.5 billion years ago, the age of the oldest known fossil, may seem to be just a number, albeit a very big number. Periods of time long enough to allow us to understand Earth's history—periods on the scale we call **geologic time**—are inherently difficult for us to comprehend. This section reviews how scientists have come to determine both the relative and absolute ages of events in Earth's history. These ages in turn help us understand the rates of geologic processes and Earth-system interactions: how long it takes soils to form or ice sheets to expand or contract, for example.

RELATIVE GEOLOGIC AGES

Geologists first began to understand Earth history by observing rocks and their relationships to one another. These "field" observations identified sequences of changes that indicated whether particular rocks were older or younger than others. They indicated their **relative ages**. The major breakthroughs came from studying sedimentary rocks.

Sedimentary rocks

Sediment is material (silt, sand, and gravel, for example) carried along by currents and deposited on river, lake, or ocean bottoms. Sediments accumulate in layers and eventually become sedimentary rocks (you will learn more about sedimentary rocks and how they form in Chapter 4). The basic facts about sedimentary rocks have been known for a long time. Nicolas Steno, the anatomist who recognized fossil shark's teeth in 1666, was one of the first to realize that sediment accumulated on the bottom of water bodies and came to form layers of sedimentary rocks like the ones shown in Figure 2-26. He pointed out that these layers were originally horizontal and that in a sequence of layers, any layer is younger than layers below it and older than layers above it. This principle, which allows us to determine the *relative* age of sedimentary rock layers, is known as *Steno's Law of Superposition*.

If we could observe all the sedimentary rocks that were ever formed in one sequence of layers, then the relative ages of Earth's entire history of sedimentary rocks would be obvious. As you will learn in Chapter 3, geosphere movements have shifted continents and formed and destroyed ocean basins. In addition, geosphere movements have commonly deformed, eroded, and displaced sedimentary rocks, leading to many incomplete sedimentary sequences of many different ages exposed around the world. As a result, there is no single place where sediments have been continuously deposited throughout Earth history. How do we determine the relative ages of rock layers in these widely scattered sedimentary sequences? In some cases, the repetition of specific rock types in a specific order suggests that one sequence is similar in age to another, but it is the fossils contained in the sedimentary rocks that convincingly reveal their relative ages.

Fossil succession

A powerful breakthrough came in the early 1800s when the English surveyor William Smith recognized that particular fossils were restricted to specific sedimentary layers and sequences. The nature of the fossils changed from lower to upper parts of a sequence or from one sequence to another in regular ways. These well-defined **fossil successions** indicated changes that were taking place in the biosphere when the sediments enclosing them were deposited. | Fossils, rocks, and time: fossil succession | Distinctive fossil successions enable the relative ages of the biosphere changes to be identified. They also allow rock sequences to be correlated over long distances—even from one continent to another—based on the fossils they contain.

Distinctive fossil successions identified from around the world form the foundation for subdividing Earth's history. As scientists better defined the historical divisions and understood their relative ages, they combined them into the **geologic time scale**, depicted in Figure 2-27 on p. 46.

The geologic time scale

The geologic time scale divides Earth history into four major intervals or **eons**. The earliest eon is the poorly known *Hadean* eon (4.5 to 3.85 billion years ago), the time when Earth internally segregated, and its surface cooled. During the *Archean* eon (3.85 to 2.5 billion years ago), the simplest forms of life evolved, and continental material aggregated into larger masses. In the *Proterozoic* eon (2.5 to 0.54 billion years ago), continental masses approached their present sizes, the oceans included many shallow seas, and life gained more complexity. The *Phanerozoic* eon (542 million years ago to the present) is marked by a tremendous diversity of life, including organisms with hard skeletons or structures. Fossils are abundant, and sedimentary rocks are widespread in the Phanerozoic. The Phanerozoic eon is subdivided into many more units including *eras*, *periods*, and even smaller subdivisions (see Figure 2-27). The age of mass extinction events are particularly helpful global time markers during the Phanerozoic and mark the boundaries between several eras (compare Figures 2-23 and 2-27). Scientists have come to understand the relative ages of the biosphere and other Earth-system changes very well.

FIGURE 2-26 ▼ Sedimentary Rock
Sedimentary rock layers are deposited in sequence one after (and thus on top of) another. Any layer in a sedimentary sequence is older than layers above it and younger than layers below it. These 60-million-year-old limestone layers are in the Dixie National Forest, Utah.

Eon	Era	Period	Boundary dates (million years ago)	Epoch	Development of Plants and Animals
PHANEROZOIC	CENOZOIC	Quaternary	0.01	Holocene	Humans develop
			2.6	Pleistocene	
		Tertiary — Neogene	5.3	Pliocene	
		Tertiary — Neogene	23	Miocene	"Age of Mammals"
		Tertiary — Paleogene	33.9	Oligocene	
		Tertiary — Paleogene	55.8	Eocene	
		Tertiary — Paleogene	65.5	Paleocene	Extinction of dinosaurs and many other species
	MESOZOIC	Cretaceous	145.5	"Age of Reptiles"	First flowering plants
		Jurassic	201.6		First birds
		Triassic	251		Dinosaurs dominant
	PALEOZOIC	Permian	299	"Age of Amphibians"	Extinction of trilobites and many other marine animals
		Carboniferous — Pennsylvanian	318		First reptiles Large coal swamps
		Carboniferous — Mississippian	359		Amphibians abundant
		Devonian	416	"Age of Fishes"	First insect fossils Fishes dominant First land plants
		Silurian	444		
		Ordovician	488	"Age of Invertebrates"	First fishes Cephalopods dominant
		Cambrian	542		Trilobites dominant First organisms with shells
PRECAMBRIAN	PROTEROZOIC		2500		First multi-celled organisms
	ARCHEAN		3850		First one-celled organisms
	HADEAN		4500		Origin of Earth

FIGURE 2-27 ◀ The Geologic Time Scale
The geologic time scale summarizes Earth's geologic history as determined from studies of rocks, fossils, and their relative ages. Radiometric dating has enabled the absolute ages of geologic events to be determined and added to the time scale. (Note that the time intervals are not drawn to scale.)

ABSOLUTE GEOLOGIC AGES

Determining relative ages and the correct sequence of Earth-system changes through geologic time is accomplished by careful observation of rocks and fossils. How do geologists know how long ago, in years, these events occurred? What are the **absolute ages** of these events?

Natural clocks

Natural clocks in rocks and minerals are used to determine the ages, in years, of geosphere materials. Elements vary naturally in the number of neutrons their atoms contain, and thus in their atomic mass (the number of neutrons plus protons in an atom's nucleus). Atoms of an element that have different atom-

TABLE 2-2 PARENT, DAUGHTER, AND HALF-LIFE OF THREE ELEMENTS COMMONLY USED IN RADIOMETRIC DATING

Parent Element	Daughter Element	Half-life (years)
Uranium, U-238	Lead, Pb-206	4.5 billion
Potassium, K-40	Argon, Ar-40	1.26 billion
Carbon, C-14	Nitrogen, N-14	5730

ic masses are called **isotopes**. The different isotopes of an element are identified symbolically by combining the isotope's atomic mass number with the element's chemical symbol. The symbol for the uranium isotope with the atomic mass number of 238 is ^{238}U, for example. Less formally, this isotope would be referred to as uranium-238, or U-238 (the style we will use in this book). In some cases, an isotope has too few or too many neutrons to be stable, and this unstable *parent isotope* spontaneously changes, or decays, to a stable *daughter isotope*, often of a different element. Elements that spontaneously decay are called radioactive elements, the natural clocks that can be used to determine the absolute ages of geologic materials.

Radioactive elements can be used as clocks because the rates at which they spontaneously decay are known. These rates can be best expressed in terms of an isotope's **half-life**: the time it takes for half of the isotope's atoms to decay (Table 2-2). Any geologic material that incorporates a radioactive isotope (parent) when it forms, and subsequently retains this isotope and its daughter, can be dated. Minerals, such as those that form when magmas solidify, are commonly excellent repositories for both parent and daughter isotopes. By measuring the amounts of the parent and daughter isotopes in the mineral, scientists can calculate how much time has passed since the host mineral formed (Figure 2-28).

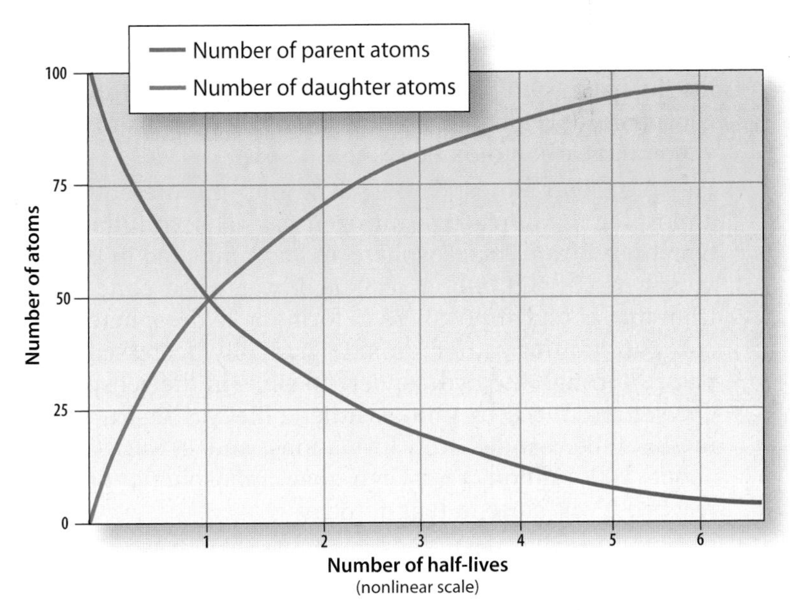

FIGURE 2-28 ▲ Radiometric Dating
Radiometric dating uses natural radioactive clocks. Because the rate at which a parent radioactive isotope decays into a daughter isotope (as reflected in its half-life) is known, determining the relative amounts of these two isotopes in a host mineral enables its absolute age to be calculated. What is the age of the mineral if 24% of its parent atoms remain?

Radiometric dating

Scientists have become very good at measuring the concentrations of parent and daughter isotopes in minerals. They can even measure these concentrations in tiny pieces of a mineral. As long as the mineral contains an appropriate radioactive parent isotope and is old enough to have measurable concentrations of its daughter isotope we can use a radioactive clock to date it. | Dating the fossil record Many radioactive elements decay so slowly that they cannot provide accurate dates for material less than 100,000 years old, but radioactive carbon can be used to date organic material from a few hundred to about 60,000 years old. This **radiometric dating** is what has provided the absolute ages on the geologic time scale.

Rates of earth-system processes

Absolute ages do more than tell us how many years ago a geologic event happened. When we know the absolute ages of geologic events, we can determine the rates at which Earth-system processes operate. For example, radiometric dating has enabled us to determine that

- Soils form in hundreds to thousands of years (Chapter 11)
- Valuable metal deposits form in several thousand to a few million years (Chapter 12)
- Oil forms and fills reservoirs in a few million to many millions of years (Chapter 13)

Quantifying the rates at which Earth-system processes produce natural resources helps us determine whether our use of them is sustainable. As we saw in Chapter 1, resources that can't be recovered in some way after their use, and are used faster than Earth's natural processes can produce them, are called *nonrenewable*. If people are careful with soils, for example, they can be considered renewable resources (Chapter 11).

Metals are natural resources that could be renewable because they are not destroyed when we use them. However, our present level of metal recycling is insufficient for use of this resource to be considered sustainable (Chapter 12). If there is a lot of a metal resource (copper, for example), its unsustainable use may not be noticed for many generations.

Oil is nonrenewable and is being used at very high rates compared to the time it takes nature to replenish it, and compared to the amount available. People's unsustainable use of oil is evident today by its increasing price and by its replacement by other energy sources. In general, people's use of nonrenewable resources is unsustainable, and we will, at some point, be faced with shortages of these resources.

PEOPLE'S PLACE IN EARTH HISTORY

In terms of absolute time, modern humans (*Homo sapiens*) have only been around a few hundred thousand years. The oldest fossils in our genus, *Homo*, date to only 2.4 million years ago. This isn't very old by Earth history standards.

Here are two analogies that help us understand people's place in Earth history. First, imagine that the combined lengths of your arms stretched out from your shoulders represents all the years of Earth history. If you scraped a fingernail file across the fingernail on your longest finger, you would remove all the time that people have been on Earth. Another analogy is to imagine that all of Earth's history is compressed into the time of one year, as shown in Figure 2-29. In this representation, modern humans have been on board for only about the last half hour. The most recent ice age ended a little over a minute ago. And the time during which people have come to be such a dominant influence on Earth, since the Industrial Revolution just a couple of hundred years ago, is less than 2 seconds. We just haven't been here very long compared to the length of Earth history.

In some respects, the success of people, even if measured only by our numbers, is amazing. There are now almost 7 billion of us. Think about this for a moment. People are at the top

FIGURE 2-29 ▼ The History of Earth as a Single Year

If we represented the entire history of Earth as a single year, the human species would not appear until the last half hour of the last day, and all of recorded history would occupy no more than about 40 seconds.

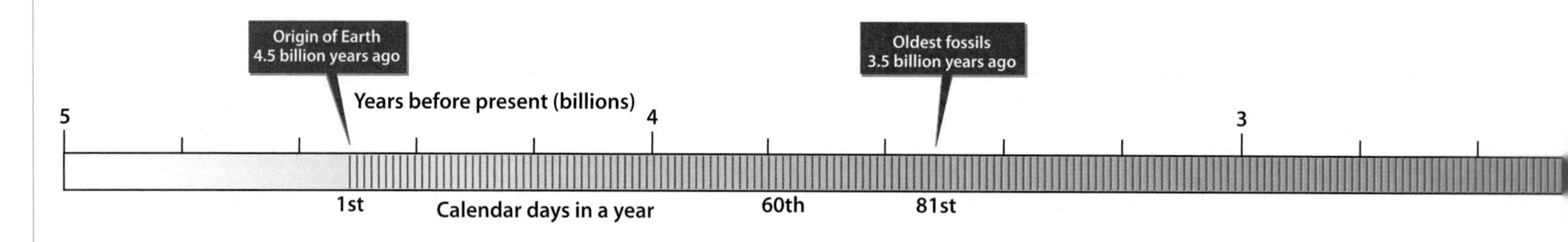

SUMMARY

Earth is just the right distance from the Sun and just the right size to evolve systems that support life. Three systems defined at the scale of the entire Earth—the geosphere, hydrosphere, and atmosphere—have evolved physical conditions that support the fourth Earth system—the biosphere. System interactions transfer and change matter and energy in ways that accommodate life, including yours (Figure 2-29).

In This Chapter You Have Learned:

- Earth's geosphere originated from the collapse of a cloud of interstellar gas 4.5 billion years ago. The initial hot Earth compositionally segregated into a still-partly molten, iron-rich core, mantle, and thin crust. There are two types of crust: continental and oceanic.
- Variations in density (estimated from changes in seismic wave velocity) define physical parts of the geosphere including (from Earth's center outward) a solid inner core, liquid outer core, lower mantle, transition zone, upper mantle, and crust. A part of the upper mantle makes up the soft, weak, pliable asthenosphere, while the top portion of the upper mantle and the crust together constitute the strong, rigid lithosphere.
- Earth's very early atmosphere (its first atmosphere) was lost to space. Outgassing of volatile components from the geosphere developed a second atmosphere rich in water vapor and carbon dioxide.
- Earth's second atmosphere evolved into the present third atmosphere. Nonreactive nitrogen gas (N_2) continued to be released from the geosphere and accumulated in the atmosphere. By 3.8 billion years ago, water had condensed from the second atmosphere to form the hydrosphere. The concentration of carbon dioxide gradually decreased as it passed through the hydrosphere to sinks in the geosphere. Oxygen produced by photosynthetic life was first trapped in sinks like banded iron formations, and it wasn't until about 500 million years ago that atmospheric oxygen reached levels close to that of today.
- Compositionally, the atmosphere has two parts. In the lower 85 kilometers (53 mi), its major constituents (N_2, O_2, and Ar) are well mixed in the homosphere. The 400 kilometers (250 mi) above the homosphere, the heterosphere, is compositionally segregated, with the heavier components more concentrated toward the bottom.
- The atmosphere has a four-part structure defined by temperature changes upward from Earth's surface: tropo-

of the biosphere's food chain, notwithstanding the occasional unlucky encounter with a bear, shark, or big cat—and the populations of these other top predators number only in the hundreds of thousands. People have accomplished this basically because of our intelligence—we have the largest brains, for our size, of any organism in the biosphere.

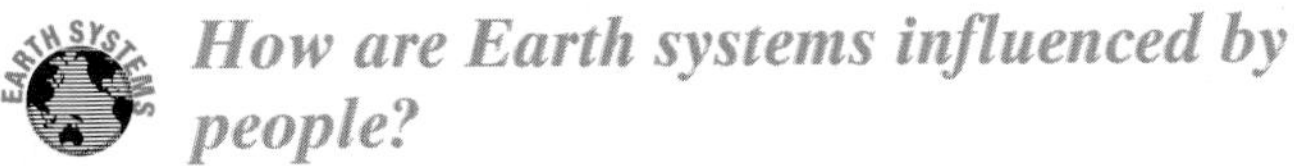

How are Earth systems influenced by people?

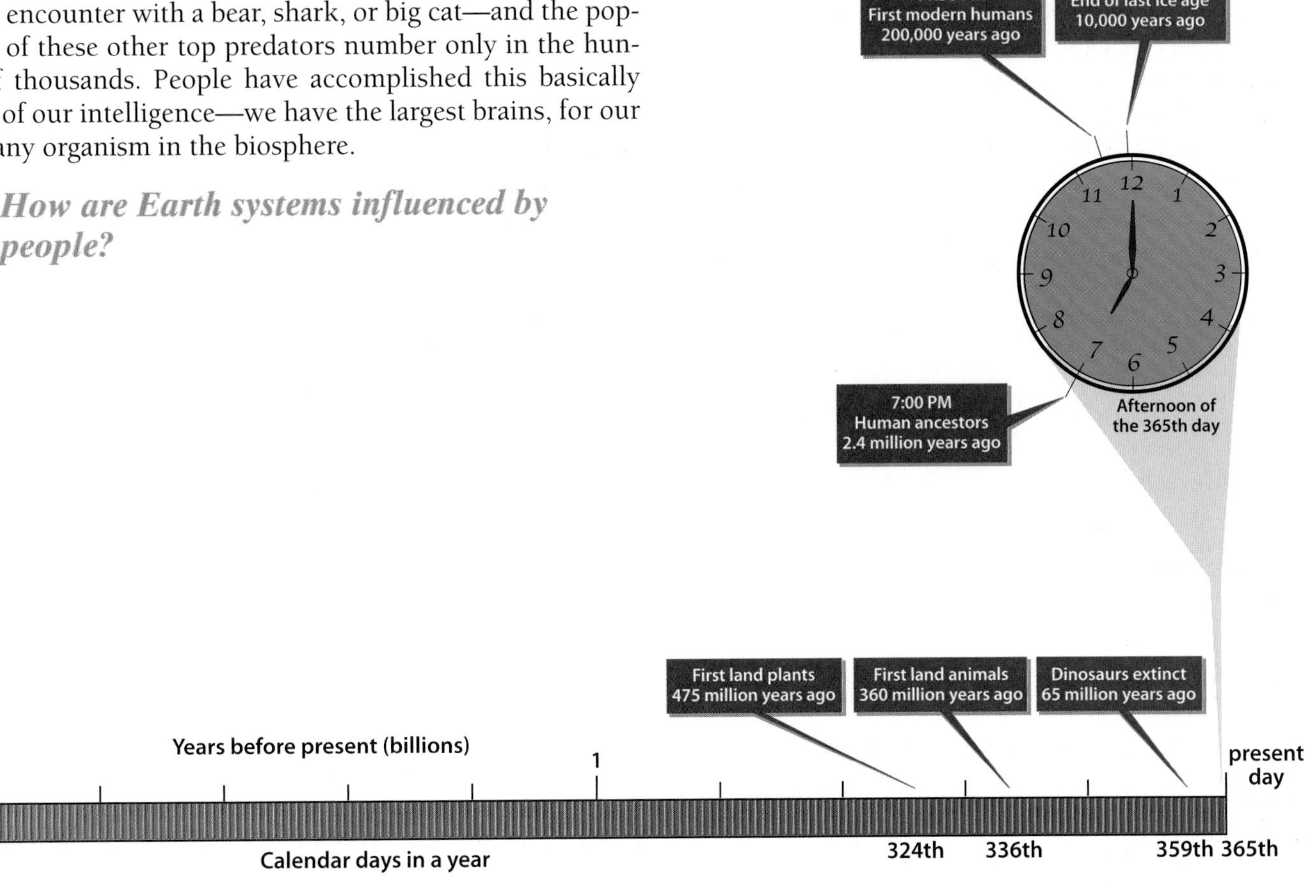

sphere, stratosphere, mesosphere, and thermosphere. The lowest part, the troposphere, contains most of the matter present in the atmosphere. The atmosphere is Earth's most dynamic system, and most of the action is in the troposphere.

- The hydrosphere's origin is tied to the evolution of Earth's second atmosphere. The age of the oldest known sedimentary rocks, 3.8 billion years, indicates that water condensed from the second atmosphere to form the oceans by this time.
- Most of the hydrosphere's water (97.2%) is in the oceans. Freshwater (nonsalty) reservoirs include the atmosphere, streams, lakes, and underground, but 99.4% of all freshwater is in glaciers, ice caps, and the two huge ice sheets centered over Greenland and Antarctica.
- Water is transferred from oceans to the atmosphere and from the atmosphere to the geosphere and biosphere in the water cycle—an excellent example of Earth-system interactions.
- Bacteria that existed at least 3.5 billion years ago are the earliest known forms of life on Earth. Since then, the biosphere has evolved incredibly diverse forms of life despite undergoing very stressful events that caused mass extinctions of species. Understanding biosphere evolution is rooted in observations that tie ancient and present life. Charles Darwin was instrumental in explaining how life evolves through natural selection. The ability of organisms to pass on traits favorable for survival and reproduction through heredity enables populations to change and species evolve.
- Modern humans are a very late addition to the biosphere, but they have come to be the major influence on the Earth's environmental health. People have caused individual species extinctions, and the current high extinction rates that the biosphere is experiencing may be caused by people.
- The evolution of Earth systems through Earth's long history, called geologic time, is understood through relative and absolute dating of geologic events. Relative ages determined through observations of sedimentary rocks and the fossils they contain define the subdivisions of Earth history in the geologic time scale. Radiometric dating, the use of radioactive elements as natural clocks, enables the absolute ages of geologic events and the rates of Earth-system interactions and processes to be determined.
- Earth systems replenish many resources at rates much slower than people use them. These are nonrenewable resources, and their use is not sustainable.

KEY TERMS

absolute age (p. 46)
asthenosphere (p. 27)
basalt (p. 25)
continental crust (p. 25)
core (p. 24)
crust (p. 24)
eon (p. 45)
fossil (p. 38)
fossil succession (p. 45)
geologic time (p. 44)
geologic time scale (p. 45)
glacier (p. 35)
granite (p. 25)
greenhouse gas (p. 28)
half-life (p. 47)
ice sheet (p. 35)
ion (p. 33)
ionosphere (p. 33)
isotope (p. 47)
lava (p. 28)
lithosphere (p. 27)
magma (p. 28)
mass extinction (p. 42)
mantle (p. 24)
mesosphere (p. 32)
Mohorovičić discontinuity (Moho) (p. 26)
natural selection (p. 41)
oceanic crust (p. 25)
ozone (p. 32)
ozone layer (p. 32)
photosynthesis (p. 30)
radiometric dating (p. 47)
relative age (p. 45)
stratosphere (p. 32)
troposphere (p. 32)
thermosphere (p. 33)

QUESTIONS FOR REVIEW

Check Your Understanding

1. Small irregularities in the early interstellar nebula produced local clusters of mass. The pull of gravity at a location is directly related to how much mass is present. How do these facts help explain the process that led to the formation of the planets?
2. Earth is divided generally into the core, mantle, and crust. What is the essential difference in composition that defines the core as very distinct from the mantle and crust? What caused these zones to form early in Earth's history?
3. The crust and mantle are both made of rocky material but are compositionally distinct. However, the upper mantle and the crust are understood to be part of one layer called the *lithosphere*. What is this layer, and how is it defined?
4. The stratosphere is located immediately above the troposphere in our atmosphere. According to the temperature profile shown in Figure 2-11, explain why the stratosphere should be relatively stable with respect to vertical air motion compared with the troposphere.
5. There is ample evidence of past and present interaction of the biosphere and atmosphere in shaping atmospheric composition. Why was the activity of the biosphere over 2 billion years ago so crucial in the development of an atmosphere conducive to modern life on Earth today?
6. Water can exist in three phases: solid, liquid, or gas. How is Earth's distance from the Sun relevant to these phases of water and to life on this planet?
7. Describe how energy and matter are transformed as they move between the atmosphere and geosphere in interactions of the water cycle. Explain how these processes are critical to all life on land.
8. Life exists in many forms today on Earth. Give an example of the oldest life on Earth. How do geologists identify early life from the rock record?
9. What is the difference between *relative* and *absolute* geologic ages? Explain how are both ways of measuring geologic age can be employed together, and why this is useful.
10. Why are fossils so crucial to the determination of reliable relative geological ages, and how are they used to construct a composite history of large regions of Earth?
11. How do radioactive isotopes found in minerals act as natural clocks to determine absolute geologic ages? What is meant by the term *half-life*?

Critical Thinking/Discussion

12. From many lines of evidence, it is clear that the biosphere has interacted extensively with the atmosphere and the hydrosphere over time. It can be argued that the current state of Earth systems has evolved to accommodate life, driven significantly *by* the actions of life. Make this argument, citing at least three examples from the text or from other systems you have studied.
13. Over time, the biosphere has grown between periods of sharp contraction—mass extinctions. Are Earth systems interactions necessary for mass extinctions to occur? Provide examples that support your answer. What is the role of evolution in the biosphere's subsequent response to a mass extinction?
14. Humans are relative newcomers to Earth. Given this, and our tremendous success as a species on this planet, why is understanding the history of Earth systems prior to our emergence as a species so crucial to shaping our own future actions and decisions?

ANSWERS TO IN-CHAPTER INSIGHT QUESTIONS

P. 30

What are some Earth systems interactions that helped make Earth's present atmosphere?

- Geosphere outgassing contributed atmosphere components such as nitrogen.
- Carbon dioxide in the atmosphere was transferred through the oceans to a long-term sink in the geosphere.
- Photosynthesis in the biosphere was the source of oxygen in the atmosphere.

P. 35

What are some potential sustainability issues associated with Earth's freshwater?

- Competition for this limited resource can affect how it is distributed and made available for use.
- The variable distribution of freshwater can prompt the creation of diversions such as dams and aqueducts that change natural drainages and habitat.
- Pollution of freshwater resources, especially groundwater, can make it unusable.

P. 49

How are Earth systems influenced by people?

- People change the hydrosphere through the many ways they transport and use water.
- People change biosphere habitat and decrease biodiversity, in some cases by causing species extinction.
- People pollute the atmosphere by their emissions of volatile wastes.
- People cause soil degradation and loss, in some cases leading to desertification.
- People establish preserves to protect ecosystems and biodiversity.
- People clean up soil, water, and air to decrease pollution.

FIGURE 3-1 ▶ Columbus Arrives in the New World
Columbus took five weeks to cross the Atlantic and reach the Bahamas. Since his journey, the distance across the Atlantic has increased over 10 meters (33 ft) as movements in Earth's geosphere have shifted the continents.

CHAPTER

3

THE DYNAMIC GEOSPHERE AND PLATE TECTONICS

Christopher Columbus made his first journey to the New World in 1492. On October 12, after crossing the Atlantic Ocean, sailors on the *Pinta* sighted land, now thought by scholars to be the island of San Salvador in the Bahamas (Figure 3-1). Four-hundred and seventy-seven years later—exploration having advanced in the interim from sailing ships to spaceships—astronaut Neil Armstrong completed another epic journey, stepping out of the lunar lander onto the Moon's surface. During those centuries, technology had clearly not stood still—but neither (less obviously) had the Bahamas. By the time Armstrong reached the Moon, Columbus' landing point in the New World was about 10 meters (33 ft) farther from his home port.

The seafloor beneath the Bahamas is in motion. Relative to Europe, it is moving due west at a rate of about 2 centimeters (0.8 in) per year. The Bahamas lie on a large piece of Earth's rigid outer layer, the lithosphere (Chapter 2). Spain and the rest of Europe lie on a different piece. Riding on their respective rafts of lithosphere, the Bahamas and Spain move farther apart each year.

Columbus would not have noticed a change in 10 meters of his sailing distance of over 3000 kilometers, but if he had made his trip 240 million years ago, he could have walked from Spain to the New World! Back then, North America and Europe were part of a single continent. This continent broke apart about 240 million years ago when the Atlantic Ocean basin began to form. Even slow movements of 2 centimeters a year add up to long distances when they continue for millions of years.

The ground under your feet moves—it is dynamic and always changing. The lithosphere, which includes the crust and uppermost mantle, is broken into a mosaic of irregularly shaped pieces called *plates* that move relative to one another, colliding head-on, sliding past each other, or sinking into Earth's interior. These dynamic geosphere processes are called **plate tectonics**. The word *tectonics* comes from the Greek *tekton*, meaning "builder." It refers to processes that sculpt Earth's major surface features, such as mountains, valleys, plateaus, ocean basins, and volcanoes. Plate tectonics builds those parts of Earth that we see and live on.

Plate tectonic processes affect people in many ways. They trigger earthquakes that destroy cities and volcanic eruptions that smother the landscape with debris. Some of these dynamic processes determine what types of mineral resources form in the crust and where they form. Without an understanding of plate tectonics, much about the geosphere's dynamic nature just wouldn't make sense.

The discovery of plate tectonics—an epic tale recounted in this chapter—is a compelling illustration of the scientific method in action. In the span of a few generations, scientists all over the world reported, debated, and finally synthesized many different types of data into a simple, coherent theory. In the process, geology was transformed into a truly global science.

IN THIS CHAPTER YOU WILL LEARN:

- The origins of plate tectonic theory—how scientists developed and tested it
- What drives geosphere movements and how plate tectonics operate
- How continents have split apart, moved, and reassembled
- How Earth system interactions at plate boundaries are related to earthquakes, volcanoes, mountains, and ocean basins
- How plate tectonic processes affect people by creating both natural hazards and vital natural resources

3.1 Early Thoughts About Moving Continents

In Chapter 1, you were introduced to the scientific method. In Chapter 2, you learned that many scientific contributions preceded Charles Darwin's work and set the stage for his understanding of evolution. In this chapter, while exploring the theory of plate tectonics, you will see how the scientific method of repeatedly questioning, hypothesizing, testing, and hypothesizing again extends across generations as many scientists investigate natural processes. The way scientists came to understand how continents move is a wonderful example of science in action. As is usual in science, the foundations of this understanding go back many generations. It started when the first global maps were made.

SETTING THE STAGE

The story of plate tectonics begins in the 16th century, when European cartographers produced maps showing the general outlines of the continents. They noticed something intriguing. South America's eastern coast fit surprisingly snugly into the crook of Africa's west coast like two pieces of a jigsaw puzzle. North America, too, seemed to fit fairly well against the west coast of Europe. But how could this be?

The Flemish cartographer Abraham Ortelius suggested in his 1596 book, *Thesaurus Geographicus*, that North and South America were once joined to Europe and Africa as a larger landmass that was torn apart by earthquakes and floods. In 1858, the French geographer Antonio Snider-Pellegrini noted that coal deposits in North America and Europe both contained fos-

FIGURE 3-2 ▲ *Glossopteris* Fossils
Early recognition that the distinctive fossil seed fern *Glossopteris* was present on all of the widely separated southern continents raised questions about how these continents may have been connected when *Glossopteris* lived (250 to 300 million years ago).

FIGURE 3-3 ▲ Alfred Wegener, Hero of Earth Science
Although his work was not appreciated by many of his peers, Alfred Wegener's hypotheses about continental drift laid the groundwork for the theory of plate tectonics.

sils of the same species of extinct plants. This observation suggested that the continents were once joined. Snider-Pellegrini proposed that continents on either side of the Atlantic had moved apart, forming the ocean between them in the process.

In the 19th century, geologists struggled to explain the origin and arrangement of the continents and oceans. Similarities in **fossil assemblages** (groups of fossil species found together), especially those containing the seed fern *Glossopteris* (shown in Figure 3-2), in India, Australia, South Africa, and South America led some to propose that the southern continents were once connected by fingers of land they called "land bridges." These continents and their connecting land bridges were called **Gondwanaland** after an area in India that contained the distinctive fossils. In this hypothesis, the southern continents didn't move; they just became separated when the land bridges sank into the oceans. It wasn't until the 20th century that global map relations, fossil data, and geology were explained by moving continents.

ALFRED WEGENER AND CONTINENTAL DRIFT

The German meteorologist Alfred Wegener (Figure 3-3) was the first to formulate a detailed, global explanation of how the continents assumed their present locations and shapes. Alfred Wegener Wegener, who received his doctorate degree in astronomy from the University of Berlin in 1904, was not a geologist by training, but as a young professor he became interested in the forces that shaped Earth's surface.

Wegener reexamined the geology of the southern continents where earlier scientists had discovered *Glossopteris* fossils. He recognized that widely separated but very similar sequences of sedimentary rocks containing the same fossil assemblages would be close to one another if the continents were moved back together, as you can see in Figure 3-4. This particularly distinctive group of sedimentary rocks, which became known as the Gondwana sequence, contained glacial deposits that clustered around Antarctica in Wegener's construction of the locations of the southern continents.

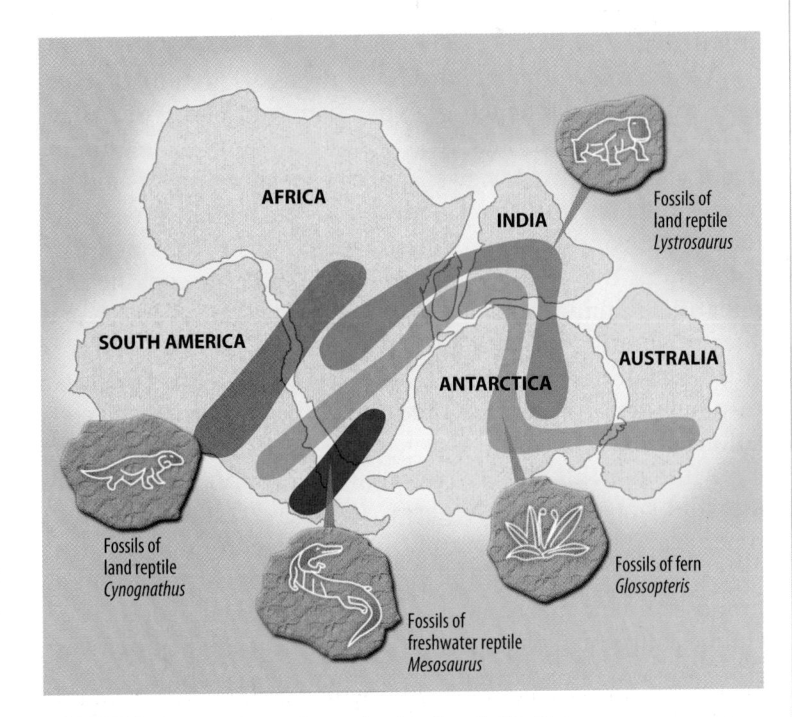

FIGURE 3-4 ▲ Fossil Evidence for Continental Drift
Cynognathus, Glossopteris, Lystrosaurus, and *Mesosaurus* are distinctive fossils now found in widely separated places on the southern continents. Wegener's reconstruction of the continents' locations at the beginning of the Mesozoic era (about 250 million years ago) connected the fossil-bearing regions.

Wegener also pointed out possible ties between North America and Europe, and proposed that all the continents—southern and northern—had once been joined in the past to form a "supercontinent." Wegener named this ancient landmass **Pangaea**, Greek for "all land." Figure 3-5 shows Pangaea and its breakup over a period of 250 million years to form the present-day continents.

To explain the breakup of Pangaea, Wegener did not invoke sinking land bridges as others had before him. Instead, he proposed a mechanism he called "continental displacement," which later came to be known as **continental drift**. Pangaea, Wegener argued, had broken apart by a process scientists call *rifting*. He pointed to the great rift valleys of Africa as examples of places where continents are splitting apart today. Subsequently, the fragments of Pangaea moved away from each other, which explained the shapes and locations of the present-day continents. Wegener first lectured publicly on his ideas in 1912 and later summarized the evidence for continental drift in his 1915 book *Die Entstehung der Kontinente und Ozeane* (*The Origin of Continents and Oceans*).

In the decades that followed, a few geologists (especially those working on the southern continents) continued studies that showed more and more geologic similarities among the continents and supported Wegener's hypothesis. Most geologists, however, particularly those in the United States and Europe, rejected Wegener's arguments for continental drift because he failed to present a viable mechanism for moving the continents. Wegener proposed that tidal forces generated by Earth's spin tugged the continents through the oceanic crust. His critics argued, rightly, that this movement was physically impossible.

Wegener's continental drift hypothesis also required geologists to reconsider much of what they had come to believe about how Earth worked. Accepting the hypothesis required a major change in worldviews, and Wegener's arguments did not have sufficient influence to dislodge the reigning ideas. Although he continued to gather evidence for continental drift throughout the 1920s, Wegener was unable to convince the scientific community of its validity. He died in 1930, shortly after celebrating his fiftieth birthday, during an expedition to cross the Greenland ice sheet, never knowing that his theory of continental drift would become one of the great scientific revelations of the 20th century.

Although the debate over continental drift subsided, the idea did not die with Wegener. During and after Wegener's time, a few scientists continued to explore continental drift. One of these scientists, the British geologist Arthur Holmes, proposed a new mechanism to drive continental drift in 1927. Holmes suggested that Earth's internal heat might drive vast, circular convection currents in the mantle that could, like the rollers beneath a conveyer belt, cause the crust to spread apart along rifts in the seafloor. But mainstream geology continued to resist the idea that the continents moved. It would be another 30 years before people would seriously reconsider mantle movements as a mechanism for continental drift.

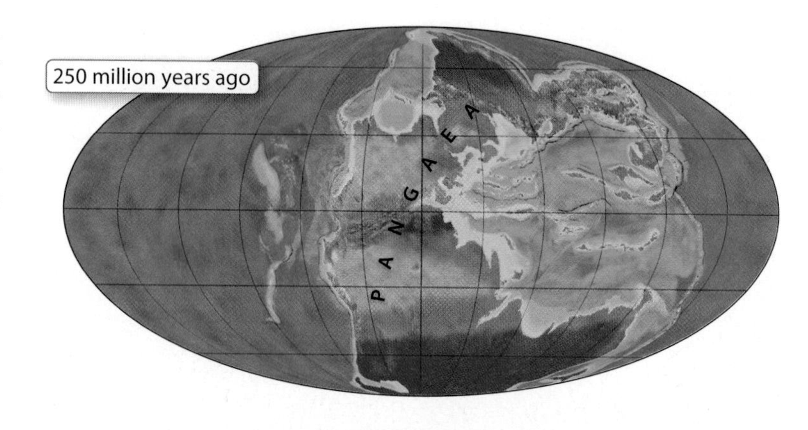

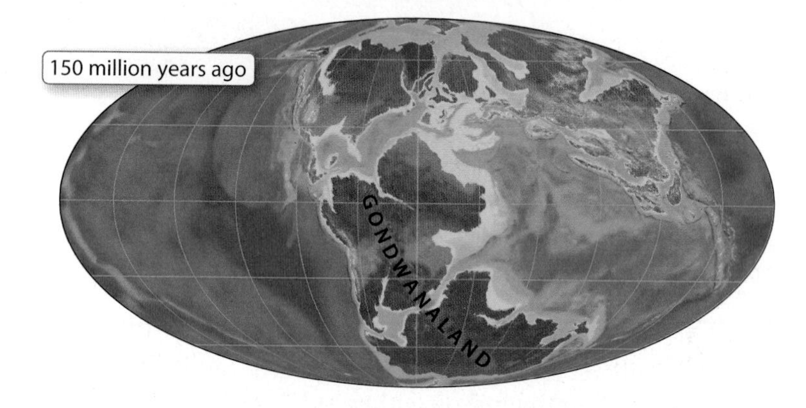

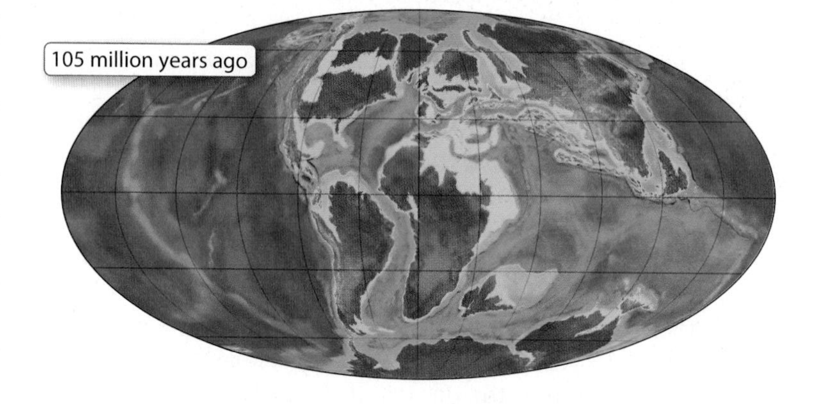

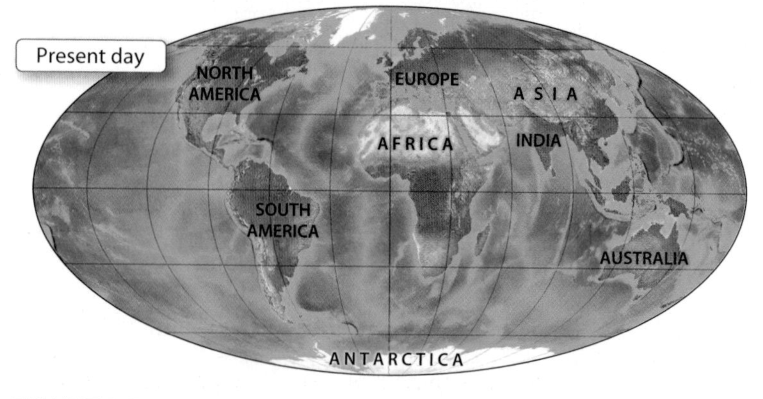

FIGURE 3-5 ▲ The Supercontinent Pangaea
Collisions combined the continents into one giant landmass about 250 million years ago. Since then, the continents have split and moved apart to their present positions.

3.2 Explaining Moving Continents—Plate Tectonics

Two areas of scientific research came together to explain how continents move. In the 1950s, studies of rock magnetism began to reveal the history of Earth's magnetic field. About the same time, studies of the ocean basins began to show what the seafloor was like. Over a period of about 40 years, from Arthur Holmes' suggestion of mantle convection to completion of the basic theory of plate tectonics in 1968, scientists united many seemingly unrelated features and characteristics of the geosphere in the comprehensive theory of plate tectonics, a theory that explains how continents move, ocean basins form, and mountains develop.

WANDERING MAGNETIC POLES

Many rocks have magnetic properties; these properties depend on the presence and composition of certain iron-bearing minerals within the rock. As we noted in Chapter 2, minerals are naturally-occurring solids with an ordered atomic structure and characteristic chemical composition. Crystals of the common iron oxide mineral magnetite have their own magnetic fields that act like little bar magnets or compass needles in the presence of Earth's magnetic field. When these minerals first form—for example, when a volcanic rock solidifies and cools—a part of the magnetic field within the crystals is aligned in an orientation determined by the direction of Earth's magnetic field at that time. This alignment—which is permanent if the rock is not subsequently reheated or chemically altered—creates a *remnant magnetism* or **paleomagnetism** in the rock that tells scientists its location relative to Earth's magnetic poles when the mineral formed.

In the 1950s, scientists measured paleomagnetism in rocks of various ages in Europe, Australia, India, and North America. Rocks of different ages from the same continent indicated different locations for Earth's magnetic poles. It was as if the location of magnetic north had moved over time. Projected onto the globe, magnetic north seemed to shift or wander with time along curved paths to its present position near Earth's geographic North Pole. These changes in the location of the magnetic poles over time are now called **apparent polar wander curves**. Figure 3-6a shows two such curves, for North America and Europe.

The word *apparent* is used because there are actually two possible, but very different, explanations for the wandering location of the magnetic poles. Either the magnetic poles moved relative to unchanging continents, or the continents moved relative to unchanging magnetic poles. If the magnetic poles

FIGURE 3-6 ▼ Apparent Polar Wander Curves
This diagram maps the changes in the apparent location of Earth's north magnetic pole at different geologic times, as recorded by magnetic rocks in North America (red) and Europe (black). **(a)** The polar wander curves for the two continents are different, although they change in similar ways at similar times. **(b)** Moving North America and Europe together (closing the Atlantic Ocean) brings the polar wander curves together. This is what would be expected if the apparent "wandering" of the pole is actually produced by continental drift, and if the two continents were once close neighbors.

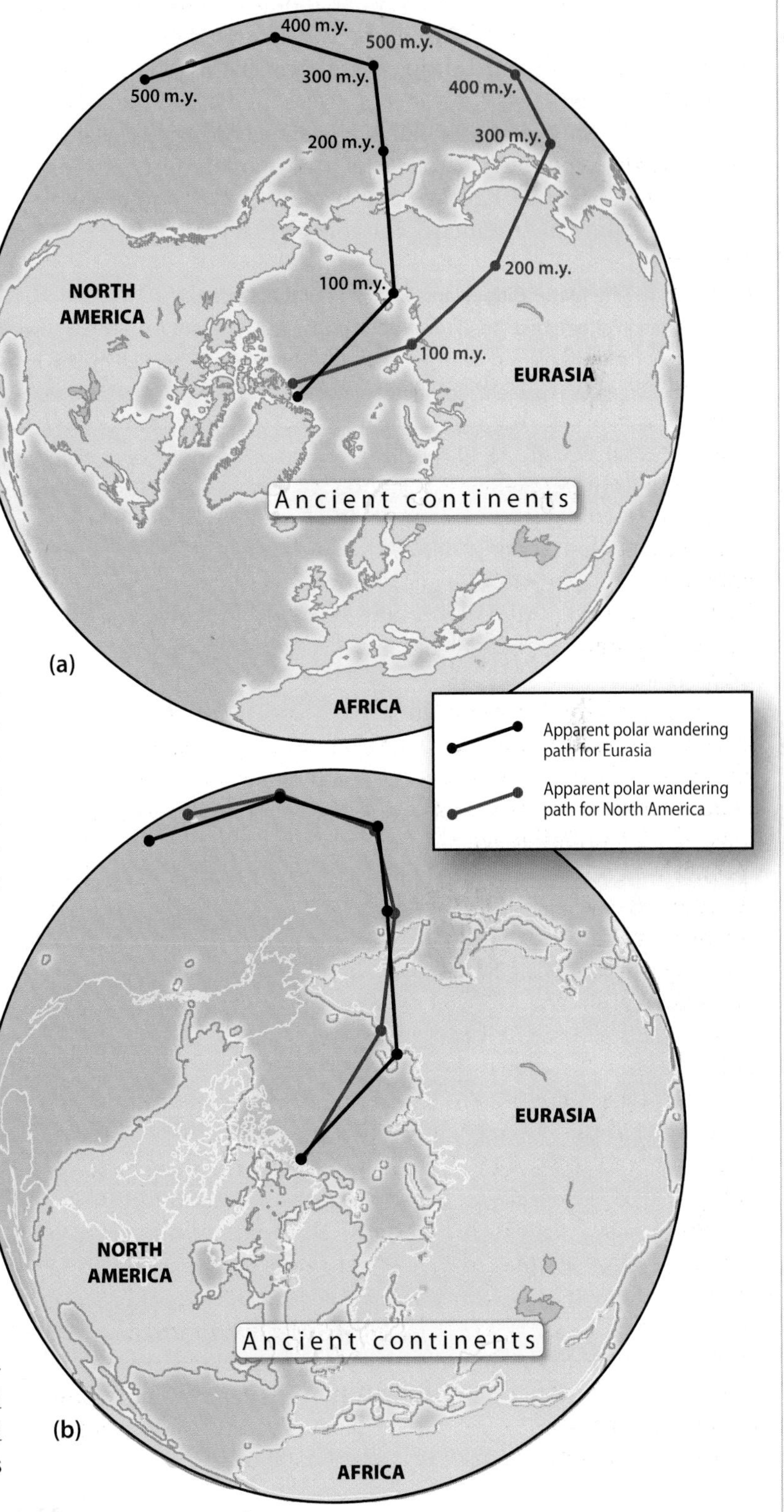

had moved and not the continents, then the polar wander curves from each continent would be the same. But if *continents* shifted position, then the polar wander curve for each landmass would be different.

In the mid-1950s, British scientists P. M. S. Blackett, Keith Runcorn, and Ted Irving reported that the polar wander curves were indeed different for each continent—which meant that the continents had in fact moved! Moreover, some of the polar wander curves were found to be roughly parallel—they changed direction in similar ways at about the same times (as you can see in Figure 3-6a). Figure 3-6b shows that the curves for North America and Europe could even be made to nearly coincide if the two continents were moved very close together—as they would once have been according to Wegener's original hypothesis. From this and similar observations, scientists recognized that the polar wander curves could represent compelling new evidence for continental drift.

EXPLORING THE OCEAN BASINS

At the same time that Blackett and his colleagues were solving the riddle of apparent polar wandering curves in Europe and North America, other scientists were beginning to unlock the mysteries of the seafloor. By the 1950s, new technologies for mapping the seafloor that had been developed during WWII had begun to show the true complexity of Earth's ocean basins. The new maps showed that, rather than being featureless abysses, the ocean basins contain huge mountain chains, deep valleys, and seamounts that rise thousands of meters above the surrounding seafloor. Extending across the floors of Earth's oceans like the seams on a baseball are raised oceanic **ridges**, while in other places, the seafloor descends into **trenches** that are more than 10 kilometers (6 mi) deep. The discovery of oceanic ridges and trenches, both of which are shown in Figure 3-7a, set the stage for the revival of the debate over continental drift.

FIGURE 3-7 ▼ Major Ridges and Trenches of the Seafloor
(a) Exploration after World War II revealed that the seafloor is not flat. Globally, the seafloor contains features such as ridges (mountain chains) and deep trenches. **(b)** A computer-generated map of part of the East Pacific Rise, a mid-ocean ridge. The red areas represent the highest elevations, the blue areas the lowest.

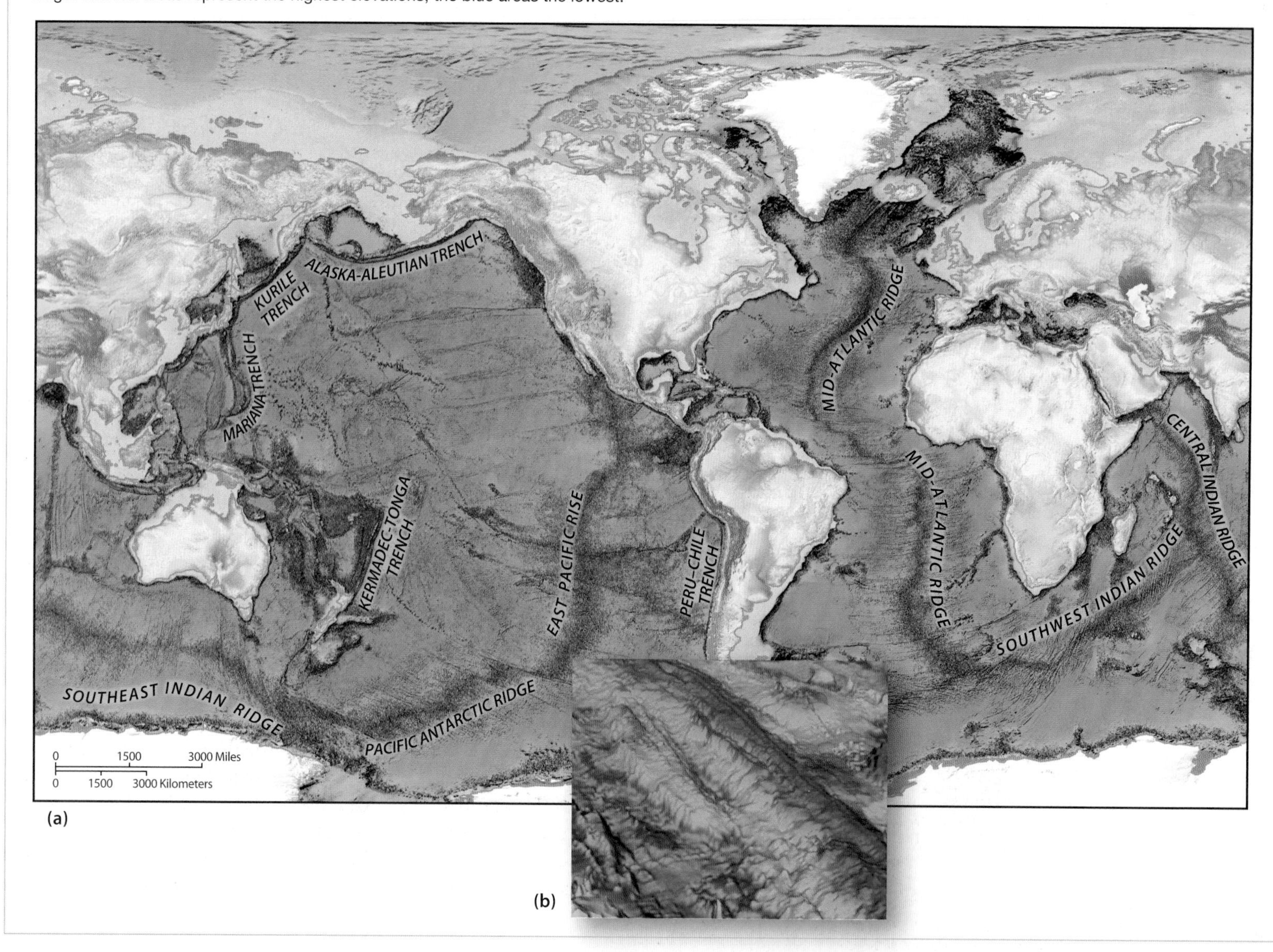

The first oceanic ridge discovered, the Mid-Atlantic Ridge, runs approximately down the middle of the Atlantic Ocean. After World War II, oceanographers began to map the ocean floor using sonar technology, which records and analyzes sound waves reflected from underwater objects. Such studies revealed that oceanic ridges form a continuous chain through all the world's ocean basins. | Exploring the global mid-ocean ridge | Scientists found ridges in the Pacific, Indian, and Antarctic Oceans. The ridges are mountains that rise several kilometers above the seafloor (Figure 3-7b). The flanks of the mountains slope outward for up to 1000 kilometers (600 mi) in some locations. Deep rift valleys run down the crests of the ridges, forming seams that divide the ridges in half. The total length of the oceanic ridge system is about 70,000 kilometers (44,000 mi).

Seafloor mapping also revealed deep trenches along the margins of continents (as in the trench off the west coast of South America) and parallel to chains of oceanic islands. The trenches mark the greatest ocean depths on Earth. The Mariana Trench, in the Pacific Ocean near Guam (see Figure 3-7), plunges to more than 11,000 meters (36,000 ft)—deep enough to swallow Mt. Everest with more than a mile to spare. | The Mariana Trench | Scientists wanted to know why the ocean floor had ridges, trenches, and seamounts. One scientist's work offered an explanation.

SEAFLOOR SPREADING

In 1962, a geologist at Princeton University, Harry Hess, published a paper titled *History of the Ocean Basins* that proposed a new hypothesis to explain continental drift. | Harry Hammond Hess: spreading the seafloor | His hypothesis connected many observations about the ocean basins and built upon the earlier ideas of Arthur Holmes about convection in the mantle. Hess's hypothesis became known as **seafloor spreading**, a fundamental part of what would become the theory of plate tectonics.

First, Hess reasoned that the ocean basins are young relative to the continents. He estimated the age of the ocean basins from the rate at which deep sea sediments are deposited. Hess estimated that seafloor sediments are about 1.3 kilometers (0.8 mi) thick on average. At an average sedimentation rate of 1 centimeter (0.4 in) per thousand years, it would only take 230 million years to deposit 1.3 kilometers of sediment. Although this was clearly a ballpark estimate, Hess concluded that the ocean basins were billions of years younger than the adjacent continents. He wondered how such an age difference could develop.

Hess also noted that the mid-ocean ridges are more than just physical features on the seafloor. High heat flow from the mantle is present at mid-ocean ridges, and many volcanoes develop there. Hess thought that flat-topped seamounts called *guyots* were volcanoes that had formed at shallow depths, where waves could erode their tops flat. He hypothesized that guyots found in deep water had moved laterally away from the shallower water where they commonly formed along the crests of mid-ocean ridges.

To explain his observations, Hess proposed that the lithosphere moves outward from mid-ocean ridges, drawn by vast convection currents in the mantle below. As the ocean floor spreads apart at the center of the ridges, **magma** (molten rock) wells up from the mantle to fill the gap and creates new seafloor, as shown in Figure 3-8. Indeed, the presence of hot, buoyant magma and rock beneath the ridges would explain their higher elevation compared to the average elevation of the seafloor. Over millions of years, Hess suggested, new seafloor spreads outward from the ridges like conveyor belts moving in opposite directions.

Once Hess and his colleagues had refined the theory of seafloor spreading, they were faced with another question. If new seafloor forms at mid-ocean ridges, then what happens to older seafloor elsewhere? While some scientists briefly considered the idea that Earth was expanding to accommodate the new lithosphere, Hess and his colleagues suggested the simple, elegant explanation that the oceanic lithosphere sinks back into the mantle at the deep ocean trenches.

Hess suggested that continents did not plow through the oceanic lithosphere on their own power, as Wegener had proposed, but instead rode passively on top of the moving

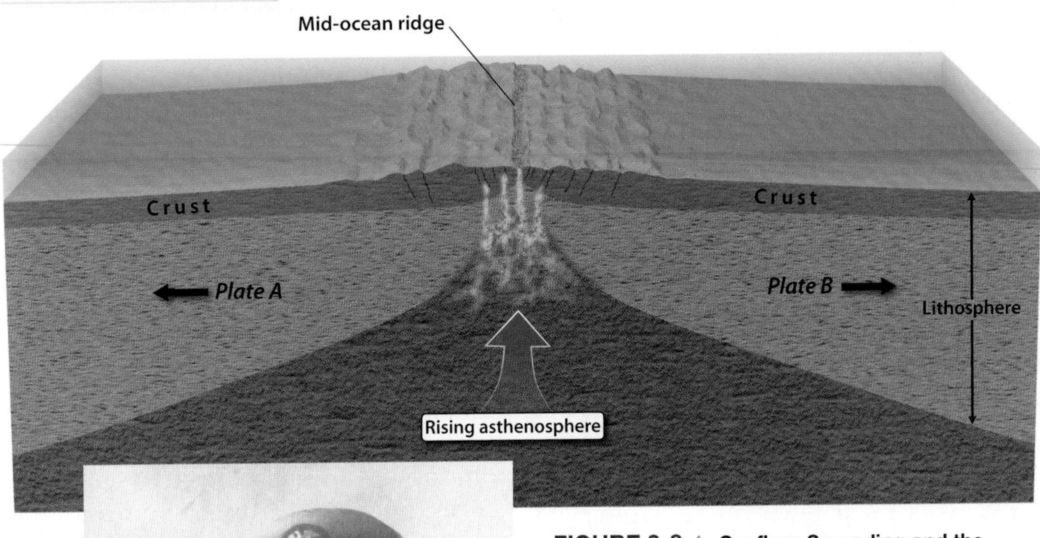

FIGURE 3-8 ▲ Seafloor Spreading and the Creation of New Ocean Crust
At mid-ocean ridges, new oceanic crust is formed as the lithosphere pulls apart and magma from the mantle wells up, cools, and solidifies. Harry Hess (inset) was the visionary geologist who developed the hypothesis of seafloor spreading to explain many fundamental characteristics of the seafloor.

mantle below. More than 50 years after Wegener first proposed continental drift, Hess's seafloor spreading hypothesis represented nothing less than a complete mechanism for continental drift. It was a remarkable synthesis and interpretation of a wide range of geologic features and data.

Recall from the discussion of the scientific method in Chapter 1 that hypotheses in science must be tested. Scientists soon put seafloor spreading to the test with new magnetic studies.

MAGNETIC STRIPES

The hypothesis of seafloor spreading was consistent with key physical features of the seafloor—oceanic ridges and trenches and the presence of guyots in deep water—but would it hold up under further testing? Researchers collected new magnetic data using shipboard instruments called *magnetometers* that were developed to detect submarines during World War II. Research ships traveled back and forth across the mid-ocean ridges and used the magnetometers to measure the strength of magnetism on the ocean floor. USGS geology in the parks plate tectonic animations The researchers discovered that the ocean floor has a distinctive magnetic character. Roughly linear regions of relatively high magnetic strength alternate with linear regions where the magnetic strength is considerably lower.

When these regions are plotted on a map, with the areas of high magnetic strength shaded dark and the areas of low magnetic strength light, they form a pattern of dark and light stripes arranged symmetrically about a mid-ocean ridge, like that shown in Figure 3-9. Because of this pattern, these ocean floor magnetic features became known as **magnetic stripes**. Magnetic stripes were a surprise discovery, and initially scientists were not certain what caused them. At first they thought they simply represented areas of stronger and weaker rock magnetism. But they were to learn this was not the case.

Scientists deciphered the true meaning of magnetic stripes when they took into account an earlier discovery—Earth's magnetic field episodically switches its polarity as if it were a giant bar magnet that periodically flips 180 degrees. Earth's magnetic field is either "normal" (pointing north as it is today) or "reversed" (pointing south). These changes, or **magnetic reversals**, have happened at irregular intervals throughout geologic time.

In 1963, Frederick Vine and Drummond Matthews, two geophysicists from Cambridge University in England, proposed that the magnetic stripes corresponded to times in the past when Earth's magnetic field was either normal or reversed. (A Canadian geophysicist, Lawrence Morley, independently thought of the same explanation for magnetic stripes in 1963.) The areas of high magnetic strength were interpreted as representing regions where Earth's present magnetic field is slightly *reinforced* by the remnant magnetic field of seafloor rocks. In these areas, the rocks were formed with their magnetic fields pointing in the same direction as that of today's Earth (normal polarity). The areas of low magnetic strength, by contrast, were regions where Earth's present magnetic field is opposed, and thus slightly *weakened*, by the remnant magnetic field of seafloor rocks. In these areas, the rocks were formed with their magnetic fields pointing in the direction opposite that of today's Earth (reversed polarity). If this was true, the stripes were the equivalent of tape recordings of past magnetic reversals.

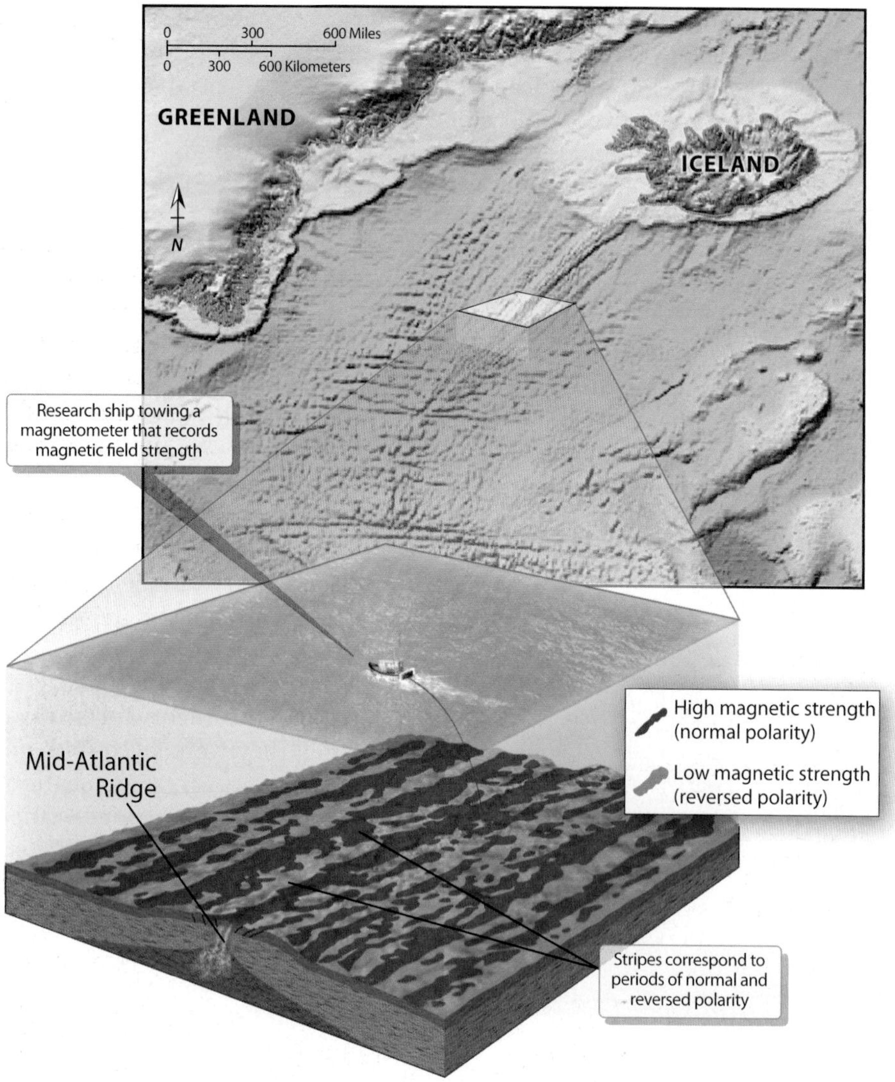

FIGURE 3-9 ◀ Magnetism of the Ocean Crust
The magnetic field of the ocean crust varies in a pattern of alternating weak and strong magnetic stripes, arranged symmetrically about the center of the mid-ocean ridges. This is part of the pattern of magnetic stripes along the Mid-Atlantic Ridge south of Iceland. Such patterns preserve a record of past reversals of Earth's magnetic field.

(a)

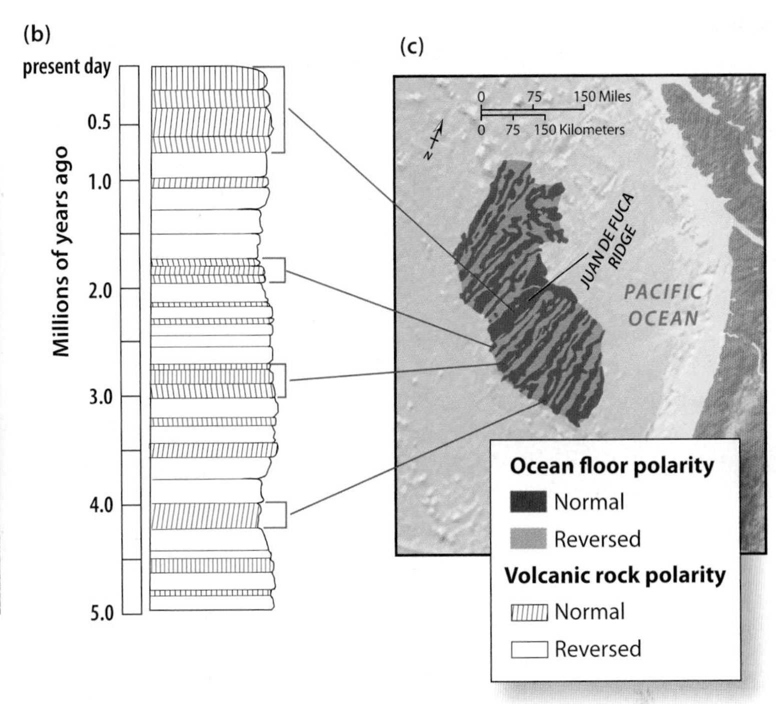

FIGURE 3-10 ▲ Volcanic Rock Sequences Were Used to Date Polarity Intervals
Analysis of well-dated rocks on land, like these 5-million-year-old basalt flows on Kauai **(a)**, made it possible to construct a timetable of magnetic polarity reversals. The timing and duration of these reversals **(b)** could be correlated with the magnetic stripe patterns on the seafloor **(c)**.

Was the Vine and Matthews proposal correct? Did the magnetic stripes represent alternating normal and reversed polarities of Earth's magnetic field? Magnetic studies of thick layered sequences of volcanic rocks on land (such as those shown in Figure 3-10a) proved critical to testing the proposal. These studies were the first to actually define and date the alternating polarity of Earth's magnetic field through geologic time. Using the radiometric dating techniques you learned about in Chapter 2, scientists could determine when the Earth's magnetic field had stayed normal or had reversed from the present back through millions of years. And with this information, they could examine magnetic stripes in the oceans to see if their alternations matched the pattern of magnetic reversals on land.

They did. The magnetic stripes showed the same alternation of normal and reversed polarity, moving laterally away from the mid-ocean ridges, as successively older volcanic rocks did on land (Figure 3-10b). This pattern is just what you would expect if the newly-formed rocks at mid-ocean ridges were split and moved in opposite directions as the seafloor spread apart. Moreover, the widths of the stripes were proportional to the durations of the polarity intervals, as would be the case if the volcanic center at the oceanic ridge produced new oceanic crust at a constant rate.

This pattern was a remarkable finding. It strongly supported Hess's proposal that new oceanic crust was formed along the mid-ocean ridges and slowly migrated away, so that the seafloor became older away from the ridges (Figure 3-11).

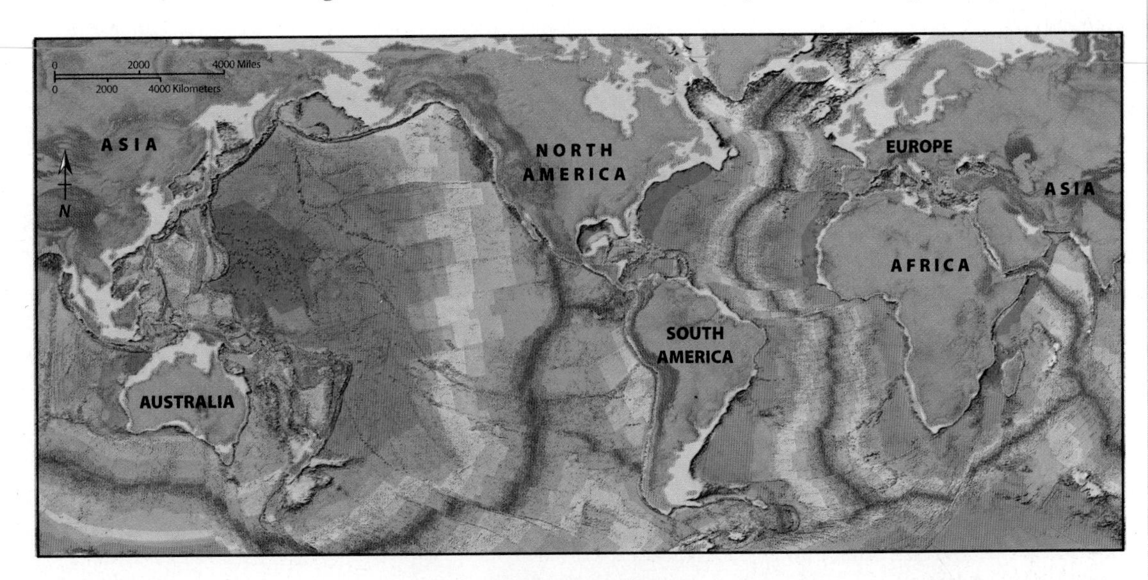

FIGURE 3-11 ▶ Age of the Seafloor
Note that the mid-ocean ridges are extremely young by geological standards—no more than 2 million years old. As you move away from the ridges, the rocks of the seafloor become progressively older.

The width of the stripes reflected the amount of new crust formed during each period between magnetic reversals. By dividing the width of each stripe by the length of time between subsequent reversals, scientists were able to calculate the rate at which the seafloor was moving away from the ridges.

EARTHQUAKES PROVIDE ANOTHER TEST

Earthquake (seismological) studies of oceanic ridges and trenches provided key evidence needed to flesh out the emerging theory of plate tectonics. In 1965, a Canadian seismologist, J. Tuzo Wilson, made a major contribution. J. Tuzo Wilson: discovering transforms and hot spots Proponents of plate tectonics already knew of two types of plate interactions: plates moved apart at oceanic ridges, and converged at oceanic trenches. Wilson proposed a third type of interaction: plates also slid past each other.

Maps of the ocean floor had revealed many **faults**, places where rock has broken and the blocks on opposite sides of the break have moved relative to each other. You can see in Figure 3-12a that many large faults cut across the ridges at right angles. Wilson recognized that the lateral sliding movement at these faults connected the spreading movements taking place at adjacent ridge segments, as shown in Figure 3-12b. He called this new type of fault **transform faults** because they convert or "transform" the motions along adjacent converging or diverging plate boundaries into lateral sliding.

Studies of earthquakes in the ocean basins confirmed the types of movement taking place at plate boundaries. The

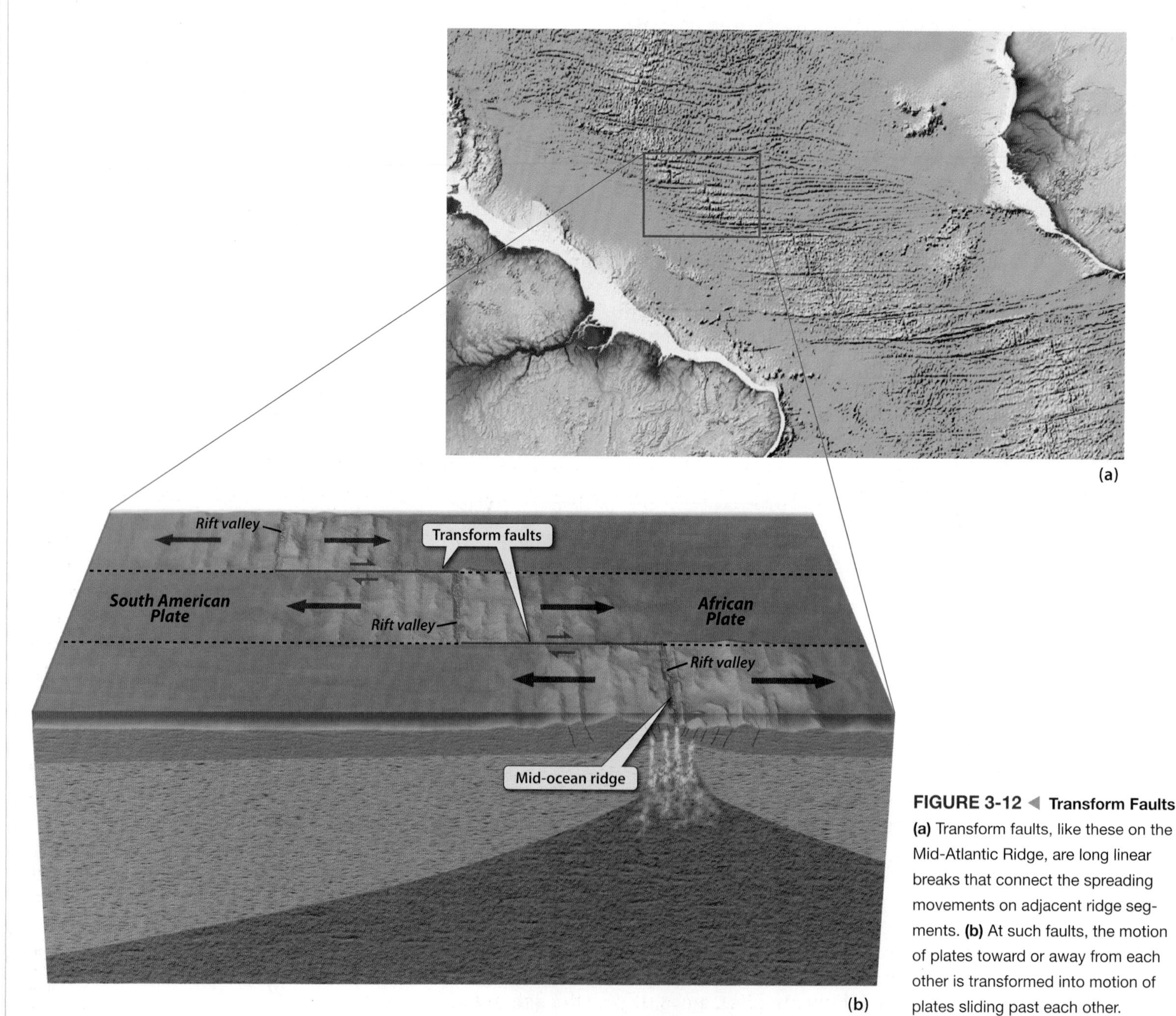

FIGURE 3-12 ◀ Transform Faults **(a)** Transform faults, like these on the Mid-Atlantic Ridge, are long linear breaks that connect the spreading movements on adjacent ridge segments. **(b)** At such faults, the motion of plates toward or away from each other is transformed into motion of plates sliding past each other.

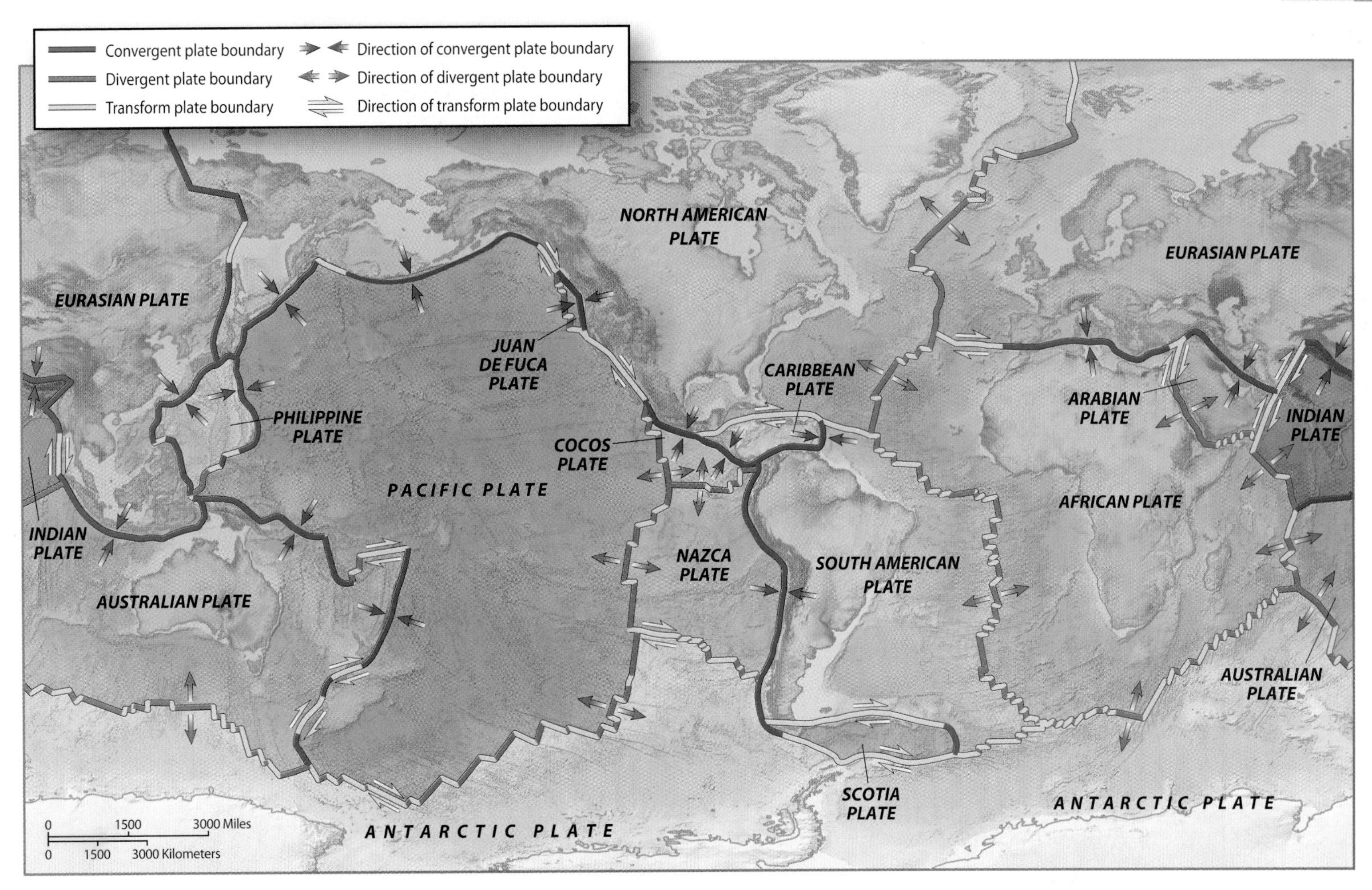

FIGURE 3-13 ▲ The Geosphere's Tectonic Plates
The tectonic plates, many of which include both oceanic and continental lithosphere, come in many sizes and shapes. Some of the motions of the major plates relative to one another are indicated by arrows.

lithospheric blocks along transform faults were found to be sliding horizontally past one another as Wilson had predicted. At deep ocean trenches, earthquakes occurred along a zone inclined 40 to 60 degrees to the seafloor. Plate convergence here was causing lithospheric blocks to slide downward into the mantle at an angle. As first proposed by Wilson, spreading at ridges, convergence at trenches, and the connecting movements along transform faults defined a global network of dynamic plate boundaries, shown in Figure 3-13. By the late 1960s, the case for plate tectonics had been made.

PLATE TECTONICS TODAY

Today plate tectonic concepts have merged into a theory that explains the major features, processes, and relationships of Earth's lithosphere, including major surface features like mountain ranges and rift valleys. In the current version of the theory, a **plate**—more formally, a *lithospheric plate* or *tectonic plate*—is a discrete piece of lithosphere that moves relative to other pieces. The plates vary in composition as well as size. Some, like the Pacific plate, are composed almost entirely of oceanic lithosphere; others, like the Eurasian plate, are mostly continental lithosphere; while still others include both. For example, the North American plate, as you can see in Figure 3-13, comprises all of North America and about half of the Atlantic Ocean.

Tectonic plates are in motion. They move slowly, at an average rate of several centimeters (a few inches) per year, or about as fast as your fingernail grows. The principal cause of this motion is temperature variations that create density differences in the lithosphere and mantle, making them unstable. Relatively warm materials rise and colder materials sink. As oceanic lithosphere moves away from a spreading center, it becomes colder, thicker, and denser. This denser oceanic lithosphere is what sinks into the mantle at oceanic trenches. As it does so, it causes plates to be pulled apart at spreading centers. Hot mantle material can then rise beneath spreading centers, partly melt, supply magma upward to the crust and seafloor, and create new oceanic lithosphere. The nature of the mantle movements involved in these processes is a major focus of research today. The movements are more complicated than the simple convection cells of the early proposals but their linking of lithosphere and mantle movements were on the right track.

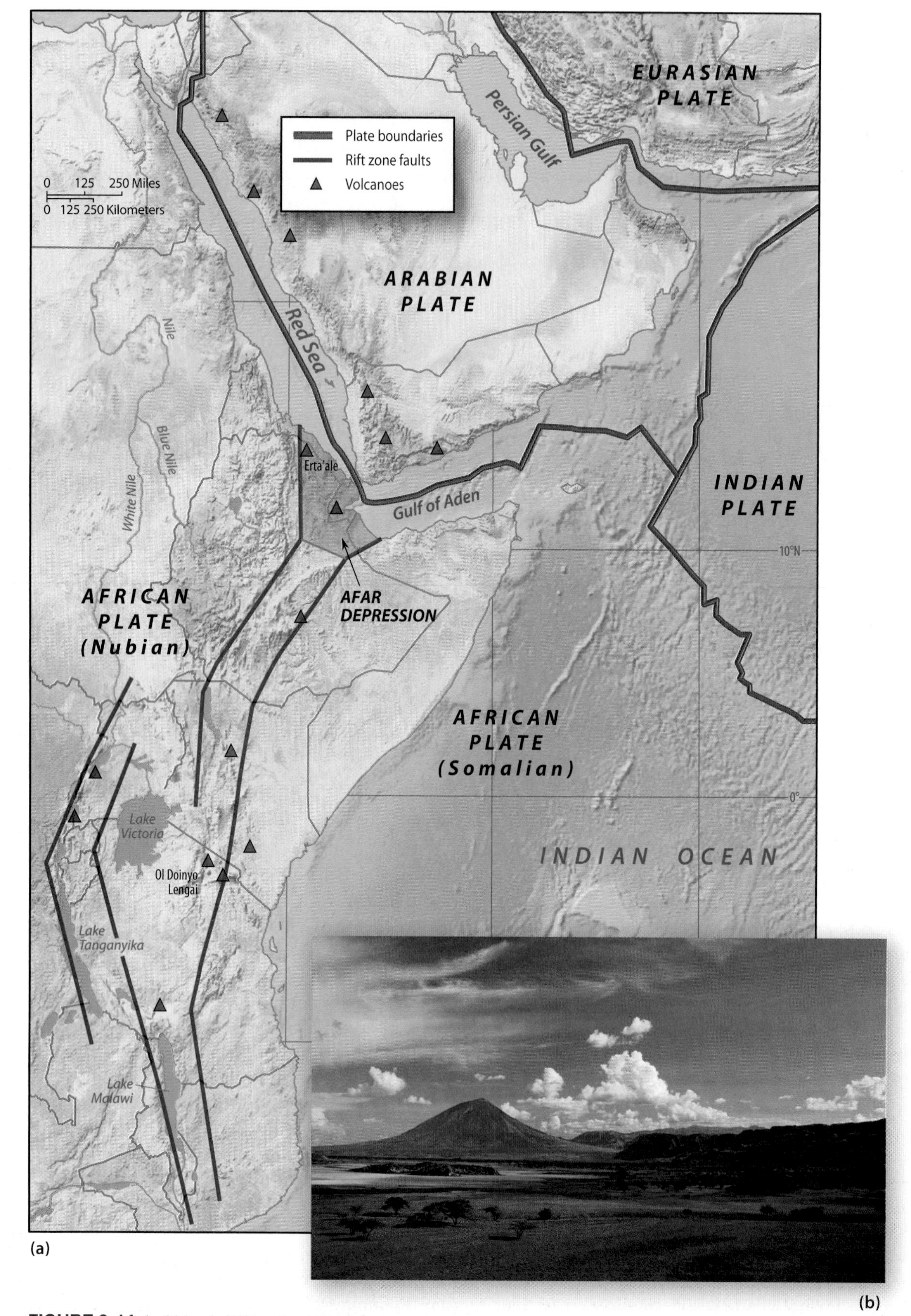

FIGURE 3-14 ▲ Africa Is Rifting Apart Today

(a) Upwelling in the asthenosphere is causing the lithosphere to thin and weaken. As a result, the African continent is being separated from Arabia along the Red Sea and split apart along the East African rift valleys **(b)**.

Let's turn the clock back to the time of Pangaea again, about 250 million years ago. It is one thing to understand how ocean basins form through seafloor spreading, but how do continents split apart? It turns out that Wegener had the right idea in explaining continental breakup.

Mantle material can rise beneath continents as well as mid-ocean ridges. Where warm material in the mantle rises, the overlying continental lithosphere thins and weakens as it is warmed and stretched. As this process continues the continental lithosphere breaks and splits apart along long linear valleys called **rifts**. One of today's best examples of this phenomenon is the Red Sea, which is forming into a new ocean basin where Africa has split off from Arabia. The rifts connecting to the Red Sea, as Figure 3-14 indicates, are now splitting a piece of East Africa away from the rest of this huge continent. Africa is splitting apart, in the same way Pangaea did, starting some 250 million years ago. Figure 3-15 summarizes how continental rifting can develop and eventually lead to splitting of continents, seafloor spreading, and creation of new oceans like the Red Sea.

You have already started to understand the importance of plate boundaries in all of these processes. Plate boundaries are where plates interact. They are places where the plates split apart, slide past one another, or converge.

3.3 Plate Boundaries—Where the Action Is

If plates move on average just several centimeters (a few inches) per year, how would a person even know it was happening? If you were standing in the middle of a plate, you probably wouldn't notice. But if you were near a plate boundary, movements would be hard to miss. If you knew what to look for, evidence of plate movements would be all around you. And once in a while, something dramatic would happen. Perhaps the ground would shake, or a volcano would erupt. Plate boundaries are where movements between plates occur and the lithosphere is broken or faulted. At the edge of a plate, rocks are stressed, and the geosphere is in action.

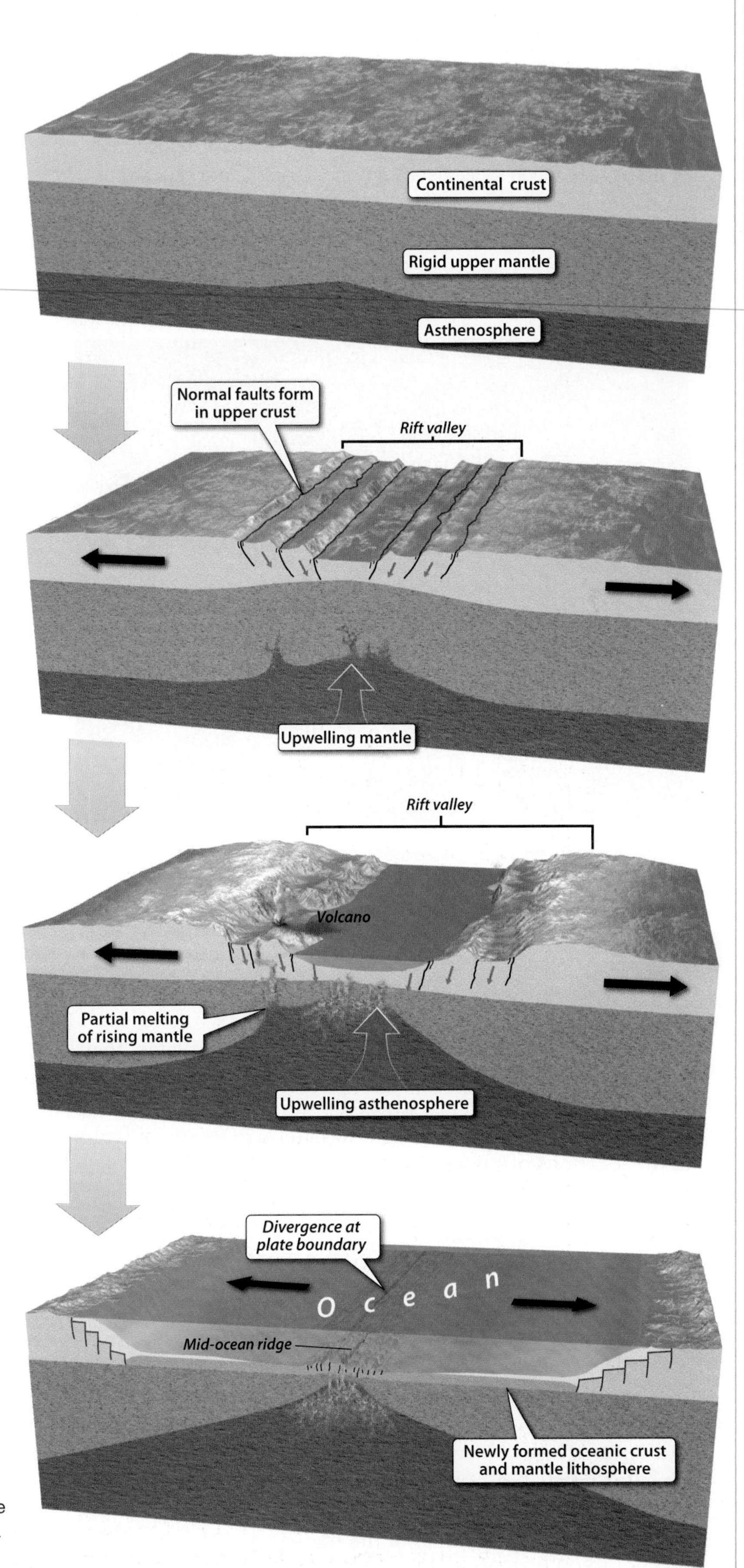

FIGURE 3-15 ▶ Rifting Splits Continents
(a) Rifting starts when hot material from the asthenosphere begins to rise. **(b)** The upwelling asthenosphere heats and weakens the crust, resulting in thinning of the lithosphere and normal faulting. **(c)** Crustal extension proceeds until the continental crust becomes thin enough to split apart. **(d)** The pieces of continent migrate away from each other as mantle upwelling continues and oceanic crust is formed where the continents once joined.

In The News

Watching Earth Move

There's more to observe from space than the geographic features that astronauts like Jack Schmidt admire on their trips to the Moon or the space station. Scientists are using a satellite network and Global Positioning System (GPS) instruments located across the Andes Mountains to monitor the convergence of the Nazca and South American plates.

At the convergent boundary where the Nazca plate sinks below the South American plate, the Andes mountains are being pushed upward at measurable rates.

GPS technology, which is the same technology you can use to locate yourself on a hike or in a car as you travel across the country (albeit a more accurate version), can detect changes of only a few millimeters (less than half an inch) in a site's location. Rates of plate movement are in the range of 10 to 200 millimeters (0.4 to 8 in) per year. At the faster rates, GPS receivers can detect changes in a site's location within just a few weeks. Over a period of a year, they can measure changes along even the slowest-moving plate boundaries.

GPS measurements show that the Nazca plate is moving relative to South America at a rate of 78 millimeters (3 in) per year. But on land, this motion can be separated into three components; (1) 35 millimeters (1.4 in) of continuous motion along the boundary (called the subduction zone) of the sinking Nazca plate, (2) 35 millimeters of episodic motion (stopping and starting), accompanied by earthquakes, along this plate boundary, and (3) 10 millimeters (0.4 in) of compressive deformation within the South American plate. This compressive deformation causes the Andes Mountains to rise—in some places by as much as a millimeter or two each year. Literally, scientists are watching the Andes Mountains grow.

Real-time measurement of Earth's surface movements with GPS technology is revolutionizing the study of many Earth systems processes. Watching mountains grow is just one example. GPS technology now enables scientists to watch volcanoes change before they erupt, to watch fault movement that helps them understand earthquakes, and to watch slowly but continuously moving landslides (Chapter 8). Using the Global Positioning System (GPS) to measure volcano deformation Don't be surprised to see the results of GPS Earth monitoring in the news. It's one way in which scientists are getting better at predicting the geosphere movements that affect people.

Scientists have made major efforts to identify plates, their boundaries, and the history of their movement since the 1960s. Plate boundaries are where the lithosphere breaks and they have been identified through the careful mapping of faults and the locations of earthquakes. There are three types of plate boundaries: divergent, convergent, and transform.

DIVERGENT PLATE BOUNDARIES

Divergent plate boundaries are regions where two plates move away from each other. They are located at the major mid-ocean ridges and continental rift zones of the world, where the upwelling of hot mantle material thins the lithosphere and extends the crust. As shown in Figures 3-8 and 3-15, magma flows upward into the crust along divergent plate boundaries. This helps warm and expand the lithosphere to form the giant underwater mountain ranges of the mid-ocean ridge system. In some places, the crests (tops) of the mid-ocean ridges are 4500 meters (14,800 ft) higher than the deep seafloor. For comparison, only a few mountains in North America have this much change in elevation (also called topographic relief) between their summits and surrounding lowlands.

As noted previously, plates move apart, or spread, at rates that range from 10 to 200 millimeters (0.4 to 8 in) per year, and the mid-ocean ridge surface features (called physiography) reflect the rate of separation. Exploring the global mid-ocean ridge There are slow-spreading and fast-spreading centers. At a slow spreading rate of 2 to 5 centimeters (0.8 to 2 in) per year, like that across the Mid-Atlantic Ridge, the crest is marked by linear, steep-walled depressions up to 10 kilometers (6 mi) long and 3 kilometers (2 mi) deep, visible in Figure 3-8a. These are fault-bounded underwater *rift valleys* or *axial rifts*. In width and depth, they are as big as the Grand Canyon.

Inclined breaks in the lithosphere where the upper block of rock has moved down relative to the lower block, as shown in Figure 3-16 on p. 67, are called **normal faults**. Geologists say that movement takes place "across" a fault. They call the motion across a fault "relative" because it is not always clear whether both blocks moved in the same direction but by different amounts, whether the lower block moved up and the upper

FIGURE 3-16 ▲ Normal Faults
Normal faults occur when an upper block (at right here) moves down relative to a lower block (left). They are common at divergent plate boundaries. (Note that the terms "upper" and "lower" refer to the position of each block relative to the fault plane.)

block moved down, or whether some other combination of movement produced the observed displacement. In normal fault displacement, the upper block of rock always moves down relative to the lower block. Normal faults are common at divergent plate boundaries.

Normal faults are especially well developed across slow-spreading centers. The two plates are being pulled apart or extended and gravity causes the blocks of rock between the normal faults to drop down. Volcanoes are common, too, both as long, linear upwellings of lava along the axial rift and as small, conical seamounts scattered across and along the ridge crest. Because spreading is slow, the amount of magma rising to the crust in places like the Mid-Atlantic Ridge is less than that along fast-spreading ridges.

Fast-spreading centers, like the one in the East Pacific Ocean, are broad and less rugged than slow-spreading centers like the Mid-Atlantic Ridge (see Figure 3-7). Normal faulting is less common here, and the crest mostly lacks well-defined axial valleys. Magma rising along fast-spreading centers is voluminous and commonly occurs along volcanic fissures rather than seamount-type volcanoes. The crust stays warm along fast-spreading centers, so they are less prone to faulting and earthquake generation than slow-spreading centers.

Volcanoes at oceanic divergent plate boundaries do not have particularly explosive eruptions. Because about 90 percent of these volcanoes are underwater and not violently explosive, we do not notice them much. We also fail to notice the underwater hot springs or *hydrothermal vents* that are on the deep ocean floor at divergent plate boundaries. Seawater seeps into the fractured and faulted hot rocks that form at ridges—the new oceanic crust—and becomes very hot. The hot water can reach 400° C (750° F). This water escapes upward to the seafloor, forming hot springs. Because of their dark, cloudy appearance, such ocean floor hot springs (shown in Figure 3-17) are called "black smokers." The hot waters contain dissolved minerals from the volcanic rocks with which they have come into contact. When the water is released onto the seafloor, new metal-bearing minerals form and accumulate on the seafloor. Specially adapted microbes use chemical energy from these minerals (instead of oxygen) and become the base of the food chain in an ecosystem where many organisms, such as the tube worms visible in the lower left portion of the photo, can flourish. Exploring the deep ocean floor: hot springs and strange creatures

What are some of the Earth systems interactions that occur at a hydrothermal vent on the ocean floor?

FIGURE 3-17 ▶ Hot Springs Form on the Seafloor Near Spreading Centers
Mineral-rich hot waters from hydrothermal vents form dark plumes called "black smokers." Organisms such as the tube worms at lower left have adapted to the very hot (up to 80° C or 176° F) environments.

We know that Earth is not expanding; therefore, if new lithosphere is created at divergent plate boundaries, then it must be consumed somewhere else. It is consumed at convergent plate boundaries.

CONVERGENT PLATE BOUNDARIES

Convergent plate boundaries are regions where two plates move toward each other. These boundaries can be between two oceanic plates, an oceanic plate and a continental plate, or two continental plates. In the first two cases, the result is similar—as shown in Figure 3-18, oceanic lithosphere sinks back into the mantle.

What makes plates sink? There are two reasons—density differences and gravity. As oceanic lithosphere moves away from a spreading center, it cools and becomes thicker and denser. Where it moves toward another plate—at a convergent plate boundary—the denser of the two plates sinks into the mantle below the other. Continental plates are less dense than oceanic plates. Therefore, where the two converge (as Figure 3-18 indicates), the oceanic plate sinks and is overridden by the continental plate.

This process of the lithosphere sinking into the mantle is called **subduction**. Subduction involves movements between two plates along complex fault systems that extend from the surface deep into the mantle. The earthquakes that these movements generate reveal the subsurface zone of faulting, called a **subduction zone**, between the two plates. A deep depression or trench on the seafloor develops where the two plates come together and the denser lithosphere first begins to sink.

The faults commonly associated with subduction are called **reverse faults**. These are inclined breaks in the lithosphere where the upper block of rock has moved *up* relative to the lower block, as shown in Figure 3-19. The displacement, therefore, is opposite to that of normal faults, and is caused by the opposite type of stress—compression rather than extension. Compression occurs where forces in the lithosphere are directed toward each other. The leading cause of compression in Earth's lithosphere is plate convergence—the movement of two plates toward each other. This is why reverse faults are characteristic of convergent plate boundaries.

FIGURE 3-19 ▲ Reverse Faults
The upper block moves up relative to the lower block on reverse faults.

What happens when two continental plates converge? Neither can sink into the mantle, so a collision occurs. Continent-to-continent collisions create broad regions of rock deformation and faulting called thrust belts, characterized by low-angle reverse faults called **thrust faults**. The collisions stack up the continental crust into mountain ranges, some of

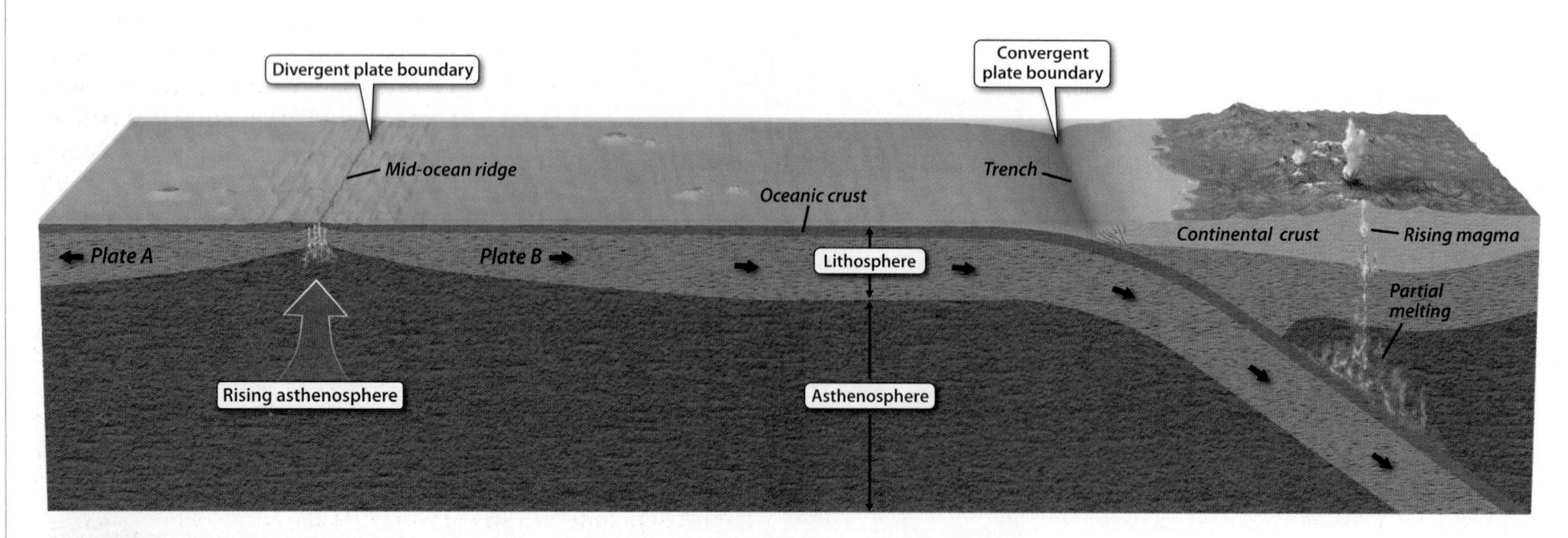

FIGURE 3-18 ▲ Consuming Lithosphere at Convergent Plate Boundaries
The sinking plate forms a seafloor trench where it begins to descend into the mantle along a subduction zone. Subduction zones, the boundaries between overriding and sinking plates, are characterized by reverse faults that can cause very large earthquakes.

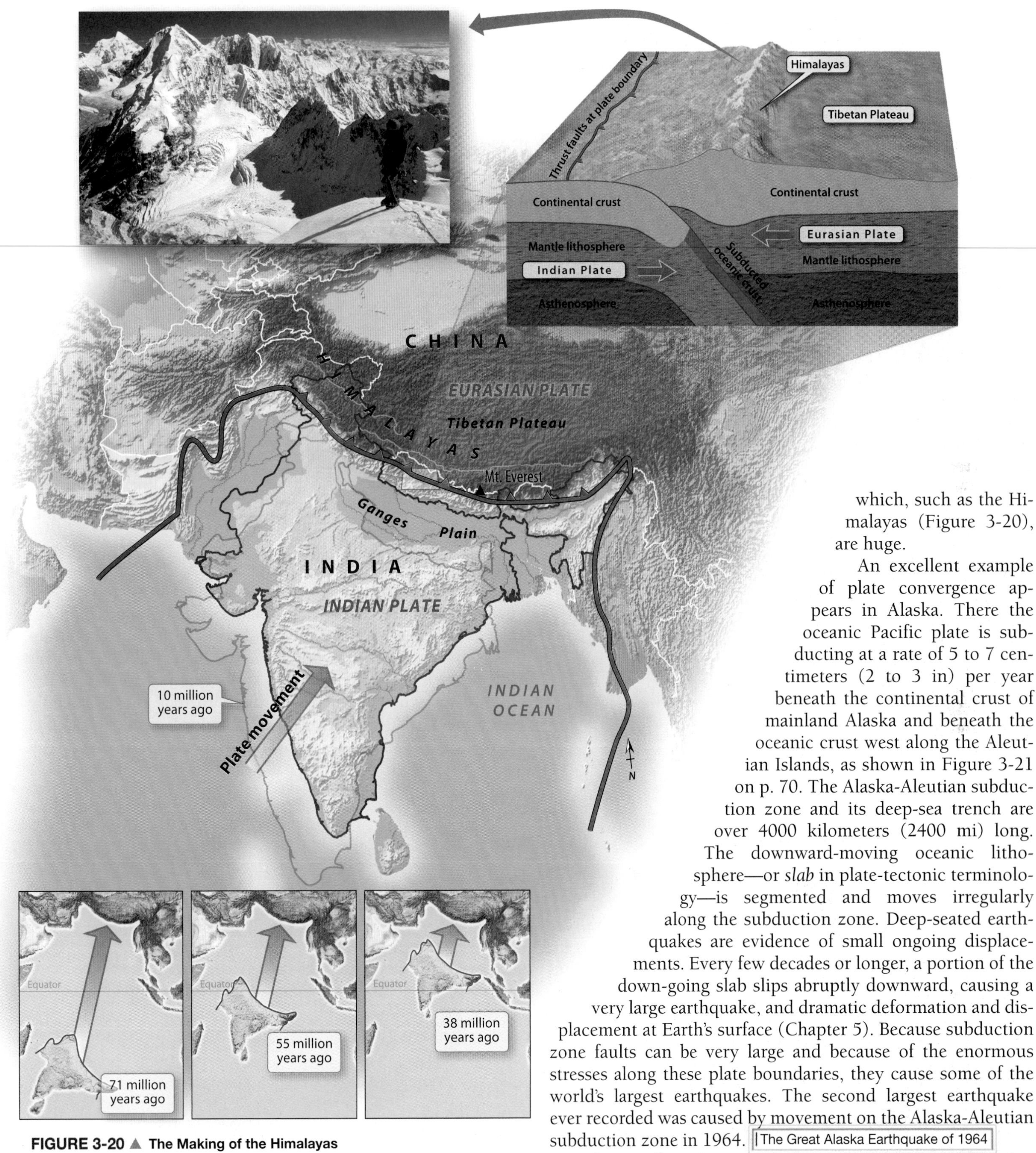

FIGURE 3-20 ▲ The Making of the Himalayas
Where continents converge, neither plate is dense enough to sink into the mantle. The two continents come together in great collisions that form the world's highest mountains. The collision between the Indian and Eurasian plates has produced the Himalayas.

which, such as the Himalayas (Figure 3-20), are huge.

An excellent example of plate convergence appears in Alaska. There the oceanic Pacific plate is subducting at a rate of 5 to 7 centimeters (2 to 3 in) per year beneath the continental crust of mainland Alaska and beneath the oceanic crust west along the Aleutian Islands, as shown in Figure 3-21 on p. 70. The Alaska-Aleutian subduction zone and its deep-sea trench are over 4000 kilometers (2400 mi) long. The downward-moving oceanic lithosphere—or *slab* in plate-tectonic terminology—is segmented and moves irregularly along the subduction zone. Deep-seated earthquakes are evidence of small ongoing displacements. Every few decades or longer, a portion of the down-going slab slips abruptly downward, causing a very large earthquake, and dramatic deformation and displacement at Earth's surface (Chapter 5). Because subduction zone faults can be very large and because of the enormous stresses along these plate boundaries, they cause some of the world's largest earthquakes. The second largest earthquake ever recorded was caused by movement on the Alaska-Aleutian subduction zone in 1964. The Great Alaska Earthquake of 1964

Chains of explosive volcanoes form above subduction zones (Chapter 6). Because these volcanic chains are commonly curved, we call them *arcs*. **Island arcs** form in the oceans, and **volcanic arcs** form on continents (see Figure 3-18). As you

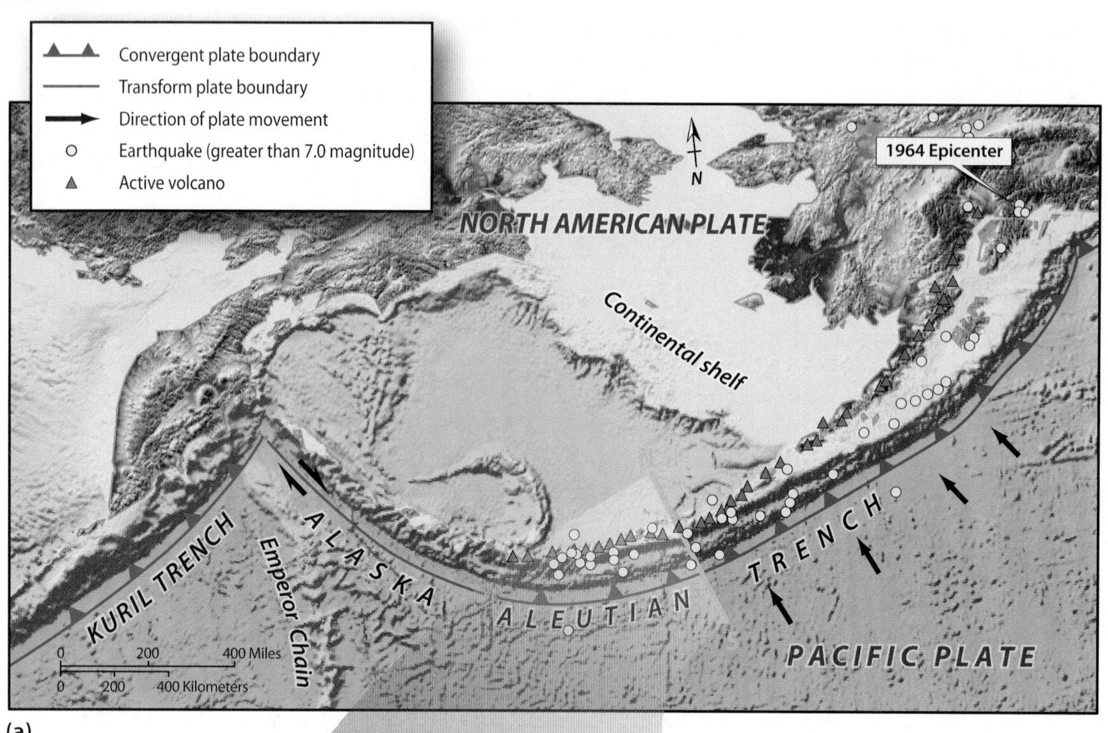

(a)

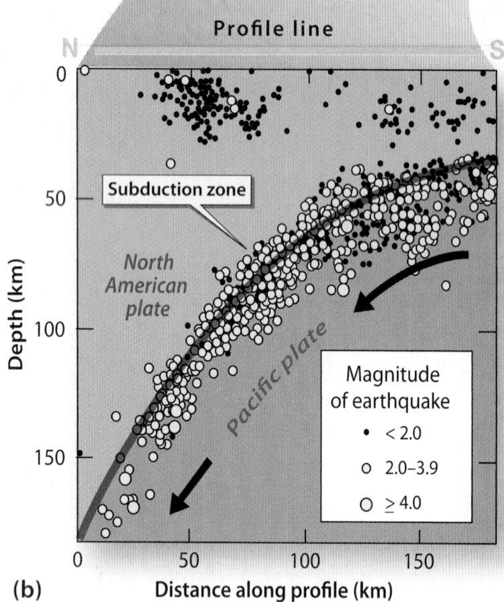

(b)

FIGURE 3-21 ◀ The Alaska-Aleutian Convergent Plate Boundary Plate convergence gives rise to deep-sea trenches, long volcanic arcs, island arcs, and earthquakes. At the Alaska-Aleutian convergent boundary **(a)**, the oceanic Pacific plate is sinking into the mantle beneath the North American plate. Earthquakes originate in the inclined subduction zone **(b)**.

will learn in Chapter 6, subduction-related volcanoes are very explosive compared to those that form along oceanic divergent plate boundaries. It is the explosive nature of subduction-related volcanoes that creates many hazards for people.

TRANSFORM PLATE BOUNDARIES

If two plates are not diverging or converging, they are sliding past one another along a **transform plate boundary**. The movement is therefore parallel to the direction, or *strike*, of the plate boundary. Where these plate boundaries connect divergent or convergent plate boundaries, they enable the type of movement to change or be "transformed," as described earlier (see Figure 3-12b).

Transform plate boundaries occur offshore, where they connect offset segments of mid-ocean ridges (as shown in Figure 3-12), but they also occur on land. Onshore transform plate boundaries are spectacular examples of the moving, dynamic geosphere. This type of plate boundary is a steeply inclined or vertical zone of interconnected faults and broken-up and pulverized rock ranging from 100 meters (330 ft) to over 1.5 kilometers (1 mi) wide. The faults common in these zones have displacements where one block of rock moves laterally (sideways) past the other. Because this displacement is parallel to the trend or strike of the fault, as shown in Figure 3-22, they

FIGURE 3-22 ▲ Strike-Slip Faults Strike-slip faults occur where blocks slide horizontally past each other. These faults are common at transform plate boundaries.

are called **strike-slip faults**. (If you are standing on one side of a strike-slip fault and the other side appears to move to your left, the fault is called a left-lateral fault. If the apparent displacement is to the right, it is a right-lateral fault.)

Transform plate boundaries and the faults developed along them can be over 1000 kilometers (600 mi) long and have rates of movement that average several centimeters (a few inches) per year. In places, the plates move imperceptibly past one another (creep), but it is more common for the movements to be intermittent and to occur in virtually instantaneous jolts that cause earthquakes.

One of the best-studied transform plate boundaries in the world is the San Andreas Fault in California (Figure 3-23). This complex fault zone, 800 kilometers (500 mi) long, connects a spreading center in the Gulf of California with a transform fault and spreading center off the coast of northern California. Movements on the San Andreas have commonly displaced the ground surface. The photo in Figure 3-23 dramatically shows how a fence north of San Francisco was offset 2.5 meters (8.2 ft) during the famous 1906 San Francisco earthquake. In 1857, surface displacement on another part of this plate boundary totaled almost 9 meters (30 ft). If you were standing near this part of the San Andreas Fault back then, you would have had little doubt about how dynamic the geosphere can be.

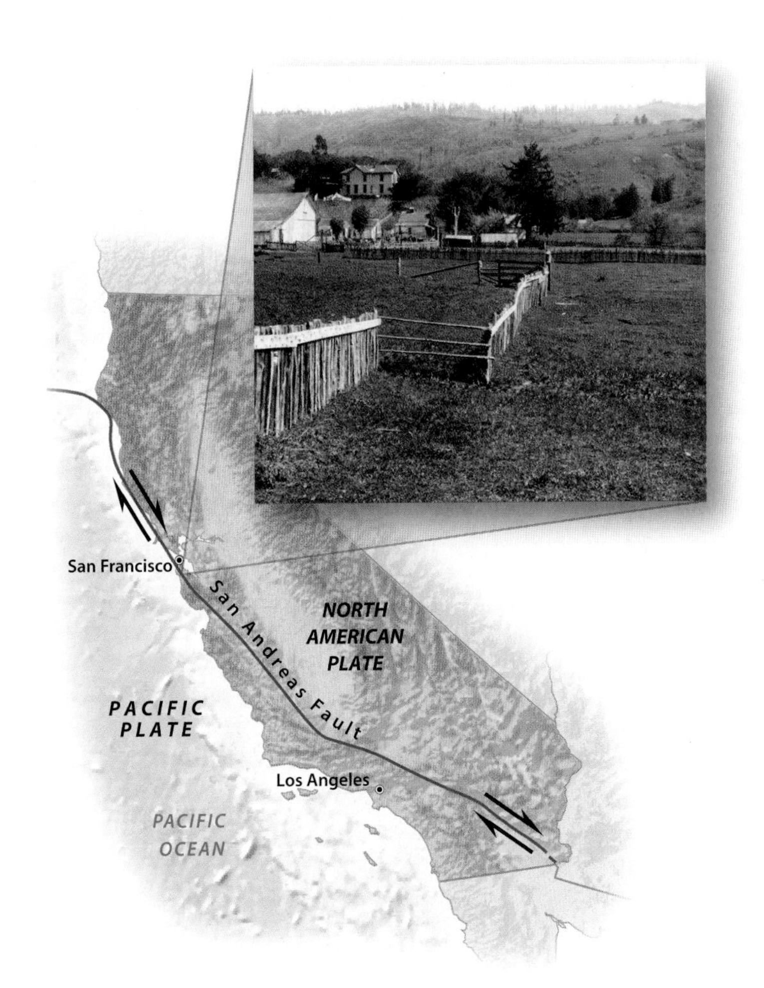

FIGURE 3-23 ▲ Surface Offset on the San Andreas Fault
The nearly instantaneous movement of the land's surface during earthquakes is convincing evidence of the dynamic nature of plate boundaries. This fence line was displaced dramatically during the Great San Francisco Earthquake of 1906, which took place along the San Andreas Fault.

You Make the Call

Living on a Plate Boundary

Imagine that you live in San Francisco. Regardless of what neighborhood you call home, living in San Francisco puts you little more than a stone's throw from an active plate boundary. For some, it's even closer than that. As the photo shows, the San Andreas Fault—the strike-slip plate boundary between the North American and Pacific plates—goes right through your city.

You now know that the San Andreas Fault is an active, earthquake-generating plate boundary. Bay Area earthquake probabilities The shaking, deformation, and fire that accompanied the 1906 earthquake killed about 700 people. Surely, local authorities have made plans for dealing with future earthquakes. But look again at the photo. Whatever the city's plans, people living very close to the San Andreas Fault are potentially in disaster's path.

- Do you think people should be allowed to live so close to an earthquake-generating fault?
- If another earthquake causes extensive damage to San Francisco, who do you think will pay to rebuild the city?
- If San Francisco is demolished by a future earthquake, do you think it should be rebuilt?

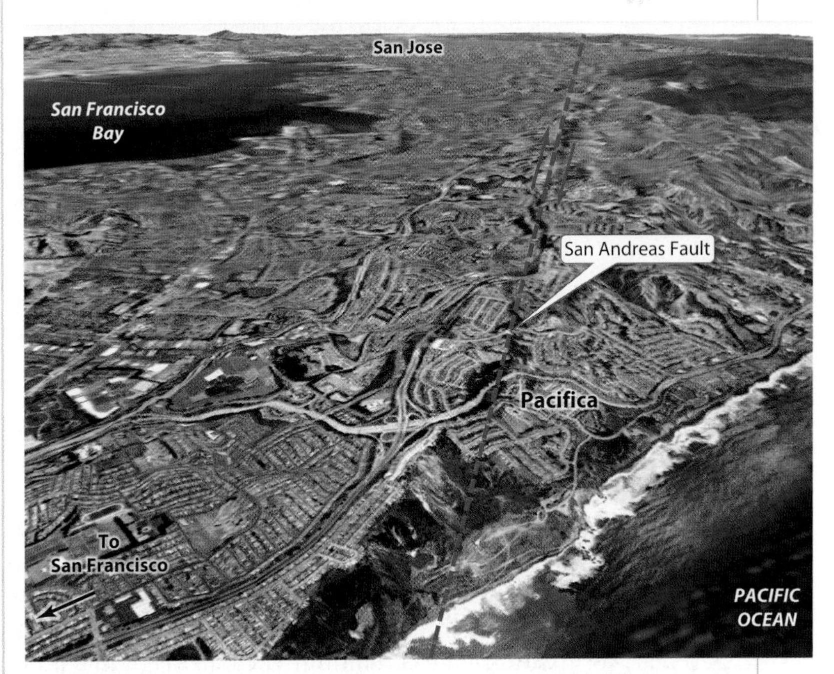

This photograph shows the approximate location of the San Andreas Fault in San Francisco.

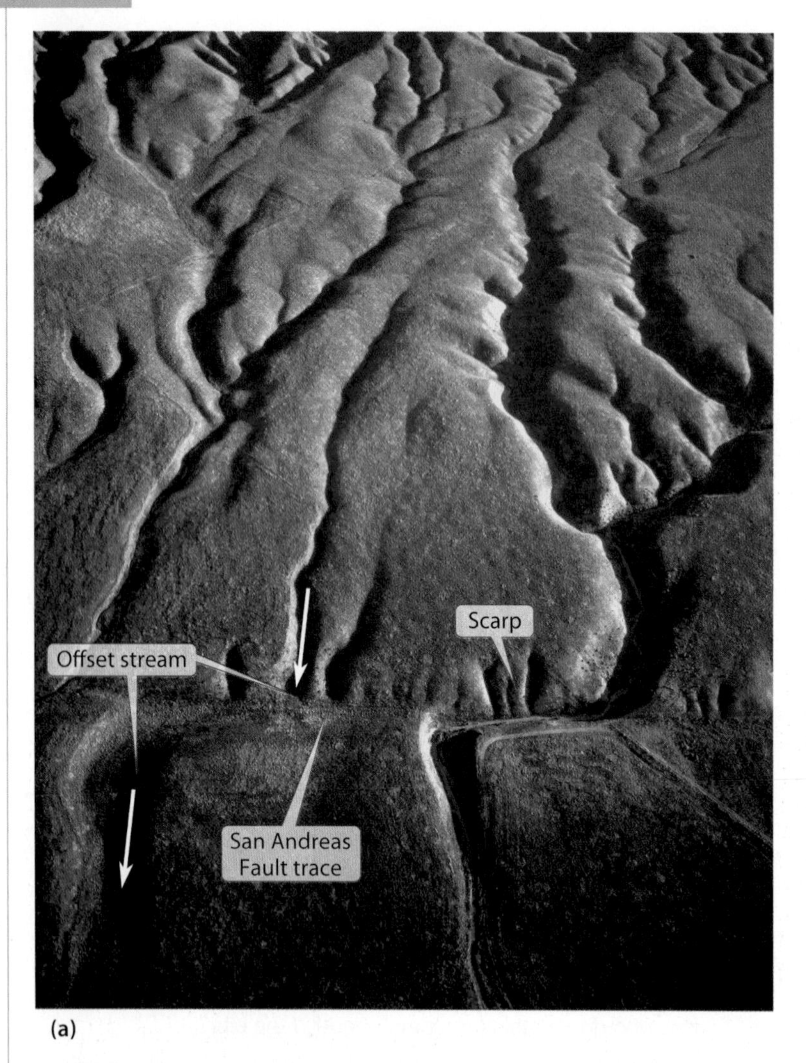

(a)

(b)

FIGURE 3-24 ▲ Effects of Earthquakes and Fault Movements at Plate Boundaries
Earthquakes and fault movements at plate boundaries can affect the land surface in many ways. Surface features along the San Andreas Fault include scarps and offset stream drainages, like these in the Wallace Creek region **(a)**, as well as linear valleys and lakes, like the Crystal Springs reservoir shown here **(b)**. (San Francisco and the Pacific Ocean can be seen in the distance.)

These types of movements create many kinds of surface features in the landscape. The San Andreas Fault and the San Francisco Bay area Surface movements offset streams, create steep banks called **scarps** (Figure 3-24a) and dam drainages to create ponds. The ground-up rock along the San Andreas Fault zone is easily eroded compared to rocks on either side of the fault zone. Over time, the erosion of these unstable rocks produces linear troughs along the plate boundary, places where lakes and valleys form, like the one shown in Figure 3-24b.

The San Andreas Fault has been moving for 20 million years or more and the total amount of displacement on this plate boundary may be as much as 560 kilometers (350 mi). If this movement continues, the land where Los Angeles is now located will eventually be moved adjacent to the land where San Francisco is today. Does this seem dramatic? Remember that the rate of movement is several centimeters (a few inches) per year, so it would take 30 million years to accomplish this change. California will not be ripped away and fall into the sea, as is the common myth.

What are some of the Earth systems interactions that occur when Earth's surface is moved along an onshore transform plate boundary?

3.4 Plate Tectonics—The Big Picture

Plate tectonics connects the geosphere to all other Earth systems. As plate tectonic processes move, change, and deform the geosphere, they influence and, in many cases, control aspects of the atmosphere, biosphere, and hydrosphere. Plate tectonic processes operate at a global scale; they are ongoing, and people cannot control them. They have both positive and negative consequences, giving rise to earthquakes and volcanoes as well as mountains and mineral resources.

PLATE TECTONICS AND EARTHQUAKES

Earthquakes are vibrations caused when rocks are broken by sudden movements on faults (Chapter 5). Each year there are 500,000 earthquakes that instruments detect, 100,000 that people feel, and 100 that cause damage. The great majority of these earthquakes are caused by movements along plate boundaries.

Examine Figure 3-25. This global map shows distinct zones where earthquakes are concentrated. The earthquake concentrations form a continuous, interconnected network around the entire Earth. The distribution of earthquakes is a very good guide to the location of tectonic plate boundaries.

Most earthquakes occur along convergent and transform plate boundaries. Along these types of plate boundaries, displacement occurs along fault zones that can be hundreds or even thousands of kilometers (hundreds of miles) long and that only intermittently releases the stresses that build up along them (a process that will be described more fully in Chapter 5).

Many of the devastating earthquakes that people experience are located around the Pacific Ocean where convergent and transform plate boundaries are common. These earthquakes shake and displace the land, cause buildings and other infrastructure to collapse, and—if they occur offshore—displace ocean waters that become dangerous seismic sea waves, or *tsunamis* (sometimes erroneously called "tidal waves"), like the devastating Indonesian tsunami that killed a quarter of a million people in 2004.

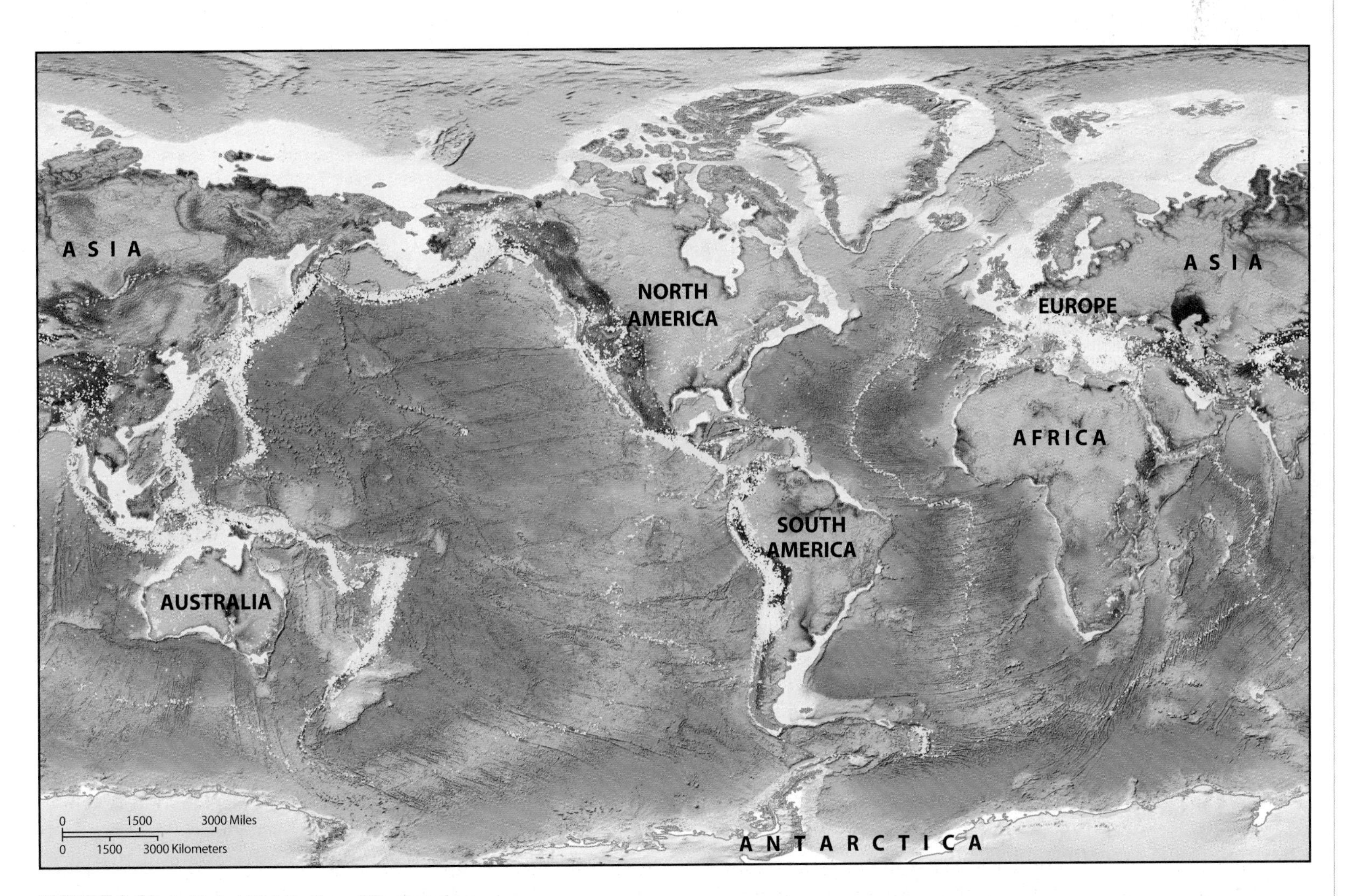

FIGURE 3-25 ▲ Global Distribution of Earthquakes

The yellow dots indicating where earthquakes have occurred reveal the world's major plate boundaries (compare this map to the plate boundary map in Figure 3-13).

What You Can Do

Keep Track of Earthquakes

Earthquakes don't have to be something that you learn about only in the news. The U.S. Geological Survey provides an almost real-time earthquake notification service that will electronically send you a report about new earthquakes around the world. The free reports arrive automatically in your email inbox within about 5 minutes of earthquakes occurring in the United States and within about 30 minutes if they occur elsewhere. The information includes the time, size, and location of the earthquake.

This is a good way to keep an eye out for really big earthquakes—you will learn about them before they get on the news. You can also use this service to keep track of earthquake activity in a region of interest to you, perhaps where you live, where a friend or relative lives, or where you plan to travel. To subscribe to this service, go to the Earthquake Notification Service. If you would rather just check in once in a while for recent earthquake news, go to Latest earthquakes in the world – past 7 days.

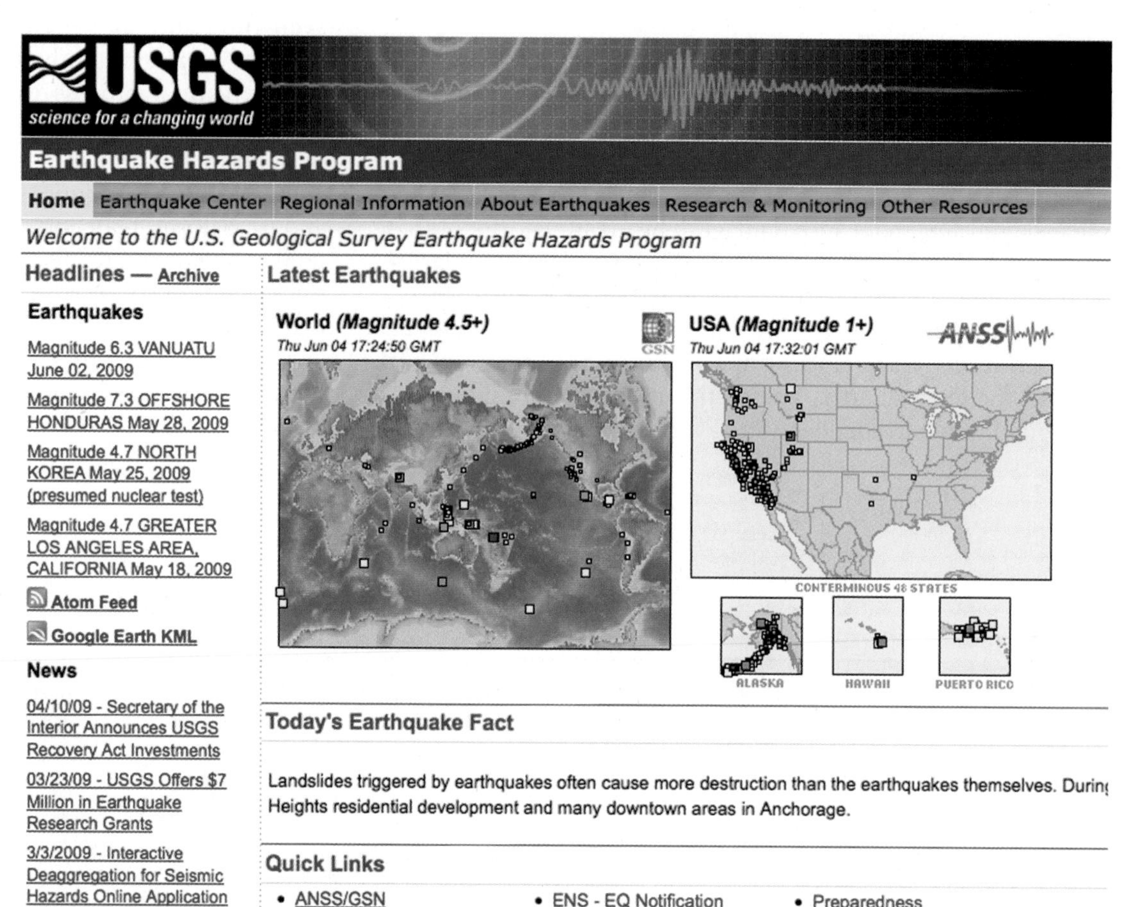

PLATE TECTONICS AND VOLCANOES

Figure 3-26 shows the distribution of most of the world's volcanoes. You will learn in Chapter 6 that significant volcanism can occur within plates, but overall, volcanoes reveal the location of plate boundaries almost as well as earthquakes. They are characteristic of both divergent and convergent plate boundaries.

In general, volcanism at divergent plate boundaries is not catastrophically dangerous, but it does involve large amounts of very hot lava (magma that erupts onto Earth's surface). It also releases gases such as sulfur dioxide. In 1783, a very large eruption in Iceland, a country composed entirely from the products of volcanic eruptions along the Mid-Atlantic Ridge, caused the deaths of 10,000 people—one-fourth of the nation's entire population at the time.

Volcanism at convergent plate boundaries is very different from that at divergent plate boundaries (Chapter 6). Explosive volcanoes in the arcs above subduction zones frequently release gases and large amounts of tiny rock fragments called volcanic "ash" into the atmosphere. (A typical ash cloud is shown in Figure 3-27.) These materials are blasted so high that winds can carry them completely around Earth. Some of the eruptions have been so large that they have temporarily changed global climate. Because erupted gases and volcanic ash absorb energy from the Sun before it reaches Earth's surface, they can cause global cooling for a few years.

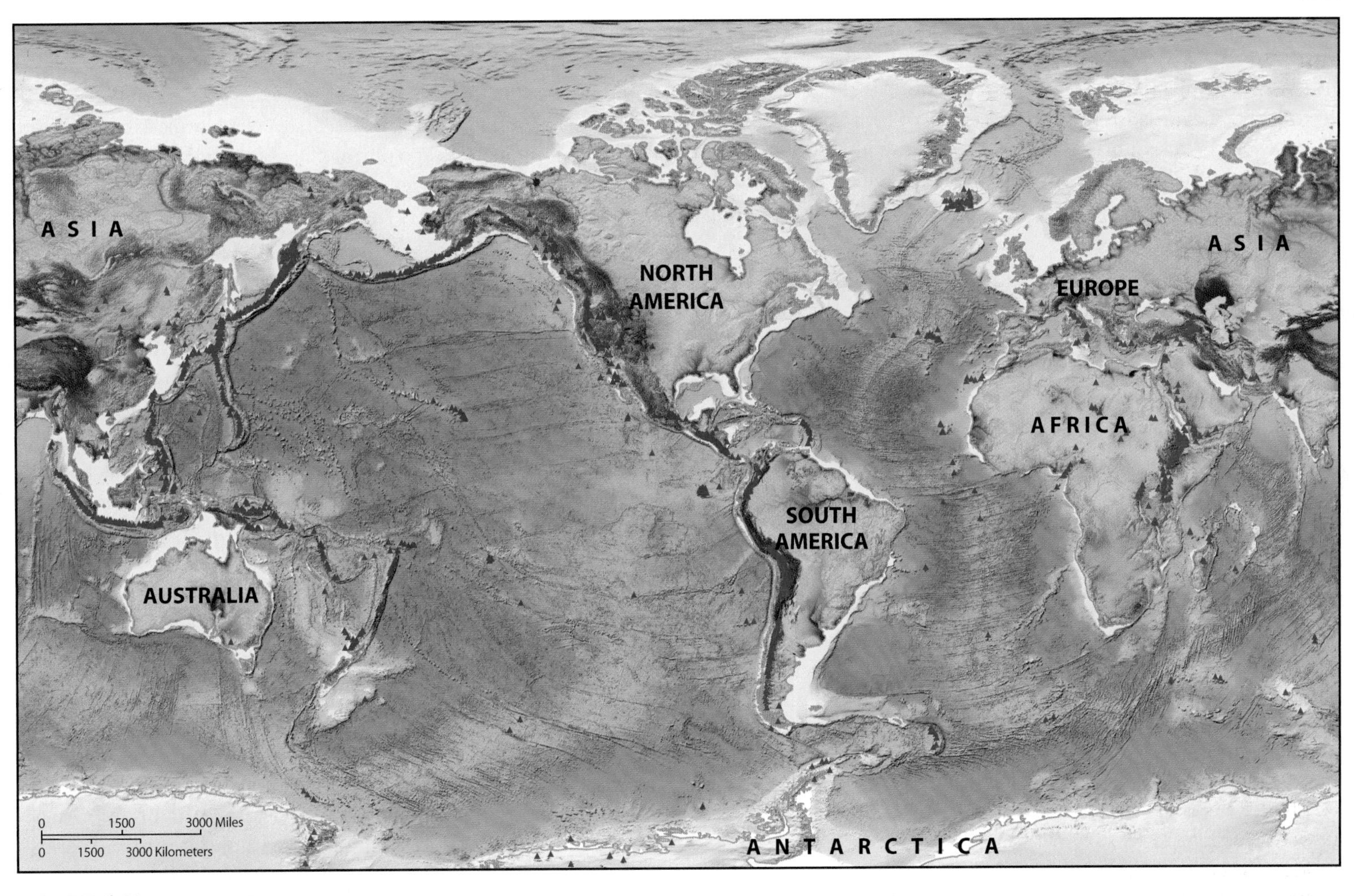

FIGURE 3-26 ▲ Global Distribution of Volcanoes
The red triangles showing where many of the world's volcanoes are located also reveal where plate boundaries are located, especially convergent boundaries. The abundant subduction-related volcanoes around the Pacific have been called the "ring of fire." (Volcanic activity is also common along the mid-ocean ridges.)

One historic example of this type of global climate change followed a massive eruption from Mt. Tambora in Indonesia. The eruption occurred in April 1815, and the following year became known as "the year without a summer." The effects were especially strong in the American Northeast, the Canadian Maritimes, and northern Europe. Killer frosts occurred in May, and snowstorms came in June. Ice formed on lakes in July and August. The effect on crops was devastating, and many people moved from the Northeast to the farmlands of the Midwest as a result. Food riots occurred in Europe. It is also said that the dark, rainy, and cold summer kept people like author Mary Shelley indoors. Perhaps the gloomy weather played a role in inspiring the 18-year-old Shelley to write her masterpiece: *Frankenstein; or, the Modern Prometheus*. It appears that global climate change can have very diverse and unexpected effects.

FIGURE 3-27 ◄ Mt. Pinatubo: A Volcano at a Convergent Plate Boundary
Volcanoes above subduction zones have explosive eruptions. Mt. Pinatubo in the Philippines, part of the Pacific "ring of fire," erupted violently in 1991, producing huge volumes of ash that blanketed fields and towns many kilometers distant.

Though some eruptions have the type of global effects caused by the 1815 Tambora event, most hazards associated with subduction-related volcanoes are more localized. Examples of these hazards include the creation of lava and masses of hot gases and rock debris that flow down the volcano's flanks. People need to learn how to live with these volcanoes because, like earthquakes, they are a fact of life along convergent plate boundaries.

What are some of the Earth systems interactions that occur when an explosive volcano erupts?

PLATE TECTONICS AND MOUNTAIN BUILDING

Mountains, the rugged topographic features that we can walk or drive to but may find very challenging to climb, characteristically form where rocks are deformed along convergent plate boundaries. This process creates not just a mountain here and there, but whole regions of mountains, thousands of kilometers (miles) of mountains that parallel the converging plate boundaries. The compressive forces that accompany plate convergence do the work as rocks are folded and thrust-faulted to create vast mountain ranges. There are several ways in which this happens.

1. **Accretion along the continental margin.** Trenches on the seafloor at converging ocean and continent plate boundaries can collect very large amounts of sediment shed from the nearby continent. In these places, the continental plate can become a giant sediment plow as sediments accumulated in the trench are scraped off the sinking oceanic plate and folded and faulted against the continent. In other words, they *accrete* or build up in quantity along the **continental margin**. The Chugach Mountains in Alaska (shown in Figure 3-28), over 600 kilometers (370 mi) long and full of ice fields and glaciers that flow from their summits to the nearby coast, are mountains formed from accreted ocean sediments.

2. **Compression at convergent plate boundaries.** The compressive forces at convergent plate boundaries can deform the lithosphere and create mountains hundreds of miles away from the plate boundary. The Rocky Mountains are a good example. About 70 million years ago, compressive forces from plate convergence along the western continental margin of North America folded and faulted the upper crust throughout the Rocky Mountain region. This deformation and uplift helped form the Rocky Mountains that we see today.

3. **Collision of continents at convergent plate boundaries.** Earth's highest mountains form where plate convergence causes continents to collide with one another. The collision of the Indian continent with Eurasia has produced the majestic Himalaya Mountains (see Figure 3-20). The Himalayas are home to 96 of the world's 109 mountain peaks whose summits are at 7300 meters (24,000 ft) or higher. Such high, steep, and rugged mountains are very young. In fact, they are actively growing as the plate collisions causing them continue. Streams, glaciers, rock falls, and landslides rapidly erode young, steep mountains. If the mountains were not growing, they would not be able to maintain their great heights. When the Appalachians in the eastern United States were formed in a collision about 300 million years ago during the assembly of Pangaea, they had towering Himalayan-style peaks. Those peaks have since eroded away, leaving the lower, less rugged mountains we know today.

4. **The role of magma.** *Magmatic processes* also help form large mountains and long mountain ranges where oceanic plates converge and subduct beneath continental margins. The magmas produced by subduction rise into Earth's crust above the subduction zone. Some of these magmas produce majestic mountains in the form of volcanoes on Earth's surface (such as the Cascades of the Pacific Northwest), but many of them get trapped and solidify within the crust. The increased amount of buoyant crustal rocks helps to form mountains, too. These processes were important in the formation of the Sierra Nevada Mountains in California.

Once mountains form, they strongly influence the climate and biosphere around them. Mountains create barriers to moisture-laden air masses that come ashore from the oceans. As this air rises and cools along the mountains, it precipitates much of its moisture. This can create rainforests on the ocean side of the mountains and a dry area, even a desert, on the other side. The phenomenon of a drier side of a mountain range is called the mountain's "rainshadow effect" and can happen at many scales. Figure 3-29 clearly shows how the mighty Andes cause a rainshadow effect over a large part of the South American continent, and the Himalayas have a similar influence in Asia.

FIGURE 3-28 ▼ The Chugach Mountains of Alaska

The offscraping (accretion) of seafloor sediments against the overriding plate at a convergent plate boundary can produce mountains like these in southern Alaska.

FIGURE 3-29 ▲ The Andes Mountains Rainshadow
The rainshadow effect is dramatically evident in this satellite photograph of South America. Moist air from the Pacific Ocean moves eastward up the slopes of the Andes and cools as it climbs the mountains. The cooling causes the moisture to condense into clouds, rain, or snow. Thus the western slopes of the Andes in Chile (left) receive most of the moisture, keeping them lush and green, while the eastern slopes in Argentina (right) are in the rainshadow and remain relatively dry.

The rainshadow effect is a major influence on climate and the type of vegetation that develops in the affected regions—what ecologists call *biomes*. The western forests and the interior grasslands of America are examples of this impact of plate tectonics.

Identify some of the Earth systems interactions that result from mountain building in the geosphere.

PLATE TECTONICS AND MINERAL RESOURCES

You will learn more about valuable mineral deposits in the geosphere and mineral resources in Chapter 12. For the present, we can make two connections between plate tectonics and mineral resources. Remember the seafloor hot springs (hydrothermal vents) that are common along spreading centers at divergent plate boundaries—the "black smokers" (see Figure 3-17a)? These springs deposit minerals on the seafloor that contain copper, lead, zinc, gold, and silver. In some places, the deposits may eventually become accreted to a continent or island arc during subduction at a convergent plate boundary. In these places, people can recover valuable metals through mining. There is a well-known example of this type of deposit on the island of Cyprus; people began recovering metals from it thousands of years ago. In fact, the word *copper* comes from the Greek word for Cyprus, *kipris*.

Plate tectonics create valuable mineral deposits along convergent plate boundaries, too. In fact, the world's largest deposits of copper (commonly accompanied by some lead, zinc, silver, and gold) form here. These deposits develop within subduction-related volcanoes. Magmas that solidify beneath the surface release fluids carrying metals. As these fluids migrate upward and cool, they deposit metals in fractures within the enclosing rock. In some places, these deposits eventually become exposed by uplift and erosion, and then people are able to mine the valuable minerals (a typical example is shown in Figure 3-30).

Magmatic processes at convergent plate boundaries help people in other ways. In some places, the hot rocks and underground water can be used to generate electricity (geothermal power, discussed in Chapter 13), and ash deposited downwind from subduction-related volcanoes helps to create fertile soils for farming (Chapters 6 and 11). Magmatic processes at convergent plate boundaries do more than simply create volcanic hazards.

FIGURE 3-30 ▲ A Copper Mine in Utah
Mineral deposits form in settings shaped by plate tectonics. Large copper deposits, such as those at the Bingham Canyon mine in Utah, form within subduction-related volcanoes.

SUMMARY

Persistent research, applications of new technology, and syntheses of diverse observations by more than three generations of scientists during the 20th century developed the theory of plate tectonics. The theory of plate tectonics explains geosphere movements and how energy and matter are transferred to Earth's lithosphere and surface from deep within Earth. Plate tectonic processes affect the hydrosphere, atmosphere, and biosphere, and scientists study these processes in order to understand and prepare for the many natural hazards they present. These natural hazards will be examined in more depth in Chapters 5–9. Resources that people use are also developed through plate tectonic processes. You will learn more about mineral, soil and energy resources of the geosphere in Chapters 11–13. Without plate tectonics, we would not understand how the geosphere works. It is an excellent example of the application of the scientific method, especially the testing of hypotheses and their related predictions.

In This Chapter You Have Learned:

- Earth's rigid lithosphere is broken into pieces called tectonic plates.
- Plates diverge, converge, or slide horizontally past one another. Density differences in lithosphere and mantle cause them to be unstable. The sinking of colder, thicker and denser lithosphere pulls plates apart and allows hot mantle material to rise where plates separate. Lithospheric plates move on the weak rocks of the asthenosphere.
- There are three types of plate boundaries, each characterized by the type of motion along the boundary. At divergent boundaries the plates move away from each other. The two types of divergent boundaries are oceanic seafloor spreading centers and continental rift zones. Plates move toward each other at convergent boundaries—subduction zones develop, where at least one of the plates is oceanic, and collision zones develop, where two continental plates converge. At transform boundaries, two plates slide horizontally relative to each other.
- Oceanic crust and lithosphere forms at divergent plate boundaries and is consumed at convergent plate boundaries. Continental crust is too buoyant to sink like oceanic crust at convergent plate boundaries. This is why the continents are geologically much older than the ocean basins, which are continuously recycled into the mantle at subduction zones.
- Movements on faults at plate boundaries cause earthquakes—the larger the movements, the larger the earthquake. The world's largest faults (and movements along them) are along convergent and transform plate boundaries.
- Plate boundaries localize volcanism. Mantle upwelling at divergent plate boundaries enables magma to rise and form oceanic crust. Magmas produced above subduction zones at convergent plate boundaries rise to form volcanoes characterized by explosive eruptions.
- The dynamic geosphere's plate tectonic processes create mountains, influence climate at local to global scales, control the distribution of biomes, and create resources that people can use.

KEY TERMS

apparent polar wander curve (p. 57)
continental drift (p. 56)
continental margin (p. 76)
convergent plate boundary (p. 68)
divergent plate boundary (p. 66)
fault (p. 62)
fossil assemblage (p. 55)
Gondwanaland (p. 55)
island arc (p. 69)
magma (p. 59)
magnetic stripes (p. 60)
magnetic reversal (p. 60)
normal fault (p. 66)
paleomagnetism (p. 57)
Pangaea (p. 56)
plate (lithospheric plate, tectonic plate) (p. 63)
plate tectonics (p. 54)
reverse fault (p. 68)
ridge (oceanic ridge, mid-ocean ridge) (p. 58)
rift (p. 65)
scarp (p. 72)
seafloor spreading (p. 59)
strike-slip fault (p. 71)
subduction (p. 68)
subduction zone (p. 68)
thrust fault (p. 68)
transform fault (p. 62)
transform plate boundary (p. 70)
trench (oceanic trench) (p. 58)
volcanic arc (p. 69)

QUESTIONS FOR REVIEW

Check Your Understanding

1. How do Earth scientists define what a *plate* is?
2. Why was the discovery of matching fossil assemblages in rocks of identical age on widely separated continents an important clue in developing the hypothesis of continental drift?
3. When Alfred Wegener proposed his hypothesis of continental drift, he could not convince other scientists that his tidal forces mechanism was correct. Why was the absence of a plausible mechanism an important problem for other Earth scientists?
4. What was the significance of apparent polar wander curves, and how did plate tectonics remove the confusion over these observations?
5. Explain how the discovery that the Earth's magnetic field has reversed itself irregularly through time allowed scientists to test the hypothesis of seafloor spreading, and how confirmation of that hypothesis in turn helped support plate tectonic theory.
6. As earthquakes became better located globally through the mid-1960s, Earth scientists recognized that earthquakes provided more evidence in support of the developing theory of plate tectonics. What insight was offered by the study of earthquakes?

7. This chapter discusses the Red Sea and the East African rift systems as features related to the current splitting apart of the African continent. Given enough geologic time, what will these features become?
8. Why do we characterize fault types and fault movement by *relative* motion *across* a fault?
9. Scientists have discovered that life flourishes around hydrothermal vents on the ocean floor. Where does the energy for life in this ecosystem come from?
10. Why does the lithosphere sink into the mantle at subduction zones?
11. What type of plate boundary is California's San Andreas Fault, and why is it a highly unusual example of this type of boundary? Where are most examples of such plate boundaries located?
12. Explosive volcanoes ring the Pacific Ocean, and are often referred to as the Ring of Fire. What kinds of plate boundaries are these explosive volcanoes associated with?
13. How are oceanic trenches at convergent margins involved in mountain building at these plate boundaries?
14. When continental crust on one plate is carried on lithosphere that is subducting beneath another continent, a massive collision results, driving up huge mountain ranges like the Himalayas. Why does this collision take place rather than subduction of the converging continent?
15. How can you explain the distance between North America and Africa today if the rate of seafloor spreading is only centimeters per year?

Critical Thinking/Discussion

16. Historians and philosophers of science recognize the advent of plate tectonics to have been a major *paradigm shift*, or change in the underlying explanatory framework for the Earth sciences. This process took one to two human generations to accomplish, despite the fact that plate tectonics provided an explanation for observed phenomena that was far superior to past ideas. Why might the scientific community have been so resistant to this new idea? Why is it generally a good thing that new ideas involving radical departures from the current understanding need to be supported by an overwhelming amount of evidence before they are accepted?
17. The development of plate tectonic theory was accelerated in the 1950s after the conclusion of World War II, when scientists gained access to new tools to map the seafloor and measure the magnetic signals of ocean floor rocks. Science and technology often develop together: Advances in science usually produce advances in technology, and new technologies often allow for the investigation of new scientific questions. Why would the technologies that led to advances in the science that supported plate tectonic theory have been developed during World War II?
18. It is understood today that motion of lithospheric plates is accommodated by flow in the largely solid asthenosphere. If this portion of the mantle is still solid, how can it flow? Can you think of any other examples of solids that flow while still solid and can provide an analogy for flow in the solid upper mantle?
19. Why is it important that modern GPS measurements of plate motion are consistent with geologic observations of separated fossil assemblages, magnetic stripes on the ocean floor, and other related features?

ANSWERS TO IN-CHAPTER INSIGHT QUESTIONS

P. 67

What are some of the Earth systems interactions that occur at a hydrothermal vent on the ocean floor?

- Parts of the hydrosphere are heated by the geosphere (energy transfer).
- Hot water dissolves material from rocks in the geosphere (material transfer)
- Vent water precipitates material on the seafloor (material transfer)
- Biosphere eats precipitated material (energy transfer).

P. 72

What are some of the Earth systems interactions that occur when Earth's surface is moved along an onshore transform plate boundary?

- Hydrosphere drainages are shifted or blocked and reservoirs (ponds or lakes) created.
- Habitat for the biosphere is changed.

P. 76

What are some of the Earth systems interactions that can occur when an explosive volcano erupts?

- Atmosphere composition can be changed (material transfer); the volcanic eruption can affect global climate.
- Groundwater in the hydrosphere can be heated by shallow magma (energy transfer).
- Biosphere habitat can be destroyed or changed.
- Hydrosphere reservoirs can be contaminated (material transfer).
- Hydrosphere drainages and reservoirs can be changed.

P. 77

Identify some of the Earth systems interactions that can result from mountain building in the geosphere.

- Mountains change atmosphere circulation.
- Mountains influence the distribution of the hydrosphere.
- Mountains influence biosphere habitat.
- Mountains affect the sediment load of the hydrosphere (Chapter 7).

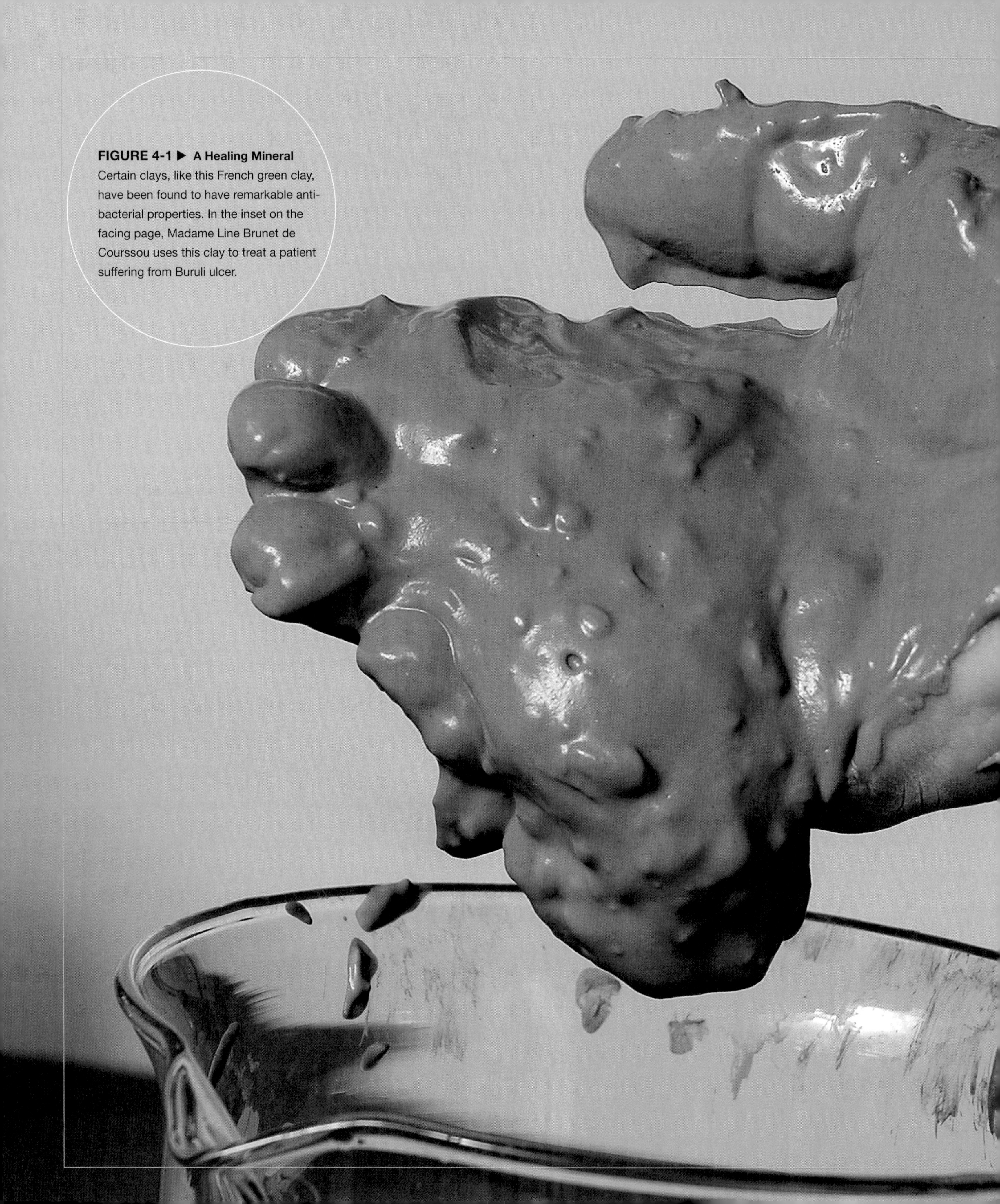

FIGURE 4-1 ▶ A Healing Mineral Certain clays, like this French green clay, have been found to have remarkable anti-bacterial properties. In the inset on the facing page, Madame Line Brunet de Courssou uses this clay to treat a patient suffering from Buruli ulcer.

CHAPTER

4

GEOSPHERE MATERIALS

Have you ever heard of Buruli ulcer? How about flesh-eating disease? Buruli ulcer, which is also known as "flesh-eating disease," is a bacterial disease related to leprosy and tuberculosis. The bacteria causing this disease first grow in people's skin, initially producing a painless swelling or a pimple-like nodule. As the bacteria multiply, the skin dies and an open ulcer develops. At this stage, the infection can move into flesh and bone. Since the only treatment has been to surgically remove the diseased area, loss of limbs and disfigurement are common results. The disease is endemic to west and central Africa, where access to adequate medical assistance is limited.

For years, Madame Line Brunet de Courssou worked at her family's health clinics in the Ivory Coast and Guinea. She knew that certain *clays*—especially French green clay—had been used in poultices to heal skin infections for hundreds, even thousands of years. Brunet de Courssou imported clays from France and used them to treat her patients with Buruli ulcer (Figure 4-1). By 2002, she had 50 carefully documented cases of healing Buruli ulcer patients with clay treatments. The World Health Organization recognized that her work with clays to treat Buruli ulcer was impressive, but funding to expand the use of this treatment was not approved because scientific studies to understand how the clay treatments worked were needed.

The effort to understand how healing clays work is now under way. Healing clay ASU research Scientists are studying how clay minerals interact with bacteria. They are finding that clays with a certain composition and structure facilitate a chemical transfer between the clay and water that releases antibacterial components. The healing clays in effect control the chemistry of the water used to make clay poultices. The bacteria-killing effectiveness of the water is startling—it even kills methicillin-resistant *Staphylococcus aureus* (MRSA), an especially virulent and deadly strain of staph that has evolved resistance to most antibiotics now in use. Searching in clay to treat disease-causing bacteria

Understanding how minerals like clays can help protect human health is a new direction for scientific research. The initial results are very promising, and much more work is being done. The potential payoff is tremendous—drug-resistant bacteria are an increasingly serious problem today. MRSA resources)

Clays, other minerals, and rocks are geosphere materials. You may have thought of geosphere materials as just rocks. Everyone has some understanding of rocks. They are hard substances that we can pick up and throw, massive formations that some of us love to climb (Figure 4-2a), materials that people have used for thousands of years to build their cities and carve their statues (Figure 4-2b), or formidable barriers that we must be careful not to drive cars, boats, or even airplanes into. In our day-to-day experience, rocks are solid, seemingly unchanging materials of the geosphere. But now we know that rocks move, break, and melt in the geosphere (Chapter 3). To understand how the geosphere can be so dynamic and how it readily interacts with the atmosphere, hydrosphere, and biosphere, we need to understand rocks and the minerals in them.

(a)

(b)

FIGURE 4-2 ▲ Rocks Are the Most Common Geosphere Material
Rocks can be fun, dangerous, beautiful, and useful.

IN THIS CHAPTER YOU WILL LEARN:

- The chemical composition of the geosphere, especially oceanic and continental crust
- What minerals are and how they are built up from chemical elements
- Why physical properties of minerals vary and how these properties can be useful or harmful
- How minerals change through interactions with the atmosphere and hydrosphere
- How minerals form rocks and how rocks change in the rock cycle
- How people use rocks

4.1 The Geosphere's Chemical Composition

Understanding the geosphere's chemical composition starts with the basics—the chemical elements. A chemical **element** is a substance that can't be broken down into other substances. There are about 90 naturally occurring elements, each with a unique name and chemical symbol. You are very familiar with at least one element, oxygen (O), that your body needs for respiration. Although you haven't actually seen oxygen, take a deep breath and think about it. Luckily for us, the atmosphere readily supplies the oxygen that we need to live.

The smallest possible division of an element is an **atom**. Atoms, made up of a nucleus of positively charged protons and neutral neutrons surrounded by one or more negatively charged electrons (as shown schematically in Figure 4-3), define the different elements. Atoms with different numbers of protons in their nuclei are different elements, ranging from hydrogen (H) with one proton to uranium (U) with 92. The number of neutrons in an atom of a specific element can vary, but the number of protons is distinctive; all oxygen (O) atoms have 8 protons, aluminum (Al) atoms have 13 protons, lithium (Li) atoms have 3 protons, and copper (Cu) atoms have 29, for example.

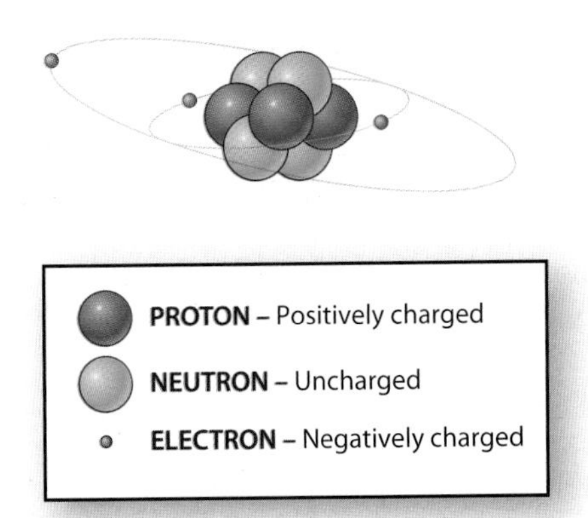

FIGURE 4-3 ▲ Diagram of the Atomic Structure of Lithium
Atoms consist of a small, dense nucleus of protons and neutrons surrounded by moving electrons. The number of protons is distinctive for different chemical elements. Lithium, for example, always contains three protons in its nucleus. An atom is normally electrically neutral because it has as many electrons as protons.

THE COMPOSITION OF THE GEOSPHERE

You learned about the compositional structure of the geosphere in Chapter 2. The crust (and some pieces of the upper mantle that find their way to Earth's surface) can be directly observed and sampled. The compositions of deeper parts of the geosphere are estimated from models based on the density of Earth as a whole, by internal density variations indicated by seismic wave velocities, by the composition of meteorites, and by assumptions about how Earth and our solar system formed. Scientists will continue to refine their models and estimates over time but the results so far are reasonably consistent and sufficient to characterize the proportion of the geosphere's more abundant elements. The major element composition of the geosphere listed in Figure 4-4a on p. 84 is fairly typical of the available estimates. Chemical composition of Earth, Venus, and Mercury

You can see from the figure that eight elements make up over 98% of the geosphere's mass, with just four of them—iron (Fe), oxygen (O), silicon (Si), and magnesium (Mg)—accounting for 91%. Many elements are only minor constituents of Earth's mass. Think about some familiar elements that are not listed in Figure 4-4a. Sodium (Na), which is in the salt we eat, and nitrogen (N), the most abundant element in the atmosphere (Chapter 2), are examples. Earth is essentially a big hunk of iron, oxygen, silicon, and magnesium.

Because the crust is such a small part of the geosphere, the four most abundant elements are largely in the core and mantle. The core is mostly iron (with some nickel) and the mantle is mainly (about 94%) a mixture of oxygen, silicon, magnesium, and iron.

Even though the proportion of crust is very small, it is the part of the geosphere that interacts most with other Earth systems. And people live on and use resources from the crust. Just eight elements make up about 99% of both oceanic and continental crust, but as Figure 4-4b shows, the composition of the crust differs significantly from that of the geosphere as a whole.

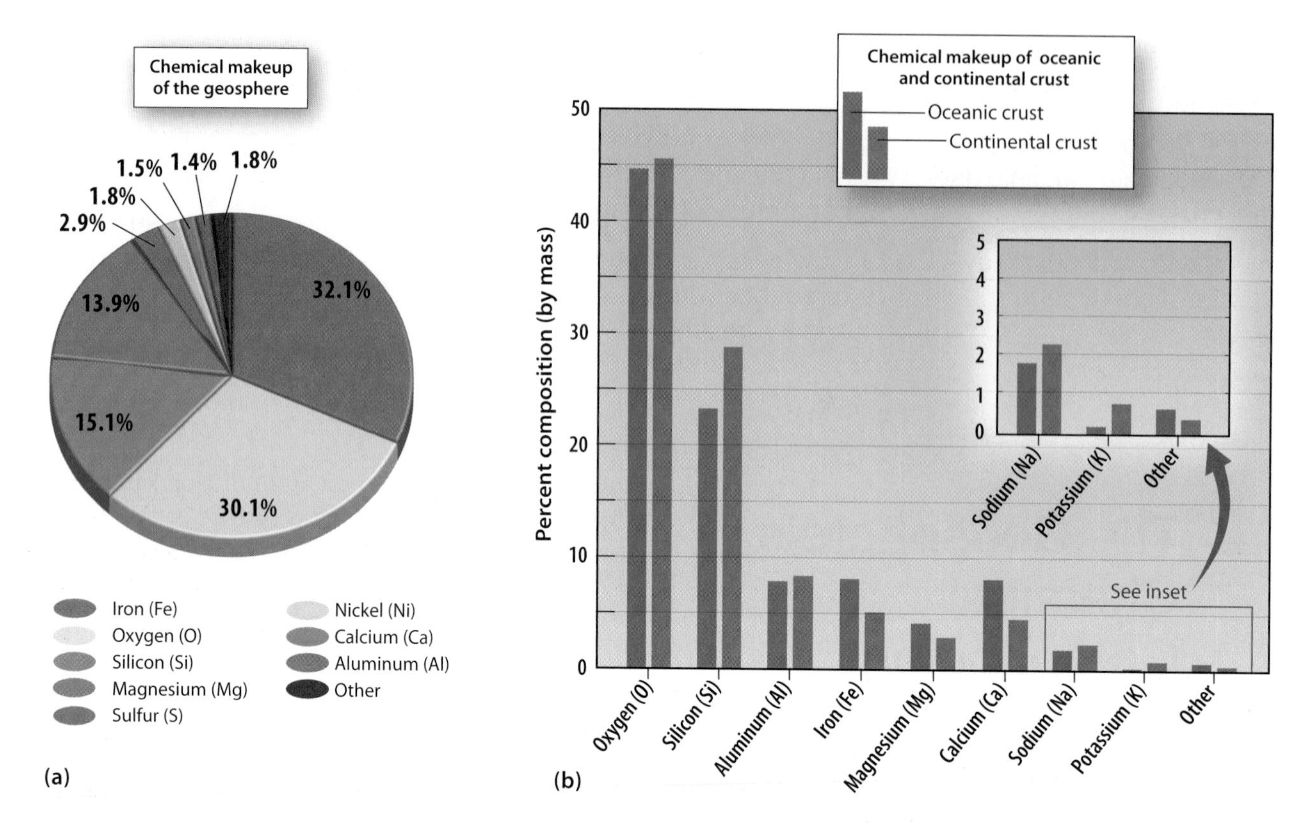

FIGURE 4-4 ▲ Average Chemical Composition of the Geosphere and Crust
(a) The chemical makeup of the geosphere as a whole. **(b)** The chemical makeup of oceanic and continental crust. These eight elements constitute some 99% of the crust. Those present in smaller amounts include titanium, manganese, and phosphorus.

3.01 Composition of continental crust

THE COMPOSITION OF THE CRUST

Earth's crust, being the part of the geosphere that people can directly observe and sample, has been extensively studied. Although there is still much to learn about its character and history, the crust's basic structure and composition are known. As we saw in Chapter 2, thinner and denser crust underlies the oceans, and thicker and less-dense crust underlies the continents.

Continental crust is compositionally diverse at local to regional scales due to its long and complicated geologic history. Overall, though, it has a three-fold compositional structure: a denser lower crust, a middle crust of intermediate density, and a less dense upper crust. All of these contribute to the average composition of continental crust shown in Figure 4-4b. Not shown in Figure 4-4b, however, are all the other elements characteristic of continental crust. Even though present in small amounts, less dense elements and elements not readily incorporated in mantle minerals (lithium, rubidium, barium, lead, thorium, and uranium, for example) are relatively concentrated in continental crust. Although these elements are useful and significant in a variety of ways, our focus here is on the elements that are most abundant in continental crust.

Although oceanic crust is largely underwater, some has become exposed on land, particularly where plates converge and the geosphere is deformed (Section 3.4). Studies of these exposures and of seafloor oceanic crust through drilling and other methods show that oceanic crust has a relatively simple internal structure and is compositionally more homogeneous than continental crust. Essentially, it is equivalent in composition to the basalt that forms at mid-ocean ridges. The average chemical composition of oceanic crust shown in Figure 4-4b is the average of many analyses of *mid-ocean ridge basalt* (also known as MORB).

Figure 4-4b shows that just eight elements make up about 99% of both oceanic and continental crust: silicon, oxygen, iron, and magnesium (the most abundant elements in the geosphere, as you can see in Figure 4-4a, along with aluminum (Al), calcium (Ca), sodium (Na), and potassium (K). However, there are significant differences in the proportions of these elements in the two types of crust.

- The combined percentage of iron and magnesium in oceanic crust (12.7%) is about 60% greater than that in continental crust (8.0%).
- The percentage of calcium in oceanic crust (8.2%) is nearly twice that in continental crust (4.6%).
- The percentage of potassium in continental crust (1.5%) is 15 times that in oceanic crust (0.1%).
- The percentage of silicon in continental crust (28.3%) is 20% greater than that in oceanic crust (23.6%).

These are significant differences that explain why continental crust is thicker and more buoyant than oceanic crust. The greater amounts of less dense elements, such as silicon, aluminum, potassium, and sodium, make it impossible for thick continental crust to sink into the mantle as oceanic crust (along with adjacent upper mantle) can. This also explains why continental crust is generally older than oceanic crust. Once it is formed it is pretty much here to stay.

4.2 Minerals—Where Elements Reside

Chemical elements are not just scattered through the geosphere as separate atoms. The atoms combine with one another to form minerals. **Minerals** are naturally occurring inorganic solids made up of an element or a combination of elements (a compound) that has an ordered arrangement of atoms and a characteristic chemical composition. People have identified and described over 4000 minerals. Rising magma carries some minerals to the surface from deep within the geosphere, but most of the described minerals formed in the crust. Because of their composition and internal atomic arrangement, each mineral has distinctive characteristics and physical properties.

MAKING MINERALS

When atoms contain equal numbers of protons and electrons, they are electrically neutral. Atoms can also gain or lose electrons, forming charged particles called ions that have different numbers of protons and electrons. Atoms with more protons than electrons are positively charged, while those with fewer protons than electrons are negatively charged. Positively charged ions attract negatively charged ions to form neutral combinations, a process called *ionic bonding*, as shown in Figure 4-5a. Neutral atoms can combine, too, simply by sharing electrons, forming molecules by a process called *covalent bonding* (Figure 4-5b).

The attraction between ions, or the sharing of electrons between atoms, forms bonds that hold them closely together, creating *chemical compounds*. As more and more ions or molecules are included in the combination, an orderly, three-dimensional arrangement develops. This orderly arrangement of bonded ions or molecules, called a **crystal**, defines the internal structure of minerals. All minerals occur as crystals, and the process of combining elements to form a solid with an orderly internal atomic structure is called *crystallization*.

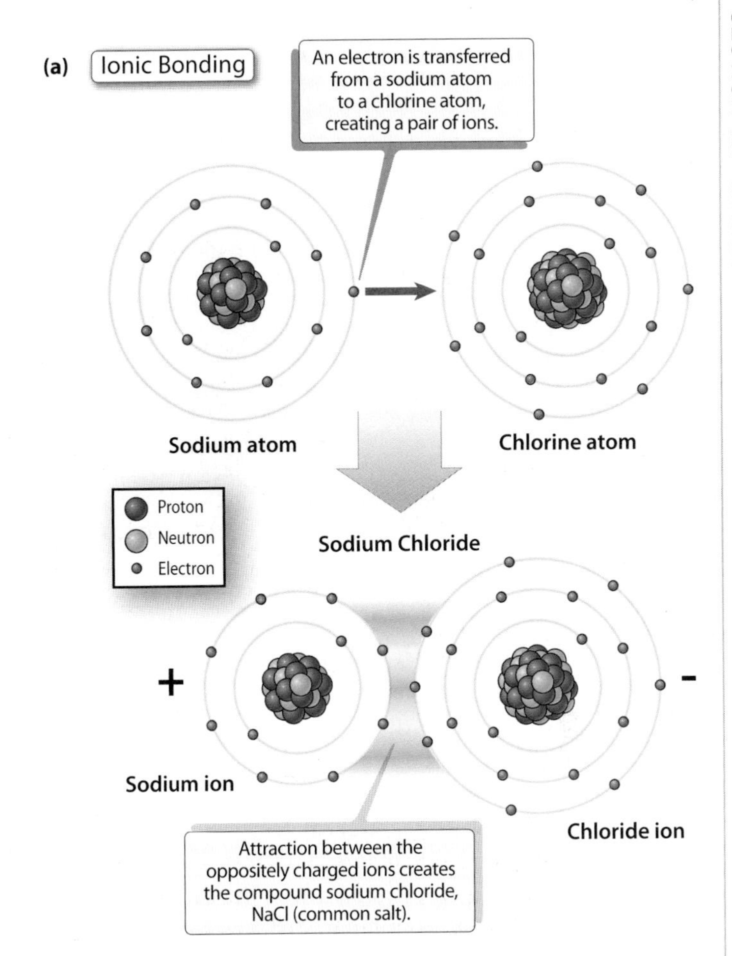

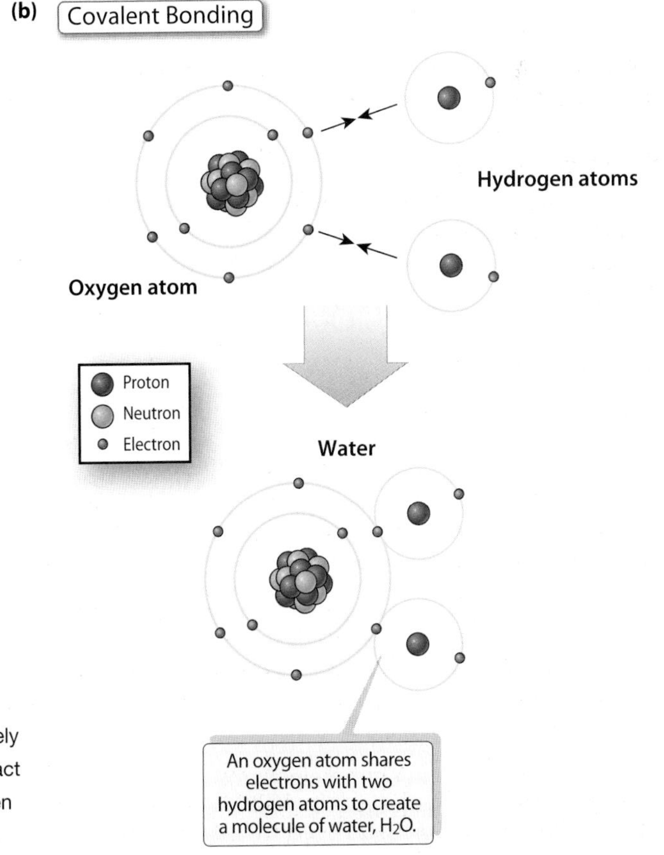

FIGURE 4-5 ▶ Elements Bond Together by Transferring or Sharing Electrons
(a) Ionic bonding: A sodium atom transfers an electron to a chlorine atom, creating a positively charged sodium ion and a negatively charged chloride ion. The oppositely charged ions attract one another, forming the compound sodium chloride or salt. **(b)** Covalent bonding: An oxygen atom shares electrons with two hydrogen atoms to form a water molecule, H_2O.

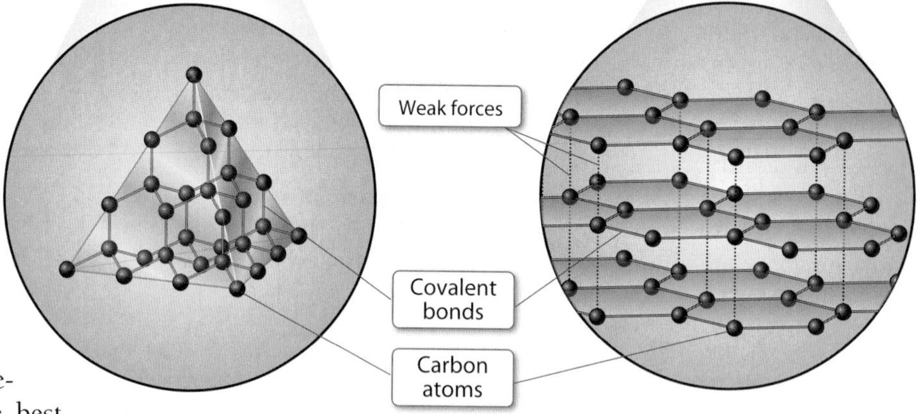

FIGURE 4-6 ▲ Diamond and Graphite Consist Only of Carbon Atoms
(a) Diamond is very hard and commonly translucent. **(b)** Graphite is dark, soft, and flaky. Their internal atomic structures explain why these two minerals have different crystal forms and physical properties, even though they are both made solely of carbon. Why is graphite soft and diamond hard if both are pure carbon?

The internal atomic structure of minerals is what gives crystals their shape and physical properties. An example of this relationship is shown in Figure 4-6. If a crystal is not physically constrained (by the presence of adjacent crystals, for example) it will develop two-dimensional, smooth, or *planar*, surfaces as it adds ions or atoms to its structure—in other words, as it grows. These planar surfaces reflect the mineral's internal atomic structure. Well-formed crystals are fairly uncommon and can become collector's items when found; they are the best examples of a mineral's geometric form.

The geosphere's most abundant elements combine to form its most abundant minerals. These are the *rock-forming minerals*, which include *quartz*, *feldspars*, and *ferromagnesium minerals*. We will discuss each of these types in turn, along with a few other types that do not fall into these three main classes.

QUARTZ—THE SILICON AND OXYGEN MINERAL

If the two most abundant elements in the crust are silicon and oxygen, it probably doesn't surprise you that these two elements form a very common mineral, **quartz** (the compound SiO_2, which contains one atom of silicon for every two atoms of oxygen). The SiO_2 compound is such an important constituent of igneous rocks (rocks formed from magma) that it is used to characterize their chemistry. In these cases, the SiO_2 compound is called **silica**.

Silicon ions are small and positively charged; they have four fewer electrons than protons. Oxygen ions, on the other hand, are several times bigger than silicon ions and negatively charged; they have two more electrons than protons. When silicon and oxygen ions combine, one small silicon ion fits nicely among four oxygen ions. This arrangement of silicon and oxygen atoms has a distinctive pyramid-like shape called a *tetrahedron* (plural *tetrahedra*), shown in Figures 4-7a and b.

The silicon and oxygen bonded in tetrahedra both transfer and share electrons, but overall this combination is negatively charged because the four oxygen ions with two negative charges each (eight total negatives) are combined with one silicon ion with a positive charge of four. To achieve electrical neutrality—which is necessary for any chemical compound—SiO_4 ions can combine with nearby positively charged ions. Moreover, if other SiO_4 ions are abundant nearby, they can share oxygen atoms at each corner of their tetrahedron and combine to form orderly three-dimensional arrangements, like those in Figure 4-7c. If a tetrahedron shares all four of its oxygens in this way, it creates the internal atomic structure of quartz, one of the most common minerals in continental crust, shown in Figure 4-8a. The quartz page: quartz structure

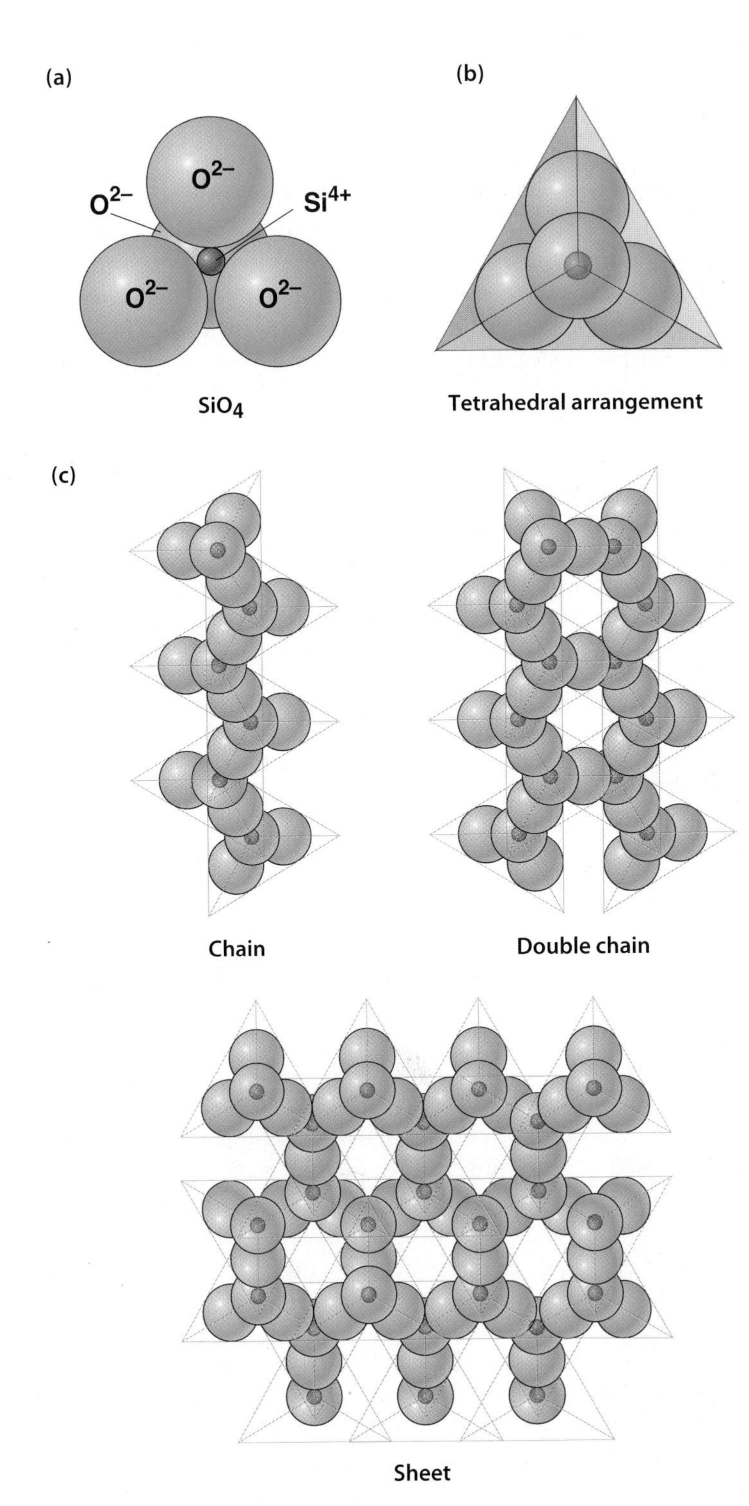

FIGURE 4-7 ▲ Silicon and Oxygen Combine to Form a Tetrahedral-Shaped Ion

(a) Small silicon ions fit between four oxygen ions to form an ion with a net charge of –4. **(b)** Because of the pyramid-like, or tetrahedral, shape of this arrangement, the ions are commonly represented in a pyramidal form. **(c)** Tetrahedral ions can share oxygen atoms to create larger structures, including chains, double chains, and sheets.

FIGURE 4-8 ▲ The Atomic Structure of Quartz

(a) The quartz structure is a three-dimensional network of silicon and oxygen tetrahedra that share oxygen atoms. **(b)** This arrangement gives quartz crystals their distinctive six sides and pointed terminations. **(c)** Quartz crystals produce curved, or conchoidal, surfaces when they fracture.

TABLE 4-1 PHYSICAL PROPERTIES OF SOME COMMON MINERALS

Mineral	Composition	Color	Luster	Hardness (1–10)	Density (g/cm^3)	Photo
Quartz	SiO_2	Colorless (if pure)	Glassy	7.0	2.62	Fig. 4-8
Plagioclase	$(Na,Ca)(Si,Al)_4O_8$	White	Glassy	6–6.5	2.68	Fig. 4-11
Biotite	$K(Mg,Fe)_3AlSi_3O_{10}(OH)_2$	Black	Pearly	2.5–3	2.8–3.4	Fig. 4-13
Olivine	$(Mg,Fe)_2SiO_4$	Olive green	Glassy	6.5–7	3.3	Fig. 4-15
Pyrite	FeS_2	Brassy yellow	Metallic	6.5	5.0	Fig. 4-19

Physical properties and occurrence

The atomic structure and composition of quartz are responsible for its physical properties such as color, luster, hardness, crystal structure, fracture (the appearance of a broken surface), and density. (A summary of some of these properties for quartz and several of the other minerals discussed in this chapter can be found in Table 4-1.) Although commonly clear, white, or gray, trace amounts of impurities give quartz many colors. For example, small amounts of titanium give it a rose color. The hardness of quartz, 7 on a commonly used scale of 1 to 10, makes it a good abrasive and it is commonly used in sandblasting. The Mohs scale of mineral hardness If quartz crystals grow into open voids, they have well-formed planar surfaces. These surfaces form the distinctive prismatic crystals with pointed terminations that can be seen in Figure 4-8b. However, broken quartz crystals have curved (*conchoidal*) surfaces, like those on broken pieces of a glass bottle, as shown in Figure 4-8c.

Quartz commonly crystallizes from magmas that originate in continental crust. It is an abundant mineral in granite, shown in Figure 4-9. Because it is a hard, tough mineral that is not chemically reactive, it survives the breakdown of rocks on Earth's surface. Individual quartz crystals, or *grains*, are major components of beach sand. The last time you were basking in the sun on a river, lake, or ocean beach, you were most likely lying on quartz-rich sand.

FIGURE 4-9 ▲ Granite: a Common Rock in Continental Crust
Granite—the material from which the faces on Mt. Rushmore were carved—is an aggregate of quartz, potassium feldspar, plagioclase, and biotite.

Silicosis

Quartz is generally not a dangerous mineral. Sharp broken edges can cut your fingers, but its chemical stability keeps it from readily reacting with other materials. However, there is a significant environmental problem associated with quartz. Very fine particles of quartz in dust (what many health reports call silica or crystalline silica) are dangerous to breathe. Of course, dust in general is not good to breathe, but quartz-rich dust is a special problem.

Because quartz is a component of many rocks, concrete, masonry, and other building materials, mechanically grinding and breaking these materials can create quartz-rich dust. When the tiny, broken pieces (grains) of quartz enter people's lungs, they get stuck in openings and cause inflammation and nodular lesions. This lung deterioration creates shortness of breath, fever, and diminished oxygen supply to the body.

This disease is called *silicosis* (the complete name is *pneumonoultramicroscopicsilicovolcanokoniosis*, the longest word in the English language!). There is no cure for silicosis and it can be deadly. In the 1930s, a few thousand workers dug a tunnel in quartz-rich rock at Hawk's Nest, West Virginia (Figure 4-10). Over 700 of these workers died from silicosis. YouTube—Hawk's Nest tunnel tragedy About the Hawk's Nest incident—background to *Book of the Dead*

FIGURE 4-10 ▲ The Hawk's Nest Tunnel Under Construction
Thousands of workers, eager for jobs during the Great Depression, helped dig this tunnel through quartz-rich rock. More than 700 eventually died of silicosis.

Because the cause of silicosis is well-known, it can be prevented. Safety procedures such as the use of water sprays, dust-collecting exhaust systems, and respirators have decreased the number of silicosis deaths in the United States from over 1,000 per year in the 1960s to fewer than 200 per year in the late 1990s. But silicosis can still be a problem where its causes are ignored or not recognized.

What are the Earth systems interactions that lead to silicosis?

THE FELDSPARS

As Figure 4-4 shows, sodium, potassium, calcium, and aluminum are four of the eight most abundant elements in both oceanic and continental crust. Ions of these elements are positive, and they readily combine with negatively charged SiO_4 ions to form a variety of **silicate minerals**. Minerals III: the silicates The most common of these are the **feldspars**.

In the feldspars, aluminum atoms combine with oxygen to form tetrahedral groups, as silicon does. The positive sodium, potassium, and calcium ions bond negative silicon-oxygen and aluminum-oxygen tetrahedra into an orderly three-dimensional arrangement.

The two most common feldspars are *potassium feldspar* and *plagioclase*, shown in Figure 4-11. There are several varieties of potassium feldspar ($KAlSi_3O_8$) with slightly varying atomic structures, depending on the temperature at which they crystallize. In plagioclase $(Na,Ca)(Si,Al)_4O_8$, sodium and calcium readily substitute for one another (the sodium and calcium substitution is accompanied by aluminum and silicon substitution to maintain electrical neutrality). The result is a continuous range in composition from the most sodium-rich ($NaAlSi_3O_8$) to the most calcium-rich ($CaAl_2Si_2O_8$) of plagioclase.

Physical properties and occurrence

Feldspars are light-colored, commonly tabular prismatic crystals that break apart along smooth planes called **cleavages**. Their hardness (6 on a scale of 1 to 10) and structure make feldspar crystals moderately resistant to mechanical breaking. Pieces or grains of feldspar commonly accompany quartz in sand. The crystal forms of the feldspars, and the various ways in which their atomic structures are repeated or duplicated in different orientations within them (*twinned*), are distinctive.

FIGURE 4-11 ▼ Feldspars
Potassium feldspar **(a)** and plagioclase **(b)** are very common rock-forming minerals. The specimens have broken along cleavages to form the smooth planar surfaces.

(a)

(b)

Potassium averages 1.5% of continental crust (see Figure 4-4b), and potassium feldspar is where much of this potassium resides. Potassium feldspar commonly crystallizes from magmas that originate in continental crust, especially those that form granite (see Figure 4-9).

From Figure 4-4, you recall that the average oceanic crust contains more calcium (8.2%) and less sodium (1.9%) and potassium (0.1%) than continental crust. Where do you think the calcium in oceanic crust resides? Much of it is present in plagioclase. Basalt, the most abundant rock in oceanic crust, is about half plagioclase by volume.

FIGURE 4-12 ▲ Building on Clay
Many clays expand and shrink with changes in wetness, making construction of buildings and roads on them very challenging.

Changing feldspars

Many minerals that form at high temperatures and pressures, as when magma cools and crystallizes in the crust, are not stable on Earth's surface. Here, lower surface temperatures and pressures combine with interactions with the hydrosphere, atmosphere, and biosphere (in soils, Chapter 11) to cause minerals to change into others that are more stable at surface conditions. The mineral changes at the surface that involve chemical reactions are called *chemical weathering*.

A key chemical weathering process is hydrolysis. *Hydrolysis* is the reaction of water (H_2O) with minerals to form new minerals that have water in their atomic structure. Feldspars undergo hydrolysis in several situations. But under many surface conditions, the hydrolysis of feldspars is a very slow process. The process is more effective in warm climates with abundant rainfall or where groundwater is hot (as within volcanoes).

If water continues to react with feldspar, some of the original constituent atoms in the feldspar can be dissolved and removed. Such *dissolution* is another chemical weathering process. In the case of feldspar, extensive reaction with water removes sodium, potassium, and calcium and leaves behind water-bearing aluminum silicate minerals called clays.

Clays are the common product of feldspar chemical weathering. They have a layered atomic structure but the layers or sheets are not strongly bonded and water can occupy spaces between the layers. As a result, some clays expand when wet and contract when dry. Soils containing this type of clay are unstable; as Figure 4-12 dramatically demonstrates, they make poor foundation material for buildings and other infrastructure.

Clays aren't all bad, however. You have already learned that clays can be useful in fighting disease. Did you know that a variety of clay is in toothpaste? Clays are also major components of cosmetics and, of course, the key ingredients in ceramics. Clays are a very valuable mineral resource.

Chemical weathering of feldspar by hydrolysis and dissolution is generally a slow process. However, if the water reacting with feldspar is hot, changing feldspar to clay is much more rapid and effective. Hot, water-rich fluids are called *hydrothermal fluids* and the mineral changes they cause are called *hydrothermal alteration*.

Feldspar hydrolysis at higher temperatures commonly produces light-colored minerals called *micas* with a strongly layered atomic structure. Muscovite is the most common light-colored mica. Its composition is very similar to that of potassium feldspar ($KAlSi_3O_8$) but with water added. The bonds between atomic layers in micas are weak, and these minerals easily break or cleave into thin flakes or sheets, as you can see in Figure 4-13. The micas formed from feldspars by hydrothermal alteration can in turn be converted to clays as temperatures cool but alteration continues.

The hydrothermal alteration of feldspars to micas and clays is extensive within volcanoes. This alteration can make large

FIGURE 4-13 ▼ Micas
Micas are silicate minerals whose internal structures include a plane with very weak bonds. They easily break (cleave) into sheets along this plane. Biotite, the dark mineral at the upper left, is an iron-bearing mica. Muscovite, the silvery mineral in the foreground, is a potassium-rich mica.

FIGURE 4-14 ▲ Mt. Rainier
Mt. Rainier, a volcano near Seattle, Washington, is extensively hydrothermally altered. The soft clays formed by hydrolysis in the volcano help make its flanks unstable. If the slopes become saturated with water they can collapse, sending muddy debris flows for long distances down the flanking valleys.

parts of volcanoes weak and susceptible to collapse. An example is Mt. Rainier in Washington state, shown in Figure 4-14. A large part of this beautiful mountain is extensively altered to micas and clays and is just waiting to collapse. This has happened many times over the last several thousand years, especially when heat from new magma warms the volcano and melts its snow and ice accumulations. The resulting collapse of muddy wet debris can flow far down the volcano's flanks and adjacent valleys, destroying everything in its path. Several communities face this type of volcanic hazard around Mt. Rainier—hydrothermal alteration is part of the reason why.

What are some Earth systems interactions associated with muddy, wet debris flows on the flanks of volcanoes?

THE FERROMAGNESIUM MINERALS

The two remaining abundant elements in oceanic and continental crust that are also associated with common minerals are iron and magnesium. These metals form positive ions, so they can combine with silicon-oxygen tetrahedra to form silicate minerals. Because iron and magnesium ions are similar in size and positive charge (+2), they readily substitute for each other in mineral structures. Silicate minerals that contain iron and magnesium in their structures are called **ferromagnesium minerals**.

Olivine and pyroxene

The simplest combination of iron, magnesium, silicon, and oxygen is the mineral *olivine*—$(Mg,Fe)_2SiO_4$. There is a continuous range in olivine composition, from samples that contain only magnesium to those that contain only iron. Olivine (Figure 4-15a) forms dark greenish-gray crystals, equal in all dimensions, that have a hardness similar to that of quartz and, like quartz, exhibit conchoidal fracture. However, olivine is much denser than quartz because of its iron and magnesium content.

(a)

(b)

FIGURE 4-15 ◄ Olivine and Pyroxene
Olivine **(a)** and pyroxene **(b)** are ferromagnesium minerals that are common in mafic rocks like basalt and gabbro.

Pyroxenes are a group of minerals that have a more complex atomic structure than olivine. They range in composition from magnesium silicates to iron-, magnesium-, and calcium-bearing silicates. Pyroxenes are dark-colored, dense, somewhat short and rectangular crystals (Figure 4-15b). They have two poorly developed cleavages, approximately at right angles to each other.

Olivine and pyroxene form at high temperatures and pressures within the mantle; they are the principal minerals in upper mantle rocks. They are also common minerals in basalt erupted at seafloor spreading centers and elsewhere. Olivine, pyroxene, and calcium-rich plagioclase are the dominant minerals in oceanic crust.

Changing olivine and pyroxene to serpentine

Olivine and pyroxene are less stable at Earth's surface than feldspars. Where they come in contact with the hydrosphere and atmosphere, they commonly undergo chemical weathering and hydrolyze to form *serpentine minerals*. Some serpentine is soft and flaky—it is smooth and slippery to the touch. Surface slopes underlain by this type of serpentine are unstable and easily collapse.

Chrysotile, shown in Figure 4-16, is a serpentine mineral that occurs as elongate fibrous crystals that can be bent or woven. Because it is very heat resistant, chrysotile has been used as fireproofing insulation in buildings. In fact, it is the principal mineral in asbestos.

The fibrous nature of chrysotile makes it dangerous. It can cause pulmonary diseases and cancer if people inhale it. For this reason, asbestos is not used in building insulation now, and government regulations have led to its removal in many older public buildings, such as schools.

FIGURE 4-16 ▲ Chrysotile
Olivine and pyroxene hydrolyze to form many soft serpentine minerals. One of these, the fibrous serpentine mineral chrysotile, is the principal mineral in asbestos. As you can see in the photograph, chrysotile breaks into fluffy fibers.

In The News

9/11 Dust

The suffocating clouds of dust that engulfed lower Manhattan as the World Trade Center collapsed on September 11, 2001 are hard to forget. People were literally covered with choking, light gray dust. It covered everything exposed and crept into buildings through broken windows and doors. It seemed to be everywhere within several blocks of the site of the tragedy.

In November and December of 2001, New York City and federal agencies sampled remnant dust in 30 residential buildings of lower Manhattan. The dust was analyzed for asbestos, fiberglass, mineral components of concrete (quartz, calcite, and portlandite), and mineral components of building wallboard (gypsum, mica, and halite). The sampled dust contained all of these components. The asbestos mineral was chrysotile.

People who were exposed extensively to the 9/11 dust have suffered serious ongoing consequences. First responders and workers who helped remove wreckage and clean up the devastated area experienced the greatest exposure and have been the most vulnerable. This group is experiencing a high incidence of respiratory illnesses, some causing death. The dust at ground zero, Katie Couric reports The 9/11 tragedy continues as illnesses traced to inhaling dust affect people. Their struggles will be in the news for years to come.

Tiny particles of many minerals, including chrysotile, were found in the dust created when the World Trade Center towers collapsed. People who worked to clean up ground zero in the days after the attack have suffered ongoing respiratory problems.

Biotite and amphibole

The other rock-forming ferromagnesium minerals are biotite and amphibole. Both of these minerals include water in their atomic structure—that is, they are *hydrated. Biotite* is the mica mineral that contains iron and magnesium. This black mineral has the typical layered-sheet structure of the micas and easily breaks into thin flakes along one prominent cleavage (see Figure 4-13). *Amphiboles* are elongate prismatic crystals that parallel the pyroxenes in composition except for the water in their atomic structure.

One common amphibole is *hornblende* (Figure 4-17). It has a complex composition that includes all the most abundant elements in oceanic and continental crust (see Figure 4-4b). The two planes along which amphiboles preferentially break (cleave) are at oblique angles to one another. This helps distinguish them from pyroxenes, which have two cleavages approximately at right angles to each other.

Both biotite and iron- and magnesium-bearing amphibole are moderately dense minerals. Typically they crystallize from water-bearing magmas that solidify within continental crust. Biotite is the dark mineral common in granite (see Figure 4-9).

Living (and dying) with fibrous amphibole

Some amphiboles have very elongate, fibrous forms like that of chrysotile. One, called tremolite (Figure 4-18a), is a minor constituent of a material that was mined for home insulation at Libby, Montana for many decades (Figure 4-18b). Dust at the mining operation and in the wind that blew across the town exposed workers and residents to this very fibrous mineral.

As you might expect, breathing needles of tremolite is not good for your health. Cases of pulmonary disease and lung cancer developed in the community. The mine closed in 1990, but residents are still being studied to determine how exposure to tremolite-bearing dust has affected their health. Asbestos and Libby health

FIGURE 4-17 ▲ Amphiboles: Common Ferromagnesium Minerals in Continental Crust
Amphibole crystals have different cleavages and tend to be more elongate than pyroxene crystals.

What are the Earth systems interactions associated with fibrous amphibole and health problems in Libby, Montana?

OTHER MINERALS—THE SULFIDES, OXIDES, AND CARBONATES

The ferromagnesium minerals, feldspars, and quartz, the common rock-forming minerals, are the most abundant minerals in Earth's crust. But the crust contains many other minerals, too. Three other mineral groups that are especially relevant to environmental issues are sulfides, oxides, and carbonates.

(a)

(b)

FIGURE 4-18 ◀ Tremolite: A Dangerous Amphibole
The elongate fibrous crystals of the amphibole mineral tremolite **(a)** have caused serious health problems for people in Libby, Montana **(b)**.

FIGURE 4-19 ▲ Pyrite
The sulfide mineral pyrite combines iron with sulfur. The cubic form of pyrite is distinctive. Its brassy yellow color and density are why it is sometimes called "fool's gold."

FIGURE 4-20 ▲ Rusty Rocks
The oxidation of pyrite and other iron-bearing minerals gives rocks and soils a rusty orange color.

Sulfides

Sulfides combine elements, especially metals, with sulfur. The common sulfur ion has a minus-2 charge and it is readily attracted to positive metal ions, such as those of copper, lead, zinc, iron, and silver. These combinations are called *sulfide minerals*. Examples are galena (lead sulfide, PbS), sphalerite (zinc sulfide, ZnS), and pyrite (iron sulfide, FeS_2, shown in Figure 4-19). Although they mostly occur in small amounts, some Earth systems interactions concentrate sulfide minerals into metal-rich deposits. These deposits are the principal source of metals mined by people. A lot of the environmental issues associated with mining (Chapter 12) stem from the processing and disposal of sulfide minerals, especially pyrite.

Oxides

Elements that combine directly with oxygen form *oxide minerals*. Positive metal ions and negative oxygen ions bond to form rutile (titanium oxide, TiO_2), cassiterite (tin oxide, SnO_2), and magnetite (iron oxide, Fe_3O_4), for example. Oxide minerals tend to be more stable and less chemically reactive on Earth's surface than other minerals. Like quartz, they commonly survive weathering and can become components of sand.

Several minerals that occur at Earth's surface react with oxygen in the atmosphere or hydrosphere. The positive ions in these minerals combine with oxygen to form oxide minerals. This chemical reaction is called **oxidation**, another chemical-weathering process.

A common example is the oxidation of iron in ferromagnesium or sulfide minerals, which produces iron oxides such as hematite (Fe_2O_3) or goethite (an iron oxide that includes some water in its structure). Hematite and goethite are the minerals that make up common rust, the reddish coatings that form on exposed steel, for example. The next time you are standing next to a friend's car, look around for a deep scratch or bent fender. Can you find some reddish stains? Just tell your friend that the car is turning to hematite before your very eyes.

Pyrite (iron sulfide) is a common iron-bearing mineral, sometimes called "fool's gold" because of its density and brassy yellow color (see Figure 4-19). It forms in many crustal settings where oxygen is not present. The iron in pyrite has a stronger affinity for oxygen than for sulfur. When pyrite comes in contact with oxygen in the atmosphere or hydrosphere, therefore, an oxidation reaction takes place and iron oxide is formed. In general, this reaction proceeds slowly, but because pyrite is widely distributed, evidence of this process is visible in many places.

The minerals produced by the oxidation of pyrite are hematite and goethite, the principal minerals in rust. Where they are produced, the surrounding rocks and soils are discolored or "stained" rusty red or orange. The next time you are driving through the countryside, keep an eye out for rusty-colored road cuts or distant hill slopes such as the one shown in Figure 4-20. If you see them, you are probably driving by a place where iron-bearing minerals, perhaps pyrite, have been oxidized.

What happens to the sulfur when pyrite reacts with oxygen? It commonly combines with water to form sulfuric acid, a form of what is called "acid rock drainage." You will learn more about the environmental issues associated with acid rock drainage in Chapter 12.

What are some Earth systems interactions associated with oxidizing pyrite?

Carbonates

The *carbonate minerals* combine positively charged ions such as calcium (+2 charge), magnesium (+2 charge), and iron (+3 charge, commonly) with negatively charged carbonate (CO_3, −2 charge). There are approximately 60 known carbonate minerals but when it comes to forming rocks, two are the most common: calcite, $CaCO_3$, and dolomite, $[Ca,Mg]CO_3$, shown in Figure 4-21. (The calcium and magnesium substitute for each other in dolomite). These commonly light-colored

(a) (b)

FIGURE 4-21 ◀ Carbonates Calcite **(a)** and dolomite **(b)** are the most common carbonate minerals.

minerals are soft (3 on a scale of 1 to 10) and easily broken along cleavages.

Large amounts of carbonate minerals form in oceans. Where the concentration of calcium and carbonate ions is high enough, they combine and fall out of the seawater as a solid, or *precipitate*. The precipitated material is a *chemical sediment*. Animals also precipitate carbonate minerals from seawater to make shells or other structures such as coral reefs. When the animals die, the carbonate minerals in these materials can also accumulate on the seafloor (see Figure 2-9). Accumulations of calcite on the seafloor eventually become a rock called **limestone**.

Carbonate minerals are prone to dissolve in acid. The most easily dissolved carbonate mineral is calcite, but even dolomite will dissolve in weak acids over long intervals of time. Rainwater is slightly acidic because it reacts with carbon dioxide in the atmosphere to form small amounts of carbonic acid (H_2CO_3). Rainwater flowing over the land and sinking into the ground along fractures will dissolve carbonate minerals. This dissolution completely removes calcite and enlarges fractures into larger and larger openings. Over long periods of time, even large caves can form this way (Figure 4-22a).

Landscapes underlain by calcite-rich rocks have many features produced by dissolution, some of which are shown in Figure 4-22b. These landscapes are called *karst terrain*, and *sinkholes* are one of their characteristic features (see Figure 4-22c). Sinkholes are a natural hazard and you will learn more about them and karst terrain in Chapter 8.

What Earth systems interactions are associated with karst terrain?

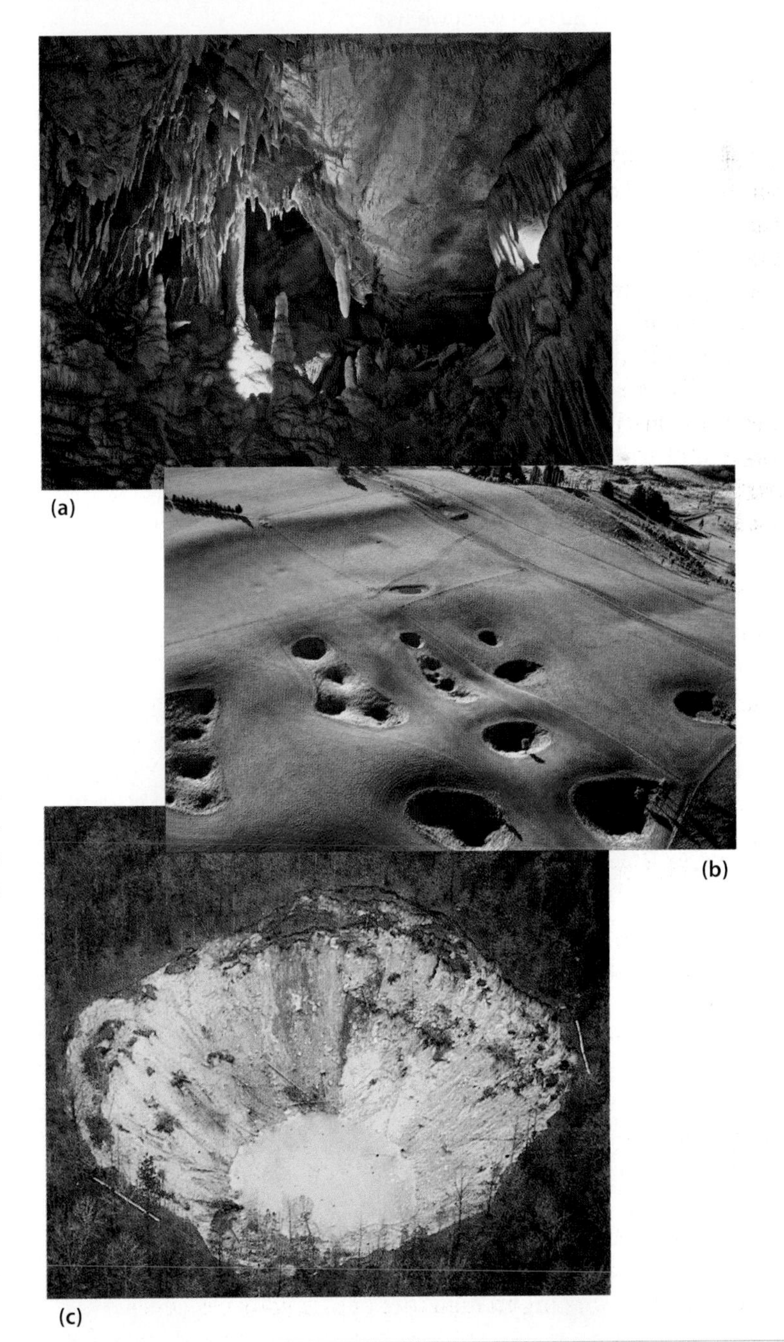

(a) (b) (c)

FIGURE 4-22 ▶ Dissolution of the Geosphere Produces Karst Terrain
Dissolution of carbonate minerals is the main process leading to the formation of caves such as Mammoth in Kentucky **(a)**, hummocky landscapes **(b)**, and sinkholes that suddenly collapse **(c)**.

What You Can Do

Investigate Mineral Use

Do you know what a mineral commodity is? It is a mineral or an element derived from a mineral that people use. People recover mineral commodities from the geosphere by mining, and they buy and sell them around the world.

The U.S. Geological Survey (USGS) keeps track of over 80 different mineral commodities—who produces them, what their uses are, who consumes them, and how much they cost, for example. These commodities include everything from aluminum to zirconium and from clays to gold. You might be surprised to learn that we use 16 mineral commodities that are not produced in the United States. We import 100% of these commodities from other countries. In fact, there are over 40 mineral commodities for which the United States imports at least 50% of what we use.

If you want to learn which countries these commodities come from, go to the USGS website that has mineral commodity information. | Mineral commodity summaries | Scroll down to "Individual Commodity Data Sheets" and pick your favorite. You can learn who is producing this commodity and how much people are paying for it.

International Minerals Statistics and Information

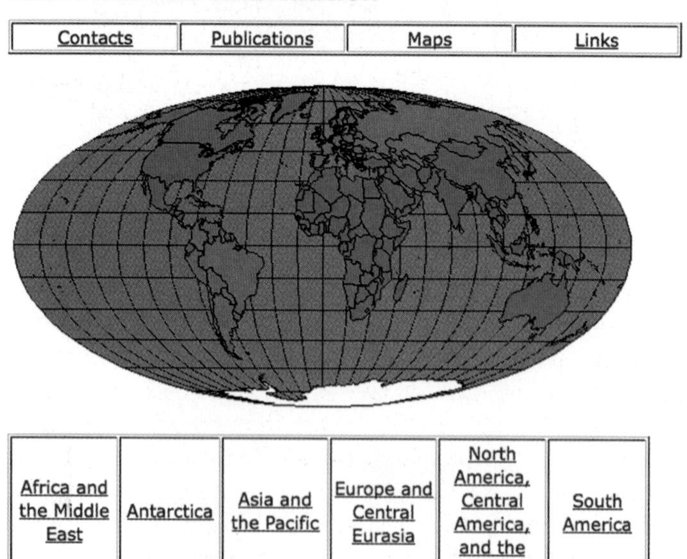

- **Contacts**
  - Country Specialists Directory
- **Publications**

4.3 Rocks—Where Minerals Reside

Minerals are distributed through the geosphere in **rocks**. Rocks can be made up of other things that are not technically minerals—volcanic glass (obsidian) and coal, for example—but most rocks are just mineral aggregates.

Rocks that form from magma are called **igneous rocks**. Igneous rocks that contain significant amounts of feldspar and quartz are said to have a **felsic** composition. These rocks have lots of silicon, sodium, and potassium and not much iron and magnesium. Chemically, they contain more than 63% silica (SiO_2). Felsic rocks are most characteristic of continental crust.

Igneous rocks that do not contain quartz but have abundant ferromagnesium minerals have a **mafic** composition. Mafic rocks are rich in iron, magnesium, and calcium, and are most characteristic of oceanic crust. Chemically, they contain 45 to 55% silica (SiO_2).

Intermediate igneous rocks have compositions between those of mafic and felsic rocks—they contain 55 to 63% silica (SiO_2). Such rocks characterize the magmas produced by subduction (as described in Chapter 3).

MAKING OCEANIC CRUST

In Chapter 3 you learned that oceanic crust is created at spreading centers along mid-ocean ridges. Magma from the mantle rises to the surface and forms the most abundant volcanic rock on Earth, basalt. Most basalt, like the sample shown in Figure 4-23a, is a mixture of olivine, pyroxene, and plagioclase. Mafic rocks like basalt commonly contain calcium-rich plagioclase. If they don't contain plagioclase, we call them *ultramafic rocks*, which are the characteristic rocks of the mantle.

Basalt crystallizes rapidly when it erupts onto the seafloor. As a result, the individual minerals do not have time to grow large. It usually takes a microscope to see the individual crystals in basalt. If rising basalt magma pools within oceanic crust, it cools more slowly and the individual minerals have time to grow large. The resulting rocks, like the one shown in Figure 4-23b, are called **gabbro**. Gabbro and basalt have the same mafic composition, but because of the very different rates at which they crystallize, they have very different mineral grain sizes.

Figure 4-24 shows the general structure of oceanic crust. There is a thin skin of sediment, but oceanic crust is predominantly igneous rocks crystallized from basalt magma. There are upper layers of basalt, a zone of tabular bodies of basalt injected below the seafloor (called *dikes*), and a lower zone with abundant gabbro. In places, ultramafic rocks form where olivine and pyroxene crystals sink to the bottom of crystallizing gabbro bod-

FIGURE 4-23 ▲ Basalt and Gabbro
(a) This sample of basalt is about 15 centimeters (6 in) across. In the breakout—a very thin slice of basalt photographed through a microscope using transmitted polarized light—plagioclase appears as rectangular crystals in various shades of gray and olivine as variably colored irregular crystals. **(b)** This gabbro specimen, also about 15 centimeters across, shows white plagioclase crystals and dark pyroxene crystals. You can see that gabbro is coarser grained than basalt—it has the same composition but is formed by much slower crystallization.

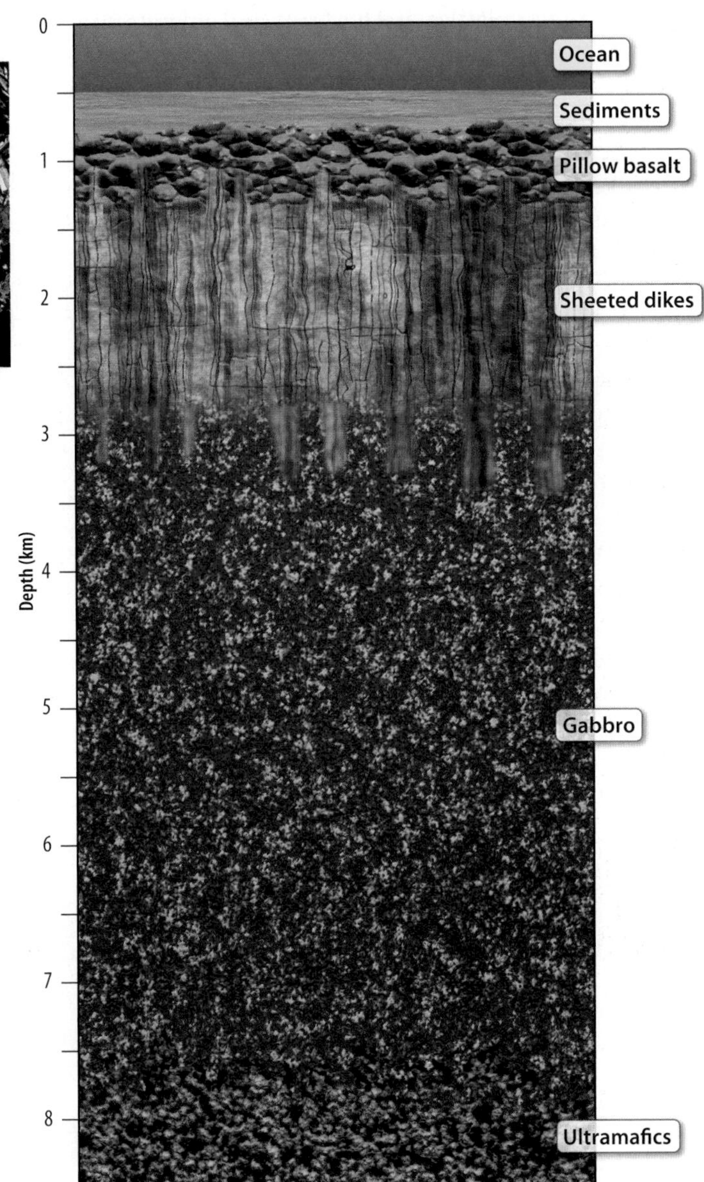

FIGURE 4-24 ▲ Oceanic Crust Has a Well-Defined Internal Structure
From the top down, oceanic crust includes deep sea sediments, basalt lava flows, sheeted basalt dikes, gabbro intrusions, and some ultramafic rocks.

ies. This section of mafic and some ultramafic rocks, 5 to 8 kilometers (3 to 5 mi) in thickness, is typical oceanic crust.

MAKING CONTINENTAL CRUST

Continental crust is more diverse compositionally than oceanic crust. It ranges from largely mafic in the lower crust to felsic, particularly in the upper crust. The average composition of continental crust is intermediate between mafic and felsic. The age and compositional complexity of continental crust makes its origin more difficult to understand than that of oceanic crust.

Earth's first crust must have been mostly basaltic (like the Moon's crust). When basalt partially melts, the magma that forms is intermediate in composition. In the hot early Earth, basalt could melt wherever it was thick enough or it was being subducted into the mantle. The intermediate-composition magmas developed by this melting could then rise and crystallize to form the initial continental crust. Over Earth's first few billion years, this process developed much of Earth's continental crust.

Today Earth is cooler, and subduction does not cause melting of basaltic oceanic crust as it sinks into the mantle at convergent plate boundaries. Instead, oceanic crust, altered by hydrolysis reactions with seawater (especially along the mid-ocean ridges, where it is warm and first encounters seawater), *recrystallizes* as it is subducted and encounters higher temperatures and pressures in the mantle. This recrystallization releases water and other volatile compounds, such as carbon dioxide, that were incorporated in the altered oceanic crust.

The separated volatiles are buoyant and migrate upward into the mantle overlying the subduction zone. Here they lower the mantle's melting point and new magmas form. These are the magmas that rise to the surface along the island and volcanic arcs described in Section 3.3.

Subduction-related volcanoes commonly have explosive eruptions that release tremendous amounts of volatiles, including water, carbon dioxide, and hydrogen sulfide. In this way, the volatiles originally incorporated into oceanic crust by hydrolysis at spreading centers get cycled back to the surface and released to the atmosphere and hydrosphere.

Magmas erupted or pooled below subduction-related volcanoes are different from those that form along mid-ocean ridge spreading centers. Erupted subduction-related magmas commonly produce *andesite*. Plagioclase, hornblende, and pyroxene are the common minerals in andesite. Andesite's chemical composition is similar to the average composition of continental crust (see Figure 4-4b).

Today there is considerable discussion among scientists about how andesitic magmas form. It's possible that the primary magma developed above subduction zones is more basaltic than andesitic. In this case, andesites could form by mixing basaltic magma with felsic magma produced by melting in the crust above subduction zones. Some andesites could also form by settling out of early-formed olivine and pyroxene crystals in basalt magma, a process called **differentiation**. This leaves the remaining magma more intermediate in composition. Andesites probably form in several ways but they are the characteristic rock formed in subduction-related volcanoes.

There is also scientific discussion today about the rates of continental crust growth over Earth's history. Subduction-related volcanoes add material and grow continental crust but at the same time, some is eroded into the oceans. Overall, it appears that the amount of new continental crust being created today is small compared to that created in the past.

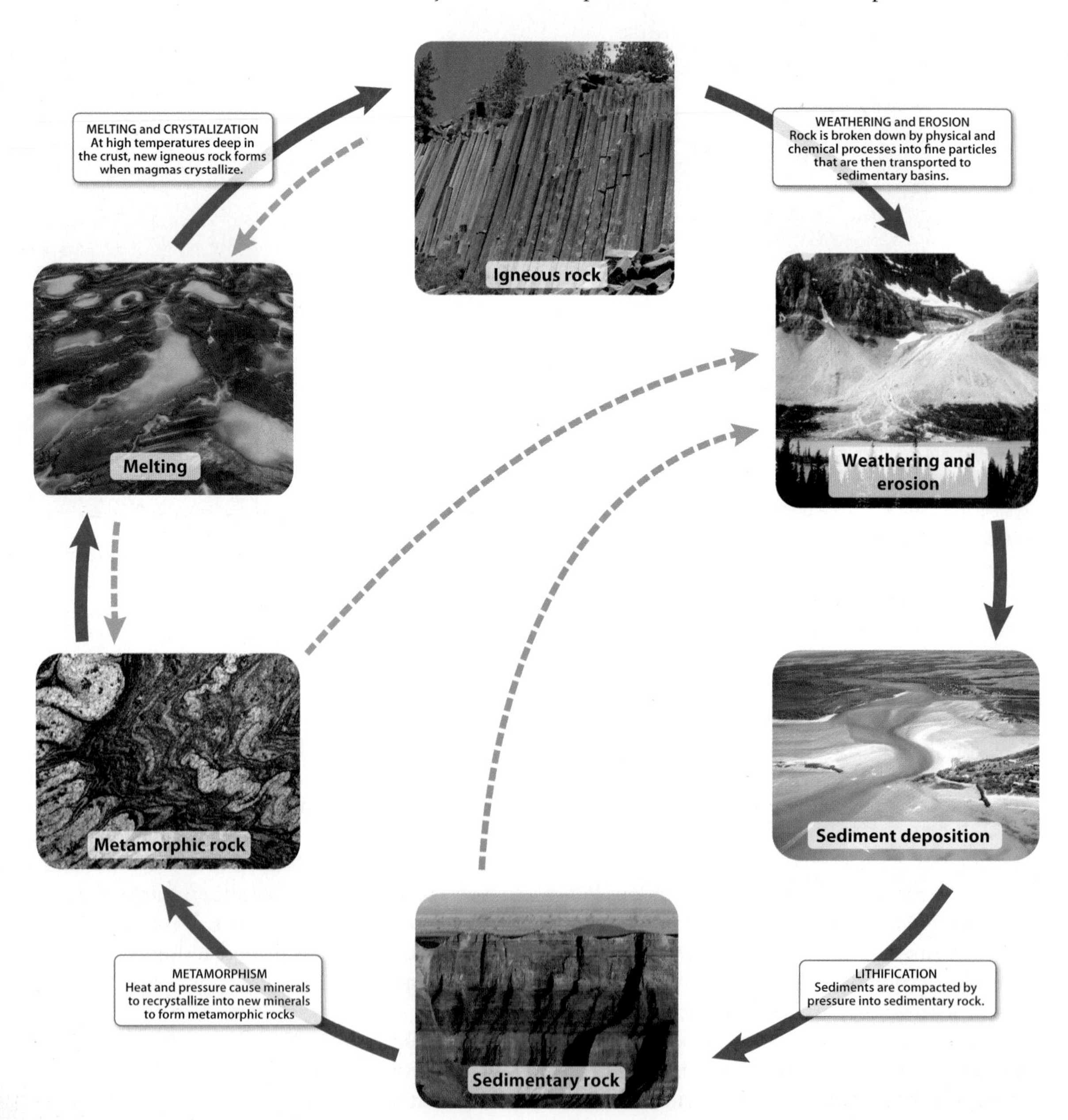

FIGURE 4-25 ▶ The Rock Cycle: An Overview
The rock cycle shows how igneous, metamorphic, and sedimentary rocks are related in continental crust and transformed into one another by geological processes. Heavy red arrows outline the main processes. The dashed arrows indicate that sedimentary and metamorphic rock can also be weathered to form sediments, and igneous rock can also undergo metamorphism.

CHANGING ROCKS IN THE ROCK CYCLE

Although continental crust hasn't grown much for a long time, this doesn't mean it hasn't changed. Over long periods of time, dynamic plate tectonic processes can greatly alter rocks in a sequence of changes called the **rock cycle**. Although we have so far been concerned mainly with igneous rocks, these are just one of three main classes of rocks in the continental crust, along with *sedimentary rocks* and *metamorphic rocks*. The rock cycle, outlined in Figure 4-25, shows how these classes of rocks are created, how they are related in the continental crust, and how they are transformed one into another. Let's begin our examination of the rock cycle with the formation of igneous rocks.

Igneous rocks and crustal melting

Igneous rocks are those that form from the solidification of magma. Magma is molten rock within the geosphere—we call it *lava* where it erupts onto Earth's surface. You are already familiar with several common igneous rocks. Rocks that form from lava, like the basalt that makes up so much of oceanic crust, are **volcanic igneous rocks**. Igneous rocks that crystallize within the crust, like gabbro and granite, are **plutonic igneous rocks**.

Crustal melting can occur where the crust becomes thick (rocks become buried to a depth where temperatures are above their melting point) or heat is added. Continental crust is thickest beneath high mountains, and heat is added where magmas from the mantle rise into the crust or asthenosphere upwelling occurs (see Figure 3-15). Crustal melting produces different types of magma, depending on how much melting occurs and the composition of the melting material. The most common magma formed from melting of continental crust is felsic; felsic rocks have lower melting points than other rocks. Felsic magma forms granite if it cools slowly within the crust and the volcanic rock *rhyolite* if it is erupted.

Crustal melts commonly crystallize to form plutonic rocks within mountain belts. These rocks form from a few kilometers to tens of kilometers (several miles) below the surface. But as the deformation that creates the mountains continues and moves them upward (geologists say they are *uplifted*) the plutonic rocks are eventually exposed at the surface or **outcrop**, as in Figure 4-9. Rocks exposed on Earth's surface, especially those in mountains, are affected by another part of the rock cycle, weathering and erosion.

Weathering and erosion

Rock exposures in youthful mountain ranges can be massive or highly fractured. Some massive and texturally homogeneous rocks, like granite, that form under high pressures deep in the crust naturally expand as they are uplifted and exposed at Earth's surface. The expansion breaks the massive rock along fractures called *joints*. Joints are common in uplifted rocks of all kinds, but they are very well developed in granite, as you can see in Figure 4-26. Jointing is commonly the first physical change that rocks experience as they are uplifted to the surface.

Once rocks are at the surface, weathering starts to break them down and change them even more. **Weathering** is a set of physical and chemical processes that change rocks at Earth's surface. *Physical weathering* breaks rocks into smaller and smaller pieces, principally through the action of freezing water. Water is one of the few substances that increases in volume when it freezes (becomes a solid). Because it commonly fills cracks and fractures in rocks, even tiny fractures along the boundaries between minerals, its expansion upon freezing exerts strong forces that push mineral grains and rock fragments apart. The results of this process, called *frost wedging*, can be seen in Figure 4-27a on p. 100. As it occurs over and over—perhaps every spring and fall night that temperatures fall below freezing—rocks are gradually broken down into smaller pieces.

Another cause of physical weathering is the growth of plant roots along fractures, known as *root wedging* (Figure 4-27b). In arid climates, minerals that precipitate along rock fractures when water evaporates can also wedge rocks apart. Frost wedging, however, is the most prevalent way rocks are physically broken into smaller and smaller pieces at Earth's surface. This process increases the surface area of rocks exposed to the atmosphere and is a big assist to chemical weathering.

FIGURE 4-26 ▼ A Jointed Granite Outcrop
Joints, sets of regular fractures, form when rocks expand as they are uplifted to the surface.

(a)

(b)

(c)

FIGURE 4-27 ▲ Weathering
(a) Frost wedging—the repeated expansion of water as it freezes in cracks and fissures—breaks rocks apart. **(b)** Root wedging can have similar effects. **(c)** Chemical weathering processes decompose rocks.

You have already been introduced to some *chemical weathering* processes—mineral hydrolysis, oxidation, and dissolution. These processes not only break down and change minerals, they also help "disaggregate" rocks. Weathering of granite will cause this hard "tombstone" rock to become a crumbly mass of loose mineral grains. Deeply weathered granite, like that shown in Figure 4-27c, is well on its way to becoming sand.

Once physical and chemical weathering have changed rocks into small pieces and perhaps even individual mineral grains, the weathered material can move. **Erosion** is the transportation of geosphere materials from one place to another by natural movements of water, wind, and ice (glaciers). Surface rock materials first move downslope by gravity. In youthful mountain ranges, rock fragments literally fall down steep slopes (Figure 4-28a). A person visiting steep, snowy mountains on a sunny summer day would probably be surprised at how noisy the surroundings are. The scattered and nearly constant fall of rock debris as surface materials thaw creates a racket.

The fallen material lands on glaciers or the sides of valleys. Glacial movement of rock debris is significant in high or northern mountains. A glacier scrapes up rocks along its base, carries the rocks along on or within the moving ice, and pushes rocks along its margins (Figure 4-28b). The bodies of rock debris carried and moved by glaciers are called **moraines**.

Elsewhere, streams are the principal movers of rock debris. A fast, deep stream that flows down steep surface gradients can move big boulders, abrading and plucking rocks from its channel bottom (Figure 4-28c). A lot of erosion occurs where streams wash against and undercut their banks. Once rock debris is added to a stream channel, the debris is rolled and bounced along the bottom as *bedload*. This is how streams move gravel. Smaller debris suspended in turbulent currents and carried along in the water makes streams muddy.

Sedimentation and lithification

Rocks and mineral grains transported by streams make up *sediment*. Wind can move sediment, too, but most is carried away from its original source by moving water in streams. What happens to sediment moved by wind and water, what we call *clastic sediment*? It becomes deposited in another part of the rock cycle.

The sediment created by weathering is transported by erosion to places where it is deposited. Depositing sediment is called **sedimentation**. Clastic sediment, like that shown in Figure 4-29a, ranges in size from tiny clay particles that make up mud to silt, sand, and gravel. In general, the farther sediment is transported, the smaller or finer in size it gets. This is partly because sediment is broken down mechanically as it is moved, but it also becomes finer downstream as streams become more sluggish, losing their ability to carry coarse sediment. Streams eventually deposit sediment—they stop carrying it as bedload or in suspension—when their rate of flow slows sufficiently. This occurs where they flow from narrow to broader valleys, into lakes, or into the ocean. Rivers, like the one shown in Figure 4-29b, are the principal way clastic sediment gets delivered to the oceans.

(a)

(b)

(c)

FIGURE 4-28 ▲ Erosion
(a) Sediment transport starts with rocks falling in steep terrain. **(b)** Glaciers deeply erode mountains. **(c)** Fast-moving streams carry and abrade larger rocks.

(a)

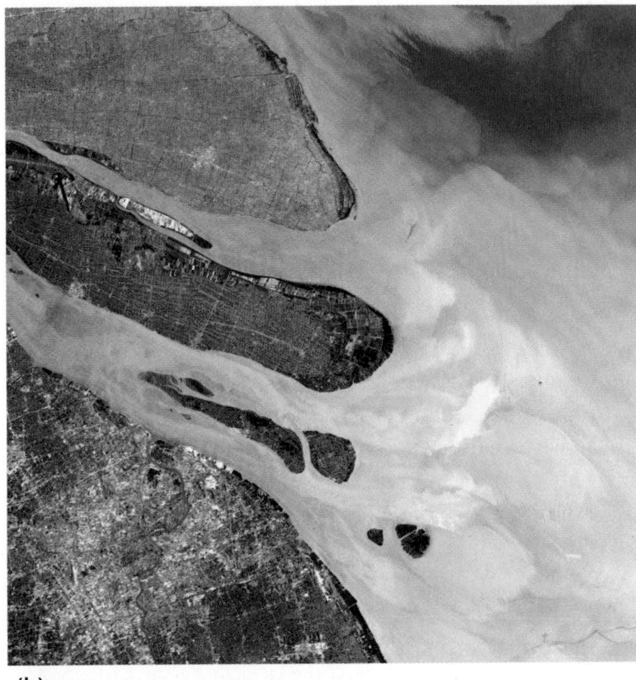

(b)

(c)

FIGURE 4-29 ▶ Sedimentation
(a) Clastic sediments range in size from tiny clay particles to sand and gravel. **(b)** Rivers carry sediment to the coast. **(c)** Thick sedimentary sections develop in basins.

(a)

(b)

(c)

FIGURE 4-30 ▶ Clastic Sedimentary Rocks
When sediments become deeply buried, they lithify. **(a)** Muddy sediment turns into shale. **(b)** Sand becomes sandstone. (The rusty stains in this specimen define individual layers or *beds*.) **(c)** Gravel forms rocks called conglomerate.

Large areas where sediment accumulates are called **sedimentary basins**. These are places where the crust subsides within or along the margins of continents. Crustal thinning—for example, in places where rifting occurs (see Figures 3-14 and 3-15)—develops basins as well. Many basins are places where the top of the crust is below sea level.

The biggest sediment-collecting basins in the world are the oceans. Because the deep oceans are so far from sources of sediment (rivers draining continents), the muddy sediments deposited in the deep sea are relatively thin—less than a kilometer (0.6 mi) thick over most of the seafloor. NGDC/MGG-Total Sediment Thickness of the World's Oceans & Marginal Sea Basins within and along continental margins, however, can develop very thick sediment deposits. In these basins, clastic sediments accumulate in layer after layer, called **beds**.

A sequence of sediment beds is called a *section*; a spectacular one is shown in Figure 4-29c. Sedimentary sections can be over 10 kilometers (6 mi) thick. The alternating beds of finer or coarser sediment reflect changes in the physical conditions under which the beds were deposited—especially changes in the ability of water currents to move sediment. The presence of finer sediments implies the existence of weaker currents when they were laid down.

Changes take place in the deeper parts of sedimentary sections. The originally deposited sediments—loose mud, silt, sand, and gravel (all saturated with water)—become compacted. In some places, the clastic mineral grains also become cemented together by minerals (such as carbonates) precipitated from water that passes through the sediment. As compaction progresses, the sediment changes from its loose "unconsolidated" state to become solid rock. We say it is **lithified**. The high pressures in deep parts of sedimentary basins and the resulting compaction turn loose sediment into **sedimentary rocks**, the second of the three major classes of rocks. Muddy sediment turns into shale, silt into siltstone, sand into sandstone, and gravel into conglomerate. These rocks, several of which are, shown in Figure 4-30, are **clastic sedimentary rocks**.

There is another general kind of sedimentary rock. You have learned how carbonate minerals form chemical sediment in oceans. The most common chemical sediment is calcium carbonate, the mineral calcite ($CaCO_3$). Where calcite is buried and lithified it becomes limestone (Figure 4-31). Other examples of minerals precipitated on the seafloor to become sedimentary rocks are gypsum (calcium sulfate, $CaSO_4$), halite (common salt, NaCl), and sylvite (potassium chloride, KCl). These minerals generally precipitate where evaporation has increased their concentration in seawater, so the rocks they form are called *evaporites*.

Sedimentary rocks form deep in sedimentary sections. What do you think happens to sedimentary rocks as they get buried deeper and deeper? If they get deep enough, where temperatures and pressures are high, they begin to change. The minerals in them recrystallize and the rocks become "metamorphosed."

Metamorphism

Metamorphic rocks, rocks changed at high temperature and pressure, make up the third major class of rocks. Any rock can become a metamorphic rock—igneous, sedimentary, and even other metamorphic rocks. All it takes is for them to recrystallize under temperature and pressure conditions different from those under which they originally formed.

At deeper levels of the crust, below 10 kilometers (6 mi), temperatures and pressures are high enough to destabilize minerals formed at the surface, such as clays and serpentine minerals. As a result, they recrystallize to form minerals stable at higher temperatures and pressures. This recrystallization largely involves dehydration reactions. In other words, recrystallization drives out water and other volatiles such as carbon dioxide incorporated in mineral structures. As these reactions progress, the new minerals grow in orientations influenced by

FIGURE 4-31 ▲ Limestone: A Marine Sedimentary Rock
Limestone is derived from the mineral calcite (Figure 4-21a), a form of calcium carbonate. Calcite that precipitates on the seafloor, or accumulates there in the remains of marine organisms, becomes limestone when it is lithified (see Figure 2-09).

pressure. They become aligned and produce "oriented fabrics" in the rocks.

The two most typical metamorphic rocks are schist and gneiss. *Schist* characteristically contains flaky mica minerals, such as biotite, that are oriented parallel to one another (Figure 4-32a). This gives the rock a strongly two-dimensional sheeted or leaf-like structure—a **foliation**—that in this case is termed *schistosity*. Sedimentary rocks like shale, siltstone, and sandstone are examples of rocks that become schist during metamorphism.

Gneiss (pronounced "nice") contains discontinuous layers (called *lenses*) of larger minerals (generally quartz and feldspar) separated by finer-grained, schistose layers (Figure 4-32b). In general, gneiss forms at higher temperatures and pressures than many schists. A sequence of schists can become

(a)

(b)

(c)

(d)

FIGURE 4-32 ▶ Metamorphic Rocks
In schist **(a)**, generally formed from sedimentary rocks like shale or sandstone, crystals of new minerals such as biotite are oriented parallel to each other. This arrangement gives the rock a well-developed foliation. In gneiss **(b)**, typically formed at higher temperatures and pressures than schist, new minerals tend to be coarser and segregated into discontinuous layers and lenses. The folded layers in this gneiss are a few to several centimeters across (about 1–10 in). Marble **(c)** is metamorphosed limestone in which the original calcite has been recrystallized. **(d)** Metamorphism can lead to melting. In this gneiss, the light-colored swirled layers were once molten.

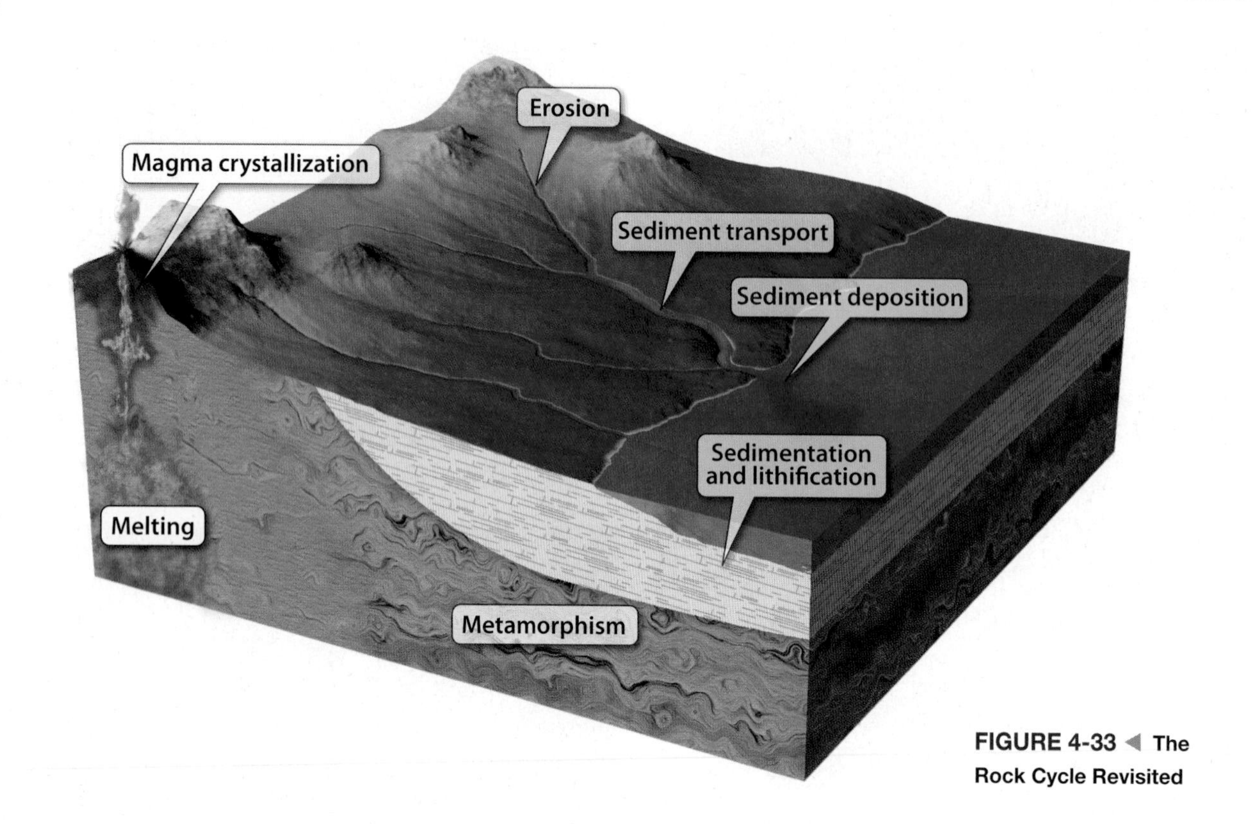

FIGURE 4-33 ◀ **The Rock Cycle Revisited**

gneiss as they recrystallize at higher and higher temperatures and pressures, what geologists call increasing *metamorphic grade*. Gneiss can form from any rock, even granite, but many form from clastic sedimentary rocks.

Another metamorphic rock is *marble* (Figure 4-32c), which is metamorphosed limestone. Marble is composed of calcite, like the limestone from which it was formed. However, in marble, the original minerals have been recrystallized. As a result, the new calcite crystals in marble tend to be larger than the original ones in limestone.

Metamorphic conditions can increase to very high temperatures and pressures. This happens especially in the deep parts of thick crust in mountain ranges. The continental crust can be up to 70 kilometers (43 mi) thick where deformation, especially along convergent plate boundaries, compresses and stacks it up in mountain ranges. In deep crustal settings, rocks can become so hot that they melt. In very high-grade gneiss, for example, some of the coarser lenses and layers were once melted (Figure 4-32d). If enough melting occurs, the molten material accumulates into a buoyant mass that can move up in the crust. This newly formed magma generally rises to intermediate or shallow levels of the crust, where it crystallizes and forms new igneous rocks. This is how most granite is formed.

A continuum of processes changes rocks in the rock cycle. Surface processes change rocks through weathering and erosion, sedimentary processes accumulate products of weathering and erosion, metamorphic processes change rocks deep in the crust, and igneous processes connect the deep crust to the surface. The entire cycle is shown in greater detail in Figure 4-33. However, keep in mind that it is not always a one-way progression through these overlapping processes. Parts of the rock cycle can be reversed or even skipped, depending on how geosphere movements affect the crust.

4.4 Using Rocks

People use rocks and minerals in many ways. Chapter 12 explains how people use minerals. Here we will examine how people use rocks.

People use rocks to build things they need—commercial and residential buildings; highways, bridges, sidewalks, and parking lots; factories and power generation facilities; water storage, filtration, and delivery systems; and wastewater collection and treatment systems. Developed countries cannot sustain their high levels of productivity, and the economies of developing nations cannot expand, without the extensive use of rocks.

Examples of using rocks in buildings are all around you. The "granite" used for countertops in homes or for flooring in public buildings such as airports, are examples. Limestone from Indiana was used to build the National Cathedral in Washington, DC, the Pentagon in Virginia, and Rockefeller Center in New York City (Figure 4-34).

FIGURE 4-34 ▲ Rock Solid
Indiana limestone was used to build many monumental structures, such as the National Cathedral in Washington, DC, shown here.

What You Can Do

Investigate Rock Use

You can investigate how rocks are used around you. Take a trip to examine and identify rocks used in your community. A good place to start is a local construction material supplier. Here you will find a wide variety of rocks available for use, especially for countertops and flooring in homes. Construction people call all of these rocks "granite," but many types of rocks are used for these purposes—limestone, gneiss, and gabbro, for example.

The cut and polished surfaces of these materials make it easy to see individual mineral grains and their textural relations. See if you can identify quartz, feldspar, and some ferromagnesium minerals. The textures and structures in gneiss are distinctive—can you find examples of gneiss? Perhaps some of the rocks have orange streaks and patches, evidence of iron oxidation. Now you probably know more about rocks and minerals than the people who sell these materials.

In many communities you can conduct similar investigations by visiting the larger, older buildings. Ask yourself, how many different rocks have been used in their construction?

AGGREGATE

A lot of rock used in construction is not easy to examine. This rock is incorporated into materials such as cement and asphalt or spread on roadways and building sites to help stabilize the ground. This type of rock is called **aggregate**; examples include sand, gravel, and crushed stone.

In 2007, the total amount of mined aggregate, 2.8 billion tonnes (3.1 billion tons), was about 8600 kilograms (19,000 lbs) for every person in the United States. This is a staggering amount of material. How can each of us need so much? For a family of four, this amount of material would make a pile in front of their residence that is 3 meters by 3 meters (10 ft by 10 ft) in area and 2 meters (6 ft) high. Of course most families didn't have this amount of material delivered directly to them in 2007. This is the amount that was used to repair, improve, and expand the basic facilities that we all use and depend on.

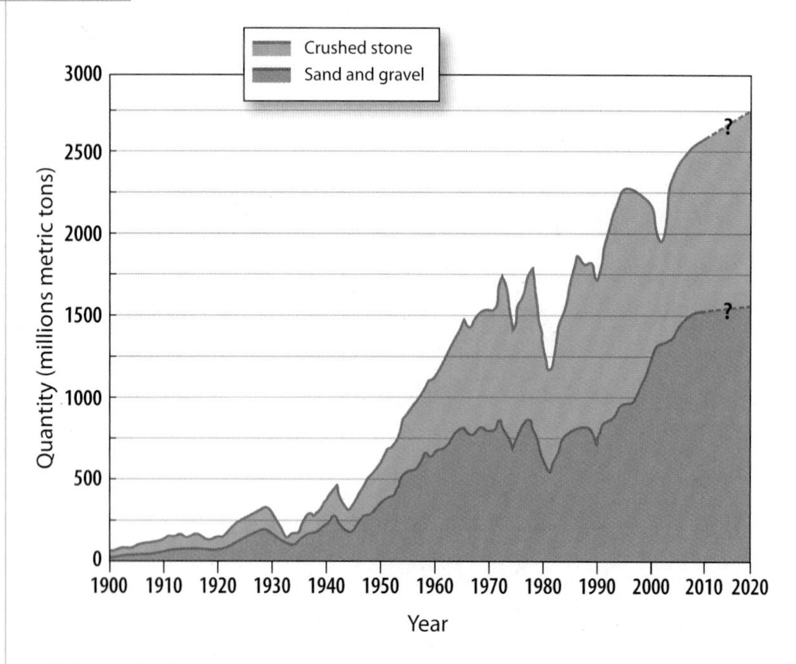

FIGURE 4-35 ▲ Past and Future Use of Aggregate
Our need for aggregate, especially crushed stone, is increasing rapidly. It is estimated that by 2020 we will be using more than twice as much as we did in 1960.

Our need for aggregate is increasing. Figure 4-35 shows the historical and estimated future use, to the year 2020, of construction aggregate in the United States. If the projections in Figure 4-35 are correct, we will use almost as much construction aggregate in the next 25 years as we used in the entire previous century.

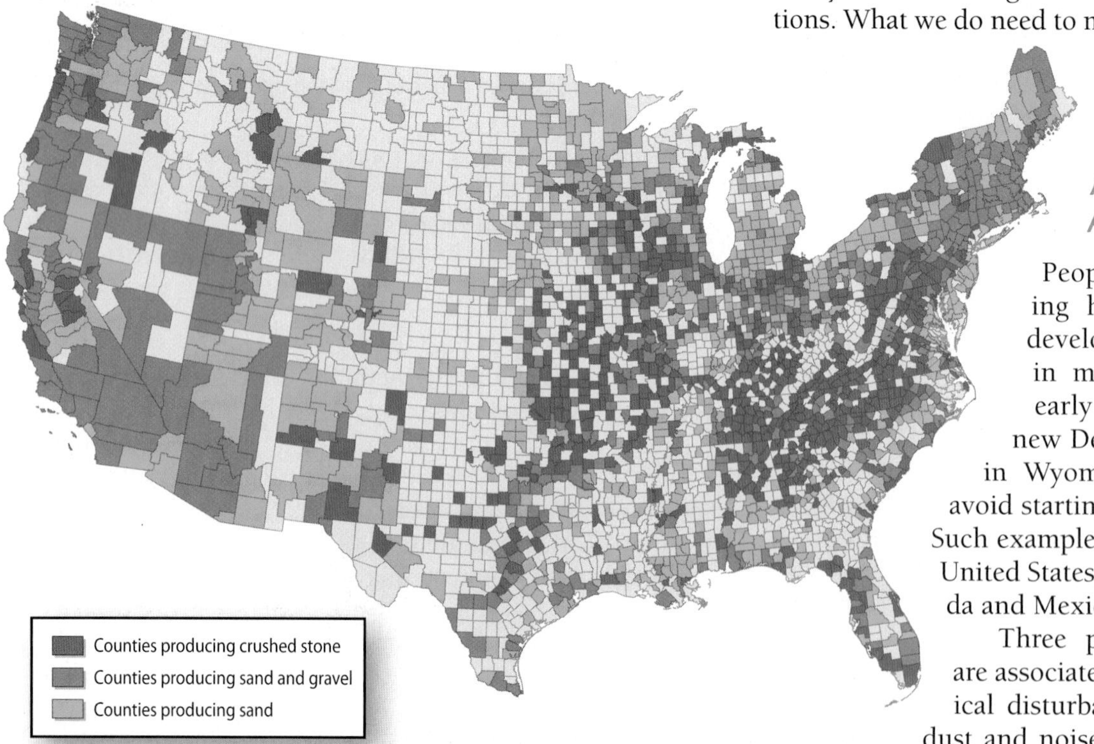

Counties producing crushed stone
Counties producing sand and gravel
Counties producing sand

FIGURE 4-36 ▲ Distribution of Aggregate Mining
Aggregate mining operations can be found near most communities in the United States.

Part of this increased need accompanies population growth, but some also reflects an increase in per capita (per person) needs. Per capita needs are increasing because, in many places, the basic infrastructure of the United States needs repair. For example, the Transportation Equity Act of the 21st Century, which was signed into law by President Clinton in 1998, initiated a program to spend $270 billion, by 2005, on the repair and improvement of U.S. highways and bridges. On August 10, 2005, President Bush signed a new law to spend another $286.4 billion over six years to continue highway and transit construction and maintenance. You probably use a street, road, or highway that you hope is addressed by this or related programs.

Construction aggregate is relatively cheap to mine and process, but transporting it is costly. This is why mining operations that supply this material are so widespread, as Figure 4-36 indicates; the operations must be close to the sites that use their products in order to keep the cost of transporting the material low.

Since aggregate is very common natural rock, sand, and gravel, we can usually find a local source to meet nearby needs. This means that you can probably find an example of this type of mining operation in your neighborhood, or at least in your general area. Perhaps your county has recently faced decisions about adding or expanding construction aggregate operations; these operations are as close as most of us get to seeing or experiencing mining activities.

Even though the amount of needed aggregate is tremendous, there is so much useful rock that we don't commonly worry about making sure there is enough for future generations. What we do need to make sure of is that the environment is not irreparably harmed by aggregate mining operations.

AGGREGATE MINING AND THE ENVIRONMENT

People's concerns about aggregate mining have been sufficient to stop the development of new aggregate resources in many places. For example, in the early 1990s, aggregate for building a new Denver, Colorado airport was mined in Wyoming and transported by train to avoid starting new operations closer to Denver. Such examples are becoming more common; the United States now imports aggregate from Canada and Mexico.

Three principal environmental concerns are associated with aggregate mining: the physical disturbances produced by quarrying, the dust and noise that may accompany the mining operations, and the congestion and safety concerns that may accompany transporting materials from mines to the sites where it is used.

Physical disturbances

Aggregate must be physically removed from its source; in other words, it must be mined. This operation produces quarries or pits where the original materials were located. These are physical disturbances to the landscape that actually look like what they are during operations—big holes in the ground, like the one shown in Figure 4-37. In addition, mining commonly removes the natural vegetation from nearby areas and necessitates building industrial facilities such as shops and processing plants. People usually consider the net result aesthetically challenged if not an eyesore.

What are the choices? Stop using aggregate? Use only recycled construction materials? Perhaps we can use smaller amounts to some extent in our future, but a future without need for new aggregate is very difficult to imagine. If we are to use aggregate, then we will need to mine it, and mining operations inherently disrupt the landscape.

FIGURE 4-37 ▲ Aggregate Mining
Aggregate mines are commonly quarries where blasting, hauling, and crushing occur.

We can mitigate the aesthetic effects or make them less severe by developing or preserving visibility barriers, ensuring well-planned and orderly operations, and reclaiming areas as operations proceed. However, mining requires some degree of disruption. What is *not* required is having the mine site remain an eyesore after operations have ceased.

All mine sites can be reclaimed to some degree. We cannot completely restore them to their original conditions but we can place pits, quarries, and disturbed surfaces back in harmony with their surroundings. Slope contouring, placement of new soil, and revegetation through a process called *reclamation* are key remedies for the negative aesthetic results of aggregate mining. Aggregate pits and quarries can become lakes for recreation and wildlife habitat. Disturbed terrain can become parks, gardens (like the one shown in Figure 4-38), or even golf courses in some cases. Golf courses spring from reclaimed mines

Aggregate is a benign natural material. Through creative reclamation, many aggregate mines may acquire new uses and

(a)

(b)

FIGURE 4-38 ▲ Reclaiming a Rock Quarry
There are many possible ways in which aggregate mines and quarries can be turned into new community assets. **(a)** Over 100 years ago this quarry on Vancouver Island, Canada, was a source of limestone for a cement factory. **(b)** Today the only surviving portion of the cement factory is the tall chimney seen in the distance. The quarry is the site of Butchart Gardens, famous for its year-round display of flowering plants and visited by close to a million people each year.

values after mining operations cease. A community may consider the eventual closure of a nearby aggregate mine as an opportunity to create new community assets.

Dust and noise

Aggregate mining, especially the production of stone, involves breaking up rocks. This is commonly accomplished by blasting, hauling, and crushing, all of which are dusty operations. Dust can cause respiratory illnesses and otherwise harm workers and others who inadvertently breathe it. All mine sites that can produce dust as part of their operations must have active air quality mitigation programs to ensure the safety of workers and the nearby community. In fact, air quality monitoring and compliance with regulatory guidelines are required at all mining sites, not just those for aggregate. Fortunately, the design of processing equipment, along with mining techniques, can provide effective dust controls. Examples include water sprayers, sweepers, and vacuum systems.

Noise from blasting and equipment operations can be a significant concern at some aggregate mines if they are near residential areas. The significance depends heavily on local conditions and needs, which are evaluated on a case-by-case basis. Controlling noise through the location and design of the equipment and the timing of operations can mitigate some of these concerns.

Congestion and safety

Perhaps one of the biggest concerns a community may have with aggregate mining is the perceived impact of transporting the mined products. If truck transport is to take place on a local road previously used only by residents, then safety concerns are easy to understand (Figure 4-39a). However, many aggregate mines can be located so that they use infrastructure, such as railroads and highways, specifically designed for transporting bulk materials (Figure 4-39b). This is another aspect of aggregate mining that people should evaluate on a case-by-case basis and can address through careful planning and infrastructure design.

(a)

(b)

FIGURE 4-39 ▲ Aggregate Transport
Aggregate transport over local roads **(a)** raises safety concerns. It is often preferable to situate mines in places that allow transport by ship or rail **(b)**.

You Make the Call

Aggregate Mining in Your Neighborhood

The example of aggregate use, future need, and related environmental concerns provides a backdrop for reflecting on the appropriateness of mining and your role in contributing to the associated problems and solutions. In this regard, how would you answer the following questions?

- Do you think that the annual and per capita consumption of aggregate accurately reflects your use of these resources?
- What do you think the future aggregate needs of the United States are going to be? What are the more important factors that influence these needs?
- Where do you think new aggregate resources should come from?
- Do you think aggregate mining should take place in your community?
- Do you think that mitigation measures such as reclamation can satisfactorily address the environmental concerns associated with aggregate mining?

SUMMARY

Geosphere materials are minerals and rocks. Although we commonly think of these solid materials as unchanging, in reality the opposite is the case. They change through interactions with the atmosphere and hydrosphere and through deeper crustal and mantle processes in the dynamic geosphere. Geosphere materials are a foundation of our economy and some even affect human health. Scientists are now learning how minerals like clays can fight deadly disease. Our use of geosphere materials can raise questions about sustainability if the materials are not abundant, but in all efforts to obtain and use geosphere materials, protection of the environment is needed.

In This Chapter You Have Learned:

- Minerals and rocks reflect the geosphere's chemical composition. The geosphere's core, mantle, and crust were created by physical segregation of the four most abundant elements—iron, oxygen, silicon, and magnesium.
- The thin, chemically distinct crust is the part of the geosphere that most directly interacts with the atmosphere, hydrosphere, and biosphere. Just eight elements make up about 99% of Earth's crust.
- The most abundant elements in crust—silicon, oxygen, aluminum, iron, magnesium, potassium, sodium, and calcium—combine in various ways to make minerals. Silicon and oxygen combine to form silica tetrahedra (SiO_4 ions), fundamental building blocks of many rock-forming minerals called silicates. The arrangement of atoms within minerals determines their physical properties, such as hardness and crystal form.
- The common rock-forming minerals are quartz, feldspars, and ferromagnesium minerals. Other minerals, including oxides, sulfides, and carbonates, are useful to people and are involved in interactions that affect the environment.
- Minerals can affect human health. Minerals that form needle-like crystals, such as chrysotile and tremolite, cause lung damage and cancer if inhaled. Some clay minerals are capable of helping kill dangerous bacteria, even some very resistant to antibiotic drugs.
- Oceanic crust is, overall, mafic in chemical composition and composed of iron-, magnesium-, and calcium-bearing silicate minerals, especially olivine, pyroxene, and plagioclase. These minerals crystallize from magmas to form the volcanic igneous rock basalt and its plutonic equivalent gabbro, characteristic rocks of oceanic crust.
- Continental crust originally formed and grew from magmas produced by melting of mafic material. Continental crust is generally intermediate in chemical composition between mafic and felsic. It is composed of sedimentary, metamorphic, and igneous rocks. The minerals in these rocks include many silicates, including potassium feldspar and plagioclase. Continental crust contains an abundance of silicon and oxygen, and as a result, rocks typical of continental crust, such as granite, contain the mineral quartz.
- Rocks are aggregates of minerals. Rocks that crystallize from magma (or lava) are igneous rocks. If the rocks crystallize slowly, as when plutonic rocks crystallize in the crust, the minerals in them can grow larger. If not, as when volcanic rocks crystallize on Earth's surface, the mineral grains are small. Magma that solidifies very quickly may become glass instead of crystals.
- Sedimentary rocks form from sediment (such as silt, sand, and gravel) that is moved from the land (usually by water) and deposited in basins. Some chemical sediments deposited directly from seawater onto the seafloor become sedimentary rocks, too. Burial of sediments causes them to become rocks, or lithify.
- Metamorphic rocks are those changed at high temperature and pressure. Any rock can be metamorphosed if it recrystallizes at temperature and pressure conditions that are different from those under which it originally formed.
- Rocks of the continental crust constantly change in the rock cycle, the overlapping effects of weathering, erosion, sedimentation, metamorphism, and igneous processes.
- Weathering is the physical and chemical breakdown of rocks into fragments and mineral grains. Physical rock breakdown is largely caused by expanding water as it freezes in rock cracks. Chemical weathering involves hydrolysis, oxidation, and dissolution of minerals.
- Erosion is the movement of weathered rock and mineral debris—sediment—by wind, water, or ice (glaciers). Erosion carries sediment from mountains to floodplains, lakes, and oceans, where it is deposited.
- Rocks and minerals are valuable resources that people use in enormous quantities and many ways. Natural aggregate—sand, gravel, and crushed stone—is a geosphere resource that is essential for infrastructure construction.

KEY TERMS

aggregate (p. 105)
atom (p. 83)
bed (p. 102)
clastic sedimentary rock (p. 102)
clay (p. 90)
cleavage (p. 89)
crystal (p. 85)
differentiation (p. 98)
element (p. 83)
erosion (p. 100)
feldspar (p. 89)
felsic (p. 96)
ferromagnesium mineral (p. 91)
foliation (p. 103)
gabbro (p. 96)
igneous rock (p. 96)
limestone (p. 95)
lithified (p. 102)
mafic (p. 96)
metamorphic rock (p. 102)
mineral (p. 85)
moraine (p. 100)
outcrop (p. 99)
oxidation (p. 94)
plutonic igneous rock (p. 99)
quartz (p. 86)
rock (p. 96)
rock cycle (p. 99)
sedimentary basin (p. 102)
sedimentary rock (p. 102)
sedimentation (p. 100)
silica (p. 86)
silicate mineral (p. 89)
volcanic igneous rock (p. 99)
weathering (p. 99)

QUESTIONS FOR REVIEW

Check Your Understanding

1. What is the relationship between a chemical element and an atom?
2. Roughly 99% of Earth's mass is made up of only eight elements. What are they, and where are they largely found?
3. The continental crust has a greater concentration, by volume, of light elements like silicon, oxygen, aluminum, potassium, and sodium than does the underlying mantle. How does this fact help explain why continental crust does not subduct into the mantle at convergent plate boundaries?
4. What are crystals, and how are they related to minerals?
5. Silicon-oxygen ions combine to form the mineral quartz. Feldspars result from the addition of a few other constituents to silicon-oxygen ions. What are these constituents, and what distinguishes feldspars from quartz?
6. When feldspars are exposed to water at high temperatures or at Earth's surface, various forms of hydrolysis create new minerals. What are the products of feldspar hydrolysis at (a) high temperatures and (b) surface conditions?
7. Which valuable mineral deposits are associated with sulfide minerals and what is the reason for this association?
8. Carbonate minerals are commonly deposited on the seafloor. What is the role of the biosphere in this process, and what are its connections to the carbon cycle (Chapter 1)?
9. What is the difference between felsic and mafic rocks, and where is each most likely to occur?
10. Why are subduction-related volcanoes more explosive than those at mid-ocean ridges?
11. What are the two main classes of igneous rocks, and how does the environment in which they form affect the differences between them?
12. What is the difference between physical and chemical weathering, and how do they work together to break down rock material?
13. How does the character of clastic material tell us something about the environmental conditions where and when it was deposited?
14. What is the process of lithification? How is it different from deposition?
15. What is the main difference between metamorphism and igneous processes, and what are the common features that result from metamorphism?

Critical Thinking/Discussion

16. The mantle and core are dominated by relatively few minerals. However, we have discovered over 4,000 minerals. Where are most of these minerals located and why?
17. Many respiratory disorders arise from the inhalation of fine mineral powders, including silica dust as well as fibrous minerals such as chrysotile and tremolite. These materials are not chemically toxic, so why are they so dangerous to inhale, especially in large, sustained doses?
18. Iron readily combines with oxygen to create oxides, mineralogically known as hematite and goethite, that make up ordinary rust. These minerals are common in red-colored soils and rocks. What are some of the implications of rust-stained soil or rock exposures?
19. Why are the oldest known rocks on Earth part of continental crust? Relate your answer to composition, origin, and density.
20. Sedimentary rocks are regarded as one of the best recorders of past environments. Geologists use these rocks to investigate conditions at the surface through time. How do they do this?

ANSWERS TO IN-CHAPTER INSIGHT QUESTIONS

P. 89

What are the Earth systems interactions that lead to silicosis?

- Quartz is transferred from the geosphere to the atmosphere in dust.
- Quartz-bearing dust is inhaled by people.

P. 91

What are some Earth systems interactions associated with muddy, wet debris flows on the flanks of volcanoes?

- Shallow magma in the geosphere heats groundwater of the hydrosphere (energy transfer).
- Heated groundwater changes minerals and weakens ground (material transfer).
- Heat from geosphere magma melts snow and ice of the hydrosphere (energy transfer).
- Surface water of the hydrosphere mixes with weakened materials of the geosphere, destabilizes slopes, and produces muddy debris flows.
- Muddy debris flows move downslope and destroy biosphere habitat and organisms (including people in some cases).

P. 93

What are the Earth systems interactions associated with fibrous amphibole and health problems in Libby, Montana?

- Tremolite is transferred from the geosphere to the atmosphere by mining operations.
- Tremolite in the atmosphere is inhaled by people (material transfer to the biosphere).
- Inhaled tremolite physically interacts with lung tissue to cause disease.

P. 94

What are some Earth systems interactions associated with oxidizing pyrite?

- Pyrite in the geosphere reacts with oxygen in the atmosphere or hydrosphere (pyrite oxidation; material changes and transfers).
- Sulfuric acid produced by pyrite oxidation can contaminate the hydrosphere (and atmosphere in some cases; material transfers).
- Acidic surface and groundwater produced by pyrite oxidation can negatively impact biosphere habitat and organisms.

P. 95

What Earth systems interactions are associated with karst terrain?

- Atmosphere and hydrosphere interactions produce acidic rain (material transfer).
- Acidic surface and groundwater produced by acidic rain dissolve the geosphere (material transfers).
- Dissolution of the geosphere leads to physical controls on surface and groundwater movements.
- Changes in the geosphere caused by dissolution influence habitat on the surface and underground.
- Unstable land can result from geosphere dissolution and physical interactions with surface and groundwater.

FIGURE 5-1 ▶ Plate Tectonics Comes Calling
The Loma Prieta earthquake of October 17, 1989 collapsed the Cypress Street Viaduct in Oakland, California, killing 42 people. The cause was a sudden movement along one of the branches of the San Andreas fault zone, a transform plate boundary.

CHAPTER

5

EARTHQUAKES

5:04 p.m., October 17, 1989. Players were gearing up for the third game of the World Series at San Francisco's Candlestick Park. Across the bay on the Cypress Street Viaduct, a double-decker elevated portion of the Nimitz Freeway, traffic was light as many people prepared to watch the game between cross-bay rivals, the Oakland Athletics and the San Francisco Giants. Steel-reinforced concrete columns supported the viaduct's upper and lower decks, each able to carry five lanes of traffic. Suddenly, an earthquake jolted the viaduct, followed by 10 to 15 seconds of strong shaking. The supports cracked, buckled, and then collapsed along a 1.4 kilometer (0.9 mi) stretch of the roadway, sandwiching the upper and lower decks (Figure 5-1). Vehicles were crushed by tons of concrete and steel. Forty-two people died on the Cypress Viaduct, representing the majority of the 67 lives lost in the earthquake.

The Loma Prieta earthquake, named for a nearby peak in the Santa Cruz Mountains, was the largest earthquake to strike a major urban center in California since the Great 1906 San Francisco earthquake. The October 17, 1989 Loma Prieta, California, earthquake The event was a catalyst for renewed attention to seismic hazards in California. The tragedy at the Cypress Viaduct repeated some lessons, too. The highway was originally built in the 1950s, but later partially reinforced to resist earthquake shaking. Despite these efforts, part of the viaduct failed catastrophically. Studies revealed that soft sediment under the failed supports had amplified the ground motions during the earthquake. The stronger shaking combined with some construction design problems led to the viaduct's collapse. ENGINEERING.com Cypress Street viaduct As with many earthquakes, the lessons learned—and relearned—came at great human and economic cost.

The late Thomas P. (Tip) O'Neill, Jr., longtime Speaker of the U.S. House of Representatives, once famously observed, "all politics is local." Similarly, to the individual caught up in an earthquake, seismic hazards are always local. To someone trapped in a car under the Nimitz Freeway, it mattered little that global forces were at work.

Most earthquakes happen because the jigsaw of tectonic plates that forms Earth's outermost shell is in constant motion. Where plates meet at plate boundaries, the lithosphere is under stress and the rocks bend and break. Some of this stress is stored as *elastic strain,* like the tension in a coiled spring, within the brittle crustal rock on which we live. The "local" consequence of global tectonics is that, as we saw in Chapter 3, the crust ruptures along faults releasing the pent-up stress in the rock in the form of seismic vibrations.

Earthquakes shake the ground beneath our feet; topple buildings and other structures; change moist soils into mush; and trigger powerful tsunamis (mistakenly called "tidal waves" by some) that inundate shorelines. It is because plate tectonics unleashes such mighty forces with little or no warning that earthquakes can be so devastating. Earthquakes may be thought of as plate tectonics knocking at our door.

Given that earthquake hazards are inevitable as long as plates continue to move, what can we do about them? It is not practical to abandon densely populated earthquake-prone areas. Nor is it possible to prevent earthquakes or to predict them with enough precision to allow timely evacuations of threatened cities. Today, our best hope is to forecast the probable locations and magnitudes of earthquakes and make sure people, buildings, and emergency response systems are prepared for the next large earthquake, hopefully minimizing loss of life and destruction of property.

Most people have a basic idea of what earthquake magnitude is and what earthquakes do to cities. However, stop and ask yourself how much you *really* understand about earthquakes and their hazards. If you moved to an earthquake-prone part of the world, would you know how to assess the hazards you would face during an earthquake? Do you know what triggers earthquakes and how much faith you can place in forecasts of future earthquakes? Do you know what areas and structures are safe if an earthquake occurs? This chapter will enable you to answer these questions.

IN THIS CHAPTER YOU WILL LEARN:

- What earthquakes are
- Where earthquakes occur
- How fault ruptures generate earthquakes
- How we measure and study earthquakes
- What earthquake hazards are
- How scientists attempt to predict earthquakes
- How people mitigate earthquake hazards

5.1 Earthquake Basics

Almost everyone has some understanding of earthquakes. We know that they move and shake the ground, sometimes violently. But technically, what are earthquakes? In this section we get down to the basics of earthquake science, or **seismology**.

WHAT EARTHQUAKES ARE

For thousands of years, people have speculated about why the ground sometimes shakes, attributing it to everything from the tantrums of angry gods to blasts of air blowing through subterranean caverns. Ancient Japanese people told of a giant catfish underground, whose lurching and wriggling caused earthquakes. | NIPPONIA catfish and earthquakes Today we know what earthquakes are and what causes them.

The simplest definition of an **earthquake** is a trembling, shaking, or vibration of Earth. Tectonic movements within the lithosphere cause most earthquakes and these movements mostly occur along preexisting faults. The movements generate vibrations or **seismic waves** that travel outward in all directions from the ruptured fault, as depicted in Figure 5-2. These seismic energy waves shake the ground. The point underground where the rock first ruptures is called the earthquake's **focus** (also sometimes known as the *hypocenter*). The point on Earth's surface directly above the focus is the **epicenter**. The epicenter is as close to where an earthquake begins as a person can get (see Figure 5-2).

Anything that causes sudden movements within the Earth can cause earthquakes. Magma movements within volcanoes commonly cause small earthquakes. In fact, as you will learn in Chapter 6, earthquake monitoring can be used to help predict volcanic eruptions. People cause earthquakes, too. Underground nuclear test explosions generate earthquakes strong enough to be detected all over the world, which is why countries use earthquake monitoring to help enforce nuclear test ban agreements. The U.S. Geological Survey (USGS) estimates that several million earthquakes occur each year. Most of these do not shake the ground strongly enough for people to feel them.

WHERE EARTHQUAKES OCCUR

When we plot the locations of earthquakes over an extended period, an unmistakable pattern emerges. Recall from Chapter 3 that earthquakes cluster densely along the edges of Earth's tectonic plate boundaries. (See the global map of earthquake epicenters in Figure 3-25.) The movement along plate boundaries generates elastic strain in the crust, most of which is then released as seismic energy. Some earthquakes, including a few large ones, occur "between the cracks," far from presently active plate boundaries. These are known as *intraplate earthquakes*.

Plate tectonics is the key to explaining the cause of most earthquakes as well as anticipating the location of future earthquakes. As is the case with many aspects of geology, little about earthquakes makes sense without seeing them in the larger picture of global plate tectonics.

Transform plate boundaries

Motions along transform plate boundaries cause many large earthquakes around the world. A famous example of this type of plate boundary is the San Andreas fault zone, which runs 1,300 kilometers (800 mi) through northern and southern California. As you can see in Figure 5-3a on p. 116, the San Andreas is the boundary along which the North American and Pacific plates slide laterally past each other. On the west side of the fault zone, a sliver of California is moving to the northwest along with the floor of the Pacific Ocean.

The San Andreas fault zone is one of the most extensively studied plate boundaries in the world, with thousands of maps, studies, and reports of various types published since the Great 1906 San Francisco earthquake. Such studies reveal that this plate boundary, like others of the same type, is not simple. The San Andreas fault zone is a system of many connected faults that allow the plates to move past each other. The main San Andreas fault cuts to the base of the crust and extends from the divergent plate boundary in the Gulf of California northward to near Cape Mendocino, where it merges with an offshore transform fault.

As Figure 5-3b indicates, however, dozens of major and minor faults lie within a zone 50 to 200 kilometers (31 to 124 mi) wide. Among these many branches or subsidiary faults are the Hayward and Calaveras faults in the San Francisco Bay area and numerous others in southern California.

As we saw in Section 3.3, many faults at transform plate boundaries are strike-slip faults (Figure 5-3c). Movements of crust along these faults account for about 80% of the relative motion of the Pacific and North American plates, which

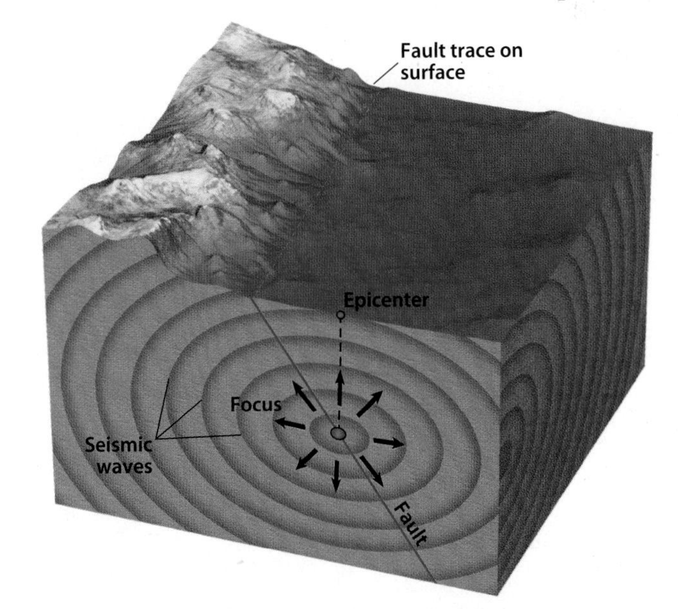

FIGURE 5-2 ▲ Movement on Faults Causes Earthquakes
Movement on a fault causes vibrations called seismic waves that originate at the focus and migrate outward. The point on the surface directly above the focus is the epicenter. Most of the earthquakes that occur around the world are too small to be felt by people, but larger ones in populous areas can be devastating.

FIGURE 5-3 ▼ The San Andreas Fault Zone

This large fault system is a transform plate boundary. **(a)** The main fault connects a divergent plate boundary in the Gulf of California to the Cascadia subduction zone, a convergent plate boundary in the Pacific Ocean off the coast of northern California, Oregon, Washington, and Canada. **(b)** The San Andreas system includes many branches and subsidiary faults. The locations and dates of major earthquakes (magnitude 7 or greater) associated with fault movement are shown. **(c)** Most of the faults at such a boundary are strike-slip faults, which allow blocks of crust to move laterally past each other (Section 3.3). When such movement occurs abruptly, the result is typically an earthquake.

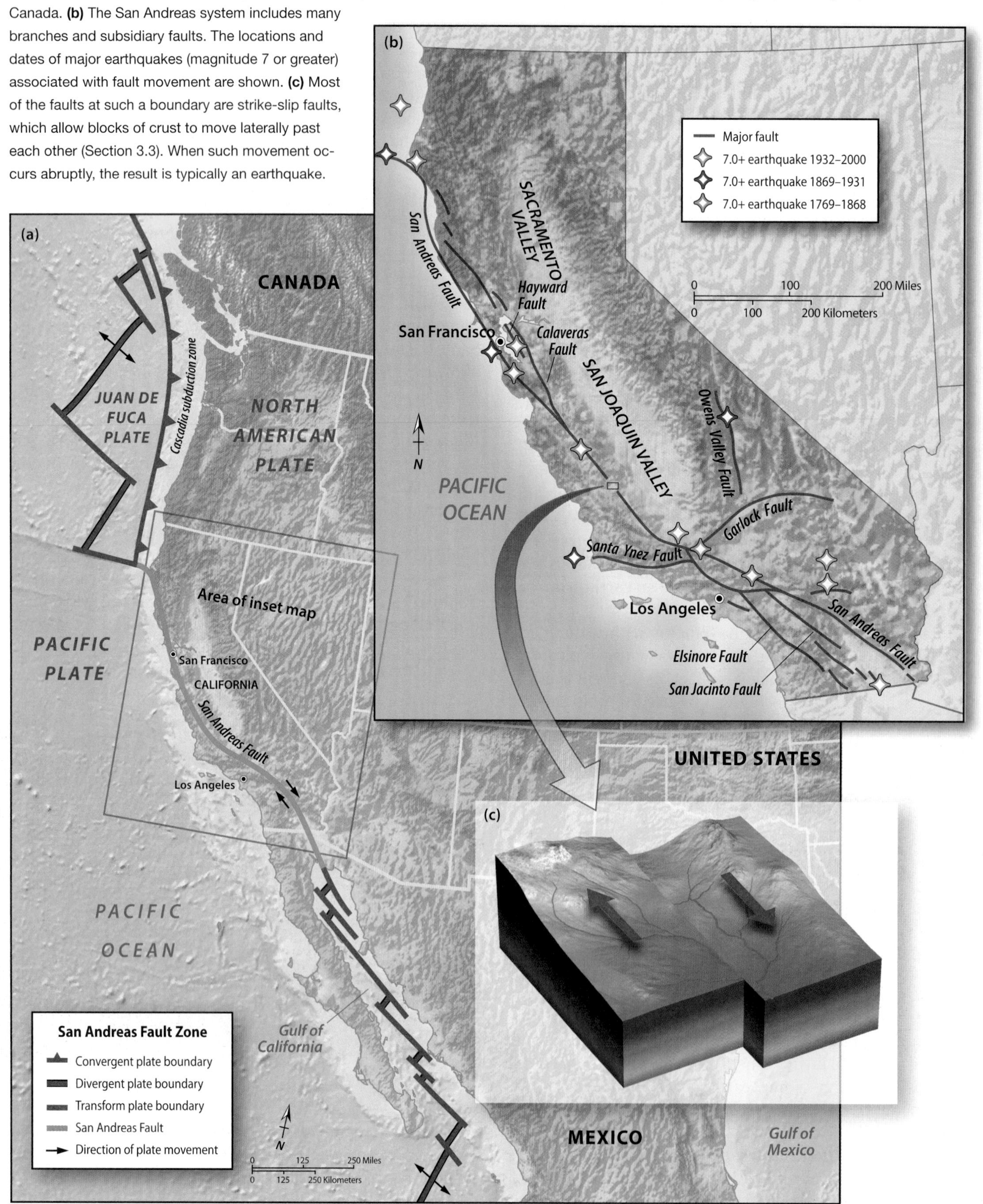

amounts to about 5 centimeters (2 in) per year. Typically these faults are "locked" until a segment gives way and triggers an earthquake. It was just such a sudden jerk along a 430 kilometer (270 mi) segment of the fault that caused the Great 1906 San Francisco earthquake. Some segments may also continuously slip by moving slowly without triggering large earthquakes, a phenomenon called *fault creep* (discussed more fully later in this chapter).

Over the approximately 20-million-year history of the San Andreas fault zone, western California has moved an estimated 560 kilometers (350 mi) relative to the North American Plate. If this movement continues for another 20 million years, rocks underlying Los Angeles on the west side of the fault will be moved next to those underlying San Francisco on the east side.

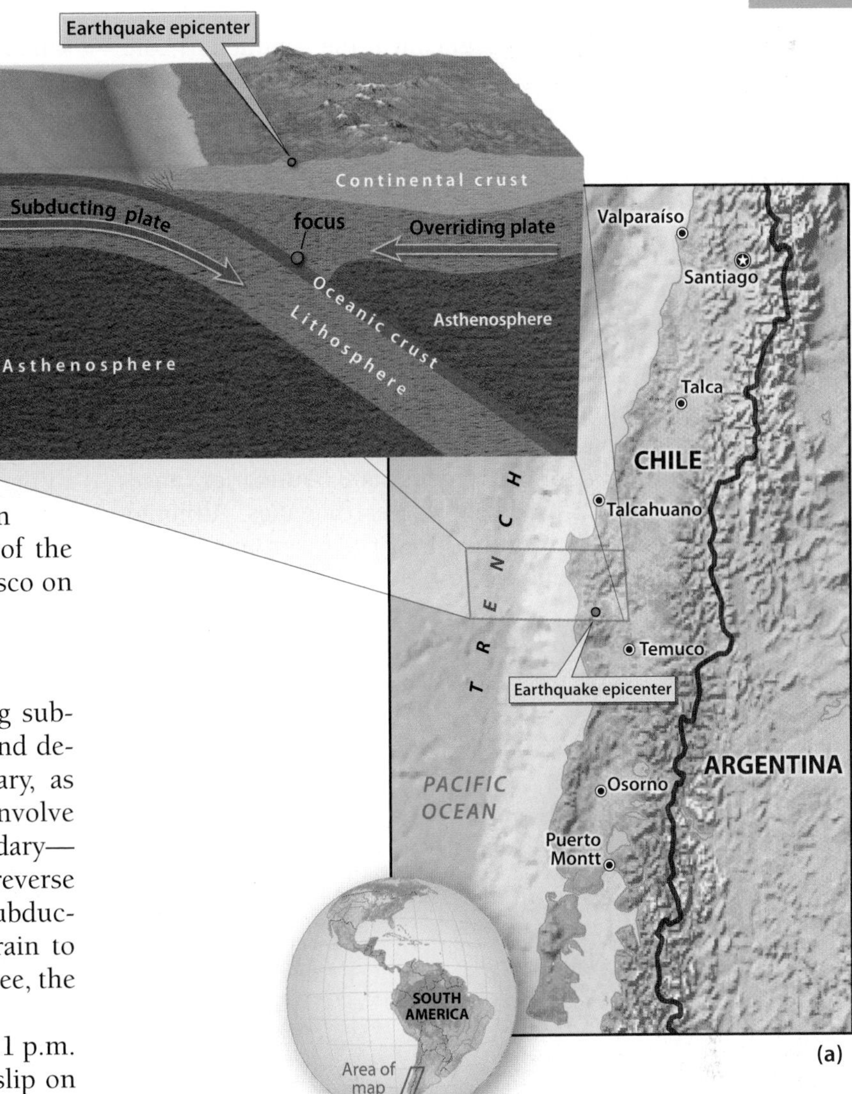

Convergent plate boundaries

The largest earthquakes on record have occurred along subduction zones, where one plate slips beneath another and descends into the mantle at a convergent plate boundary, as described in Section 3.3. Subduction-zone earthquakes involve sudden slips along segments of a convergent plate boundary—a very large system of shallow to moderately dipping reverse faults. These earthquakes occur where a portion of the subduction zone becomes stuck or locked, causing elastic strain to build up. When the down-going plate suddenly breaks free, the result is a subduction-zone earthquake.

The largest earthquake ever recorded occurred at 2:11 p.m. local time on May 22, 1960, as the result of a sudden slip on part of the Nazca Plate beneath South America (Figure 5-4).

(b)

(c)

FIGURE 5-4 ▲ The Largest Earthquakes Occur at Convergent Plate Boundaries

Earthquakes occur when sudden movement occurs on reverse faults in subduction zones. The largest earthquake ever recorded occurred along the convergent plate boundary off the coast of Chile on May 22, 1960 **(a)**. This magnitude 9.5 earthquake killed more than 1600 people, injured 3000, left 2,000,000 homeless, and caused $550 million of damage in southern Chile **(b)**. It also generated a tsunami that took 22 hours to reach Japan, but still damaged property and killed 138 people there **(c)**.

Historic earthquakes Chile 1960 A segment of the subduction zone about 1,000 kilometers (620 mi) long suddenly broke, generating intense shaking and a giant wave (tsunami) resulting from the sudden movement of the seafloor. The earthquake killed over 1,600 people in Chile, and the tsunami killed 138 more in Japan. Wave heights on the coasts of Chile and Peru exceeded 25 meters (82 ft).

Divergent plate boundaries

As you learned in Chapter 3, mapping earthquakes is very helpful in recognizing and understanding mid-ocean ridges. The pulling apart or *extension* of the lithosphere and normal faulting along these oceanic divergent plate boundaries causes many shallow, small-to-moderate-size earthquakes. Although frequent and scientifically interesting, these earthquakes are commonly not large enough or close enough to people to cause much concern.

Extension of the lithosphere, normal faulting, and related earthquakes occur on the continents, too. Recall from Chapter 3 that new divergent plate boundaries are developing where rifting is breaking up northeast Africa. As Figure 5-5a shows, moderate-size earthquakes are common along the normal faults of the East African rift valleys, and occasional large quakes have caused significant damage. One such earthquake toppled buildings, caused power outages, and killed several people in Mozambique in February 2006 (Figure 5-5b).

Although continental breakup is not occurring in the United States today, extension, normal faulting, and related earthquake generation are. The largest region being actively extended is the Basin and Range province, which includes a large part of Nevada and its neighboring states, as shown in Figure 5-6a. Geologic provinces of the United States: Basin and Range province In this region, high heat flow from the mantle is thinning the lithosphere and causing the crust to extend. A result is

(a)

(b)

FIGURE 5-5 ◀ Earthquakes at Divergent Plate Boundaries
(a) Earthquakes are common in the East African rift zone, where normal faulting is common as new divergent plate boundaries develop. **(b)** The 2006 Mozambique earthquake produced vertical dislocations of more than a meter (3.3 ft) in some places.

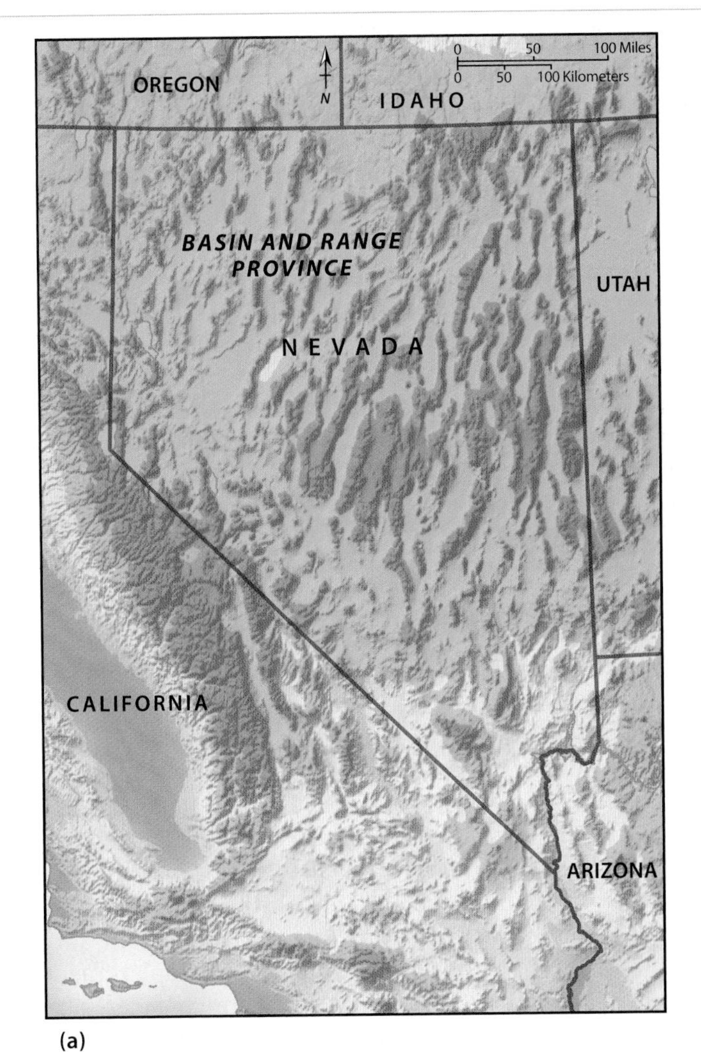

(a)

the distinctive topography of the Basin and Range province, characterized by long narrow mountain ranges separated by wide, flat valleys (Figure 5-6b). The mountain ranges are bound by normal faults that cause the mountains to be uplifted relative to the basins that are dropped down between them (Figure 5-6c). About 150 of these faults have had displacements capable of generating earthquakes within the last 15,000 years. | Summary of the Late Quaternary tectonics of the Basin and Range province in Nevada, Eastern California, and Utah | The most recent earthquakes have mostly been on the western and eastern margins of the province.

What are some Earth system changes or interactions associated with earthquakes in the Basin and Range province?

Intraplate earthquakes

In the winter of 1811–1812, strong earthquakes rocked the central Mississippi Valley. The crust shook strongly across a region from Memphis, Tennessee, north to Cairo, Illinois. The shaking first hit the town of New Madrid, Missouri, on December 16, 1811. At New Madrid, seismic waves made the ground crack and roll like a stormy ocean. The crust buckled and changes in the land surface caused flooding and permanent changes in the Mississippi River channel, even making the Mississippi River flow backward for a short while. The New Madrid earthquakes were **intraplate earthquakes**, earthquakes that originate far from an active tectonic plate boundary. | Historic earthquakes New Madrid |

(b)

(c)

Up

Down

Fault scarp

FIGURE 5-6 ▲ The Basin and Range Province Is Extending
(a) The Basin and Range province of the American west is centered on Nevada. **(b)** Normal faults are located at the boundaries between the long mountain ranges and the intervening valleys, or basins. **(c)** Displacement on normal faults generates many small-to-moderate-size earthquakes and creates surface scarps.

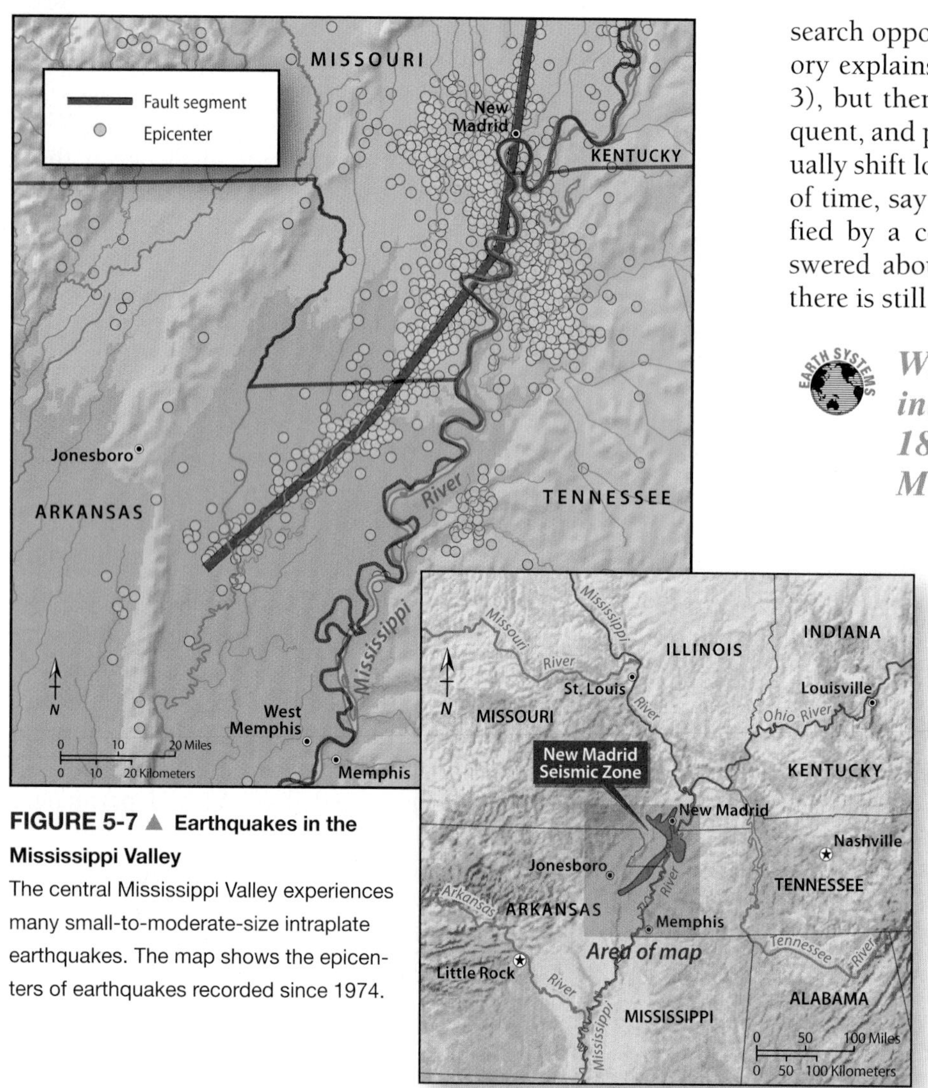

FIGURE 5-7 ▲ Earthquakes in the Mississippi Valley
The central Mississippi Valley experiences many small-to-moderate-size intraplate earthquakes. The map shows the epicenters of earthquakes recorded since 1974.

People in the United States hear so much about western U.S. earthquakes that it's easy to overlook seismic risks in other parts of the country. The New Madrid seismic zone is associated with a set of faults deep in the crust along an ancient rift, and as Figure 5-7 indicates, it remains an active threat. Since the shocks of the early 19th century, the population of New Madrid County has increased from about 400 to more than 20,000 people. There are now some 12 million people living in the affected zone, including densely populated cities like St. Louis and Memphis. The lower Mississippi is also a key shipping lane, so earthquake damage here could have national repercussions. Catastrophic earthquakes like those of 1811–1812 may recur every 500 years or more, but there is a much higher probability of moderate shocks. The Mississippi Valley-"Whole Lotta Shakin' Goin' On"

Intraplate earthquakes, such as those of the New Madrid area, can be large and cause widespread damage. In continental interiors, the crust is old, thick, and strong. When it breaks, it can unleash large amounts of seismic energy and transmit it long distances.

The cause of most large intraplate earthquakes is not well understood. As a result they are perplexing but exciting research opportunities for scientists. After all, plate tectonic theory explains so much about the dynamic geosphere (Chapter 3), but then along come intraplate earthquakes—large, infrequent, and poorly understood. What if these earthquakes gradually shift location over time? If they reoccur over long periods of time, say a few thousand years, what preparations are justified by a community? There are many questions to be answered about intraplate earthquakes. They serve notice that there is still much to learn about the geosphere.

EARTH SYSTEMS

What are some Earth systems changes or interactions that accompanied the 1811–1812 earthquakes in the central Mississippi Valley?

EARTHQUAKES AND FAULTS

Historically, geologists located earthquake-generating faults by looking for particular features in the landscape. If a fault breaks the surface, it's surface features such as scarps form a sinuous or linear **fault trace** across the land, like the one shown in Figure 5-6c. The age of the surface materials that are offset by the fault can be used to constrain the timing of surface displacement. Faults that break the surface or show other evidence of displacement within the last 10,000 years are considered to be *active* and capable of generating earthquakes. But what actually happens when a fault moves and generates earthquakes?

The elastic rebound theory of earthquakes

Until the early 20th century, geologists and seismologists lacked a simple and consistent mechanical explanation of what actually goes on in the lithosphere before, during, and after earthquakes. It was not entirely clear whether earthquakes caused faulting, or faulting caused earthquakes. In the wake of the Great 1906 San Francisco earthquake, a remarkably simple but powerful explanation emerged, thanks to Henry Fielding Reid, a professor of geology at Johns Hopkins University. He is largely credited with developing the **elastic rebound theory**, outlined in Figure 5-8.

In 1910, Reid proposed that before an earthquake occurs, rocks under tectonic stress are increasingly bent and deformed, a condition known as **elastic strain** (Figure 5-8b). At some point, the rock can't bend any further and suddenly breaks, most of the time slipping along a preexisting fault (Figure 5-8c). The rocks on either side of the fault unbend as they snap to a new position (a process known as *elastic rebound*), causing the vibrations we experience as an earthquake. This movement displaces rocks on opposite sides of the fault from their original positions (Figure 5-8d). The cycle then starts over with elastic strain building along the fault until it slips again. This pattern is sometimes called the **earthquake cycle**.

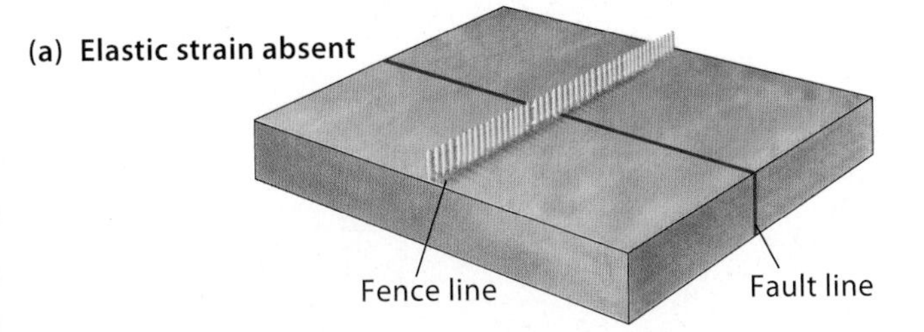

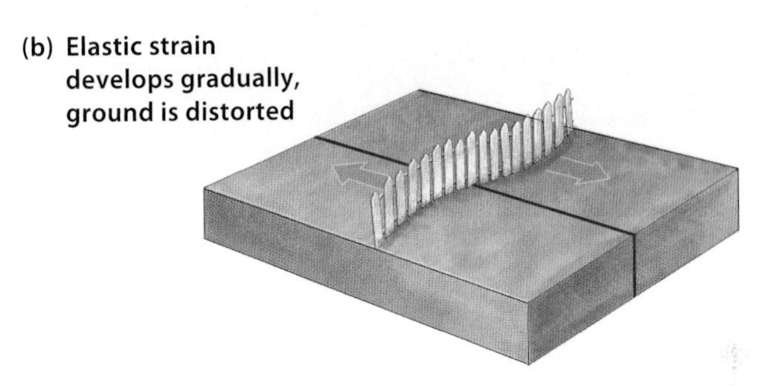

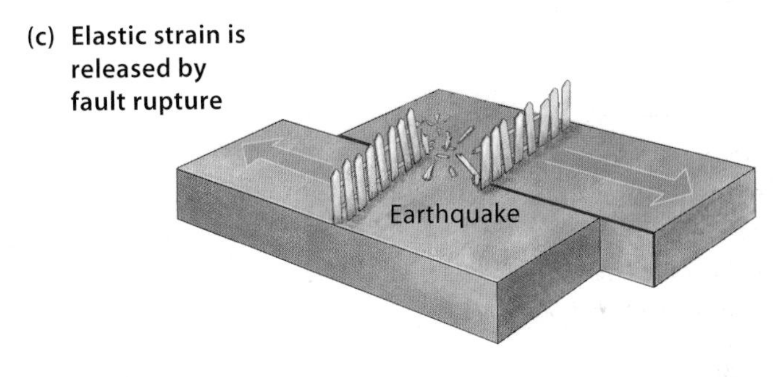

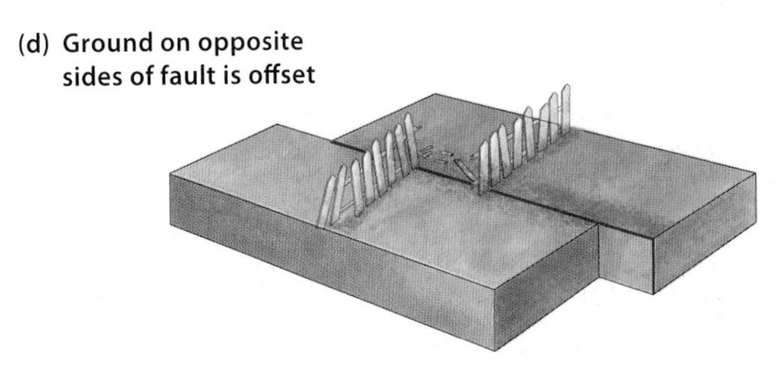

FIGURE 5-8 ▲ The Elastic Rebound Theory Explains Fault Rupture and Earthquake Generation

The sudden release of elastic strain that builds up along a fault releases seismic energy and causes earthquakes, as shown at right. The theory was developed from observations of displacements on the San Andreas fault in 1906 such as the offset fence line in the photo above.

You can simulate the behavior of faults by holding two blocks of semi-rigid foam together and then forcing them to move in opposite directions. For a while, the friction between the blocks holds them together as the foam slightly deforms. When you overcome the friction between the blocks, they quickly move to new positions.

Faults have irregular surfaces. They have rough spots, bumps, and in some places, prominent bends. These irregularities are what help keep the fault stuck, or locked, between earthquakes. In the common case, the elastic strain is not evenly distributed along the fault and it does not unlock uniformly when it slips. | YouTube asperities on earthquake fault The slip, or rupture, begins at a restricted point (the focus—see Figure 5-2) and propagates away from this point along the fault. The larger the area of rock that ruptures, the more seismic energy is released and the longer the shaking lasts.

Creepy faults

There is one more key player in the earthquake cycle: **fault creep**, the slow and gradual movement along a fault that doesn't cause significant earthquakes. The big difference between displacements that cause earthquakes and those that don't is the rate of fault slip. During a large earthquake, slip along a fault of several meters may occur in less than a minute, whereas the rate of fault creep is commonly less than a few tens of millimeters *per year*. The creep rate may slow down or speed up at different times.

Fault creep poses little risk unless your house happens to be right on top of a creeping fault trace. There are a few celebrated cases of this in California, notably Memorial Stadium at the University of California at Berkeley. This sports arena straddles a section of the active Hayward Fault that creeps at a rate of about 1 centimeter (0.4 in) per year. Thus, plate tectonics is slowly tearing the stadium in two (Figure 5-9 on p. 122).

(a)

(b)

FIGURE 5-9 ▲ The Hayward Fault Creeps
Part of the San Andreas Fault system, the Hayward Fault goes from goalpost to goalpost across the UC Berkeley football stadium **(a)**. The creeping displacement on the fault is very slowly tearing the stadium apart **(b)**.

EARTHQUAKE WAVES

Sound and light are forms of energy that move as waves. Seismic energy travels as waves, too. In the case of sound, the waves move through the medium of air. Seismic energy moves through the medium of rock, although some types of earthquake waves can pass through liquid and air. Seismic waves transfer the energy of a fault rupture away from the focus.

Sound and seismic energy both begin as a mechanical disturbance. If you strike a gong, the metal surface of the gong vibrates and sets the air around it in motion. That motion then propagates through the air, eventually reaching your eardrums. The sudden movement of rock along a fault is also a mechanical disturbance, like the strike of the gong. Part of the energy is converted into heat through friction; another portion of the energy cracks and pulverizes rock. The rest is converted into seismic waves. Seismic waves are distinguished by the different effects they have on the medium (rock) through which they move, and the different kinds of ground motions they produce. There are two broad categories of seismic waves: body waves and surface waves.

Body waves

Body waves travel through Earth's interior. There are two types, P waves and S waves. **P waves**, or *primary waves*, are compressional. Like sound waves, they alternatively push and pull (that is, compress and expand) rocks along their direction of travel (Figure 5-10a). P waves can propagate through a solid or a liquid; on the surface, they can even travel through a body of water, such as a lake or ocean, or the air. People sometimes report hearing low rumbling during an earthquake because part of the P wave energy may be converted into sound as the surface of the ground vibrates up and down like a stereo speaker cone.

The other class of body waves, **S waves**, vibrate rock perpendicular to the direction of wave movement—that is, up and down or side to side, as in waves on a shaken rope (Figure 5-10b). Such rock deformation is termed *shearing*, so S waves are sometimes referred to as *shear waves*. They are also known as *secondary waves* because they travel through Earth more slowly than P waves.

When an earthquake occurs, instruments first register the arrival of the faster P waves and then detect the S waves. As you will see, recording the different arrival times of P and S waves is essential to determining the location of an earthquake.

Surface waves

When body waves approach the Earth's surface, a part of their energy is converted into **surface waves**. Surface waves travel along Earth's outer edges rather than through its interior. *Love waves* (Figure 5-10c), named after the British mathematician A. E. H. Love, who studied them, are surface waves that vibrate the ground back and forth but not vertically. *Rayleigh waves* (Figure 5-10d), in contrast, propagate across Earth's surface in a rolling motion similar to that of ocean waves. Surface waves propagate more slowly than body waves, and therefore arrive later.

5.2 Investigating Earthquakes

Fleets of satellites orbit Earth hundreds to thousands of kilometers overhead, tracking weather patterns, checking the upper atmosphere for pollutants, measuring the minutest changes in the ocean surface, and performing myriad other scientific, commercial, and military functions. The technology of remote sensing allows us to gather all kinds of information that would be too dangerous, expensive, or time-consuming to collect in person. Modern-day earthquake scientists use

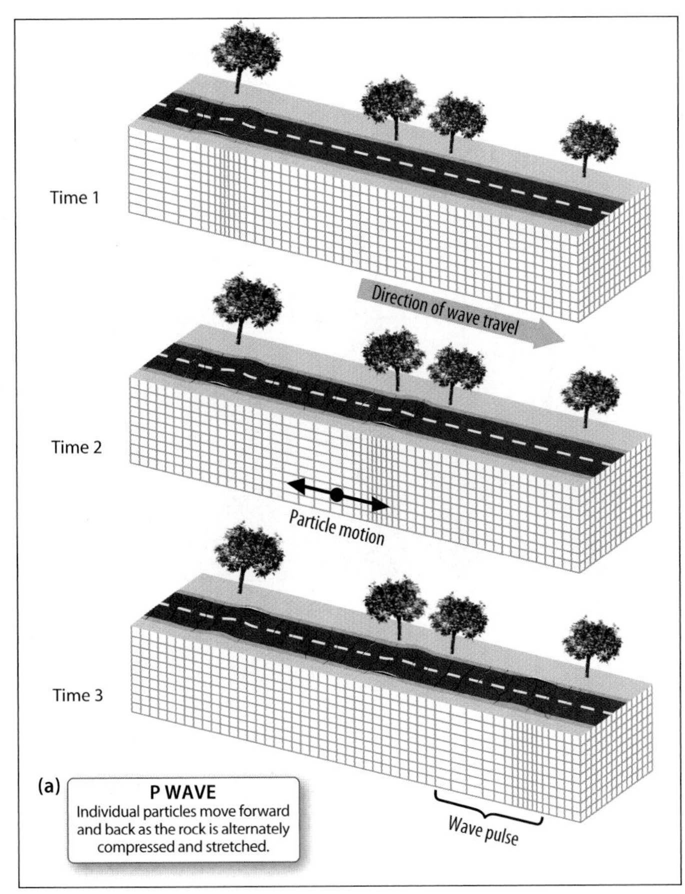

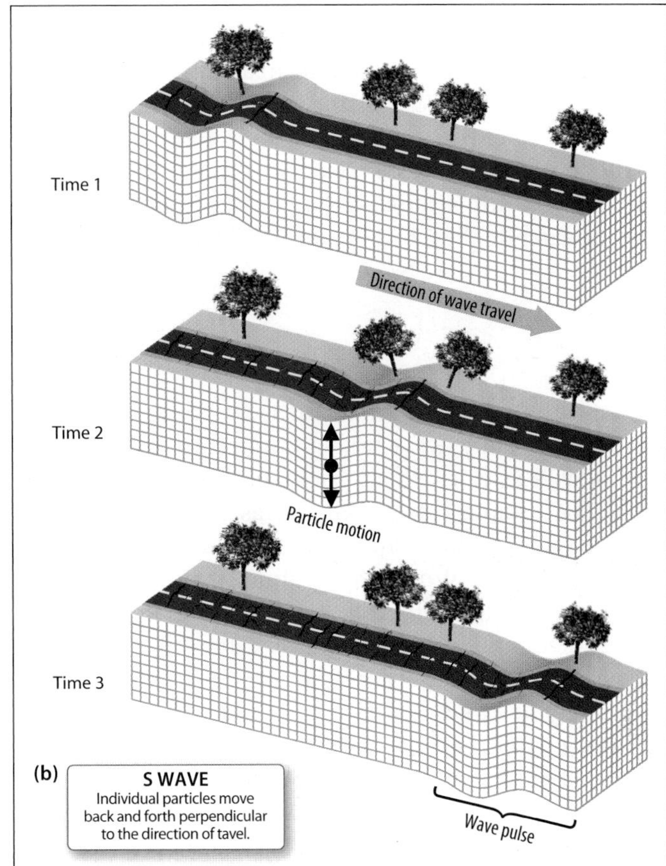

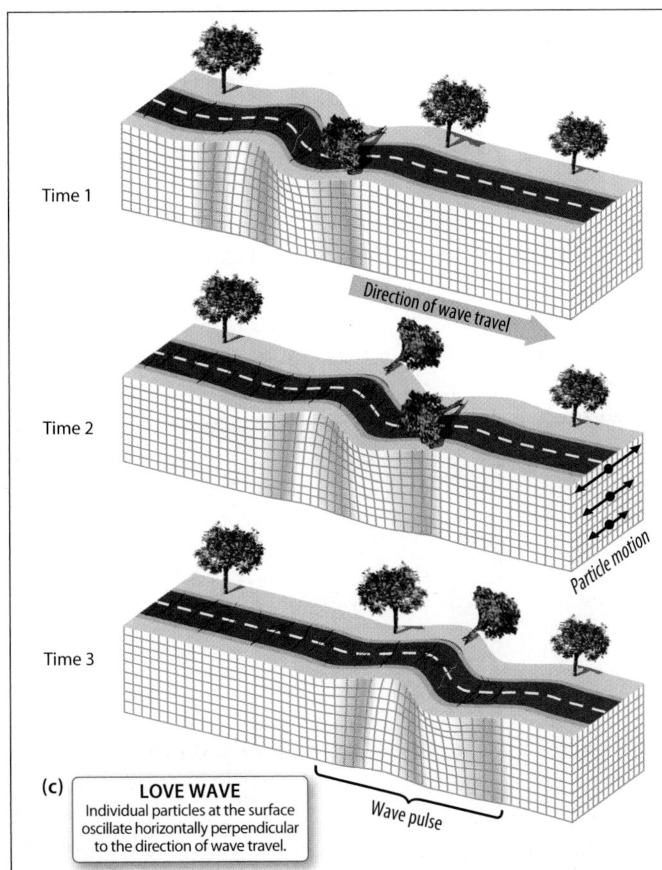

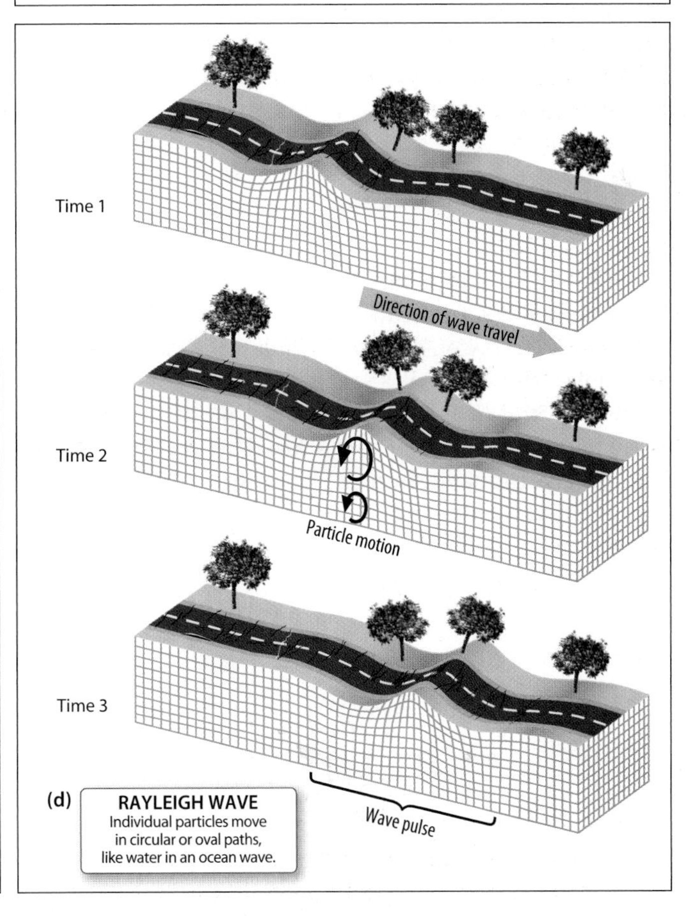

FIGURE 5-10 ▲ How Seismic Waves Pass Through Rock

Primary, or P waves **(a)** and secondary, or S waves **(b)** are body waves, which travel through Earth's interior. Love waves **(c)** and Rayleigh waves **(d)** are surface waves.

satellites to measure changes in the land surface in fault zones and to collect and transmit data from instruments, but they are not newcomers to remote sensing. Since the late 19th century, earthquake scientists have had their own device for recording the motions that occur deep within Earth's interior: the seismometer. With seismometers, we can measure the magnitudes of earthquakes and locate the fault ruptures that cause them.

MEASURING EARTHQUAKES

Seismometers are instruments that measure ground motions caused by passing seismic waves. They can record vertical (up-and-down) as well as horizontal (back-and-forth) motions of the ground. The simplest seismometer, as shown in Figure 5-11a, consists of a weight mounted so that it can move independently of its frame. When the ground shakes, the mass tends to stay put due to its inertia, but the frame moves with the ground. A pen connected to the mass via mechanical linkages traces a permanent record of the ground motion in the form of a squiggly line on a rotating paper drum. This record is called a **seismogram** (Figure 5-11b).

Modern seismometers record the vibration electronically and very accurately, note the time, and store the data in digital form, but the principle behind them is the same. They are distributed around the world in a network that can detect earthquakes virtually anywhere on Earth. Special types of seismometers are designed to accurately detect ultra-faint or very strong ground motions.

Strong-motion seismometers

The signals seismometers detect from a distant earthquake, perhaps thousands of kilometers away, are so faint that most need to be amplified electronically many times to create a readable record. But this sensitivity means that seismograms recorded near the epicenter of a strong earthquake quickly go off the scale. For this reason, a different type of instrument, called a *strong-motion seismometer,* is used to record ground motions near an earthquake source. The motion is measured as ground acceleration, or the rate of change in the velocity of the ground motion as the seismic waves vibrate it.

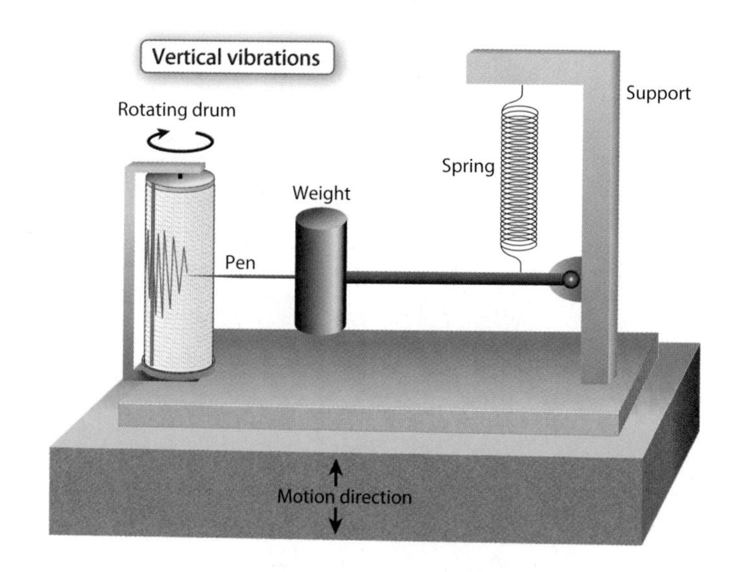

(a)

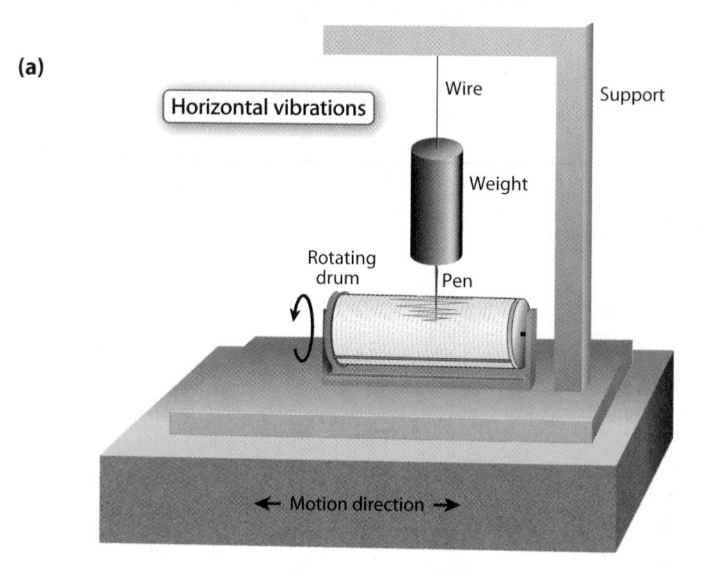

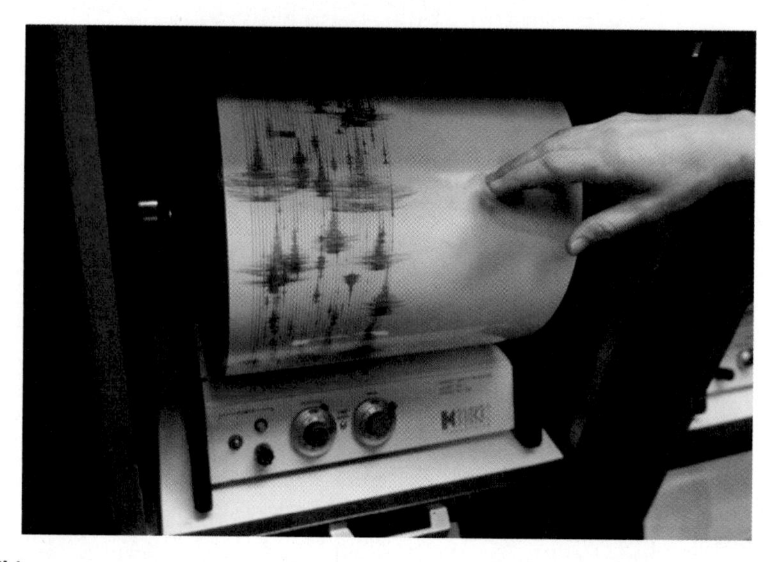

(b)

FIGURE 5-11 ▶ Earthquake Vibrations Are Recorded by a Seismometer

(a) Seismometers are sensitive instruments that mechanically measure (1) vertical and (2) horizontal earth vibrations. There are many types but all have a base that is firmly attached to the ground and a suspended weight that stays still relative to the shaking bedrock. (You can see a home-made seismometer by searching for the following phrase.) Fred's junk box seismometer **(b)** A seismogram is a recording of ground motions caused by seismic waves.

FIGURE 5-12 ▲ Buildings Are Particularly Vulnerable to Horizontal Motion

Because most buildings are strongest in the vertical dimension, horizontal motion can be more destructive than vertical shaking.

The data provided by strong-motion seismometers are critical for designing buildings that can withstand earthquakes without collapsing on their occupants. (Structural collapse is the main cause of death from earthquakes.) The damage an earthquake does to buildings and other objects is related to the magnitude of ground motion accelerations. Normally, gravity holds you, your home, and everything else securely on the ground. If the ground below you suddenly lurched upward with an acceleration greater than that of gravity you would be tossed into the air. Buildings can be knocked off their foundations or have their supports dislodged by vertical ground accelerations. Horizontal ground motion accelerations are even more likely to damage buildings. Buildings tend to be more strongly constructed vertically than horizontally. This makes it easier for horizontal accelerations to shake them apart (Figure 5-12).

Experiencing what seismometers measure

On June 12, 1897, a great earthquake struck the province of Assam in northeastern India, killing at least 1542 people and injuring thousands of others. This region is under tremendous convergent plate boundary stresses as the Indian subcontinent collides with Asia, raising the Himalayas in the process. Recent research suggests that fault displacements caused the Shillong Plateau to move almost 15 meters (49 ft) in just 3 seconds. The ground motions that accompanied this event, known as the Great Assam earthquake, were spectacular.

Farmers reported seeing fields of rice rolling like ocean swells as seismic surface waves passed through them. Indeed, in one field the soil preserved an impression of the waves, reportedly having a peak-to-trough height of up to 3 meters (10 ft). Most alarming, the ground motions were strong enough to overcome the force of gravity, tossing boulders, tombstones, and people into the air.

EARTHQUAKE MAGNITUDE

When an earthquake happens, one thing people want to know immediately is how powerful the event was. Seismometers produce seismograms that provide the data needed to answer this question. Seismograms are composed of many jagged peaks and troughs, reflecting the complex motions of the ground during an earthquake. They record the broad spectrum of seismic waves reaching the surface, just as a microphone records the spectrum of tones in the human voice. The time it takes for one complete peak to peak vibration to pass the seismometer is the **period** of that particular wave. The number of vibrations that the wave completes in one second, measured in cycles per second (hertz, Hz), is the wave's **frequency**. The height of the peaks, or **amplitude**, reflects the strength of the shaking. Scientists use the wave measurements recorded on seismograms to estimate the size, or **magnitude**, of an earthquake.

Earthquake scientists have devised a variety of ways to categorize the size of earthquakes. The California geophysicist, Charles F. Richter, developed the first version in the 1930s. It estimates the size of an earthquake based on the strongest seismic wave amplitude recorded at a standard distance of 100 kilometers (62 mi) from the epicenter. The **Richter magnitude scale** (also called the Local Magnitude Scale, M_L) is given as a whole number with one decimal place—for example a magnitude 6.4 earthquake. The scale is logarithmic, meaning that a change in one whole number represents a factor of 10. For example, the seismic wave amplitude of a magnitude 6.0 earthquake is 10 times greater than that of a magnitude 5.0 earthquake; a magnitude 7.0 earthquake is 100 times greater than a 5.0; and so on.

The original Richter magnitude scale was devised to compare California earthquakes that were recorded by a specific type of seismometer. Alternatives to the Richter scale, each tied to measurements of a specific type of seismic wave, have been developed. However, specific body and surface wave magnitudes sometimes give different estimates of the same earthquake's size, and each is accurate only within a certain range of earthquake magnitudes. One result is that they may sharply underestimate the true size of large earthquakes. To get around these problems, seismologists use a scale called moment magnitude.

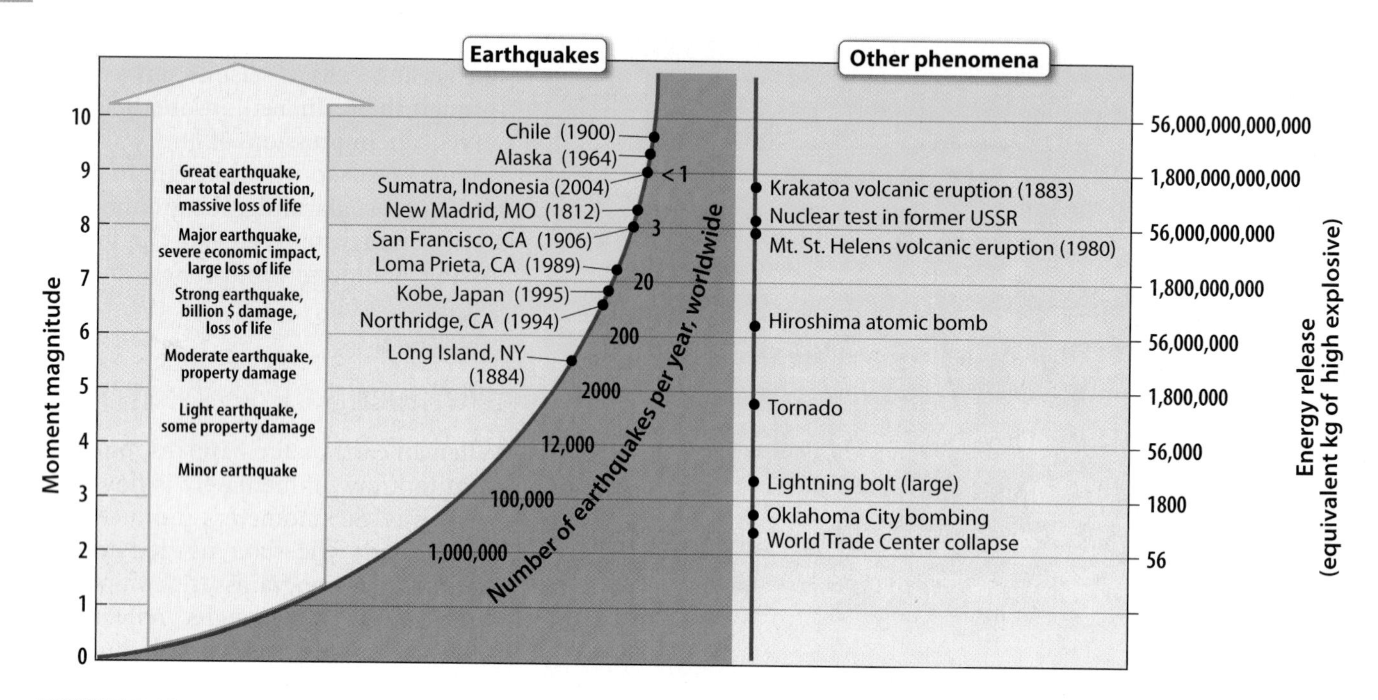

FIGURE 5-13 ▲ The Amount of Energy Released by Earthquakes of Different Magnitudes
Comparing the amount of energy released by earthquakes to the energy associated with other natural and human-produced phenomena helps us understand how incredibly powerful they can be. The figure also gives the energy equivalents of moment magnitudes in terms of kilograms of high explosive. Note that each step on the magnitude scale represents a nearly 32-fold increase in energy.

Moment magnitude, M_W (the "w" stands for "work"), is a numerical scale of the amount of energy released by an earthquake. It is calculated on the basis of the total area of the fault rupture, how far the rocks move along the fault during the earthquake, and the strength of the rock that ruptures. These aspects of an earthquake are related to long-period seismic waves recorded on seismograms. The resulting magnitudes are a good estimate of the seismic energy released by earthquakes, especially very large ones. On the moment magnitude numerical scale, a change of one represents a change of $10^{1.5}$, or 31.6 times the amount of energy released. Thus, a M_W 6 earthquake releases 31.6 times more energy than a M_W 5 earthquake, and 31.6×31.6, or about 1,000 times (10^3) more than a M_W 4 earthquake. (The power of large earthquakes is staggering; Figure 5-13 will give you an idea of the energy associated with earthquakes of various magnitudes compared to other phenomena.) Although not always identified as such, moment magnitudes are now the most commonly reported measure of earthquake size. Unless we indicate otherwise, earthquake magnitudes provided in this book are moment magnitudes.

EARTHQUAKE INTENSITY

Earthquake magnitude tells us the overall size of an earthquake but not the degree of shaking or its effects at a particular location. This is known as the **intensity**, and is measured in terms of its effect on people and structures. The widely used **Modified Mercalli Scale** ranks earthquake intensities on a 12-point scale, expressed as Roman numerals I to XII (Table 5-1). Each increase on the scale represents greater intensity and therefore more damaging or violent effects on people, structures, and the ground itself. The intensities are typically mapped on the basis of eyewitness accounts or post-earthquake surveys of damage in the area around the earthquake's epicenter. Category I shaking is detectable to very few people under the most favorable conditions. At category XII, most structures are in ruins and the ground lurches violently enough to toss objects into the air and knock people off their feet. The highest intensities tend to be near the epicenter of the earthquake, but certain conditions, like the type of rock or sediment and the shape of the fault system, also determine the areas of greatest shaking intensity. Mapping Mercalli intensities using historical descriptions of earthquake effects is the only way we can estimate the size of earthquakes that predate the widespread use of seismometers.

Figure 5-14 on p. 128 shows Mercalli intensity maps that compare the shaking and destruction from the 1989 Loma Prieta earthquake to that caused by the Great San Francisco earthquake of 1906. (You read about both of these San Francisco Bay area earthquakes earlier in the chapter.)

What You Can Do

Map Earthquake Intensity

You can actually help make an earthquake intensity map. If you experience an earthquake, go to the USGS Earthquake Hazards Program website and submit information about the earthquake.

"Did you feel it?" The USGS uses such information to generate a Community Internet Intensity Map that is updated every few minutes following a major earthquake. In this way, you can be involved in real-time earthquake studies. The intensity map you help make will evolve before your eyes as your neighbors and others submit information in the aftermath of an earthquake.

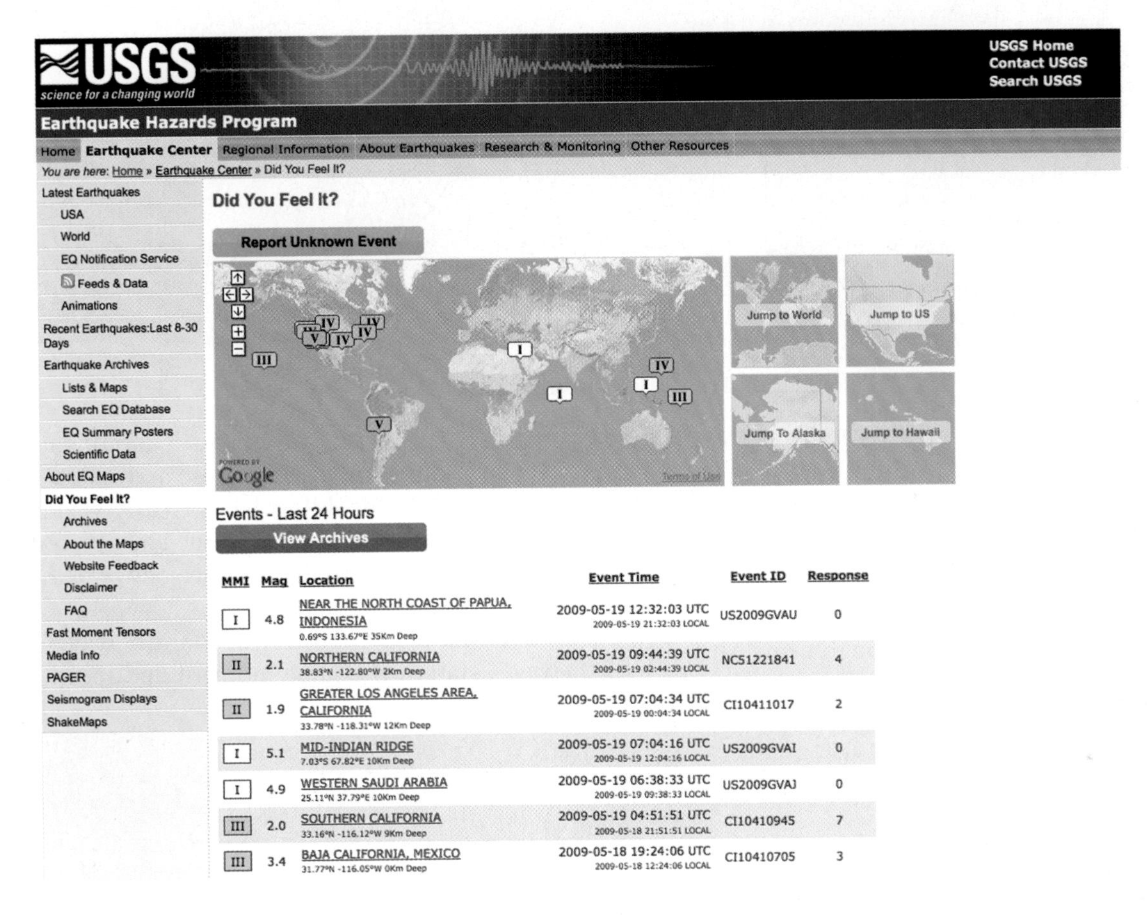

TABLE 5-1 THE MODIFIED MERCALLI INTENSITY SCALE

Mercalli Intensity (at epicenter)	Magnitude	Witness Observations
I	1 to 2	Felt by very few people; barely noticeable.
II	2 to 3	Felt by a few people, especially on upper floors.
III	3 to 4	Noticeable indoors, especially on upper floors, but may not be recognized as an earthquake.
IV	4	Felt by many indoors, few outdoors. May feel like heavy truck passing by.
V	4 to 5	Felt by almost everyone, some people awakened. Small objects moved. Trees and poles may shake.
VI	5 to 6	Felt by everyone. Difficult to stand. Some heavy furniture moved, some plaster falls. Chimneys may be slightly damaged.
VII	6	Slight to moderate damage in well-built, ordinary structures. Considerable damage to poorly built structures. Some walls may fall.
VIII	6 to 7	Little damage in specially built structures. Considerable damage to ordinary buildings, severe damage to poorly built structures. Some walls collapse.
IX	7	Considerable damage to specially built structures, buildings shifted off foundations. Ground cracked noticeably. Wholesale destruction. Landslides.
X	7 to 8	Most masonry and frame structures and their foundations destroyed. Ground badly cracked. Landslides. Wholesale destruction.
XI	8	Total damage. Few, if any, structures standing. Bridges destroyed. Wide cracks in ground. Waves seen on ground.
XII	8 or greater	Total damage. Waves seen on ground. Objects thrown up into air.

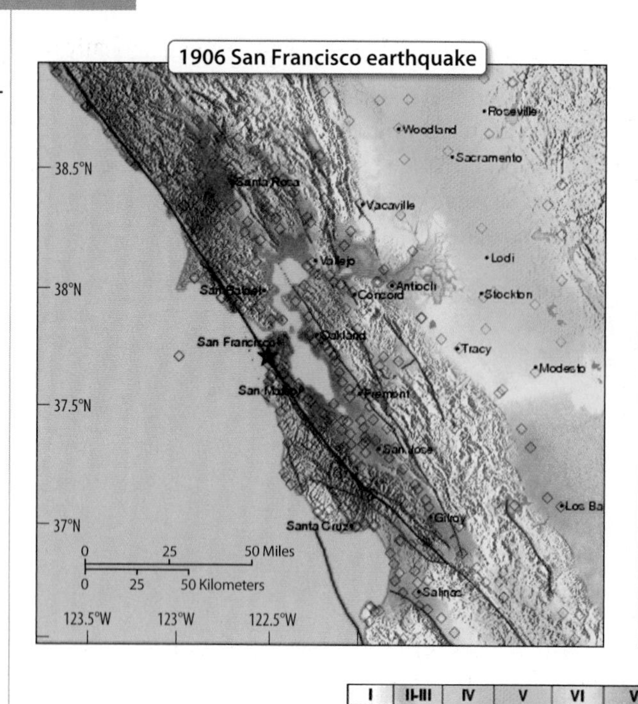

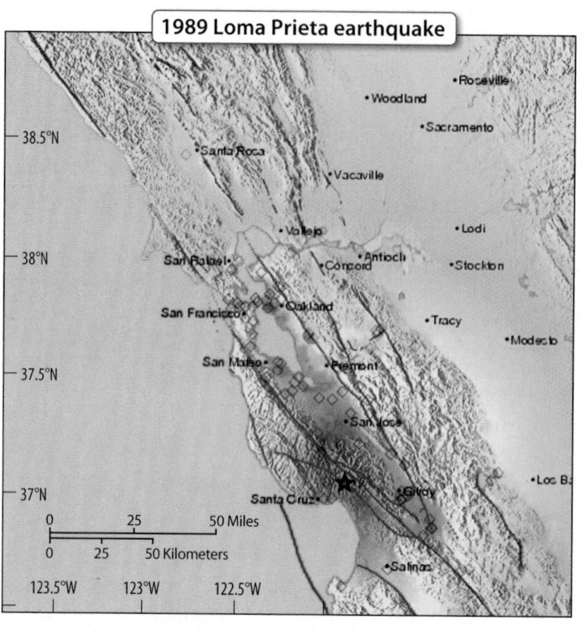

I | II-III | IV | V | VI | VII | VIII | IX | X+

Mercalli intensity

FIGURE 5-14 ◀ Mercalli Intensity Maps of Two Famous Bay Area Earthquakes
These maps show that the 1906 earthquake was much more extensive and destructive than that of 1989.

LOCATING EARTHQUAKES

Scientific organizations such as the USGS maintain a global network of instruments to detect earthquakes, measure their strengths, and locate their sources. Although many earthquakes occur around the world, only a few cause large numbers of deaths and widespread, severe damage. Scientists locate the epicenter and depth of larger earthquakes using data recorded by seismometers around the world.

Locating the epicenter

Seismologists estimate the location of an earthquake's epicenter by analyzing the difference in arrival times of P and S waves at several different seismometers. Recall that P (primary) waves and S (secondary) waves travel at different speeds and so arrive at a seismometer at different times. This difference is clearly visible on a seismogram like that shown in Figure 5-15a. The difference in arrival times is directly related to the distance between the seismometer and the earthquake's epicenter.

Think of the journey of P and S waves as a car race. The waves start at the same moment and location (the earthquake

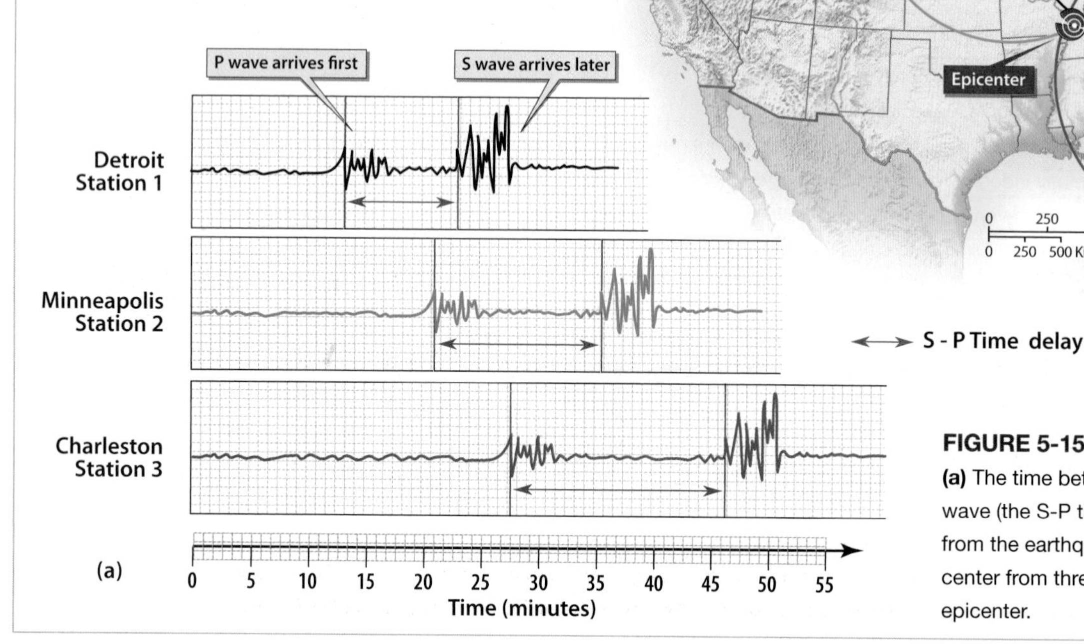

FIGURE 5-15 ▲ Locating the Epicenter of an Earthquake
(a) The time between the arrival of the first P wave and the first S wave (the S-P time delay) is greater the farther the seismometer is from the earthquake. **(b)** Triangulation of the distance to the epicenter from three or more seismometers gives the location of the epicenter.

focus). The P waves take an early lead, and get farther and farther ahead of the S waves as time goes by. Let's say the waves reach a seismometer station only 20 or 30 kilometers (12 or 19 mi) from the epicenter. At this point, the P waves may be only a few seconds ahead of the S waves. At a second station, several hundred kilometers away from the epicenter, the P waves have a larger lead. The greater the lead that the P waves gain over the S waves, the greater the distance between the seismometer and the epicenter of the earthquake.

The distance of any seismometer from the epicenter can be quickly calculated using the difference in arrival times and the known velocities of the seismic waves. However, this data does not tell seismologists the location of the epicenter—the P and S waves could have come from any direction on the compass. The solution to this dilemma is the age-old method of *triangulation*. Using records from three or more seismometers, triangulation locates the epicenter by mapping circles with centers at the stations and radii equal to the stations' distance from the epicenter, as shown in Figure 5-15b. The point at which all the circles intersect is the earthquake's epicenter. Seismologists today use computer programs and digital data from many seismograms rather than compasses and maps for the triangulation, but the concept is the same.

Determining earthquake depth

The depth of an earthquake's focus is also calculated with data from multiple seismometer stations. The method is more complicated than using triangulation to locate the epicenter because it uses distances to the earthquake (based on arrival time differences) to calculate three-dimensional spheres instead of two-dimensional circles around the seismometer location. Where the calculated spheres intersect in the geosphere is the earthquake's focus.

Seismologists have classified earthquake depths in three categories. Shallow earthquakes occur within the first 70 kilometers (43 mi) of the lithosphere. Intermediate-depth earthquakes occur from 71 to 300 kilometers (186 mi). Finally, deep earthquakes occur 301 to 700 kilometers (435 mi) below the surface. All other things being equal, the deeper the earthquake is, the lower its potential hazard to people on the surface because a seismic wave loses strength as it passes through Earth.

The deepest earthquakes ever recorded are associated with subduction zones, where oceanic plates sink into the mantle. In fact, part of the evidence for plate tectonics is the distinctive patterns of earthquakes that occur at convergent plate boundaries, as discussed in Section 3.3 (see Figures 3-17 and 5-4). Deep earthquakes are valuable for understanding the conditions deep in Earth's interior, but may produce barely a quiver of shaking on the surface. Earthquakes that occur within the crust, as is common in California, present a far greater hazard.

5.3 Earthquake Hazards

At 5:26 a.m. on December 26, 2003, a magnitude 6.6 earthquake devastated the ancient city of Bam in southeastern Iran. At least 41,000 people perished, most after being crushed beneath the rubble of their mud brick homes (Figure 5-16a).

(a)

(b)

FIGURE 5-16 ▲ Earthquakes Don't Kill People—Collapsing Buildings Do

(a) The city of Bam before the earthquake of December 26, 2003 (top) and afterward (below), reduced to piles of mud bricks. This magnitude 6.6 earthquake caused over 41,000 deaths. **(b)** The Northridge, California earthquake of 1994, though equally strong, claimed only 57 lives.

Compare this death total to the impact of the 6.7 magnitude Northridge earthquake that struck a densely populated part of southern California on January 17, 1994. The Northridge earthquake caused $20 billion dollars worth of damage (Figure 5-16b), but killed only 57 people.

The hazards faced by people in Iran and other places not prepared for earthquakes are sobering. Since 856 A.D., there have been some 60 earthquakes that killed at least 5000 people each—including a dozen that killed over 80,000 each. Among these was the May 12, 2008 Sichuan, China earthquake that left 87,600 dead or missing. China was also the site of the most deadly earthquake ever recorded, which took 830,000 lives in 1556 (see Table 5-2). |Earthquake lists and maps

As the shocking body counts from the Bam disaster and numerous other earthquakes demonstrate, earthquakes don't kill people—collapsing buildings do. Most of the people at Bam would not have died if their homes had been built to withstand strong shaking. In addition, weak, water-saturated soils turn into a soft mush when shaken vigorously, causing buildings to sink or topple over; landslides destroy homes; fires rage unchecked through devastated cities; and sudden lurching of the seafloor stirs up giant, destructive tsunamis. Scientists help us anticipate these and other hazards of earthquakes, providing an opportunity to reduce destruction and loss of life.

GROUND SHAKING

Earthquakes, by definition, shake Earth's surface and everything on it—including homes, offices, roads, bridges (like the Cypress Street Viaduct), and people. Whether the shaking in-

TABLE 5-2 THE MOST DEADLY EARTHQUAKES

Date	Location	Deaths	Magnitude	Comments
Jan. 23, 1556	Shaanxi, China	830,000	~8	Buildings and city walls collapsed. Damage as far as 430 km (270 mi) from epicenter, shocks felt up to 800 km (500 mi) away. Reports of ground fissures, uplift, subsidence, liquefaction, and landslides.
July 27, 1976	Tangshan, China	255,000 (official)	7.5	Official death toll has been questioned; estimates run as high as 655,000.
Aug. 9, 1138	Aleppo, Syria	230,000		
Dec. 26, 2004	Sumatra	228,000	9.1	Earthquake-generated tsunami, which caused more casualties than any other in history, was recorded nearly world-wide. Approx. 1.7 million people displaced in 14 countries of South Asia and East Africa.
Dec. 22, 856	Damghan, Iran	200,000		
Dec. 16, 1920	Haiyuan/Ningxia, China	200,000	7.8	About 200 km (125 mi) of surface faulting; many landslides and ground cracks in region of epicenter. Some rivers dammed, others changed course.
Mar. 23, 893	Ardabil, Iran	150,000		
Sept. 1, 1923	Kwanto, Japan	143,000	7.9	Extreme destruction in Tokyo-Yokohama area. Firestorms burned about 380,000 of nearly 700,000 houses destroyed or damaged. Tsunami with maximum height of 12 m (39 ft).
Oct. 5, 1948	Ashgabat, Turkmenistan	110,000	7.3	
Sept. 27, 1290	Chihli, China	100,000		
May 12, 2008	Eastern Sichuan, China	87,600	7.9	More than 45 million people affected, at least 15 million evacuated, more than 5 million left homeless. Over 5 million buildings collapsed and 21 million damaged. Total economic loss estimated at $86 billion.
Oct. 8, 2005	Pakistan	86,000	7.6	
Nov. 1667	Shemakha, Caucasia	80,000		
Nov. 18, 1727	Tabriz, Iran	77,000		
Dec. 28, 1908	Messina, Italy	72,000	7.2	Over 40% of the city population and more than 25% in the region killed by earthquake, fires, and tsunami as high as 12 m (39 ft). Aftershocks continued for more than 4 years.
May 31, 1970	Chimbote, Peru	70,000	7.9	A debris avalanche of rock, ice, and mud buried the town of Yungay, which had a population of about 20,000.
Nov. 1, 1755	Lisbon, Portugal	70,000	8.7	Earthquake occurred on All Saint's Day while many of the 250,000 inhabitants were in church. Stone buildings swayed violently and then collapsed on the population. Many who sought safety on the riverfront were drowned by a large tsunami. Fire ravaged the city.
Jan 11, 1693	Silicia, Asia Minor	60,000	7.5	
1268	Sicily, Italy	60,000		
Jun. 20, 1990	Western Iran	40,000–50,000	7.4	
Feb. 4, 1783	Calabria, Italy	50,000		

tensity will be mild and relatively harmless or violent and destructive depends on a variety of natural factors, including the earthquake's magnitude, the site's distance from the earthquake focus, and the site's geologic characteristics.

Magnitude

The overall size of the earthquake, of course, influences the intensity of ground shaking. The larger the rupture, the more seismic energy is released. The size of the rupture—essentially, how large an area slips on the fault—determines the duration of shaking. For example, the 1994 magnitude 6.7 Northridge earthquake involved slip along an estimated 12.5 kilometers (8 mi) of a fault. It took about 8 seconds for the slip to be complete, producing about 10 seconds of strong ground shaking. | Northridge earthquake rupture model Now compare this to the Great 1964 Alaska earthquake, involving a rupture about 800 kilometers (500 mi) in length. Strong shaking from this magnitude 9.2 subduction-zone earthquake lasted for several minutes. Here is one anonymous person's experience during the Alaska earthquake.

> "This was a vicious earthquake, and we were pretty well shaken up. I'm glad we were all home, because I would have hated to have been downtown and watch the streets and buildings dissolve! We had just finished dinner when the quake hit, and at first it started out kind of gentle like one of our earth tremors up here, but it very quickly built up to a strength we'd never felt before. Claudia made it out the back door and promptly sat down in the snow. It was impossible to stand up without some sort of support and she grabbed the earth as though she could hold it still by brute force. I was in the back door and being batted back and forth by the door jams as the house wobbled around, and tried to keep an eye on Claudia, the kitchen, and Gloria all at the same time. Gloria rattled around the kitchen for a while and finally grabbed something to hold on to in the kitchen doorway, and we just stood and watched our cupboards empty themselves onto the floor.
>
> It seemed to go on forever. I guess it lasted five or six minutes but I was afraid our house would go, it was shaking and creaking and groaning so much and I thought of the car in the garage. I started around the side of the house to get it out in case the house collapsed, but I couldn't stand up. The earth moved so violently that it was impossible to walk, and I got as far as the fence and held on. It was almost enough to make a person seasick."

Distance from the focus

Other things being equal (magnitude, for example), the deeper the earthquake's focus, the less severe is the ground shaking. As noted earlier, most earthquakes on the San Andreas fault system occur 10 to 16 kilometers (6 to 10 mi) beneath the surface. At these shallow depths even a moderate earthquake can cause severe damage. Now consider an earthquake that occurred in 1994 some 640 kilometers (400 mi) beneath Bolivia. The ground shook for 40 seconds from the magnitude 8.3 quake. It was the most powerful deep earthquake ever detected, and people felt the shaking as far away as Canada. However, it resulted in little or no damage because the seismic waves had to travel so far to reach the surface that they lost most of their energy along the way.

Shaking is expected to be stronger closer to the epicenter, the closest surface location to the focus. Near the epicenter, both short-period and long-period waves are significant but their effects change as they migrate away. Short-period body waves are like high-pitched musical notes that weaken quickly with distance. They cause strong shaking close to the epicenter. Long-period surface waves are like bass notes, which maintain more of their strength over distance than do high pitched notes. They can continue to cause strong shaking farther from the epicenter. Body waves dissipate, but the surface waves persist even at considerable distance from an epicenter.

Site geology

After an earthquake, seismologists often discover that the ground shaking was concentrated and more intense in some areas than in others. Local geology commonly causes amplification of shaking from seismic waves in some places and dampening in others. Geologic factors that play a role may be in the bedrock or in the surface materials that overlie the bedrock.

Seismic waves change as they pass through different materials, such as rock or sediment layers with different densities. These physical discontinuities can reflect seismic waves or change their velocities as they pass through. This type of velocity change causes the waves to bend (refract) and propagate in a different direction. Reflection or refraction can focus seismic energy toward a particular area on the surface and increase shaking there.

Loose, unconsolidated surface materials on bedrock can be a significant factor in amplifying ground shaking. These materials, commonly soft sediment or soil, act somewhat like a hunk of Jell-O sitting on a table. Jar the table and the Jell-O shakes more and longer than the table. S waves in particular change as they pass through unconsolidated materials. They slow down, their amplitudes build, and the material shakes more as a result. | Soil type and shaking hazard in the San Francisco Bay area This is the type of earthquake amplification that helped collapse the Cypress Street Viaduct.

GROUND DISPLACEMENT AND FAILURE

Shaking is a hazard itself but it is even more of a problem when it leads to ground failure. All buildings and structures rest on foundations of some sort, and those foundations rest on the ground. Where earthquakes cause permanent surface displacement or the surface materials to lose their strength, the result can be both deadly and costly. Examples of ground displacement or failure include liquefaction, slope failure, surface rupture, and crustal deformation.

Liquefaction

Unconsolidated surface materials, such as silty and sandy sediments, are subject to **liquefaction** if they are poorly drained and saturated with water. Seismic waves that vigorously vibrate the sediment can cause it to lose its cohesiveness. The sediment grains are repeatedly separated with each vibration, so that eventually they are completely surrounded and supported only by water. At this point, the materials become slurry and can no longer bear weight. Objects sitting on them, like houses, sink or shift. In some places, the liquefied material flows out from beneath structural supports or foundations. When the shaking stops, the ground becomes firm again.

Urban areas with low-lying coastal land are vulnerable to damage from liquefaction. It is also common where ports, harbor facilities and even residential areas rest on loose fill, such as dredged silt or sand, dumped there to extend the land surface. The magnitude 6.9 earthquake that devastated Kobe, Japan, in 1995 killed 5502 people, most in building collapses. This earthquake was a direct hit, as Kobe was located over the fault rupture, but significant damage was also traced to liquefaction under the city's extensive harbor facilities. As the ground settled, levees heaved and cracked and docks and seawalls failed (Figure 5-17). Strong shaking during the Great 1964 Alaska earthquake triggered liquefaction and slope failure in a housing development on the bluffs overlooking Cook Inlet in Anchorage. Soft, silt-rich sediments liquefied and flowed during the earthquake, taking many houses with it.

Slope failure

Earthquakes may trigger landslides and other forms of *slope failure* (discussed more fully in Chapter 8). Slope failures include everything from small rock falls to massive muddy debris flows that have buried entire towns and killed thousands of people. The 1994 Northridge earthquake, for example, caused more than 11,000 slides of various kinds. A 1970 earthquake in the Andes Mountains of Peru triggered a massive mud and rock flow that buried the towns of Ranrahirca and Yungay (Figure 5-18). An estimated 18,000 people lost their lives.

FIGURE 5-18 ▲ Earthquakes May Trigger Landslides
Ranrahirco and Yungay, Peru were devastated by a large mud and rock flow caused by a 1970 earthquake in the Andes.

FIGURE 5-17 ▲ Kobe, Japan after the 1995 Earthquake
Liquefaction of soft fill material along the Kobe waterfront destroyed levees and other port facilities during the 1995 magnitude 6.9 earthquake.

Surface ruptures

Among the direct effects of earthquakes are fault ruptures that permanently displace the ground surface (Figure 5-19a). These movements may create fault scarps (see Figures 5-5b and 5-6b) or other surface features that can do a great deal of harm to any structure resting along the fault trace. Surface ruptures may damage roads and buildings or cause expensive damage to

(a) (b)

FIGURE 5-19 ▲ Surface Rupture

(a) Surface rupture can permanently scar the landscape, as in this example from New Zealand. **(b)** It can also cause great damage to urban infrastructure, as this photo of Anchorage, Alaska after the great earthquake of 1964 shows.

buried utility pipes and cables (Figure 5-19b). (Although Hollywood scriptwriters are fond of having earthquake ruptures swallow people and deposit them in the bowels of the Earth, in reality that does not happen.)

Crustal deformation

Large changes in elevation, caused by shifts in crustal blocks, have occurred during earthquakes. When blocks of the crust shift, some areas may rise (uplift) whereas others may sink (subside). In coastal areas, whole regions may subside below sea level. Some of the most dramatic occurrences of uplift and subsidence in recent history occurred during the Great 1964 Alaska earthquake. Some segments of the coastline rose, leaving barnacle-encrusted seafloor high and dry. In the Pacific Northwest, drowned and buried forests testify to major coastal subsidence due to past earthquakes along the Cascadia subduction zone.

What are some Earth systems changes or interactions that accompany earthquake-induced ground failure?

TSUNAMIS

Large ocean waves, or **tsunamis**, are a potentially devastating earthquake hazard. *Tsunami* means "big harbor wave" in Japanese. Some people refer to them as tidal waves, but they have nothing to do with tides. As Figure 5-20 on p. 134 shows, most tsunamis are generated by large movements of the seafloor during subduction-zone earthquakes. Displacement of the seafloor in turn displaces a large volume of water above it. | What causes a tsunami? | This sudden displacement, like that created by a stone tossed into a pond but much more massive, sends waves in all directions. The waves are tsunamis. Each represents a large mass of seawater in motion.

In recent decades, earthquakes in subduction zones offshore Alaska, Chile, Nicaragua, Mexico, and Indonesia have launched tsunamis, taking many lives. A devastating example was the tsunami of December 26, 2004 in the Indian Ocean. It was caused by a magnitude 9.1 earthquake off the coast of northern Sumatra, Indonesia. This tsunami killed more than 225,000 people (Figure 5-21, p. 135)—the most of any in recorded history.

In the open ocean, tsunamis travel extremely fast (hundreds of kilometers per hour), but have long periods and low amplitudes—that is, the wave crests are widely spaced and not very high. They are therefore fairly inconspicuous, and can easily be mistaken for ordinary ocean swells. As the waves enter shallower water, however, friction with the seafloor causes them to slow down, their period to shorten, and their amplitude to increase (see the tsunami animation at the SEED website, The Earth A Living Planet). | The Earth: a living planet—tsunami | Witnesses frequently describe an approaching tsunami not as a breaking surf wave, but as a wall of water. A tsunami that struck Okushiri, Japan, in 1993 reached a height of 32 meters (105 ft), or as tall as an eight-story building.

Tsunamis make big waves, but the first thing people may notice as a tsunami approaches is the withdrawal of water along the shoreline. This occurs when the trough of the tsunami approaches shore before its crest. The receding water is drawn toward the building tsunami offshore and exposes the seafloor as it recedes. Receding shorelines soon after an earthquake warn that a tsunami is coming. If people quickly

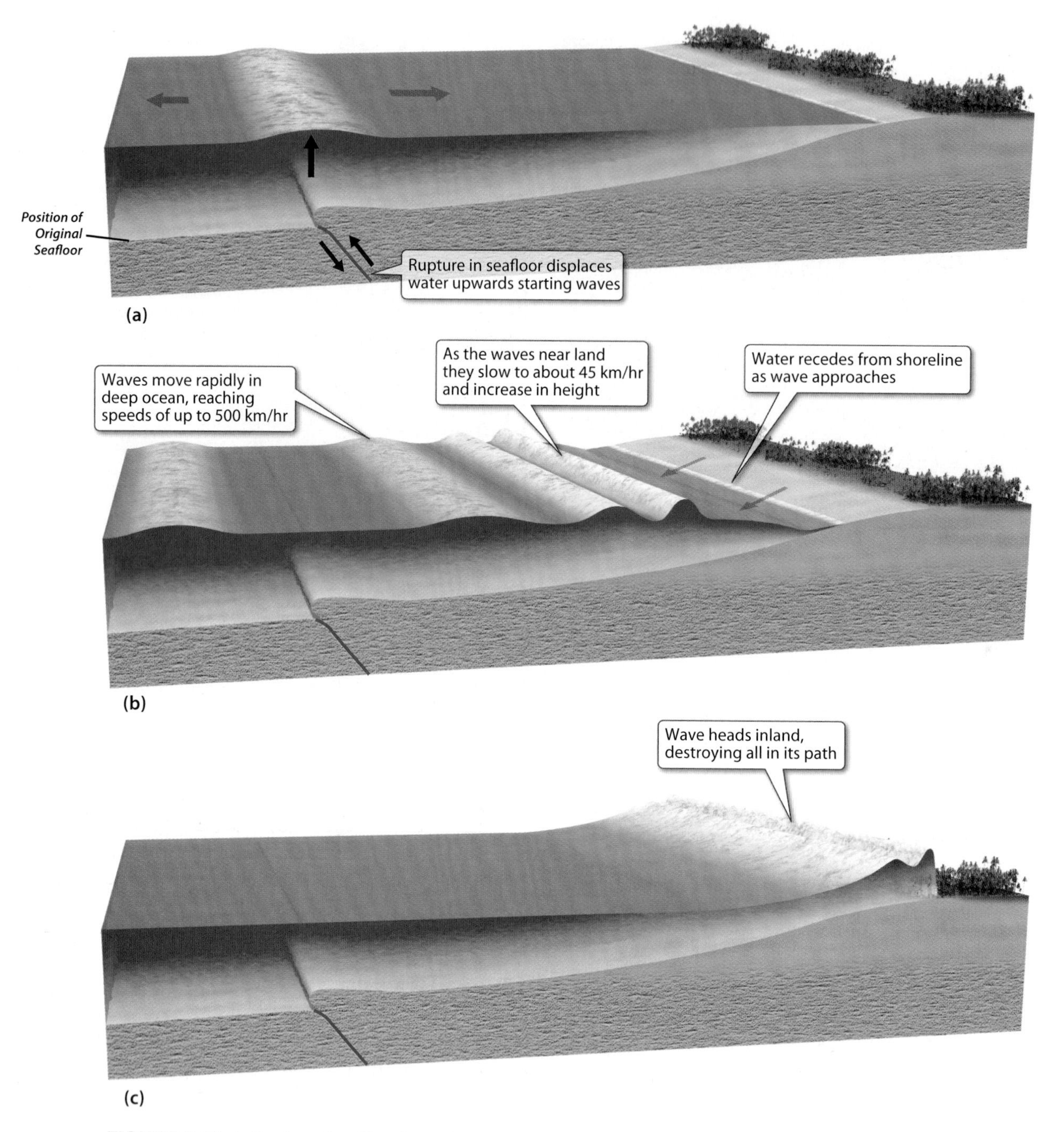

FIGURE 5-20 ▲ Earthquakes May Cause Tsunami
(a) Tsunami form when water is displaced above a sudden offset on the seafloor. **(b)** As they enter shallow water, tsunami slow down and gain height.

recognize this warning, they may have time to escape to higher ground.

Unfortunately, tsunamis can travel for great distances in the open ocean with minimal loss of energy. For this reason, large subduction-zone earthquakes generate tsunamis that may threaten people far removed from the worst shaking. Such was certainly the case in the Great 1964 Alaska earthquake. The resulting tsunami killed 106 of the 115 people who lost their lives in the Alaskan quake, as well as four people in Oregon and 13 in California. The physical damage to boats and harbors was extensive, reaching as far as Hawaii. The 2004 Indian Ocean tsunami was much worse, causing many thousands of deaths as far away as Sri Lanka and India, thousands of kilometers distant.

What are some Earth systems changes or interactions that accompany earthquake-induced tsunamis?

(a)

FIGURE 5-21 ◀

Tsunami Devastation

On December 26, 2004 a subduction-zone earthquake of M_w 9.1—the third-largest ever recorded—occurred off the coast of Sumatra, Indonesia. The earthquake generated a tsunami **(a)** that killed over 225,000 people in 11 countries. The tsunami surprised nearby communities, and some were completely wiped out **(b)**.

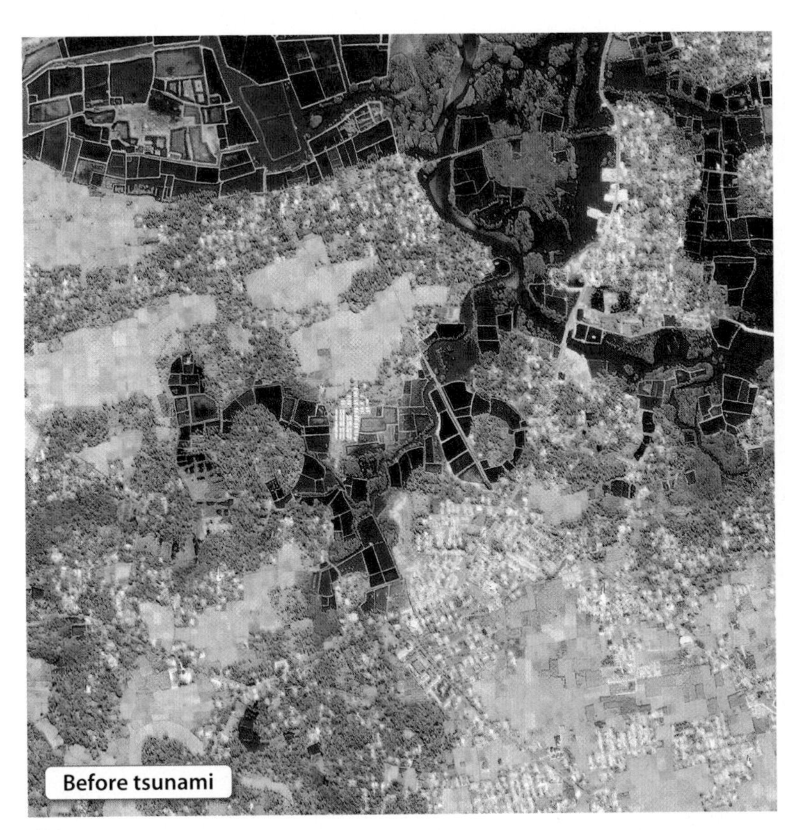

(b)

(a)

(b)

FIGURE 5-22 ▲ Fires Can Become Major Secondary Effects of Earthquakes
(a) Large parts of San Francisco burned as a result of the Great 1906 San Francisco earthquake. **(b)** In 1995, the Japanese city of Kobe suffered extensive fire damage after a powerful earthquake.

FIRES

One of the greatest hazards to urban areas stricken by earthquakes is fire. Severed gas mains and electrical lines are commonly where such fires start. Damage to the water supply, blocked roadways, and overwhelmed emergency crews make the fires hard to subdue. Many of the estimated 28,188 buildings lost in the Great 1906 San Francisco earthquake fell to the fires that burned for three days, devastating 12 square kilometers (4.6 mi^2) of the city (Figure 5-22a). More recently, during the 1995 Kobe earthquake, fires struck the crowded central city district where wood construction was common (Figure 5-22b). Earthquake damage cut off the water supply for fighting fires for several hours. Once the water was flowing, the rubble of numerous building collapses blocked access to some fires. The Japanese have faced this hazard before, and on an even larger scale. In the Great 1923 Tokyo earthquake, tens of thousands of people perished in fires. Altogether, a total of 140,000 are estimated to have died in Tokyo and the neighboring industrial city of Yokohama.

CONSTRUCTION DESIGN

Remember the phrase, "earthquakes don't kill people—collapsing buildings do" and the contrast in the number of deaths caused by similar-magnitude earthquakes at Bam, Iran, (41,000 deaths) and Northridge, California, (57 deaths)? Other things being equal (such as population density), construction design is a key reason the number of deaths can be so different for similar-size earthquakes. Buildings that cannot withstand earthquake shaking become tombs for people as they collapse. Even in California, collapsing buildings or debris falling off buildings cause many of the earthquake-related deaths and injuries.

The construction design problem is one that people inherit in older communities like Bam, Iran. The traditional building techniques in these places use materials and designs that predate our understanding of how to make structures safer during earthquakes. Even in places like California, older buildings may not meet contemporary earthquake-resistant design standards. In other places, the cost of "reengineering" or constructing new earthquake-resistant structures may be a barrier to their widespread use.

As you will see, the good news is that engineering can provide construction designs that withstand strong earthquake shaking for everything from homes to natural gas pipelines. Many governments in earthquake-prone areas are now requiring, through construction permits and regulations, new structures to be earthquake resistant.

In The News

The Great 1906 San Francisco Earthquake

> ". . . about 5:15 a.m., Wednesday morning, I was thrown out of bed and in a twinkling of an eye the side of our house . . . was dashed to the ground. How we got into the street I will never be able to tell, as I fell and crawled down the stairs amid flying glass and timber and plaster. When the dust cleared away I saw nothing but a ruin of a house and home that it had taken twenty years to build. I saw the fires from the city arising in great clouds . . ."

This is part of a letter written by E. H. Adams recounting his experience during the Great 1906 Earthquake in San Francisco. The strong shaking from this earthquake lasted about a minute and was felt from Oregon to southern California. Most of the buildings still standing after the shaking had stopped were subsequently destroyed by fire (see Figure 5-22). When it was over, about 3,000 people were dead and 225,000 left homeless. But that was more than a century ago; why is this earthquake still in the news?

The 100th anniversary of this earthquake was widely noted in 2006. You may wonder why a natural disaster is commemorated. Of course, the many ways people help each other and rebuild after a disaster are always worth remembering, but for Earth scientists, the Great 1906 San Francisco earthquake created a revolution in how earthquakes are studied and understood.

The Great 1906 San Francisco Earthquake—What was learned scientifically from 1906

The very first comprehensive scientific investigation of earthquakes in the United States was commissioned by the State of California three days after the 1906 earthquake. The State Earthquake Investigation Commission, under the leadership of Professor Andrew C. Lawson, chairman of the geology department at the University of California, Berkeley, integrated and unified the investigations under way by scientists from many institutions.

Some of the fundamental contributions made by the commission to the scientific understanding of earthquakes and their impact, contributions that remain foundations of today's earthquake studies, include the following:

- The studies showed that engineering designs of buildings and other infrastructure had a major effect on earthquake impact. These were the initial foundations for the building codes that now guide construction designs to withstand ground shaking.
- Maps of ground shaking intensity showed that the underlying geology influenced the severity of shaking. Areas on poorly consolidated sediment and other loose and soft material were much more strongly shaken than other areas. Such materials amplified ground shaking. This recognition is the foundation of modern maps that predict ground shaking severity—maps that guide land-use zoning decisions on such issues as the location of schools, hospitals, homes, and other facilities.
- A member of the commission, Professor H. F. Reid of Johns Hopkins University, developed the theory of elastic rebound as a result of his studies of the San Andreas fault rupture. This theory is the foundation of modern seismology. It explains how earth movements build and release elastic strain along faults and cause earthquakes.

You encountered the challenges of living on a plate boundary in Chapter 3 (*You Make the Call*, p. 71). The plate boundary was the San Andreas fault and the challenging location was San Francisco. You can expect this fault zone and the earthquakes it generates to be in the news many times during your lifetime.

5.4 Earthquake Prediction

On February 4, 1975, authorities in China issued an extraordinary statement: A major earthquake was imminent, and residents of the northeastern province of Liaoning should take immediate precautions—up to and including abandoning their homes temporarily. What happened next that same day was even more extraordinary: A magnitude 7.0 earthquake struck, reportedly damaging or destroying 90% of buildings in the city of Haicheng and many in the city of Yingkow. Precise death statistics are not available, but it is widely believed that thousands of lives were saved by the government's action.

The prediction was based on a variety of possible warning signs, or **precursors**. The key precursor turned out to be an increase in regional earthquakes that scientists interpreted as foreshocks of a much larger event to come. People had also noted uncharacteristic animal behavior in the months leading up to the prediction, meaning perhaps that the animals sensed something going on underground associated with increasing elastic strain. The afterglow of this episode, hailed as a successful earthquake prediction, didn't last for long. The following year, a magnitude 7.5 earthquake devastated the city of Tangshan, killing at least 240,000 (and possibly many more)—the largest earthquake death toll of the 20th century, and probably one of the greatest in history. The warning signs observed before the Liaoning earthquake were absent.

To this day, the ability to accurately predict the size, location, and time of earthquakes remains elusive. The dream of prediction has been superseded by a more achievable goal of making long-term, less-specific earthquake forecasts. A typical forecast states the odds of an earthquake within a certain magnitude range occurring on known faults over a specific range of time. A forecast could span months, years, or decades. Forecasts are intended not to trigger evacuations, but instead to guide decisions on how to prepare for a large earthquake. In response to a forecast, communities at risk can take steps such as beefing up building codes, planning emergency response, and steering dense development away from areas likely to experience very strong ground shaking, displacement, or failure.

SHORT-TERM PREDICTIONS

The phrase "short-term prediction" is open to interpretation, but is commonly taken to mean a specific statement that an earthquake of a given size is imminent within a stated number of hours, days, or weeks. This would provide time to take precautions, such as putting emergency services on alert or perhaps taking the extreme step of evacuating a town or city.

Scientists interested in short-term earthquake prediction have examined a variety of potential precursors. The general theory behind these efforts is straightforward: As elastic strain builds up in the crust, it should have secondary effects that we can detect before the main earthquake. Some of the precursors studied include:

- An increase in the number of small earthquakes on a fault
- A sharp increase (or decrease) in the rate of fault creep
- Fluctuations in groundwater levels
- A decrease in the resistance of the ground to electrical current
- An increase in the ground's emission of radon gas or methane
- Tilting or uplifting of the ground surface near a fault

Despite a great deal of effort to monitor these phenomena in places like California and Japan, none of these precursors have proven to be reliable enough to allow accurate short-term earthquake predictions. For example, some earthquakes are preceded by foreshocks, but many are not. Some scientists have argued persuasively that the processes that trigger earthquakes may be inherently unpredictable, or at least not predictable enough to allow a statement that an earthquake will occur at a specific location at a specific time. Others still hold out hope that further research will make short-term predictions more reliable.

FORECASTS

An alternative to short-term earthquake prediction is making less specific, longer-term statements called forecasts. These forecasts are more like general hazard assessments, and they address such questions as:

- Where is the next damaging earthquake likely to occur?
- Is it likely to happen in the next decade, or much later?
- How large could the next earthquake be?

By studying faults and earthquake history, scientists can determine the average amount of time between successive earthquakes on a fault (the *recurrence interval*). They can then identify places on an active fault where expected earthquakes have not occurred (*seismic gaps*), and use this information to make forecasts.

Seismic gaps

Large, continuous faults don't move along their entire length at once. Instead, they tend to rupture in discrete segments. The segments themselves are commonly bounded by bends in the trend of the fault that may concentrate elastic strain. Any segment of the fault that has not ruptured recently, in comparison to neighboring segments, is said to be a **seismic gap**. Seismic gaps are primed for an earthquake. The gap represents a fault segment that has been locked for a long period of time and has presumably stored up more elastic strain than others, and is therefore more likely to rupture.

Identifying seismic gaps requires long-term records of past earthquakes to determine which parts of the fault have been quiet and for how long. Where direct reports of earthquakes are lacking, geologists study displaced surface features along a fault to determine its long-term earthquake history.

The North Anatolian fault system in Turkey is an example of a place where we can see the seismic gap model in action. It is about 1500 kilometers (930 mi) in length, stretching across much of northern Turkey. As on the San Andreas fault zone, movement along the North Anatolian is predominately transform motion along strike-slip faults. As the Arabian Plate pushes against Eurasian Plate, the North Anatolian Fault accommodates part of the strain. Turkey moves westward along the fault at a rate of 10 to 20 millimeters (0.4 to 0.8 in) per year. As you can see on the map shown in Figure 5-23, the main trace of the fault splits into two strands in the west—one of which extends close to the city of Izmit in northwestern Turkey—that were identified as seismic gaps.

Starting with a magnitude 7.9 shock in 1939 that ruptured 350 kilometers (220 mi) of the eastern end of the fault, there have been 11 large earthquakes along the North Anatolian fault. What is extraordinary is that, with a few exceptions, different segments of the fault have ruptured in sequence from east to west. This trend culminated in a magnitude 7.6 earthquake on August 17, 1999, whose epicenter was near the city of Izmit. Officially known as the Kocaeli earthquake, for the Turkish province where it occurred, it happened at 3:02 a.m., when most people were asleep in their homes. Due to many catastrophic building collapses, the earthquake killed more than 17,000 people, injured nearly 44,000, and left more than a quarter of a million homeless (Figure 5-24). This earthquake occurred on the western strand of the fault previously identified as a seismic gap. A large earthquake here was anticipated. But anticipation did not save lives in this case as it was not possible to move or rebuild the city to modern earthquake-resistant standards.

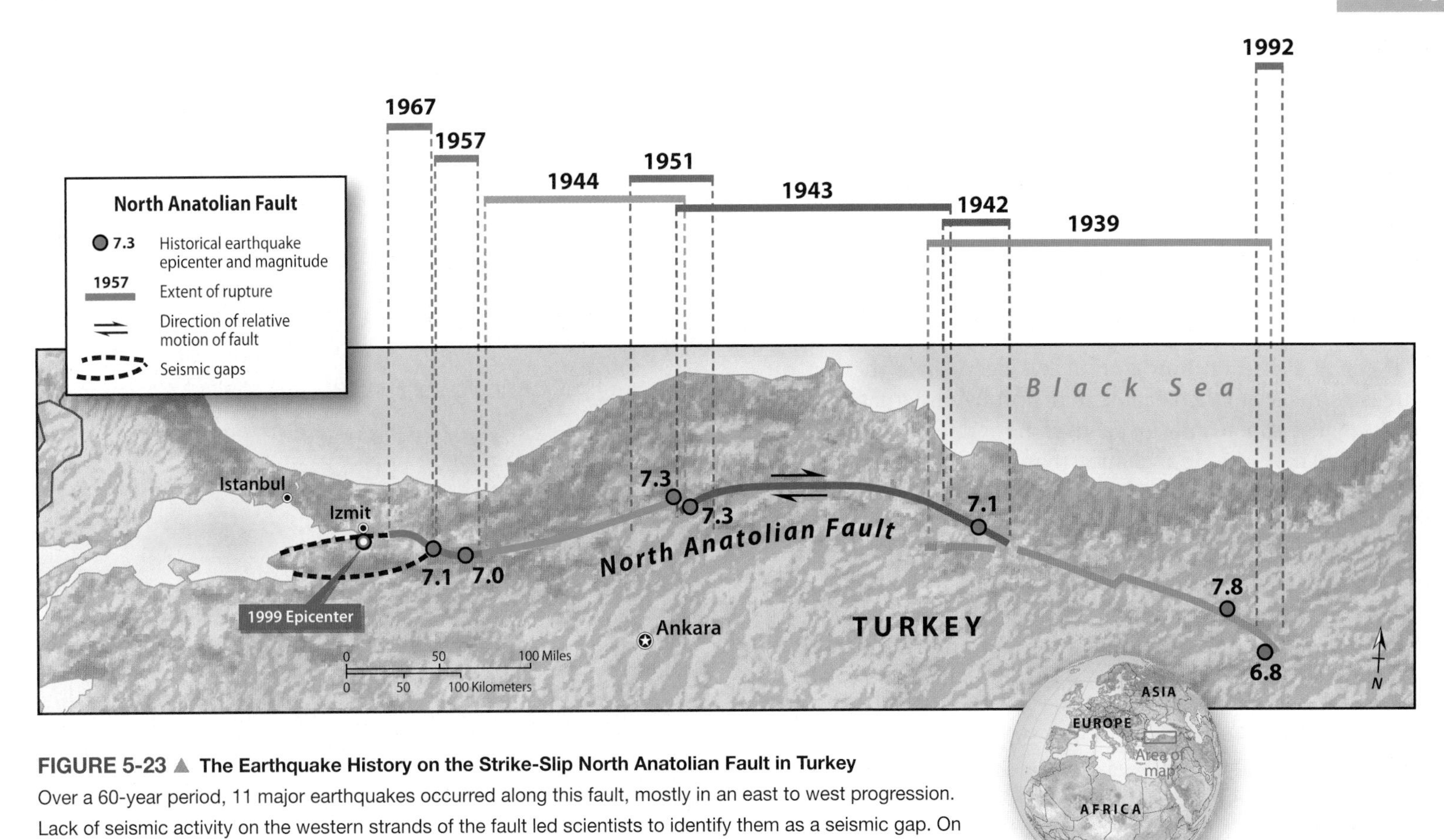

FIGURE 5-23 ▲ The Earthquake History on the Strike-Slip North Anatolian Fault in Turkey
Over a 60-year period, 11 major earthquakes occurred along this fault, mostly in an east to west progression. Lack of seismic activity on the western strands of the fault led scientists to identify them as a seismic gap. On August 17, 1999 a M_w 7.6 earthquake occurred there.

FIGURE 5-24 ◀ Damage in the City of Izmit, Turkey from the August 17, 1999 Earthquake
A few of the sturdier buildings, including the old mosque in the foreground, survived the earthquake, but many structures collapsed. Over 17,000 people died, and hundreds of thousands were left homeless.

Recurrence intervals

Records of past earthquakes, including the size of the rupture, allow us to identify where earthquakes are likely to occur in the future. Another question, of course, is "when?" Past activity on fault segments should allow us to recognize what a typical or "characteristic earthquake" has been in the past, including its size. And long-term records establish the average time, or **recurrence interval**, between characteristic earthquakes. The more time that passes after the last earthquake, the more likely it becomes that the characteristic earthquake will occur. Based on how long it has been since the last rupture, scientists can then calculate the probability that the earthquake will happen in a given time period. For example they can say, "There is a 30% chance that an earthquake of magnitude 5.0 to 6.0 will occur on this fault in the next decade."

Scientists with the USGS and other institutions launched a test of this model in the 1980s. Records indicated that moderate earthquakes have occurred on a segment of the San Andreas fault near Parkfield, California, at a fairly consistent rate since the mid-1800s. Earthquakes of around magnitude 6.0 struck in 1857, 1881, 1901, 1922, 1934, and 1966. With the exception of the 1934 earthquake, which was "early," earthquakes occurred on this part of the San Andreas fault approximately every 22 years. Seismic records suggest that the ruptures happened on the same area of the fault, explaining why the earthquakes were all roughly the same size. The scientists estimated that there was a 95% chance that another characteristic earthquake would occur sometime between 1985 and 1993. They wired the area with a network of seismometers and other instruments designed to "capture" the earthquake. Presumably, the sensors would document earthquake precursors and provide some clues about the processes that trigger earthquakes. As Figure 5-25 suggests, everyone was ready.

The earthquake, however, didn't happen in the predicted time period. (A 6.0 magnitude earthquake occurred on September 28, 2004, about 15 years "late".) Although the experiment has generated a lot of new information about the fault segment, it did not provide the hoped-for demonstration of the usefulness of recurrence interval determinations.

Making forecasts

In California, a panel of earthquake scientists and civil planners periodically reviews the latest information on the San Andreas fault system and issues long-range forecasts. It is one of the few places on Earth where this is possible, owing to the large volume of data on California faults and the dense network of instruments monitoring them. The forecasts take into account how much elastic strain is imposed on the faults due to regional plate motion, how much of this strain is removed each year by slow fault creep, and how much of the elastic strain was relieved by previous earthquakes. Each fault has its own forecast, including the magnitude range of the expected earthquake and the probability it may occur within a given time period.

FIGURE 5-25 ▼ Waiting for the Big One
The people of Parkfield, California were ready for the predicted earthquake, but it arrived 15 years late.

For example, the forecast published in 2003 for the San Francisco Bay area, shown in Figure 5-26a, states that there is a 62% probability that one or more magnitude 6.7 earthquakes will occur by 2032 on a fault within the region. | Bay area earthquake probabilities summary of main results | This warning, of course, means that there is a 38% chance that there won't be any such earthquake. Each fault has its own probability attached. The highest-risk fault in this forecast is the Hayward Fault, which cuts through Oakland on the east shore of San Francisco Bay. (This is the fault that runs under Memorial Stadium at the University of California at Berkeley—see Figures 5-3 and 5-9.)

This forecast, and others like it, contains uncertainties. Scientists have to make many assumptions about how the faults will behave. Indeed, the 62% probability value represents the midpoint between two extreme estimates of how the faults in the Bay area will adjust to the stresses along the plate boundary. On the low end, the probability is 37%, and the high end estimate is an 87% chance of a large earthquake by 2032. The probabilities change depending on the assumptions one makes about the faults' physical characteristics and how elastic strain builds up along them.

However uncertain, long-range forecasts allow emergency managers to prepare for the most likely scenarios. Forecast

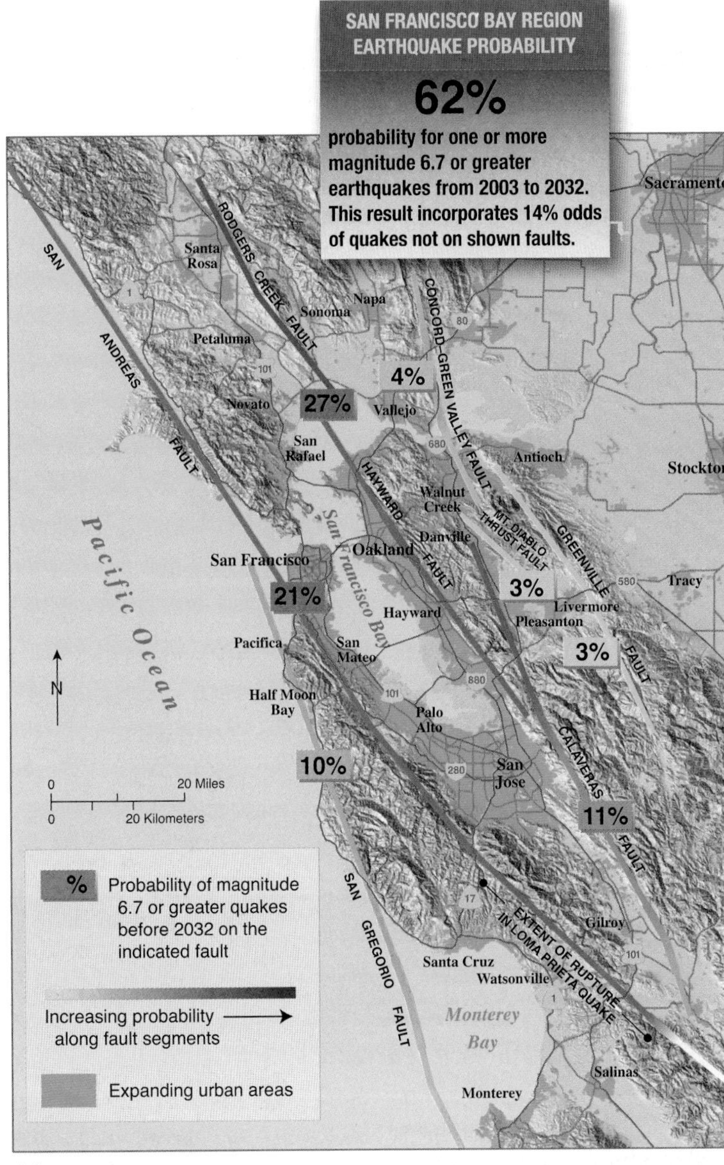

(a)

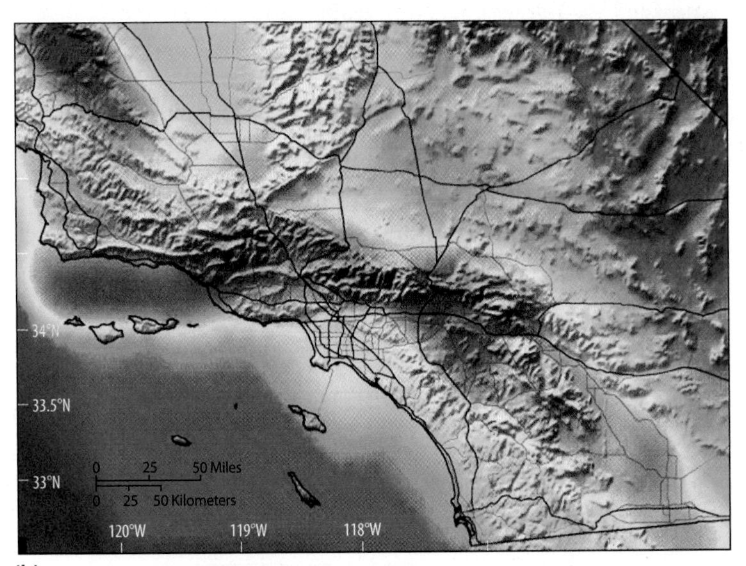

(b)

FIGURE 5-26 ◀ **Looking to the Future**
Earthquake forecasts give the probability that an earthquake will occur on a fault within a specified period of time. Only a few areas in the world are well enough studied to justify an earthquake forecast, however. The San Francisco Bay Region **(a)** is one of them. Another way in which such forecast information can be displayed is shown in this map of the Los Angeles area **(b).**

maps like the one shown in Figure 5-26b can also guide municipal decisions about building codes and land-use choices. Individuals, too, can decide where or how to build their homes and what they need to do to prepare for a large earthquake. This is how people diminish or mitigate earthquake hazards.

5.5 Mitigating Earthquake Hazards

When emergency managers in California have nightmares, they may be dreaming about something like this: It's 2:00 p.m., Monday, March 12, 2013. Office buildings in the San Francisco Bay area are packed with workers. A segment of the San Andreas fault spanning much of Northern California ruptures, unleashing a magnitude 7.9 earthquake. Some modern earthquake-hardened buildings sustain major damage but survive relatively intact, protecting their occupants. Others collapse. Fires ignited by gas leaks also inflict heavy damage. In the days after the disaster, the number of confirmed deaths climbs into the thousands and estimates of damages balloon upward. Repairing the Bay area's water supply alone is projected to cost tens of billions of dollars. More than 150,000 homes and apartment buildings have been rendered uninhabitable. Reporters begin to refer to the event as "Kobe in California," an allusion to the major earthquake that devastated Kobe, Japan, in 1995 and killed 5502 people.

This fictional account of a catastrophic earthquake probably sounds familiar to you. We've all heard about the "Big One" that could hit California sometime in our lifetimes. However, people have learned a lot about living with earthquakes. They have learned how to map earthquake hazards, build earthquake-resistant structures, prepare for—and respond to—earthquakes, and warn people about impending tsunamis.

EARTHQUAKE HAZARDS MAPPING

The USGS maintains national maps showing the peak horizontal ground motion accelerations expected from earthquakes. The maps are part of the National Earthquake Hazard Reduction program. A quick glance at these maps shows the major seismic hazard zones, the western states, the New Madrid region, and a few other smaller regions. In one version of the map (shown in

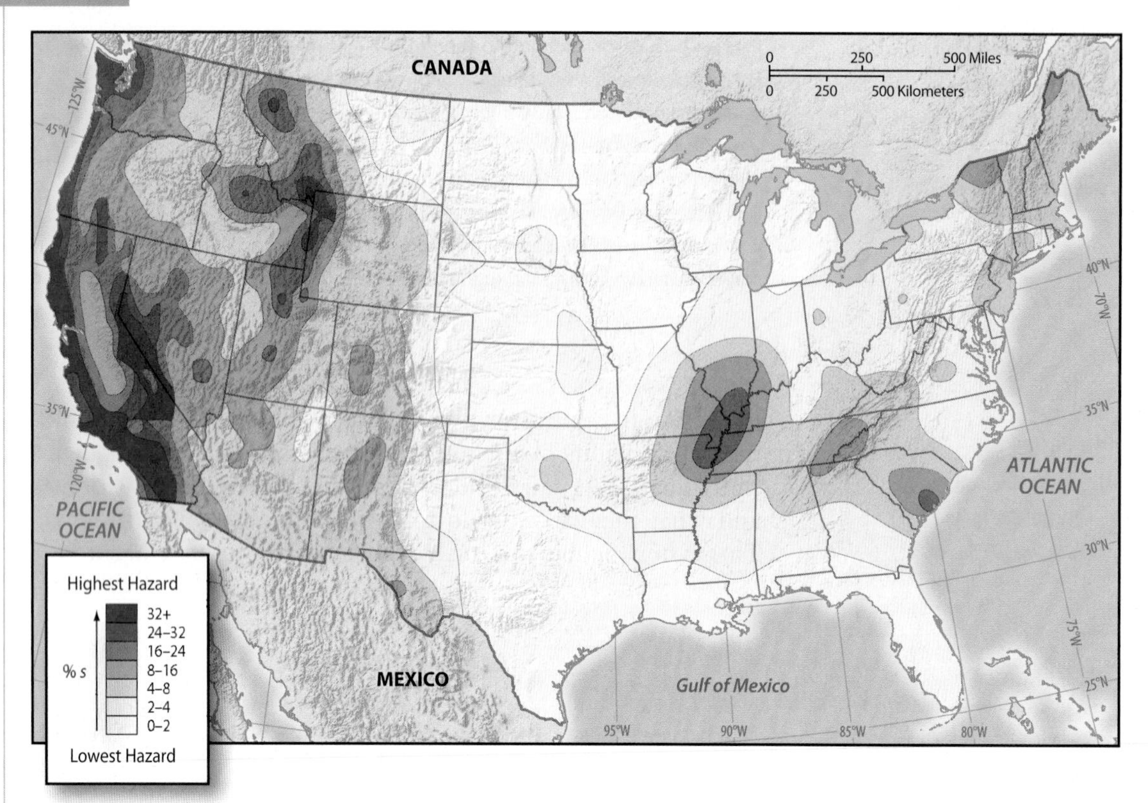

FIGURE 5-27 ◀ The USGS National Shaking Hazard Map
Colors on this map show the levels of horizontal shaking that have a 1-in-10 chance of being exceeded in a 50-year period. Shaking is expressed as a percentage of the acceleration of a falling object due to gravity (g).

Figure 5-27), shaded contours indicate levels of ground shaking that have a 10% chance of being exceeded during an earthquake in the next 50 years.

People use the USGS maps in many ways. Insurance companies use them to set rates. They are also used to estimate the risk of earthquake-triggered landslides and by the Federal Emergency Management Agency (FEMA) to target funding for earthquake safety education.

Scientists map hazards at the local level of towns and cities, too. They map not only shaking intensities, but the likelihood of hazards such as liquefaction or tsunami inundation. This kind of information can be used to guide future land use. A town could decide, for example, not to let people build certain types of buildings in low-lying areas prone to liquefaction or tsunami inundation. In this way, earthquake hazard maps become key influences on a community's land-use zoning: Future development is steered away from hazardous areas.

What You Can Do

Investigate Earthquake Hazards

Do you know what an earthquake's effects could be in your state? You can investigate earthquake shaking potential across the United States at USGS websites. At the Earthquake Hazards 101 site, you can learn about hazards maps, how they are made, and how you can use them. | Earthquake hazards 101—the basics | At the Earthquake Hazard Program Regional Information site, you can learn about the earthquake history of your state and the likelihood of certain levels of shaking intensity there. | U S earthquake information by state |

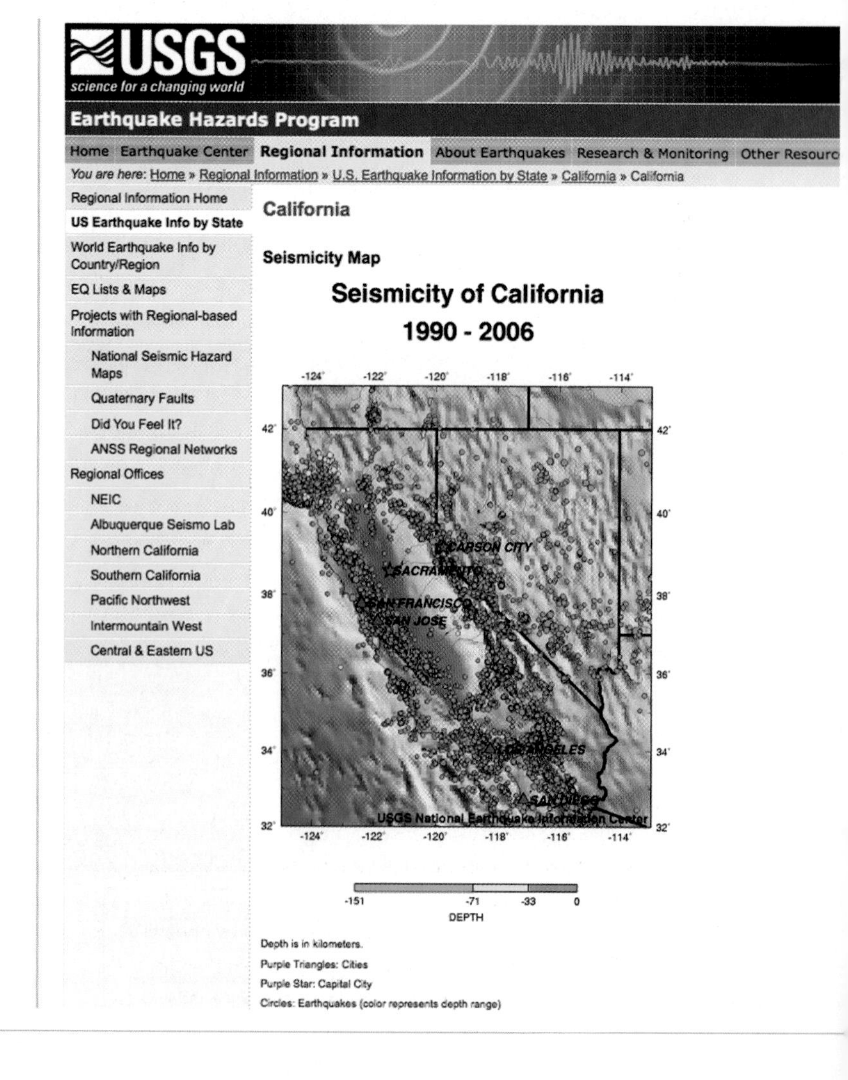

ENGINEERING FOR EARTHQUAKES

The tremendous cost of earthquakes in urban areas stems from damage to buildings, bridges, highways, rail lines, dams, and other infrastructure. In designing structures to resist earthquake damage, engineers must take into account not only the likely intensity of shaking, but also the direction of shaking. During an earthquake, a building might shake in three directions simultaneously—up and down, left to right, and forward and backward. Shearing ground motions, those that shake buildings sideways, are especially damaging to structures.

Structures of almost any type, including buildings, bridges, roads, and dams, can be engineered to resist strong ground shaking. The word *resist* is key since there are physical and economic limits to how strong a structure we can build. The goal is to prevent a catastrophic collapse and keep structures repairable. Seismic engineering is an advanced field, and a variety of construction practices can make a structure more shake-resistant.

Buildings often fail because cracks develop at corners where vertical and horizontal beams or supports are joined. With continued shaking, the cracks evolve into complete breaks. Buildings made from unreinforced masonry, stone, or cement block may be particularly hazardous because the materials lack an important earthquake-resistant quality, *ductility*, which allows the structure to bend instead of crack. Extra bracing at the corners of buildings, either included in the original construction or added later as a retrofit, can reduce the danger of failure. The building also has to have enough "give" so that the supporting beams, columns, or walls themselves can ride out horizontal shaking without crumbling. These are the main reasons that wood frame homes tend to withstand earthquake shaking better than brick or masonry homes.

Another approach to designing earthquake-resistant structures is to add features that reduce the motion imposed on the building. In one strategy, *base isolation*, the entire building rests on flexible supports that act as a buffer between the building and the ground. One common type of base isolation "pad" is constructed of alternating layers of steel and rubber, with a post of lead down the center. The building is anchored firmly to its foundations but is allowed to move horizontally during an earthquake. The lead post helps to dampen, or reduce, the seismic vibrations. Such pads can decrease the accelerations experienced by a building up to 75%. Multistory commercial and apartment buildings in earthquake-prone cities have been constructed using base isolation technology.

In the United States, where wood-frame construction is common for single-family homes, relatively simple features can reduce damage during an earthquake. One common feature of seismic building codes is a requirement that walls be firmly anchored to their foundations with bolts. This helps prevent the building from being jarred off its foundations.

The Trans-Alaska Pipeline System (TAPS) is a remarkable example of successful earthquake engineering. This pipeline has carried up to 2 million barrels of oil a day, right across the earthquake-generating Denali Fault. The Denali Fault is a 1600 kilometer (1000 mi) strike-slip fault capable of causing large earthquakes. Geologists determined that the eastern 350 kilometers (220 mi) segment of the fault, which is the part TAPS crosses, was most likely to rupture next.

Engineers planned for a large earthquake occurring under TAPS when it was constructed in the 1970s. The pipeline's design included supports made of Teflon that allow it to slide laterally (sliders) and bends that give it extra length, both of which can be seen in Figure 5-28.

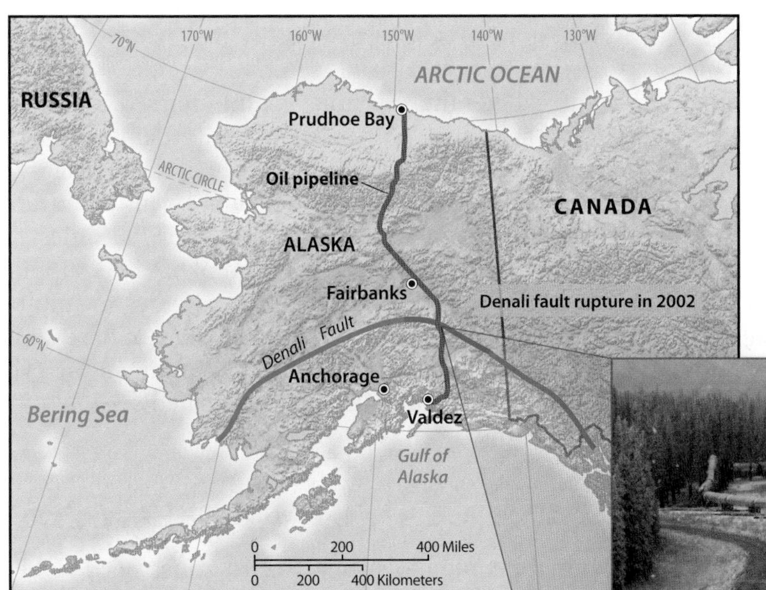

FIGURE 5-28 ▲ The Trans-Alaska Pipeline System (TAPS) Crosses the Active Strike-Slip Denali Fault

With information provided by geologists about the location of the fault and the size of expected earthquakes on it, engineers developed a special pipeline design for the fault crossing. The design worked. The pipeline very successfully withstood 5.5 meters (18 ft) of horizontal and 1 meter (3 ft) of vertical surface displacement caused by the November 3, 2002 M_W 7.9 earthquake on the Denali fault.

The pipeline was designed to move without breaking when shaking occurred.

On November 3, 2002, a 7.9 magnitude earthquake ruptured the Denali Fault directly beneath the pipeline. In addition to tremendous shaking, ground offset beneath the pipeline was 5.5 meters (18 ft) horizontally and 1 meter (3.3 ft) vertically. But the pipeline didn't spill a drop of oil and was back in full service, after inspections, just three days later.

EMERGENCY RESPONSE

After a damaging earthquake, many lives can be saved if the community is equipped to respond quickly to rescue people from damaged or collapsed buildings, fight fires, and restore utilities. The value of emergency response was seen during the earthquake that struck Kobe, Japan, in 1995. Although the government had constructed an emergency water supply, it didn't work in the aftermath of the quake because of heavy damage to underground piping. About 100 fires broke out in the city in the first few minutes after the earthquake, but firefighters were hampered for several hours by lack of water and large parts of the city burned as a result. Earthquake effects in Kobe, Japan

One critical task emergency managers have is allocating resources. They have only so many crews available to fight fires and rescue people, so where should they go first? In the first few hours after the 1994 Northridge quake, municipal leaders used a map of shaking intensity (Figure 5-29), prepared by seismologists and engineers, to manage emergency services. Southern California had already put in place a system, called TriNet, that generates maps to help managers determine *within minutes* where the strongest shaking occurred and, therefore, where the greatest damage was likely.

TriNet consists of a network of seismometers and ground-motion instruments connected to a central computer and communication system. When an earthquake occurs, TriNet rapidly determines the magnitude, location, and time of the earthquake. It also generates "shake maps" automatically, within minutes, showing where the strongest ground motions are likely to be. These help emergency managers decide where first to send help, such as medical assistance, and to deliver food and water to displaced people. The maps can also help utility repair crews target the most heavily damaged services and restore them sooner.

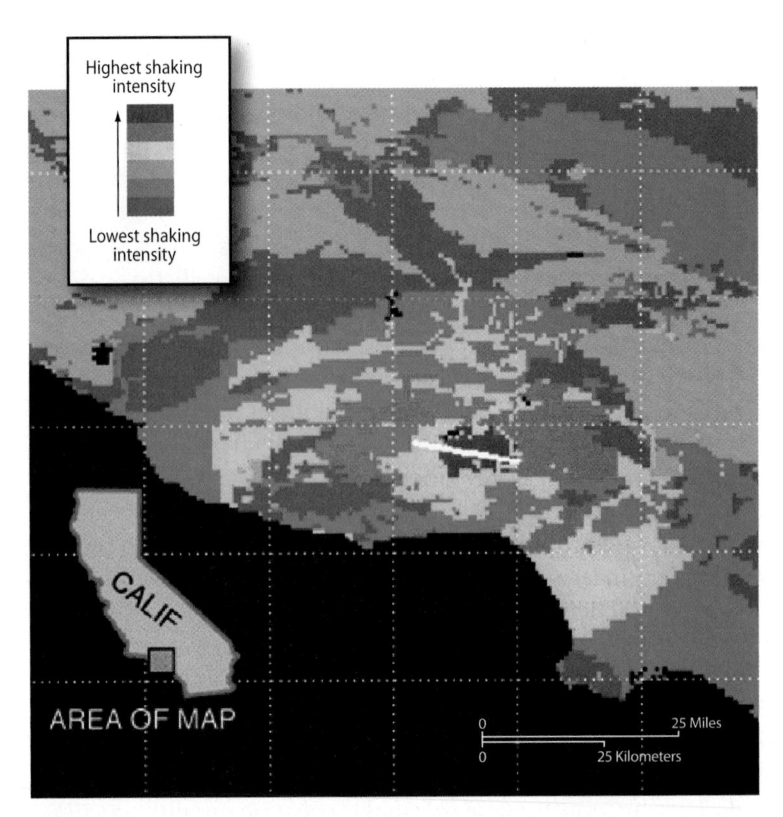

FIGURE 5-29 ▲ Map of Shaking Intensity Created Immediately after the Northridge Earthquake
This rapidly prepared map guided the allocation of emergency services to tens of thousands of people in the Northridge area.

EARTHQUAKE EARLY WARNING SYSTEMS

Because P waves move faster than other types of seismic waves, they are being used in some instances to provide a brief but potentially lifesaving warning to people in harm's way. An early warning system in Mexico City, for example, can detect an initial shock from the subduction zone off Mexico's west coast soon enough to provide 60 seconds of warning to authorities before strong surface waves arrive. In Southern California, seismologists are using TriNet to test an earthquake warning system called the Seismic Computerized Alert Network (SCAN) that can identify a potentially damaging earthquake seconds to tens of seconds in advance of the strongest shaking, depending on how far away it occurs. Even this little warning might allow people to "duck and cover," utilities to secure their power grids and shut off flammable fuel supplies, businesses to preserve their databases, and cities to stop trains before the strongest shaking takes place. Japan, for example, uses seismic sensors to stop its high-speed trains in the event of an earthquake, and maintains a sophisticated seismic monitoring system (Figure 5-30).

PUBLIC EDUCATION AND PREPAREDNESS

Individual people in seismic hazard zones need to be well informed about what they can do before, during, and after an earthquake to reduce their personal risk and to help their community respond. Individuals often can't do much to eliminate an earthquake hazard, but they can do a lot to prepare for a seismic event and therefore to reduce their personal risk. Public education that starts in elementary school has helped Californians live more safely with earthquakes. In California, as in many other places, even children know where to go (in doorways and under tables) and where not go (by windows and next to buildings) during earthquakes (Figure 5-31). Public education also helps people live more safely along coasts prone to tsunamis. When the sea suddenly recedes, whether one feels an earthquake or not, it's time to head for higher ground along marked emergency evacuation routes.

In The News

Tsunami Education Saves Lives

Tilly Smith remembered her lessons well. This English schoolgirl had learned about tsunamis in her geography class, including their warning signs such as a suddenly receding sea. She and her family were vacationing on a beautiful Thailand beach the morning of December 26, 2004. She noticed the water withdrawing from the shoreline and warned her family and about 100 others on the beach. Tilly is credited with saving the lives of these people as they retreated to higher ground before a devastating tsunami came ashore.

FIGURE 5-30 ▲ Preparedness on the National Level
At the Earthquake, Tsunami and Volcanic Activity Data Management Center operated by the Japan Meteorological Agency, a minimum of five staff members are on duty at all times, ready to set in motion measures to protect the population from a destructive earthquake.

TSUNAMI WARNING SYSTEMS

Tsunamis are a serious hazard for countries located around the Pacific Ocean, and many operate tsunami warning and response centers. When a large earthquake occurs, seismologists determine its location, its magnitude, and whether it has the potential to generate a tsunami. The typical "tsunamigenic" earthquake occurs on a subduction zone, such as those that ring the Pacific, where sudden lurching of the ocean floor sets the waves in motion.

The Pacific Tsunami Warning Center (PTWC), in Hawaii, was founded in 1948. It was a direct response to the loss of life and extensive property damage caused by a tsunami in 1946 that was triggered by an earthquake in the Aleutian Islands, Alaska. The center issues a *tsunami watch* when seismographs detect a large earthquake capable of displacing the seafloor and spawning a tsunami. If ocean floor sensors or tide gauges confirm a tsunami, the center issues a tsunami warning, along with predicted arrival times at various locations. This gives coastal communities minutes to hours to respond, depending on their distances from the earthquake.

FIGURE 5-31 ▼ Preparedness on the Local Level
In many countries, schoolchildren are taught what to do when an earthquake strikes.

FIGURE 5-32 ▲ Road Signs Used to Identify Tsunami Hazard Zones and Evacuation Routes

In the United States and Canada, the region including northern California, Oregon, Washington, British Columbia, and Alaska are all at risk of tsunami inundation after a large earthquake in the Cascadia subduction zone. Many communities have their own tsunami warning sirens and post signs (such as those shown in Figure 5-32) instructing people to head for higher ground if they feel the earth shake, see the sea recede or hear warning sirens.

You Make the Call

Who Is Responsible for Tsunami Warning Systems?

On December 26, 2004, a fault rupture along about 1000 kilometers (620 mi) of the subduction zone between the Indian and Burmese plates caused a magnitude 9.1 earthquake off the coast of northern Sumatra. Movement along seafloor faults during the earthquake displaced tremendous volumes of water, creating a devastating tsunami that killed more than 225,000 people in eleven countries (see Figure 5-21). Nearby coastal communities would have had about 20 minutes to prepare from the time of the earthquake until the deadly tsunami came ashore, if they had known it was coming, and it took hours or more for the waves to travel across the Indian Ocean and strike places like India and Sri Lanka. Tragically, the tsunami's arrival was a surprise, and hundreds of thousands of people died.

The Indian Ocean is bordered by Africa, India, and the southern Asian countries of Myanmar, Thailand, Indonesia, and Australia. The convergent plate boundary that generates large earthquakes in this region is on the east side of the ocean. Although tsunamis in the Indian Ocean are less frequent than in the Pacific Ocean, seven have been recorded around this ocean basin since 1762. The tsunami on August 26, 1883, was caused by a volcanic eruption, but the others were related to earthquakes.

The December 26, 2004, tsunami seriously affected twelve countries located on both sides of the Indian Ocean. NSF after the tsunami introduction An Indian Ocean tsunami warning system was not in place at the time. Waverly Person of the USGS National Earthquake Information Center has concluded that most of the lives that were lost could have been saved by a tsunami warning system. The tsunami warning system for the Pacific Ocean, including community education programs that teach people how to recognize and respond to potential tsunamis, has worked effectively for many years. After the December 26, 2004, Indian Ocean tsunami, it was painfully apparent that the people living around the Indian Ocean need a tsunami warning system.

At least 27 countries located around the Indian Ocean are concerned about tsunamis. But devastating tsunamis like that in 2004 affect the entire world. Individuals, organizations, and governments around the world responded to help the people affected by this disaster. From this perspective, there are many challenging questions. If a tsunami warning system is to be developed in the Indian Ocean:

- Who should participate in developing it?

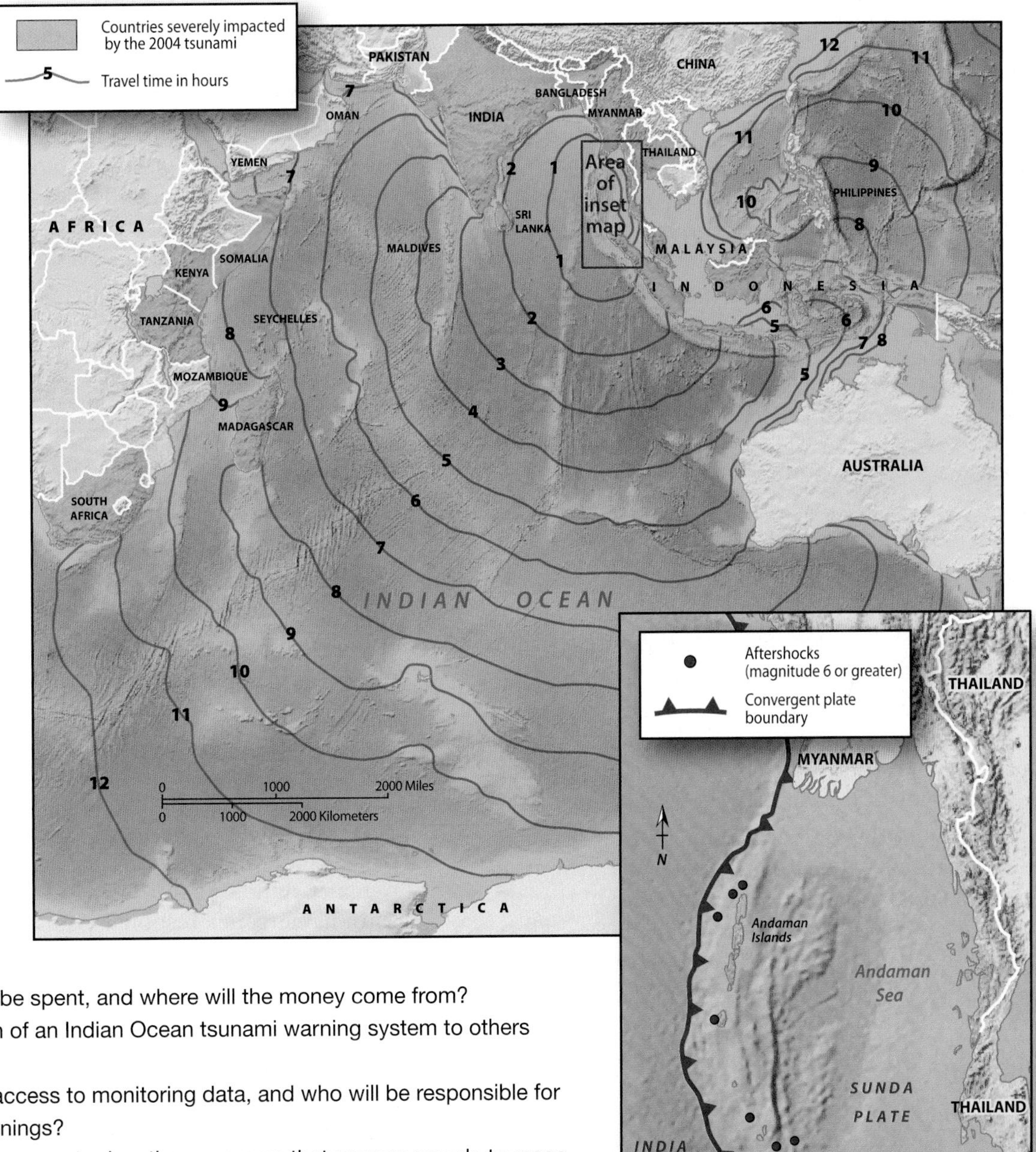

The December 26, 2004 tsunami severely affected many countries around the Indian Ocean and killed over 225,000 people, more than any other tsunami in recorded history.

- How much should be spent, and where will the money come from?
- What is the relation of an Indian Ocean tsunami warning system to others around the world?
- Who should have access to monitoring data, and who will be responsible for timely tsunami warnings?
- Who will plan and carry out education programs that prepare people to recognize and respond to potential tsunamis?

In some ways, setting up a tsunami warning system seems straightforward. After all, the Pacific tsunami warning system has worked very well. But this system benefits by a having wealthy countries located along large segments of its earthquake-prone regions. How do 27 countries, with a wide range of resources and capabilities, develop an effective tsunami warning system? The United Nations Educational, Scientific, and Cultural Organization (UNESCO) began coordinating such an effort in 2005.

SUMMARY

Movements in the dynamic geosphere cause earthquakes. Earthquakes are common globally although most of them are too small to be noticed. People definitely feel larger earthquakes, and some of these cause death and destruction. Earthquakes present many hazards, even very far from where they occur if the hydrosphere's oceans are affected and a tsunami develops. Earthquake destruction affects human habitats, and their damage to infrastructure is a continuing strain on people's resources around the world. Thus, diminishing the economic and human costs of future earthquakes can be a step toward achieving sustainability. Scientists and engineers are helping people to understand and live more safely with earthquakes. Because earthquakes will be in the news and in your future whether you live in earthquake country or not, it is useful for you to understand earthquakes.

In This Chapter You Have Learned:

- Most earthquakes occur on or near boundaries between tectonic plates. Subduction zones, where tectonic plates sink into the mantle, generate extremely powerful earthquakes. Transform plate boundaries and places where continents collide are also sites of many earthquakes. The poorly understood earthquakes that occur in the interiors of continents, intraplate earthquakes, are less common but can still be large and destructive.
- Earthquakes are vibrations caused by movements on faults. The elastic rebound theory explains the relationship between earthquakes, tectonic stresses, and faulting. Elastic strain builds up on a fault, bending or deforming rock. The elastic strain is released and the rock rebounds as fault rupture occurs.
- Seismic energy travels in the form of waves. Body waves move through Earth's interior, whereas surface waves move along the surface.
- Surface waves travel more slowly than body waves and remain hazardous for a greater distance; they are a major cause of shaking and damage to structures.
- Seismometers record ground motions due to seismic waves. They enable us to estimate the size (magnitude) of an earthquake, its location (epicenter), and the point beneath the surface where the earthquake began (focus).
- The Richter magnitude scale was the first quantitative method developed for calculating the size of earthquakes based on seismogram readings. A newer system, the moment magnitude scale (M_W), ranks earthquakes based on the total amount of seismic energy they release, and is a more accurate method of measuring the size of very large earthquakes.
- Earthquake intensity refers to the effects of ground shaking at a specific point on the surface. People's observations of damage are used to determine earthquake intensity using the Modified Mercalli Scale.
- Strong ground motion is the most deadly hazard of earthquakes because it frequently leads to catastrophic collapse of buildings. Other major hazards include liquefaction, permanent surface displacements, landslides, tsunamis, and fires.
- An understanding of where active faults are and their earthquake history helps us to identify seismic hazards, but it is not yet possible to exactly predict the location, size, and time of earthquakes.
- Less-precise long-term forecasts of earthquake occurrence are presently more useful than short-term predictions. They state the statistical probability that an earthquake of a certain magnitude or range of magnitudes will occur within a given period of years on a specific fault or fault segment.
- Long-term forecasts are frequently based on identifying seismic gaps and earthquake recurrence intervals on active faults. Seismic gap theory holds that the longer a segment of an active fault goes without an earthquake, the more likely it is to have major slip. The recurrence interval of a characteristic earthquake on a fault is the average amount of time between these earthquakes.
- Communities in seismic and tsunami hazard zones can reduce loss of life and property by developing seismic hazard maps, tightening building codes, engineering buildings to remain standing during earthquakes, and developing sound emergency response plans and warning systems.
- Residents of seismic hazard zones should be well-informed about the risks they face and what to do during an earthquake.

KEY TERMS

amplitude (p. 125)
body waves (p. 122)
earthquake (p. 115)
earthquake cycle (p. 120)
elastic rebound theory (p. 120)
elastic strain (p. 120)
epicenter (p. 115)
fault creep (p. 121)
fault trace (p. 120)
focus (p. 115)
frequency (p. 125)
intensity (p. 126)
intraplate earthquake (p. 119)
liquefaction (p. 132)
magnitude (p. 125)
Modified Mercalli Scale (p. 126)
moment magnitude (p. 126)
period (p. 125)
precursor (p. 137)
P (primary) waves (p. 122)
recurrence interval (p. 140)
Richter magnitude scale (p. 125)
S (secondary or shear) waves (p. 122)
seismic gap (p. 138)
seismic waves (p. 115)
seismogram (p. 124)
seismology (p. 115)
seismometer (p. 124)
surface waves (p. 122)
tsunami (p. 133)

QUESTIONS FOR REVIEW

Check Your Understanding

1. How does elastic rebound theory relate elastic strain and its release to earthquakes?
2. What is the relationship between the focus and the epicenter for a given earthquake? Why is it important to locate both in order to understand earthquake effects at the surface?
3. When studying transform faults on land (such as the San Andreas fault) and working to understand seismic hazard, why is it important to realize that plate boundaries usually comprise complex systems of interacting smaller faults?
4. Why and how do earthquakes at offshore subduction zones often pose a tsunami risk?

5. The sudden release of stored elastic strain energy in earthquakes takes many forms. In what ways is this stored energy released into the surrounding rock around a fault?
6. Why do surface waves cause more damage to structures than P and S waves?
7. What is the relationship between the amplitude of earthquake body waves and the magnitude of an earthquake measured by the Richter scale?
8. What is the basic difference between earthquake intensity and earthquake magnitude? Can a given earthquake have more than one intensity value? Can a given earthquake have more than one moment magnitude value?
9. Why do seismologists require at least three seismic records to determine the location of an epicenter from the time difference between the arrival of P and S waves?
10. Why do structures built directly on hard granite rock tend to survive earthquakes better than those built on soft sediments?
11. Why are fires a major problem associated with big earthquakes in large cities?
12. Why has short-term seismic prediction proven frustratingly elusive for earthquake scientists? How does longer term probability-based earthquake forecasting help protect people and property from future earthquake destruction?
13. What fact about the velocities of P, S, and surface waves do earthquake early warning systems exploit to our advantage? How do these systems help communities respond during an earthquake?
14. How does the engineering technique called base isolation protect tall buildings from earthquake damage?

Critical Thinking/Discussion

15. Intraplate earthquakes pose an interesting geological challenge in that it is not immediately possible to relate faults at these locations to active plate boundaries. Based on what you've learned in Chapters 3 and 5, provide three possible explanations for the accumulation of elastic strain that can lead to earthquakes in intraplate regions.
16. What factors account for the remarkably low loss of life in the Loma Prieta and Northridge earthquakes in California, despite very high damage costs? Discuss these factors in relation to the much higher loss of life in the Bam (Iran) and Mexico City earthquakes, which were comparable in size to the California earthquakes.
17. Identifying seismic gaps on fault systems and determining when and how often a fault has moved in the past (recurrence intervals) for fault segments are key tools in assessing the seismic hazard in a tectonically active region. While both provide valuable insights into seismic risk, they are not foolproof. What are the difficulties associated with these methods, and what are the resulting uncertainties in risk assessments.

ANSWERS TO IN-CHAPTER INSIGHT QUESTIONS

P. 119

What are some Earth systems changes or interactions that accompany earthquakes in the Basin and Range province?

- Geosphere movements that cause earthquakes create mountains and valleys.
- Mountains and valleys are an influence on biosphere habitat.
- Mountains and valleys influence stream flow and lake development.

P. 120

What are some Earth systems changes or interactions that accompanied the large 1811–1812 earthquakes in the central Mississippi Valley?

- The geosphere's surface moved significantly.
- The geosphere movements affected the course of the Mississippi River and flooding occurred.

P. 133

What are some Earth system changes or interactions that accompany earthquake-induced ground displacement or failure?

- Surface displacements can change hydrosphere shorelines and drainages.
- Ground failure can change hydrosphere shorelines, drainages, and biosphere habitat.
- Ground failure has destroyed property and killed thousands of people.

P. 134

What are some Earth systems changes or interactions that accompany earthquake-induced tsunamis?

- Geosphere movements that displace ocean water generate tsunamis.
- Tsunamis inundate coasts and disrupt, change, or destroy biosphere habitat.
- Some tsunamis damage or destroy property and kill people.

FIGURE 6-1 ▶ Redoubt Volcano
Redoubt Volcano in Alaska erupted from December 1989 through April 1990. Ash clouds, blasted as high as 12 kilometers (39,000 ft), damaged aircraft, disrupted air traffic, and caused local power outages and school closures.

CHAPTER

6

VOLCANOES

On December 14, 1989, a KLM 747-400 jumbo jet high over the Talkeetna Mountains of Alaska starts its final approach to Anchorage International Airport after an uneventful over-the-pole journey from Amsterdam. With 231 passengers and 14 crew members on board, KLM-867 is a routine flight. The pilot at the controls radios Anchorage Air Traffic Control (ATC):

"KLM-867 heavy is reaching level 250 heading 140."

Heavy indeed. The aircraft, just recently off the assembly line, has a maximum takeoff weight of 438 tons. It is a wide-body jumbo jet, a "heavy" in the jargon of commercial aviation. It is one of the most advanced and safest flying machines ever built. But KLM-867 is flying into a problem. Mount Redoubt, a volcano 176 kilometers (109 mi) southwest of Anchorage, is erupting (Figure 6-1).

ATC: "Do you have good sight of the ash plume?"

Mount Redoubt is one of the 40 historically active volcanoes strung along the Aleutian Islands chain. It has blasted a plume of fine pulverized lava and hot gases more than 8 kilometers (5 mi) into the atmosphere. The prevailing winds have blown this mass of ash into the path of KLM-867.

Previously, the crew had requested a flight path they assumed would steer them clear of the ash plume, a hazard to aircraft that many other crews before them had learned to respect. In 1982, for example, a British Airways 747 cruising at 11,300 meters (37,000 ft) with 241 passengers flew into an ash cloud from Galunggung volcano in Indonesia. All four engines flamed out and the aircraft glided for 16 minutes without power, falling to an altitude of 3800 meters (12,500 ft) before the crew was able to restart the engines. A month later, another 747 lost power after encountering ash from the same volcano. It landed with only one engine still operating.

KLM-867: "It's just cloudy, it could be ashes. It's just a little browner than a normal cloud."

The ash cloud is not visible on the aircraft's onboard radar. It looks like a dark cumulus cloud. If it were nighttime, the crew would not have seen anything at all. But they begin to notice fine brown ash and the smell of sulfur in the cabin.

KLM-867: "We have to go left now . . . it's smoky in the cockpit at the moment, sir."

ATC: "KLM-867 heavy, roger, left at your discretion."

KLM-867: "Climbing to level 390, we're in the black cloud, heading 130."

The aircraft is in a dangerous place and the pilots want to get out of it as quickly as possible. They push the throttles to full power and start to climb out of the cloud. But in hindsight, throttling up is the worst thing they could do. Temperatures in the engine, normally in the range of 700° C (1300° F), begin to rise. Tiny shards of volcanic glass in the ash plume melt onto the engine's whirling compressor blades. Fine ash clogs vent holes and other narrow openings in the engines. Within 58 seconds of starting the climb, the engines stop turning: flameout. The aircraft goes into a glide. The pilot sounds remarkably composed over the radio as she updates Anchorage ATC of the situation.

KLM-867: "We have flameout all engines and we are descending now."

In the cockpit, alarms are sounding. Readouts for airspeed and other critical information are not functioning. A warning light indicates, incorrectly, that there is a fire in one of the forward cargo compartments.

ATC: "KLM heavy . . . Anchorage."

KLM-867: "We are descending now . . . *We are in a fall!*"

Years of training and experience kick in. The first officer uses the 747's inertial navigation system to maintain the stricken craft's airspeed. The captain and second officer repeatedly attempt to restart the engines. Below, the mountain peaks, themselves nearly 3350 meters (11,000 ft) high, get closer. The plane falls for nearly seven minutes without power. Two engines come to life at 5240 meters (17,200 ft); the remaining two come back online at 4055 meters (13,300 ft). As the crew regains full power, they are less than two minutes from possible impact with the mountains. They land safely in Anchorage, but it will cost $80 million to repair the damage done to the aircraft by Mount Redoubt's ash plume.

Volcanoes may seem a remote concern to many people reading this book. However, as the KLM-867 incident illustrates, the hazards of volcanoes may be closer than you think. The encounter with Mount Redoubt's ash plume was a close call for the passengers and crew of KLM-867. Anyone who has flown along the rim of the Pacific Ocean has flown over potentially dangerous volcanoes (Figure 6-2).

This chapter starts with some basics about volcanoes:

FIGURE 6-2 ◀ Over the Volcano
If you fly over the Pacific rim, you will probably find yourself looking down on volcanoes. This erupting volcano, in the Aleutian Islands of Alaska, is on the main flight path between Asia and North America.

what they are, where they are, and why they form. But what's important to people are volcanic hazards. When volcanoes erupt, they have been known to threaten the lives and property of people living hundreds of kilometers distant—even, in some cases, to affect atmospheric conditions and weather all over the world. Volcanic hazards, such as explosions of debris, clouds of ash, and cascading lava flows, can be devastating. Volcanoes are a product of the dynamic geosphere and no one can prevent or stop their eruptions. What people can do is recognize the risks, stay out of the way, and make decisions that minimize death and loss of property from volcanic hazards.

IN THIS CHAPTER YOU WILL LEARN:

- What volcanoes are, where they occur, and why they form
- The different types of volcanoes and the hazards associated with each type
- How scientists study volcanoes and come to understand and predict their eruptions
- How people can lower the risks posed by volcanoes

6.1 Volcano Basics

Most people have an idea of what volcanoes are. You would probably describe a volcano as a cone-shaped, snow-capped mountain with a dark cloud billowing from its peak, perhaps with glowing lava flowing down its slopes. This image has been planted in our collective imagination by innumerable paintings, films, and cartoons—not to mention grade-school science projects resulting in papier-mâché volcanoes and baking-soda-and-vinegar eruptions. This image of the volcano is accurate for many places on Earth, especially the chains of volcanoes like Mount Redoubt that ring the Pacific Ocean. But there are different types of volcanoes, and some are much more hazardous than others.

WHAT VOLCANOES ARE

In the most general sense, **volcanoes** are places where molten rock, or magma, rises from great depths to the uppermost levels of the crust and onto the surface. When magma erupts onto the surface it is called lava. Magma is much less dense, and therefore lighter, than the solid rock of its surroundings, like a hot air balloon released on a cool fall day. Because magma is hot and buoyant, it rises upward through the crust; when it breaks through to the surface, an eruption results. As Figure 6-3 shows, magma quickly cools and solidifies, either on the surface or at relatively shallow depths in the crust, making volcanoes igneous rock factories.

Volcanoes also emit hot gases when they erupt. Magma contains dissolved gases, mostly water vapor (steam), carbon dioxide, and sulfur dioxide. As magma rises toward the surface and the confining pressure of the overlying rocks decreases, dissolved gases bubble out of solution, like the fizz in a bottle

FIGURE 6-3 ▲ Volcanoes Are Rock Factories
Magma quickly cools and solidifies into igneous rock. The surface of this lava flow in Hawaii is changing to a gray color as it solidifies.

of soda when you open it and release the pressure. As you will see, the amount of dissolved gases in magmas helps determine how explosive an eruption will be. The more dissolved gas, the more dangerous the magma.

DEFINING AND COUNTING VOLCANOES

What if a volcano has not erupted during human memory? Is it still a volcano people should be concerned about? Determining the total number of "active" volcanoes in the world, and therefore how many people may be at risk, depends on how you define *active*. Virtually every day, a volcano is exhibiting some sort of threatening activity somewhere on Earth. However, "volcanic activity" is not synonymous with "volcanic eruption." A new eruption occurs only when fresh magma reaches the surface. Lesser events, such as the release of steam, do not necessarily indicate that a large, sustained eruption is happening or will happen soon. Massive, destructive eruptions like those of Mount St. Helens in Washington (1980) and Mount Pinatubo in the Philippines (1991) occur less frequently than routine, day-to-day volcanic unrest. For every major volcanic eruption like the one that blasted 400 meters (1300 ft) off of Mount St. Helens in 1980, there are scores of smaller events at other volcanoes in the same year.

Scientists at the Smithsonian Institution keep a running count of volcanoes that have been active in the last 10,000 years. The Smithsonian team has tallied at least 1300 volcanoes on land with probable eruptions in that period, and perhaps as many as 1500. Most of these are evidenced in the geologic record, as only about 550 volcanic eruptions were witnessed by someone who left a written account. (Some of the more notable ones are listed in Table 6-1.) As you read this book, about 20 volcanoes are probably erupting somewhere on land; between 50 and 70 volcanoes erupt in an average year. And as you learned in Chapter 3, many more volcanic eruptions occur in the deep oceans.

Even volcanoes that last erupted thousands of years ago are still considered capable of erupting in the future. People commonly refer to these as *dormant* or *extinct* volcanoes. However, to volcano scientists, called *volcanologists*, any volcano that has erupted in the last 10,000 years could do so again.

Because of the difference in scale between human lifetimes and geological time (Chapter 2), volcanoes sometimes catch local populations off guard. Volcanologists had established by the 1970s that Mount St. Helens was the most active volcano in the Cascade Range (Figure 6-4) and could erupt again before the end of the century. However, when Mount St. Helens essentially blew up in March 1980, many people in the region seemed to have forgotten that they were living near an active and dangerous volcano.

TYPES OF MAGMAS

Some volcanoes erupt in spectacular explosions and some just ooze red-hot lava onto the surface. This difference reflects the fact that not all magmas are the same. They have different temperatures, compositions, and dissolved-gas contents.

Temperature and composition control magma's resistance to flow, or its **viscosity**. Viscosity may be thought of as the "stickiness" of the magma. For example, consider how butter at room temperature is more viscous than, say, molasses. The butter does not flow freely, so you must spread it with a knife.

TABLE 6-1 SOME NOTABLE VOLCANIC ERUPTIONS

Date	Name/Location	Fatalities	Comments
79	Vesuvius, Italy	15,000 +	Pompeii completely buried, not rediscovered until 1700s.
1783	Laki, Iceland	10,000	Enormous lava flows and gas releases destroyed crops and livestock; many people died of famine.
1792	Unzen, Japan	14,000	Debris avalanche, tsunamis.
1815	Tambora, Indonesia	92,000	Most violent historical eruption. Atmospheric ash caused worldwide cooling.
1883	Krakatoa, Indonesia	36,000	Collapse of caldera produced huge tsunami. Explosion heard thousands of miles away.
1902	Mount Pelee, West Indies	30,000 +	Pyroclastic flows totally destroyed town of St. Pierre.
1951	Mount Lamington, Papua New Guinea	6000	Ash flows.
1980	Mount St. Helens, Washington, USA	54	Evacuation warnings prevented higher death toll.
1982	El Chichon, Mexico	2000	Ash flows.
1983–present	Kilauea, Hawaii, USA	—	Longest recorded continuing eruption.
1985	Nevado del Ruiz, Colombia	22,000	Eruption unleashed massive mudflows.
1991	Mount Pinatubo, Philippines	800	Violent eruption; ash flows, lahars.
1991	Mount Unzen, Japan	41	Three volcanologists killed.

FIGURE 6-4 ◀ **Potentially Active Volcanoes of the Western United States**
Many people do not realize that even in the continental United States there are several potentially active volcanoes dangerously close to large population centers.

If you heat the butter, however, it becomes less viscous and you can pour it from a container. Contrast this quality of butter to molasses. Molasses flows at room temperature, but when cooled it will become thick enough to spread with a knife. Temperature affects magma viscosity in the same way. The hotter the magma, the less viscous it is and the more easily it flows.

Composition has a greater influence on magma viscosity than temperature does. The key compositional factor is silica content—the same silica that characterizes the composition of igneous rocks, as described in Section 4.3. Silica (SiO_2) makes magma more viscous because of the strong tendency of silicon and oxygen atoms to combine together even before magma solidifies. And because magmas with more silica solidify at lower temperatures, their viscosity increases even more.

The primary determinant of magma's silica content is the composition of the rock that melts to produce the magma. If a rock melts completely, the magma will have the same composition as the rock. But rocks commonly don't melt completely. As a rule, the magma that forms when rocks partially melt has more silica than the parent rock. Thus, partial melting of ultramafic rocks produces mafic magma, partial melting of mafic rocks produces intermediate magma, and partial melting of intermediate rocks produces felsic magma. As Table 6-2 indicates, mafic magma, with the least silica, is the least viscous; it

TABLE 6-2 MAGMA TYPES

Composition	Silica Content (%)	Viscosity	Eruption Temperatures (°C)	Eruption Style
Mafic	45–52	Low	Up to 1300	Flows
Intermediate	53–65	Intermediate	About 1000	Flows and explosions
Felsic	Greater than 65	High	Less than 900	Domes and explosions

tends to flow easily, while the more viscous intermediate to felsic magmas do not. This is important because, as we shall see, magma viscosity is a major factor in how volcanoes erupt.

Besides silica content, the compositional factor that most influences eruptions is the gas content of the magma. The most abundant gas dissolved in magma is water, but magmas also contain carbon dioxide, sulfur dioxide, and smaller amounts of other gases. | Volcanic gases These gases (especially water) help lower viscosity (by reducing the tendency of SiO_4 ions to combine, as discussed in Section 4.2), and allow magma to migrate more easily toward the surface.

FIGURE 6-5 ▼ Shield Volcano
The repeated eruption of less-viscous mafic magma produces broad volcanoes with gentle slopes that resemble a shield in shape.

As magma rises, changes happen that can separate gases from the magma. The confining pressure probably decreases, the magma may partially crystallize (and thus hold less liquid), and the temperature may go down. All of these changes can cause gases to leave the magma. As gases are released, they can bubble through a less-viscous magma like fizz (carbon dioxide) in a bottle of soda. However, in viscous magma they may be trapped. The gases accumulate and build up gas (vapor) pressure. The gases that build up in magmas can be released rapidly, especially when the magma gets near the surface and the confining pressure of the surrounding rocks becomes much lower than the vapor pressure of the gases. Rapid release of gases from magma causes violent eruptions.

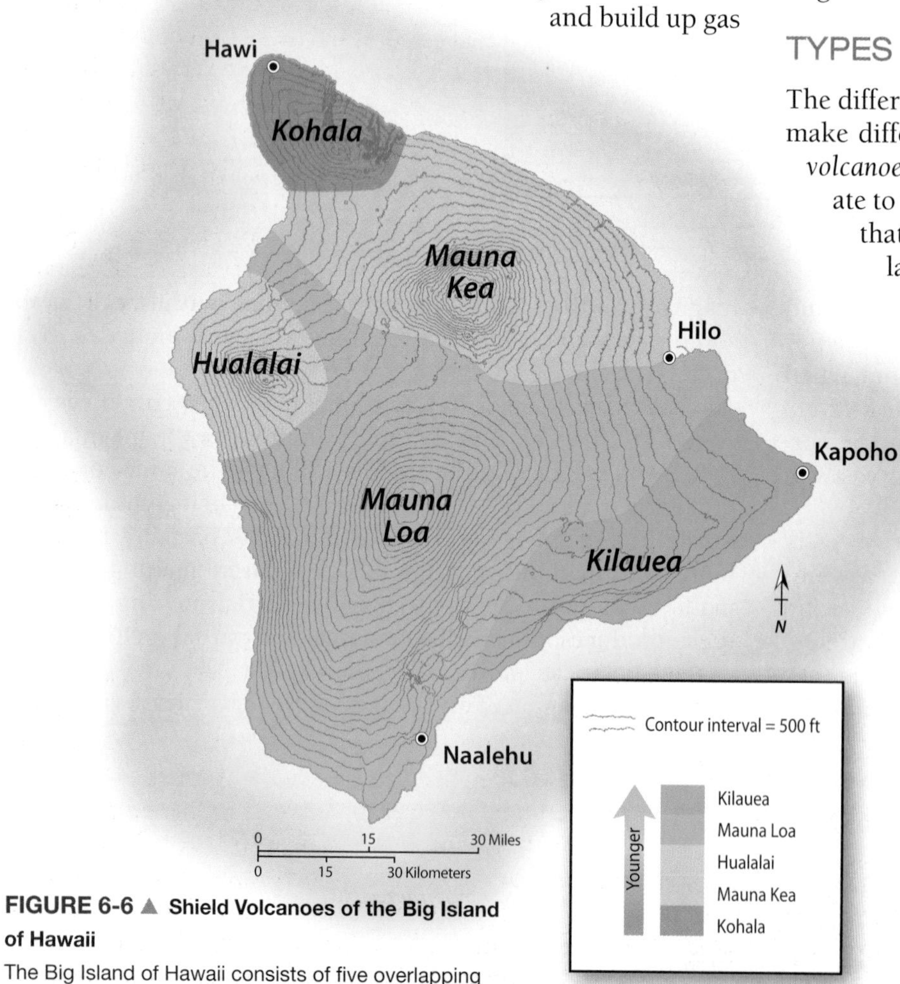

FIGURE 6-6 ▲ Shield Volcanoes of the Big Island of Hawaii
The Big Island of Hawaii consists of five overlapping shield volcanoes formed at different times over the last few hundred thousand years. Kilauea is the currently active volcano on the island.

TYPES OF VOLCANOES

The different types of magma and the different ways they erupt make different types of volcanoes. Mafic magmas form *shield volcanoes* and extensive areas flooded with basalt. Intermediate to felsic magmas form *stratovolcanoes* and a related type that has very large (caldera-forming) eruptions. The landforms developed by the different types of volcanoes are some of their most distinguishing features.

Shield volcanoes

Shield volcanoes, as you can see in Figure 6-5, are wide and gently sloping, with a shape similar to an upturned warrior's shield. They look the way they do because they are almost entirely composed of layer upon layer of solidified mafic (basaltic) lava flows. Recall that mafic magma is hot and less viscous, so it flows easily over long distances before cooling and solidifying. These volcanoes are therefore the largest on Earth.

As they gradually build over hundreds of thousands of years, shield volcanoes become enormous. The classic example is the island of Hawaii. It consists of five successively younger, overlapping shield volcanoes (shown in Figure 6-6). Measured from its submerged base on the seafloor, the island of Hawaii is the largest mountain on Earth. Its highest point, the volcano Mauna Kea, is nearly 10,200 meters (33,500 ft) above the seafloor, higher than the elevation above sea level of Mount

FIGURE 6-7 ▲ Lava Tube
Lava can flow long distances underground through lava tubes. This lava tube is near the summit of Kilauea Volcano on the Big Island of Hawaii.

Everest (8850 meters, or 29,035 ft). Hawaii is so massive that the seafloor crust actually sags under its weight. Iceland, like Hawaii, is composed of shield volcanoes.

Lava may erupt at shield volcanoes from long cracks, or **fissures**, as is the case with the ongoing eruptions at the East Rift Zone of Hawaii's Kilauea volcano. The construction of a shield volcano is also aided by the formation of **lava tubes** like the one shown in Figure 6-7. A mafic lava flow crusts over as it cools (see Figure 6-3). This crust acts as an insulating roof that allows the interior of the flow to remain hot and mobile. As a result, the lava can continue to flow for much longer distances before solidifying. When the eruption eventually stops, the lava in the interior of the tube keeps flowing and drains out, leaving a hollow tube behind. Lava tubes can run out for many kilometers from their source. Kazumura cave, an old lava tube on the flanks of Kilauea volcano on the island of Hawaii, is nearly 12 kilometers (7.5 miles) long.

Flood basalts

Mafic magmas that erupt from fissures on land can cover extensive areas with basalt. The mafic flows fill in low areas and valleys and gradually stack up over many thousands of years to form thick accumulations of **flood basalt**. Flood basalt doesn't build great mounds like shield volcanoes, but the amount of magma erupted can be tremendous. The largest in the United States is the Columbia River flood basalt, shown in Figure 6-8, which covers 130,000 square kilometers (50,000 mi^2) in parts of Washington, Idaho, and Oregon. Columbia River flood basalt province Some 100,000 cubic kilometers (24,000 mi^3) of basalt was erupted here about 15 million years ago.

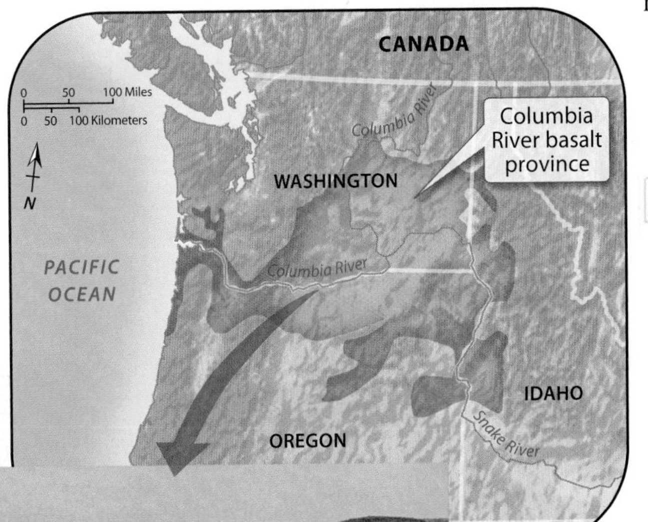

FIGURE 6-8 ▶ Columbia River Flood Basalt
Many lava flows erupted from fissures formed in the extensive Columbia River flood basalt province in adjacent parts of Washington, Oregon, and Idaho. Over 95% of these lavas were erupted between 17 and 15 million years ago. The photograph shows horizontal basalt flows along the Columbia River.

Cinder cones

Cinder cones are a common volcanic feature where mafic magma erupts. They form from mafic magma that is rich in gas and they erupt at shield volcanoes, flood basalt areas, and even stratovolcanoes.

The gas-rich mafic magma spews bits and pieces hundreds of meters (feet) up into the air. Sunset crater volcano lava flow trail, stop 5 The pieces quickly solidify and rain down as chunks of rock known as *cinders* and larger blocks called *lava bombs*. This loose debris falls down around the vent to form cone-shaped piles like the one shown in Figure 6-9 on p. 158. Cinder cone eruptions are not large or explosive enough to send magma flying farther than the immediate vicinity of the vent. And cinder cones are typically short-lived. They develop around a new vent, exhaust their magma supply, and shut down. The whole process may play out in a matter of weeks or a few years.

(a)

(b)

FIGURE 6-9 ▲ Cinder Cones
(a) Gaseous basalt magma spews lava short distances into the air. The quickly solidified magma (cinders) falls to the ground nearby to form cinder cones.
(b) This cinder cone in Lassen Volcano National Park, California partly covers older lava flows.

Paricutín is a cinder cone that erupted in a Mexican cornfield in 1943. The eruption began on February 20 in Mexico's Itzicuaro Valley, a flood basalt province that contains approximately 1000 cinder cones. Amazingly, local inhabitants witnessed the actual birth of Paricutín. A farmer, Dionisio Pulido, at first saw steam and ash rising from a hole in his cornfield. The ground swelled and later began to spew glowing rock into the sky. Within a day, the cone grew to a height of about 40 meters (130 ft). Within a week, it had piled up debris to a height of 167 meters (548 ft); after a year, 336 meters (1100 ft). Finally, by 1952, the cone reached a height of 424 meters (1390 ft), with a base diameter of about 1000 meters (3280 ft).

Meanwhile, lava had begun to erupt from Paricutín's base. Over time, an estimated 700 million cubic meters (900 million yd^3) of lava engulfed approximately 25 square kilometers (nearly 10 mi^2) of the countryside, including the nearby village of San Juan Parangaricutiro. In the end, only the town's church steeple peeked above the flows (as you can see in Figure 6-10). Paricutín has not erupted since 1952.

Stratovolcanoes

Stratovolcanoes like the one in Figure 6-11 take the shape of towering, steep-sloped, and frequently symmetrical mountains. They are also called "composite" volcanoes because they are constructed of complexly alternating layers of lava and other volcanic debris, as shown in Figure 6-12. The magma that builds stratovolcanoes is typically intermediate to felsic in composition but includes mafic magma, as well. Because of their height, many stratovolcanoes commonly develop a cap of glacial ice and snow and can be quite scenic. Japan's Mount Fuji, for example, depicted in an old Japanese wood-block print on p. 151, is a veritable icon of the snow-capped stratovolcano.

FIGURE 6-10 ▼ Burial of the Village of San Juan Parangaricutiro, Mexico
By the time basalt stopped erupting at Mexico's Paracutin volcano, the only thing visible in the village of San Juan Parangaricutiro was the church steeple.

FIGURE 6-11 ◀ **Beautiful but Dangerous**
Arenal Volcano in Costa Rica shows the classic stratovolcano profile. Stratovolcanoes are picturesque but potentially deadly, because they tend to erupt explosively.

However, the beauty of these volcanoes belies their danger. Because the magmas erupted at stratovolcanoes are viscous and gas-rich, they can erupt explosively. As the magma nears the surface, the dissolved gases come out of solution and form a froth of molten rock and trapped gas bubbles. The expanding bubbles provide the explosive force that ejects magma out of the vent. Stratovolcanoes can send a column of debris into the stratosphere, spreading a choking layer over thousands of square kilometers (miles), as shown in the opening photo of Mount Redoubt. Because of their potential for violent eruptions, stratovolcanoes are newsmakers. In the United States, they include Mount St. Helens and Mount Redoubt, as well as many others in the western states and Alaska.

Volcanic materials erupted into the atmosphere are called **pyroclastics**, from the Greek words for "fire" (*pyro*) and "broken" (*klastos*). (Another general term for material erupted into the atmosphere is **tephra**.) Pyroclastics consist of fragments of various sizes, including **ash** (small crystals, rock fragments, and bits of glassy frozen magma, Figure 6-13a on p. 160), cinders, **pumice** (glassy solidified magma that contains abundant gas bubbles, Figure 6-13b), and great blocks of rock that may, in the largest eruptions, be the size of automobiles or houses (Figure 6-13c). Explosive eruptions also hurl the glowing, semi-molten blobs that form cinders and lava bombs (Figure 6-13d)—an impressive sight at night as their glowing bodies trace graceful ballistic arcs through the sky (Figure 6-13e). The heaviest pyroclastics fall closest to the vent.

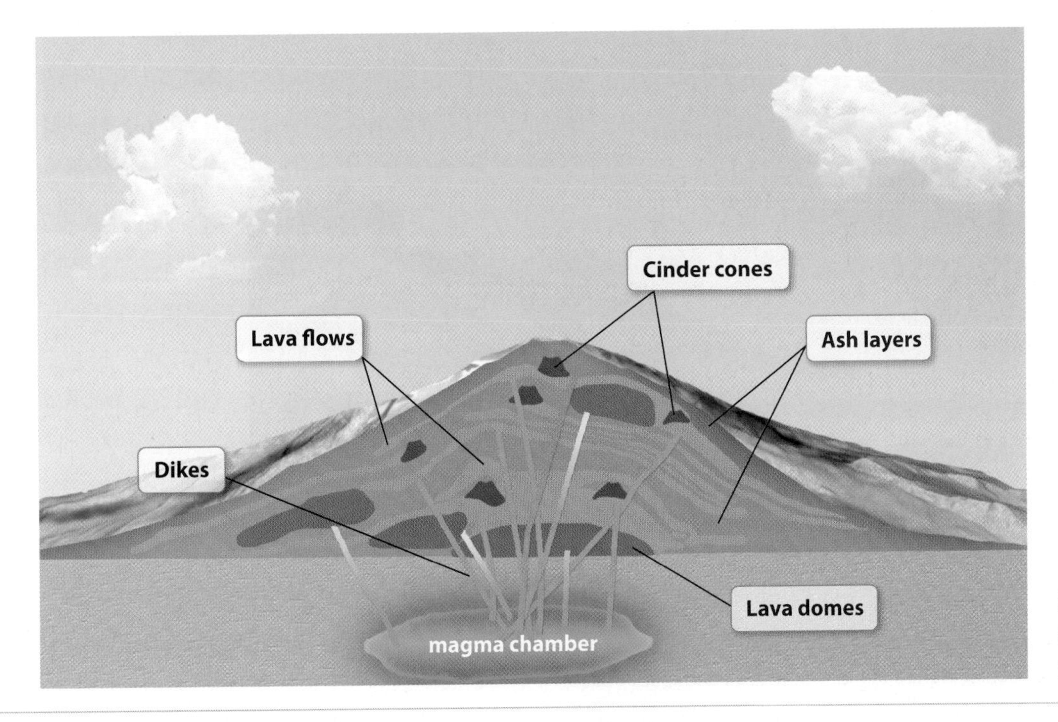

FIGURE 6-12 ◀ **The Anatomy of a Stratovolcano**
Stratovolcanoes are built up from alternating layers of lava, ash, and other volcanic debris.

FIGURE 6-13 ▲ Pyroclastics

Pyroclastic debris ranges from fine ash **(a)** to solidified magma with many holes created by gas bubbles, called pumice **(b)** to large blocks of rock **(c)**. Glowing lava bombs **(d)**, are a spectacular feature of many stratovolcano eruptions **(e)**.

FIGURE 6-14 ▲ Lava Dome
Lava domes are viscous and plug up vents. This lava dome formed on the floor of the Mt. St. Helens crater in 1981.

FIGURE 6-15 ▲ Crater Lake, Oregon, Fills a Caldera
A caldera-forming eruption created Oregon's Crater Lake about 7700 years ago.

Felsic magmas are too viscous to flow easily, so they tend to form mounds or masses, called **domes**, that plug volcanic vents. A typical example is shown in Figure 6-14. A dome grows by internal inflation as magma intrudes into its hot core, causing it to slowly expand upward and outward. After the explosive eruption of Mount St. Helens in 1980, a dome formed in the ruins of the volcano's summit. The dome reached a height of about 267 meters (876 ft) above the crater floor. If magma continues to inflate and build a dome, the dome may become unstable and collapse. Collapsing domes have triggered many violent eruptions as the vent "unplugs" and gas-rich magma is allowed to explode to the surface.

Large calderas

Majestic stratovolcanoes can be completely obliterated by *caldera-forming eruptions*, as occurred in the formation of Crater Lake in southern Oregon (Figure 6-15). **Calderas** are large, circular-to-oblong depressions that form when magma chambers erupt their contents and the volcanic mountain above them collapses into the empty **magma chamber**.

Caldera-forming eruptions can be huge. They expel tens to thousands of cubic kilometers (cubic miles) of ash that completely blankets whole regions. (Volcanic ash that becomes lithified is called **tuff**.) As you can see from Figure 6-16 on p. 162, caldera-forming eruptions can make the 1980 Mount St. Helens eruption look like a hiccup.

The most recent really big caldera-forming eruption in the United States was in the area of Yellowstone National Park. The 45 by 75 kilometers (28 by 47 mi) caldera formed here 640,000 years ago when the 1000 cubic kilometers (240 mi^3) Lava Creek Tuff erupted. As enormous as this eruption was, other caldera-forming eruptions have been even bigger. The Toba caldera on the island of Sumatra in Indonesia erupted 2800 cubic kilometers (670 mi^3) of tuff 75,000 years ago making it the largest volcanic eruption during the last 2 million years. Twenty-eight million years ago, the La Garita caldera in New Mexico erupted 5000 cubic kilometers (1200 mi^3) of tuff.

Caldera-forming eruptions must be devastating. They are capable of killing every living thing over large areas around them. Fortunately, the larger the eruption, the less frequently they occur. There hasn't been a really big caldera-forming eruption in recorded history. However, three large eruptions occurred at Yellowstone 2, 1.2, and 0.6 million years ago, about 600,000 to 800,000 years apart. It's been 640,000 years since the last one. Is another one brewing? In 2001, Yellowstone became a national volcano observatory. Scientists study and monitor the area to gain a better understanding of the relationships between surface features, such as its famous geysers and hot springs, and subsurface magma. Steam explosions, earthquakes, and volcanic eruptions—what's in Yellowstone's future? The next eruption may not be soon, but there's no reason to think it won't happen.

(a)

La Garita, Colorado
Ejected material = 5000 km³

Toba, Indonesia
Ejected material = 2800 km³

Yellowstone Huckleberry Ridge Tuff
Ejected material = 2500 km³

Yellowstone Lava Creek Tuff
Ejected material = 1000 km³

Mount St. Helens
Ejected material = 1 km³

(b)

Mount St. Helens Ash

Yellowstone Caldera

UNITED STATES

Huckleberry Ridge Tuff

0 250 500 Miles
0 250 500 Kilometers

FIGURE 6-16 ▲ Caldera-Forming Eruptions Can Be Huge
(a) The eruption at La Garita caldera, Colorado, 28 million years ago is the largest volcanic eruption known. **(b)** Ash from Yellowstone's 2.1-million-year-old Huckleberry Ridge eruption covered about a third of the United States.

The major types of volcanoes and their chief characteristics are summarized in Table 6-3.

ERUPTION MAGNITUDE

Eruptions unleash a tremendous amount of energy. One way to get a handle on the size of an eruption is to estimate how much material, including ash and solid debris, the volcano ejected during the event. The largest eruption in recorded history is the 1815 eruption of Tambora in Indonesia, which expelled an estimated 50 cubic kilometers (12 mi^3) of magma. CVO website—Tambora volcano, Indonesia We now know that there have been even bigger eruptions, in the range of hundreds to thousands of cubic kilometers (cubic miles) of erupted material, as shown in Figure 6-16.

The total volume of an eruption does not tell the whole story. How long it takes to erupt the material, and in what manner it is expelled, are other key factors. A single explosive event that lasts a few days is qualitatively different from a longer-lasting eruption that expels a much greater volume of material over a period of months or years. The violent destruction of Krakatau Island, Indonesia, in 1883 is a good example of a short-lived but devastating event. Before the eruption, the island consisted of a line of overlapping volcanic cones (Figure 6-17), with the high-

FIGURE 6-17 ▼ A Violent Eruption: Krakatau
In August 1883, a series of explosions blasted much of Krakatau island to bits, demolishing the 800- meter (2600-ft) Rakata volcano and creating a caldera 300 meters (1000 ft) below sea level **(a).** The shock waves from this cataclysm were still detectable after traveling seven times around the globe, and the sound was heard 4800 kilometers (nearly 3000 mi) away **(b).**

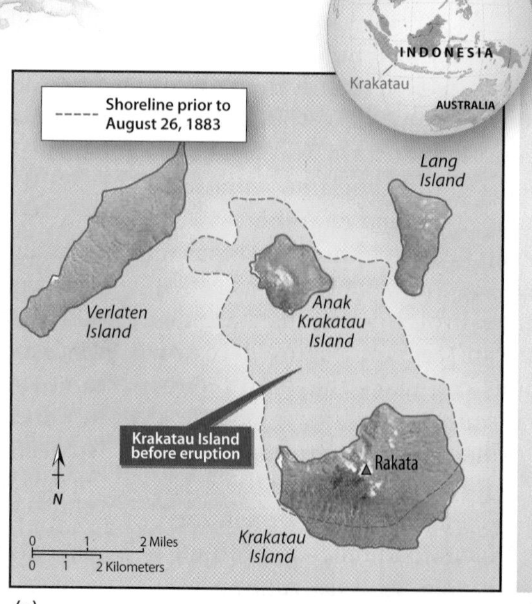

(a)

THE ILLUSTRATED LONDON NEWS

(b)

TABLE 6-3 TYPES OF VOLCANOES

Increasing Violence / Increasing Viscosity (arrow pointing down the table)

Volcano Type	Characteristics	Examples	Simplified Diagram
Flood Basalt	Lava; flows very widespread; emitted from fractures	Columbia, River Plateau	1 Mile
Shield Volcano	Lava emitted from a central vent; large; sometimes has a collapse caldera	Larch Mountain, Mount Sylvania, Highland Butte, Hawaiian volcanoes	1 Mile
Cinder Cone	Explosive lava; small; emitted from a central vent;	Mount Tabor, Mount Zion, Chamberlain Hill, Pilot Butte, Lava Butte, Craters of the Moon	1 Mile
Composite or Stratovolcano	More viscous lavas, much explosive (pyroclastic) debris; large, emitted from a central vent	Mount Baker, Mount Rainier, Mount St. Helens, Mount Hood, Mount Shasta	1 Mile
Volcanic Dome	Very viscous lava; relatively small; can be explosive; commonly occurs adjacent to craters	Novarupta, Mount St. Helens Lava Dome, Mount Lassen, Shastina, Mono Craters	1 Mile
Caldera	Very large composite volcano collapse after an explosive period	Crater Lake, Newberry, Kilauea, Long Valley, Medicine Lake, Yellowstone	1 Mile

est being Rakata at nearly 800 meters (2600 ft). After a period of escalating unrest, the island was virtually destroyed in a few days by a series of explosions and a caldera-forming collapse that triggered a tsunami responsible for killing about 36,000 people. |CVO website—1883 eruption of Krakatoa| In contrast, the volcanoes of Hawaii have gradually brought enormous amounts of magma to the surface in the form of lava flows. In recent history, these flows have destroyed valuable property but have caused relatively few deaths.

Given the diversity of eruptions, how can we compare them directly? Scientists have developed a relative scale, called the **Volcanic Explosivity Index** (VEI), to compare one eruption to another. |VHP photo glossary VEI| The VEI (Table 6-4, p. 164) assigns a volcano an index number from 0 to 8 based on the volume of material ejected by past eruptions, the height of the eruption column, the style of the eruption (lava flows versus explosive eruptions, for example), and how long the eruption lasted. However, some scientists think that using the mass (not the volume) of erupted material is the key to adequate comparisons. |How to measure the size of volcanic eruptions| Stay tuned: comparing volcanic eruptions is a work-in-progress.

THE BENEFITS OF VOLCANOES

Before discussing specific volcanic hazards, consider a few benefits of volcanoes. Their benefits are what lure people to live near or even on volcanoes. For example, volcanic activity is the ultimate source of valuable natural resources. Many valuable metal deposits (Chapter 12) are found in the roots of extinct volcanoes. People mine the solid debris that erupts from volcanoes, such as pumice and ash, for use as construction materials and abrasives.

Even to nearby communities threatened by hazards, volcanoes have value. Towering cone-shaped volcanoes such as Mount Fuji in Japan create breathtaking mountain scenery that draws tourists. Snow-covered volcanic slopes are the centerpieces of many ski resort areas, such as Mount Hood in Oregon. Geothermal springs in volcanically active areas are used as spas

TABLE 6-4 THE VOLCANIC EXPLOSIVITY INDEX

VEI	Eruption Frequency	Plume Height	Volume of Erupted Material	Comparable Volume (approximate)	Example
0	Daily	<100 m	<10,000 m^3	Large private house 2000 m^3	Mauna Loa (many) Hawaii, USA
1	Daily	100–1000 m	10,000–1,000,000 m^3	Lincoln Memorial 76,700 m^3	Stromboli (many) Italy
2	Weekly	1–5 km	1,000,000–10,000,000 m^3	New Orleans Superdome 3,500,000 m^3	Galeras (1993) Columbia
3	Yearly	3–15 km	10,000,000–100,000,000 m^3	Three Gorges Dam, China 27,200,000 m^3	Cordon Caulle (1921) Chile
4	≥10 Years	10–25 km	100,000,000 m^3–1 km^3	Mount Rushmore 200,000,000 m^3	Mount Peleé (1902)
5	≥50 Years	>25 km	1–10 km^3	Midtown Manhattan (to skyscraper height) 4 km^3	Mount St. Helens (1980) Washington, USA
6	≥100 Years	>25 km	10–100 km^3	Mount Everest (above local terrain) 20 km^3	Mount Pinatubo (1991) Philippines
7	≥1000 Years	>25 km	100–1000 km^3	Water in Lake Erie 545 km^3	Tambora (1815) Indonesia
8	≥10,000 Years	>25 km	>1000 km^3	Mauna Loa volcano, Hawaii 40,000 km^3	Toba (73,000 years ago) Indonesia

(a)

(b)

FIGURE 6-18 ◀ The Upside of Volcanism

(a) Yellowstone park's hot springs and geysers are heated by underground magma. **(b)** Soils derived from volcanic rock, typically rich and fertile, support crops such as wine grapes in the region around Italy's Mount Vesuvius and coffee in Guatemala (shown here).

in many places, and Yellowstone National Park's famous geysers are heated by an underlying magma chamber (Figure 6-18a). Hot rocks are sources of steam and electricity generation in volcanically active areas such as Iceland (Chapter 13).

Perhaps most valuable, weathering transforms ash and lava into fertile soil that supports agriculture in many lands. CVO: the plus side of volcanoes—fertile soil Rich soils on the volcanic flanks of Mount Vesuvius in Italy support flourishing gardens and vineyards compared to other parts of the region where much poorer soils are developed on limestone bedrock. The combination of rich volcanic soils and high elevations in the Andes Mountains of South America are ideal for growing coffee, as are the slopes of Central American volcanoes (Figure 6-18b). For proof that volcanoes are not all bad, look no further than your morning cup of coffee.

For all the hazards volcanoes present, people can be thankful for them. In a way, they are a renewable source of energy, useful materials, and rich soils. From this perspective, volcanoes are an important piece of the puzzle when it comes to developing a sustainable future.

What are some Earth systems interactions associated with the benefits of volcanoes?

6.2 Volcanoes: Where and Why

You could say it was a dark day for volcanology in cinema when the feature film "Volcano" was released in 1997. In keeping with the advertising tagline for the film, "The Coast Is Toast," this action-adventure extravaganza chronicled a volcanic eruption in downtown Los Angeles, complete with glowing lava flows, exploding lava bombs, and various other forms of destruction and mayhem (Figure 6-19, p. 166). However, the geologists in the audience probably spent a good bit of the film quietly chuckling. The destruction of Los Angeles by a volcano is not going to happen.

Why can't volcanoes form in Los Angeles? Plate tectonics provides a large part of the answer. In fact, most volcanoes, notably the numerous volcanoes ringing the Pacific Ocean (Chapter 3), directly result from the processes that occur at plate boundaries. In addition, some very significant volcanoes develop within plates, far from plate boundaries.

VOLCANOES AT DIVERGENT PLATE BOUNDARIES

You learned in Chapter 3 how scientists used volcanoes to help identify divergent plate boundaries and spreading centers at mid-ocean ridges. In terms of volume, divergent plate boundaries are where most of Earth's volcanism occurs. About 62% of

FIGURE 6-19 ▲ Hollywood Volcanism
In this 1997 film, a volcano erupts in Los Angeles. (The creative team working on the project probably didn't include a geologist.)

FIGURE 6-20 ▲ Volcanism at a Divergent Plate Boundary
A fissure eruption of basalt lava in Iceland. Iceland's shield volcanoes are part of a huge volcanic center that has formed near and along the Mid-Atlantic Ridge.

the magma erupted on Earth is at mid-ocean ridges. This magma is generated because mantle material decompresses as it rises below divergent plate boundaries. As the mantle material rises (remember that parts of the mantle can slowly flow due to density variations), the pressure decreases and so does the melting point. At a depth of 100 kilometers (62 mi), the temperature of the mantle is between 1200° and 1400° C (2200°–2550° F). That's pretty hot, but melting doesn't occur because the enormous pressures at these depths keep the melting point high. Eventually the material rises to lower-pressure regions, where the melting point is lower than the temperature of the mantle material, and melting begins. The resulting mafic magma, derived from partially melted ultramafic mantle rock (as previously described), is buoyant and wells up along the spreading plate boundary above.

There may be as many as 1 million submarine volcanoes, many of them scattered along mid-ocean ridges. These submarine volcanoes are for the most part "out of sight, out of mind." Iceland, a vast volcanic plateau astride the Mid-Atlantic Ridge, is covered with shield volcanoes whose eruptions frequently issue from great fissures (Figure 6-20). The large volcanic province here breaches the surface to form islands, but this is not typical of divergent plate boundaries.

VOLCANOES AT CONVERGENT PLATE BOUNDARIES

About 26% of the total magma erupted on Earth is generated at subduction zones, which are the seams along which oceanic lithosphere sinks into the mantle at convergent plate boundaries (see Chapters 3 and 5). As explained in Section 4.3, the descending plate carries with it water trapped in altered oceanic crust and sediments deposited on the seafloor. In the mantle, the oceanic plate recrystallizes and releases water and other volatiles, such as carbon dioxide, that rise into the overlying mantle and lower its melting point. As a result, the overlying mantle begins to melt. This process primarily produces mafic (basaltic) magma.

You also learned in Section 4.3 that the magma that erupts at subduction-zone volcanoes is primarily intermediate in composition (60% silica), making it an andesite. How does rising mafic (basaltic) magma become intermediate (andesitic), or even felsic, in composition? This probably occurs in several ways. The mafic magma may *differentiate* as silica-poor minerals like olivine and pyroxene crystallize first, leaving a greater proportion of silica behind in the melt. Alternatively, the heat of the rising magma may partly melt the surrounding crust, forming more felsic magmas with low melting points that may then be assimilated into the mafic magma. Either way or perhaps in both ways, these processes create viscous, gas-rich magmas that fuel explosive eruptions.

The processes that change magmas in the roots of stratovolcanoes also involve interaction with groundwater. A lot of the steam that accompanies explosive eruptions can originally be from groundwater that becomes heated and mobilized by near-surface magma. Groundwater is also the source of water in hot springs and geysers around volcanoes.

What are some Earth systems interactions associated with development of subduction-related volcanoes?

VOLCANOES WITHIN PLATES

Although most volcanic activity in the world occurs on or near plate boundaries, the globe is also peppered with volcanoes that are far from plate boundaries. The Hawaiian Islands, for example, lie within the Pacific Plate, far from any boundary. Yellowstone is within the North American Plate, far inland from the convergent boundary and Cascade Range volcanic arc of the Pacific Northwest. **Intraplate volcanoes** make up for their lack of numbers with size. They have erupted enormous amounts of magma.

The origin of intraplate volcanism is less well understood than that at divergent and convergent plate boundaries. One hypothesis is that intraplate volcanism is caused by local anomalies in the mantle called **hot spots**. Hot spots are places where voluminous mantle material rises and melts to form mafic magma. (For this reason, Iceland's huge volcanic plateau is likely due to hot spot processes rather than typical divergent plate boundary volcanism). Crust that becomes heated above the hot spots melts to produce intermediate and felsic magma.

A hot spot origin has been proposed for a very long chain of volcanoes that includes the Hawaiian Islands. These volcanoes, called the Emperor-Hawaiian Seamount Chain (Figure 6-21a), extend from near the Aleutian Trench in Alaska (Chapter 3) south to the Hawaiian Islands. The more than 80 volcanoes in this chain are progressively younger moving from north to south—from 80 million years old next to the Aleutian Trench to those that are active and growing in the Hawaiian Islands.

The change in age along the chain has been attributed to movement of the Pacific Plate over a fixed mantle hot spot, as shown in Figure 6-21b. In this interpretation, the dogleg in the chain provides evidence that the direction of Pacific Plate motion changed while the volcanoes were forming. The mantle hot spot just stayed put, as the lithosphere above migrated first northerly, then more westerly, above it.

Some scientists are now wondering if mantle hot spots are really that stationary and whether they represent deep or shallow mantle processes. Melting and magma movement in the mantle can be influenced by things other than the mantle's thermal structure. For example, mantle composition can influence melting. Some parts of the mantle may have larger amounts of lower-melting-point material (probably cycled into the mantle at subduction zones). These materials can melt even in the upper mantle, and they can produce large amounts of mafic magma. Scientists call this type of mantle "fertile." The thermal, compositional, and even structural character of the mantle may all play roles in developing hot spots.

Scientists will make major progress in understanding the mantle and how it works when they figure out hot spots. It's an

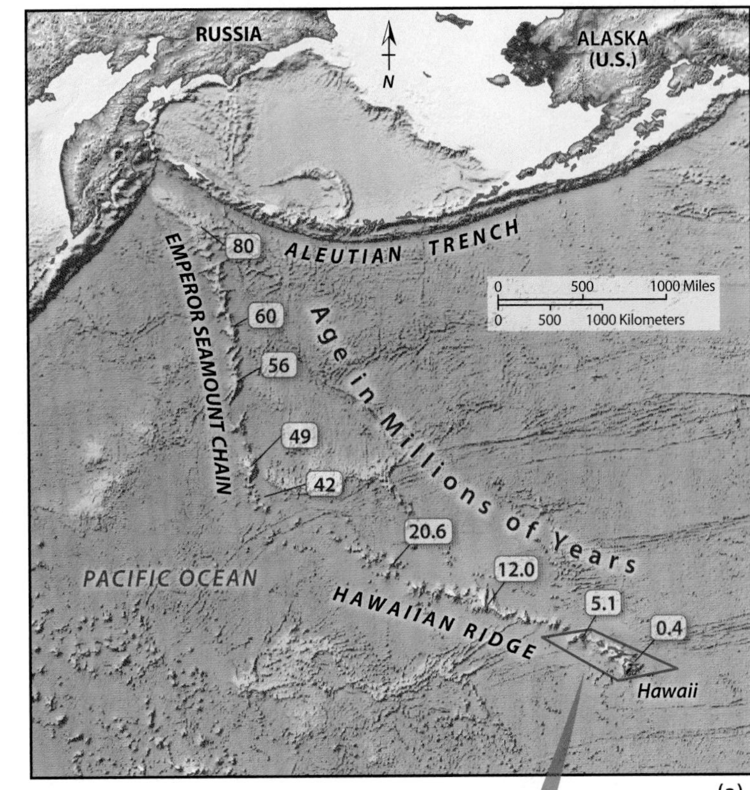

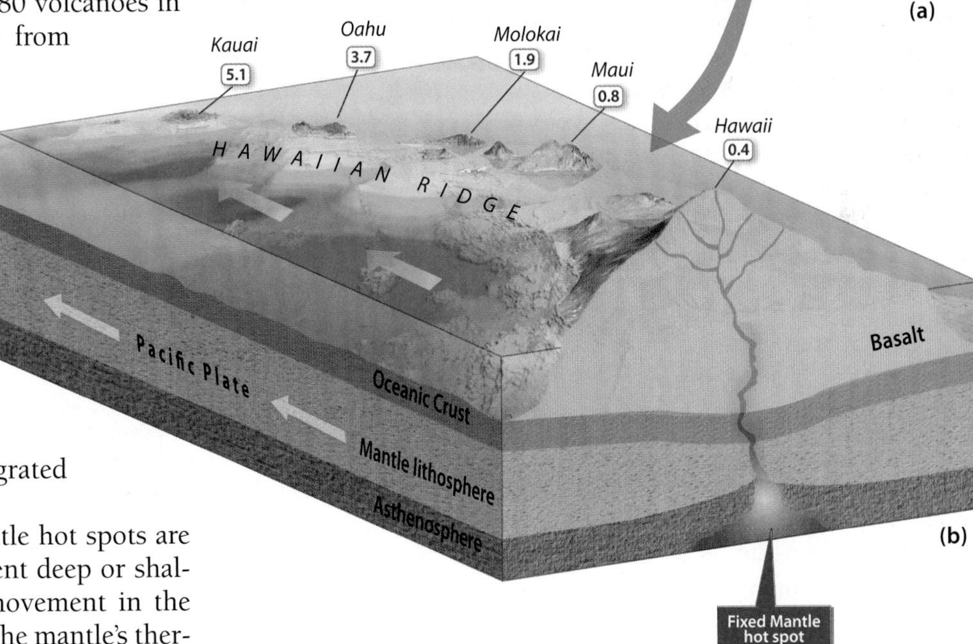

FIGURE 6-21 ▲ The Hawaiian Islands Are the Youngest in a Long Chain of Intraplate Volcanoes

(a) The progressive change in age of the Emperor-Hawaiian seamount chain—getting younger from north to south—has been interpreted to reflect the movement of the Pacific plate over a "fixed" mantle hotspot, as shown in part **(b)**. Recent research is suggesting that hotspots may not all be stationary.

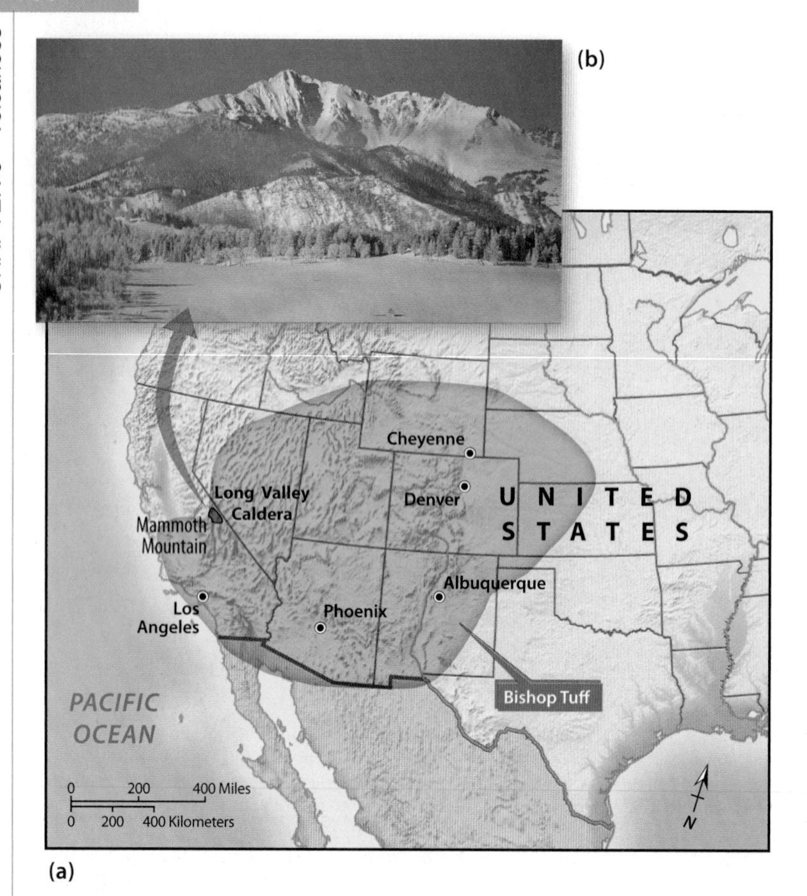

FIGURE 6-22 ▲ The Long Valley Caldera Erupted the Bishop Ash 760,000 Years Ago
(a) This volcanic eruption covered most of the southwestern United States, including Los Angeles, with ash. The continued volcanic activity, which produced Mammoth Mountain **(b)**, with its popular ski slopes, is still active and is being closely monitored by the U.S. Geological Survey.

exciting area of contemporary research. For us, the take-home message about hot spots is that they cause some of the world's largest volcanoes, from mafic ones like Hawaii to felsic ones like Yellowstone, and they mostly occur within tectonic plates, far from divergent or convergent plate boundaries.

The coast might be toast (someday)

After considering the processes that create magma, it should be clear why a major eruption in Los Angeles is unlikely. That city does not lie near a subduction zone or mid-ocean ridge. However, there is an intraplate volcano—a large caldera-forming volcano—along the California-Nevada border. The 15 by 30 kilometer (9 by 19 mi) Long Valley Caldera formed 760,000 years ago. As you can see in Figure 6-22a, the erupted ash covered most of the southwest United States, including the area that is now Los Angeles. Since then, continued volcanic activity has built Mammoth Mountain, a popular ski area for southern Californians (Figure 6-22b). The most recent small eruptions were 250 years ago, and earthquake swarms, hot springs, and doming within the caldera are evidence of continued volcanic activity. The USGS monitors the Long Valley Caldera because of its potential volcanic hazards. They don't want it to make anyone toast, even in the distant future.

What You Can Do

Investigate Volcanism in Your State

If you live in Nebraska, for example, you probably don't think volcanoes are very likely to disturb you. You're basically right when it comes to being threatened by active volcanoes, but this hasn't always been the case in your state. Volcanism has affected every part of America at some time in its geologic history. Do you want to learn when and how? Go to the USGS website "America's Volcanic Past." This site provides interesting summaries and descriptions of the volcanic events that have occurred in each state. America's volcanic past

This is a great way to start your own research about volcanoes. For one thing, you might be able to take a Sunday drive to check out volcanic rocks near where you live. You may be surprised by how widespread they are and how common volcanism has been in Earth's history.

USGS/Cascades Volcano Observatory, Vancouver, Washington

America's Volcanic Past
Cascade Range

"Though few people in the United States may actually experience an erupting volcano, the evidence for earlier volcanism is preserved in many rocks of North America. Features seen in volcanic rocks only hours old are also present in ancient volcanic rocks, both at the surface and buried beneath younger deposits." -- *Excerpt from: Brantley, 1994*

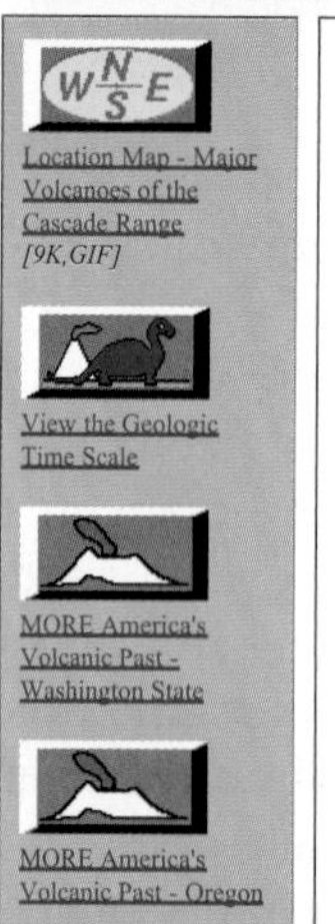

Volcanic Highlights and Features:

[NOTE: This list is just a sample of various Cascade Range features or events and is by no means inclusive. All information presented here was gathered from other online websites and each excerpt is attributed back to the original source. Please use those sources in referencing any information on this webpage, and please visit those websites for more information on the Geology of the Cascade Range.]

- Cascade Range
- Pacific Mountain System
- Cascade Range Vulcanism
- Major Cascade Range Volcanoes
- Cascade Range National Parks

Cascade Range

The Cascades arose through the plate collisions that have enlarged the western continent in Tertiary-to-Holocene time. This mountain range contains such large and geologically recent active volcanoes as Rainier, Hood, and Shasta.

Major Volcanoes of the Cascade Range
[Map,10K,InlineGIF]

http://vulcan.wr.usgs.gov/LivingWith/VolcanicPast/Places/volcanic_past_cascade_range.html

6.3 Volcanic Hazards

It began on June 8, 1783, like some biblical cataclysm. The ground near Laki, Iceland, cracked open along a 32-kilometer (20-mi) fissure and lava poured forth. Lava continued to erupt through a number of fissures for eight months, covering an area of 565 square kilometers (218 mi^2). The eruption released toxic gases that caused crop failures and livestock deaths, triggering a famine that killed nearly 10,000 people (about one-quarter of Iceland's population at the time). The Laki fissure erupted nearly 15 cubic kilometers (3.6 mi^3) of lava and ash. That's about the size of 15,000 Empire State Buildings.

The destructive potential of major eruptions is sobering. According to a study by the volcanologist R. J. Blong, 564,612 people died in volcanic eruptions from 1600 to 1986. This includes eruptions at Laki, Iceland. The 1813 eruption of Tambora volcano in Indonesia took an even worse toll, around 80,000 lives lost to starvation and disease when ash from the eruption killed off forests. More than 23,000 people perished in 1985 from a mammoth volcanic debris flow unleashed by Nevado del Ruiz volcano in Colombia. About 300 people died in the 1991 eruption of Mount Pinatubo in the Philippines.

High body counts don't tell the whole story, however. Volcanic eruptions can also inflict great damage to the expensive infrastructure of society, destroying roads, bridges, agricultural fields, and homes. Learning to live more safely with volcanoes is a step toward living more sustainably by protecting against devastation and preserving valuable human and economic resources. Volcanoes will continue to erupt but if people learn their lessons well, death and destruction will diminish.

Our lessons come from past eruptions. The goal is to untangle all of the strands of evidence and build up a coherent picture of what the volcano did in past eruptions so that we can anticipate what it may do in the future. As noted earlier, we can trace the general level of hazard posed by a volcano to the composition of its magma. Contrasting the typical hazards of shield and stratovolcanoes illustrates this idea very clearly.

HAZARDS OF STRATOVOLCANOES

When a stratovolcano erupts explosively, the driving force is the gas dissolved in the magma. A typical eruption sequence might unfold this way:

- Hot, gassy magma rises into the roots of the volcano.
- A blockage below the vent prevents gas boiling out of the magma from escaping. The pressure builds and the volcano adjusts; small earthquakes and surface doming are evidence of the trapped magma.
- The building pressure overcomes the strength of the confining rocks and in a violent, throat-clearing explosion, opens the vent and discharges a mass of debris and ash.
- A mix of hot gases and ash rises; its internal heat enabling it to climb higher and higher. Riding high-altitude winds, ash in this eruption column can travel for hundreds or even thousands of kilometers.

The hazards of such an eruption are many. They put all living things near the volcano at risk, can affect large parts of continents, and can even change global climate for a while.

Ash hazards on the ground

Ash can smother whole communities. The eruption of Italy's Vesuvius volcano in 79 A.D. covered the city of Pompeii to a depth of 3 meters (10 ft) in ash and pumice (Figure 6-23a). The buried city and its ruined surroundings were lost until the 1700s, when they were rediscovered along with casts of its citizens entombed by the ash (Figure 6-23b).

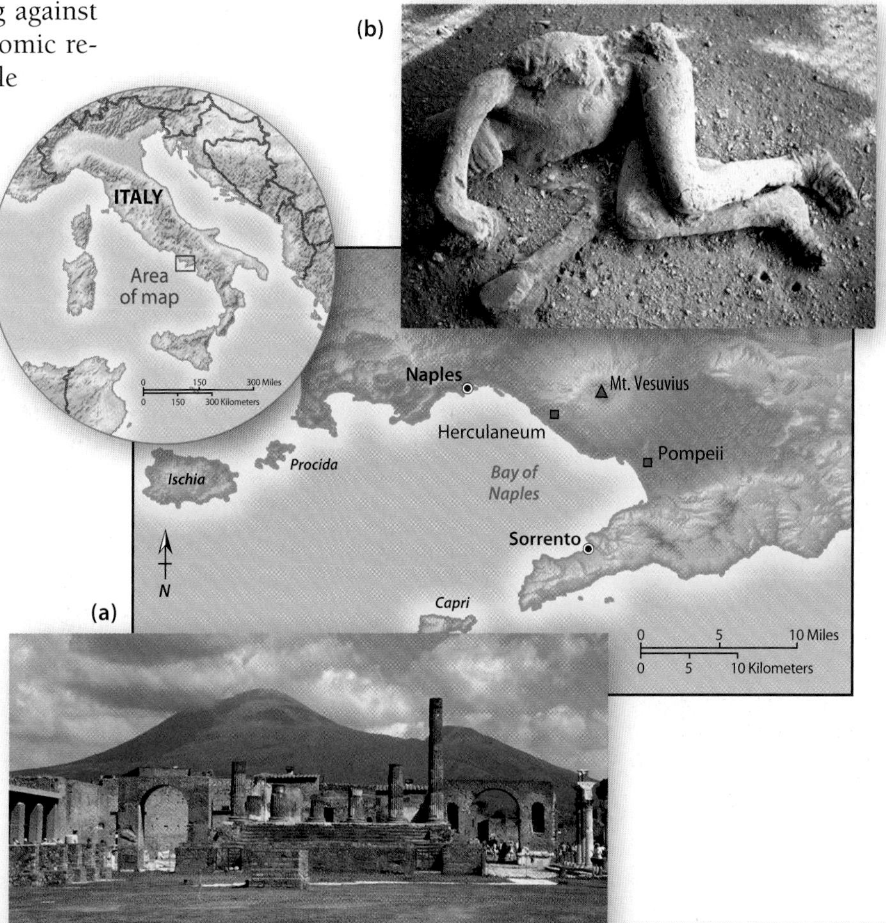

FIGURE 6-23 ▶ Ghosts of Pompeii
(a) The eruption of Mount Vesuvius completely buried the Roman town of Pompeii and its inhabitants in 79 A.D. **(b)** When the bodies of the victims decayed, they left cavities in the deep layers of ash. By pouring plaster into these spaces, investigators were able to produce lifelike casts of the dead Pompeiians.

FIGURE 6-24 ▲ The Eruption of Mount St. Helens, Washington
Ash, steam, water, and debris were blasted to a height of 18,300 meters (60,000 ft) during the May 18, 1980 eruption of Mt. St. Helens.

In more recent history, people in the United States experienced the far-flung effects of volcanic ash. | Volcanic hazards: Tephra, including volcanic ash | When Mount St. Helens erupted on May 18, 1980, an ash column rose to an altitude of more than 18 kilometers (11 mi) (Figure 6-24). It blanketed downwind areas in Washington with several centimeters (a few inches) of ash (Figure 6-25a), and the next day some had reached as far as the Dakotas. The ash turned the sky dark, and when mixed with rain it made roadways treacherous.

Volcanic ash can kill even after it settles on the ground. After the 1991 eruption of Mount Pinatubo in the Philippines, hundreds died when rain-soaked ash caused the roofs of many houses and other structures to collapse (Figure 6-25b).

Ash hazards in the atmosphere

Erupted ash can travel through the atmosphere for days and journey around the world. It's not precisely clear when such long-lived plumes cease to be hazardous to aircraft. We only know that pilots should avoid them.

For people in North America, the North Pacific rim is where ash is most hazardous to aircraft. Air routes there ferry cargo and people across the ocean or over the pole to Europe. The region is a veritable shooting gallery. Batteries of active volcanoes sit on all sides: those of Russia's Kamchatka Peninsula on the west, those of the Aleutian Islands to the north, and those of the Cascade Range to the east (Figure 6-26). It is not unusual for several volcanoes to erupt every year in this very large region.

Aircraft engines aren't the only things that shouldn't breathe volcanic ash; people shouldn't either. Ash is full of glass shards (quickly solidified magma) and tiny, sharp crystals (Figure 6-27). People generally know that it's not healthful to stand around in a cloud of falling volcanic ash. What is less well understood are the hazards associated with cleaning up after an ash fall. The abrasive ash gets stirred up by sweeping, shoveling, gutter cleanout, or just by driving a car along the road. This "secondary" ash is just as potentially harmful to people and machines as the original ash fall.

(a)

(b)

FIGURE 6-25 ▲ Volcanic Ash
(a) Ash from the eruption of Mount St. Helens blanketed areas downwind of the volcano. **(b)** After the eruption of Mount Pinatubo, roofs collapsed under the weight of rain-soaked ash, killing many people.

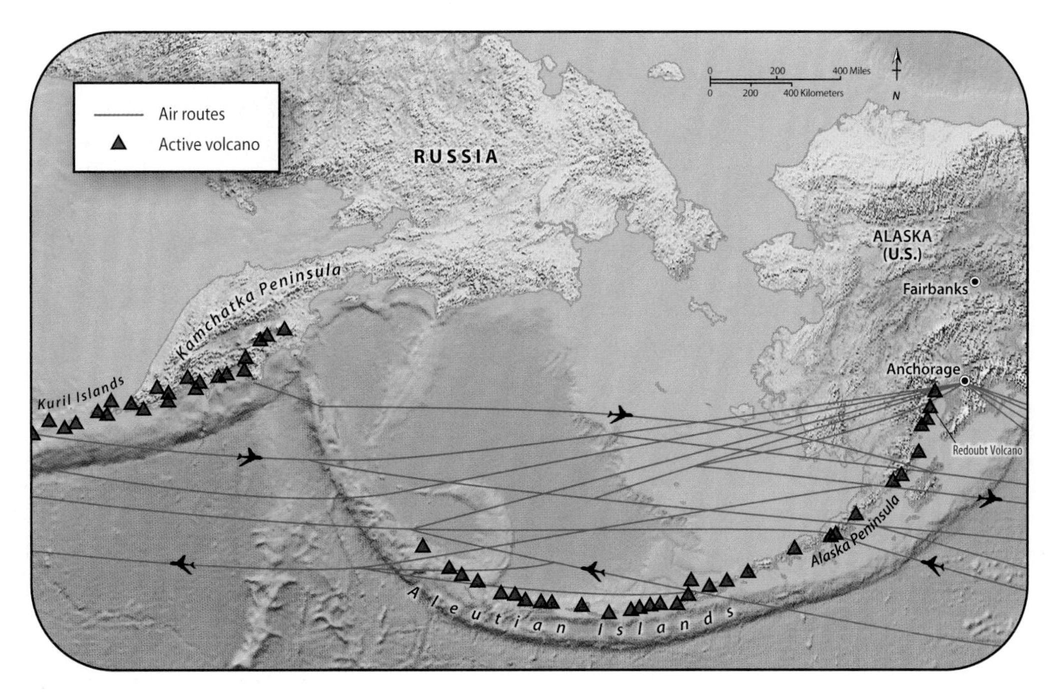

FIGURE 6-26 ▲ The Volcano Shooting Gallery
Aircraft flying North Pacific and Russian Far East air routes, carrying more than 10,000 passengers and millions of dollars of cargo each day, pass over or near more than a hundred potentially active volcanoes.

Pyroclastic flows

Pyroclastic flows are among the most deadly and destructive hazards of stratovolcanoes. Volcanic hazards: pyroclastic flows and surges In a pyroclastic flow, like the one shown in Figure 6-28, hot ash and other solid debris cascade down the flanks of the volcano, smashing or incinerating everything in its path. This deadly material moves very rapidly, up to 160 kilometers/hour (100 mph), and quickly overwhelms its victims before they can flee. Pyroclastic flows tend to follow the low areas and valleys around the volcano. The largest single loss of life in the 20th century during an eruption was caused by a 1902 pyroclastic flow at Mount Pelée, a volcano on the Caribbean island of Martinique. In a matter of minutes, the flow virtually destroyed the city of St. Pierre and killed an estimated 28,000 people (Figure 6-29 on p. 172).

A common cause of pyroclastic flows is gravitational collapse of the column of ash, debris, and hot gas blasted into the air above a stratovolcano. The hot ash and gas in the column rise until they reach cooler altitudes and lose buoyancy. Then the column falls back in on itself and spreads out on the ground as a pyroclastic flow.

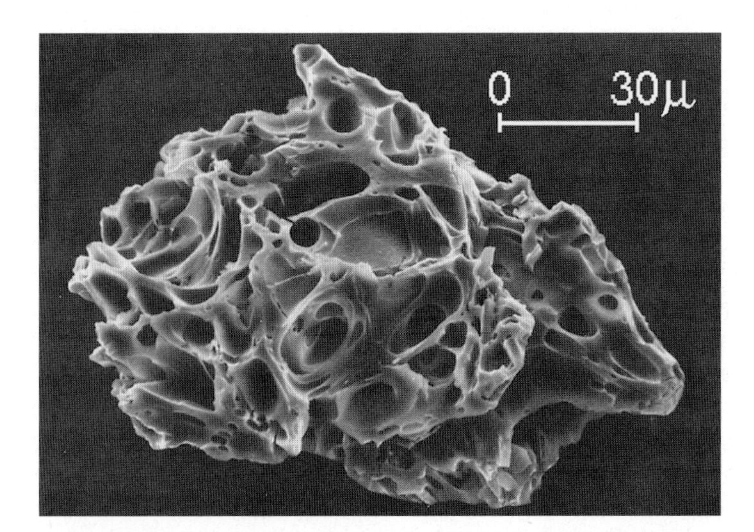

FIGURE 6-27 ▲ Unhealthy to Breathe
Volcanic ash contains tiny particles of glassy solidified magma (shown here in a scanning electron micrograph) and small rock and crystal fragments. The size of this particle is about 0.01 centimeters (0.004 in).

FIGURE 6-28 ▲ Pyroclastic Flow
Pyroclastic flows, like this one produced by the eruption of the Soufriere Hills volcano on Montserrat Island in the Caribbean, consist of hot gases, ash, and debris. Extremely destructive, they can move with deadly speed.

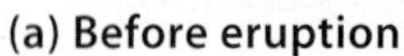

(a) Before eruption

(b) After eruption

FIGURE 6-29 ▲ Before and After at St. Pierre, Martinique
(a) St. Pierre was a bustling seaside town on the Caribbean island of Martinique before the 1902 eruption of nearby Mount Pelee. **(b)** The 1902 eruption generated a pyroclastic flow and surge that essentially obliterated the town, killing all but two of its 28,000 inhabitants. (One of the known survivors was a man lucky enough to be imprisoned in a deep, dungeon-like jail cell.)

Pyroclastic flows also occur when the parts of a volcano confining a shallow magma chamber fail. In these cases, the side of a volcano may collapse, releasing pressure on magma and initiating a lateral (sideways) pyroclastic eruption. The 1980 Mount St. Helens eruption is an example. Collapse of a dome plugging a vent can trigger pyroclastic flows in a similar way. In some cases, the volcano simply spills ash and gas over the rim of the volcanic vent, like a pot of rice boiling over.

Pyroclastic flows are heavy. Large flows can snap trees like twigs and smash masonry walls up to a meter thick. A pyroclastic flow unleashed during the 1991 eruption of Mount Unzen, Japan, threw a car 120 meters (390 ft). The added hazard in a pyroclastic flow is heat. Indeed, an older term for a pyroclastic flow is *nuée ardente*, French for "glowing cloud." Material within the flow can be hotter than 800° C (1500° F). Pyroclastic flows kill their victims chiefly by burning and asphyxiation even before burying them in debris.

Pyroclastic flows are accompanied by an envelope of hot gas and ash called a *pyroclastic surge*. The heaviest material in the flow settles to the bottom, and because of its weight it tends to hug the ground along valley floors and other low areas. This dense, ground-hugging mass of ash, cinders, and blocks of rock is called the *base flow*. The pyroclastic surge enveloping the base flow is lighter and behaves more like a fluid.

Surges are especially dangerous because they can detach from the main flow and continue to advance, flowing up—and sometimes over—ridges and valley walls. It was, in fact, a pyroclastic surge that destroyed St. Pierre and killed all but two of its 28,000 inhabitants (see Figure 6-29). In 1991, a pyroclastic surge killed 43 people—including volcanologists—observing the eruption of Mount Unzen in Japan.

Lahars

Another common and extremely deadly hazard of stratovolcanoes is a **lahar**, an Indonesian word for a wet debris flow (like the one shown in Figure 6-30) originating on the flanks of a volcano. The flows are a slurry of ash, lava debris, and water, as well as any other material such as soil, rocks, and trees that the flow picks up on the way down the slope. Gravity is the source of the destructive force of lahars, and because stratovolcanoes are commonly high in elevation, lahars pick up speed and momentum as they flow down their flanks.

Lahars flow when water mixed with loose pyroclastic material becomes unstable. To understand a lahar, imagine a muddy avalanche as thick as wet concrete, containing a jumble of rock, ash, soil, and trees (Figure 6-31). Because they are so dense and massive, lahars can bulldoze and bury entire towns. In 1985, a lahar at Nevado del Ruiz volcano in Colombia took a terrible toll of life. Deadly lahars from Nevada del Ruiz, November 13, 1985 A relatively small eruption melted snow and ice around the peak of this volcano. The water that was released saturated loose flank debris and triggered lahars that destroyed more than 5000 homes and killed more than 23,000 people. Worst hit was the town of Armero, 72 kilometers (45 mi) from the volcano. Much of the town was buried and approximately three-quarters of its residents lost their lives. Lahars can kill long after a volcanic eruption. All that is needed is loose flank debris and water, even water from heavy rains. After the 1991 eruption, lahars continued to flow down the sides of Mount Pinatubo for years, burying whole villages.

The hazards associated with stratovolcanoes are summarized in Figure 6-32.

FIGURE 6-30 ▲ A Lahar
Snow and ice melted by an eruption—even heavy rain—can send wet volcanic ash and rock cascading down the slopes of a volcano. Such lahars can carry with them soil, tree trunks, even boulders.

FIGURE 6-31 ▼ Lahar Damage
Lahars, which can be as dense as wet cement, buried this car on the slopes of Mount St. Helens.

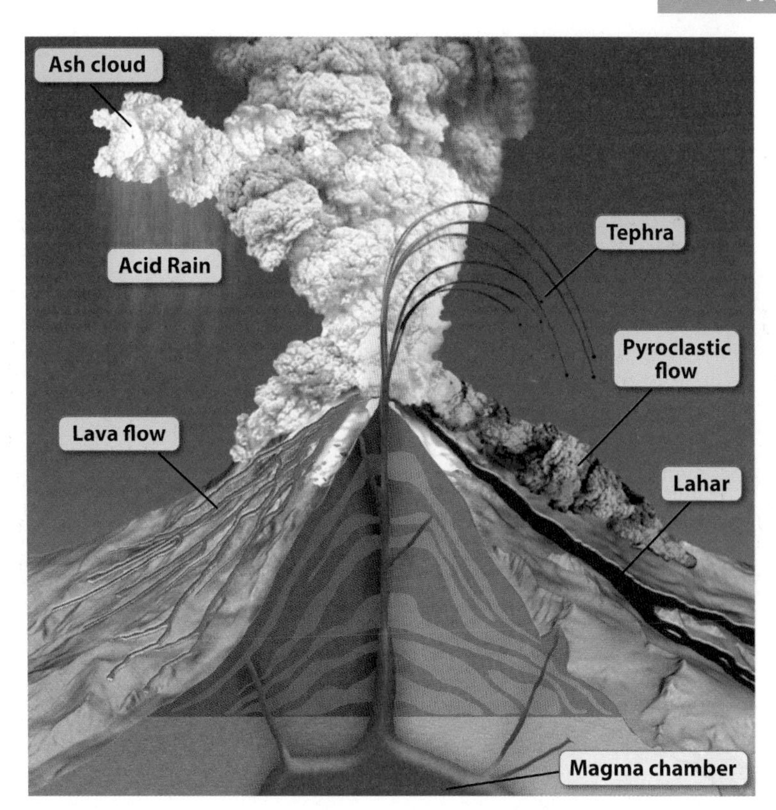

FIGURE 6-32 ▲ Stratovolcano Eruptions Can Create Many Hazards

What are some Earth systems interactions associated with lahars?

A stratovolcano in action: Mount St. Helens

The eruption of Mount St. Helens was the first major eruption in the Cascade Range since a series of explosions rocked California's Mount Lassen in 1914–1917. Even though Mount St. Helens was the worst volcanic disaster in U.S. history, the volcano's hazards were well understood and scientists and officials prevented much loss of life.

Some Native American tribes in the Pacific Northwest call Mount St. Helens Louwala-Clough, or "smoking mountain." Its symmetrical, snow-capped cone was nicknamed America's Mount Fuji. Canadian artist Paul Kane painted Mount St. Helens erupting a plume of ash in 1847, during a period of intermittent unrest that ceased in 1857. The volcano remained inactive for 123 years, until a magnitude 4.2 earthquake on March 20, 1980, signaled that the mountain had awakened. The earthquakes meant that new magma was refilling the volcano's plumbing system.

Within two weeks of the start of unrest at Mount St. Helens, scientists from the USGS and the University of Washington were closely monitoring the volcano and working to determine the potential hazards. On March 26, officials from the U.S. Forest Service and the county and state governments agreed to close access to the area surrounding the volcano to the general public. The following day, at 8 a.m., the USGS issued an official hazard watch. At 12:36 p.m., the volcano obliged by hurling ash and steam into the air as a crater was blasted into the summit. These small eruptions continued intermittently as hot rock in the volcano flashed groundwater into steam, chipping away at the dome of lava plugging the throat of the volcano.

Through April, the mountain shook with hundreds of small earthquakes caused by magma movements in the volcano's roots. The unrest stopped, started again on May 7, quieted down, then resumed May 16. By this time the north face of the volcano had developed a clearly visible bulge, distended hundreds of meters outward from the original slope. Measurements taken with laser beams showed that the bulge was growing at a rate of 1.5 meters (4.9 ft) per day, a clue that pressure was building inside the volcano. Small steam and ash eruptions continued to evidence the volcano's unrest. Scientists discussed the possibility that the bulge could collapse, initiating a violent eruption. However, they could not be sure it would actually happen, or how soon. As a precaution, officials set up roadblocks to control access to the hazard zone.

On the morning of May 18, USGS volcanologist David Johnston was at an observation station on Coldwater Ridge about 10 kilometers (6 mi) north of Mount St. Helens when the distended north side finally collapsed. The last his colleagues heard from Coldwater Ridge was Johnston's voice on the radio, "Vancouver, Vancouver, this is it!"

The bulge collapsed and the resulting landslide uncorked the volcano's pent-up pressure (Figure 6-33). A powerful lateral blast of ash and debris surged north. This "stone wind" fanned out as far as 30 kilometers (19 mi), devastating 590 square kilometers (230 mi^2) of the surrounding forest as it snapped stout old-growth trees like matchsticks (Figure 6-34a). The blast took the lives of David Johnston and several others. Before the eruption was over, 56 people and nearly 7000 deer, elk, and bear perished. In all, about 1.25 square kilometers (0.3 mi^3) of material was erupted by the volcano.

Within 10 minutes of the blast, an ash column rose more than 19 kilometers (12 mi) into the atmosphere (Figure 6-24). The cloud darkened the skies downwind and traces of it reached the East Coast within two days. During nine hours of eruption, about 500 billion kilograms (550,000,000 tons) of ash fell on an area of 56,400 square kilometers (21,800 mi^2). When the most violent phase of the eruption ceased the next day, an amphitheater-shaped crater was all that remained of the mountain's formerly majestic peak. A full 400 meters (1300 ft) was missing from the summit. Fluidized by melted snow and ice, portions of the landslide debris flowed down the flanks of the volcano as dense lahars (Figure 6-34b). One lahar advanced 21 kilometers (13 mi) down the North Fork of the Toutle River.

Right up to the moment of the eruption, it remained unclear when or if the volcano would erupt. Before the May 18 blast, the rate of earthquakes, bulging of the slope, and release of sulfur dioxide gas all increased steadily. However, scientists did not see a sudden or unusually sharp upturn in these precursors, which would have indicated an imminent eruption. Still, the warnings and restricted access to the volcano undoubtedly saved many lives.

Luck played a significant role, too. Had the eruption occurred the next day, rather than on Sunday, many more people would have been in the kill zone, including several hundred loggers. The direction of the wind, which blew much of the blast debris away from the most populous areas of Vancouver and Portland, also helped mitigate the loss of life and damage. Nevertheless, the economic impact of the eruption was pegged at more than $1.1 billion (in 1980 dollars).

(b)

(a)

FIGURE 6-33 ◀ The 1980 Eruption of Mount St. Helens
(a) Mount St. Helens is located in the southwestern corner of Washington state. **(b)** The volcano's north flank collapsed in less than a minute to trigger the May 18, 1980 eruption.

(a)

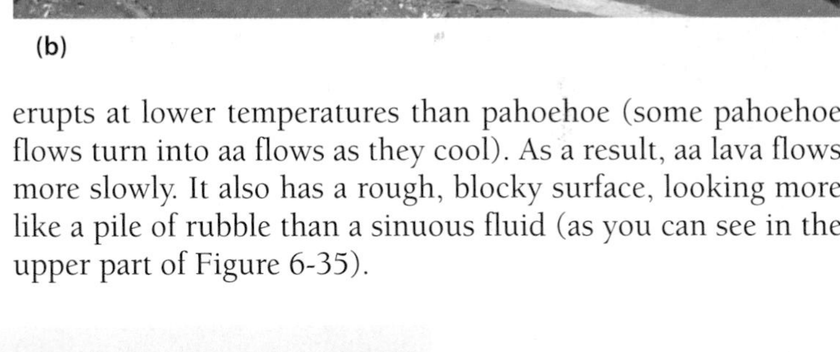

(b)

FIGURE 6-34 ◀ After the Eruption
(a) Surrounding forests were devastated by the 1980 eruption of Mount St. Helens. Over four billion board feet of usable timber, enough to build 150,000 homes, was damaged or destroyed. **(b)** Lahars severely damaged property and stream drainages. The lahar that devastated the Muddy River drainage carried this huge boulder along with it.

HAZARDS OF SHIELD VOLCANOES

Compared with stratovolcanoes, shield volcanoes and other places mafic lava erupts seem relatively harmless, but they present hazards as well. Lava flows are the main hazard of shield volcanoes.

Lava flows

Lava flows downhill, but how quickly it flows depends on its viscosity. The hottest mafic lavas flow the fastest. Lava of this kind is referred to as **pahoehoe** (pronounced pa-hoy-hoy). When they crust over and harden, pahoehoe lava flows assume a variety of forms. They may be smooth, ropey, or wrinkled (as in the lower part of Figure 6-35). Another type of mafic lava is called **aa** (pronounced ah-ah). This lava is more viscous and erupts at lower temperatures than pahoehoe (some pahoehoe flows turn into aa flows as they cool). As a result, aa lava flows more slowly. It also has a rough, blocky surface, looking more like a pile of rubble than a sinuous fluid (as you can see in the upper part of Figure 6-35).

FIGURE 6-35 ◀ Pahoehoe and Aa Lava Flows
Pahoehoe flows are fluid and develop smooth, hummocky, or ropy surfaces when they solidify. Aa flows, which are lower in temperature and more viscous than pahoehoe, develop a rough rubbly surface of broken lava blocks. Here an aa flow is advancing over a previously solidified pahoehoe flow. (See also Figure 6-3 for another view of a pahoehoe flow.)

Lava flows are hazardous because they are extremely hot and almost unstoppable as they move downslope. | Volcano hazards: lava flows | Thus, lava flows burn and engulf objects in their path. In many cases, lava moves slowly enough for people to move out of the way. Flows may creep along at a snail's pace of a meter or so per hour. If hot fluid lava is flowing from a volcano at a high enough rate, however, it might move faster than a person can run. During an eruption at Hawaii's Mauna Loa volcano in 1950, lava flows were clocked at about 16 kilometers per hour (10 mph) through thick forest. Fast lava flows can be deadly. In 1977, a lava flow moving down the slope of Nyiragongo volcano in the Congo at 15 kilometers per hour (9.3 mph) engulfed a town, killing 70 people.

Shield volcanoes in action: Kilauea

Kilauea is one of the five overlapping shield volcanoes that make up Hawaii's "Big Island." It occupies the southeast corner of the island, overlapping the much larger and older Mauna Loa volcano (see Figure 6-6). The ongoing eruption of Kilauea began in January 1983. As of September 2002, new flows covered 110 square kilometers (42 mi^2) of the southeast side of the volcano, extending all the way to the ocean. The flows destroyed 189 houses and other structures, including a Volcano National Park visitor's center, but caused no deaths.

The current unrest is typical of the volcanic activity that has built Hawaii for hundreds of thousands of years. Kilauea, like the other volcanoes of the island, consists of stacks of lava flows interlayered with small amounts of chunky debris blasted into the air by more explosive gas-rich eruptions. Magma rises from deep sources into a shallow reservoir inside Kilauea, breaks out at various vents for a time, and then subsides until the magma supply is replenished. There have been dozens of these eruptive episodes since 1983.

Eruptions have occurred at three locations, shown in Figure 6-36. One, the summit crater, is a round depression at the highest point on the Kilauea shield, although not the highest point on the island. Sweeping downslope and away from the summit are two rift zones. The East Rift Zone, dotted by about a dozen active or formerly active craters and cones, has been the focus of most of the activity since 1983. The other area of activity is the Southwest Rift Zone.

One of the craters of the East Rift Zone, Pu'u 'O'o, Hawaiian for "Hill of the O'o Bird," staged spectacular displays of lava fountaining for the first three years of the eruption. For a day at a time, lava sprayed as high as 470 meters (1540 ft) into the air. This gradually built a cone of cinders and splattered lava some 255 meters (837 ft) high. To add to the drama, Mauna Loa volcano erupted in March 1984 with a large lava flow that headed for, but stopped short of, the city of Hilo.

The flows originating from the Pu'u 'O'o crater were the more viscous and blocky aa type (as opposed to more fluid pahoehoe). They moved downhill at maximum speeds of 500 meters (1640 ft) per hour, a slow crawl at best. However, houses and roads, unlike people, cannot step out of the way of even the slowest-moving flow. In 1983 and 1984, the lava flows cut a dark swath through a housing development built 6 kilometers (4 mi) from the active vent, the Royal Gardens subdivision (Figure 6-37, p. 178). This flow destroyed 16 homes and wiped out the only road to the subdivision.

In the summer of 1986, the eruption shifted to a vent, Kupaianaha, 3 kilometers (2 mi) northeast of Pu'u 'O'o. Under the protection of a lava tube, rivers of pahoehoe lava were able to flow 11 kilometers (7 mi) to the sea, quenched in billowing clouds of steam. The flows closed the coastal highway and, in 1990, residents of the village of Kalapana could only stand by and watch as lava burned their homes and buried the remains under up to 25 meters (82 ft) of basalt. The destruction wrought by Kilauea since 1983 has not been massive, but it reminds us that all we can really do is get out of the way.

VOLCANIC GASES

Volcanoes of all types emit vast amounts of gases, some of them hazardous. | Volcanic hazards: gases | Among the most abundant are water vapor (steam), carbon dioxide (CO_2), and sulfur dioxide (SO_2). The more caustic gases directly harm lung tissue if inhaled, and some people straying too close to active craters have been harmed.

Hazards to people

As violent as volcanoes can be, sometimes they kill with barely a whisper. An incident in 1986 in Cameroon, West Africa, revealed the rare but dangerous hazard created by gas buildup in a lake-filled crater of a mafic volcano.

In August 1984, a geological disturbance in a volcanically active area of Cameroon churned up a deadly cloud of colorless, odorless carbon dioxide (CO_2) gas from the lower depths of Lake Monoun. Volcanic vents below the lake leak this and other gases, which because of the great pressure at that depth normally remain dissolved in the water, like the fizz in a bottle of soda. Something—possibly an earthquake—disturbed the lake, causing a huge volume of water saturated with CO_2 to rise toward the surface. At shallow depths, the lower pressure allowed the gas to bubble out of solution and escape. Being heavier than air, the carbon dioxide hugged the ground and flowed downhill into nearby villages, displacing the oxygen needed to sustain life. Thirty-nine people from the village of Njindoun were asphyxiated.

This event was a preamble to a much worse incident in the same area—this time originating from Lake Nyos (Figure 6-38a on p. 178). In August 1986, carbon dioxide erupted from the lake and spread over an area of 300 square kilometers (116 mi^2). The human victims were estimated at 1700 in addition to numerous wildlife and livestock (Figure 6-38b, p. 178). When the gas burst to the surface of the lake, perhaps released when an underwater landslide started the gassy water on the bottom flowing upward, it created a wave up to 75 meters (246 ft) high.

Seeping CO_2 also created a hazard at California's Mammoth Mountain, on the southwest rim of the volcanically active

FIGURE 6-36 ◀ **A Shield Volcano in Action**
Kilauea is the active volcano on the Big Island of Hawaii (see Figure 6-6). The current eruption started in 1982. Episodic eruptions have sent many lava flows to the sea, some through the Royal Gardens subdivision.

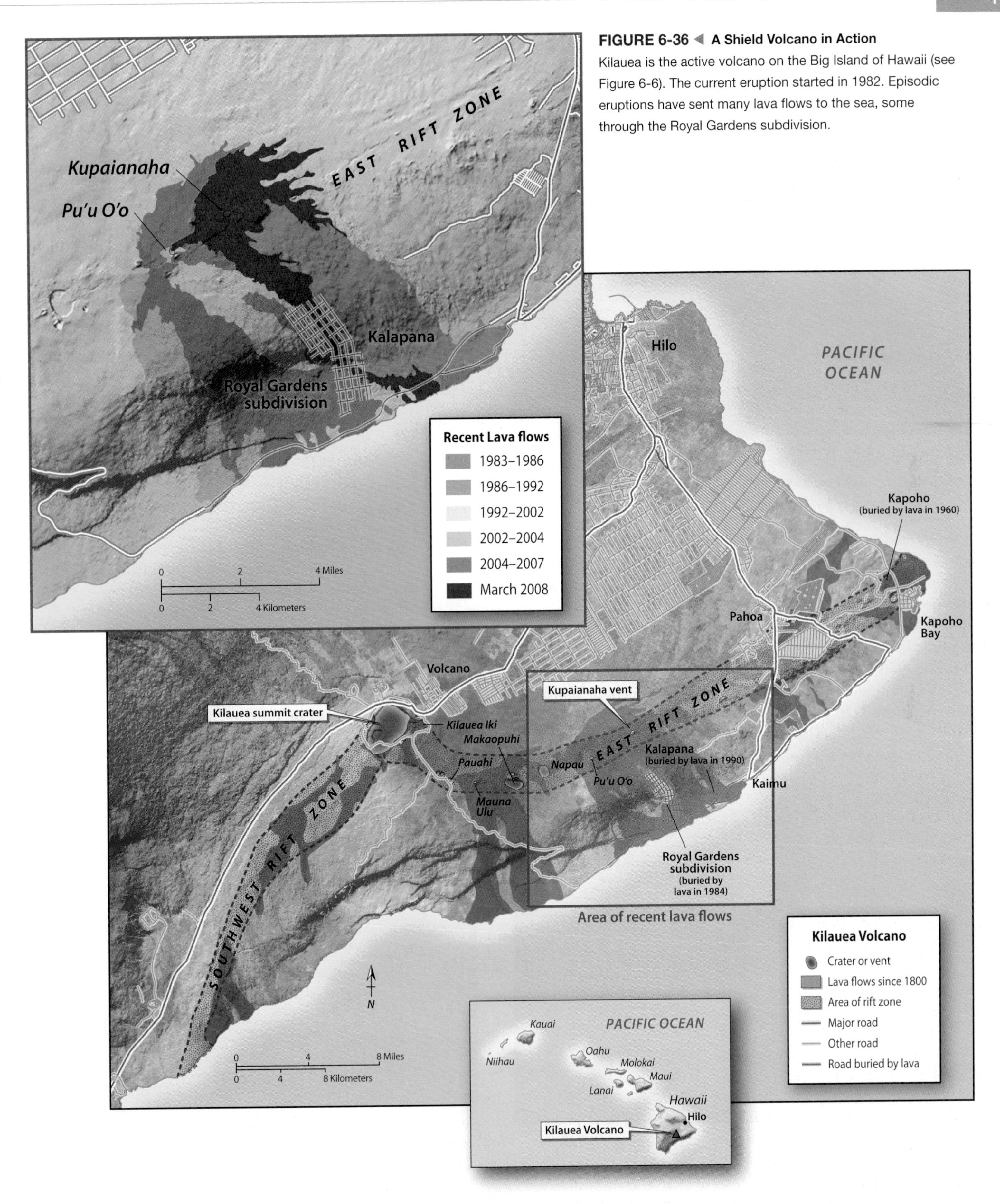

FIGURE 6-37 ▲ People, but not Subdivisions, Can Get Out of the Way of Lava Flows
A 1983 eruption sent flows from Kilauea's Pu'u 'O'o vent into the Royal Gardens subdivision. (Can you identify the kind of lava in this photo?)

Long Valley caldera (Figure 6-22). Starting in 1990, CO_2 gas seeping into soil killed trees in a small patch of forest. Because CO_2 tends to settle into low spots and linger, some people entering enclosed, snow-covered cabins in the area were nearly overcome.

As a result of lava eruptions at Hawaii's Kilauea Volcano, sulfur dioxide (SO_2) and other gases react chemically with moisture and oxygen to form a type of air pollution called *volcanic fog*, or *VOG*. On some days VOG forms a visible haze within Hawaii Volcanoes National Park and other areas. It irritates the skin and sensitive tissue lining the eyes, nose, and throat; can cause breathing trouble in people with respiratory problems, such as emphysema and asthma; and has even led to a few deaths.

Hazards to plant life

Sulfur gases can cause problems for the health of vegetation, too. This problem is occurring today at Masaya Volcano in Nicaragua. Large amounts of SO_2 have vented from Masaya since the 19th century. When the gas combines with moisture in the air, it generates tiny droplets (aerosols) of sulfuric acid. This makes rains in the area acidic. Masaya's acid rain is strong enough to corrode iron gates, but what is really most damaging is its destruction of local coffee crops. Various heroic remedies have been tried in the past, including attempts to cap or divert the gas, or explode a bomb in the crater to seal it. Masaya continues to release SO_2 in large volumes, as much as 2.3 million kilograms (2500 tons per day).

Gases released during the 1793 fissure eruptions near Laki, Iceland, contributed to the famine that killed more than 9000 people. Grasses contaminated by fluorine gas poisoned most of Iceland's livestock, and SO_2 generated acid rain that caused crop failures.

Climate changes

Volcanic gases can affect the global climate. Although not always directly hazardous in all cases, some eruptions have had significant impacts on people. The 1991 eruption of Mount Pinatubo injected an estimated 22 million tons of SO_2 into the

(a)

(b)

FIGURE 6-38 ▲ Tragedy at Lake Nyos
In August 1986, an imperceptible cloud of heavier-than-air carbon dioxide emerged from Lake Nyos **(a)**, a crater lake in Cameroon. The gas flowed silently across lowlands, suffocating 1700 people and their animals **(b)**.

stratosphere, where winds carried it around the world. The gas combined with water to form aerosols of sulfuric acid. The aerosols acted like a global sunscreen, reducing the sunlight that reached Earth's surface. The result was cooling of up to 0.5° C (0.9° F) in some areas.

Sulfur aerosols that formed after the 1815 eruption of Tambora volcano, Indonesia, had an even more dramatic effect on world climate. Because of dramatic cooling in the Northern Hemisphere, people in Europe and North America referred to 1816 as "the year without a summer" (see Chapter 3).

What are some Earth systems interactions associated with eruption of volcanic gases?

In The News

Getting Your Own Weekly Volcano Report

If a violent eruption somewhere in the world threatens people and property, it will probably be in the news. If people are killed in a volcanic eruption, then it may even be in your evening TV news. But volcanoes are erupting somewhere on Earth every day, and most eruptions won't be reported in your local paper or TV news program. There is a way for you to keep up with the volcano news in case you're traveling, or just interested in the subject. The Smithsonian Institution (SI) and the USGS provide a weekly online report of volcano activity around the world. | SI/USGS weekly volcanic activity report | These reports include information about new as well as ongoing volcanic activity if:

- A volcano observatory raises or lowers the alert level at a volcano.
- A volcanic ash advisory has been released.
- A verifiable news report of new activity or a change in activity at a volcano has been issued.
- Observers have reported a significant change in volcanic activity.

The reports give geologic summaries of the volcano, descriptions of the current activity, and a map showing the volcano's location. They provide web links to the sources of information. With the "Weekly Volcanic Activity Report" you can not only keep up with, but in some cases stay ahead of, the news.

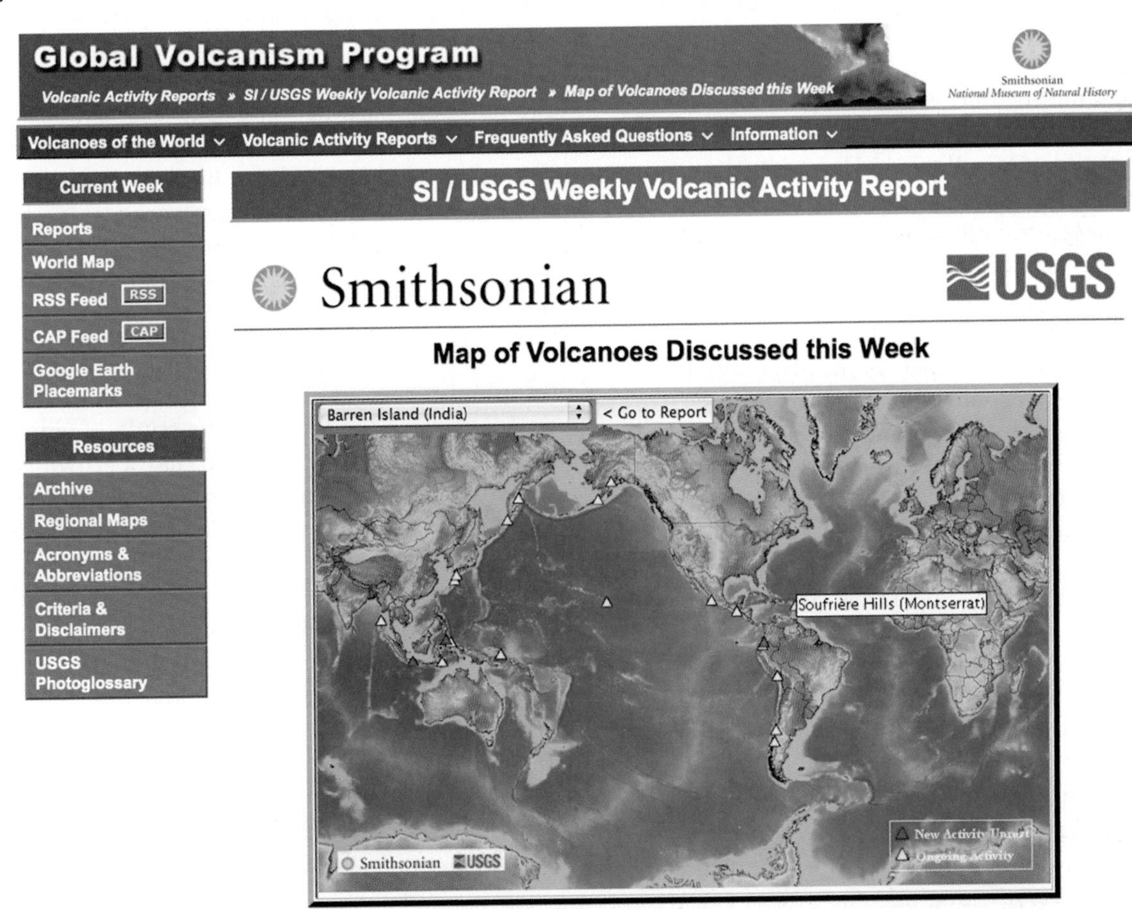

6.4 Living with Volcanoes

Worldwide, an estimated 500 million people live in volcanic hazard zones. Many of these zones are in the developing world, where land is precious and people are reluctant or unable to move away from a potentially active volcano. Population growth increases volcanic risks in many parts of the world. UNESCO has found that the average number of deaths per year in the 20th century from volcanic hazards actually rose, even as our understanding of volcanism improved dramatically. Part of the reason for this trend may be the increasing number of people living near active volcanoes. About 1 million people live within 40 kilometers (25 mi) of Mexico's Popocatepetl volcano. More than 530,000 people live within 20 kilometers (12 mi) of Japan's Sakurajima. About 2.7 million people live within 15 kilometers (9 mi) of Italy's Vesuvius volcano.

FIGURE 6-39 ▲ Going to War with Lava Flows in Iceland
In 1973, the community of Vestmannaeyjar, Iceland was threatened by an erupting volcano, and fought the advance of lava flows by cooling them with seawater.

GOING TO WAR WITH PELE

In Hawaiian mythology, Pele, the volcano goddess, is a force of nature not to be trifled with. When angry, she unleashes torrents of lava from her home, Kilauea volcano. To most people, the appropriate response to eruptions is to flee. However, some have attempted from time to time to stand and fight—sometimes quite literally. In 1935, 1942, and again in 1975–1976, attempts were made to divert lava flows in Hawaii by dropping bombs on them from aircraft.

People have sometimes tried to divert lava flows in less violent ways. Seeing access to their critical seaport threatened by lava flows, the citizens of Heimaey, Iceland, chose to make a stand. Using a network of pipes, the defenders pumped 32,400 tonnes (36,000 tons) of seawater per hour onto the flows (Figure 6-39). The idea was to cool and harden the outer skin of the flow front, causing it to divert away from the harbor. The effort continued for five months. In the end, the flow did change course and did not block the harbor completely, although it's not entirely certain that the water spray was ultimately responsible.

At Mount Etna, an active volcano on the island of Sicily, there is a long tradition of battling lava flows. The first known attempt to do so at any volcano occurred there in 1669. In 1983, an effort was mounted to divert lava from the towns of Belpasso and Nicolosi by blasting through the walls of a channel feeding the flow. It was partially successful. During the same eruption, people built earthen barriers in an attempt to prevent a lava flow from spreading outward and destroying buildings. In 1992, they tried several methods to control a lava flow. An earthen barrier slowed the flow's advance long enough to dig a channel and blast an opening into the wall of a nearby lava tube. The entire flow poured into the channel and down the tube. Most recently, in 2001, earthen barriers were again constructed to divert lava into artificial channels and protect buildings.

Is it realistic to go to war with an erupting volcano? Some efforts may help temporarily, but in general, it is better to understand and avoid volcanic hazards than to confront them. How do communities near active volcanoes effectively respond to the potential hazards? This is a complex issue, but the fundamentals can be reduced to two parts: (1) understanding the magnitude and nature of the hazards, and (2) applying this knowledge so that the risk of death and destruction is reduced. Volcanism is a force of nature that cannot be controlled or contained, but people can learn to live more safely with volcanoes.

HOW SCIENCE HELPS

Volcanologists are geologists, geophysicists, seismologists, and other specialists who focus on the study of volcanoes. Starting in the 20th century, they began to play a growing role in the response to volcanic hazards. With every eruption that occurs around the world, field volcanologists scramble to the scene to observe, to study, to test hypotheses, and to try out new methods of monitoring volcano behavior. Their hope is to reduce the number of deaths from eruptions, and there have been some successes as well as disappointments. Ironically, volcanoes are sufficiently unpredictable that a number of volcanologists have lost their lives to volcanic hazards, despite all of their expertise.

Volcanology—a hazardous profession

Field volcanologists who work on active or erupting volcanoes are a small group. There are only several hundred in the world. One of their fundamental tasks is to figure out what is going on inside hazardous volcanoes. Like doctors with stethoscopes, their work sometimes involves getting up close and personal with potentially deadly volcanoes. In the 25-year period from 1975 to 2000, 29 volcano scientists died in the line of duty.

In 1991, for example, a volcanologist, two French photographers, and 40 Japanese journalists were observing pyroclastic flows on the flanks of Mount Unzen in Japan (see Figure 6-30). The scientist was Harry Glicken, who in 1980 had barely escaped death at the eruption of Mount St. Helens. (He had been staffing the Coldwater Ridge observation post until a day before the blast. The scientist who took over for him, David Johnston, ultimately perished instead of Glicken.) But Glicken's luck ran out on June 3, 1991, when a surge of hot gases unexpectedly separated from a pyroclastic flow and reached an area that Glicken thought was safe for observations. Among the other victims were Maurice and Katia Krafft, who had spent their lives filming eruptions under extremely dangerous conditions.

The most notorious incident in recent history occurred barely two years later, on January 14, 1993. The location was Galeras volcano, an active stratovolcano that looms over the town of Pasto, Colombia, and its 350,000 residents. The volcano had been active for several years, and had erupted as recently as July 1992. During a scientific meeting about Galeras hosted by Colombian volcanologists, an international group of scientists went to Galeras's active crater to measure gravity levels and collect gas samples. This kind of data is potentially helpful for determining if a volcano is gearing up for an eruption. Fifteen people were near the active vent, including three local residents who hiked in for sightseeing, when a short but explosive eruption occurred. Six scientists and the three tourists were killed, while others were gravely injured.

The incident provoked a lot of public soul-searching about whether the Galeras party had been careless. Most, for example, were not wearing the most basic piece of safety equipment, a hardhat, which might have prevented some of the head injuries suffered that afternoon as the scientists tried to flee the rain of lava debris, grapefruit-sized and larger, exploding from the crater. Most volcanologists agree that visits to active volcanic vents are necessary, but should be conducted with great caution and adequate protective clothing and gear—not to mention a healthy respect for the inherent unpredictability and violence of volcanoes.

Hazard assessments

When confronting an active volcano, communities need to know the degree of hazard they are facing. Scientists provide this information in the form of a *hazard assessment*. Geologic mapping and dating of lava and ash from past eruptions can reveal approximately how often a volcano has large eruptions. As a rough rule of thumb, the larger the last eruption, the longer it will be until the next one. This is because a volcano needs a certain amount of time to replenish its supply of magma.

Hazard mapping involves locating and describing geological deposits near the volcano to identify them as lava, ash, pyroclastic flows, lahars, and so forth. Laboratory work can date these deposits, sometimes precisely, down to individual lava flows and the calendar year they erupted. The volume or thickness of individual deposits provides clues to the magnitude of past eruptions, which in turn puts some constraints on how large a future eruption could be.

If they know enough about the geologic history and behavior of a volcano, scientists may be able to make long-range forecasts. These are not specific predictions of where and when the volcano will erupt. Instead, forecasts state the probability that an eruption of a certain size will occur within a certain time period. For example, a forecast might state that an eruption large enough to devastate everything within 50 kilometers (31 mi) could occur within the following 10 to 20 years.

Monitoring volcanic activity

To respond to an eruption, you obviously have to know whether or not one may be starting. Another role that scientists play is monitoring volcanoes for signs of unrest, such as increased earthquake activity or an upturn in gas emissions from the vent (Figure 6-40). In the United States, the USGS operates observatories to monitor activity in Hawaii, the Cascade Range, Alaska, Long Valley (California), and Yellowstone (Wyoming). If there are signs of unrest, scientists at the observatories can issue alerts. Numerous other countries with active volcanoes, such as Japan, Colombia, Mexico, and Indonesia, also maintain volcano observatories. At these often-remote outposts, scientists use seismometers (see Chapter 5) and other instruments to monitor changes in the volcano's level of activity.

FIGURE 6-40 ▲ Monitoring Volcanic Activity Is an Important but Sometimes Dangerous Job

Volcanologists and other scientists frequently monitor volcanoes to determine when an eruption may be imminent. They may visit active volcanoes to gather data collected by sensors, take samples of magma or volcanic gases, and make close-up observations of ongoing activity. These investigations sometimes put them in harm's way, and have cost a few scientists their lives.

In remote areas where no staffed observatory exists, instruments can be set up to electronically send information back to a centralized location. In the Aleutian Islands and Cook Inlet region of Alaska, for example, a number of remote volcanoes are "wired" with seismometers that transmit data back to the Alaska Volcano Observatory in Anchorage via cell phone. Satellites that can detect concentrations of heat or ash plumes are increasingly used to monitor remote or inaccessible volcanoes.

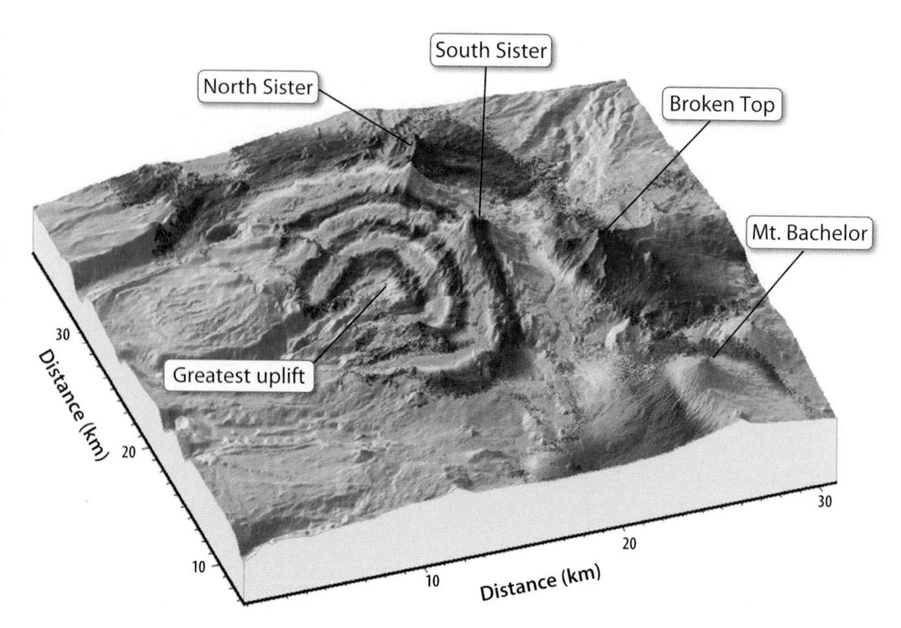

FIGURE 6-41 ▲ **Ground Deformation Can Be Detected from Satellites Using Radar** A satellite radar system collected the data used to make this map of ground uplift near South Sister volcano in Oregon. This uplift occurred between 1997 and 2001 and suggests a new area of magma intrusion. A flurry of earthquakes, located within the area of uplift, occurred in 2004. (Each complete sequence of colors represents 28.3 millimeters, or 1.1 in, of uplift, so the maximum uplift is about 100 millimeters, or 4 in, centered in middle of the circular feature.)

Monitoring eruption precursors

The foundations of short-term forecasting are patterns of volcanic activity, or precursors that might indicate an eruption is imminent. No precursor has been identified yet that reliably predicts eruptions every time. Through a combination of techniques, however, scientists have scored a few successes in making short-term predictions that an eruption is likely to occur in days to hours. More than anything, the scientists need to determine what may be happening within the volcano and its plumbing system. For example, is a large amount of fresh, gas-charged magma recharging the system? Scientists become like doctors trying to make a diagnosis, listening to the volcano, feeling its pulse, measuring its temperature, looking for swelling on its flanks. As in medicine, this is as much craft as science and depends on experience and judgment.

To help diagnose volcanoes, scientists use a variety of monitoring methods—their equivalents of the stethoscope, tongue depressor, X-ray, and ultrasound. They especially monitor earthquakes, ground deformation, and the release of volcanic gases, all of which are influenced by movement of magma within a volcano.

As magma flows through a volcano's plumbing system, the ground shakes. Very often, an increase in small- to medium-sized earthquakes at shallow depths beneath the volcano is the first sign that an eruption may be brewing. This was certainly the case at Mount St. Helens in March 1980. Specific patterns of earthquakes are often associated with certain changes in the volcano's plumbing.

Monitoring ground deformation has been put to effective use at Kilauea volcano in Hawaii. Using mechanical tilt meters, Global Positioning System (GPS) receivers, or field surveying, scientists monitor volcanoes for changes in their surface elevation. At Kilauea and Mauna Loa, swelling or "inflation" of the volcano's slopes indicates that the system may be recharging with fresh magma. (Recall that the north slope of Mount St. Helens swelled dramatically in the weeks leading up to its eruption, a sign that magma was inflating the mountain—see Figure 6-33.) Remote sensing techniques, such as satellite radar mapping, can also detect such telltale bulges, as Figure 6-41 demonstrates.

As magma rises, a series of gases bubble out of solution at different depths. One that is commonly monitored is sulfur dioxide (SO_2). This monitoring can, fortunately, be done from a distance with a device called a correlation spectrometer (COSPEC), which was originally developed to monitor sulfur pollutants from industrial smokestacks. Any abrupt change can be significant. A large increase in SO_2 emissions from an active volcano, for example, may indicate that magma is rising higher in the system and is therefore more likely to erupt. If gas emissions plummet, on the other hand, it could indicate that the volcanic vent is plugged and at higher risk of an explosion.

Numerous other precursors may come into play, such as minute changes in the force of gravity on a volcano's slopes or fluctuations in electrical conductivity that reflect changes in the magma. Scientists often combine these with other precursors to make educated guesses on the movement of magma within a volcano.

Volcanic crisis response

If a volcano shows signs of a pending eruption, scientists take on the role of advisors to communities at risk. A volcanic crisis begins with the initial start of unrest and spans the buildup to an eruption, the eruption itself, and the immediate aftermath. At these times, local officials, the public, and the media rely on volcanologists and other experts to answer the critical questions:

Is a major eruption coming?

If so, when?

How big will it be?

What areas will be in the most danger?

FIGURE 6-42 ▲ Mount Pinatubo: An Eruption Foreseen
(a) Mount Pinatubo, in the Philippines, erupted on June 12, 1991, but scientists had been closely monitoring the volcano since April. In response to advance warnings, Clark Air Base and other nearby regions were evacuated before being covered by a thick blanket of ash. A combination of ash and heavy rain destroyed houses many kilometers from the volcano **(b)**, but evacuations ordered before the eruption probably saved thousands of lives.

Unfortunately, these questions do not have simple answers, and short-term eruption forecasting is fraught with uncertainties. A typical short-term forecast or prediction is a statement that an eruption of a given size will probably occur in the near future, typically in days to weeks. These statements may also be called warnings or alerts. USGS volcano-warning scheme for the United States It's common to issue alerts on a multilevel scale of increasing severity, from no alert (normal background), to the first level (signs of unrest), up to the top level (eruption in progress). The alert levels are lettered, numbered, or color-coded (as shown in Table 6-5), and commonly tied to responses such as evacuations and other measures to protect people at risk.

Pinatubo: a successful crisis response

The eruption of Mount Pinatubo in June 1991 was among the largest in the 20th century. About 300 people died, but few died from the direct effects of the climactic blast on June 30 that sent ash rocketing 35 kilometers (22 mi) into the sky (Figure 6-42a).

TABLE 6-5 VOLCANO ALERT LEVELS

Alert Level	Color	Symbol	Description
Normal	Green	N	Volcano is in typical background, noneruptive state; *or, after a change from a higher level*, volcanic activity has ceased and volcano has returned to noneruptive background state.
Advisory	Yellow	A	Volcano is exhibiting signs of elevated unrest above known background level; *or, after a change from a higher level*, volcanic activity has decreased significantly but continues to be closely monitored for possible renewed increase.
Watch	Orange		Volcano is exhibiting heightened or escalating unrest with increased potential of eruption, timeframe uncertain; *or*, eruption is underway with no or minor volcanic-ash emissions [ash-plume height specified, if possible].
Warning	Red	!	Eruption is imminent with significant emission of volcanic ash into the atmosphere likely; *or*, eruption is under way or suspected with significant emission of volcanic ash into the atmosphere [ash-plume height specified, if possible].

On April 2, 1991, steam explosions on the northeast flank of Mount Pinatubo announced the beginning of the eruption. In response, scientists at the Philippine Institute of Volcanology and Seismology set up seismometers around the volcano. Hundreds of small earthquakes beneath the volcano suggested that magma was on the move. The scientists recommended evacuating rural villagers within 10 kilometers (6 mi) of the volcano.

The Philippine scientists requested assistance from their colleagues at the USGS, who arrived on April 23 and began closely monitoring the volcano. Geologists also surveyed past eruption deposits, discovering that pyroclastic flows threatened areas up to 24 kilometers (15 mi) from the volcano. This included the U.S. military's huge Clark Air Base, home to 18,000 military personnel and their families.

On May 23, the scientists distributed a hazard map to local Philippine authorities. An alert system was established on a scale of 1 to 5.

By June 5, the alert level was at 3, meaning that an eruption could occur within two weeks. Efforts began to evacuate areas that could be reached by pyroclastic flows from a large eruption.

On June 7, as thousands of earthquakes popped off beneath the volcano and a lava dome grew in the vent, the alert was raised to 4: eruption possible within 24 hours. Scientists recommended widening the evacuation zone to 20 kilometers (12 mi).

On June 10, Clark Air Base was evacuated. Personnel removed military aircraft and other expensive hardware.

By June 12, explosive eruptions and pyroclastic flows had begun. The alert level was 5: eruption under way.

On June 14, explosions continued and earthquake activity beneath the volcano increased even further. Evacuation was recommended out to 30 kilometers (19 mi).

By June 15, the day of the climactic eruption that blew a huge caldera in the body of the volcano, a total of about 78,000 people had been evacuated from the flanks and valleys around Mount Pinatubo and the air base. Most of the deaths due to the eruption occurred when the thick ash blanket spewed by the volcano (Figure 6-42b) mixed with rain from a passing typhoon and collapsed people's homes (see Figure 6-25b).

What made the difference?

The success at Mount Pinatubo has been attributed to a combination of decisive, effective action and a bit of luck. On the scientific side, volcanologists responded quickly, set up a high-quality seismic monitoring network, instituted a simple five-level warning system, and accurately predicted the hazards posed and who was at risk. They were also successful at convincing most local officials and residents of the risks they faced. The local officials, to their great credit, acted on the warnings and took the dramatic step of evacuating large numbers of people. The scientists and officials were in close daily contact. The scientists were willing to stick their necks out and issue eruption forecasts, even though they were not and *could not be* totally certain of what the volcano would do.

At the same time, there were skeptics. Right up to the climactic eruption of June 15, there were people who doubted that an eruption was imminent. Most notably, the mayor of Angeles City, population 300,000, was openly dismissive of the scientists' warnings right up to the point when the eruption began. Scientists had to work hard to convince people they were in danger, and everyone involved was justifiably anxious about the consequences of issuing a false alarm. Despite these complications, they pulled off what has been called the most successful volcanic crisis response in history. They were also fortunate that the pre-eruption phase lasted long enough to give people time to do thorough studies and start to educate people about the hazards. A video about volcanic hazards was particularly helpful, proving the old axiom that a picture is worth a thousand words.

HOW COMMUNITIES RESPOND TO VOLCANIC HAZARDS

During a volcanic crisis, the many variables associated with forecasting eruptions places scientists in a difficult position. They may only be able to say that an eruption is likely to occur within a certain time period, not that it will definitely occur in an hour, a day, or a week. The fear of the "cry wolf" syndrome is ever-present in these tough decisions. If scientists recommend an evacuation and an eruption doesn't happen, people will be skeptical of the next forecast, or warnings may not be heeded at all.

In part, the loss of more than 23,000 lives to a lahar in 1985 at Nevado del Ruiz volcano (Section 6.3) occurred because local officials didn't understand, trust, or act on warnings by geologists. During the eruption of Mount St. Helens, the closure of areas close to the volcano met with resistance, but the warnings and closures ultimately saved many lives. In the end, volcanologists can only act as advisors; they cannot "order" anyone to do anything. Only the communities facing hazards, specifically local decision makers, can make the final call on how to respond to hazards scientists identify, translating scientific knowledge into policies and actions that save lives.

You Make the Call

Living in the Shadow of Mount Rainier

Mount Rainier is a beautiful stratovolcano in the Cascade Range of the western United States. On clear days it can be seen throughout the Seattle, Washington, area. River valleys draining Mount Rainier's flanks have become densely populated, especially near Puget Sound and the port of Tacoma—over 3.4 million people live in the greater Seattle area. Cities such as Puyallup and smaller towns like Ortig are located in the lowlands of the Puyallup River valley. The Puyallup River's headwaters are on the flanks of Mount Rainier and the people living in this river valley are in the path of lahars.

Mount Rainier Is a Sleeping Giant
Many communities around Mount Rainier, Washington, face volcanic hazards from this beautiful but dangerously close stratovolcano.

Scientists have mapped and dated lahar deposits throughout the Puyallup River valley. They have learned that lahars have cascaded down the valley over 100 kilometers (62 mi) to the shores of Puget Sound. And these catastrophic mud flows have occurred at least every 500 to 1000 years. Lahars are commonly caused by melting of snow and ice high on the summits of stratovolcanoes as they erupt. However, scientists worry that many lahars in the Puyallup River valley may not have formed during major eruptions but instead have been caused by ongoing low levels of activity, especially steam eruptions that melt ice and snow and destabilize the volcano's flanks. A specific 5700-year-old lahar is thought to have formed this way and scientists developed a video called "Perilous Beauty—the Hidden Dangers of Mount Rainier" to help us understand how this event occurred. In these cases, precursors of an eruption would not be noticeable. How can the people living in the Puyallup River valley be protected from such a stealthy lahar?

Think about the lahar dangers in the Puyallup River valley. Mount Rainier is an active volcano and lahars will again flow down its flanks and along the Puyallup River.

If you lived there, what would you do to protect yourself and your family from lahars? What should your community do?

These are tough questions for people and communities that have been in this valley for generations. Compare your thoughts to those of the people living there now. Are they taking adequate steps in light of the risks to life and property they face? Geotimes—paths of destruction: the hidden threat of Mount Rainier

Mount Rainier: living safely with a volcano in your backyard

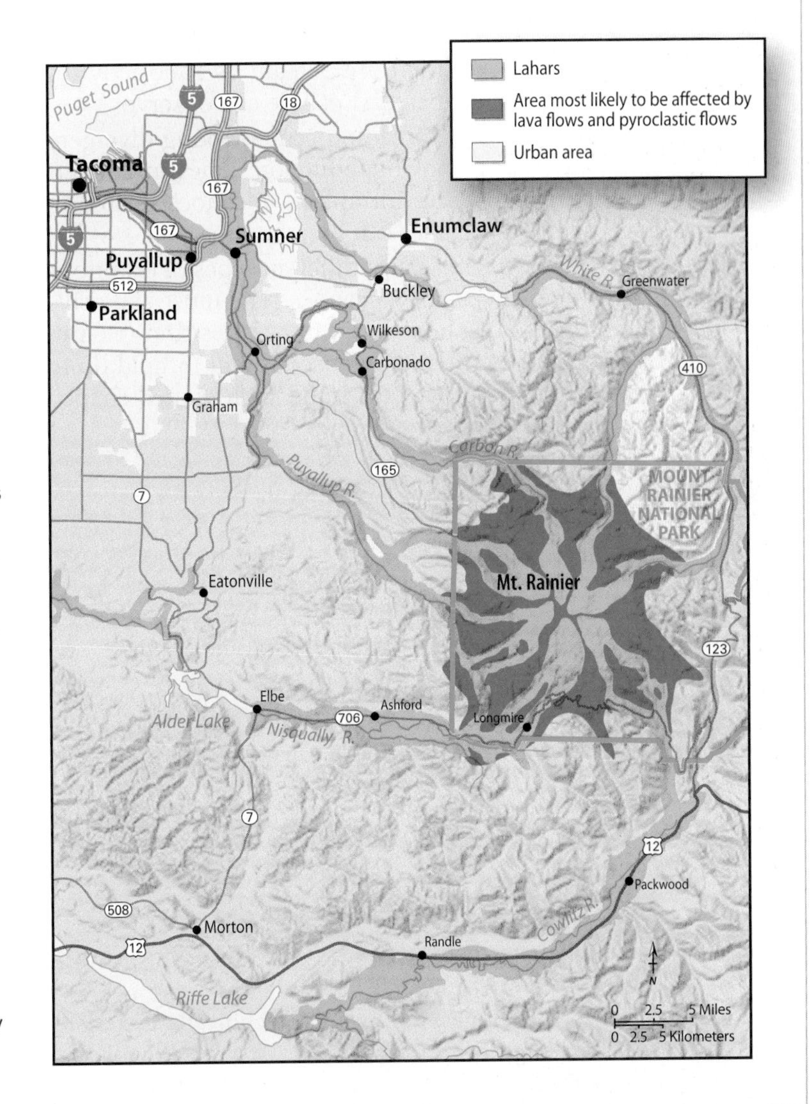

Too Close for Comfort?
Geological studies have shown that lahars have flowed down river valleys flanking Mount Rainier many times, through areas that are now heavily populated.

SUMMARY

The dynamic geosphere transfers energy and materials to Earth's surface at volcanoes. Most people imagine volcanoes as mountains erupting lava, but volcanism takes many forms. At bare minimum, a volcano is an opening, or vent, in the crust that allows molten rock (magma) to reach the surface from deep sources. Wherever volcanic eruptions occur, they can impact the atmosphere, hydrosphere, and biosphere, and are especially devastating to people and their property. Despite their destructive potential, volcanoes are beneficial to people. They are essentially a renewable source of energy and material at Earth's surface—material that develops fertile soils or is used for construction and industrial purposes. By studying volcanoes, scientists can help people prepare for and avoid many of the disastrous consequences of volcanic eruptions. By applying lessons learned from studying volcanoes, people can diminish the harm done by eruptions and preserve both human and economic resources for other purposes. In this way, living more safely with volcanoes can contribute to a sustainable future. Consequently, everyone has a stake in understanding volcanoes.

In This Chapter You Have Learned:

- Volcanic activity is commonplace on Earth. Annually, 50 to 70 volcanoes erupt, and there are as many as 1500 active or potentially active volcanoes that people can observe.
- Most volcanoes form at spreading centers along divergent plate boundaries and above subduction zones at convergent plate boundaries. But intraplate volcanoes related to more local mantle processes can be very large.
- Different types of magma cause different types of eruptions and form different types of volcanoes. The most explosive magmas are intermediate to felsic in composition (rich in silica), which causes them to be viscous and to hold on to dissolved gases. These magmas build high, coned-shaped stratovolcanoes and large calderas.
- Mafic magmas poor in silica content, such as those that fuel Hawaiian-type volcanoes, are less viscous, and tend to erupt as lava flows. They build broad, gently sloping shield volcanoes and vast areas of flood basalt.
- Stratovolcanoes are the most explosive and the most dangerous. They create avalanches of hot pyroclastic material (pyroclastic flows), volcanic mudflows (lahars), and ash plumes in the atmosphere that can be blown around the world. Stratovolcanoes have killed tens of thousands of people in the past century alone.
- Lava flows from shield volcanoes destroy property, but are associated with less loss of life than the more explosive eruptions at stratovolcanoes.
- Volcanic gases, released by all types of volcanoes, can be extremely hazardous, and in large eruptions can influence the global climate.
- Volcanic eruptions vary tremendously in size. The largest, those that form calderas, erupt hundreds to thousands of cubic kilometers of material. Fortunately, large eruptions are much less frequent than small ones. Hundreds of thousands of years may pass between caldera-forming eruptions.
- Scientists study the eruption history of volcanoes to determine where and how large future eruptions may take place. Mapping the distribution and character of erupted material allows them to identify hazards and predict where future eruptions are likely to affect people. They have also found that changes in the nature of earthquakes, ground deformation, and released gases can be precursors to eruptions.
- Communities potentially threatened by volcanoes can live more safely with them by working closely with scientists and translating scientific understanding into sound policies and actions. People in volcanic hazard zones can help protect themselves by being well informed about volcanic hazards and what to do if a volcano erupts.

KEY TERMS

aa (p. 175)
ash (p. 159)
caldera (p. 161)
cinder cone (p. 157)
dome (p. 161)
fissure (p. 157)
flood basalt (p. 157)
hot spot (p. 167)
intraplate volcano (p. 167)
lahar (p. 172)
lava tube (p. 157)
magma chamber (p. 161)
pahoehoe (p. 175)
pumice (p. 159)
pyroclastic flow (p. 171)
pyroclastics (p. 159)
shield volcano (p. 156)
stratovolcano (p. 158)
tephra (p. 159)
tuff (p. 161)
viscosity (p. 154)
volcanic explosivity index (VEI) (p. 163)
volcano (p. 153)

QUESTIONS FOR REVIEW

Check Your Understanding

1. What is the difference between an "active" and a "dormant," or "extinct," volcano?
2. Viscosity, or the "stickiness" of magma (technically its resistance to flow) changes with temperature. Describe how the viscosity of a magma or lava changes with temperature.
3. Magma viscosity is also related to silica content. How and why is this true?
4. How does partial melting of a rock increase the silica content of a magma relative to the rock from which it originated?

5. Why are volcanoes that are built from more silica-rich magmas inherently more explosive than mafic volcanoes?
6. Why do Hawaiian-style shield volcanoes grow to cover such a large area, especially compared to their much smaller and more explosive stratovolcano counterparts?
7. Why is falling volcanic ash so much more dangerous to people and built structures than falling wood ash from a forest fire?
8. What are calderas? Why is the discovery of these geologic features so significant?
9. Volcanoes are not without their real social and economic benefits. Name and describe three natural resources that we owe to volcanoes.
10. What are hot spots? Where do they occur, thought to be and why?
11. Why are pyroclastic flows around stratovolcanoes so dangerous?
12. Why are mudflows, or lahars, a major volcanic hazard? What is the source of the water in lahars?
13. Why are the volcanic hazards related to lava flows on the big island of Hawaii so different, and ultimately much less frightening and life-threatening, than the volcanic hazards of Mount St. Helens in Washington?
14. Why is carbon dioxide (CO_2) from volcanoes such a deadly volcanic hazard to people and livestock?
15. Volcanic hazards can be anticipated and predicted to some extent. What kinds of things might volcanologists measure to help predict a volcanic eruption?

Critical Thinking/Discussion

16. The Yellowstone Caldera was something of a geologic mystery for many years. Geoscientists recognized many volcanic features, but failed to truly recognize the scale of the volcanic structure present in the national park until the advent of satellite imagery. Why do you think that was the case, and what significance would the discovery of such large volcanic structures have for the assessment of volcanic hazards?
17. Human efforts in Iceland, Hawaii, and Sicily to directly combat or minimize the impact of ongoing volcanic eruptions on human communities usually meet with only modest success at best, and often fail. Why is this the case when we understand the nature of these volcanic hazards relatively well?
18. The scientific efforts to predict the eruption of Mount Pinatubo in the Philippines are often cited as among the most successful in protecting human life. However, good science—while essential—was not the only factor involved in preventing loss of life and resources in the Philippines during this event. What other entities had to cooperate actively with scientists to achieve this evacuation? Explain how good communication of science played a major role in the success of this response to an imminent volcanic hazard.

ANSWERS TO IN-CHAPTER INSIGHT QUESTIONS

What are some Earth systems interactions associated with the benefits of volcanoes?

P. 165

- Groundwater heated by magma from the geosphere is a source of energy.
- Erupted material dispersed across the land by the atmosphere can develop into rich soils.
- Development of rich soils involves interaction of all Earth systems (Chapter 10).
- A variety of biosphere habitats develop on volcanic mountains.

What are some Earth systems interactions associated with development of subduction-related volcanoes?

P. 166

- Oceanic crust, hydrated by interaction with seawater, releases water and other volatiles during subduction that facilitates melting of the geosphere.
- Erupted magma releases water and other volatiles to the atmosphere.
- Magma heats groundwater that can form hot springs and contribute to steam emissions to the atmosphere.

What are some Earth systems interactions associated with lahars?

P. 173

- Rain from the atmosphere saturates ash and other erupted debris with water that makes it unstable and prone to slope failure.
- Heat from magma within a volcano can melt snow and ice, causing slope materials to be saturated with water and unstable.
- Lahars flow down stream valleys, destroying habitat and parts of the biosphere.

What are some Earth systems interactions associated with eruption of volcanic gases?

P. 179

- Volcanic gases can change habitat and destroy parts of the biosphere.
- Volcanic sulfur dioxide erupted to high altitudes can cause temporary global cooling that affects the biosphere and the water cycle.

FIGURE 7-1 ▶ Illinois: Summer 1993
The Mississippi River floods of 1993 submerged many towns, including Valmeyer and nearby Kaskaskia (shown here). In places the depth of the water reached 6.5 meters (21 ft).

CHAPTER

7

RIVERS AND FLOODING

For generations, communities all along the Mississippi River have experienced inundation when levees and other flood-control measures failed. Valmeyer, Illinois, located 40 kilometers (25 mi) south of St. Louis on the Mississippi's east bank, was flooded in 1910, 1943, and 1944, but the levees had protected the town from the great Mississippi flood of 1973. During the Mississippi River flood of 1993, however, Valmeyer and nearby communities were submerged to the rooftops (Figure 7-1). Essentially, the whole town was in the river.

In September of that year, Valmeyer's 900 citizens decided that they'd had enough and made a radical decision: Pick up and leave the lowland for higher ground. With funds from several government programs, the town bought 334 homes and properties so that their owners could abandon them. With the money from selling their homes and other sources, the people of Valmeyer rebuilt the entire town on bluffs 3 kilometers (2 mi) away, 120 meters (390 ft) above the river (Figure 7-2, p. 190). The decision brought a variety of hardships. Even with government assistance, many people had to start over with new long-term mortgages after making payments on their former, now-ruined homes for decades. It took time, too. The first business opened in May of 1994, and the first people did not move into the new Valmeyer until April of the following year.

This was a reasonably happy ending for the flood-struck town of Valmeyer, but what about the 12 million people who live in 125 counties and parishes bordering the Mississippi? It is simply not possible to evacuate all flood-threatened towns, and the fundamental question remains: What can and should we do about flood hazards?

Throughout history, people have been drawn to rivers by their many benefits. Rivers contain water we need for drinking and for irrigating crops. They have been a valuable means of transportation from ancient times to the present—they were our first highways. When the Industrial Revolution began, the waterwheels along rivers drove

FIGURE 7-2 ◀ Valmeyer's New Location
The 1993 flood at Valmeyer, Illinois was the last straw for this community. The town of 900 citizens moved Valmeyer to the high bluffs of Salt Lick Point. Here the town's mayor introduces the new community plan.

the bellows, spinning wheels, and mills of the developing technological societies in Europe and North America. Industrial towns built along rivers and on the lands adjacent to them were convenient locations for housing, train tracks, and roads. Rivers are key links in the water cycle (Chapter 2) and rock cycle (Chapter 4). Even floods are beneficial when they deposit sediments that sustain soils. And rivers are usually pleasant to live by—at least until a flood occurs.

Measured by lives lost and property damaged, floods were the leading natural disasters in the United States during the 20th century. They can happen in any part of the country, at any time of the year. Floods are part of the natural and necessary behavior of rivers, and people encounter these hazards because they live on flood-prone lands. Periodic floods are a price we pay for the benefits of living near rivers.

Technology and engineering have allowed us to reduce flood hazards to some extent. Historically, society has built barriers to stop the waters or to rechannel them safely around populated areas. However, efforts to control flooding have consequences. Dams, for example, can regulate the flow of water during times of heavy rain and reduce flooding but they may also alter erosion and sedimentation patterns, harm fish and other wildlife, and lead to worse flooding under certain circumstances. Maintaining natural river processes and decreasing the costs of flooding will be a part of what people must do to achieve a sustainable future.

IN THIS CHAPTER YOU WILL LEARN:

- What watersheds are and the role they play in the natural history of rivers
- How scientists describe and measure rivers, including their discharge, flow, base level, and longitudinal profile
- How rivers erode, carry, and deposit sediment
- What causes floods and how rivers develop floodplains
- How people measure and forecast floods
- How people attempt to control or mitigate flooding

7.1 River Basics

The Amazon River is the mightiest on Earth in terms of the amount of water it carries and the size of the area it drains. Starting its run in the Andes Mountains of Peru, this river travels some 6400 kilometers (4000 mi) to its mouth on the Atlantic Ocean (Figure 7-3). The volume of water flowing downstream in an Amazon flood can exceed 175,000 cubic meters per second (cms), or over 6 million cubic feet per second (cfs). That's enough water to fill the volume of a large football stadium more than 10 times every second. And with this water comes sediment—the Amazon dumps about a billion tonnes (tons) of sediment into the sea every year.

Despite its grand size and power, the Amazon has much in common with the humblest stream burbling through a park near you. Both are parts of the water cycle (Chapter 2). Fed by rainwater, snowmelt, and groundwater, both flow to lower elevations such as lakes or oceans, and change their courses or flood their banks from time to time. All flows of surface water, from tiny streams to mighty rivers, share basic characteristics and follow the same physical rules.

WATERSHEDS

A **watershed** is an area where the surface runoff from precipitation onto the land flows together toward lower areas such as lakes or oceans. As you can see in Figure 7-4 on p. 192, watersheds have a connected system of streams that work like the veins in your circulatory system; small veins collect blood in

FIGURE 7-3 ▲ The Amazon River Watershed

The Amazon and its many tributaries flow from headwaters in the Andes Mountains east to the Atlantic Ocean. This huge watershed covers nearly 7 million square kilometers (2,670,000 mi^2), or some 40% of South America. Depending on where one starts measuring, the main trunk of the Amazon is between 6,300 and 6,800 kilometers (3900 and 4200 mi) long. One-fifth of all the freshwater entering the world's oceans comes from the Amazon.

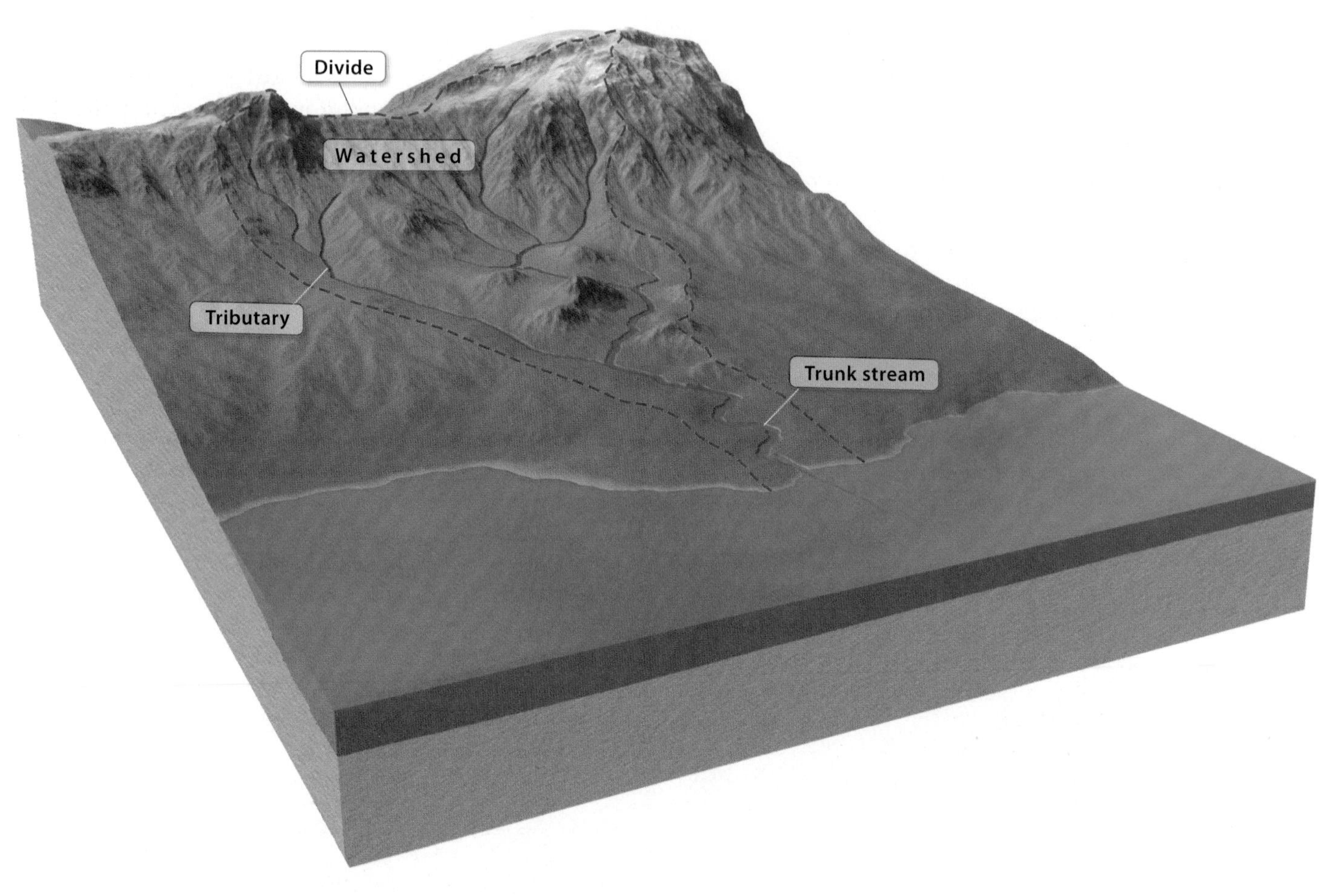

FIGURE 7-4 ▲ Rivers Are a Product of Their Watersheds
Rivers develop, flow, and change through their watersheds—the area that contributes water to a river. All Earth systems interact in watersheds.

your body's extremities, the small veins flow together and merge into larger ones, and the largest veins deliver blood back to your heart. Surface water is collected and moved through a watershed in a similar way. Small streams in the watershed's extremities flow together and form larger and larger streams. Streams that flow into another are called **tributaries**, and tributaries merge into the watershed's main **trunk river**. The watershed is all the land that contributes water to a trunk river, including all the tributary streams. The boundaries between watersheds of all sizes are elevated areas called **divides**. Watersheds vary tremendously in size and shape, from small mountain valleys to vast continental plains.

A watershed's area and the amount of precipitation within it are key determinants of the amount of water available to rivers. However, precipitation in watersheds doesn't all flow overland and feed into streams. Some precipitation infiltrates the ground. The amount of infiltration depends on ground conditions and how much water is already stored there. If the ground is saturated with water, then overland flow, or **runoff**, will be the main way in which water gets into rivers. However, water that infiltrates the ground is not necessarily lost forever to rivers. Some may eventually make its way to springs and contribute to stream flow, since springs are fairly common on the bottom of rivers.

The amount of infiltration and runoff in a watershed is influenced by people. For example,

- Some of the runoff stored in ponds or dams infiltrates ground.
- Removal of vegetation, as when timber or crops are harvested, can increase runoff.
- Urbanization that covers areas with buildings and pavement decreases infiltration and increases runoff.

The way to understand rivers is to understand their watersheds. It's essentially impossible to examine a river at a single place, or even study a considerable stretch of river, and really understand why and how it will change its flow characteristics. But a watershed perspective includes all the factors that affect a river, from the natural ones like precipitation to the people-related ones like urbanization.

What are some Earth systems interactions in watersheds?

What You Can Do

Investigate Your Watershed

Everyone lives in a watershed. What do you know about yours? The Environmental Protection Agency (EPA) has a website where you can learn about your watershed. Surf your watershed Would you like to know where your watershed is just by entering your zip code? Do you want to know how much water is flowing in the major streams in your watershed? Surf Your Watershed links to real-time USGS streamflow data. Other things you can learn about include the citizens' groups working in your watershed, any river corridor and wetland restoration projects that are under way, and the science projects that have been completed in your watershed.

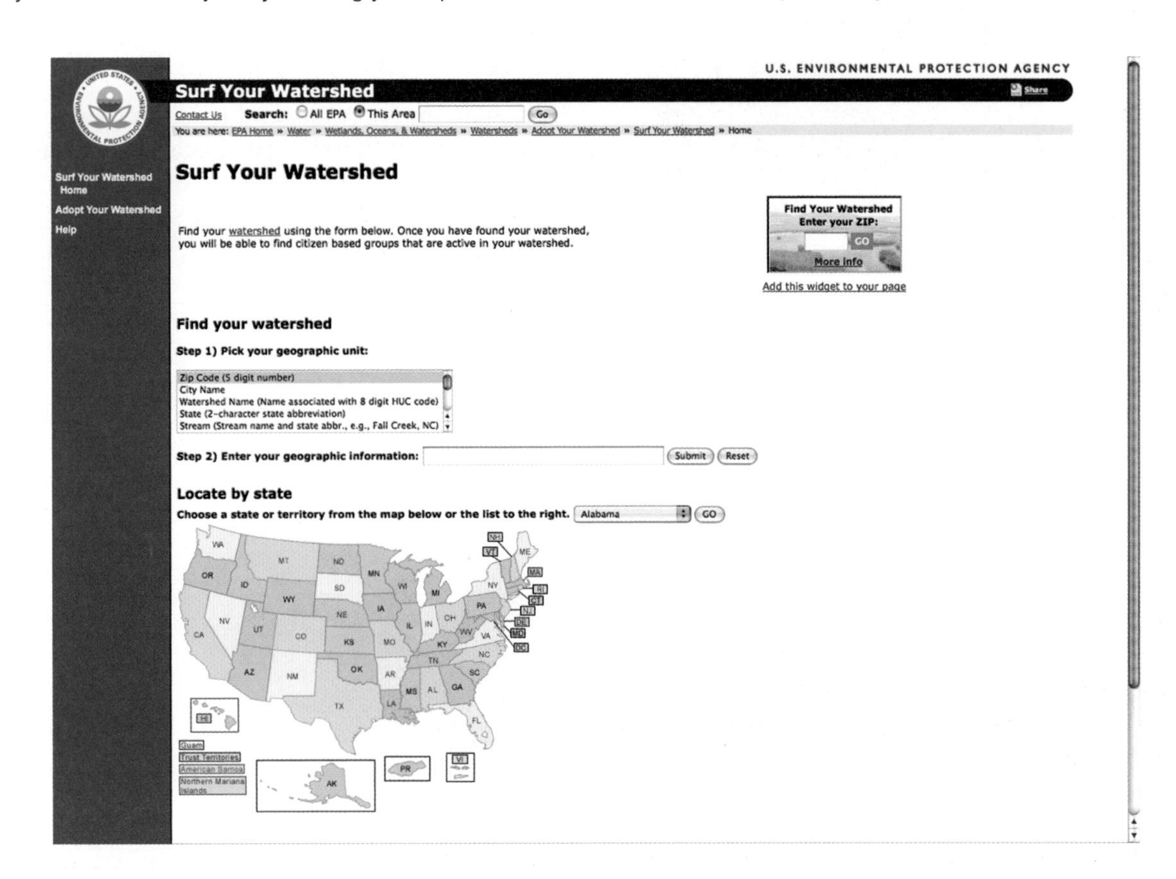

FLOW, DISCHARGE, AND CHANNELS

The amount of water flowing in a river at any given time is its **discharge**. It is the volume of water flowing past a specific point on the river per second. The discharge of a river depends mainly on two factors: the size of its watershed and the amount of precipitation (both rain and snow) the watershed receives.

We calculate discharge by multiplying the velocity of the water (in meters or feet per second, m/s or ft/s) by the river's cross-sectional area (width multiplied by depth). The relationship between a river's flow velocity, cross-sectional area, and discharge is:

$$\text{Discharge } (m^3/s) = \text{Area } (m^2) \times \text{velocity } (m/s)$$

Scientists use this fundamental relationship to understand how streams respond to changes in the amount of water they carry. For example, this relation explains why, if discharge increases but a river's cross-sectional area stays the same, its flow velocity must increase.

A river flows in a **channel**, the place where water is more or less continuously present and the main current flows. Channels vary in shape from narrow and deep to broad and shallow. However, Figure 7-5 on p. 194 shows that different channel shapes can have the same cross-sectional area. Channel cross sectional area is the key characteristic that determines how much water a river can carry. When discharge is unusually high, such as during spring thaws or winter storms, it may be more than a river's channel can hold. Water then overtops the channel's banks and flooding occurs.

BASE LEVEL

Rivers flow downhill until they get to a place where they can't flow any farther. These places are areas of lower elevation, such as lakes and enclosed surface basins, but the most common ones are ocean shores. The elevation at which a river cannot flow farther or erode deeper into the ground is its **base level**. The place that defines base level can be thousands of kilometers

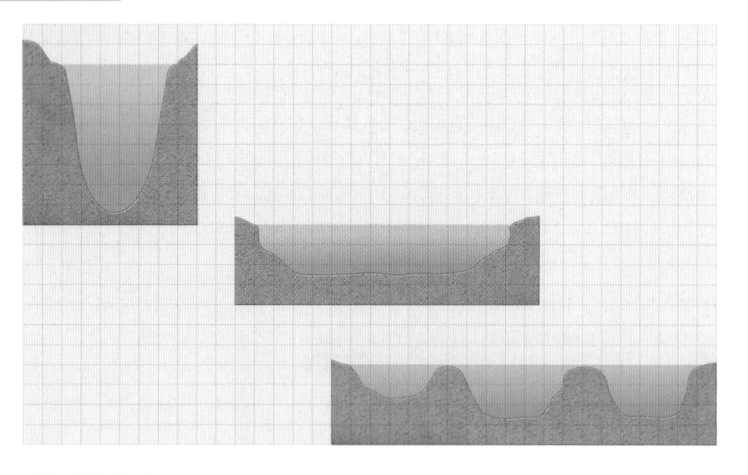

FIGURE 7-5 ▲ Channel Cross-Sectional Area
River channels of different shapes can have the same cross-sectional area.

(miles) from a river's source. The Mississippi River's base level is sea level at the Gulf of Mexico shoreline. The base level for a high mountain stream might be a nearby lake. Generally speaking, the ultimate base level for all rivers is sea level. Even where a river empties into a lake—a local base level—the water is residing in a temporary reservoir. Although it may take awhile, the water will eventually return to the ocean through one pathway or another of the water cycle (see Figure 2-17).

LONGITUDINAL PROFILES AND GRADIENT

Rivers take many twists and turns as they follow the local topography of the landscape. Regardless of all the twists and turns, we can show a river's overall vertical drop with a **longitudinal profile** like the one shown in Figure 7-6. A river's longitudinal profile shows the elevation of its channel bottom

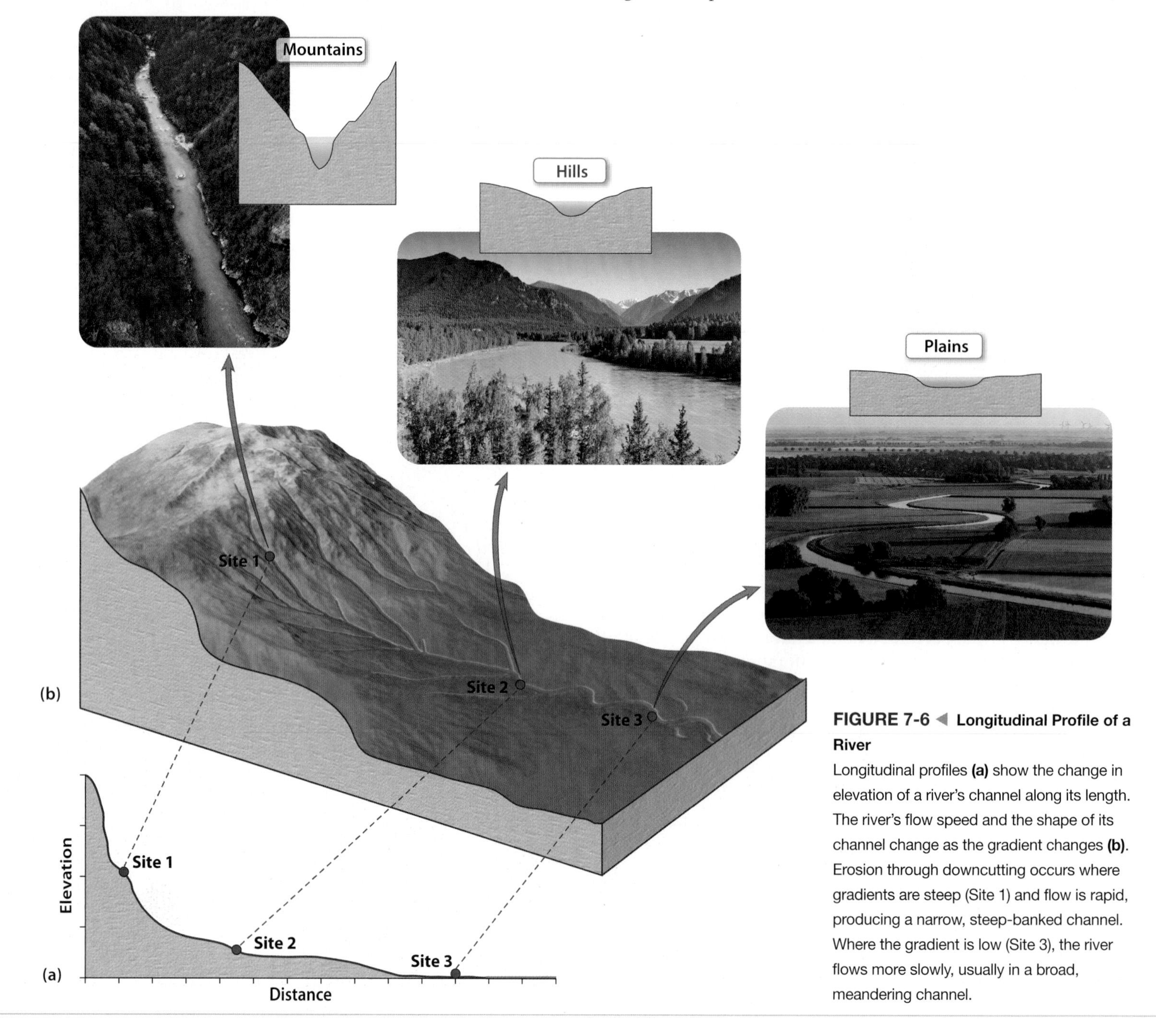

FIGURE 7-6 ◄ Longitudinal Profile of a River
Longitudinal profiles **(a)** show the change in elevation of a river's channel along its length. The river's flow speed and the shape of its channel change as the gradient changes **(b)**. Erosion through downcutting occurs where gradients are steep (Site 1) and flow is rapid, producing a narrow, steep-banked channel. Where the gradient is low (Site 3), the river flows more slowly, usually in a broad, meandering channel.

from the headwater areas to its downstream termination—the river's **mouth**. The loss of elevation along a river channel is its **gradient** (measured in meters per kilometer or feet per mile). In general, the steeper a river's gradient, the faster it flows, the more effectively it can erode, and the more sediment it can carry. These factors, in turn, influence a river's channel geometry. As Figure 7-6 indicates, steep gradients and fast flow rates lead to narrow, steep-sided channels, whereas gentle gradients and slow flow rates lead to broad, meandering channels.

EROSION

A river can erode in two directions. It can cut its channel down to lower and lower elevations or cut laterally into the channel's sides, the river's *banks*. Streams that flow across solid rock of the geosphere—bedrock—cut downward by plucking out loose rock fragments (Figure 7-7). Sand, gravel, and boulders carried along on the stream bottom wear down or abrade the bedrock, too. Over time, the result can be spectacular, as Figure 7-8 shows. The Colorado River has cut over 1500 meters (4900 ft) through bedrock to form the Grand Canyon. In this case, the river maintained its elevation by channel cutting as the surrounding region—the Colorado Plateau—was gradually uplifted by tectonic forces over millions of years.

If a river flows over sediment and not bedrock, it must pick up and carry sediment particles in order to erode downward. Sediment on the channel bottom will not move unless the river exerts a force greater than the sediment's resistance to movement, which is generally a function of its density as well as the friction and cohesion within the sediment. A river's erosive force depends on the depth of the water (the greater the depth, the greater the weight of the water on the sediment) and the river's gradient. A steeper gradient does two things to help erode sediment on a river bottom. If discharge is constant, a steeper gradient increases the flow velocity across the sediment on the channel bottom. A steeper gradient also means that the river bottom sediment is easier to move downstream: It is resting on a steeper slope and its resistance to movement is less than on a shallower gradient.

Where rivers bend, they can erode laterally into their banks. This is because a river's velocity tends to be greater and its channel deeper along the outside parts of bends. The greater the velocity and channel depth, the more force the river exerts against its bank. As it removes bank material, the river commonly develops a bare, steep, or overhanging bank called a **cutbank**. As erosion into this bank continues, it gets steeper and becomes unstable. Collapse of cutbanks is a common way for sediment to enter a river. Keep your eyes out for cutbanks on your next

FIGURE 7-7 ▲ A Steep-Gradient Stream
In headwaters where gradients are steep, rivers and streams flow fast, cut steep-sided channels, erode vigorously, and carry coarse sediment.

FIGURE 7-8 ▲ Rivers Erode the Geosphere
The Grand Canyon is a spectacular example of how effective river erosion can be.

FIGURE 7-9 ▲ Erosion Along River Cutbanks Can Be Hazardous

cross-country trip. As you can see from Figure 7-9, they are definitely not places where people should build homes or other structures. Cutbanks clearly show ongoing lateral river erosion.

Broad loops and curves are common along the lower parts of rivers, near their base levels. These bends, or **meanders**, which you can see in Figure 7-10a, change as a result of erosion and deposition working hand in hand. Erosion enables them to move laterally as much as hundreds of meters (yards) per year in some places, but lower rates are most common. Over time, rivers develop more and more pronounced meanders that may swoop back on themselves, leaving only a narrow strip of land and very little gradient difference between them. During floods, channel erosion may cause the water to take the shortcut across the narrow part of the loop. Sedimentation along the sides of the shortcut channel may then seal off the former loop, creating an **oxbow lake** like the one shown in Figure 7-10b. As the river continues to meander, the entire channel may shift position, leaving the oxbow lake a considerable distance from its river of origin.

What are some Earth systems interactions associated with stream erosion?

SEDIMENT TRANSPORT

If you've looked at a variety of streams and rivers, you probably noticed that some are clear and others are muddy or silty. The color varies, too, from coffee-colored to gray or even yellowish. The color reflects the fact that river water does not travel alone on its journey downstream. It also carries a cargo of sediment, dissolved minerals, and organic matter eroded from upstream sources in the watershed. All material that the river carries is collectively known as its **load**.

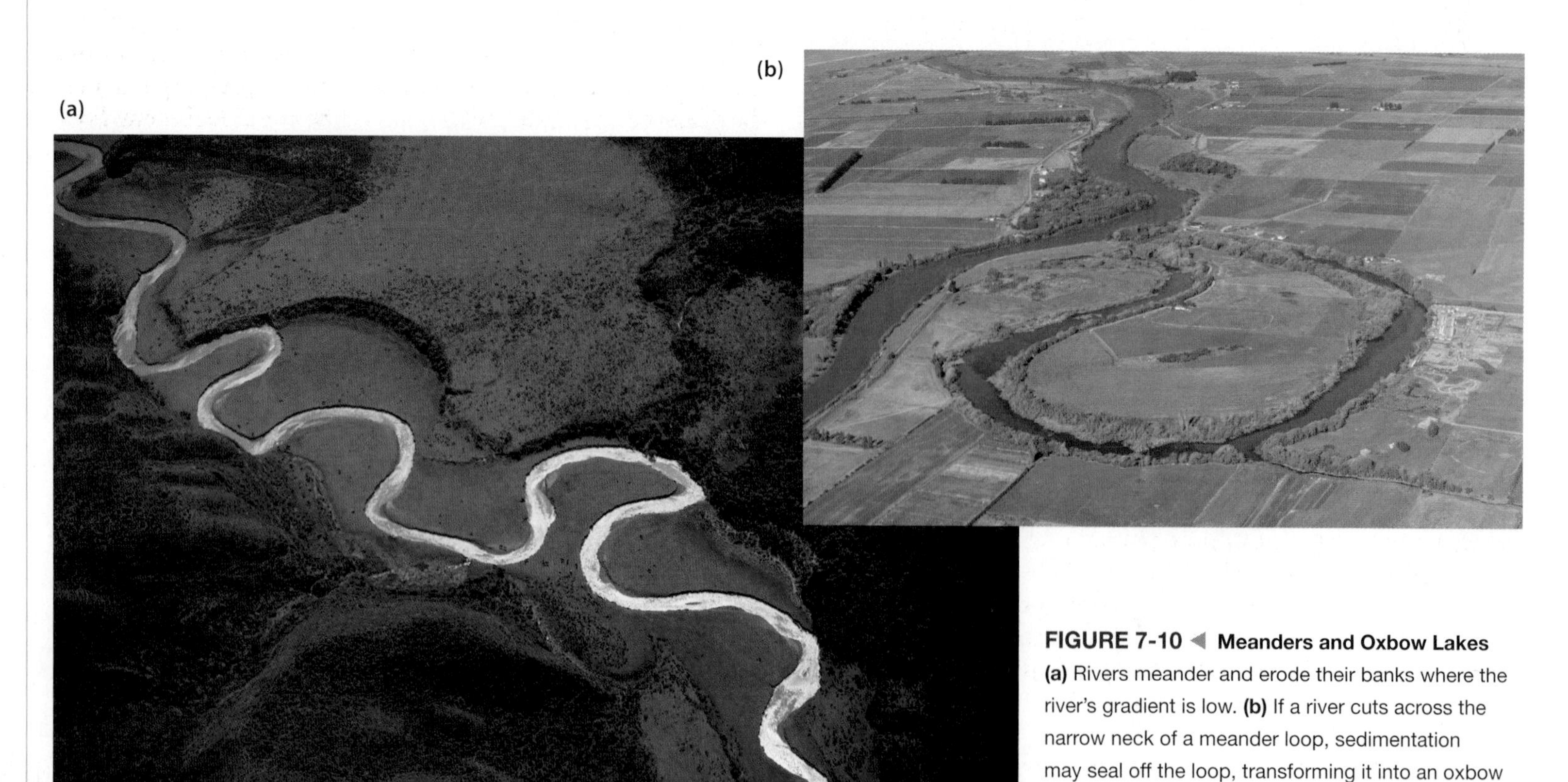

FIGURE 7-10 ◀ Meanders and Oxbow Lakes
(a) Rivers meander and erode their banks where the river's gradient is low. **(b)** If a river cuts across the narrow neck of a meander loop, sedimentation may seal off the loop, transforming it into an oxbow lake.

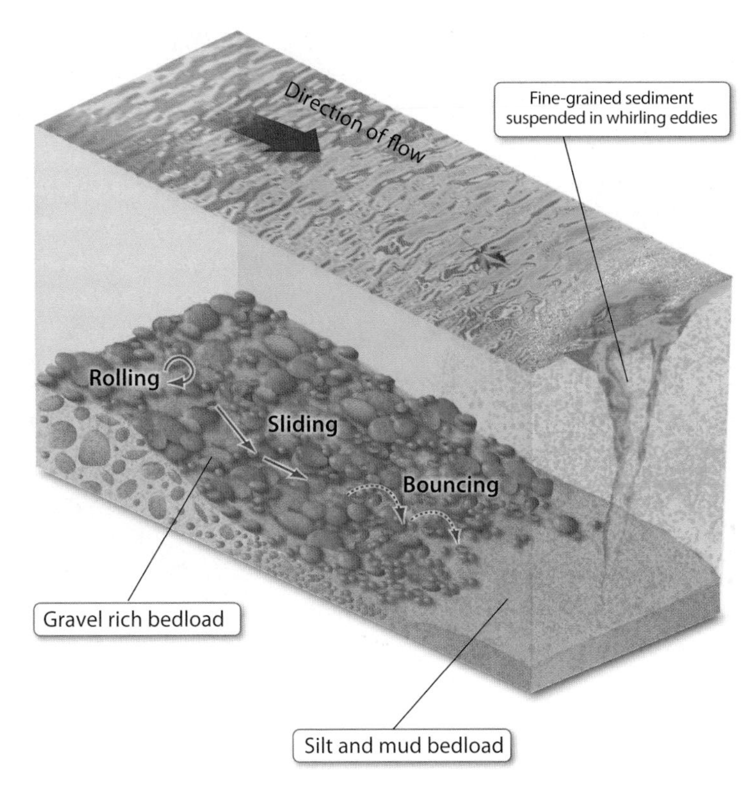

FIGURE 7-11 ▲ Mechanisms of Sediment Transport
Gravel-rich bedload is abundant along steeper river gradients, whereas suspended and dissolved loads are common at lower gradients.

FIGURE 7-12 ▲ The World's Muddiest River
China's Huang Ho (Yellow River) carries a tremendous suspended load of silt—up to 70% of the river's volume.

The sediment carried by a river may include gravel, sand, silt, and mud. The faster the river's velocity, the larger the sediment material it can carry. At any specific velocity there is a maximum sediment size that can be transported. The larger sediment, the *bedload*, is dragged, rolled, skipped or bounced over the river bottom—a process called *saltation* (Figure 7-11). The finer particles carried along in the river water itself are the *suspended load*. Dissolved minerals and some organic materials leached from the soil and rock in the river's watershed make up the *dissolved load*.

At a specific velocity and discharge, a stream has a maximum amount of bed and suspended load it can carry, called its **capacity**. A stream's capacity increases with increasing velocity and discharge, which is why much more erosion and sediment transport takes place at times of flooding. In general, most of the material transported in large rivers consists of suspended load, such as fine silt.

China's Huang Ho River has an impressive suspended load—it is the world's muddiest river. Its headwaters lie in mountainous Tibet, and it flows a total of nearly 5500 kilometers (3400 mi) to the Yellow Sea. In English translation, it is called the "Yellow River," a reference to the color of the sediment it carries (see Figure 7-12). Its middle course cuts through loose, easily eroded deposits of windblown silt called *loess*. During flood conditions, the river may carry as much as 34 kilograms (75 lb) of silt per cubic meter of water, so that the silt makes up 70% of the river's volume. For comparison, consider that the Nile River carries approximately 1 kilogram (2.2 lb) of sediment per cubic meter. Annually, the Yellow River moves an estimated 1.5 billion tonnes (1.7 billion tons) of sediment to the sea.

How can the Yellow River carry such an enormous load of sediment? As in any river, the swirling, turbulent motion of the water carries the suspended load. Have you ever stirred a pot of soup? As you swirl the spoon around in the mix, the liquid becomes turbulent. Whorls and eddies in the broth pick up the heavier ingredients and carry them along. If you stop stirring, the ingredients begin settling to the bottom. Turbulent flow keeps ingredients in a soup or sediment in a river suspended. As in your soup pot, river water carries more and bigger sediment the faster it flows and the more turbulent it is.

SEDIMENT DEPOSITION

Wherever a river loses its ability to transport sediment, it deposits sediment. Where is this likely? A river deposits sediment where its velocity decreases. Velocity decreases along the inside of bends, for example, where sand and gravel bars form.

In rivers with abundant sediment, lower-velocity zones within the river's channel can be places of sediment deposition. Over time, within-channel deposits may become islands and a **braided channel** forms. Braided channels, as you can see in Figure 7-13 on p. 198, are places where many small channels interconnect between areas of sediment deposition. There is just too much sediment for the river to carry under normal flow conditions. At times of higher flow, the within-channel sediment can be moved and redeposited, causing changes in the channel network. Braided channels are common where glaciers flow into valleys and release meltwater and abundant sediment.

FIGURE 7-13 ▲ Braided Channels
Channels become braided where sediment is abundant.

FIGURE 7-14 ▼ An Alluvial Fan
Tributaries can deposit sediment in alluvial fans where they reach local base levels.

FIGURE 7-15 ▼ The Mississippi River Delta
Many rivers deposit sediment in deltas when they reach base level at their mouths. You can explore the features of the Mississippi River delta shown in this NASA photograph on the internet. Mississippi River delta

However, the largest amounts of sediment are deposited where rivers approach their base level, even local base levels such as valleys and lakes. A stream rushing down a steep mountain canyon to a nearby valley commonly carries large amounts of sand and gravel, or *alluvium*. Where the stream enters the valley its gradient abruptly diminishes, the discharge spreads out over several channels, and its sediment is deposited in an **alluvial fan** like the one seen in Figure 7-14.

If the base level is defined by a body of water such as a lake or the ocean, the sediment is deposited at the mouth of the stream in a fan-shaped **delta**. In deltas, the river's velocity diminishes so much that most of the sediment load is quickly deposited. This clogs the river channel with sediment and causes new channels, *distributaries*, to branch off to its sides. Distributaries disperse flow rather than bring it together, as tributaries do. Distributary channels commonly have a bird's-foot pattern and give deltas their characteristic fanlike shape, like that of the Mississippi River's delta, shown in a satellite photo in Figure 7-15.

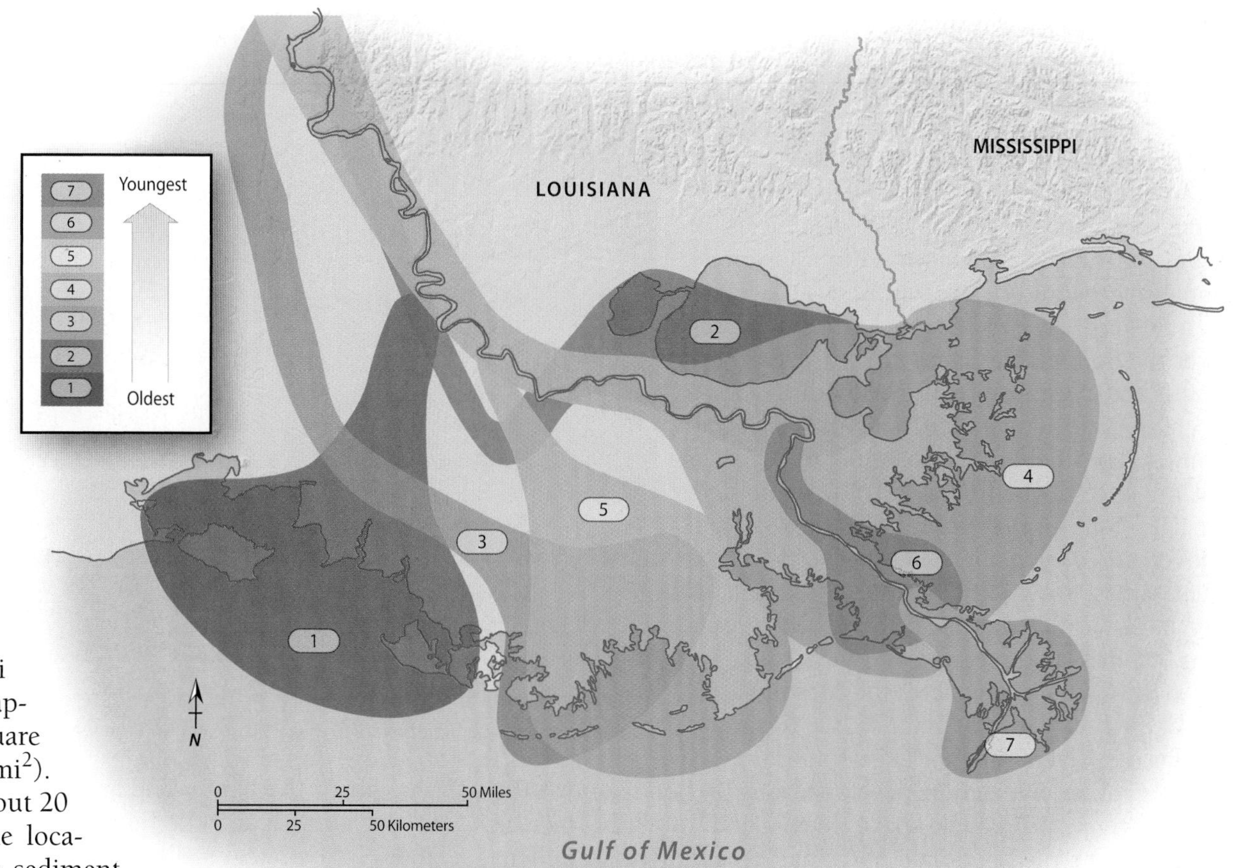

FIGURE 7-16 ▶ Mississippi Delta Sedimentation
Sediment deposition at the mouth of the Mississippi River has caused the river channel to shift and form several different delta lobes over the last 5 to 6 thousand years.

Today's Mississippi delta covers an area of approximately 40,000 square kilometers (15,000 mi^2). Since the last ice age (about 20 thousand years ago), the location of the river's active sediment deposition has shifted several times as the river slightly moved its course east or west. Each time, the sediments would form a new lobe—a bulbous platform of sediment connected to the shore by narrower channel deposits, as you can see in Figure 7-16. The older sediment lobes have compacted and subsided below sea level.

Through deposition processes, rivers sort sediment by size. The larger, heavier (coarser) sediment is deposited first, so these materials, bouldery gravel for example, will tend to be deposited in faster, upstream parts of the river. Finer sediment can be carried farther. This factor, combined with the breaking down of sediment grains as they are carried along in the bedload, means the sediment in lower parts of a river tends to be sand-, silt-, and mud-rich. The farther from a river's headwaters you are, the finer the river's sediment becomes. Through erosion, sediment transport, and sediment deposition, rivers play a key role in the rock cycle (Chapter 4).

What are some Earth systems interactions associated with stream sediment deposition?

FLOODPLAINS

Have you noticed that the land along a river tends to be low and level—essentially flat? And have you also noticed that when the river leaves its channel and floods, this low area is where waters cover the land first? This is the river's **floodplain**, a distinctive feature along rivers (a typical example is shown in Figure 7-17). Floodplains along a river's upper *reaches* (specific sections of a river) may be narrow parts of valleys, but along lower reaches they can be very broad and extend for many tens of kilometers (miles) from its banks.

FIGURE 7-17 ▲ A Typical Floodplain
Floodplains are relatively level areas adjacent to rivers that floodwaters commonly inundate.

Floodplains form because the river changes the nearby land, especially during flooding. The river's flow during flooding helps erode and level the land. Flooding also carries fine suspended sediment onto the land and deposits it as floodwaters recede. Sediment deposits on floodplains are sheets of mud, silt, sand, or gravel that fill in low areas and leave behind a level ground surface. As floodwaters leave the channel and spread out over the floodplain, the maximum size of sediment and the capacity of the flooding waters decrease and larger (coarser) material becomes deposited along the banks of the river (at the edge of the floodplain) as natural levees. In some places, lateral and downstream movement of meanders helps level the land near rivers, too. Although some floodplains are narrow, all rivers have them, and people need to cross them to get to a river's bank.

FIGURE 7-18 ▲ Brisbane, Australia
People crowd the banks of rivers in many places around the world.

The ancient city-states of Mesopotamia developed on the floodplain of the Tigris and Euphrates Rivers in what is present-day Iraq. The Egyptian civilization developed on the floodplain of the Nile River. And people have lived—and drowned by the millions—on the floodplain of China's Yellow River for millennia. Throughout history, rivers and their floodplains have been magnets for human society.

Today, populations are still concentrated on floodplains. (Figure 7-18 shows Brisbane, Australia, but you can probably think of many other examples.) In the United States, millions of people live on them. They include the vast floodplains of the Mississippi River and innumerable smaller creeks, streams, and rivers. The agricultural activity, industrial activity, and shipping along rivers are an important part of the national economy. This puts us in a bit of a bind, however. Living on floodplains may put the interests of people in conflict with nature. Our interest is to prevent floods that destroy property and threaten public health and safety. Rivers, on the other hand, will periodically overtop their banks and spill onto the floodplain when there is too much discharge for their channels to hold.

What are some Earth systems interactions associated with floodplains?

7.2 Floods

The ancient Egyptians worshipped the Nile River as a god. This is not surprising when you consider that this vast waterway flowing north from the interior of Africa was largely responsible for the survival and flourishing of Egyptian civilization for 3000 years. Indeed, Egypt is sometimes called "the gift of the Nile." The annual spring flooding of the Nile was welcomed, and the Egyptians abetted inundation of their lands. In a strategy called "basin irrigation," farmers parceled the fields into areas covering up to 20,000 hectares (50,000 acres) surrounded by earthen levees. When the Nile flooded, the water lingered within the levees and deposited a layer of rich silt. This provided fresh, nutritious soil for crops of wheat, barley, and fruit.

Since the completion of the Nile's Aswan High Dam in 1959 and other control measures, Egypt's river god has been put under the yoke, and the relationship of the people to the river has changed. Reengineering the Nile provided a steady water supply to Egypt and Sudan, hydroelectric power, and some protection from large floods. But trapping sediment behind the dam has led to delta erosion at the river's mouth and stopped the annual soil replenishment that came with flooding—the floodplain's rich agricultural lands are no longer sustained by the river's natural processes.

WHAT FLOODS ARE

Flooding occurs when the discharge of a river is so great that the water rises and overtops the river's natural or artificial banks (**levees**). River levels naturally fluctuate up and down within their channels but once they overtop their banks and endanger people or property, they have reached *flood stage*. The extent of the flood is then given in terms of how many feet or meters the river is above flood stage. The term **flood crest** refers to the highest level above flood stage that a river achieves during a flood.

Flooding is not rare or unnatural. In river systems, flooding is inevitable and necessary. To understand why this is so, recall the relationship between a river's discharge, its cross-sectional area, and its flow velocity. If the amount of water entering the channel increases, the river can respond in two ways. One way is for the velocity of the water to increase. The water moves through the channel more quickly because the river must convey more water through the same cross-sectional area. (The same thing occurs if the amount of water in the river remains the same but the channel narrows. Now the river must move the same amount of water through a smaller space, so it speeds up.)

There are limits to the water's velocity, however. Water encounters frictional resistance as it flows down its channel. Every rock on the riverbed or bend in the channel adds friction, impeding the flow. The flow becomes more turbulent and the water begins to back up. At a certain threshold, the only way that the river can accommodate the volume of water in the channel is to increase its cross-sectional area. A river can increase its cross-sectional area during times of high discharge by removing sediment from and scouring its channel bottom. More commonly, though, it first rises and completely fills the channel, then overtops its banks and floods.

PRECIPITATION AND FLOODING

In most cases, the source of the extra discharge that results in flooding is precipitation within the river's watershed. It certainly seems intuitively logical that the more precipitation within a watershed, the greater the likelihood of a flood. However, the relationship between precipitation and flooding can be complex, and factors such as the intensity, duration, and timing of the precipitation play important roles.

Intensity

If enough rain falls in a short enough period of time, flooding can occur. This was surely the case on June 9, 1972 when Rapid Creek surged through Rapid City, South Dakota, killing 238 people and causing over $600 million in damage, some of which you can see in Figure 7-19a. The 1972 Black Hills-Rapid City Flood revisited

On the afternoon of June 9, a group of thunderstorms parked over the Rapid Creek watershed, and within one 6-hour period, up to 38 centimeters (15 in) of rain fell near the town of Nemo. Overall, an area of 155 square kilometers (60 mi^2) received more than 25 centimeters (10 in) of rain in just a few hours. The flood struck Rapid City at 9:30 p.m., peaked at midnight, and receded to channel levels by 5:00 a.m. on June 10. The exceptionally intense rainfall created the largest flood ever experienced in the Black Hills area.

Duration

Precipitation that continues for lengthy periods can deliver amounts of water that a watershed's stream channels cannot hold. The March 17–19, 2008 floods in the central Mississippi valley are an example. Flooding occurred over parts of several states but was particularly extensive in Missouri. The cause was a stationary weather front that stalled moisture-laden air masses moving north from the Gulf of Mexico. Rain fell for two days along the weather front. The amount of rain varied from 2.5 to 25 centimeters (1–10 in) through the region. In some places two to three inches fell in four hours, but overall it was the cumulative rainfall over two days that led to extensive flooding along the major rivers (Figure 7-19b). This precipitation was

(a)

(b)

FIGURE 7-19 ▲ Both Intense and Prolonged Rainfall Can Cause Flooding
(a) In June 1972, thunderstorms dumped intense rains in the Rapid Creek watershed. Enough rain fell in just a few hours to cause a flash flood that devastated Rapid City, South Dakota and killed 238 people. **(b)** Sustained rainfall over two days in March 2008 produced extensive flooding along the Mississippi River.

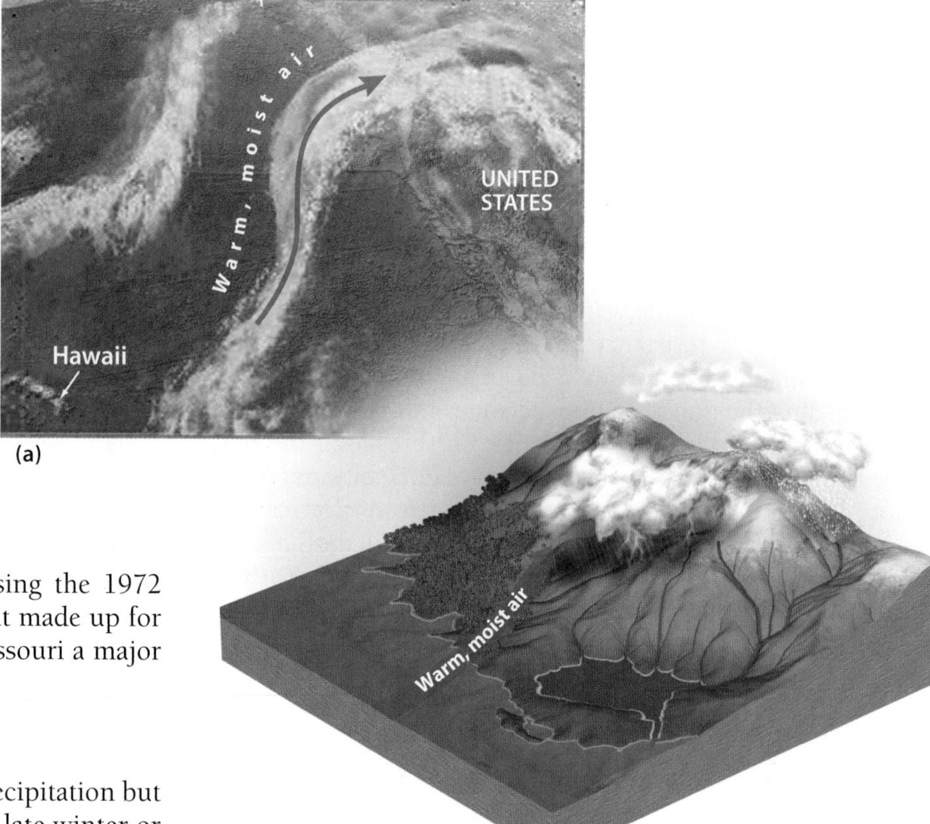

FIGURE 7-20 ▶ The "Pineapple Express"
Warm, moist air from the Pacific Ocean is carried by high-altitude winds to the U.S. Pacific Coast **(a)**. When it reaches the snow-covered western slopes of the Cascades and Sierra Nevada mountains, it releases its moisture as rain **(b)**. The combination of rain and melting snow can cause severe flooding.

mild in intensity compared to the rainfall causing the 1972 Rapid City flood but what it lacked in intensity it made up for in duration. President Bush declared parts of Missouri a major disaster area on the evening of March 19, 2008.

Timing

Sometimes it's not the intensity or duration of precipitation but its timing that leads to flooding. Flooding in the late winter or early spring can occur if what are otherwise normal rainstorms quickly melt the winter snow pack. This situation occurs in Washington, Oregon, and Northern California due to a weather pattern called the Pineapple Express. During the winter, when snow drapes the mountains of the Pacific Northwest, high-altitude jet-stream winds carry moist tropical air from the vicinity of Hawaii toward the U.S. coast (Figure 7-20a). If this air drops its rain on accumulated snow, the meltwater it creates can set off floods on the west slopes of the Cascades and the Sierra Nevada Mountains (Figure 7-20b).

Seasonal flooding commonly results from events that coincide in time. The Red River of the North's watershed covers the eastern third of North Dakota, the northwest corner of Minnesota, and parts of adjacent Manitoba, Canada. In many past floods, the underlying cause was spring snowmelt, in combination with factors such as saturation of the soil by fall rains, abnormally heavy winter snows, and frozen soil conditions that increased surface runoff. Flooding in the river's basin in 1997 followed unusually heavy winter snows and a late spring blizzard. This flood was especially devastating to Grand Forks, North Dakota, where an uncontrollable fire added to the destruction.

In 2002, the cause of flooding on the Red River of the North was quite different. Two spells of torrential rain fell in June, less than two weeks apart. It was bad timing. The first rainstorm so completely saturated the watershed that rain from the second storm had no place to go but across the land and downstream. Among the casualties was the town of Roseau, Minnesota, where 95% of the town was inundated (Figure 7-21). The flood stage recorded on June 24 by a gauge on the Wild Rice River, which lies within the Red River of the North watershed, indicated a flood discharge of a size not seen more than once every 500 years.

FIGURE 7-21 ▲ The Timing of Rainfall Can Also Be Crucial
In June 2002, two periods of heavy rain led to the inundation of Roseau, Minnesota. The first heavy rain saturated the ground, causing large amounts that fell during the second period to run off into the Red River of the North. The resulting flood inundated almost the whole town of Roseau, Minnesota.

FAILING DAMS AND FLOODING

Dams have far-reaching effects on drainages. They block channels and create reservoirs that store large amounts of water. If they catastrophically fail, a flood of water is released downstream. An interesting fact about dams is that they are not all man-made. Floods have been caused by the failure of natural dams created by landslides that blocked drainages and the break-up of natural ice dams as well as the failure of man-made dams.

Failure of landslide dams

Some landslides are large enough to extend completely across valleys, damming river channels and creating lakes upstream. On June 23, 1925 such a landslide blocked the Gros Ventre River near Jackson Hole, Wyoming. A lake soon formed upstream from the landslide. This natural dam was examined and thought safe, but within two years it failed. On May 18, 1927, after two weeks of heavy rains, the lake overflowed the landslide. Erosion by the escaping water caused the natural dam to collapse, sending a flood of water down the Gros Ventre River. Vigilant observers discovered the eroding dam and rushed to warn those downstream. The warning saved many lives but six people died and the town of Kelly was devastated. | Grand Tetons, the Gros Ventre Landslide – Wyoming tales and trails | Landslide dams are not that common but they are clearly not constructed to engineering standards. Those living on the floodplains downstream from such dams are assuming that they will hold and keep them safe from catastrophic flooding.

Failure of ice dams

Ice dams come in a few varieties. Many are related to the seasonal breakup of frozen rivers, but the advance of glaciers creates dams in some places, too.

In many northern states, rivers commonly freeze over each winter and thaw each spring. When the rivers start to thaw, blocks of ice break free and begin floating downstream. It's common, though, for the ice blocks to get jumbled together and hung up in *ice jams*. Unlike dams, ice jams do not completely block a river channel because some water continues to flow underneath them. They do, however, decrease the channel's cross-sectional area and partially block the river's flow. Interestingly, ice jams can lead to flooding both upstream and downstream. While the ice jam is intact, it can trap water on the upstream side. Growing lakes upstream from ice jams have flooded many towns.

Ice jams are always temporary features, as the warm days of spring continue to melt the river ice. Melting ice jams have been known to fail within hours. When they do, the surge of water that heads downstream can cause flooding. Flooding related to spring breakup of frozen rivers has occurred in 36 states and causes $125 million in damages each year. They are so common in some places that communities have begun to invest in ways to control these nearly annual events. | Ice dams: taming an icy river |

Where glaciers send rivers of ice down valleys, they can block tributary drainages and create lakes upstream from the ice dam. If these dams fail due to glacier retreat, melting, or physical breakup, the lake water can be released within hours to cause downstream flooding. This happens in Alaska today. | 2007 ice jam floods | But an amazing example occurred 13 to 15 thousand years ago in the Columbia River watershed of the Pacific Northwest.

Tremendous volumes of water literally roared across large parts of Idaho, Washington, and Oregon when the glacier damming ancient Lake Missoula failed. | CVO website – glacial Lake Missoula | Former Lake Missoula was 320 kilometers (200 mi) long and held over 2000 cubic kilometers (500 mi^3) of water. It formed repeatedly upstream from ice dams 1250 meters (4100 ft) high that blocked the Clark Fork River Valley in western Montana. The advancing glacier that created the ice dams was at the edge of the extensive ice sheet that covered large parts of North America at the time.

The glacier dam failed and then reformed many times during a 2000-year period. For anything trying to live downstream, each failure was a catastrophe. A volume of water ten times the flow of all the rivers in the world rushed down the Columbia River drainage, scouring the land. Many of the surface features of the region are directly attributable to the massive floods, and the scoured land became known as the *channeled scablands* (Figure 7-22, p. 204). | Hugefloods | Geologist J. Harlen Bretz coined this term in the 1920s. He was the first to recognize the tremendous size, extent, and effect of the flooding. But because such huge floods were previously unknown, he had a struggle convincing some others that such vast floods had actually occurred. After geologist Joseph Pardee provided a source for the floodwaters by identifying where ice dam failures had catastrophically released the waters of ancient Lake Missoula, the controversy diminished and Bretz was shown to be correct. The repeated failure of the glacier dam to ancient Lake Missoula did indeed create tremendous floods that scoured the land. | USGS: the channeled scablands of eastern Washington |

Failure of constructed dams

Most dams on rivers are constructed by people. There are about 75,000 constructed dams in the United States alone. Most of these dams are soundly constructed and have successfully withstood the wide range of river flows that accompany severe storms. However, from time to time problems develop. Some dams are over 100 years old, their original designs may have been flawed, and their physical conditions may have deteriorated. The Kelly Barnes Dam in Georgia is an example. It failed at 1:30 a.m. on November 6, 1977 and the resulting flood killed 39 people. | USGS Atlas HA-613: Kelly Barnes Dam flood |

In some cases, earthquake shaking has led to dam collapse or landslides into dam reservoirs have caused dam failure. More commonly, though, failure is caused by unusually heavy precipitation and river flow that stress dams beyond their

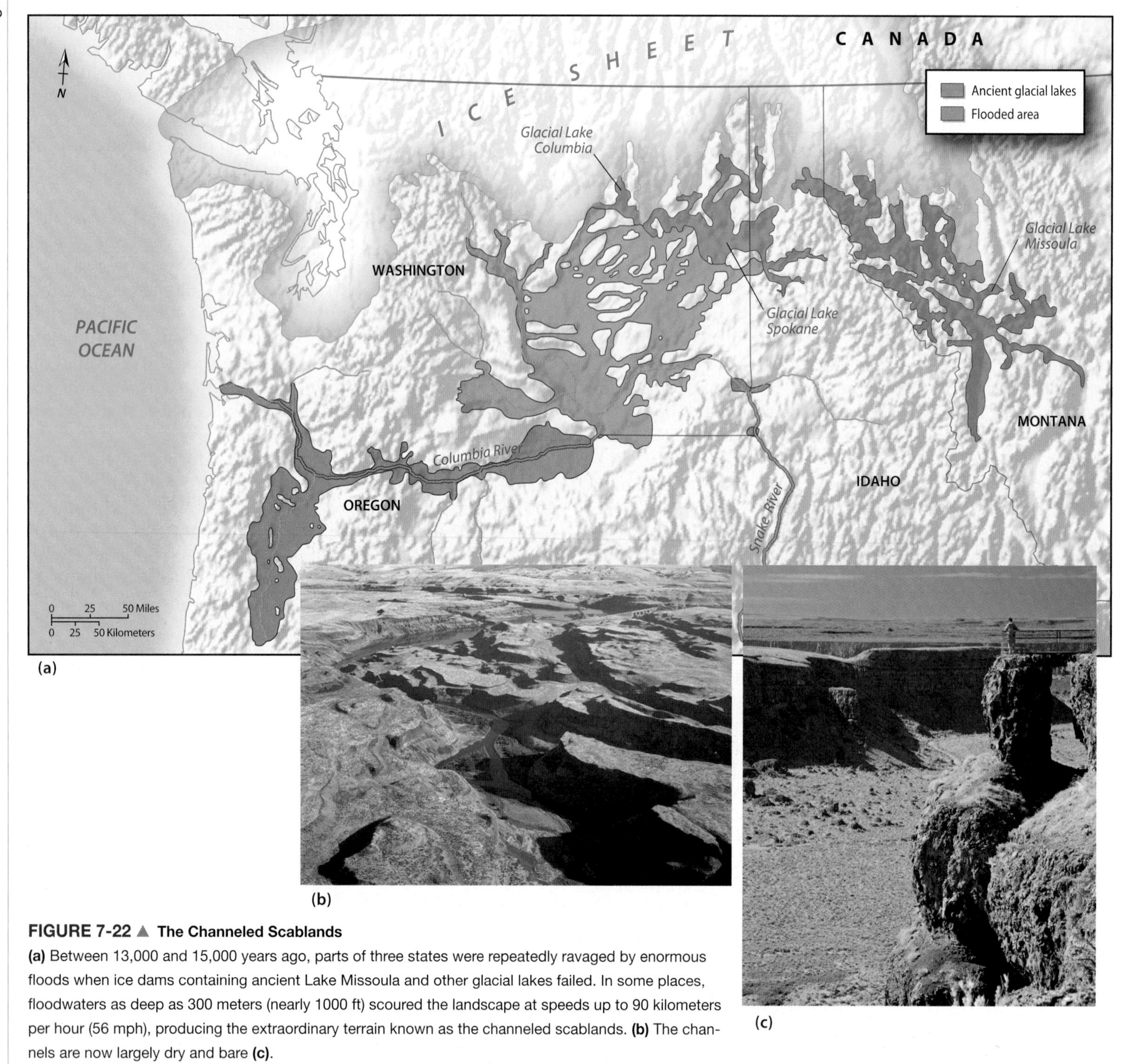

(a)

(b)

(c)

FIGURE 7-22 ▲ The Channeled Scablands

(a) Between 13,000 and 15,000 years ago, parts of three states were repeatedly ravaged by enormous floods when ice dams containing ancient Lake Missoula and other glacial lakes failed. In some places, floodwaters as deep as 300 meters (nearly 1000 ft) scoured the landscape at speeds up to 90 kilometers per hour (56 mph), producing the extraordinary terrain known as the channeled scablands. **(b)** The channels are now largely dry and bare **(c)**.

design capacity. This occurred in 1975 when Typhoon Nina brought extremely heavy rainfall to the Ru River drainage of southern China. Bangiao Reservoir Dam was one of many dams in the region. It could not adequately release the water pooling upstream, the reservoir overtopped the dam, and the dam failed on the evening of August 7, 1975. The resulting flooding led to more dam collapses and when the storm was over, 26,000 people had been killed directly by flooding, 145,000 died from subsequent disease and famine, and nearly 6 million buildings were destroyed. Overall, 11 million people were impacted by the flooding.

LAND USE AND FLOODING

Human activities that affect soil conditions and the flow of water through a watershed can contribute to flooding. Everything from how fields are tilled to where land is paved over for streets and parking lots plays a role. Here are some examples.

Effects of cultivation

Bare soils are susceptible to erosion. Tilling for agriculture that leaves bare soils for rainfall to wash away, as in Figure 7-23, can increase the amount of sediment load in a river. If this sed-

FIGURE 7-23 ▲ Farming Practices Influence Runoff and Sedimentation
Bare fields contribute much sediment to surface runoff and rivers—sediment that was valuable soil.

iment is deposited in the river channel, it decreases its cross-sectional area, and therefore its maximum discharge. Unless the river can carry the sediment away, the volume it can accommodate during times of high water will be reduced, and it will be more likely to spill over its banks.

Loss of wetlands

Think of a **wetland** as a combination of sponge and reservoir. Wetlands, like the one shown in Figure 7-24, impound water and also allow it to soak into the groundwater system. Converting land to agricultural use may involve draining or filling wetlands. The result is that water runs off into the river instead of remaining in wetlands. This may raise the level of the river during floods, making floods larger.

Urbanization

With increased development, buildings, roads, and paved parking lots may sprawl across the formerly open lands or agricultural fields in a river's watershed. Towns construct drainage ditches and storm sewers to funnel runoff from the pavement and roofs to nearby rivers. Where urbanization transforms forests, fields, and wetlands into buildings and pavement, it reduces the amount of water that can soak into the ground or be absorbed by plants. (Plants may absorb 50% or more of the rainwater that falls in a watershed.) Instead, the water runs off the pavements and roofs and pours into the river in a matter of minutes or hours. Effects of urban development on floods This has two effects: During floods, the stage of the river rises higher and faster. Meanwhile, in the populated areas, local flooding occurs if the volume of runoff exceeds the capacity of the storm sewers and surface runoff channels.

FIGURE 7-24 ▼ Wetlands
Wetlands like this one are a natural flood-control mechanism along rivers.

In The News

The Eastern Deluge of 2006

It was a deluge, a deluge that lasted for days. In late June 2006 officials placed 15 eastern states and the District of Columbia on flood alert. And the record rains came, a foot per day in many places. Two moisture-rich weather systems, one from the Gulf of Mexico and one from the Atlantic Ocean, met and stalled over the region, and the deluge began. Before it was over, 12 people died, hundreds of thousands had to evacuate their homes, people had to be rescued from rooftops, and drinking water became scarce. Across the nation, scenes of the flooding—along rivers, in subdivisions, and across parking lots—were in news broadcasts for days. One of the stark images captured during the flooding was of private airplanes sitting in a field of water.

This event brought home to anyone watching the news unfold how urbanization can affect flooding. The parking garage at the Department of Justice in Washington, DC, flooded—too much water with no place to go. Parking lots and streets became lakes and streams across the highly urbanized eastern states. It was a record rainfall event and flooding was likely even without urbanization, but examples of overflowing storm water drains were everywhere. Urbanization decreases the amount of water that the land can adsorb, decreases the pathways runoff can take across the land, and leads to flooding in unlikely places—like the Department of Justice parking lot. Urbanization can make even regular storms hazardous enough to be newsworthy.

MAN-MADE FLOODS

Although the word *flood* doesn't have positive associations for most people, ecologists have discovered that floods actually have a variety of benefits. Natural flood cycles help sustain and renew plant and animal habitat. It was with such considerations in mind that environmental managers have begun releasing periodic floods at the foot of the Glen Canyon Dam. | Controlled flooding on the Colorado River in the Grand Canyon | The dam interfered with the natural cycle of spring flooding. Much of the sediment that would otherwise have been deposited on the river's bank remained in Lake Powell, the reservoir created by damming the river. One result was a decline in the amount of sandbar space. Sandbars provide living space for wildlife and space for campers to enjoy the river. Hatchling fish live in the backwaters shielded by sandbars, which sometimes evolve into nutrient-rich wetlands. In a free river, these sandbars and backwater areas may be scoured during floods, but when the flow of the river slows down again, the sandbars reform. Trapping sediment behind the dam meant that erosion exceeded deposition of sediment downstream from the dam. The result was loss of habitat.

The first flood to reverse this impact of the dam was an experiment in 1996. | Geotimes Grand Canyon floods | The turbulent water scoured sand from the riverbed and deposited it on the margins of the channel. As you can see in Figure 7-25, the redeposited sand formed beaches large enough for camping. Animals benefited, too. Numerous backwater channels were established, creating potential habitat for humpbacked chub and other endangered fish. Carefully designed, man-made floods are now a means of partially restoring ecologically beneficial flood conditions below dams while preserving the benefits of dams to people. They are a way of sustaining river habitat. The most recent man-made flood below the Glen Canyon Dam was on March 5, 2008.

TYPES OF RIVER FLOODS

Floods vary significantly in how they develop. Some rush down rivers and surprise people, while others slowly build over days. These are examples of flash floods and riverine floods.

Flash floods

Flash floods commonly occur when a storm dumps a large amount of rain over a watershed in a relatively short period of time. Flash floods tend to crest rapidly and recede rapidly. They are essentially a large mass of water flowing down a constricted space, with a correspondingly large peak discharge. The danger to people in the immediate vicinity can be very grave and the damage severe, though not necessarily widespread. The 1972 flood at Rapid City, South Dakota is an example.

Another factor that can contribute to flash flooding is the **permeability** of the ground in the drainage basin, defined as the capacity of rock, sediment, or soil to allow water to pass through. In a climate in which the ground is rocky or dry and hard—conditions common in the western United States—rain from a sudden downpour may run off the land instead of soaking in.

In narrow drainages with steep slopes, such as mountain canyons, the topography promotes flash flooding by funneling a large volume of water into a restricted space. The flash flood may arrive as a rumbling wall of debris-choked water. This multiplies the water's destructive force. The most deadly floods

FIGURE 7-25 ◀ Man-Made Floods Reestablish Habitat in the Grand Canyon
A sandbar downstream of Glen Canyon Dam, at the 30-mile mark on the Colorado River, gained area and volume after an experimental flood in 1996.

are commonly flash floods. The 1976 Big Thompson Canyon flood is an example.

The Thompson Canyon flash flood occurred on July 31, 1976, on the Front Range of the Rocky Mountains, northwest of Denver. The canyon and its stream, the Big Thompson River, are popular recreation destinations, and the road up Thompson Canyon heads to Rocky Mountain National Park. Many people were in the canyon that weekend for camping and other activities. Capping a day of beautiful, sunny weather, winds aloft pushed moist humid air up the mountain slopes (Figure 7-26a)—one of the key ingredients for a summer thunderstorm.

FIGURE 7-26 ◀ The 1976 Flash Flood in Big Thompson Canyon
(a) On July 31, 1976, a thunderstorm dropped torrential rains over the headwaters of the Big Thompson river. A wall of water up to 6 meters (20 ft) high rushed down the canyon **(b)**, destroying everything in its path (including parts of U.S. Highway 34) and killing 145 people.

Most often, high-level winds blow developing thunderstorms away from the Front Range. However, on this particular day, the winds were weak. The storm parked over the drainage of the Big Thompson River, and starting in the early evening it began to pour. As much rain fell that night as typically falls in an entire year. The discharge of the Big Thompson River rose from 3.9 cms (140 cfs) at 6 p.m. to 884 cms (31,200 cfs) by 9 p.m. At the same time, the rocky walls of the canyon and steepness of its slope funneled the water rapidly toward people and towns downstream. Tragically, hundreds of motorists were still making their way down the canyon on U.S. Highway 34 when a wall of water as high as 6 meters (20 ft) descended on them, choked with mud and debris (Figure 7-26b). The death toll for the night was 145, including six people whose bodies were never found. Hundreds of homes were destroyed, and scores of automobiles were heaped up in piles.

What are some Earth systems interactions that accompany flash floods?

Riverine floods

Riverine or regional floods commonly involve larger volumes of water than flash floods because they develop in larger watersheds and draw on more widespread sources. The effects are correspondingly greater in extent, but the flood stage takes more time to build and to recede. Riverine floods are typical on rivers with large watersheds, multiple tributaries, and extensive floodplains.

Riverine floods commonly result from the effects of multiple rainstorms as well as spring runoff from melting snow. The flood may develop over a period of days to weeks, and it may take months for the floodwaters scattered across a floodplain to completely recede. Riverine floods cause more property damage than flash floods, but fewer deaths, as people have more warning and more time to evacuate the floodplain ahead of the flood.

A large riverine flood that lasts for weeks can have multiple flood crests. After the river channel fills initially and reaches flood stage, it is like a bathtub filled to the brim: Any additional water that enters the tub sloshes over the side. New rainwater may enter the river's upstream tributaries at different times and generate additional flood crests as they merge downstream. This is what leads to flooding that lasts for days or weeks. It is only when the rain abates throughout the watershed that flood levels will begin to gradually recede.

Take the 1993 Mississippi flood, the last straw for Valmeyer, Illinois, as an example. The Great USA Flood of 1993 The watershed of the Mississippi River includes six major subwatersheds that feed water and sediment into the main channel of the Mississippi River at various points from Minnesota to Louisiana. Roughly midway down the length of the river, two major tributaries empty into the main trunk of the Mississippi. These are the Missouri River, which enters north of St. Louis, and the Ohio River, which enters at Cairo, Illinois. Tributaries of the upper Mississippi include the Illinois, Rock, and Wisconsin Rivers. On the lower Mississippi River, south of Cairo, the Arkansas River is a major tributary.

Conditions on any of the tributaries of the Mississippi can contribute to flooding. During the 1993 flood, the worst in the river's history in terms of property damage, economic costs, and disruption of people's lives, rainfall in the upper Mississippi as early as November 1992 began to saturate the soils. This ensured that a large part of any subsequent rainfall or snowmelt would run off into the river. Then, in the spring and summer, regional weather patterns combined to create a "rain machine" over the upper Midwest. As shown in Figure 7-27, an alignment of high-altitude winds drew warm, moist air from the

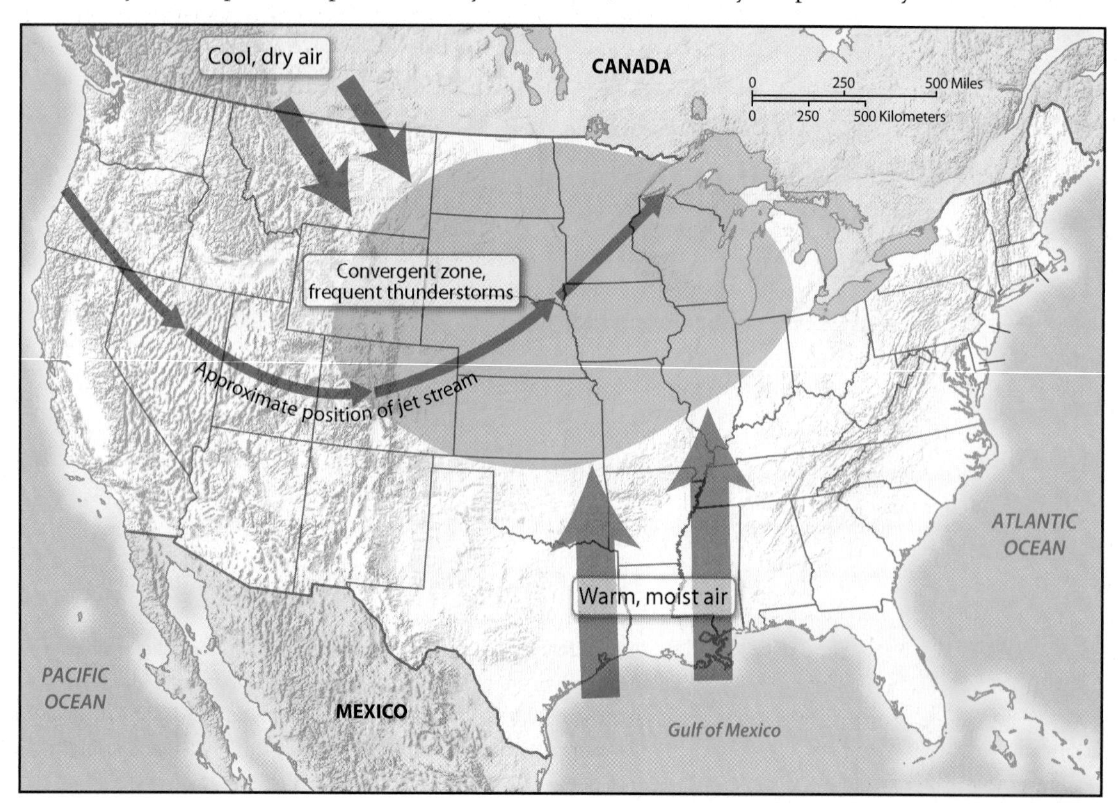

FIGURE 7-27 ◀ The 1993 "Rain Machine" over the North-Central United States
Regional weather patterns created a "rain-machine" over the upper Midwest that led to the great 1993 Mississippi River flood.

FIGURE 7-28 ▲ 1993 Mississippi River Flooding
Valmeyer, Illinois wasn't the only community inundated by the great 1993 Mississippi River flood. This photograph shows flooding in Missouri in July of that year.

Gulf of Mexico and put it on a collision course with cool air from Canada. A late snowmelt, due to the lack of the usual January thaw, was also a contributing factor.

As a result, drenching thunderstorms formed repeatedly over the upper Mississippi watershed from June through August in a pattern known as a train effect. The river overtopped or breached many levees, and flooding was widespread in the northern Mississippi River and its tributaries (Figure 7-28). The Mississippi was closed to shipping for weeks, and many channels experienced severe erosion. The final cost of the damage was estimated at $20 billion.

7.3 Measuring and Forecasting Floods

Given the damaging effects of floods, it's valuable to have a way of estimating the largest floods that are likely on a river and how often such events might occur. This allows planning for floods. Fortunately, there are ways to use historical records to measure floods, calculate their recurrence intervals, and forecast future floods.

HYDROGRAPHS

Hydrographs are plots of river discharge (or height in some cases) over time. Scientists routinely monitor rivers with gages (or gauges) that record the water level above a standard reference point that represents the height of the riverbed. The measured water level at any time is called the river's **stage**. A one-year hydrograph like that shown in Figure 7-29 records the river level over a 12-month period, with measurements taken at a set time interval (hourly, daily, weekly, monthly). The USGS operates more than 7000 stream-gaging stations in the United States, Puerto Rico, and the Virgin Islands.

| Stream gages – measuring the pulse of our nation's rivers

Many stations have at least 30 years of discharge records, and some date back to the late 1800s.

| USGS real-time water data for the nation

Hydrograph records commonly identify the normal **bank-full stage** of the river—the level at which the channel is completely full. The bank-full stage marks the water level at which people need to be concerned about flooding. Any discharge on the hydrograph above this level can indicate flooding, but the bank-full stage and flood stage are not always the same level. In some cases, water levels can exceed the bank-full stage but the resulting flooding is local and inconsequential. At flood stage, water levels can significantly impact people and property. The highest point on the curve above flood stage represents the flood crest, or the peak discharge of that particular flood.

Hydrographs reveal what time of year we can expect a river to reach its peak flow. In addition, the slope (or steepness) of a particular peak on a hydrograph provides useful information about the river's flooding history. If the curve rises steeply, the river built quickly to flood stage (as Figure 7-29 shows occurred in June of 2001). A more gradual rise indicates that the river built to peak discharge over a longer period. In practical terms, slower flood development means more time for people to prepare. It can also mean that the flooded area will probably remain under water longer.

FIGURE 7-29 ▼ A Stream Hydrograph
This hydrograph records the discharge (daily mean stream flow) of South Boulder Creek, Colorado, for the period of a year. More than 4,000 USGS stream gages in the United States have over 30 years of hydrograph recordings.

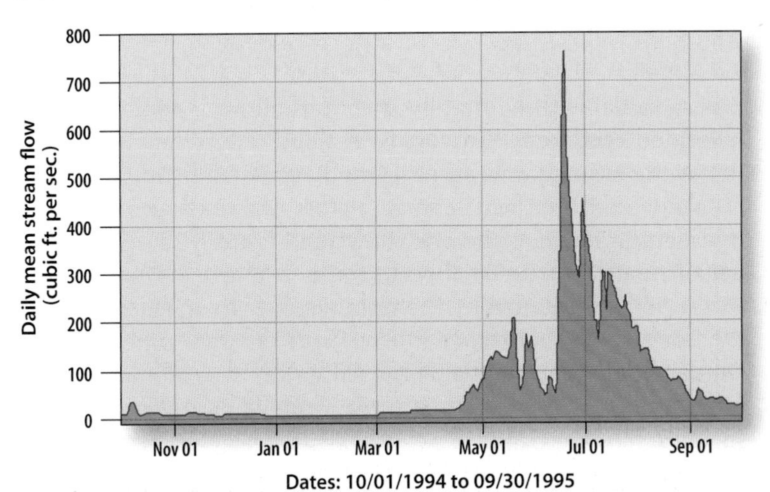

What You Can Do

Investigate Real-Time Stream Gauge Data

Want to see where flooding is occurring right now? Perhaps you want to keep track of the stage of a river near you? Go to NOAA – National Weather Service – water. This website links you to real-time stream gauge data throughout the United States and identifies those stream gauge stations that are near or above flood stage. The data for each stream gauge are displayed on a hydrograph that shows several days of recordings of the stream's stage. You can literally watch a flood develop—with hydrograph data—at this site.

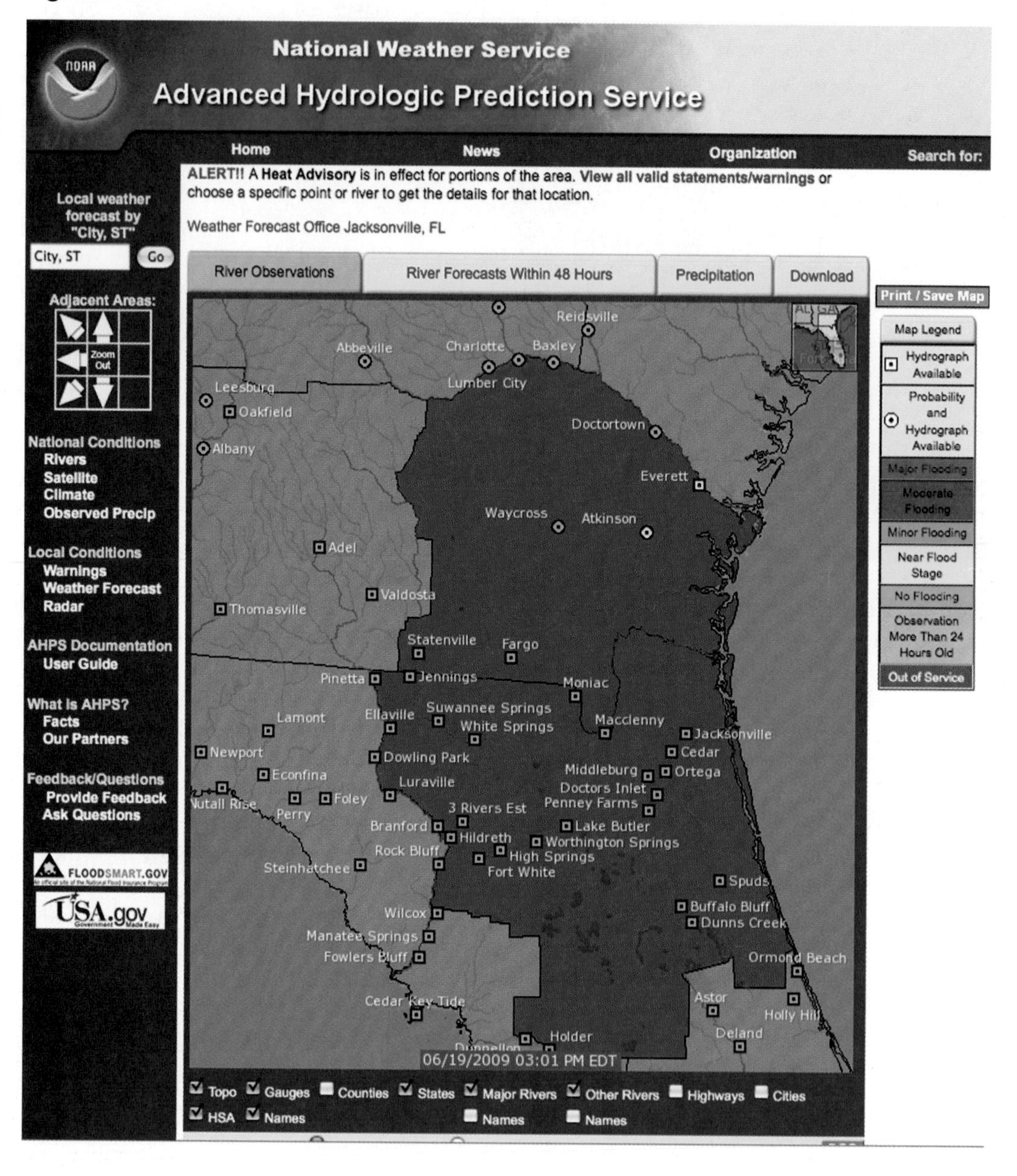

FLOOD RECURRENCE INTERVALS

A very valuable thing that hydrographs allow us to do is calculate flood **recurrence intervals**. A flood recurrence interval is the average time between past flood events of a similar size. It is calculated from the discharge history of a river recorded on a hydrograph (such as the one shown in Figure 7-29). The number of times a particular flood level is reached during the historical record, divided by the number of years in the record, is the flood levels recurrence interval—the average time between similar flood events. This is the same way we calculate recurrence intervals for other natural events such as earthquakes (Section 5.4).

Scientists have found that a very useful way to characterize a river's flooding history over a period of years is to focus on its peak discharge each year—the river's **annual flood**. The size of the annual flood varies over time; in many years it is very low, in a few it is very high. These annual flood discharges can be ranked according to their size, starting with the largest on record as number one. If the historical record covers ten years, the annual flood ranked number ten has the smallest discharge—this flood level was reached or exceeded every year in the historical record.

An especially useful way to understand a river's flooding history is to determine the recurrence interval for all its annual flood ranks. The recurrence interval (RI) for a specific annual flood rank is the number of years in the historical record divided by the annual flood rank (r):

$$RI = [(T_2 - T_1) + 1]/r$$

where T_1 is the first year of the record and T_2 is the last year. The term $(T_2 - T_1) + 1$ is the number of years data are recorded. (We need to make sure to count both the first and last years of recordings, as just subtracting T_1 from T_2 leaves us one year short—a record from 1997 to 2006, for example, covers 10 years of data, not 9.)

The recurrence interval for an annual flood ranked 5 in the example 1997 to 2006 period is 2 years—this flood level is reached or exceeded every 2 years on average. On the other hand, the recurrence interval for an annual flood ranked 1 in this record is 10 years—this flood level is reached or exceeded every 10 years on average. This is a very short record but even this simplified example readily shows a fundamental characteristic of flood history—large annual floods are less frequent than small ones—they have longer recurrence intervals.

THE 100-YEAR FLOOD

Have you heard about *100-year floods*? Has your community experienced a 100-year flood? Perhaps some of your neighbors are expecting a 100-year flood soon? So what is a 100-year flood? A 100-year flood is an estimate of the size of a flood having a recurrence interval of 100 years based on extrapolations of historical annual flood and recurrence interval data.

A plot of annual flood discharges and their recurrence intervals, like that shown in Figure 7-30, is called a **flood frequency curve**. Rivers with many years (commonly 30 years or more) of annual flood records show a systematic relationship between peak discharge (annual flood level) and recurrence interval. This relationship enables the peak discharge of a flood having a recurrence interval of 100 years to be estimated. A 100-year flood's estimated size is a benchmark for evaluating a river's flooding capability.

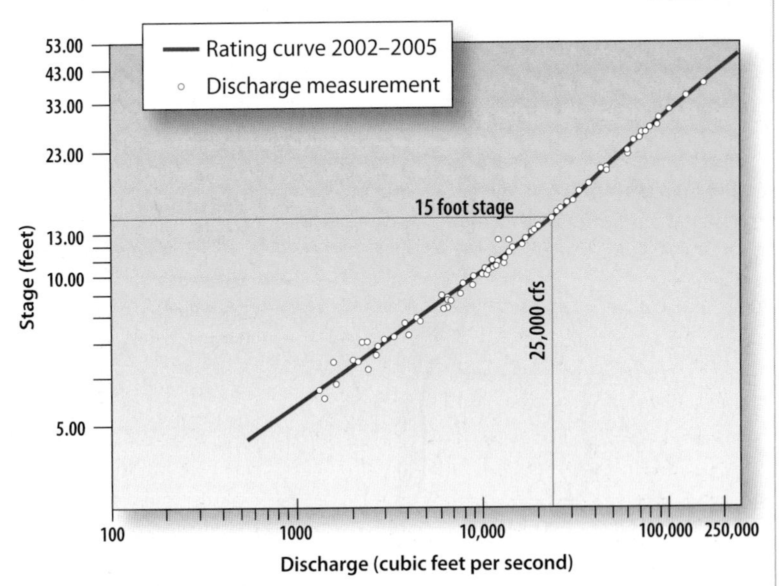

FIGURE 7-31 ▲ A Rating Curve

Rating curves show the relationship between a river's discharge and the corresponding stage (water level) over a particular period of time. This rating curve is for the Connecticut River at Montague City.

Communities can use the estimated size of a 100-year flood to evaluate flood hazards and guide planning decisions. For example, the estimated discharge for a 100-year flood (see Figure 7-30) can be plotted on a *rating curve* for the river. A rating curve, like the one shown in Figure 7-31, is a graph that shows what stage (that is, depth) the river reached during various discharges over a period of historical time. This enables the stage for a 100-year flood to be estimated and used in community planning. Areas potentially inundated by the stage reached in a 100-year flood can be identified on maps showing land elevation (topographic maps) and new development guided away from where flooding would occur in such an event. For example, if a community wants to make sure that a new school is at low risk from future flooding, it can require the school to be constructed outside the area that a 100-year flood would inundate (Figure 7-32, p. 212).

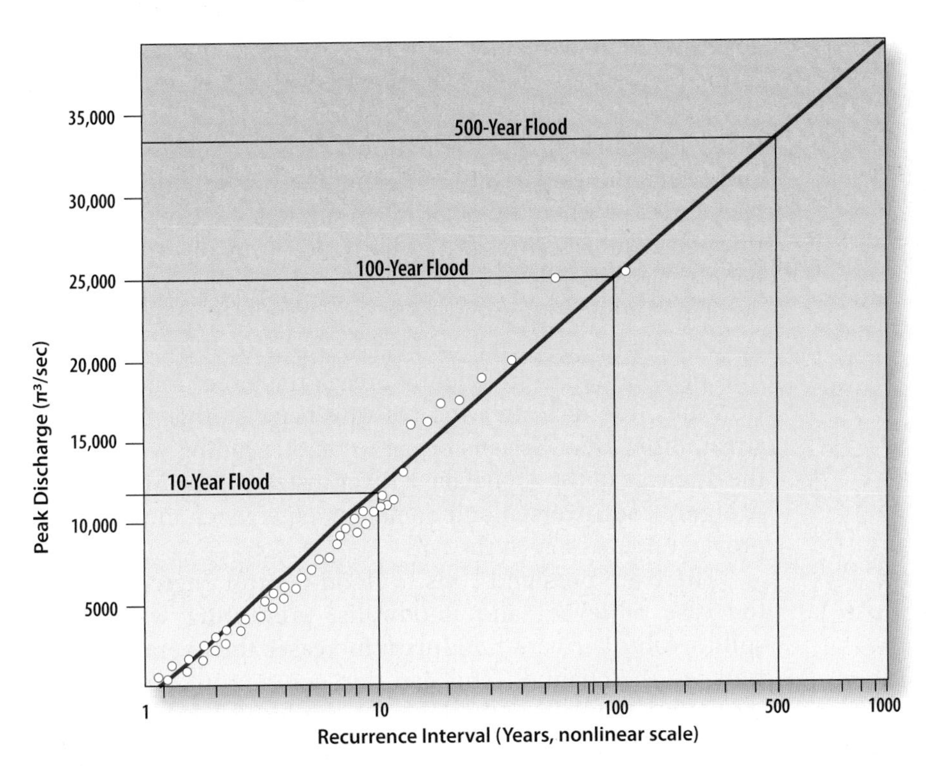

FIGURE 7-30 ▲ A Flood Frequency Curve

The flood frequency curve for the Red River of the North is based on hydrograph records from 1882 to 1994 at Fargo, North Dakota.

FLOOD PROBABILITY

Just how risky is it to be within the reach of a 100-year flood? Fortunately, recurrence intervals can be used to calculate the probability that a future flood will meet or exceed a specified size (magnitude). The probability (P) that a specific-size flood will occur is the inverse of its recurrence interval (RI):

$$P = 1/RI$$

A 100-year flood has 1 in 100 chance ($P = 0.01$ or 1 percent probability) of occurring in any year. A "one-year flood,"

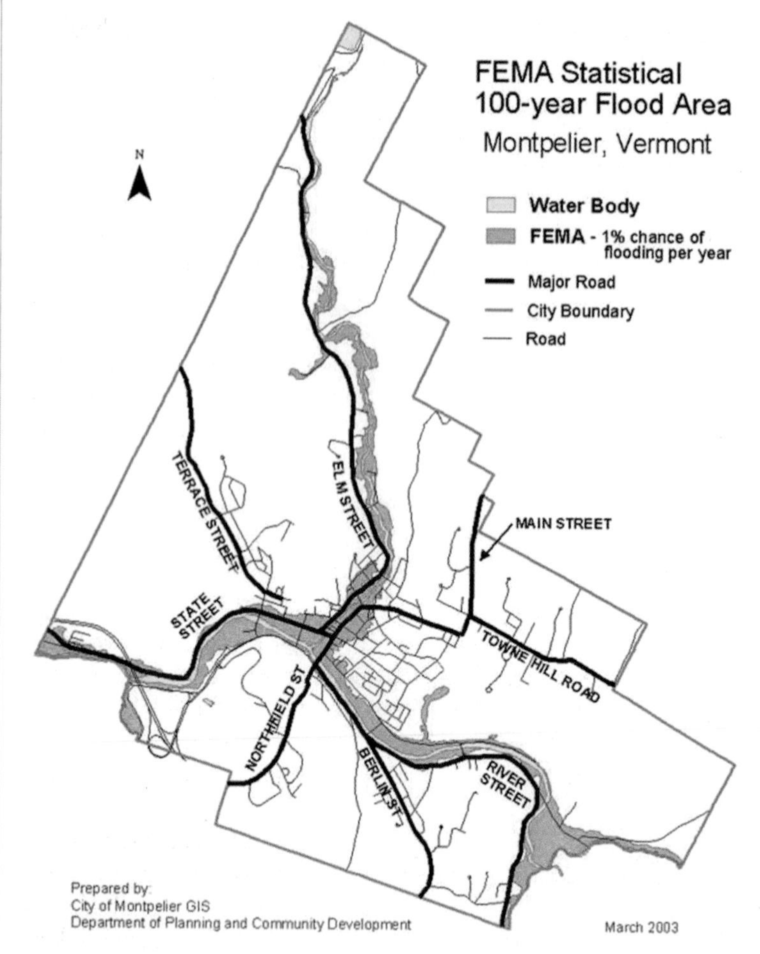

FIGURE 7-32 ▲ A 100-Year Flood Map of Montpelier, Vermont
Maps like this one can serve as guides to safe development.

the smallest annual flood in a record, has a 1 in 1 chance (P = 1 or 100 percent probability) of occurring in any given year. (This does not mean that such a flood will invariably occur once a year, every single year; rather, it means that the *average* occurrence will be once a year.) Assigning probabilities to different flood sizes helps clarify how much risk people are exposed to when they live on a floodplain.

Flood probabilities do not change from one year to the next, regardless of what the recent flooding history has been. If a 100-year flood occurred last year, the probability is still 1 in 100 (P = 0.01) that such a flood will occur this year. If a 100-year flood hasn't occurred for 100 years, the probability is still 1 in 100 (P = 0.01) that it will occur this year. A 100-year flood is never "due." Since the probability of a 100-year flood is based on its recurrence interval, the best we can say is that over a long period of time, say a thousand years, the discharge levels of a 100-year flood will be reached or exceeded about ten times.

LIMITATIONS OF HISTORICAL DATA

Using historical stream discharge data to characterize future floods requires making some assumptions. First of all, we assume that flood events are independent of one another—that what happened in the past will not influence what happens in the future. This is usually the case for floods, but not always. We also assume that the past physical conditions of a watershed will continue into the future—that precipitation patterns will be similar, for example. This may not be the case, as longer term precipitation patterns have changed in some places. Stream-flow trends in the United States You have already learned how urbanization can change watershed characteristics. Roads, forestry operations, and agricultural practices can also affect how much runoff water and sediment enter the river system. Even flood control measures such as dams and levees can limit the usefulness of historical stream discharge data. But as long as the future conditions are reasonably similar to those of the past, recurrence interval and flood probability determinations can be powerful sources of insight into a river's future flooding capabilities.

7.4 Living with Floods

During the great Mississippi flood of 1927, engineers found themselves trapped in a web of their own making. As chronicled by John M. Barry's book *Rising Tide*, a decision had been made in the 19th century to contain floods with a system of constructed earthen levees along the river's main channel as well as its major tributaries. But the 1927 flood proved too great for the levees to withstand. As the waters rose, exerting greater and greater pressure on the flood control system, the levees began to break.

The leaders of the commercial capital of New Orleans, Louisiana, grimly contemplated the consequences of a major flood in the city. A decision was made, backed by the federal government, to dynamite the levee at Poydras, on the east bank of the Mississippi. The Poydras levee was at a bend where the river turned sharply west. Breaching the levee here, it was thought, would allow water to flow south into rural St. Bernard and Plaquemines parishes, relieving the pressure on levees in New Orleans. There was a problem, though—letting the water through would destroy the homes and farms of some 10,000 people. Nonetheless, the project went ahead.

The "public execution" of these communities, as one local leader called it, occurred on Friday, April 29, 1927. It took 39 tons of dynamite to sufficiently breach the levee (Figure 7-33). Ironically, the day after the blasting began, other levees upstream from New Orleans began to burst, sending water down the drainage of the Atchafalaya River instead of the Mississippi. A highly controversial and damaging emergency measure had proved unnecessary in the end.

There are pros and cons to flood-control engineering. In the case of levees and floodwalls, preventing water from spilling onto a river's floodplain increases the overall level of water in the channel. This can trap communities in an ever-escalating struggle with the river, building larger and ever more expensive barriers as each "great flood" proves bigger than the last. However, in areas that have not completely sprawled to the banks of the river, other approaches are possible. For example, communities can adopt policies that dis-

FIGURE 7-33 ▼ Blowing the Levee
New Orleans thought it would be protected by blowing up the levee at Poydras, Mississippi in 1927.

courage development in areas prone to damaging floods; they can mandate sufficient flood insurance to cover damages; or they can use flood-prone lands for public parks, which are less costly to rebuild than commercial properties or housing. Some of these alternatives, especially those that leave floodplains for purposes other than urbanization, can contribute to sustaining valuable soil resources and biodiversity.

FLOOD HAZARDS

Floods are a significant hazard throughout the world. In the United States, floods on average kill 140 people each year and cause $6 billion in damage. | Natural hazards – floods In the 20th century, flooding was the leading type of natural disaster in terms of deaths and destruction of property in the United States, as Table 7-1 shows. | Significant floods in the United States during the 20th century In countries with higher population densities on floodplains and less developed flood-control measures, the human toll can be far greater. During the 20th century, there were 12 floods that killed 18,000 or more people; ten of these were in China, where death tolls from specific floods have reached the millions.

The sight of a house torn from its foundations, tumbling down a raging river, or of automobiles heaped into piles like children's toys, captures the tremendous force of flowing water. Flood-driven erosion can undermine bridges, leading to their collapse. Even if structures survive the deluge intact, inundation leaves basements flooded, walls and foundations waterlogged, and personal possessions ruined—all of which must be repaired or replaced.

Flood impacts on communities include direct and indirect economic losses. There is always the huge cost of rebuilding or repairing public facilities such as sewage treatment plants, roads, bridges, power transmission facilities, and other infrastructure. Every business disrupted or damaged results in lost wages and reduced tax revenue. Lost crops and drowned livestock mean less profit after years of investment. These costs all add up and tax dollars help pay them. Billions of dollars in flood damage losses are paid through government funds (taxes) and private insurance companies.

Another hazard of flooding is the impact on human health. Sewage, the carcasses of drowned livestock, and pollutants such as paint, pesticides, and petroleum can be released into the local environment by floodwaters, creating a lingering public health threat. It always seems difficult to get clean drinking water when it floods.

Avoiding economic and human losses due to flooding can be part of achieving sustainability. The money, time, and effort saved by avoiding or limiting flood damage could be used for other purposes, such as developing more renewable energy sources, for example. So far, people's efforts to avoid flood losses have focused largely on engineering structures to control flooding, but as we will see, other approaches are also being explored.

MITIGATING FLOODS: THE STRUCTURAL APPROACH

The structural approach to flood control emphasizes engineering. It involves physically restructuring the river to control water flow. It can be likened to a vast replumbing project meant to lower flood levels, reduce choke points in the river where water may back up, and divert high water before it can inundate valuable property.

Remember the relationship between a river's discharge, cross-sectional area, and velocity? Structural approaches try to (1) decrease discharge, (2) increase a channel's cross-sectional area to allow more discharge without flooding, or (3) allow a river's flow velocity to increase in response to increased discharge.

Channel alteration

Channel alterations are designed to make channel cross-sectional area greater or to increase flow velocity. The most fundamental way channel cross-sectional area is increased is to raise the bank height with earthen levees or concrete dikes and floodwalls. The modified channel can then hold a greater volume of water before flooding. Special *fuse-plug levees*, built lower than the main levee, allow rising water to flow out of the channel at strategic points where it will not be damaging, thus decreasing the discharge continuing downstream.

Any measure that straightens, deepens, widens, or clears a channel can allow the water to flow faster. Straightening channels, for example, reduces the resistance to flow that bends create. The riverbed may be dredged deeper or cleared of large stones and sand bars, increasing the channel's cross-sectional area. In addition, many cities with rivers flowing through their heavily developed centers have chosen to pave the entire course of the river with concrete and to raise its banks with vertical floodwalls. The Los Angeles River provides a classic example of this approach.

The Los Angeles River has made cameo appearances in many films. In the 1950s science fiction movie *Them*, giant mutant ants colonized it. The marauding liquid-metal assassin from the future in *Terminator II* chased his human prey down its bed. However, many moviegoers probably failed to recognize

TABLE 7-1 SIGNIFICANT FLOODS IN THE UNITED STATES DURING THE 20TH CENTURY

Flood Type	Date	Area or Stream with Flooding	Reported Deaths	Estimated Cost (M = million, B = billion)	Comments
Regional flood	Mar.–Apr. 1913	Ohio, statewide	467	$143M	Excessive regional rain
	Apr.–May 1927	Mississippi River from Missouri to Louisiana	unknown	$230M	Record discharge downstream from Cairo, Illinois
	Mar. 1936	New England	150+	$300M	Excessive rainfall on snow
	July 1951	Kansas and Neosho River Basins, Kansas	15	$800M	Excessive regional rain
	Dec. 1964–Jan. 1965	Pacific Northwest	47	$430M	Excessive rainfall on snow
	June 1965	South Platte and Arkansas Rivers, Colorado	24	$570M	14 inches of rain in a few hours in eastern Colorado
	June 1972	Northeastern United States	117	$3.2B	Extratropical remnants of Hurricane Agnes
	Apr.–June 1983 June 1983–1986	Shoreline of Great Salt Lake, Utah	unknown	$620M	In June 1986, the Great Salt Lake reached its highest elevation
	May 1983	Central and northeast Mississippi	1	$500M	Excessive regional rain
	Nov. 1985	Shenandoah, James, and Roanoke Rivers, Virginia and West Virginia	69	$1.25B	Excessive regional rain
	Apr. 1990	Trinity, Arkansas, and Red Rivers, Texas, Arkansas, and Oklahoma	17	$1B	Recurring intense thunderstorms
	Jan. 1993	Gila, Salt, and Santa Cruz Rivers, Arizona	unknown	$400M	Persistent winter precipitation
	May–Sept. 1993	Mississippi River Basin, central United States	48	$20B	Long period of excessive rainfall
	May 1995	South-central United States	32	$5–6B	Rain from recurring thunderstorms
	Jan.–Mar. 1995	California	27	$3B	Frequent winter storms
	Feb. 1996	Pacific Northwest and Montana	9	$1B	Torrential rains and snowmelt
	Dec. 1996–Jan. 1997	Pacific Northwest and Montana	36	$2–3B	Torrential rains and snowmelt
	Mar. 1997	Ohio River and tributaries	50+	$500M	Slow-moving frontal system
	Apr.–May 1997	Red River of the North, North Dakota and Minnesota	8	$2B	Very rapid snowmelt
	Sept. 1999	Eastern North Carolina	42	$6B	Slow-moving Hurricane Floyd
Flash flood	June 14, 1903	Willow Creek, Oregon	225	unknown	City of Heppner, Oregon, destroyed
	June 9–10, 1972	Rapid City, South Dakota	238	$600M	15 inches of rain in 5 hours
	July 31, 1976	Big Thompson River, Colorado	145	$39M	Flash flood in canyon after excessive rainfall
	July 19–20, 1977	Conemaugh River, Pennsylvania	78	$300M	12 inches of rain in 6–8 hours
Ice-jam flood	May 1992	Yukon River, Alaska	0	unknown	100-year flood on Yukon River
Storm-surge flood	Sept. 1900	Galveston, Texas	6,000+	unknown	Hurricane
	Sept. 1938	Northeast United States	494	$306M	Hurricane
	Aug. 1969	Gulf Coast, Mississippi and Louisiana	259	$1.4B	Hurricane Camille
Dam-failure flood	Feb. 2, 1972	Buffalo Creek, West Virginia	125	$60M	Dam failure after excessive rainfall
	June 5, 1976	Teton River in Idaho	11	$400M	Earthen dam breached
	Nov. 8, 1977	Toccoa Creek, Georgia	39	$2.8M	Dam failure after excessive rainfall

the Los Angeles River as a river at all. From its headwaters in Canoga Park to its mouth at San Pedro Bay, the Los Angeles River's 82-kilometer (51-mi) course is virtually all concrete pavement, like the section shown in Figure 7-34.

The Los Angeles River once meandered through cottonwood and alders. In the 18th century, the first European settlers in what is today urban Los Angeles tapped the river to irrigate fields of corn, wheat, and grapes. Steelhead trout populated its waters, pursued hungrily by grizzly bears. The river also had the power to destroy, as water rushed down its steep gradient from the nearby mountains. In 1815, a flood went right through downtown Los Angeles. After a 1938 flood that killed 87 people, the city decided to turn the river—by then heavily polluted by local industry—into a straight and armored drainage ditch. Sloping, vegetated banks and a natural floodplain have given way to concrete walls.

In recent years, community members have advocated restoring Los Angeles County rivers. The river project However,

FIGURE 7-34 ▲ The Los Angeles River

The Los Angeles River channel is now all concrete. The skyline of downtown Los Angeles is visible in the distance.

the 2260 square kilometer (872 mi^2) Los Angeles River watershed is probably one of the most urbanized and structurally modified in the world. Although it is doubtful that all of the concrete will ever be jackhammered out of the channel, there is broad-based support for planting trees along parts of the river, clearing the channel of copious amounts of trash and debris, and establishing new parks and paths to create a more natural-looking river corridor.

Flood-control dams

A dam captures river discharge in the reservoir behind it. Dams can have several purposes: to provide irrigation and drinking water (Chapter 10), to generate hydroelectric power (Chapter 13), to create a recreational area for boating and fishing, to control water levels for commercial shipping, and to control flooding. So-called multipurpose dams may provide several or all of these benefits. As a flood-control measure, dams allow us to store excess discharge (say, from a spring flood) in the reservoir and release it gradually during times of lower water. The largest dam ever built was recently completed on the Yangtze River in China.

China's Three Gorges Dam, shown in Figure 7-35, soars 175 meters (574 ft) into the air and is wider than 100 Washington Monuments laid end to end. It stretches 2.4 kilometers (1.5 mi) across the upper Yangtze River and will create a reservoir 600 kilometers (373 mi) long in the famous Three Gorges region of China. Its 34 generators will ultimately produce 22,500 megawatts (million watts) of power. Overall, the final cost will approach \$70 billion for this massive project. In short, it's a project on a scale that previous dam-building nations, such as the United States, would be unlikely even to contemplate.

The flow of water that must be tamed by the Three Gorges Dam is just as enormous. The discharge of the river after passing through the Three Gorges region is 14,800 cms (523,000 cfs). The river also carries a large load of sediment—over a billion tons per year are delivered to the sea. The huge discharge of the river, building to even greater flows as various tributaries empty into the Yangtze, inevitably brings flooding to the densely populated agricultural lowlands downstream.

Each year, monsoon rains make the river swell. The average fluctuation in river level is 20 meters (66 ft). Flooding begins in the spring and peaks in August. Catastrophic floods occur on average twice per century. A 1931 summer flood on the Yangtze killed an astounding 1.4 million people. In 1996, floods killed more than 3000 people and caused \$24 billion in damage. By regulating the flow of water into the middle and lower Yangtze, China hopes to reduce the flooding that has plagued the population of the Yangtze River valley for thousands of years.

FIGURE 7-35 ▼ The Three Gorges Dam

The Three Gorges Dam on China's Yangtze River is the largest in the world.

However, numerous critics of the project are asking: Flood control at what cost? Much of the opposition to the project has focused on the ecological and social drawbacks to large dams. To make way for filling the reservoir, more than 1.3 million people have been relocated, while cultural, scenic, and archeological resources have been inundated. There are concerns that industrial pollutants from inland ports now under construction, in addition to sewage and agricultural runoff, will build up in the waters of the reservoir, creating a health risk and harming wildlife. Downstream, the dam could deprive wetlands and coastal areas of silt and nutrients. Sediment gradually collecting behind the dam will eventually decrease the reservoir's capacity to produce hydropower.

The Chinese government is steadfast in its belief that regardless of these potential disadvantages, the project, on the whole, is vital for China's future prosperity. Some "dam refugees" are prospering with thousands of construction jobs, while others struggle to make it through the transition. It may not be clear for decades to come whether the dam will be a net gain or loss, and whether it adequately prevents flooding of the river valley that holds more than one-third of China's population.

You Make the Call

Would You Build the Three Gorges Dam?

The Three Gorges Dam will help protect people and property and supply much needed power for a fast-growing Chinese economy. But the negative consequences of this dam may be significant, too. River habitat has been changed over large areas, people have been dislocated, and there is uncertainty about what sedimentation in the dam's reservoir will lead to.

- What would you do to understand the pros and cons of the Three Gorges Dam project?
- If you were in charge, would you build the Three Gorges Dam?

Diversion channels/floodways

Diversions locally decrease a river's discharge. Many cities with flood-prone rivers running through heavily developed residential and commercial areas have adopted this strategy. Upstream from the city, a dam is constructed with floodgates that can be opened or closed. At a time of high water, the gates open and divert a portion of the river's flow to a channel (the *floodway*) that flows around the city and deposits the water either back into the river, downstream from the flood-prone area, or into a lake or other nearby reservoir. For example, within the Mississippi River flood-control system, Lake Pontchartrain in New Orleans serves this purpose.

Detention ponds

Detention ponds are locally constructed reservoirs that decrease the rate at which surface water runs off into rivers. They collect runoff water and then slowly release it back into a stream or let it seep into the ground, as opposed to allowing the water to enter the stream or river immediately through its natural drainage or through storm drains and sewers. Developers often incorporate retention ponds into new subdivisions or other paved developments that might otherwise increase runoff into local streams.

THE MISSISSIPPI RIVER FLOOD-CONTROL SYSTEM

In the 20th century, flood control engineering on the Mississippi River reached a whole new scale. To more than 2560 kilometers (1590 mi) of levees on the main trunk river, the Army Corps of Engineers added several control structures intended to divert water from the main channel. The purpose was to lower flood stages in the lower Mississippi River.

The Birds Point–New Madrid Floodway sits just below Cairo, Illinois, where the Mississippi and Ohio rivers join. The floodway is an irregularly shaped lowland area, bordered by levees, on the west side of the Mississippi River. It stretches 56 kilometers (35 miles) from north to south and ranges from 6 to 19 kilometers (4 to 12 mi) in width. When the river reaches a certain stage, water diverts into the floodway. Sprawling across an area of 520 square kilometers (200 mi^2), it functions as a temporary floodplain. The floodway can accommodate 15,400 cms (544,000 cfs) of river water, lowering the flood stage in the area of Cairo by as much as 2 meters (7 ft). The diverted water reenters the river at New Madrid, Missouri.

The Old River Control Complex lies on the west bank of the Mississippi between Natchez and Baton Rouge. Louisiana River control

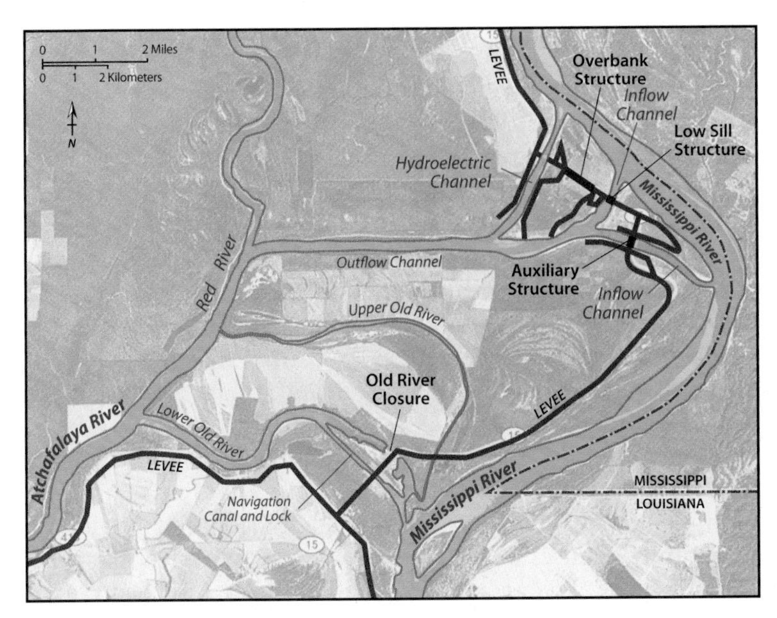

FIGURE 7-36 ▲ The Old River Flood Control Structures
These structures divert high waters on the Mississippi River to the Atchafalaya River basin, a straighter and steeper path to the Gulf of Mexico.

As Figure 7-36 indicates, it consists of three structures that act as valves to regulate the flow of water between the Mississippi and Atchafalaya Rivers. During a flood, the Old River Control Complex can be used to lower Mississippi flood stages as it diverts water to the Atchafalaya basin.

About 48 kilometers (30 mi) above New Orleans, the Bonnet Carré Floodway diverts water from the Mississippi River into Lake Pontchartrain (Figure 7-37). The outlet, or *spillway*, 2130 meters (6990 ft) wide, directs overflow into a constructed channel that continues 9 kilometers (5.6 mi) to the lake. When in use, the system turns Lake Pontchartrain into an extension of the river's floodplain. It can accommodate 7000 cms (250,000 cfs). These diversions are just the largest engineering projects in the system. There are hundreds of smaller reservoirs and flood-control structures on tributaries of the Mississippi.

PROS AND CONS OF THE STRUCTURAL APPROACH

The main benefit of the structural approach is that it reduces the frequency of damaging floods. This saves lives and can prevent flood damages amounting to billions of dollars. However, it comes at a cost. Here are some of the drawbacks.

- The structural approach can be expensive. Large-scale construction projects, regulatory compliance, and permitting all take time and a lot of money. These are all "up-front" costs, meaning that it takes a large initial investment to prevent later damages when a flood occurs.
- Levees and floodwalls raise a river's flood stage, essentially increasing the level to which the water can rise before it overtops the barrier. If the levees fail or are overtopped, the losses and hazards may be greater than ever before. Also, artificial barriers may trap floodwaters, prolonging the time it takes for the water to recede back into the channel. Water scattered about in the floodplain may linger for months as it evaporates or soaks into the ground.
- Restricting the lower course of a river creates bottlenecks and increases the risk of flooding upstream because of the bottlenecks. On the other hand, upstream channelization can send discharge downstream more quickly, putting people and property at greater risk. For levees and floodwalls to protect everyone equally, the river and its watershed must be understood as a whole.
- Dams essentially create local base levels. As the river empties into a dam's reservoir, it deposits sediments. Even in a very large dam project, the reservoir may fill up with sediment after a few decades. This requires dredging to maintain the reservoir's flood-control capacity. The resulting lack of sediment downstream from the dam can increase erosion and significantly change habitat in the river corridor. Continuous sediment supply is also needed to maintain deltas and their wetlands at river mouths.
- Flood control is a two-edged sword. On one hand, people can occupy and use valuable floodplain land. On the other hand, this puts more and more people at risk if a flood occurs, and can lock the communities into an ever-escalating program of flood control. To use a gambling metaphor, once you get into the game, it becomes harder and harder to get out.
- The changes people make along a river have environmental consequences. Channel alteration disturbs or destroys the habitat of fish or the vegetation on riverbanks that other types of wildlife utilize. Draining and filling wetlands, which is sometimes part of flood control efforts, also eliminates wildlife habitat and removes a natural reservoir that would otherwise help absorb high water levels in a river. Dams represent barriers to fish migration and spawning, and in many areas have led to plummeting fish populations.

FIGURE 7-37 ▼ The Bonnet Carre Spillway
The spillway diverts high water on the Mississippi River to Lake Pontchartrain in the distance.

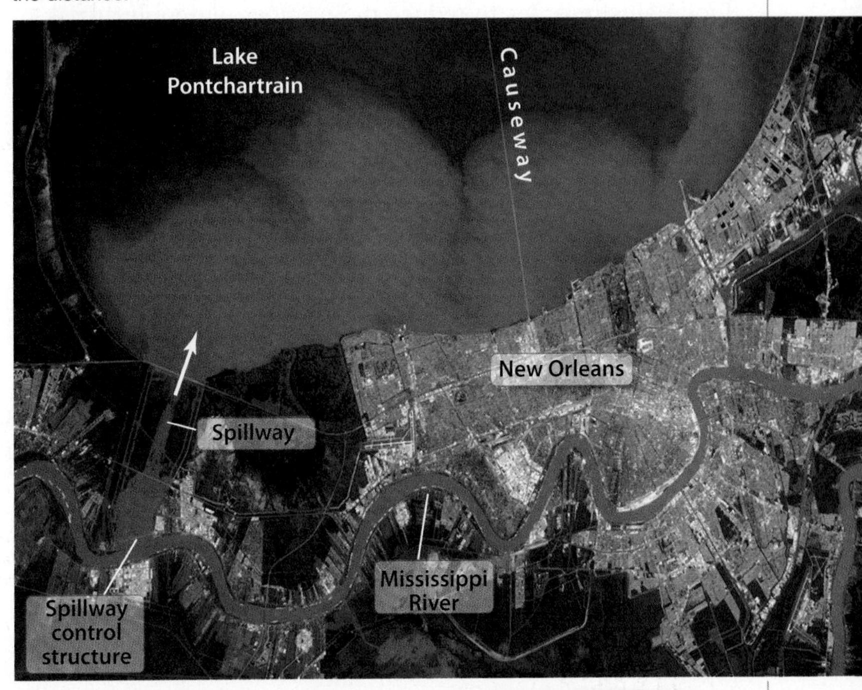

How do structural flood control measures influence Earth systems and their interactions?

MITIGATING FLOODS: THE NONSTRUCTURAL APPROACH

The Flood Control Act of 1928 inaugurated the golden age of replumbing rivers in the United States, literally in the wake of the catastrophic Mississippi River flood of 1927. This act and subsequent initiatives assumed the grand goal of reducing or even eliminating flooding on the nation's waterways. From 1927 until the mid-1960s, the federal government spent $13 billion on dams, levees, diversion projects, and other structural measures, with the Army Corps of Engineers assuming responsibility for designing, building, and maintaining the system. The structural approach did hold the waters back in many instances, but large damaging floods continued to occur. In the 1960s, national policy broadened to include other means of reducing flood damage.

The 1993 Mississippi flood catalyzed soul-searching about existing approaches to mitigating flood hazards. Continued development on floodplains meant that each time a flood overwhelmed our physical defenses, the price tag would be larger. After centuries of settlement, it's not possible to move everyone off the floodplain of the Mississippi River. However, as illustrated by the story of Valmeyer, Illinois, that opened this chapter, there are alternatives to simply rebuilding towns until the next great flood comes.

Today, the government officials, engineers, and scientists concerned with flooding usually talk in terms of floodplain management as opposed to flood prevention or control. Floodplain management still includes structural measures, but there is now greater recognition of their limitations. The overarching goal is to encourage changes in how people occupy and use floodplains in order to reduce the number of people at risk of floods and to decrease the damages associated with floods.

The U.S. National Flood Insurance Program (NFIP)

On the level of national policy in the United States, the National Flood Insurance Program (NFIP) represents the principal nonstructural approach to managing floodplains. National Flood Insurance Program It covers flood-prone areas both on rivers and on the seashore. The NFIP became law in 1968 as the National Flood Insurance Act, and was subsequently amended and expanded by further acts of Congress in 1973 and 1994. It offers government-subsidized flood insurance to communities if they adopt floodplain management policies that discourage development in the most hazardous areas and reduce flood damage to insured buildings.

For communities near rivers, the program covers structures that lie within a river's 100-year flood zone. The areas that would be inundated by a 100-year flood are identified on maps and the community chooses a "regulatory floodway," the area that is set aside to accommodate a 100-year flood. Regulations and land-use zoning then apply to development in the defined floodway.

Communities in the NFIP also have to enforce certain minimum standards for reducing damage to buildings. For new buildings and those undergoing major renovations, the minimum action would be raising the lowest floor (including the basement) above the elevation that water would reach during a 100-year flood. The building could be raised on fill material, on raised foundations, or on columns, as shown in Figure 7-38. Nearly 20,000 communities now participate in the program.

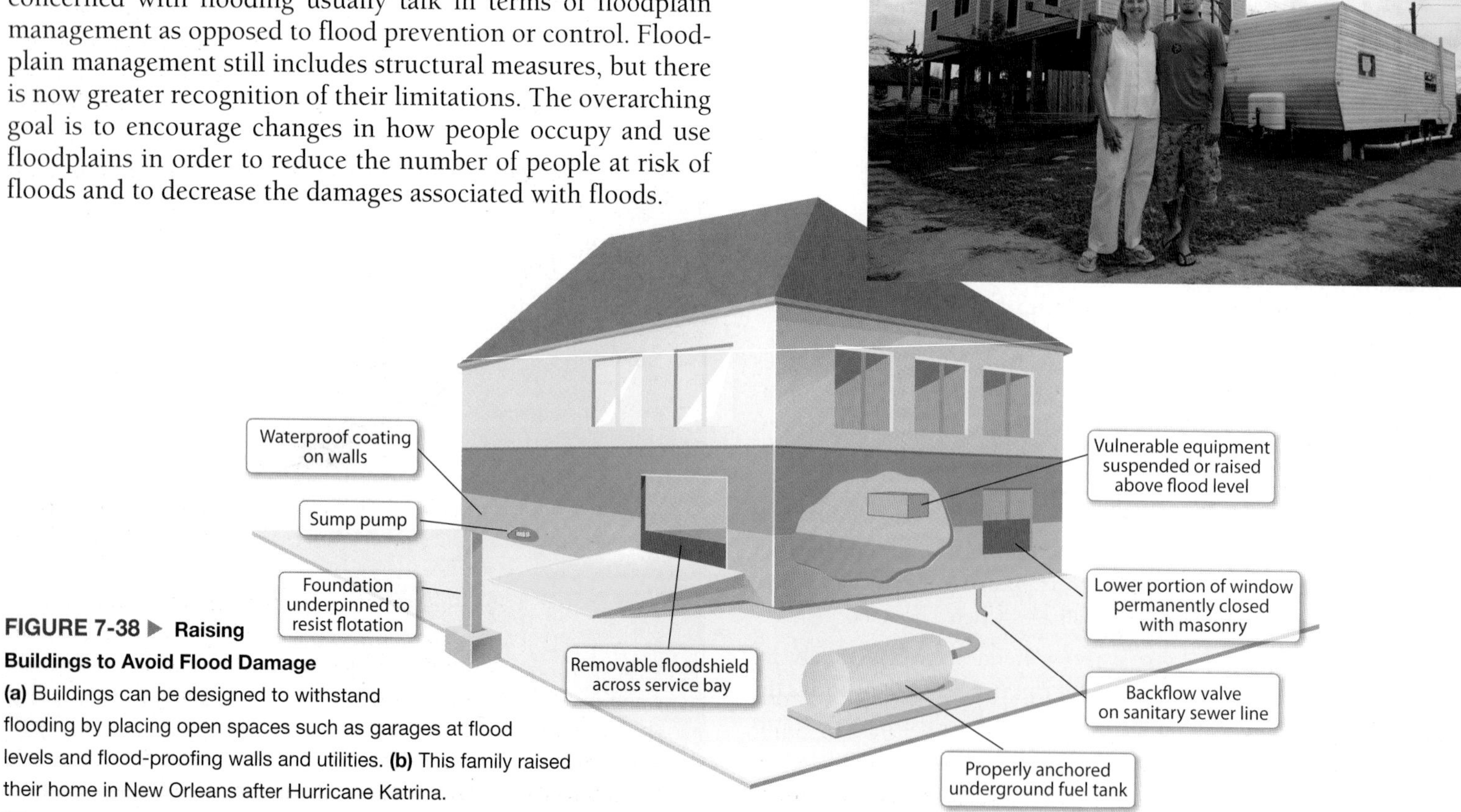

FIGURE 7-38 ▶ Raising Buildings to Avoid Flood Damage
(a) Buildings can be designed to withstand flooding by placing open spaces such as garages at flood levels and flood-proofing walls and utilities. **(b)** This family raised their home in New Orleans after Hurricane Katrina.

The NFIP has its limitations. The program is voluntary, so not all flood-prone communities have joined. Some people, unwilling or unable to pay the insurance premiums, choose not to participate. If a flood occurs, taxpayers must bear much of the cost of damage to uninsured property. This is because flooded areas are commonly declared disaster areas, qualifying them for subsidized rebuilding loans and other assistance through the Federal Emergency Management Agency (FEMA). People know this, and it serves as a disincentive for everyone to join. Also, the 100-year flood zoning can't protect everyone, since larger floods can and do occur.

In some cases, the amount of money paid out to flood-besieged homeowners has exceeded the market value of the house and property. From 1978 to 1995, 40% of insurance payments were to *repetitive loss properties*, defined as those that have sustained flood losses of $1000 or higher two or more times in the previous decade. One out of 10 of these properties had received insurance payments that exceeded the value of the property. Thus, repetitive loss properties make up a relatively small number of the communities threatened by flooding but contribute an inordinate share of the cost of flooding on a national level.

Relocations and voluntary buyouts

The ultimate solution to mitigating flood hazards is the one chosen by Valmeyer, Illinois—get out of the way of floods. Toward this goal, the 1994 amendment to the NFIP and other federal programs have provided funds for communities at risk of flooding to relocate buildings or to purchase and demolish them. Since the 1993 Mississippi flood, more than 17,000 landowners in 36 states and one territory opted for voluntary buyout. Once buildings and other structures are removed, the land can be redeveloped as parks, open space, or agricultural land—a more sustainable land use. When a flood does occur in these areas, the damage is much less severe than it would have been had the land been used for residential or commercial purposes.

Sustainable floodplain management

A more recent national policy shift in the United States has been toward sustainable floodplain management. The general idea is that when floodplains are preserved or returned to a natural state, they help mitigate floods. Federal legislation that embodies this approach is the Flood Mitigation and Riverine Restoration Program, part of the Water Resources Development Act of 1999. Flood mitigation and riverine restoration program

This program, informally known as Challenge 21, stresses nonstructural approaches and restoration of floodplain ecosystems to a state as close to natural as possible. Congress allocated funds for restorations on a variety of rivers around the United States. The projects must be on floodplains and can include relocation of threatened structures, conservation and restoration of wetlands and natural floodwater storage areas, and planning for responses to potential future floods.

Restoring or preserving a floodplain involves more than just limiting development to give the river more room to spread out when it floods. Natural floodplain ecosystems include the river channel and its surrounding floodplain lands, but also vegetation along the banks, wetlands, and forested or otherwise vegetated lands bordering the floodplain. Wetlands absorb and store floodwater and gradually return it to the river. Forests and other vegetation maintain the permeability of the soil and slow the passage of runoff water into streams and rivers. Meandering rivers are longer than straight rivers, and represent a greater volume of storage space for high water.

The hope is that once a river channel and its floodplain are preserved or restored, the system will help control flooding with minimal human intervention. In addition, floodplain ecosystems provide habitat for wildlife and may improve local water quality, since wetlands perform a filtering function (Chapter 10). Sustainable floodplain management de-emphasizes structural approaches but does not necessarily shun them. Dams, levees, and artificial channels may still be needed to protect people from flood hazards. These measures become part of an overall plan, not an end in themselves.

SUMMARY

As with many environmental issues, the question of how best to manage flooding hazards does not have a clear-cut and simple answer. Floodplains are valuable real estate, and millions of people live on them. It is not possible to move every community away from floodplains, particularly on the Mississippi and the world's other major rivers. On a smaller scale, even people living on small creeks and rivers choose to stay, despite the constant risk of financial and personal loss.

Floodplains are communities as well as real estate. Moving away from home is traumatic, particularly to people who have lived near a river all their lives. People are often tied to floodplains by their jobs, too. However, a river's flooding capability and the hazards it presents are well understood and used to guide flood-control engineering and wise land-use planning. People can strive to strike a balance between the risks and rewards of living along rivers.

In This Chapter You Have Learned:

- Rivers play a key role in the water and rock cycles—they transfer water and sediment from the surface of the continents to the oceans.
- Watersheds are the areas where water collects and flows together to form streams and rivers. They vary from small valleys in mountains to large parts of continents. The atmosphere, hydrosphere, geosphere, and biosphere interact in watersheds.
- The amount of water flowing in a river—its discharge—is related to its channel cross-sectional area and flow velocity.

A river's character changes as these parameters change, either from natural influxes of water and sediment or from people's effects on its watersheds.

- Rivers will flow and carry sediment to their base level. Local base levels are defined by places like lakes, but the most common base level is sea level.
- The more elevation difference there is between a river's channel and its base level, the steeper a river's gradient and the greater its flow velocity can be.
- Rivers can erode their banks and channels. Erosion supplies sediment that can be carried as bedload, suspended load, or dissolved load. Most sediment is deposited where rivers near or reach their base level.
- Rivers develop floodplains, low areas on either side of the main channel that accommodate high water levels during floods. Floodplains are where many people live to be close to rivers and the benefits they provide for transportation, energy, agriculture, and water resources.
- Earth systems interact on floodplains. Floods irrigate wetlands, replenish soils, and help create habitat for the biosphere.
- Flash floods commonly develop when storms precipitate large amounts of rain in a watershed in a short period of time. Because of their sudden onset and high peak flows, flash floods can inundate people with little warning.
- Riverine floods have more widespread sources and may develop over an extended period of time, from days to months. Because they affect large areas where many people live, riverine floods can be very destructive to property.
- Recurrence intervals for peak discharges each year (annual floods) are estimates of the average amount of time between annual floods of different sizes. Annual flood recurrence intervals can be useful for long-range planning in communities near rivers.
- The extrapolation of historical annual flood discharge and recurrence intervals enables the size of an annual flood having a recurrence interval of 100 years to be estimated. Such 100-year floods have an unchanging 1 in 100 chance (probability = 0.01) of occurring each year. The area that would be inundated by a 100-year flood is commonly used to guide community planning and flood mitigation efforts.
- Floods present significant hazards to lives, property, economies, and human health. Since people will continue to live on floodplains, it's important to find ways to reduce the risks of flooding.
- One flood-mitigation strategy, called the structural approach, is to build levees, dams, and otherwise reengineer a channel and surrounding land to contain and control floodwaters. The structural approach increases channel cross-sectional area, enables water to flow faster, or both. The structural approach can be expensive, lead to even greater flood losses over time, and negatively impact wetlands and other habitats.
- Another flood-mitigation strategy is floodplain management, which combines structural and nonstructural approaches, such as restricting development on floodplains and restoring river ecosystems.
- Flooding is a natural process that brings benefits to floodplains in many ways. However, flooding causes more damage than any other natural disaster. Finding ways to diminish loss of life and property and maintain natural river processes will help people achieve a more sustainable future.

KEY TERMS

alluvial fan (p. 198)
annual flood (p. 210)
bank-full stage (p. 209)
base level (p. 193)
braided channel (p. 197)
capacity (p. 197)
channel (p. 193)
cutbank (p. 195)
delta (p. 198)
discharge (p. 193)
divide (p. 192)
flash flood (p. 206)
flooding (p. 200)
flood crest (p. 200)
flood frequency curve (p. 211)
floodplain (p. 199)
gradient (p. 195)
hydrograph (p. 209)
levee (p. 200)
load (p. 196)
longitudinal profile (p. 194)
meander (p. 196)
mouth (p. 195)
oxbow lake (p. 196)
permeability (p. 206)
recurrence interval (p. 210)
runoff (p. 192)
stage (p. 209)
tributary (p. 192)
trunk river (p. 192)
watershed (p. 191)
wetland (p. 205)

QUESTIONS FOR REVIEW

Check Your Understanding

1. What is meant by a watershed?
2. How are tributaries and trunk rivers related?
3. How is discharge defined, measured, and calculated for a given portion of a river? Is it the same for all parts of a river along its length?
4. What is base level and why is this an important concept for understanding river systems?
5. In which directions (relative to its channel) can a river erode?
6. Rivers carry sediment in many different forms. What are the main types of sediment loads, what are they made of, and how are they transported?

7. A flood is defined by a river rising above its banks. What are the most common natural events leading to flooding, and how common are small floods?
8. Why are wetlands important in mitigating flooding and flood effects?
9. How does urbanization contribute to an increase in flooding frequency and severity?
10. What are the basic differences between flash floods and riverine floods?
11. What is a hydrograph? Who constructs hydrographs, how is raw data to make these plots collected, and why are they useful?
12. What is a flood recurrence interval? If it has been 99 years since a river's last 100-year flood, does this recurrence interval tell us that we can expect a 100-year flood next year? Why or why not?
13. Name and describe the function of three separate man-made structures used for flood control.
14. How can flooding be beneficial, and even essential for maintaining ecosystems along rivers?

Critical Thinking/Discussion

15. Why is a watershed an important unit and scale for analysis of human impacts on and benefits from river systems?
16. From a geologist's perspective, how can river energy in the past be determined by observations of old river deposits that are no longer part of the active channel?
17. Using what you have learned from the chapter about erosion and sediment deposition along river channels, explain how a river meander can migrate laterally with little to no net change in width of the channel through time. Can this process continue indefinitely, or can meanders become too pronounced?
18. Living on floodplains is dangerous, but also highly desirable for many people. Why? What are the benefits that drive many people to accept the tradeoff between flood hazards and potential gains?
19. Maintaining open floodplains along rivers and channelizing them with levees and concrete walls are both valid approaches to the structural management of rivers. What are the benefits and drawbacks of each approach?
20. The U.S. National Flood Insurance Program is cited in the text as the principal nonstructural approach to managing floodplains. The program provides benefits to communities that are willing to work with and around 100-year flood inundation levels. How might this strategy actually cost less than government-funded structural river management (dams, levees, etc.) in many circumstances? What are some possible drawbacks of this program?

ANSWERS TO IN-CHAPTER INSIGHT QUESTIONS

What are some Earth systems interactions in watersheds?

P. 192

- Water is transferred from the atmosphere to the hydrosphere and geosphere.
- Hydrosphere waters either infiltrate or run off the geosphere.
- Hydrosphere reservoirs and movements are influenced by geosphere and biosphere characteristics within watersheds.
- Geosphere and hydrosphere interactions influence biosphere habitat such as wetlands.

What are some Earth systems interactions associated with stream erosion?

P. 196

- Geosphere materials are dislodged and moved as energy from the hydrosphere does work on the geosphere.
- Erosion influences biosphere habitat as stream banks collapse, channels deepen or migrate, and valleys change in size or physiography.

What are some Earth systems interactions associated with stream sediment deposition?

P. 199

- Sediment deposition transfers material from the hydrosphere to the geosphere.
- Biosphere habitat such as wetlands can be created by stream sediment deposition.

What are some Earth systems interactions associated with floodplains?

P. 200

- Hydrosphere sediment transport and deposition occurs on floodplains.
- Sediment accumulations on floodplains provide materials and nutrients to geosphere soils.
- Specialized biosphere habitat develops on floodplains.

What are some Earth systems interactions that accompany flash floods?

P. 208

- Atmosphere conditions, some influenced by interactions with the geosphere, create local intense precipitation.
- Geosphere topography funnels hydrosphere runoff into restricted channels.
- The geosphere is eroded and much material is transferred downstream.
- Flash floods surprise and devastate biosphere within and along stream channels.

How do structural flood control measures influence Earth systems and their interactions?

P. 218

- Floodplain sedimentation may be prevented or dislocated.
- Aquatic habitat can be changed, fragmented, dislocated, or destroyed.
- Interactions between surface water and groundwater (Chapter 10) may be changed or prevented.
- River mouth sedimentation may be changed in ways that affect coastal stability (Chapter 9) and habitats such as wetlands.

FIGURE 8-1 ▶ The Portuguese Bend Landslide
Only a few homes are left in the area of the Portuguese Bend landslide. The main road in the area, Palos Verdes Drive, is still in use, but the hummocky, winding road is in almost constant need of repair as the slide continues to move slowly downslope to the sea.

CHAPTER

8

UNSTABLE LAND

California's Palos Verdes Peninsula overlooks the Pacific Ocean 40 kilometers (25 mi) south of central Los Angeles. Sloping to the sea, it terminates along bluffs bordering a narrow strip of beach. Early in the 20th century, cattle roamed its grassy slopes and farmers raised beans, peas, and tomatoes. Eventually land developers realized that the peninsula's beautiful ocean views and isolated rural setting so close to the urban center of Los Angeles made it a great place for homes. In the 1950s, they started to sow a new and different crop on the western slope of Palos Verdes Peninsula—houses in a subdivision called Portuguese Bend.

In the summer of 1956, people in Portuguese Bend noticed that the ground beneath them had begun to move. A 105-hectare (260-acre) section of the slope was breaking free, carrying Portuguese Bend with it. Within six months, movement reached about 2.5 centimeters (1 in) per day. By 1961, the rate had slowed to about 1 centimeter (3/8 in) per day, but by then 154 homes within the slide area were either destroyed or significantly damaged (Figure 8-1).

Some people chose to ride out the landslide—literally—by bolting their homes onto steel frames that would hold the structure together as the land shifted. Portuguese Bend resident Robert McJones came up with a different solution. After buying a home in 1975, he placed the structure on three steel cargo containers that had been welded into a triangular base. As the ground shifted, he raised the corners to level the house. Perching the house on a tripod eliminated twisting forces that had damaged other homes. The people still living at Portuguese Bend are serviced by above-ground water and gas mains with flexible joints and the city, after considerable debate, still allows cargo containers to be used as part of homes. Palos Verdes Peninsula news – city sets parameters for cargo container usage

The Portuguese Bend landslide is an example of **mass wasting,** the downslope movement of earth materials under the influence of gravity. Individual episodes of mass wasting are called **mass movements**, and many of these are what people generally call **landslides.** Mass wasting is one of a pair of gravity-driven processes that continually shapes Earth's surface. The other is **subsidence**, the settling or collapse of the surface with little or no horizontal motion.

As at Portuguese Bend, demand for living space sometimes drives people toward development on unstable land. Slopes are desirable places to live if they offer scenic views, and people will pay a premium to live on them despite their potential hazards. According to the U.S. Geological Survey (USGS), the cost of mass movements and subsidence is enormous. Landslide hazards – a national threat Annually in the United States, 25 to 50 people die in the collapse of unstable slopes, at a cost in damage to property, roads, housing, and other infrastructure of $3.5 billion. Worldwide, such events are common and sometimes shockingly deadly. In the Andes Mountains of South America, high elevations with steep slopes and unstable ice-capped peaks combine to unleash mass movements that periodically take thousands of lives in densely populated towns and cities below.

Whether the process is mass wasting or subsidence, gravity always wins in the end. Because of this fact, people must remain respectful of the forces moving materials on Earth's surface. Fortunately, many effects of unstable land are preventable with changes in land use and other measures.

IN THIS CHAPTER YOU WILL LEARN:

- How driving and resisting forces control slope stability
- What factors initiate slope failure, including weather, earthquakes, and slope steepening
- The characteristics of the different types of mass movements, including falls, slides, flows, and creep
- What land subsidence is and how people's activities can cause it
- How people attempt to protect property against unstable land and help prevent slope failure
- How science and engineering help guide land-use decisions and help mitigate the impact of unstable land

8.1 Slope Stability Basics

Slopes are not static piles of dirt and rock. In an instant of geological time, such as a human life span, a slope may remain still. Over longer periods, however, all slopes are potentially unstable and some form of downslope movement of material will eventually occur. Mountains, valley walls, hills, and ridges are in a dynamic tug of war between the internal strength of their slopes and the downward pull of gravity. Gravity is the **driving force** behind slope failure. **Resisting forces** oppose gravity and work to maintain slope stability. A slope may become unstable and fail if the driving force exceeds the resisting forces.

THE DRIVING FORCE—GRAVITY

Earth's gravity is always pulling on surface materials. For objects on a horizontal surface, gravity's entire force is directed downward, perpendicular to the surface. But if the surface is inclined—if it slopes—then gravity's force has two components. As you can see in Figure 8-2, one component is parallel to the slope and the other is perpendicular to the slope. The slope-parallel component of gravity makes anything resting on

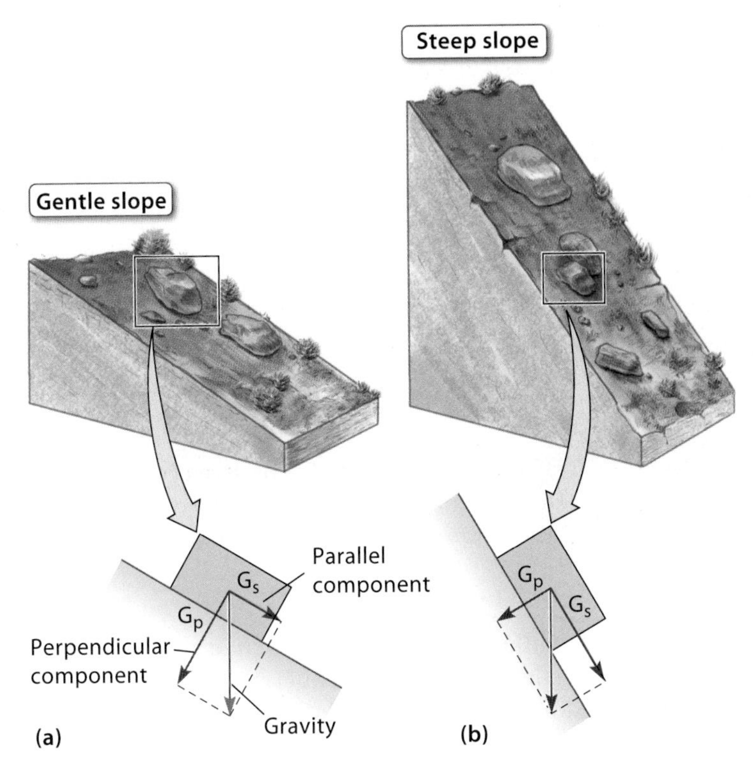

FIGURE 8-2 ▲ Gravitational Forces
(a) On slopes, the force of gravity (G_t) can be divided into two components. One component, G_p, acts perpendicular to the slope, pulling the object *against* the slope surface. The other, G_s, acts parallel to the slope, pulling the object down *along* the slope surface. **(b)** As the slope steepens, G_p diminishes and G_s increases.

an inclined surface inherently unstable. Even people standing on a slope must resist gravity's downslope pull or they will fall, slide, roll, or otherwise wind up at the bottom. If the resistance to gravity's downslope pull is insufficient, even bedrock can become unstable. In the long-term, slopes gradually diminish as a result of downslope movement of surface materials.

RESISTING GRAVITY

The existence of slopes is proof that forces do resist gravity. Unbroken, massive bedrock, such as granite, has internal strength that resists gravity due to the interlocking of its mineral grains (Chapter 4). Such rocks form cliffs like those in Figure 8-3, but wherever they are broken or where weathering diminishes their internal strength, gravity makes loose fragments fall. Piles of angular rock fragments, or **talus**, at the bases of cliffs show that even strong features are only temporary parts of the landscape.

The forces resisting gravity in *loose* materials are friction and cohesion. **Friction** is resistance to movement along a contact between two bodies such as blocks of rock or sand grains (Figure 8-4, p. 226). Materials with rough surfaces have more friction at their contacts than smooth, slippery ones. This is why hiking boots have jagged treads—they increase the friction between a hiker and the underlying slope he or she is trying to walk across.

FIGURE 8-3 ▲ Internal Strength vs. Gravitational Force
Despite its great strength, even granite can't resist the pull of gravity indefinitely. Rockfalls form talus deposits at the base of steep granite bedrock outcrops, like these.

Cohesion is the force created by attractions between grains of material—it's the force that makes grains stick together. Electrical attractions between particles at the atomic level are the principal source of cohesive forces, which vary with the material and the amount of moisture present. Dry sand, for example, has little or no cohesion, whereas clay and moist sand both have a good deal.

The interaction of gravity, friction, and cohesion in loose materials determines a slope's stability. If the resisting forces are less than the slope-parallel component of gravity, **slope failure** will result.

Slope stability can be assessed with the *Safety Factor*, the ratio of the resisting forces to the driving force. If this ratio is slightly more than 1, the slope is close to being unstable. But if

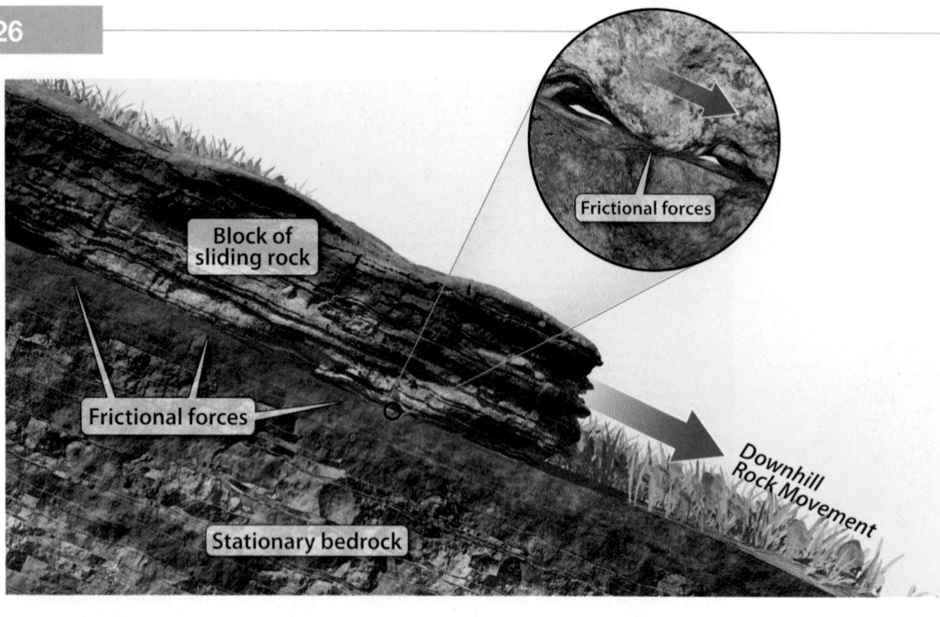

FIGURE 8-4 ◀ Friction Is a Resisting Force
Frictional forces between blocks of solid material on potential failure surfaces oppose the force of gravity.

the ratio is significantly more than 1, say 1.5 or 2, then the resisting forces are much greater than the driving force (1.5 and 2 times greater, in this example) and the slope is stable. Calculating the Safety Factor can be complicated, as it requires knowing material strength, density, and volume in addition to slope steepness and other parameters. In a general way though, just knowing which way the factor of safety changes when slope characteristics change can be very helpful. Key characteristics that influence slope stability include the slope's materials, its steepness, its water content, and its vegetation cover.

SLOPE MATERIALS

Some materials are inherently stronger or weaker than others, and materials fail in different ways for different reasons. Solid materials can form stable slopes at steeper angles due to their greater internal strength. These materials include solid bedrock, such as granite. But even the strongest bedrock may have features that can lead to slope failure. One example is a slope containing different layers of sedimentary rock, such as sandstone and shale. The boundaries between the layers—the sedimentary bedding planes—are relatively weak zones. If the planes are tilted (dip) downslope, the layers may break free along the bedding planes, as shown in Figure 8-5. Faults and fractures may also be weak zones in bedrock that enable a slope to break apart and fail.

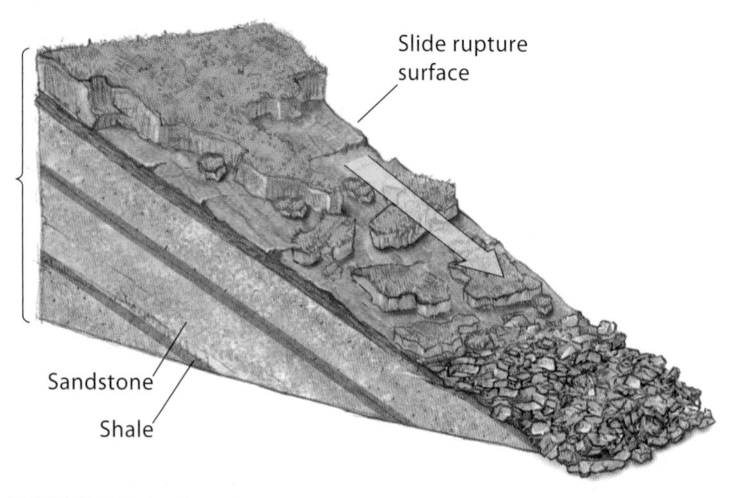

FIGURE 8-5 ▲ Slope Failure
Discontinuities in rocks that are inclined in the direction of the slope are weak zones that can fail more easily under the force of gravity.

The most unstable slopes are generally those containing loose or **unconsolidated materials** such as soil, sand, and broken rock debris. Friction between individual particles in these materials helps stabilize them. Because there is a wide range of resisting forces in these types of materials, they respond differently to gravity's driving force on slopes. Take a pile of dry sand sitting on level ground, for example. Friction between the sand grains holds the pile up; gravity tries to overcome the friction between the grains and pull the pile down. When these two forces are in balance, the pile assumes a characteristic maximum slope, or **angle of repose**. As you can see in Figure 8-6, dry sand has an angle of repose of about 35 degrees. Coarser irregular materials, such as broken rock, have steeper angles of repose; there is more friction between the jagged fragments than between smooth sand grains.

SLOPE STEEPNESS

We can see the dynamic nature of slopes in the relationship between slope steepness and slope stability. The slope-parallel driving force of gravity increases as the slope becomes steeper (see Figure 8-2). Imagine sitting on a toboggan perched at the top of a snow-covered hill. On a shallow incline, the friction between the bottom of the toboggan and the snow may prevent the toboggan from sliding if the downslope pull of gravity is unable to overcome the frictional resistance. However, if you move to a steeper hill, the downslope component of gravity may increase enough to overcome friction and send you sliding downward. Because it influences the driving force so much, slope steepness is commonly the most important single factor in determining slope stability.

(a)

(b)

FIGURE 8-6 ▲ The Angle of Repose
Unconsolidated materials have different angles of repose on slopes, depending on their internal friction. The angle of repose for dry sand **(a)** is commonly about 35 degrees and for coarse angular regolith **(b)** about 40 degrees.

WATER CONTENT

The water content of slope materials most commonly affects their stability in two ways. First, an increased amount of water in slope materials increases their weight. This increased weight increases the downslope component of gravity (the driving force) and makes it more likely that slope failure will occur. For example, heavy rains led to the 1983 landslide at Thistle, Utah that dammed the Spanish Fork River (Figure 8-7). A highway and railroad were wiped out and the town of Thistle inundated by the lake that formed upstream from the landslide. Thistle Landslide – Utah Geological Survey The direct and indirect costs, exceeding $400 million, are a record for landslide damages in the United States.

Second, increasing water content can decrease internal resisting forces. As water infiltrates a slope, or the water table rises, it fills the tiny spaces or pores between particles of the slope material. When only a small amount of water is present, its presence can actually help the particles to stick together—this is why you can build a sand castle more easily with moist sand than with dry sand (Figure 8-8a, p. 228). If the material becomes saturated, however, the pressure of the water at any particular place in the material—the **pore pressure**—increases. The pore pressure is roughly proportional to the amount of water above it (since it is the weight of the overlying connected water that produces the pore pressure). The rising pore pressure counteracts the downward force of gravity, which tends to compact particles together in the slope material. This increase in pore pressure in effect decreases the resisting forces, thereby diminishing the internal strength of the material. In bedrock, water can destabilize a slope as rising pore pressure reduces the friction on bedding planes or fractures. Note that the water is not acting like a lubricant—it is the pore pressure's opposition to gravity squeezing particles together that leads to decreased resisting forces and slope instability.

FIGURE 8-7 ◀ The Thistle, Utah Landslide
Heavy rains caused the 1983 landslide that dammed the Spanish Fork River in Utah. The lake that formed upstream of the landslide flooded the town of Thistle.

(a)

(b)

FIGURE 8-8 ▲ Water Content and Vegetation Influence Slope Stability **(a)** Up to a certain point, attraction between water molecules and solid particles, as well as between the water molecules themselves, increases cohesion in loose materials such as sand, making them more stable. Too much water, however, has the opposite effect. **(b)** Cutting of the forest on this slope in western Washington lowered the soil's internal strength, while heavy rains increased the weight of slope materials and the pore pressure within them. These factors decreased the resisting forces sufficiently to cause slope failure.

VEGETATION

Overall, vegetation tends to help stabilize slopes. The roots of plants can strengthen unconsolidated slope materials. The longer and more extensive the roots, the more likely they are to increase soil strength and resisting forces. Plants also work as pumps that remove water from the slope materials. They draw water up from their roots and evaporate it from their leaves (transpiration). By removing water, plants decrease the driving force and increase the resisting forces if the pore pressure of water is reduced. This is why stripping slopes of trees and other vegetation can often lead to landslides following heavy rains (as in Figure 8-8b).

What are some Earth systems interactions that influence slope stability?

8.2 Types of Unstable Land

Although the stability of all slopes is determined in the same way—by the interaction of driving and resisting forces, and how these change with slope steepness and other factors—slopes fail in several different ways. Land can also become unstable and fail in places where slopes are not present. In these places, the land is said to *subside*.

SLOPE FAILURES

People commonly refer to any material that moves rapidly down a slope as a "landslide," but geologists have developed more specific terms for the different types of mass movements. These are categorized according to three fundamental characteristics:

1. **How quickly the material moves down the slope.** Some mass movements, like the Portuguese Bend landslide described earlier, take place slowly over many years, whereas other types of slope failure occur in seconds.
2. **The type of earth material involved.** Mass movements may involve solid rock, rock debris, clay-rich sediment, soil, or mixtures of soil and rock debris. Depending on the type of material that makes up a slope, mass movements have different underlying causes and pose different hazards.
3. **The type of movement.** Materials moving on a slope fall, slide, flow, or creep.

The relationship between these factors and the types of mass movements that develop are summarized in Figure 8-9.

Falls

Falls are the fastest mass movements, characterized by the tumbling, rolling, or free fall of materials down a steep slope or cliff. In a rockfall, a mass of bedrock breaks free and falls down to the base of the slope. The fractured blocks of rock, or talus, collect at the base, as in Figure 8-10 (see also Figure 8-3). Rockfalls are a serious hazard along highway roadcuts and for mountain hikers and climbers. They are especially common in spring as daily freezing and thawing cycles break loose rock fragments.

In Yosemite National Park, California, the sheer granite cliffs in Yosemite Valley are irresistible attractions to climbers and sightseers. They can also be deadly, as demonstrated by an unusual rockfall early on July 11, 1996. Two massive blocks of granite detached at a height of 665 meters (2180 ft), 14 seconds apart,

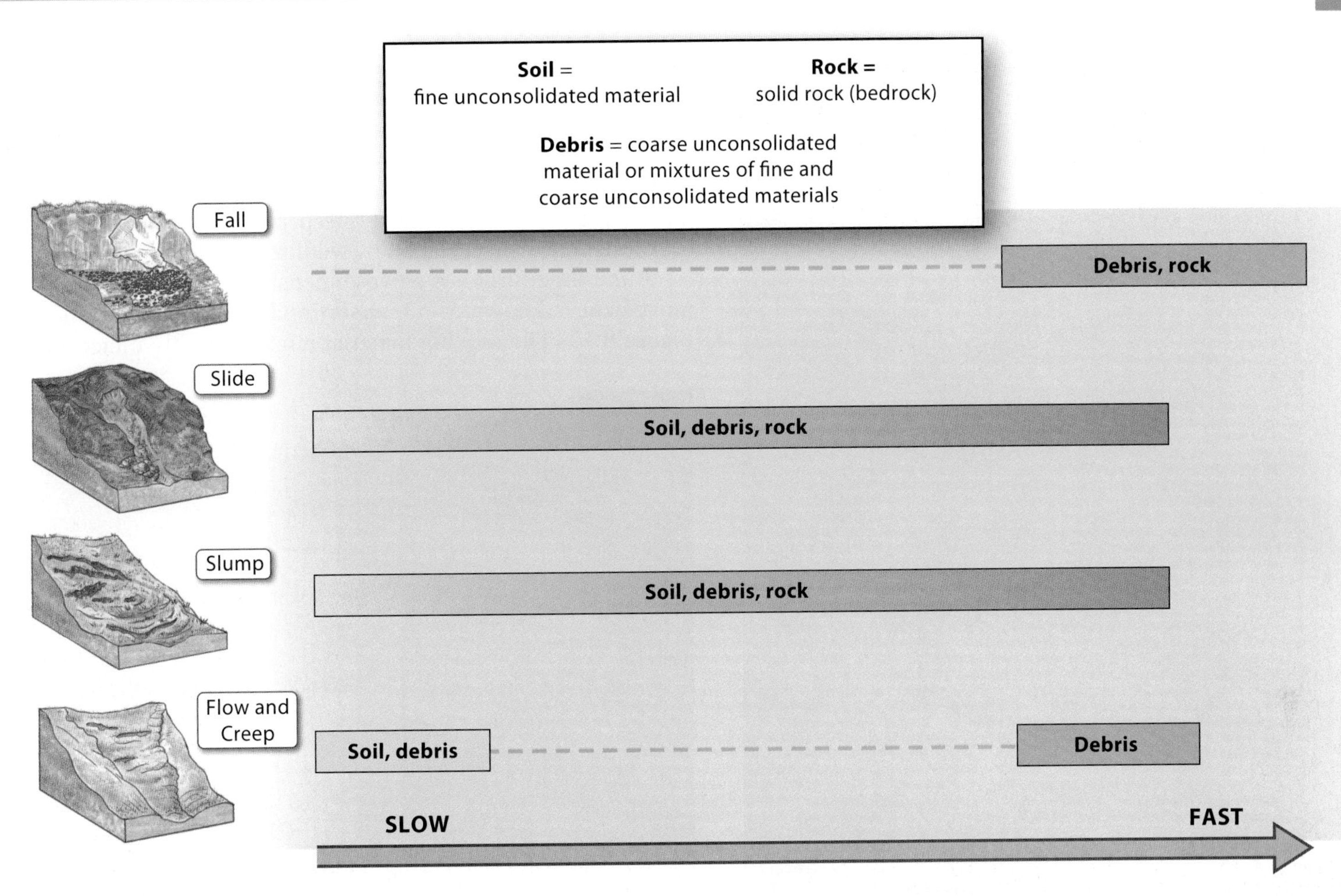

FIGURE 8-9 ▲ Classification of Mass Movements
Relationships between the type of movement, the type of material, and the rate of movement are used to classify mass movements.

FIGURE 8-10 ▼ Rockfall and Talus
Loose blocks of rock and rock fragments, as talus fall from bedrock formations and accumulate at their base as talus. (See also Figure 8.3.)

FIGURE 8-11 ▲ Rockfall, Yosemite, June 1999
This dramatic fall in Yosemite National Park, photographed in progress from across the canyon by a rock climber dangling on a rope 600 meters (2000 ft) above the valley floor, killed one climber and injured two others.

and fell to the valley floor and shattered. The combined volume of the two blocks was estimated to be as much as 38,000 m^3(50,000 yd^3). That's equivalent to a cube of rock 34 meters (112 ft) on a side. The impact generated a powerful blast of air. Moving at velocities exceeding 400 kph (250 mph), the blast flattened about 1,000 trees. The falling trees damaged the nearby Happy Isles Nature Center, destroyed a snack bar and bridge, caused one death, and injured several people. A similar fall nearly three years later, captured on film in Figure 8-11, also claimed a life. As of 2004, there were 14 documented deaths due to rockfalls in Yosemite National Park. Rockfall monitoring in Yosemite National Park Rockfalls continue in Yosemite. In October, 2008, two rockfalls damaged tent cabins at Curry Village, but luckily there were no fatalities, only minor injuries to a few campers.

Slides

In a **slide**, material moves downslope along a sloping surface, as opposed to free falling, tumbling, or bouncing. The rate of movement can vary from very slow (essentially imperceptible) to very rapid (often catastrophic) collapses. Slides are a common hazard on slopes that have been overly steepened by construction projects or denuded of vegetation by wildfires, erosion, or logging.

If the slide surface is a two-dimensional planar surface, the movement is known as a **translational slide**, like the one in Figure 8-12. The moving material remains largely intact, slid-

(a)

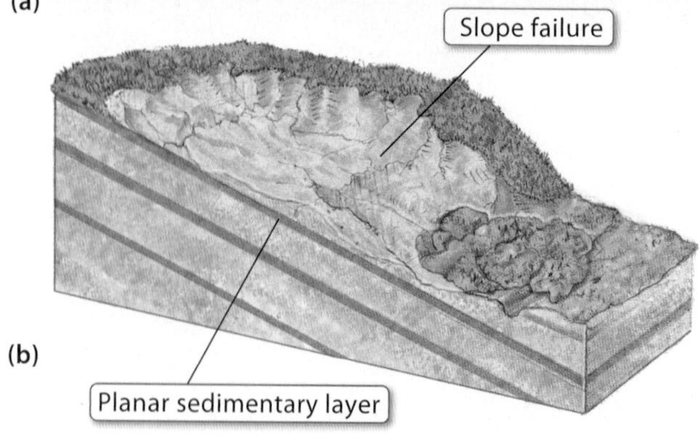

(b)

FIGURE 8-12 ▲ Translational Slide
The largest landslide in U.S. history, the 1925 Gros Ventre slide in Wyoming **(a)**, was a translational slide. The slope failed along a weak and planar sedimentary layer—the Amsden Shale **(b)**.

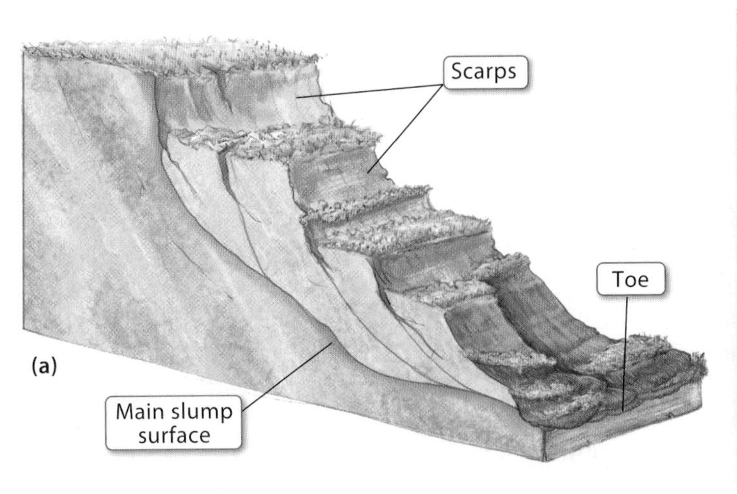

FIGURE 8-13 ▲ Rotational Slide
(a) In a rotational slide, or slump, the downslope movement is along curved surfaces at the base, and often between separate blocks of the slide material as well. **(b)** This landslide in Japan has the characteristic curved scarps and hummocky surface of a rotational slump.

ing as a single coherent mass or a group of blocks. Translational slides commonly occur along weak planes in bedrock such as sedimentary bedding planes, faults, or fractures that dip parallel to the surface slope (as shown previously in Figure 8-5).

In a **rotational slide** or **slump**, shown in Figure 8-13a, the unstable material slides downward and outward along a concave surface (like the bowl of a spoon) rather than a planar surface. Slumps can move as largely intact blocks or be broken into pieces along several concave surfaces that merge at their base with the main slump surface. Scarps form along the surfaces at the upslope head of the slump. Slumps may pile up material at the toe, where the blocks start to turn upward as the slump surfaces reverse their orientation and face upslope (Figure 8-13a). They may also transition downslope to earthflows (described below). The different movements of internal parts of a slump give them a characteristic undulating (hummocky) and scalloped surface that you can see in Figure 8-13b.

Flows

In a **flow**, rather than falling, slumping, or sliding as a block or mass, materials move more like a liquid. Particles in the material are in motion independently of each other. Some kinds of flows, such as lahars (Section 6.3), are common on the flanks of volcanoes, but flows can occur wherever unconsolidated slope materials become saturated with water.

Flows take a variety of forms, depending on the type of material involved and its water content. Most **debris flows**, like the one shown in Figure 8-14, involve relatively coarse material mixed with water. YouTube—debris flow Clear Creek County, Colorado In this case "coarse" means that more than half of the particles are larger than sand grains. Material in debris flows typically includes rock fragments, and the mixture has the consistency of wet concrete.

FIGURE 8-14 ◄ Debris Flow in Venezuela
In December 1999, days of torrential rain triggered debris flows of water, mud, boulders, and trees. As many as 30,000 people were killed. The devastation shown here is in a suburb of Caracas.

Heavy rainfall or snowmelt initiates most debris flows. The flows commonly channel into streambeds and other low areas, sweeping up boulders, broken trees, and—in developed areas—even automobiles and the wreckage of homes and bridges, as we saw in Figure 6-31. The larger the included debris, the more destructive these flows can be. Debris flow hazards in the United States Debris flows can attain speeds of 16 meters per second (36 mph) or greater. An extremely rapid type of debris flow is called a *debris avalanche*. Commonly originating on high or steep slopes, debris avalanches are a highly mobile and fast flow of debris, water, and air.

Debris flows composed of fine-grained material such as soil, sand, or silt are termed **earthflows**. The amount of water in the flow varies. *Mudflows*, for example, are wet slurries of soil and other fine material. They typically follow streambeds.

Some earthflows occur in a type of sediment known as **quick clay** that can rapidly turn into a flowing mass if disturbed. Quick clay soil is composed of silt grains surrounded by a jumble of thin, platy clay minerals, forming a chaotic "house of cards" structure filled with water. People may live, farm, and build on quick clays for many years until a jolt—perhaps from an earthquake or construction-related blasting—makes the structure collapse. Incredibly, the solid ground can turn into a soft slurry in a matter of minutes and flow away at the speed of a river.

On May 4, 1971, the town of Saint-Jean-Vianney in Quebec, Canada, was destroyed by a quick clay flow. Forty homes and 31 people were swept away. The flow occurred in clay deposits formed 9000 years ago in a shallow sea. In fact, the town was located within the depression left behind by a much larger slide that occurred some 500 years ago.

In 1993, people in the Ontario town of Lemieux were spared the same fate, although at the cost of abandoning a town some families had occupied since the 19th century. Geologists had discovered that Lemieux, a small community on the South Nation River, lay atop sensitive clays that were vulnerable to the same type of rapid ground failure that claimed Saint-Jean-Vianney. The people relocated, and just two years later, on June 30, 1993, the town's main street slid away, leaving behind a void 680 meters (2230 ft) long and 320 meters (1050 ft) wide (Figure 8-15).

What You Can Do

Tour an Earthflow

The Slumgullion Creek earthflow has been slowly moving downslope for 2000 years. This very large earthflow in southwestern Colorado tilts trees, scars the landscape, and dams a fork of the Gunnison River to form Lake San Cristobol. It continues to move at slow but variable rates, from 1 centimeter (0.4 in) to 6 meters (20 ft) per year.

You can take a virtual tour of the Slumgullion earthflow online. Virtual field trip of the Slumgullion earthflow This website enables you to visit many localities on the flow and zoom in to see particular features. You can learn about the results of the USGS's comprehensive studies of the Slumgullion flow online, too. The Slumgullion earthflow: a large-scale natural laboratory

This earthflow in Hinsdale County, Colorado has been slowly moving downslope for over 2000 years. The red dots identify localities where the USGS has photographic coverage of the earthflow online. Virtual field trip of the Slumgullion earthflow

FIGURE 8-15 ◀ Quick-Clay Failure
Scientists determined that Lemieux, Ontario was in a hazardous area and residents moved away two years before failure of clay-rich surface materials destroyed part of the former townsite.

Creep

Creep is a very slow type of earthflow. It causes soil and weathered rock to move downslope at rates as slow as 1 millimeter (0.04 in) per year. Even though the process is slow, creep can inflict significant damage over time, even on gentle slopes. It warps railroad lines, cracks roadways, dislocates underground utilities, and causes houses to shift and crack.

The most vigorous slope creep is driven by cycles of freezing and thawing. Soil and weathered rock contain many small individual particles. Upon freezing, moisture between the particles expands. This lifts the particles up, perpendicular to the slope surface, as shown in Figure 8-16. When the material thaws, gravity pulls the particles directly down. This makes every stone, pebble, or particle in a soil gradually zigzag down the slope.

Cycles of wetting and drying also contribute to creep in soils that contain certain kinds of clay minerals. These clay minerals expand when they get wet and contract when they dry. This causes the same kind of zigzag movement of surface material down a slope that freezing and thawing cycles do.

Materials closest to the surface of the slope creep more than those below. This causes objects anchored in the soil, such as trees, telephone poles, fence posts, and even tombstones, to tilt in the downward direction of the slope. Trees compensate by growing upward toward the Sun, resulting in curved trunks (Figure 8-17).

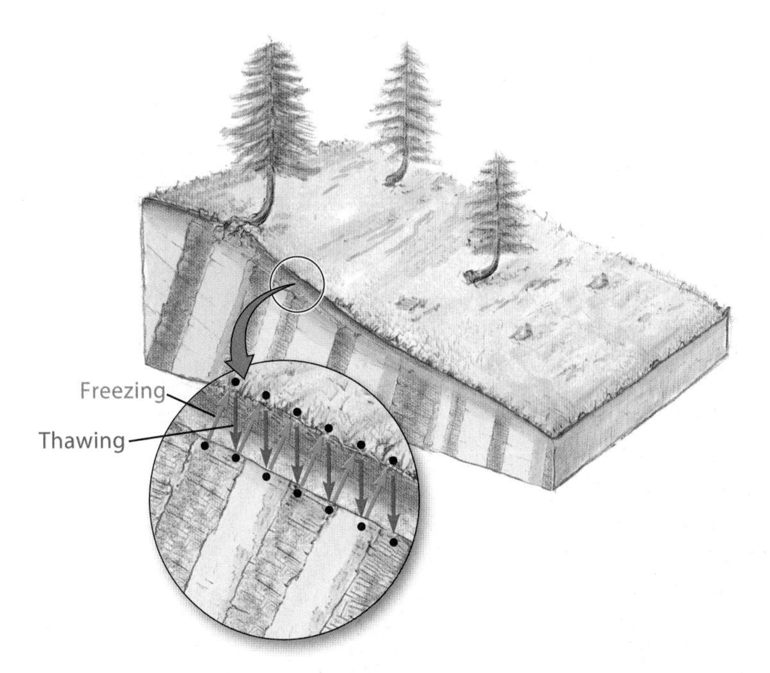

FIGURE 8-16 ▲ Creep
Creep on a slope is commonly caused by expansion of surface materials (blue arrows) during freezes (or wetting if expandable clays are present). When the surface material thaws (or dries), gravity pulls the material slightly downslope (red arrows).

FIGURE 8-17 ▲ Evidence of Creep
Curved tree trunks are evidence of creep in underlying surface materials.

Although always slow, the rates and amounts of creep within a soil vary, causing humps or bulges in the surface to form. These surface features are especially common in cold climates where the subsurface is permanently frozen (permafrost). Here, slope creep is known as *solifluction*. On a hillside in permafrost regions, only the uppermost soil thaws in the spring, leaving deeper layers frozen. Gravity draws the soggy, thawed layers downslope, making them seem to ooze or flow (Figure 8-18).

COMPLEX MASS MOVEMENTS

Mass movements, like most natural phenomena, do not always obey the simple schemes we devise to classify them. During a particular event, one type of mass movement can evolve into a different type. One such complex mass movement occurred in 1970 in Peru. It ranks as one of the deadliest mass movements on record. It began when a major earthquake jolted Nevados Huascarán, a 6770-meter-high (22,200-ft) mountain in the Andes. With an explosive boom, a mass of rock, ice, and snow fell away from the mountain's peak. At the source, the slope was a sheer 70 to 80 degrees, so the rubble rapidly accelerated, cascading down the steep slope as a swift rockfall.

As the rock fragmented and mixed with ice, snow, and air, it evolved into a debris avalanche with a volume of up to 60 million cubic meters (79 million yd^3). The debris was funneled into a steep-walled valley, aimed straight at the city of Ranrahirca, less than 16 kilometers (10 mi) away. As it rumbled down the valley at an average speed of 160 kilometers per hour (100 mph), the flow picked up rock, soil, trees, and other debris. On the way, tongues of debris diverted into side canyons, following the major drainages on the flanks of the mountain.

FIGURE 8-18 ▼ Solifluction

Creep caused by freeze/thaw cycles moves surface material downslope in solifluction lobes.

Farther on, the avalanche struck a ridge, splitting into two lobes—one aimed at the city of Yungay and the other at Ranrahirca (Figure 8-19a). First hit were a number of small mountain villages, some overrun by the debris and others devastated by a rain of boulders—some as heavy as 73 tons—which were launched into the air by the thousands as they struck obstacles in their path. Yungay was almost completely inundated by the flow. The depth of the avalanche reached 10 meters (33 ft) and at least 18,000 people died (Figure 8-19b). Some survived only because they happened to flee to a hilltop cemetery, where they watched the river of debris race past them and consume the city. In Ranrahirca, 170 perished. An avalanche and debris flow from Nevados Huascarán in 1962 had killed about 4000 people there. The devastated zone from the earlier catastrophe remained largely abandoned in 1970, so there were fewer people at risk.

SUBSIDENCE

The gradual settling or sudden collapse of level or gently sloping land is termed subsidence. Gradual regional subsidence occurs over long periods of time and can affect large areas. Subsidence in **karst terrain**—areas where underground cavities such as caves are common—produces smaller, circular depressions called **sinkholes**. The reasons for both types of subsidence are well understood, and much can be done to prevent them or reduce risks associated with them.

Regional subsidence

Regional subsidence occurs when the land gradually becomes lower over an extended area. The amount of subsidence can range from barely measurable to tens of meters (yards), and the affected area can cover thousands of square kilometers (miles). Regional subsidence is caused by the compaction of underlying porous material, and can result from either natural processes or human activities.

Natural subsidence occurs where water-saturated sediments become buried and lose the water they contain. This is common in deltas at the mouths of rivers. The loss of water lowers the supporting pore pressure in the sediments, and they compact. As compaction proceeds, the ground surface subsides. Most regional subsidence, however, is caused by people withdrawing fluids (groundwater or petroleum) from underground. We can stop regional subsidence, but it may not be possible to reverse its effects on the underground materials or the land surface.

Karst-related subsidence

On May 8, 1981, the ground opened up like a mouth and started to swallow a neighborhood at the corner of Fairbanks Avenue and Denning Drive in Winter Park, Florida. At about 8:00 p.m., one homeowner heard a curious swishing sound and observed a large sycamore tree descending into a hole. The tree was later followed into the hole by his house. Over several days, the hole enlarged into a sloping, round pit 106 meters (348 ft) across and 30 meters (100 ft) deep. This sinkhole, shown in Figure 8-20, eventually consumed numerous trees,

(a)

FIGURE 8-19 ◀ Debris Avalanche Damage in Yungay, Peru

(a) Over 18,000 people died in Yungay, Peru from an earthquake-triggered debris avalanche that passed over a ridge 100 to 200 meters (330 to 660 ft) high and buried the town. **(b)** This photograph shows the former plaza area and several palm trees buried to a depth of 5 meters (16 ft). The ridge that was overtopped by the debris flow is visible in the distance. The wreckage in the right middle ground consists of a smashed bus and truck.

(b)

FIGURE 8-20 ◀ Down a Hole

Everything from parts of a swimming pool to two Porsches collapsed into this sinkhole in Winter Park, Florida on May 8, 1981.

half of a highway, a large chunk of a public swimming pool, and portions of three businesses, including parking lots with two Porsches and a pickup camper. The initially dry hole subsequently filled with water. To Floridians, this headline-grabbing event was familiar. Florida is plagued by sinkholes. Like many others before it, the Winter Park sinkhole disrupted traffic, damaged buried utilities, and necessitated costly and protracted repairs.

Sinkholes develop when the land surface collapses into underground caverns and other cavities. Wherever bedrock is composed of rocks that groundwater can dissolve, sinkholes can form. These areas are most commonly underlain by rocks made up of carbonate minerals, mainly calcite in limestone. Other minerals that can readily dissolve in groundwater are those precipitated from seawater when it evaporates—halite (table salt), for example.

Rainfall and water percolating through soil are commonly weakly acidic. This water will gradually dissolve carbonate minerals (especially calcite) with which it comes in contact. Dissolution starts along fractures and, over time, gradually enlarges them into channels and other openings. As shown in Figure 8-21, this process produces a complex network of interconnected fissures, channels, and caves in the underground bedrock. Where underground openings come close to the surface, the land can become unstable and collapse into them.

Some sinkholes form gradually, with the land settling into bowl-shaped depressions, whereas others form suddenly with the collapse of the ground surface into a steep-walled pit. The thickness and type of material overlying underground cavities determines the type of sinkholes that form. Where cover material is soft and sandy, infiltrating water carries some material down with it into the bedrock and the ground above slowly subsides to fill the void. In areas where the cover is more rigid, the land surface maintains a solid roof over underground voids before suddenly collapsing.

Over time, the surface adjustments caused by dissolution of bedrock create distinctive karst landscapes (Figure 8-22a). ("Karst" is derived from the Slovenian word *kras* for a mountain range along the border between Slovenia and Italy.) Such landscapes are characterized by sinkholes, caves, and streams that disappear below ground or suddenly appear as springs. This type of terrain makes up about 10% of Earth's land surface. It comprises about 20% of the continental United States and is widespread in Alabama, Kentucky, Missouri, Tennessee, Texas, Virginia, and West Virginia (Figure 8-22b).

What are some Earth systems interactions associated with sinkhole formation in karst terrain?

Mining-related subsidence

Mining activity can create underground openings near the surface, too. Collapse or gradual subsidence above these openings can damage roads, buildings, and buried utilities. In a particularly dramatic example, the 1994 collapse of an active salt mine in the Genesee Valley of northwestern New York State caused widespread subsidence and threats to public health. Until the collapse, the Retsof mine was the second largest salt mine in the world. It had operated continuously for 110 years and supplied huge amounts of rock salt to the Northeast for road deicing. As is standard practice in mines of this type, salt was

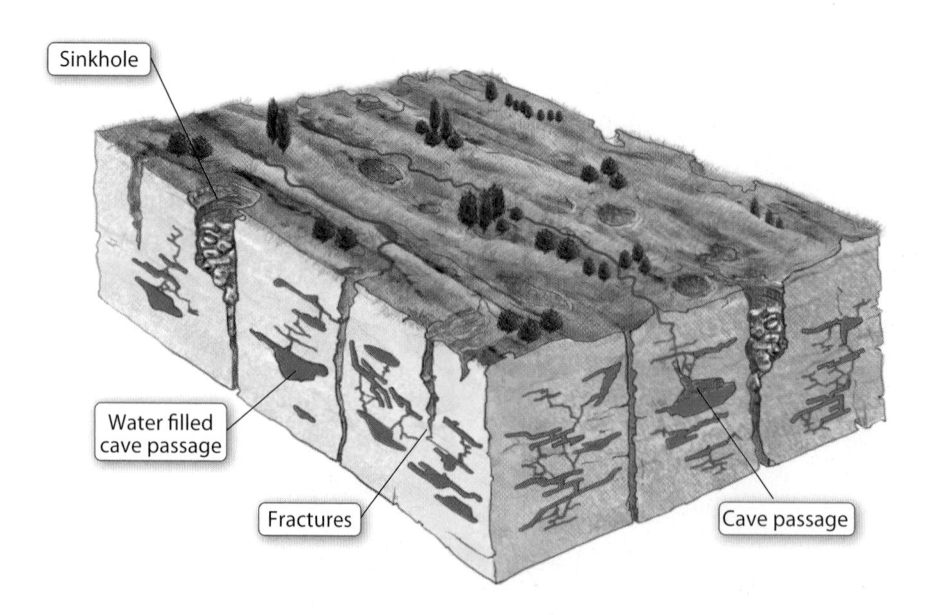

FIGURE 8-21 ▲ Formation of Karst
Interactions of the hydrosphere and geosphere produce the characteristic surface and underground features of karst.

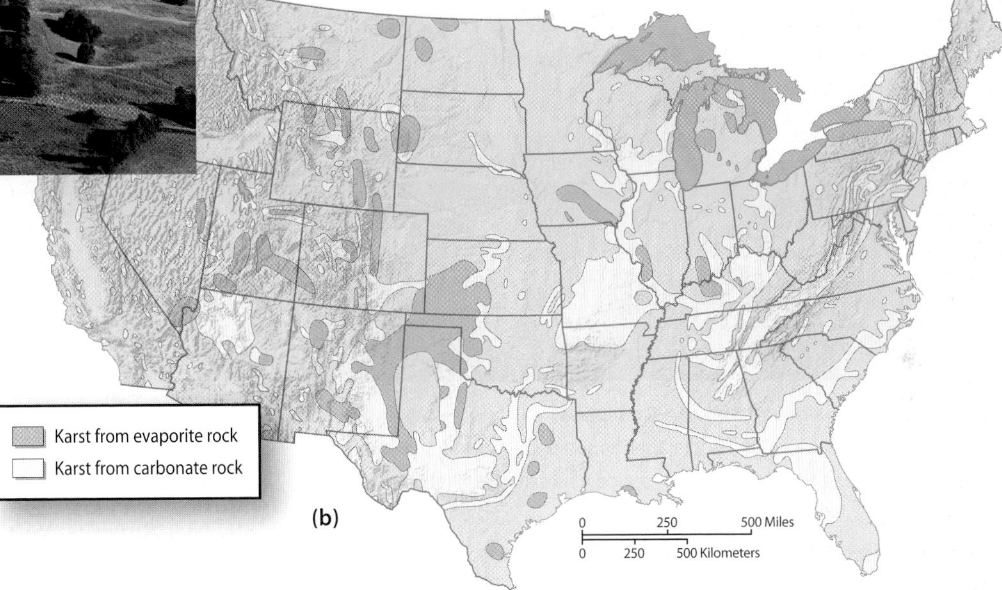

FIGURE 8-22 ◀ Karst Terrain
(a) Mass movements can occur on level land where bedrock has been dissolved and overlying material gradually or suddenly collapses into underground openings. This produces undulating land surfaces dimpled with sinkholes and surface depressions characteristic of karst. **(b)** Soluble bedrock is widespread in the United States.

excavated to form hollow chambers or "rooms," with pillars of unmined salt supporting the massive bedrock roof over the salt deposits. The mine's underground workings spread out over an area of about 26 square kilometers (10 mi^2).

On March 12, 1994, at 5:43 a.m., a room's roof collapsed at the southern end of the mine, some 365 meters (1200 ft) beneath the surface. The collapse opened a pathway for groundwater to flow down into the mine at a rate of almost 19,000 liters (5000 gal) per minute. Because the mine became shallower to the north, the entire mine did not flood at once. Instead, the water gradually filled the mine, dissolving the salt in successive rooms as it advanced. On April 6, a sinkhole opened in the ground above the initial collapse area, and gradually widened to a diameter 183 meters (600 ft). Two days later, another part of the mine roof collapsed, opening another sinkhole 244 meters (800 ft) in width on the surface (shown in Figure 8-23). Flow of groundwater into the mine tripled, to 83 million liters (22 million gallons) per day. The mine was completely flooded by December 1995.

FIGURE 8-23 ▶ The Retsof Mine Sinkhole
Surface subsidence resulting from the collapse of underground openings in the Retsof salt mine was just the visible part of the problem. Groundwater completely flooded the mine, dissolved its internal support pillars, and became contaminated.

FIGURE 8-24 ▲ Subsidence Above a Coal Mine
Areas prone to subsidence, such as this abandoned underground coal mine area in Colorado, are candidates for open-space land use designations.

In the aftermath of the initial collapses, some water wells went dry or were contaminated by salty water. The disturbance caused groundwater to give off methane and hydrogen sulfide gases. The gases migrated up through water wells, creating a nuisance (from the rotten-egg smell of hydrogen sulfide) and a potential safety and health hazard. Subsidence spread on the surface above the salt mine, causing prominent surface cracks in agricultural fields and tilting of the land surface. Drawing down of groundwater also caused the ground to compact and sink. Areas have been affected as far as 13 kilometers (8 mi) from the initial collapse zone.

Subsidence associated with underground mines is a significant problem in the Appalachian coalfields of Pennsylvania, Virginia, West Virginia, and Kentucky. Many other states share this problem: Ohio has about 6000 abandoned mines, most of them coal mines. Figure 8-24 shows an example from Colorado.

8.3 Causes of Land Failure

What caused the Portuguese Bend landslide? Residents saw their beloved country homes torn apart by natural forces they barely understood. California geologists might have offered some insight, since they had long known that the southern slope of the Palos Verdes Peninsula was scarred by the signs of multiple past landslides. A portion of one of the largest ancient slides, which draped across the entire subdivision and beyond, had been somehow reawakened. Those affected by the slide began to ask, Who is to blame?

The matter soon found its way to the courts, spawning 129 separate lawsuits, which were consolidated and settled in a nonjury trial in 1961. The developer sued the county, charging that roadwork on the upper slope of the subdivision had triggered the slide. The homeowners sued the developer for selling them homes in a hazardous area. In turn, the county sued the developer and the homeowners, charging that outflow from septic systems and lawn sprinklers had lowered the shear strength of the soils and initiated the slide.

Testimony by some engineers and geologists suggested that grading and filling the upper slope to extend a road, Crenshaw Boulevard, had increased the driving force on the old slide enough to start it moving again. Other experts testified on behalf of the county that the weight of the road fill was too small to reactivate the slide, based on calculations of the slope's overall stability. The judge decided, in the end, that the road grading was responsible. The homeowners and developers received compensation for their losses.

One thing is undisputed: Until August 1956, Portuguese Bend was not moving. Something pushed this marginally stable slope over the edge, triggering mass movement that continues to this day.

Natural causes of slope failures include above-normal rainfall, earthquakes, wildfires, and slope steepening. People's activities, too, can set the stage for slope failures as well as many examples of subsidence.

WEATHER

Many of the damaging or deadly mass movements that make headlines are debris flows, frequently reported by the media as "mudslides." The typical cause of a debris flow is a drenching rainstorm, or series of storms, that saturates unconsolidated slope materials. The water decreases the resisting forces in the slope material and also adds weight, causing the slope to fail. Not surprisingly, debris flows frequently come hand in hand with floods. A typical scenario would include a period of higher-than-normal rain for a week or two that saturates the soil, followed by a storm bringing additional precipitation that further burdens slopes, making them unstable.

A tragic example of this scenario occurred in Venezuela, when several days of torrential rain in 1999 led to flash floods and debris flows that killed an estimated 30,000 people. It was the worst natural disaster in Venezuela in the 20th century. The focus of the disaster was a narrow strip of coastline on the Caribbean Sea at the base of the coastal mountains north of Caracas. About 300,000 people inhabit this coastal zone within the state of Vargas, concentrated on alluvial fans at the mouths of valleys that cut through the mountains. During times of heavy rainfall, the valleys funnel water and debris from the flanks of the mountains directly onto towns and cities. As you can see from Figure 8-25, the residents of these communities were virtually living in the crosshairs of a loaded gun.

For the first two weeks of December, rainfall was above average. Then, on December 14–16, storms dumped 900 millimeters (35 in) of rain along the 40 kilometers (25 mi) coastal

FIGURE 8-25 ◀ Ground Zero for Landslide and Flood Hazards in Venezuela
The village of Tanaguarena, Venezuela, was one of many in the crosshairs of unstable land hazards during the heavy rains of 1999.

zone. Debris flow and flooding hazards associated with the December 1999 storm in coastal Venezuela and strategies for mitigation The resulting floods and debris flows reached the towns on the night of December 15 and continued in waves until the following afternoon. The debris flows and floods destroyed more than 8000 homes and 700 apartment buildings, causing $1.8 billion in damage. Many of the victims were never seen again, either buried under tons of mud or carried out to sea.

Hurricanes

As sources of intense, drenching rains, hurricanes rank high as causes of mass movements. Hurricane Mitch, which swept across Central America from October 27 to November 1, 1998, was one of the most powerful and damaging Atlantic hurricanes in 200 years. Heavy rains associated with the storm caused widespread flooding and landslides in Belize, Costa Rica, El Salvador, Guatemala, Honduras, and Nicaragua. In Honduras, total measured rainfall exceeded 900 millimeters (35 in), triggering more than a half *million* individual mass movements. Estimates based on satellite data suggest that local rainfall in some areas may have reached an astonishing 1900 millimeters (74 in)! The slope failures and damage to crops and homes in rural areas produced widespread misery and thousands of deaths (Figure 8-26).

Central America also has many stratovolcanoes, and mixing volcanic ash with hurricane rains is a recipe for lahars (Chapter 6). One lahar spawned by Hurricane Mitch swept down the flanks of Casita Volcano in Nicaragua, destroying several towns and killing about 1600 people.

FIGURE 8-26 ▼ Slump and Debris Flow Damage in Honduras
The heavy rains that came ashore with Hurricane Mitch in 1998 triggered mass movements on unstable slopes over large parts of Central America. The slumps and debris flows that devastated Tegucigalpa, Honduras, are just one example.

El Niño

The large-scale ocean-atmosphere circulation pattern known as the El Niño Southern Oscillation (ENSO) is often associated with destructive landslide activity on the U.S. Pacific coast. ENSO comes and goes about every three to seven years, as westerly winds episodically weaken and warmer than normal sea surface temperatures develop in the eastern equatorial Pacific. This pattern often brings unusually heavy winter and spring rains to America's Pacific coast. Heavier rainfall usually means more slope failures. During the 1997–1998 ENSO, spring storms triggered landslides throughout the ten-county San Francisco Bay region that caused about $156 million in damages.

Weather-related landslides occur year after year in California. The devastating La Conchita landslide on January 10, 2005, described in the accompanying *In the News* feature, is an example.

What are some Earth systems interactions that accompany weather-related mass movements?

In The News

Landslide Weather

One of the geologists developing a landslide warning system in California is Raymond Wilson. Here's his recollection of his experience on January 10, 2005. Geotimes—November 2005—watching for landslide weather in California

I learned of the fatal January 10 landslide from a local radio reporter, who was requesting a comment from me about "the deadly landslide at La Conchita." At first I thought she meant the massive slump that occurred there in 1995, but that had not killed anybody. No, she said, this landslide happened today, in the past hour, hadn't I heard about it?

"No" I replied, first puzzled, then with growing horror. "Oh . . . no!" I hadn't heard; I was working on a press release, the latest in a series, warning of the risk of landslides from the heavy rains that had been soaking California since a few days after Christmas.

I downloaded a news photo of the new landslide from the Internet. The scar from the 1995 landslide was still clearly visible, looming over the houses of La Conchita. Smaller than the 1995 event, but more fluid, the 2005 landslide could move much faster and, therefore, was much more dangerous. Earlier in the day, I later learned, smaller debris flows had traveled down the hill and through the streets, all the way across the railroad and Highway 1. Residents anxiously watched the hillside, but did not evacuate. Ten people died, several of them children.

The reporter asked the usual questions: "Why were these people still living there" (after 1995)? Was there a landslide warning?" "Why hadn't these people been evacuated?" I remember reciting the usual answers: Land-use policy is best addressed by local officials. Public advisories had been issued by the U.S. Geological Survey and the National Weather Service, but these were regional and not specific to La Conchita.

Meanwhile, staring at the news photo, I felt shock, like a physical blow, and profound sorrow. I couldn't deny what I was seeing, but couldn't fully assimilate it either. After watching, over many days of heavy rains, the landslide potential progress from possible to threatening, to horribly inevitable, how could I be surprised? In the face of such human tragedy, the reporter's questions—and my stock answers—began to seem weak and incomplete. I felt deeply disturbed by the La Conchita event; I still do.

The close tie between storms and landslides in California is why geologists must understand the weather to be able to predict when landslides will occur. The historical record in California enables them to closely compare precipitation patterns and the frequency of past landslides. These historical relationships help scientists identify areas where new storms—bringing specific amounts of precipitation within specific periods of time—can be expected to cause landslides and debris flows. By watching the weather and mapping where slopes have historically failed, geologists can achieve their goal of warning people of impending landslides.

You Make the Call

What Would You Do with La Conchita?

Soon after the January 10, 2005, La Conchita landslide, Governor Arnold Schwarzenegger visited the stricken community and offered his support. In fact, he told the residents that he would help them stand strong and rebuild. "We'll be back" the residents said.

Landslides destroyed homes at La Conchita in 1995 and homes and people in 2005. The area is exceptionally prone to landslides. What would you do with La Conchita:

- if you were the California governor?
- if you were a resident whose intact home was near another that was destroyed?
- if you were a geologist like Ray Wilson, working so hard to warn people of the risks they face?

FIGURE 8-27 ◀ Earthquakes Can Trigger Landslides
In January 2001, an earthquake of magnitude 7.6 caused this slump that buried part of the Pan-American Highway in El Salvador.

EARTHQUAKES

Ground shaking during earthquakes frequently triggers mass movements by temporarily reducing friction and cohesion in slope materials (Figure 8-27). On a rocky slope, earthquakes may shake loose boulders or slabs of bedrock. In some earthquakes, slope failure accounts for most of the property damage and deaths. This was certainly true of the 1964 Good Friday earthquake that struck Anchorage, Alaska, and the surrounding region (Chapter 5). The prolonged and powerful ground shaking from this magnitude 9.2 earthquake triggered local mass movements scattered over an area of about 260,000 square kilometers (100,000 mi^2).

WILDFIRES

Wildfires are common during the dry summers in parts of the American West. They consume tree cover, surface leaf litter, and organic material within the soil. This may make the soils harden and repel water, thus reducing the amount of precipitation that infiltrates the ground. Such conditions increase surface runoff and can mobilize large amounts of sediment into fast-moving debris flows.

In California, summer wildfires commonly lead to winter and spring debris flows. These do much physical damage and sometimes catch the unwary, with tragic results. In October 2003, a wildfire scorched Waterman Canyon north of San Bernardino. On Christmas day that year, 28 people were spending the holiday with the caretaker of a Greek Orthodox camp in the lower Waterman Canyon. Heavy rain touched off a debris flow in the canyon that caught the people in Camp St. Sophia by surprise. Fourteen people in the group died. The remains of the fourteenth and final victim of the slide, an 11-year-old boy, were not found until April, 24 kilometers (15 mi) downstream of the campsite.

SLOPE STEEPENING

Erosion at the base of slopes by rivers or ocean waves can trigger mass movements of various types. This is the principal way rivers erode their banks (Chapter 7). The material at the base of a slope helps support the material above. If erosion removes the base, the slope is locally steepened and the resulting increase in driving force leads to slope instability.

River erosion played a leading role in a famous slide that occurred in Wyoming's Gros Ventre River Valley in 1925 (see Figure 8-12). A bed of sandstone formed the base of the valley wall, which sloped gently down to the river. Over time, the river eroded through much of the sandstone, which lay atop a layer of weak shale. Spring snowmelt and heavy rainfall increased the driving force and decreased the resisting forces along the weak shale layer. One day a huge portion of the slope broke free along the unsupported shale layer and slid down into the valley, carrying the forest cover with it. The fast-moving slide mass filled the valley bottom and traveled more than 30 meters (100 ft) up the opposite side before settling into a natural dam across the Gros Ventre River. This is the natural dam that failed two years later, causing a flood that killed six people and devastated the town of Kelly, Wyoming (Chapter 7).

PEOPLE AND SLOPE FAILURE

People can significantly influence a slope's stability. The great desirability of a "room with a view," despite the potential hazards, continues to attract development onto hillsides (Figure 8-28).

FIGURE 8-28 ▲ Where's My Backyard?
People often build on sites that aren't as stable as they appear.

In a typical scenario, developers bulldoze a flat bench or create one by compacting fill material on the slope for house construction. Constructing access roads to new housing also changes slopes and rearranges surface materials. This can compromise stability in several ways.

- Constructed facilities, like houses, add weight to the slope, increasing driving forces.
- Excavating terraces and roads may locally steepen slopes and interrupt normal drainage patterns. Precipitation may either lead to erosion or saturate the soil and increase the driving force.
- Septic systems, lawn sprinklers, and leaking water pipes can increase the amount of water that infiltrates the hillside, thereby increasing the driving force and reducing the resisting forces.
- Clearing trees, brush, and other vegetation that has well-developed deeper roots can decrease the slope's resisting forces.

PEOPLE AND SUBSIDENCE

People cause land subsidence by withdrawing fluids from underground. In several places, the recovery of oil or natural gas from underground reservoirs produces subsidence, but the most common cause is withdrawal of groundwater | USGS groundwater information: land subsidence in the United States | In the United States, groundwater pumping accounts for more than 80% of subsidence, affecting about 44,000 square kilometers (17,000 mi^2) spread out over 45 states (Figure 8-29).

The groundwater reservoirs, or *aquifers*, most vulnerable to compaction are those composed of unconsolidated sediments or sedimentary rock containing fluid-filled pores. The pore pressure of water in the aquifer helps support the weight of the materials above. With the ebb and flow of seasonal demands on groundwater, aquifers expand and contract, imperceptibly raising and lowering the ground surface. If people pump water out too quickly or in excessive amounts, the aquifer may compact in ways that cannot be reversed, and the land surface permanently subsides.

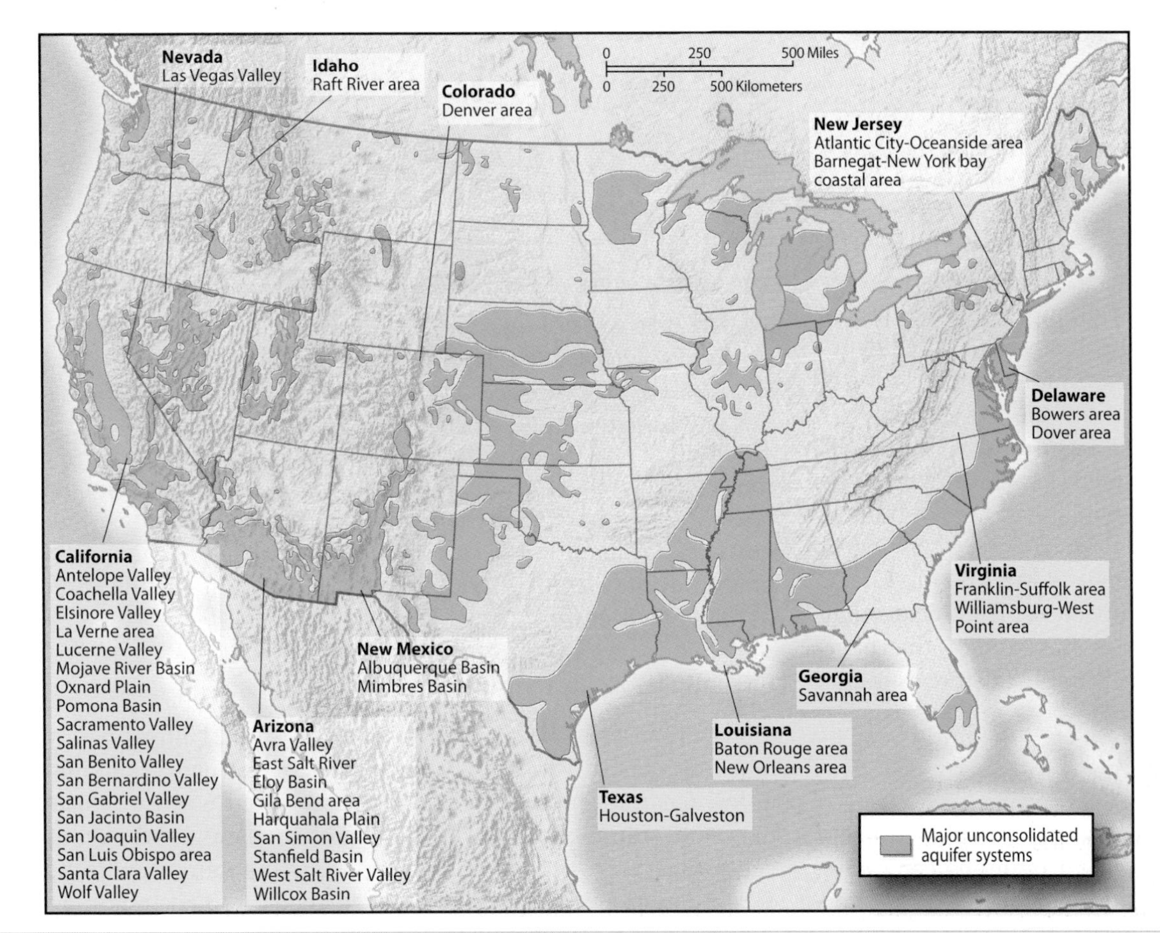

FIGURE 8-29 ◀ Groundwater Pumping and Subsidence
Subsidence caused by groundwater pumping is widespread in the United States. Some of the areas most affected are shown on this map.

Subsidence in California's San Joaquin Valley

An extreme example of regional subsidence has occurred in the San Joaquin Valley in California. It occupies the southern two-thirds of the larger Central Valley, a sediment-filled trough bordered on the west by the Coast Ranges and on the east by the Sierra Nevada Mountains. Groundwater here is annually recharged by snowmelt from the Sierra Nevada. The valley itself is arid, so people had to pump a lot of groundwater to irrigate fields and turn the region into an agricultural powerhouse.

In the 1930s, electricity became available in the valley, allowing farmers to pump water from deeper sources than before. That same decade, people first noticed that the ground surface was sinking—the aquifer settled and compacted like a drying sponge as farmers drew water from it faster than the mountains could replenish it. By the 1960s, the water table in the valley had fallen as much as 30 meters (100 ft) in some places.

By the end of the decade, canals and aqueducts supplied surface water to the fields, slowing the rate of subsidence, and by 1974, most subsidence had ceased. However, the effect of past groundwater withdrawal was dramatic (Figure 8-30). Maximum subsidence in the valley, near the town of Mendota, was 8.5 meters (28 ft).

FIGURE 8-30 ▲ San Joaquin Valley Subsidence
Groundwater withdrawal has led to major subsidence in the San Joaquin Valley of California. In 1925, the ground surface was 8.5 m (28 ft) higher than it was in 1977, the date of this photo.

Regional subsidence and cities

When significant subsidence occurs in an urban area, it can cost dearly. A case in point is the widespread subsidence around Houston, Texas, due to withdrawal of groundwater. The huge area affected overlaps Galveston Bay and has permanently altered the coastline (Figure 8-31a). Some developments, such as the Brownwood subdivision on Galveston Bay, have been flooded out and abandoned (Figure 8-31b). The elevation of the subdivision, built on a small peninsula, decreased from 3 meters (10 ft) to only 0.6 meters (2 ft) above sea level. About 40 hectares (99 acres) of the San Jacinto Battleground Historical Park on Galveston Bay were permanently flooded.

Regional subsidence causes damage to urban infrastructure, such as warped or crushed well pipes, broken water mains, cracked house foundations and roads, and altered surface drainage that can lead to flooding. In Mexico City, for example, regional subsidence followed heavy pumping of groundwater from the sediment-filled basin underlying much of the city. Among other effects, subsidence damaged buildings and disabled the city's sewer system, which had to be rebuilt at great cost.

A less-obvious impact of subsidence is the loss of storage capacity in aquifers that undergo compaction. That means that even if the rate of pumping is reduced, the aquifers that have been permanently damaged will never be able to store as much water as they did before.

Groundwater pumping and sinkholes

In areas where the groundwater level is high enough to keep bedrock cavities full of water, groundwater pumping can cause sinkholes. The water helps support the weight of the roof over the cavity. If that support is withdrawn, the roof may no longer be able to support its own weight. In one field of water wells north of Tampa, Florida, rapid pumping caused 64 new sinkholes to form within a 1.6 kilometers (1 mi) radius in the space of only one month.

Urbanization and sinkholes

Many urban areas lie partly or wholly on karst. Unfortunately, many aspects of city living can trigger the formation of sinkholes. Changes in surface drainage associated with urbanization are commonly a factor. In open land covered by vegetation, seepage is more diffuse, occurring more slowly and over a larger area. In contrast, paved roads, roofed buildings,

FIGURE 8-31 ▶ Subsidence at Galveston, Texas

Subsidence caused by withdrawal of groundwater in the Houston, Texas, metropolitan area from 1906 to 1995 permanently flooded parts of the shoreline along Galveston Bay **(a)**, including this home in the Brownwood subdivision **(b)**.

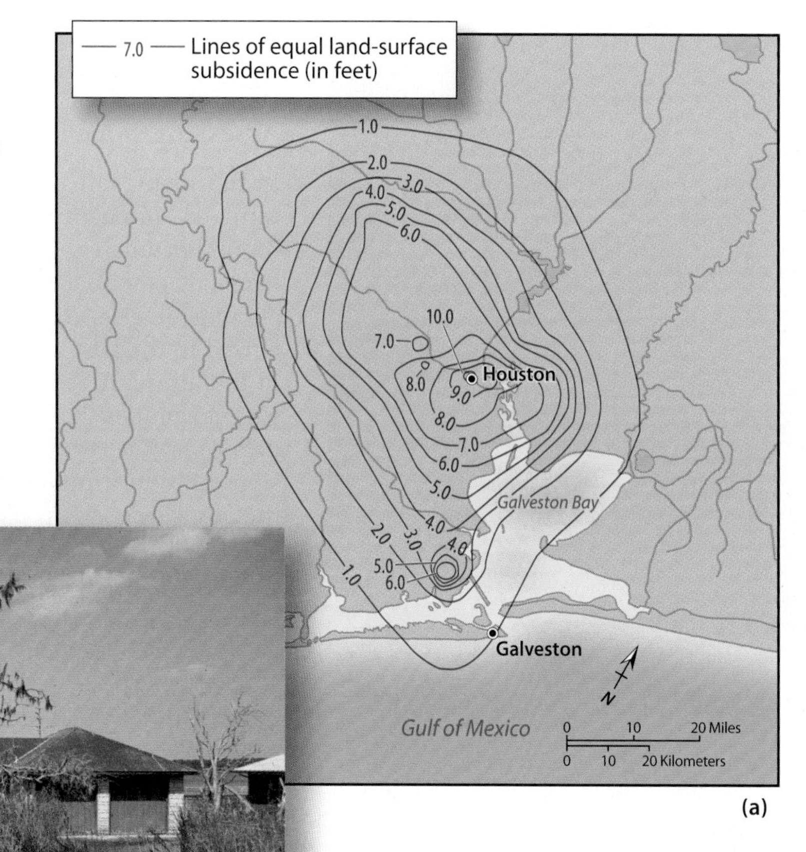

(b)

asphalt parking lots, drainage ditches, and other aspects of the built environment concentrate groundwater in ways that trigger sinkholes. Constructed facilities also add weight to the cover material that can make land unstable. Here are a few examples typical of cities built on karst.

- Along a new section of paved highway, rainwater runoff drains into a roadside ditch lined with crushed stone. During a heavy rain, the water percolates downward through the cover material and into a dry cavity in the bedrock below. The soggy cover material starts to fall into the cavity, removing material upward until the surface can't be supported and it collapses into the cavity. This undermines the road, leading to costly repairs.
- Leaky gutters and downspouts soak the sandy soil alongside a large commercial building over several weeks of heavy rains. The soil starts to subside, leaking into open fissures in the bedrock below. The soil movements in turn cause the buildings constructed on them to sag and settle unevenly, so doors begin to stick, windows don't open easily, and large cracks appear in the foundation walls.
- Water from a leaky water main permeates the loose soil beneath a town's main street and carries sediment into the bedrock below. A void develops around the water main. One hot day, the soft asphalt collapses to reveal a sinkhole 30 meters (100 ft) wide that swallows up two parked cars, a section of sidewalk, and half of a homeowner's front yard. The damage to property plus the cost of plugging the deep bedrock source of the sinkhole and rebuilding the road exceeds $1 million.

Sinkhole flooding is another significant problem in urbanized karst areas. In some communities, underground channels serve as a storm water drainage system and flooding can occur if the sinkholes and street drains discharge more water into the bedrock than the system can accommodate. Water then begins to back up into the sinkholes and nearby houses may become flooded. Sinkholes may also become clogged with debris and then fill up like a bathtub with a clogged drain. Illegal dumping of trash into sinkholes, a significant problem in rural karst areas, increases the likelihood of flooding.

8.4 Living with Unstable Land

In the winter of 1996–1997, the wooded slopes around Puget Sound in Washington began to give way under the burden of an unusually wet season. On Bainbridge Island, west of Seattle, homes were perched on a narrow beach along Rolling Bay. On the evening of January 19, 1997, the Herren family was asleep in the basement of their home: Dwight Herren; his wife, Jennifer; and two children, Skyler and Cooper. Herren was

FIGURE 8-32 ▲ Tragedy on Bainbridge Island
The Herren family died when a landslide collapsed their home on the shores of Bainbridge Island, Washington on January 19, 1997.

remodeling the house. Then, in a matter of seconds, the Herren's home was crushed by tons of soil, rocks, and uprooted trees (Figure 8-32). A debris flow originating on the steep slope behind the house killed the entire family. | The liquid Earth

Can people prevent this kind of tragedy? In many cases, the answer is yes. There are proven strategies for preventing damage and loss of life due to unstable land.

LIVING WITH UNSTABLE SLOPES

To learn to live more safely in areas with unstable slopes starts with identifying and assessing hazardous areas. Once hazards are well delineated and understood, people can take steps to guide or even prevent development in risky areas through land-use zoning and regulations. If slope failures affect an area, all may not be lost. Engineering can help mitigate the effects and decrease the risks that people face.

Assessing slope hazards

In too many cases, people are unaware of a landslide hazard until the moment a hillside comes crashing into the living room. Fortunately, geologists can determine where the greatest landslide potential exists. Aerial and field surveys, for example, allow geologists to identify slopes where landslides have occurred in the past and therefore where they are likely to occur in the future. A geologist, engineer, or other landslide investigator examines the area for telltale signs of previous slope failure. | Landslide warning signs In undeveloped areas, unstable slopes produce undulating hummocky features, bare scars in the landscape, sharp banks (scarps) or gashes in the surface, and tilted trees with curved trunks. In developed areas, sidewalks, fences, roads, foundations, and other constructed features can be cracked, tilted, or offset where the land is unstable. By making maps that show the distribution of landslide surface features, the investigators delineate the overall extent of the unstable land.

However, identifying areas with previous landslides is just the first step in identifying hazards. Combining this information with slope steepness and the nature of the underlying material enables scientists to make landslide hazard maps.

Landslide hazard maps delineate areas with different likelihoods of slope failure. Some simply label slopes that are at high, medium, or low risk of failure. Some assign unstable slopes a probability of failure, say a 10%, 50%, and 90% chance of failing. Here's an example.

Kansas doesn't have mountains but it still has landslides. In Leavenworth County, geologists combined a geologic map showing the distribution of bedrock units (Figure 8-33a) with another map showing where landslides had already occurred (Figure 8-33b). The combination showed that landslides were most common where bedrock units containing shale layers were present on steep slopes. The relationships among bedrock geology, previous landslides, and slope steepness enabled the geologists to produce a *landslide susceptibility map,* shown in Figure 8-33c. This map shows areas with different probabilities for future landslides, from very low (less than 1 in 1000) to high (greater than 1 in 10). Such maps are valuable guides to land-use decisions by land owners, planners, developers, local government officials, insurance companies, and lending institutions.

Acting on hazard information

As emphasized by Ray Wilson's experience in California, identifying a hazard is not enough; individuals and communities must then take action to reduce the risks they face from slope failures. The good news is that hazard experts have identified a number of general strategies for reducing loss of life and property damage due to slope failures.

Ultimately, the most effective way to reduce landslide hazards is to minimize the number of people living and working on

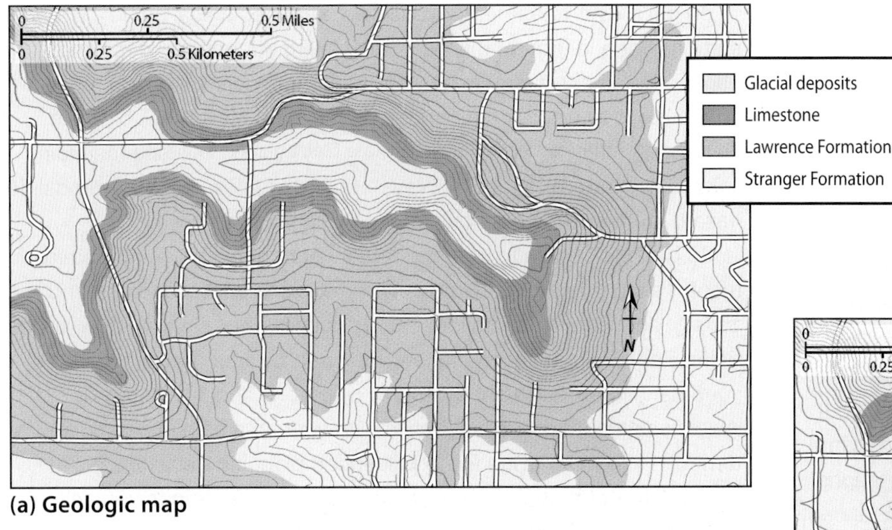

(a) Geologic map

FIGURE 8-33 ◀ Assessing Landslide Hazards in Kansas
Scientists use a geologic map **(a)** and a map showing where landslides have taken place **(b)** to make a landslide susceptibility map **(c)** of part of Leavenworth County, Kansas. Landslide susceptibility maps are guides for land-use and related decisions such as building, financing, and insuring new construction.

0 0.25 0.5 Miles
0 0.25 0.5 Kilometers
N
Landslides

(b) Previous landslides map

0 0.25 0.5 Miles
0 0.25 0.5 Kilometers
N
Probability of landslides
Less than 0.1%
0.1% – 1.0%
1.0% – 10%
Greater than 10%

(c) Landslide hazard map

or near unstable slopes. This approach is called *avoidance*. One method of avoidance is to require people to disclose the potential for slope failure when they sell their land or home. This gives the prospective buyer an opportunity to decline the purchase or obtain insurance for potential losses. Private or government-subsidized mortgage lenders may also deny loans to people who want to build in areas susceptible to mass movements.

In areas that have experienced large or repeated losses due to slope failure, officials may discover that the costs have begun to exceed the benefits. In this case, it might make more sense in the long run to abandon hazardous areas altogether. Rather than continue to repair roads and public utilities after every new slope failure, government officials could cut their losses by purchasing property and housing in hazard zones. They would then ban new construction or convert the land to open space and public parks, reducing people's exposure to the hazard.

The human population continues to grow in most places in the world. As a result, completely prohibiting development on potentially unstable slopes is not always possible—especially given the fact that people are willing to pay a premium price for hillside views. However, local governments have the power to regulate land use and construction so that it doesn't increase the risk of slope failure. Grading codes are an effective method.

Grading codes typically require builders to obtain a permit before altering a slope in any way, including excavating or filling it to create flat areas for roads or foundations, removing hillside vegetation, or doing anything that would alter the flow of surface water or groundwater. Governments may also require developers to take special steps to protect slope stability, such as installing underground drains or planting vegetation.

In 1952, Los Angeles County adopted the first grading codes in the United States. The codes were a response to major slope failures on hillsides developed during the post-World War II housing construction boom. Over time, the grading codes produced dramatic benefits. Losses after severe storms in 1968 and 1969 were far lower in developments covered under the modern grading codes than in developments built earlier. One study found that grading codes have reduced losses in Los Angeles due to slope failure by 92 to 97%.

What You Can Do

Investigate Unstable Slopes in Your Community

Do you know or suspect that landslides could occur near where you live? You can take four steps to learn more about landslides in your community. Use information from the USGS, the geological survey in your state, your local county planning department, or your own observations and mapping.

The USGS has a very active landslide hazards program. | USGS landslide hazards | This program develops landslide susceptibility maps and landslide forecasts for particularly vulnerable areas. The USGS website has much general information about landslides, but the program also maintains databases of printed information, contact information for active landslide researchers, landslide photographs, and government ordinances related to landslides. The USGS accepts calls requesting information and help at 800-654-4966. Who knows, USGS researchers may be working near where you live.

To get closer to home, try your own state geological survey. Every state has one, although this work may fall under another department's name. The quickest way to find out about geologists working for your state is to go to the American Association of State Geologists website. | American Association of State Geologists | This site maintains links to all state geological surveys. Hazard studies, including landslide hazard studies, are commonly a priority of these surveys.

To get really close to home, contact your county planning department. These departments implement decisions about land use and zoning, so wherever landslides are a concern, they should know about them. Perhaps you know of a landslide nearby. You can check up on your planning department by asking questions about the landslide. When was the landslide last active? How much area has been disturbed? How has the landslide influenced planning decisions?

Don't forget that you can always investigate landslides yourself. It's best to start with a topographic map of the area where you live. On this map, note the steepest areas where people's homes, community developments, and roads are constructed. Drive or walk through these areas looking for evidence of unstable slopes—curved trees, cracked sidewalks, and frequently repaired roads, for example. | Landslide warning signs | Bare scars on a slope surface are a clear indication that something isn't stable. You can plot the location of these features on your map and start to get a feel for how stable the slopes are around you. Perhaps most of the slopes are stable, but you might also discover some landslides.

Unstable slopes are so common that just about every cross-country trip will reveal some. Watch for hummocky slopes, mountainside scars, and failing roadcuts wherever you go. Unstable slopes will be there somewhere.

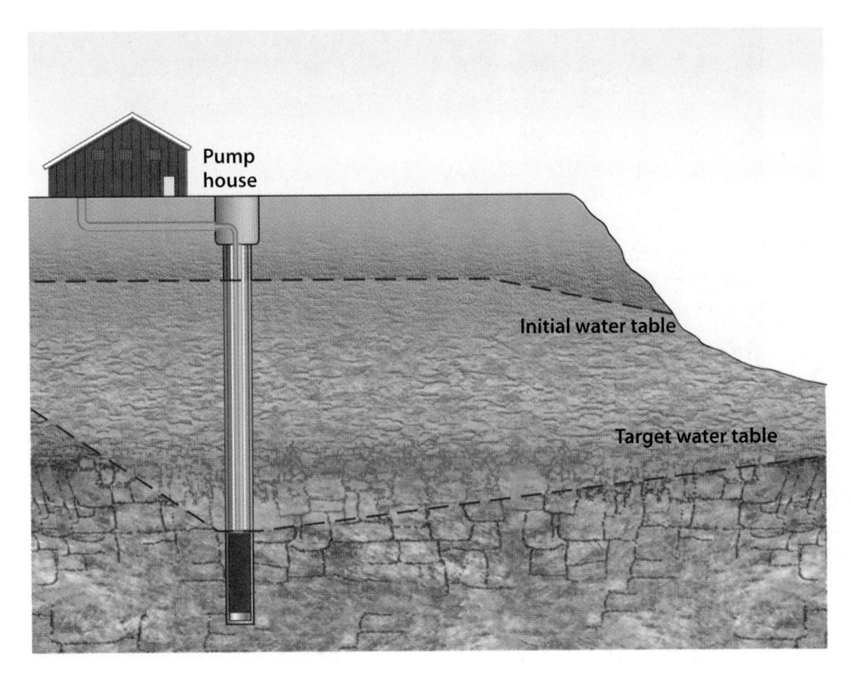

FIGURE 8-34 ▲ Dewatering Slopes
Decreasing the water content of landslide-prone materials can increase resisting forces and help prevent mass movements.

Engineering stronger slopes

Engineering approaches can help make slopes more stable. Most engineering measures aim to decrease the driving force (water content, steepness) or increase the resisting forces on the slope (installing walls, anchors, and other physical supports). The downside of slope engineering is that it doesn't always work and it can be expensive.

Controlling runoff and infiltration commonly involves installing surface drains that collect and reroute rainwater runoff or snowmelt away from the slope, preventing it from soaking into the ground. Drains can also be constructed to gather surface water and carry it away. At a roadcut, for example, perforated pipes can be installed deep enough in the slope to intercept groundwater and conduct it out to the roadway or storm sewer system. Another approach is to cover the slope with concrete, asphalt, vegetation, or plastic sheeting to prevent runoff water from infiltrating. Trenches filled with coarse gravel are also used to drain subsurface water. In places where groundwater is the main problem, wells and pumping can purposely lower the groundwater table enough to increase slope stability (Figure 8-34).

Decreasing slopes increases stability by decreasing the driving force. On slopes composed of unconsolidated materials, decreasing steepness can be very helpful. The top of the slope can be excavated and transferred to the base, reducing its overall grade. This unloads a portion of the mass bearing down on the head of the slope. Excavating a series of steps, or benches, into a slope can also enhance stability (Figure 8-35).

Supporting slopes increases their resisting forces. There are many ways to add physical support to a slope and increase its overall stability. The most popular method, buttressing, adds a stable, heavy mass of reinforced soil fill or broken stone to the base of the slope. This prevents the slope above from sliding downward. Retaining walls made of concrete, steel plates bolted together, or stacks of wire baskets filled with broken stone (gabions) can be used to buttress the base of a slope, as shown in Figure 8-36 on p. 250. Another option is driving a wall of steel or concrete piles across the base of the slope to prevent sliding or slumping.

Surface pins and anchor bolts are also used to support slopes. These are driven into the ground perpendicular to the slope, increasing the slope's overall strength. This is a common way of reinforcing roadcuts through bedrock, where masses of

FIGURE 8-35 ◀ Decreasing Slopes Increases Stability
Constructing a series of benches on a steep slope is a common method for increasing slope stability.

FIGURE 8-36 ▲ Supporting Unstable Slopes
Supporting unstable slopes is another way of increasing their resisting forces to prevent mass movements.

rock may detach along planes of weakness and fall onto the roadway below. To prevent this, workers drill holes deep into more stable layers of the slope and cement steel pins (called *soil nails*) into the holes, as shown in Figure 8-37. Erosion control soil nailing

Some areas can be protected from unstable land with barrier systems. Reducing the hazards of rockfalls is a special challenge, particularly in mountainous areas where steep, high roadcuts are often necessary and the slope lies perilously close to the roadway. In cases where the slope cannot be completely stabilized by physically covering it or installing anchors, nets of strong wire mesh can be laid across the slide area. This funnels falling debris safety to the base of the slope, where it is frequently captured by a low ditch or prevented from rolling outward by a wall like the one seen in Figure 8-38.

Dams and other physical barriers can help reduce hazards in areas where debris flows are common. *Check dams* like that shown in Figure 8-39, known as *Sabo dams* in Japan, partially block drainages in order to decrease flow velocity and capture the heavier material (typically boulders) responsible for much of the destructive power of a debris flow. The reservoir space behind the dam traps and stores the material until workers can remove it. Check dams are sometimes equipped with grates or bars to strain out larger debris. The drawbacks of check dams are that they are expensive, must be cleaned and repaired after debris flows, and are vulnerable to catastrophic failure if they are hit by a flow that exceeds their design strength. However they have proven helpful in appropriate situations. After the 1980 eruption of Mount St. Helens, for example, the U.S. Army Corps of Engineers built a large dam to capture debris flows and prevent clogging of the Lower Toutle and Cowlitz Rivers. Officials anticipate that the dam will completely fill with sediment by 2035.

FIGURE 8-37 ◀ Using Soil Nails to Stabilize Slopes
Installing soil nails—commonly, steel rods—increases the internal strength and resisting forces of surface materials. This technique is often used to help stabilize steep to vertical walls of surface excavations.

FIGURE 8-38 ▲ Barriers
Constructed barriers can provide protection from rockfalls.

FIGURE 8-39 ▲ Check Dams
There are many types of check (or Sabo) dams, but they all help decrease the impact of debris flows by decreasing their velocity and removing larger debris. Such structures typically function by locally decreasing drainage gradients, so that the speed of the flow is decreased and larger debris stops moving.

In some cases, real-time landslide monitoring alerts people to slope movements. In 1997, failing slopes along U.S. Highway 50 in California caused extensive property damage and briefly dammed the South Fork of the American River. The USGS immediately deployed an electronic system to monitor active slides along U.S. Highway 50 in northern California. The sensors monitor vibrations in the ground caused by sliding, stretching, or compression of the soil, pore-water pressure, and local rainfall. The data are transmitted by radio to computers. The system can be used to alert emergency response personnel if a catastrophic slope failure occurs. It also provides valuable information that engineers can use to design methods for stopping active slides or stabilizing slopes.

What You Can Do

Monitor the U.S. Highway 50 Landslide

You can keep track of rainfall, slope movement, groundwater pore pressure, and vibrations in the U.S. Highway 50 landslide at a USGS website. | Real-time monitoring of an active landslide above U.S. Highway 50, California | The data are collected every 15 minutes and reported on graphs. These data would be particularly interesting to monitor during stormy periods with abundant rainfall.

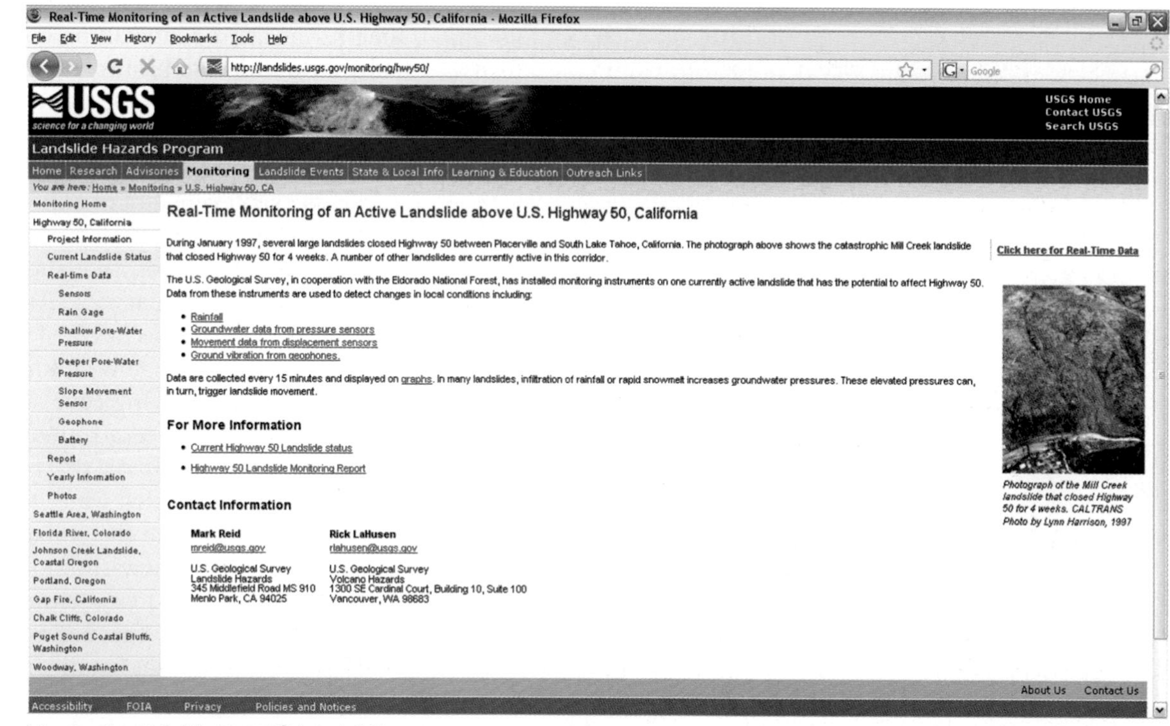

Monitor the U.S. Highway 50 Landslide

LIVING WITH SUBSIDENCE

The subsidence that affects people is mostly caused by people. Where they recognize it, people can come to understand the causes and make adjustments to living in subsidence-prone areas. This is particularly the case where regional subsidence occurs. In karst terrain, anticipating and predicting sinkhole development is more challenging, but there are still ways to live more safely with their development. In former mining areas, knowing where underground mine workings are can guide people's land-use decisions.

Regional subsidence

Preventing regional subsidence due to groundwater withdrawal requires regulating aquifer use. Aquifers, like checking accounts, must not be "overdrawn." In practice, this is not a simple task. As we will see in Chapter 10, the vital need for water in the short term may lead communities to pump too much water too quickly.

If subsidence has already occurred, it's not necessarily too late for a remedy. Regulating withdrawals, encouraging water conservation, and switching to surface supplies such as rivers can help mitigate regional subsidence. Houston initiated these measures in the 1970s when subsidence in some locations reached 3 meters (10 ft) and had caused significant and costly damage (see Figure 8-31). The switch to surface water supplies slowed or halted subsidence east of the city, although pumping from underground sources continued in other areas. In these other areas, subsidence is still taking place. However, a plan is in the works to rely more on surface water sources, such as reservoirs fed by rainwater runoff.

The response to regional subsidence in Santa Clara Valley, California, came earlier than in Houston. The Santa Clara Valley lies south of San Francisco Bay. Today it hosts the technology-driven corridor known as Silicon Valley. In earlier times, the area was an agricultural center that, like the San Joaquin Valley, drew heavily on groundwater to irrigate its fruits and vegetables. After the residents recognized that the land was subsiding, they formed the Santa Clara Valley Regional Water District to regulate water use. By using imported surface water, reservoirs and retention ponds that captured storm water runoff, and injection wells that actively recharged the depleted aquifers, people halted subsidence in the valley by 1969. From 1964 to 1995, the depth to groundwater beneath San Jose rose from a historic low of 72 meters (236 ft) to 11 meters (36 ft), reflecting the recharging of the aquifer (Figure 8-40). Conservation of water during dry periods is also part of the valley's management plan. It helps, of course, that the area is now far less reliant on agriculture than it was at the time subsidence became a problem.

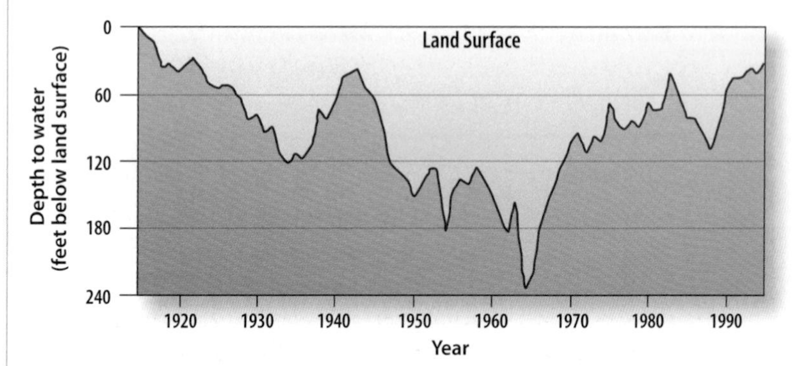

FIGURE 8-40 ▲ Groundwater Levels in Santa Clara Valley, California
Subsidence was halted in the 1960s when decreased groundwater pumping, conservation of surface water, and water injection enabled the underlying aquifer to recharge, bringing groundwater levels closer to the surface.

Karst-related subsidence

In the United States, many communities have adopted a variety of laws designed to mitigate the hazards associated with sinkholes. First, towns, cities, or counties must locate existing sinkholes and identify areas especially vulnerable to ground collapse. This allows them to steer development away from hazardous areas, or at least regulate it through building permits and zoning laws. In Bowling Green, Kentucky, for example, development in areas prone to sinkholes is restricted, and some homes that have repeatedly flooded have been purchased and demolished.

Storm water management has been largely successful in avoiding sinkhole-related problems such as flooding in Bowling Green. Its more than 320 kilometers (200 mi) of streets are served by only 37 kilometers (23 mi) of storm sewers. Numerous sinkholes in the city drain surface water into a network of underground streams that eventually discharge through springs to the Barren River. To avoid overtaxing the capacity of the underground drainage system, new parking lots and other forms of impervious cover must incorporate retention ponds that allow water to seep into the ground gradually. Another helpful measure is installing concrete boxes with grates in established sinkhole basins to prevent them from clogging with debris during storms. To prevent ponding of water on roads during heavy rains, the city has sunk numerous vertical drywells, like the one in Figure 8-41, that drain into the bedrock.

However, even a lot of experience with karst doesn't always avoid problems. Dishman Lane was a newly constructed roadway in Bowling Green that rudely interrupted 5:00 p.m. rush hour traffic on February 22, 2002, when it collapsed into State Trooper Cave (Figure 8-42). The collapse was so sudden that two cars fell into the new sinkhole, but fortunately no one was injured. The underlying cave system was well-known, but the new roadway followed a route where the cave's cover material was thin. Geotimes slip and slide in Kentucky Storm water controls along the road funneled water to a weak area and the increased weight contributed to the collapse.

In cases where new sinkholes do appear under streets and buildings, all is not lost. Before filling the sinkhole, pumping concrete grout into the bedrock may seal the cavity at the source of the collapse. Small sinkholes can be turned into surface drains by filling them with crushed stone. Engineers have also developed special types of foundations to support even

FIGURE 8-41 ◀ Controlling Surface Runoff
Bowling Green, Kentucky uses drywells to direct storm water into selected underground openings of the karst terrain the city is developed on.

very large buildings in sinkhole-prone areas. The sinkhole on Dishman Lane was repaired—at a cost of $1 million.

FIGURE 8-42 ▲ Collapse of State Trooper Cave's Roof
The sinkhole that developed on Dishman Lane in Bowling Green, Kentucky cost $1 million to repair.

Mining related subsidence

A common underground mining technique in coal mines, past and present, is the excavation of large "rooms" in coal beds, leaving behind regularly spaced pillars of unmined material to support their roofs. Operators now carefully determine the size and number of pillars needed to prevent surface subsidence. Portions of the coal bed can also be left beneath roads or developed areas to support the roof. Rooms in closed mines are sometimes reinforced or backfilled to reduce subsidence.

In 1977, the U.S. Congress passed the Surface Mining Control and Reclamation Act to ensure that coal mining is conducted in an environmentally acceptable manner. The act also established a fund to repair and restore lands altered by mining before 1977. This includes preventing and controlling subsidence associated with underground coal mining.

In areas where mining occurred years ago, maps showing underground workings and their depths below the surface can guide decisions about surface land use. These areas are candidates for open-space designations rather than community development. Where mining occurred many years ago, such maps may not be available. So, mine-related subsidence can still surprise people and cause damages.

SUMMARY

Unstable land is common wherever slopes exist or underground openings are close to the surface. Interactions of all Earth systems, but especially the hydrosphere and geosphere, can cause mass movements. The resulting slope failures and land subsidence can be devastating to people and property. Scientists and engineers can identify unstable land hazards and provide ways to live more safely with them. Although people have come to understand the causes of unstable land, they still build and live in hazardous areas. Getting better at preventing human and economic losses from unstable land is part of what people must do to develop a more sustainable future.

In This Chapter You Have Learned:

- Slope stability is determined by the interaction of driving force (gravity) and resisting forces, especially friction and cohesion in unconsolidated materials.
- Changes in slope steepness, material character, water content, and vegetation cover can change the relations between driving and resisting forces and lead to unstable land.
- When land fails, it can move down a slope or subside or collapse vertically. Falls are the fast movement of mostly coarse rock material on steep to vertical slopes. Slides are

the slow (sometimes barely perceptible) to fast (often catastrophic) movement of distinct blocks of material. Flows are the moderate to fast movement of mixtures of surface materials and water. Creep is the slow movement of soil material.

- Subsidence and collapse into underground openings (sinkholes and mines) involve vertical movements of materials and can occur on level land. Natural dissolution of rocks in karst terrain creates caves and other underground openings. Where these are close to the surface, sinkholes can gradually or suddenly develop as materials collapse into them. The same thing can happen where underground mine openings are near the surface.
- Regional subsidence is the gradual lowering of the land surface over a large area. Most of this subsidence is caused by the withdrawal of fluids, especially water, from underground reservoirs or aquifers. This withdrawal lowers the pore pressure of the fluid and allows the reservoir to compact. Compaction lowers the land surface, and can be irreparable.
- Both natural processes and people's activities initiate surface land movements. Weather brings heavy rains that increase the driving force on slopes, earthquake shaking can decrease friction and lower cohesion in surface materials, wildfires destroy slope-strengthening vegetation, and rivers and coastal waves increase slopes by erosion. People initiate slope movements in many ways, but especially by increasing the driving force through slope steepening or placing burdens such as homes on the slope.
- Living more safely with unstable land first requires delineating and assessing hazardous areas. Geologists can identify and map areas susceptible to land failure and assess the likelihood that this land will fail in the future.
- If unstable land develops in an area, several engineering approaches can change the driving and resisting forces to help stabilize the land. People can control water infiltration, lessen slopes, and place supports to strengthen materials. Barrier systems constructed to divert some types of falls, landslides, and debris flows can protect people and property from unstable land hazards. Real-time monitoring and warning systems can help in areas prone to land failure.
- How people use their understanding of unstable land—where it is and what causes it—is the biggest factor in mitigating these hazards and lowering risks associated with them. Land-use decisions are a critical part of living more safely with unstable land.

CHAPTER REVIEW

KEY TERMS

angle of repose (p. 226)
cohesion (p. 225)
creep (p. 233)
debris flow (p. 231)
driving force (p. 224)
earthflow (p. 232)
fall (p. 228)
flow (p. 231)
friction (p. 225)
karst terrain (p. 234)
landslide (p. 224)
mass movement (p. 224)
mass wasting (p. 224)
pore pressure (p. 227)
quick clay (p. 232)
regional subsidence (p. 234)
resisting force (p. 224)
rotational slide (p. 231)
sinkhole (p. 234)
slide (p. 230)
slope failure (p. 225)
slump (p. 231)
subsidence (p. 224)
talus (p. 225)
translational slide (p. 230)
unconsolidated material (p. 226)

QUESTIONS FOR REVIEW

Check Your Understanding

1. What is the primary force that makes slopes unstable? What are the factors that resist this force?
2. What is the difference between friction and cohesion, and in what kinds of materials is each the dominant factor?
3. Why is slope instability likely to be greater when sedimentary layering is inclined downslope?
4. What characteristics of granular material contribute to a steeper stable angle of repose? Provide examples with your answer.
5. Describe how the presence of water in loose material can lower resisting forces.
6. What are the fundamental characteristics by which slope failures are classified?
7. What is the basic difference between a rockfall and a rock slide?
8. In debris flows, earthflows, and mudflows, solid material has become a flowing mass. How does this happen and under what circumstances?
9. How is earth/soil creep different from an earth/soil flow? How do the hazards associated with these types of mass movements differ?
10. Regional subsidence is typically driven by the compaction of porous material when fluids like water or oil are removed from it. Why is this generally not a reversible process?
11. How can urbanization (for example, the creation of roadways, drainage channels, and storm drains) actually increase

the formation and risk of ground failure due to sinkholes? Provide examples with your answer.

12. What are the common warning signs of imminent slope failures that can be readily observed and even mapped prior to complete failure?
13. Slopes can be mechanically stabilized with a variety of engineering approaches. Pick three examples of such approaches from the text and explain why they work in terms of driving or resisting forces.

Critical Thinking/Discussion

14. Using the concepts of driving and resisting forces, explain how rapid heavy rainfall is involved in triggering different kinds of slope failures.
15. Slope failures are common along roadways at the base of steep slopes. How does the action of road crews clearing the roadway of freshly failed material now at the base of the slope actually enhance the risk of future slope failures at that same location? Think in terms of how they are altering resisting forces.
16. Local subsidence and sinkholes are sometimes formed by mine collapse. If you were in charge of land development safety in a state with many abandoned mines, what would you recommend be done about this widespread but localized hazard? What roles can state and local governments play in mitigating or avoiding hazardous building situations related to old mines?
17. Grading codes for builders have proven to be an effective way for state and local governments to avoid an increase in slope failure risk. Explain and illustrate by example or scenario how land managers can use Earth systems thinking to develop such codes.
18. What are the factors that drive development on unstable land? Is it usually necessary to develop housing in dangerous areas, or are there other factors at work that lead to people building houses in harm's way? Discuss your answer in terms of economic factors and the balance between risk and benefit.

ANSWERS TO IN-CHAPTER INSIGHT QUESTIONS

P. 228

What are some Earth systems interactions that influence slope stability?

- The amount of precipitation from the atmosphere can add weight to geosphere materials and increase the driving force.
- Biosphere vegetation can strengthen geosphere materials and increase resisting forces.
- Biosphere vegetation may enhance infiltration of atmosphere precipitation into geosphere materials increasing the driving force.

P. 236

What are some Earth systems interactions associated with sinkhole formation in karst terrain?

- Atmosphere transfers precipitation to the geosphere.
- Precipitation infiltrates and dissolves geosphere materials.
- Hydrosphere flow over and within the geosphere may destabilize cavern roofs leading to sinkhole formation.

P. 240

What are some Earth systems interactions that accompany weather-related mass movements?

- Ocean evaporation is commonly the source of water transferred from the hydrosphere to the atmosphere.
- Atmosphere precipitation transfers water to the geosphere.
- Infiltration of water into the geosphere increases the driving force and decreases the resisting forces.
- Resulting mass movements clog drainages and disrupt or destroy biosphere components and habitat.

FIGURE 9-1 ▶ Galveston, Texas after the 1900 Hurricane An estimated 8000 people died when Galveston, Texas—built on a low coastal island—was destroyed by a hurricane on September 8, 1900.

CHAPTER

9

CHANGING COASTS

People in Galveston, Texas, still refer to it as "The Storm." On September 8, 1900, a powerful hurricane swept ashore from the Gulf of Mexico and devastated the city. An estimated 8000 people died out of a total population of 37,000. It still ranks as the worst natural disaster in U.S. history in terms of the number of lives lost. There were more fatalities in Galveston that day than in all the 325 big storms that have struck U.S. coasts since then.

Galveston is on the coast of the Gulf of Mexico, southeast of metropolitan Houston. In 1900, as now, Galveston was a popular resort area. Its deepwater port was the point of departure for most of the country's cotton. The community was prosperous but tragically vulnerable to hurricanes. The reason? The highest elevations on Galveston Island, on the east side of town, were no more than 2.7 meters (9 ft) above high tide, and portions of the west side of town were not much higher than 1 meter (3 ft) above high tide.

Large, rolling waves from the southeast were crashing over the Galveston shore by the afternoon of September 7. Unrealized at the time, these waves were just the faint ripples of an enormous disturbance in the Gulf—a great hurricane. The next day, Isaac Cline, the U.S. Weather Bureau official in Galveston, sent a final telegram to Washington, DC, before the lines blew down: "Gulf rising rapidly, half the city now under water . . ." As the storm bore down on the shore, winds later estimated to be in excess of 190 kilometers per hour (120 mph) pushed the ocean to a height of nearly 5 meters (16 ft) above normal.

At this point, there was little people could do to escape, so they huddled in their homes to wait it out. The flood heaped up the debris from smashed houses ahead of it, bulldozing a swath of destruction through the city (Figure 9-1). As many as 3600 buildings were completely destroyed. Bodies lay tangled amid the splintered wreckage of the city. Lacking the open space to inter so many people, the city buried corpses at sea. Then, compounding the grief of the survivors, scores of bodies began to wash ashore. Ironically, the victims had to be burned on the beach on pyres of debris from their own destroyed homes and businesses.

The storm's aftermath epitomized the hazards of living on coasts. The response to the disaster also epitomized an approach to mitigating coastal storm hazards that prevailed for most of the 20th century. To protect the city from future inundation, the citizens decided to build a massive seawall on the south coast to hold back the waters during future storms (Figure 9-2). The concrete wall was poured with a curving face to help deflect the energy of crashing waves. The first wall, 5 meters (16 ft) high at its crest and 5.3 kilometers (3.3 mi) long, was completed in 1904. Later additions extended the wall to just over 16 kilometers (10 mi). In addition, the city's height above sea level was raised by jacking up buildings and pouring sand beneath them that had been dredged from Galveston Bay. It took 9.2 million cubic meters (12 million yd^3) to do the job, which was completed in 1910. More than 2100 buildings were jacked up by hand to make room for the fill. Galveston is safe—at least until a very large storm overtops the barrier.

FIGURE 9-2 ▲ The Galveston, Texas Seawall
The citizens of Galveston, Texas have constructed a seawall 16 kilometers (10 mi) long to protect their city from storm waves.

The coast exerts a strong attraction on people. It is estimated that half of the human population lives within 100 kilometers (60 mi) of a coastline. In the United States, 30 of the 50 states, as well as five territories, lie at least in part on a coast (if we include those of the Great Lakes, which are virtually inland seas). Even in the most developed areas, a fortunate few can still buy their own piece of the coast. That's where the trouble starts. Once people achieve the dream of living on or near the shore, they would much prefer the shore to be as unmoving as the foundations of their beach cottages, seaside luxury hotels, and cliff-top retirement homes. In short, people would prefer that the coast do the one thing that it is least likely to do: stay put.

The coast cannot "stay put" because it is where all Earth systems meet and interact (Figure 9-3). This boundary is in constant motion as waves wash back and forth across the shore. The waves along shores are caused by interactions among the atmosphere, hydrosphere, and geosphere—winds cause waves, and differential heating of the atmosphere by oceans and land causes winds. Organisms of the biosphere inhabit all parts of the coastal zone, both onshore and offshore.

FIGURE 9-3 ▲ The Coast: Where Earth Systems Meet
The coast is a constantly shifting boundary where hydrosphere, geosphere, atmosphere, and biosphere interact.

Any boundary between Earth systems is a place for their interactions, and the coast is one of the best places in the world to observe and understand them. These interactions make the coast one of the most dynamic and changing environments on Earth.

If it were not for the fact that people insist on living along the coast, its changing character would probably not be a "problem." Unfortunately, our response to coastal changes cannot be as simple as, "Stay off the coast." We are already *on* the coast in enormous numbers, so finding ways to live with this dynamic environment is unavoidable. Through decades of experience, some of it costly and bitter, we have learned much about how to do this—what works and what does not. Given enough time, virtually all attempts to resist natural coastal processes are likely to fail, so we must try to coexist with them.

IN THIS CHAPTER YOU WILL LEARN:

- The major coastal processes, including waves, currents, tides, and sea level change, and how they work
- The characteristics of the most common types of landforms along coasts, including beaches, estuaries, bays, and headlands
- What causes coastal erosion, how it happens, and what change it creates
- What causes coastal sedimentation, how it happens, and what change it creates
- The role of storms in changing coasts
- How people attempt to respond to changes caused by flooding, erosion, and sedimentation along coasts

9.1 Coastal Basics

The "extreme" sport of big-wave surfing is thrilling to watch. On January 10, 2004, surfer Pete Cabrinha broke the world record by riding a 21-meter (69-ft) monster wave that reared up from a location off the North Maui coast nicknamed "Jaws" (Figure 9-4). For beating the previous 68-ft record set in 2002—and living to tell about it—Cabrinha took home the $70,000 top prize in the annual Billabong XXL Global Big Wave Contest.

In addition to being surfers' playthings, waves are the major force sculpting coasts. It doesn't take monster waves, however. The action of innumerable smaller waves over time can erode rocky coastal promontories or cause entire sand islands to migrate toward the mainland. Waves also drive currents that move sand along shores. Tides cause other daily fluctuations of the shoreline. Overlain on the short-term processes of waves and tides are changes in sea level. The combination of these processes—waves and currents, tides, and sea level change—has produced a variety of features along coasts.

WAVES

Most waves are born where the atmosphere's wind blows across a body of water, setting the water's surface in motion. The wind essentially transfers a portion of its energy of movement (kinetic energy) to the surface and near-surface water. The water itself is a passive medium through which the energy moves.

Measuring waves

The main attributes of water waves can be described in much the same way we described earthquake waves in Section 5.2.

FIGURE 9-4 ▲ Surf's Up
A surfer riding "Jaws" off the Maui coast.

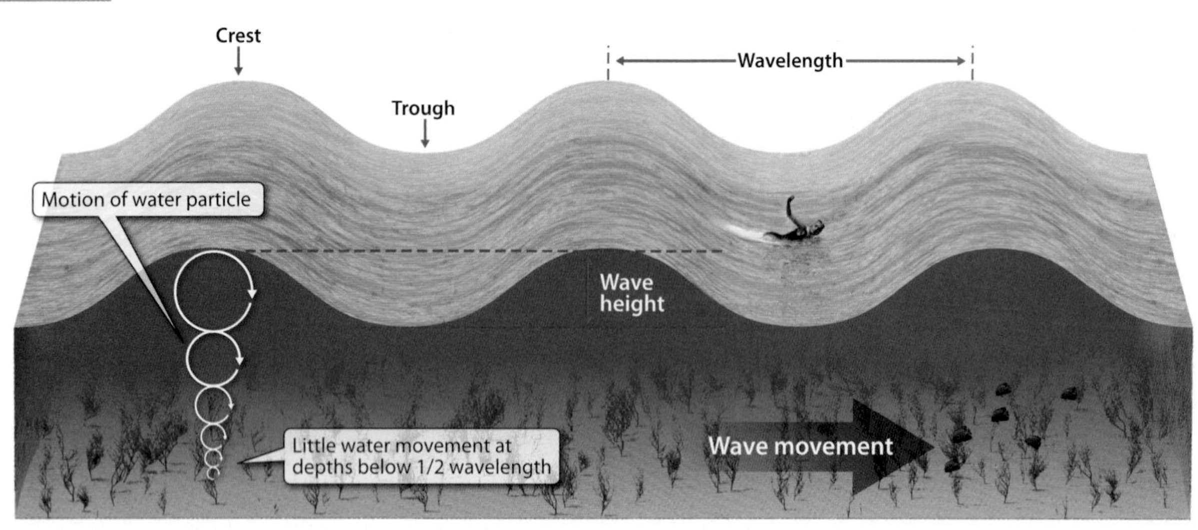

FIGURE 9-5 ◀ Wave Properties Height (the vertical distance from a crest to an adjacent trough) and wavelength (the horizontal distance from one crest or trough to the next) are fundamental properties of a wave.

As shown in Figure 9-5, the wave *height* is the vertical distance between a *crest* (peak) of the wave and an adjacent *trough* (lowest point). The *wavelength* is the horizontal distance between two successive crests. The wave *period* is the amount of time it takes two successive crests to pass by a fixed point. The wave *velocity* is the rate at which a particular wave moves through the water (it equals the wavelength divided by the period).

Three factors determine the size of waves generated on the open water: how strong the wind is, how long the wind blows in a consistent direction over the surface of the water, and the distance over which the wind blows (called the *fetch*). Large, long-lived storms in the Pacific Ocean—the largest ocean on Earth—can produce enormous waves because the potential fetch is so large. For example, winds with a fetch of more than 500 kilometers (300 mi) may produce wave heights exceeding 20 meters (66 ft), traveling at 80 kph (50 mph). In contrast, winds blowing over a smaller body of water, such as an enclosed coastal bay, would produce far smaller waves, commonly a meter (3 ft) or less high.

Waves in deep water

Ocean waves can be caused by local winds or by winds in distant storms. Waves generated by distant storms, called *swells*, have gently rounded crests and troughs. Swells can travel great distances in the open ocean. Those originating in the storm-tossed waters of the Southern Ocean off Antarctica, for example, may travel 16,000 kilometers (10,000 mi) before crashing ashore on the coast of California—much to the delight of big-wave surfers.

As Figure 9-5 shows, waves do induce motion in the water as they pass through it, even in deep water. Floating in the water as a wave passes, you would move in a forward circular motion, as if riding in a Ferris wheel. The diameter of each circle—the width of the imaginary Ferris wheel—is equal to the height of the wave.

Beneath the water's surface, things change. A scuba diver hovering 1 or 2 meters (3 to 6 ft) below you would also move in forward circles. However, the diameter of the circle would be smaller, diminishing more and more with depth until it becomes zero where the depth is about half the wavelength. Thus, most of the energy of the wave exerts its influence near the surface. For the same reason, deepwater waves experience virtually no friction with the ocean bottom. This is what allows them to keep most of their energy despite journeys of thousands of kilometers (miles).

Note, however, that just as a person riding a Ferris wheel eventually returns to the place he or she started, a person floating in the ocean as a wave passes doesn't end up any closer to shore. The motion of the water serves only to transmit energy in the direction of the wave motion. Objects in deepwater waves (including water molecules), like passengers on the Ferris wheel, are just along for the ride.

Waves in shallow water

Waves undergo a series of transformations as they encounter shallow water depths. When they reach a water depth equal to about one-half their wavelength, waves begin to "feel bottom"—the water movements within them become elliptical, as shown in Figure 9-6. For example, storm waves on the open ocean with a wavelength of 200 meters (660 ft) would start to interact with the sea bottom at a water depth of around 100 meters (330 ft). The wave slows down at this point due to friction between the moving water in the wave and the seafloor. The crests of the waves crowd together, and the wave height increases. At some point, the wave begins to collapse—break—under its own weight.

How the wave breaks depends on the geometry of the seafloor, as shown in Figure 9-7. If the seafloor gradually becomes shallower, the wave develops into *spilling breakers*. In this case, the crest is frothy and turbulent but does not hook forward prominently. On a more steeply inclined seafloor, the

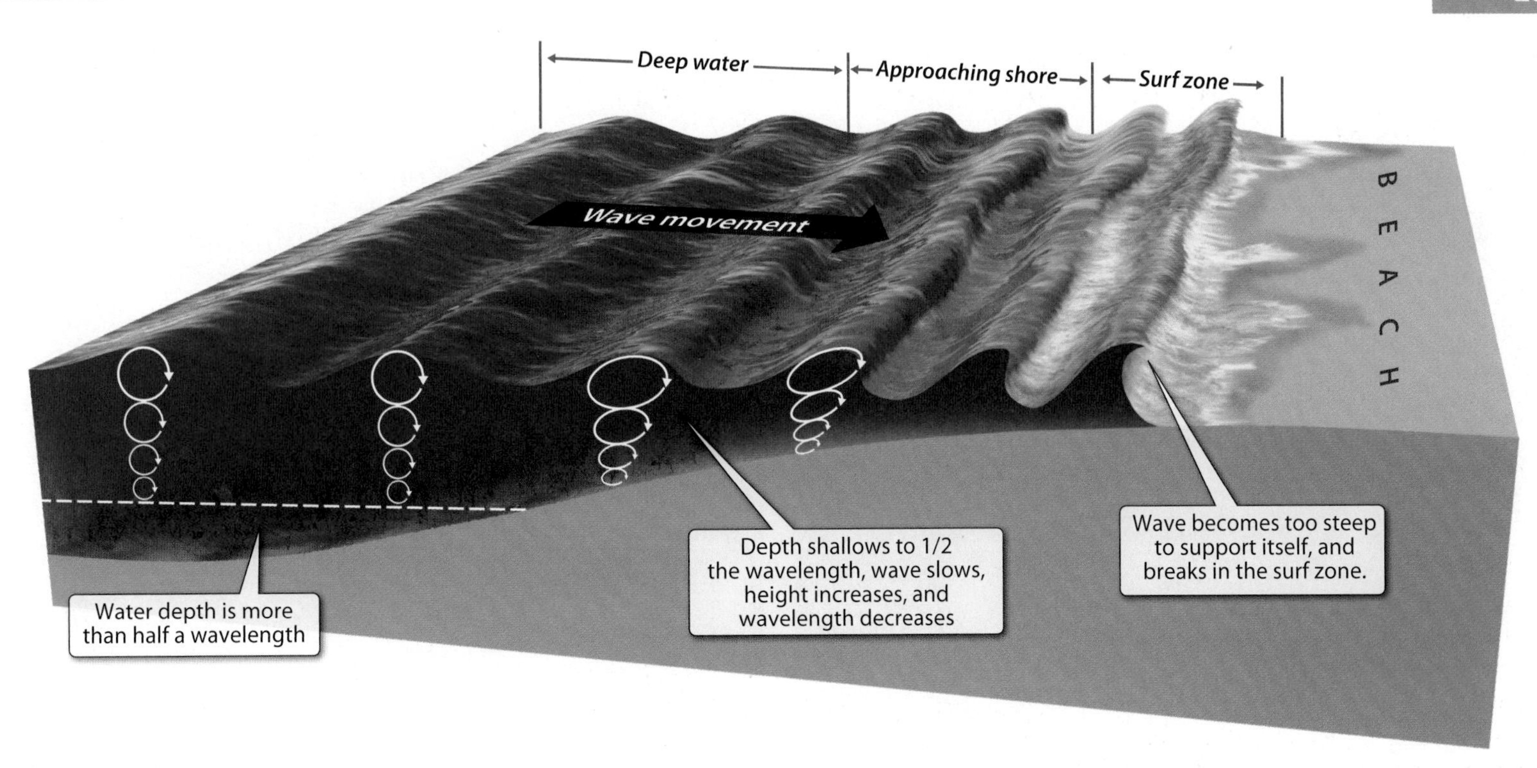

FIGURE 9-6 ▲ Waves in Deep and Shallow Water
In deep areas, movement of water within a wave is circular, but the size of the circle decreases downward to a depth of about half a wavelength. As waves approach shore and reach areas where the water depth is less than half a wavelength, movement within the wave becomes increasingly elliptical. Slowed by friction with the bottom, the troughs travel more slowly than the crests, causing the wave's height to increase and its wavelength to decrease. You can explore wave motion in online simulations at the following site: CACR linear wave velocities

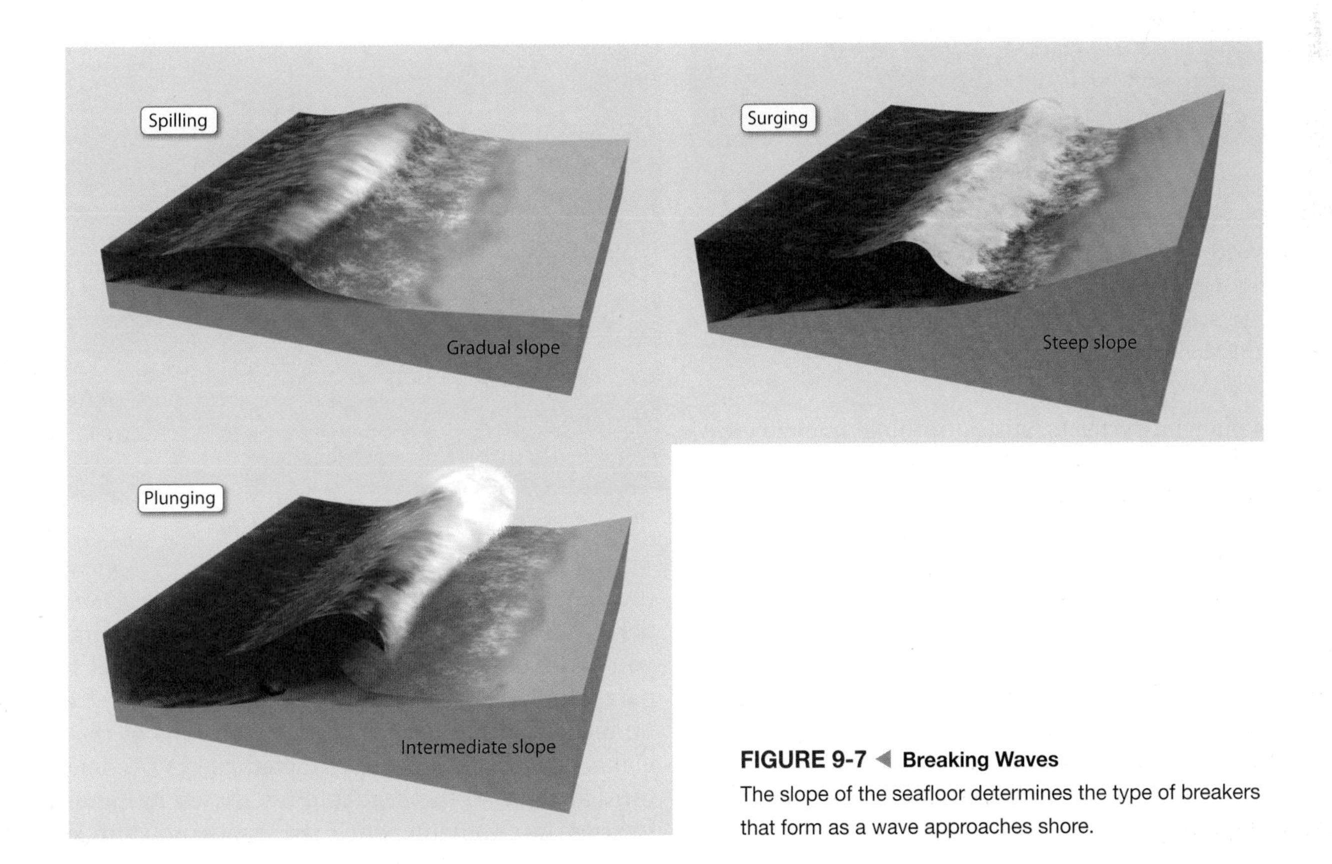

FIGURE 9-7 ◀ Breaking Waves
The slope of the seafloor determines the type of breakers that form as a wave approaches shore.

(a)

(b)

FIGURE 9-8 ▲ Wave Refraction

(a) Waves that approach shore obliquely undergo refraction, becoming more parallel to the shoreline as friction slows the parts of the waves that enter shallow water first. **(b)** Refraction focuses wave energy on headlands. The resulting erosion may eventually convert the headland into an island.

wave may form *plunging breakers,* with a frothing, hooked crest and a tube-like hollow space beneath, like the wave in Figure 9-4. If the seafloor is particularly steep, the wave may begin to crest but then just rush up the beach as a *surging breaker.* Breaking waves create a turbulent region along the coast known as the **surf zone**. The turbulence disturbs and moves sand and gravel on the bottom.

Wave refraction

Waves typically approach the shore at an angle instead of parallel to the shore. This means that part of the wave will encounter shallow water before other parts. The part that reaches shallower water first slows down while the rest keeps going at its original velocity. The change in velocity along the length of the wave causes it to bend, or undergo *refraction,* as different parts "feel bottom" at different times. Such refraction, which you can see in Figure 9-8a, explains why waves become more parallel to a shore as they approach it. Wave refraction is pronounced around local seaward extensions of the coast, called *points* or *headlands.* Here refracted waves become directed toward all sides of the headland (as shown in Figure 9-8b). This focuses wave energy onto the headland and leads to high coastal erosion rates in these settings.

FIGURE 9-9 ◀ **Longshore Drift**
Waves that approach the shore obliquely create longshore currents parallel to the beach that can transport sediment along the coast. They also move sand grains in a zigzag motion along the beach.

NEARSHORE CURRENTS

When waves break along a coast, they transfer part of their energy to currents and the movement of sediment. This transfer can give rise to *longshore drift* and *rip currents.*

Longshore drift

Even though waves are often refracted when they enter shallow water, they generally come ashore at an angle. This causes movement of sediment along the shore. As waves wash up the face of the beach at an angle, they carry grains of beach sediment with them. Where the water washes back down the slope of the beach, it carries the sediment grains directly back to the shore. Consequently, the sediment grains gradually move in a zigzag pattern along the shore in a process called **beach drift**, as shown in Figure 9-9.

Waves breaking offshore at an angle cause water to move along the shore, too. These **longshore currents** also transport sediment along the coast. If you've ever played in the surf on an ocean beach and found yourself progressively farther down the beach from your blanket with every crashing wave, you have experienced a longshore current. The movement of sediment by longshore currents is known as **longshore drift**.

Rip currents

As breaking waves encounter the shore, the water flows up the beach in a turbulent sheet called *uprush,* pauses for an instant, then flows backward as the *backwash.* The water flowing back into the surf zone may organize into plume-shaped **rip currents**, shown in Figure 9-10 on p. 264, that convey the water from breaking waves back through the surf zone.

Rip currents can be dangerous. They commonly move faster than even a strong swimmer can fight against. Panicked, people struggle in vain against the current until they are too worn out to stay afloat. Hundreds of people die every year in rip currents; thousands more are rescued from them by lifeguards. | National Weather Service rip current safety

Anyone caught in a rip current should swim across, not against, the flow. When you are carried beyond the surf zone and no longer feel the tug of the current, head back to shore or wait for assistance. The rip current will probably be no more than a few tens of meters (yards) across. To surfers, rip currents are welcome, since surfers can often hitch a quick ride on them out to where waves are breaking.

What are some Earth systems interactions associated with waves along a coast?

FIGURE 9-10 ◀ Rip Currents
(a) Where waves bring water directly ashore, it is likely to return to the sea in a rip current. **(b)** Scientists used blue dye to highlight this rip current in Australia—it took about two minutes for the dye to be carried from the water's edge to its seaward location in this photograph.

TIDES

People visiting the coast will soon notice that beaches gradually widen and the water level at docks gradually lowers for a few hours. Then these trends reverse for a few hours. This daily fluctuating rise and fall of sea level is the **tide**. Sea level fluctuates between two high and two low tides each day (actually every 24 hours and 50 minutes) on most coasts. The tides are caused by gravitational interactions among Earth, the Moon, and the Sun—but especially between Earth and the Moon. The gravitational pull of the Moon easily moves the liquid oceans. The amount of rise and fall during a tidal cycle is the *tidal range*. This range varies from day to day depending on the location of the Moon relative to Earth.

The tidal range also varies during a month depending on the location of the Moon relative to Earth and the Sun. When Earth, the Moon, and the Sun are aligned (Figure 9-11a), the gravitational pull of the Moon and Sun combine to create especially high and especially low tides. This is when the difference between a day's high and low tides, the tidal range, is greatest. The opposite happens when the Moon is at right angles to the alignment of the Sun and Earth (Figure 9-11b). The tides during the two times each lunar month that tidal ranges are greatest are called *spring tides* (these are not related to the season!). The tides during the times when tidal ranges are the least are called *neap tides*.

Although the timing of high and low tides along a coast is very predictable, determining the expected tidal range depends on local factors. This is because the shape of the coast and seafloor influences the tidal range. Where the tidal flow moves into restricted areas, such as a long narrow bay or inlet, the tidal range increases. The water flows into a smaller volume of space, where the only direction it can go is up.

The greatest tidal range in the world is in such a setting at the Bay of Fundy in Canada (shown in Figure 9-12). Here the range from high to low tide is over 12 meters (39 feet)! At low tide you would have to walk far down the bay to get to the shoreline—and you would have to hustle back to keep ahead

(a)

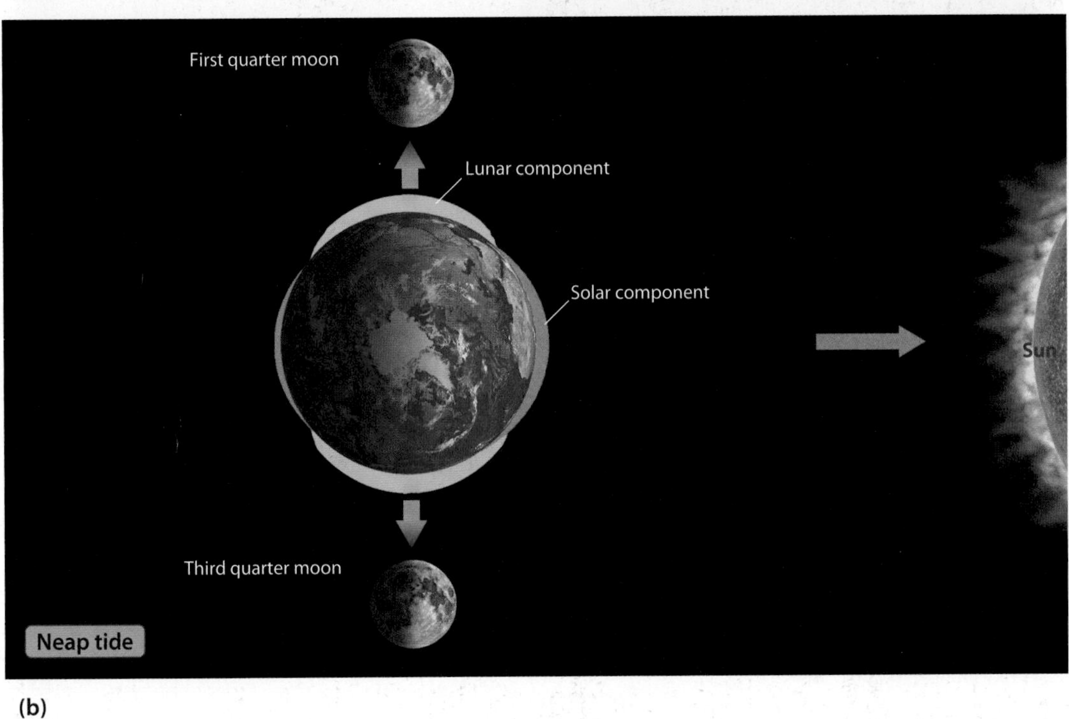

(b)

FIGURE 9-11 ◀ Making Tides
The influence of the Moon's and Sun's gravity on Earth's hydrosphere causes the daily sea level fluctuations called tides. **(a)** When Earth, the Moon, and the Sun are aligned, the combined gravitational pull of the Moon and Sun give rise to especially high and especially low tides. **(b)** When the three bodies form a right angle, the tides are smaller. (Note that the sizes of the tidal bulges in both parts are greatly exaggerated.)

FIGURE 9-12 ▼ Tidal Range
Tidal range is greatest in long, narrow bays and inlets where the flow of the incoming tide is constricted. The Bay of Fundy in Canada is a classic example: **(a)** high tide; **(b)** low tide, about 6 hours later.

(a)

(b)

FIGURE 9-13 ◀ The Intertidal Zone Habitat Organisms of the biosphere (sea stars, mussels, barnacles, and anemones, for example) have adapted to the constantly changing physical conditions of the intertidal zone.

of the tide when it turned and rose to fill the bay again. The opposite happens along open ocean coasts. For example, mid-ocean islands far from shallow coasts and bays of the continents have tidal ranges of less than 1 meter (3 ft).

The Bay of Fundy illustrates very well one of the main effects of tides, the repeated flooding followed by exposure to the atmosphere of parts of the coast. This creates unique habitat, and the biosphere has many interesting species that have adapted well to life in the tidal zone (Figure 9-13).

Another effect of tides is coastal currents. Where tides flow through restricted channels, such as between islands or the mouths of bays, the currents that develop called *tidal currents*, can be very strong. The tidal currents flowing through the Golden Gate, the mouth of San Francisco Bay (Figure 9-14), are some of the strongest tidal currents in the world, averaging 9.0 kilometers per hour (5.6 mph). These currents were a major hurdle to escapees from the prison on Alcatraz Island—the currents are too strong to swim against for long distances. As you might imagine, tidal currents can do the things that other currents do on a coast, such as erode the seafloor and transport sediment.

What are some Earth systems interactions associated with tides?

FIGURE 9-14 ◀ Tidal Currents Tidal currents flow rapidly through the opening to San Francisco Bay beneath the Golden Gate Bridge.

What You Can Do

Keep Track of Tides

Anyone playing or working along a coast can find keeping track of the tides helpful. The timing and range of tides can influence channel depths, the strength of currents, and fishing success, for example. Maybe you are just curious about how tides are affecting a coast you're interested in—where you might go on spring break, perhaps? It's easy to keep track of tides.

At the Saltwatertides website you will find predictions for tides at over 2500 locations on the coasts of the continental United States. Saltwater tides This site gives not just the time and height of tides, but also times for sunrises, sunsets, moonrises, and moonsets. Just pick the location and dates you're interested in—even months in the future—and up comes an easily printed chart showing all the relevant data.

Do you want to keep track of the tide at your favorite coastal spot in real time? The National Oceanographic and Atmospheric Administration (NOAA) provides real-time monitoring of tides at many locations along the nation's coasts. NOAA tides online Just click on the coastal state you are interested in and find the location nearest your favorite spot. You'll see how the tide is changing there throughout the day. The charts display the data for yesterday, today, and tomorrow and commonly provide water temperature and wind speed and direction, as well.

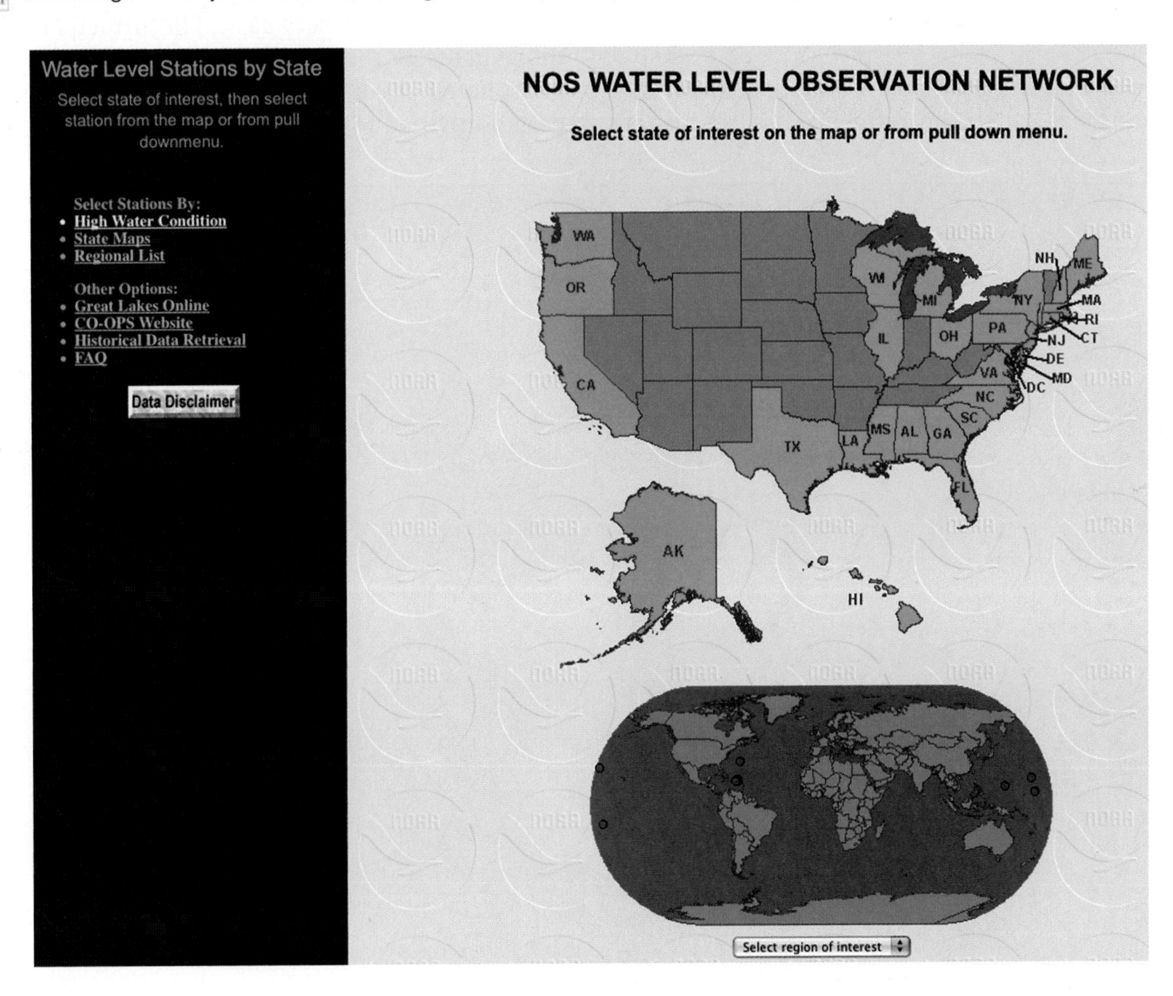

SEA LEVEL CHANGE

Tides are not the only things that cause change in sea level along a coast. Over geologic time, the biggest influence on sea level has been climate—the shifts from warm to cold (glacial) periods that have global effects. Changes of sea level along coasts also happen because of changes in the geosphere. The land locally rises or falls relative to sea level. Because a person can't always be sure which is changing—the oceans or the land (or both)—the local changes are called *relative* sea level rises or falls.

Global sea level change

Global climate has fluctuated from warm to cold many times in Earth's history. These natural changes are primarily caused by variations in how much of the Sun's radiation the Earth receives, variations controlled by changes in Earth's orbit around the Sun (discussed in Chapter 14). During global cold periods, the water cycle transfers tremendous amounts of ocean water to polar ice accumulations. As these ice accumulations expand, Earth enters a glacial period. The resulting loss of water from the oceans lowers global sea level.

At the peak of the last glaciation 20,000 years ago, when thick ice sheets covered most of Canada and extended south into the Midwest, sea level was about 110 to 125 meters (361–410 ft) lower than today. As you can see in Figure 9-15, large parts of the U.S. continental shelf were exposed as land then, and rivers flowed across these regions. Since then, melting of most of the ice has transferred much water back to the oceans, and global sea level has risen. Most of the melting concluded about 7000 years ago, leaving the shoreline near where it is today over large parts of Earth. The flooding of the continental margins, caused by melting ice and the expansion of ocean water as it warmed, is responsible for many of the major coastal features around the world—it's why the British Isles are now islands and Alaska is separated from Siberia by the Bering Sea.

As we will see in Chapter 14, global warming is under way today. This warming has caused an estimated 15 centimeters (6 in) of sea level rise during the last 100 years as ice melted and seawater warmed and expanded. Continued warming during the next 100 years could cause sea level to rise even farther. For low-lying coasts, and the people who live along them, a comparable rise in sea level would bring many changes.

Local sea level changes

Movements of the geosphere also cause changes in relative sea level. Because these are not global in scale, we call them local changes, but they can occur along a thousand kilometers (miles) of coastline or more. The geosphere's exposed surface—land—can move up or down, and this movement can be gradual or very fast. The principal cause of gradual movements is the loading and unloading of weight on the geosphere's crust. The more rapid geosphere movements are commonly tied to plate tectonic movements.

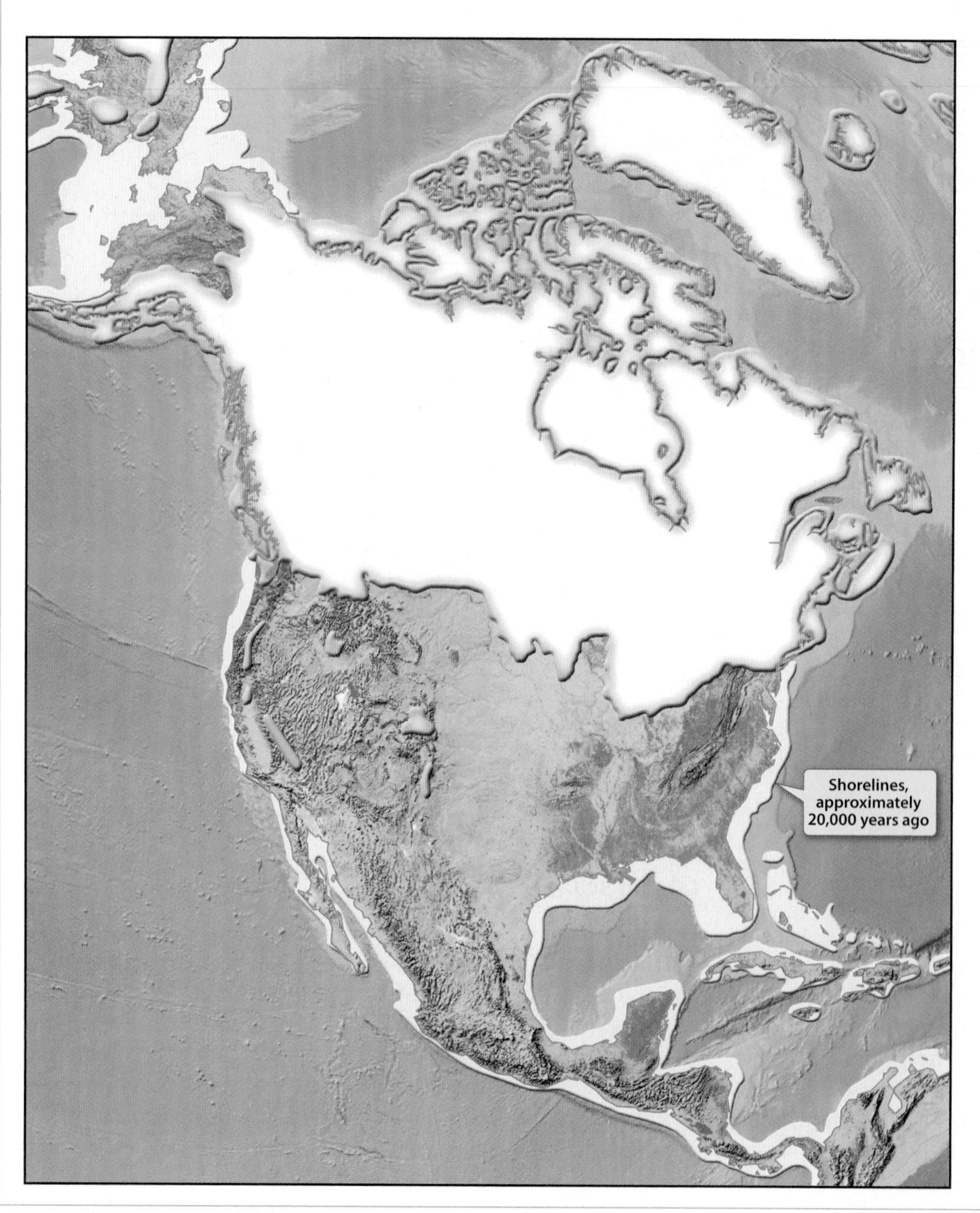

FIGURE 9-15 ◀ Sea Level during the Last Glaciation

Large ice sheets covered Canada and parts of the northern United States during the last glacial maximum 20,000 years ago. Shorelines were about 110 to 125 meters lower than today, and much of the Gulf of Mexico and Atlantic continental shelves were exposed plains with rivers and forests.

When the ice accumulations over Canada were thickest (about 2 kilometers, or 1.2 miles) 20,000 years ago, the weight of the ice pushed the crust down as much as 600 meters (2000 ft). Of course, the opposite happens as the ice melts—the crust gradually rebounds back to its original position. But if the land doesn't rise as much (or as fast) as sea level does when this melting occurs, the coast can still be flooded. The interplay of these processes can make determining the absolute sea level change at a particular location very challenging. This is especially the case where tectonic movements are also a factor.

Displacements of the coast by tectonic movements can be either up or down. During the Great Alaska earthquake in 1964, a large coastal area in the western Gulf of Alaska was uplifted several meters and another depressed in seconds (Chapter 5). In the eastern Gulf of Alaska, the ongoing collision of a small crustal block along the continental margin is raising the land relative to sea level (Figure 9-16a). The coastal plain here has sequences of beach ridges that were abandoned as the land gradually rose (and relative sea level fell) during the last few thousand years (Figure 9-16b). In other places along this coast, the land must have risen in discrete jumps, because sequences of wave-cut platforms (terraces) rather than beach ridges are present (Figure 9-16c). Active tectonic processes that displace and deform the coast are always possible wherever there are plate boundaries along continental margins.

What are some Earth systems interactions associated with sea level changes?

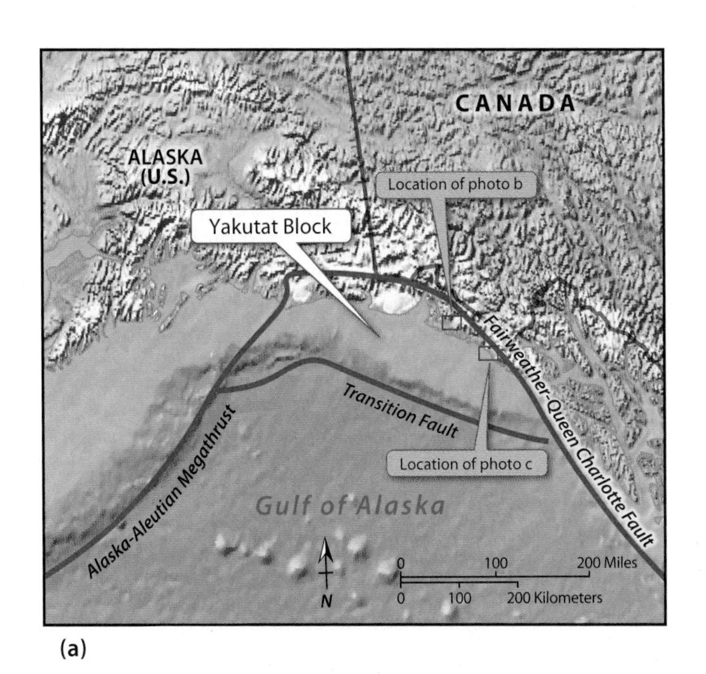

(a)

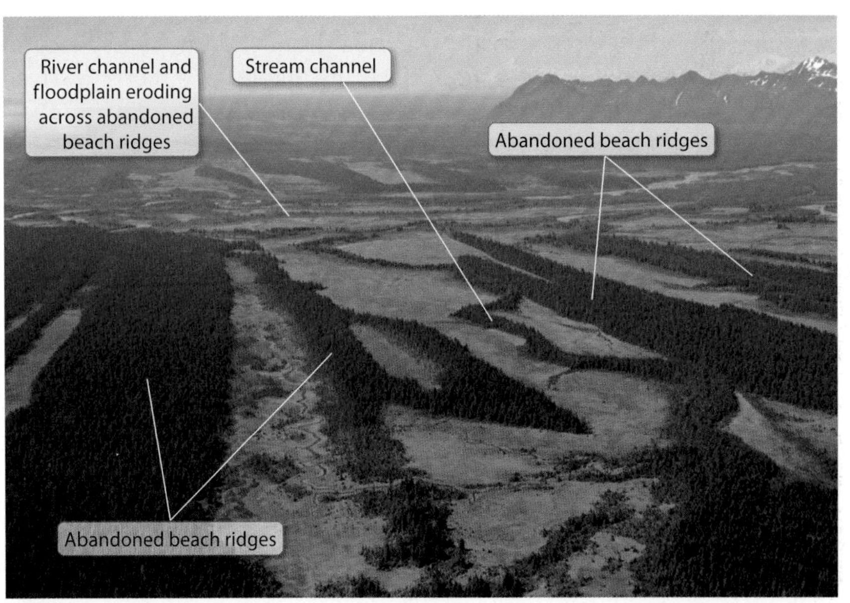

(b)

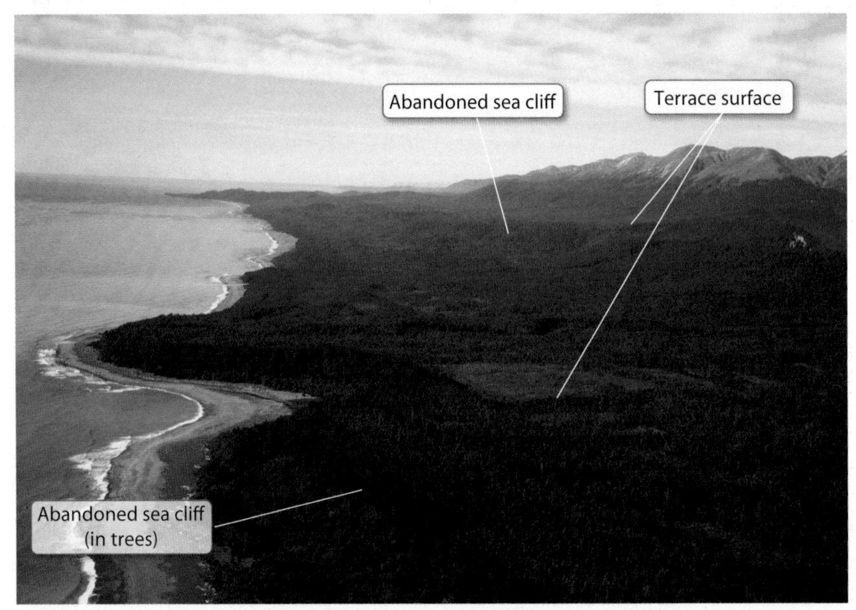

(c)

FIGURE 9-16 ▶ Plate Collision Influences the Gulf of Alaska Coast

Ongoing collision between the Yakutat block and central Alaska **(a)** is uplifting the shoreline throughout the eastern Gulf of Alaska. Where uplift is more gradual **(b)**, the shoreline recedes and leaves abandoned beach ridges behind. Where uplift is more episodic **(c)**, the shoreline retreats in distinct jumps, leaving segments of the former seafloor (wave-cut platforms called marine terraces) behind.

9.2 Coastal Features

Earth's coasts include those fringed by massive glaciers, broad white sandy beaches, forbidding rocky cliffs, and materials created by marine creatures in coral reefs. The 20,000 kilometers (12,400 mi) of U.S. coast have examples of the coastal features common around the world. | America's coastline album These features are developed along the Atlantic, Florida, Gulf of Mexico, Pacific, and Alaska coasts.

THE ATLANTIC COAST

The Atlantic coast, 5300 kilometers (3300 mi) long, extends from Maine to Florida. It is part of the North American continental margin (the part of the continent between the shoreline and the deep ocean floor) that formed when rifting and seafloor spreading split North America from Europe and opened the Atlantic Ocean, as described in Chapter 3. America's east coast is a *passive continental margin* like the one diagrammed in Figure 9-17. This type of continental margin subsides in various ways as seafloor spreading opens its bordering ocean. The subsiding continental margin is where sediment derived from the continent gradually accumulates (as shown in Figure 9-17). The coast is the inner edge of this continental margin, and much of its character is determined by the interplay of subsidence, sea level change, and sediment supply.

The major features of the Northeast Atlantic coast, from Maine to northern New Jersey, are strongly influenced by the most recent glaciation. From Maine to Rhode Island, continental ice sheets scraped the land down to bedrock. As ice melted and sea level rose between about 20,000 and 7000 years ago, the rugged bedrock was gradually submerged. The resulting features are rocky cliffs, headlands, and islands. Sediment eroded from the headlands forms narrow sand and gravel pocket beaches in the bays between headlands.

In contrast to the scouring action of the glaciers in the north, in southern New England (Rhode Island to northern New Jersey) the ice sheets were a conveyor belt that transported silt, sand, gravel, cobbles, and boulders from the continental interior and deposited them to form a thick apron of sediment. The rising sea eroded into the glacial sediment and formed bluffs with beaches strewn with sand, gravel, and large boulders at their base, like the one shown in Figure 9-18. Depending on the composition of the source material, beaches on this coast can be composed almost entirely of coarse gravel or a mix of sand and gravel.

Glaciation did not directly affect the coast from southern New Jersey to Florida. Here the land slopes gently to the sea on a coastal plain that continues offshore and merges with the sloping underwater platform of the continental shelf. This part of the continental shelf was carpeted with forests and meandering rivers back when the sea was lower, 20,000 years ago. Poking above the waves near the shoreline is a chain of about 300 sandy barrier islands stretching virtually unbroken from Fire Island, on the south shore of Long Island, to Florida. **Barrier islands**, like the one shown in Figure 9-19, are long, low-lying ridges of sand and gravel isolated from the mainland by shallow lagoons, bays, or marshes. (Similar ridges that remain connected to land at one end—peninsulas rather than islands—are called *barrier spits*.) Bays or inlets separate the individual islands or spits. They bear the name "barrier" because they buffer the mainland from the powerful waves of coastal storms.

Barrier islands are a characteristic feature of the coast from southern New Jersey to Florida, but their origins are a subject of continuing scientific research. | Barrier islands: formation and evolution They formed during the last 6000 or 7000 years,

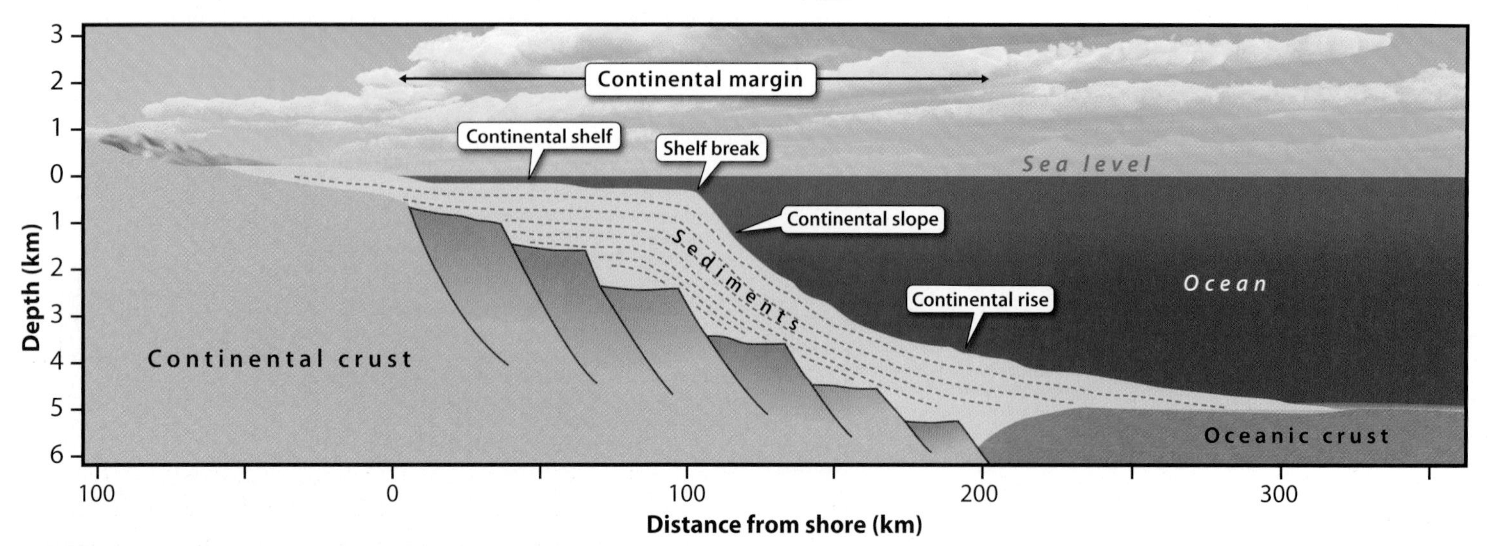

FIGURE 9-17 ▲ A Passive Continental Margin

Small changes in sea level can cause shorelines to advance (transgress) or recede significant distances on gently sloping shelves and coastal plains of passive continental margins. (Note that the slope of the continental margin is actually much more gradual than it appears here. In this representation the vertical and horizontal scales differ by a factor of about 16. If both dimensions were drawn to the same scale, this figure would be about 16 book pages wide.)

FIGURE 9-18 ▲ Bluff Erosion
Erosion of New England coastal bluffs provides sediment for beaches along this coast.

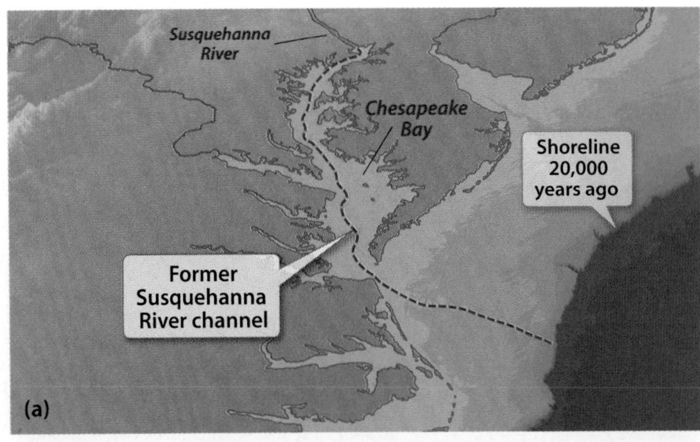

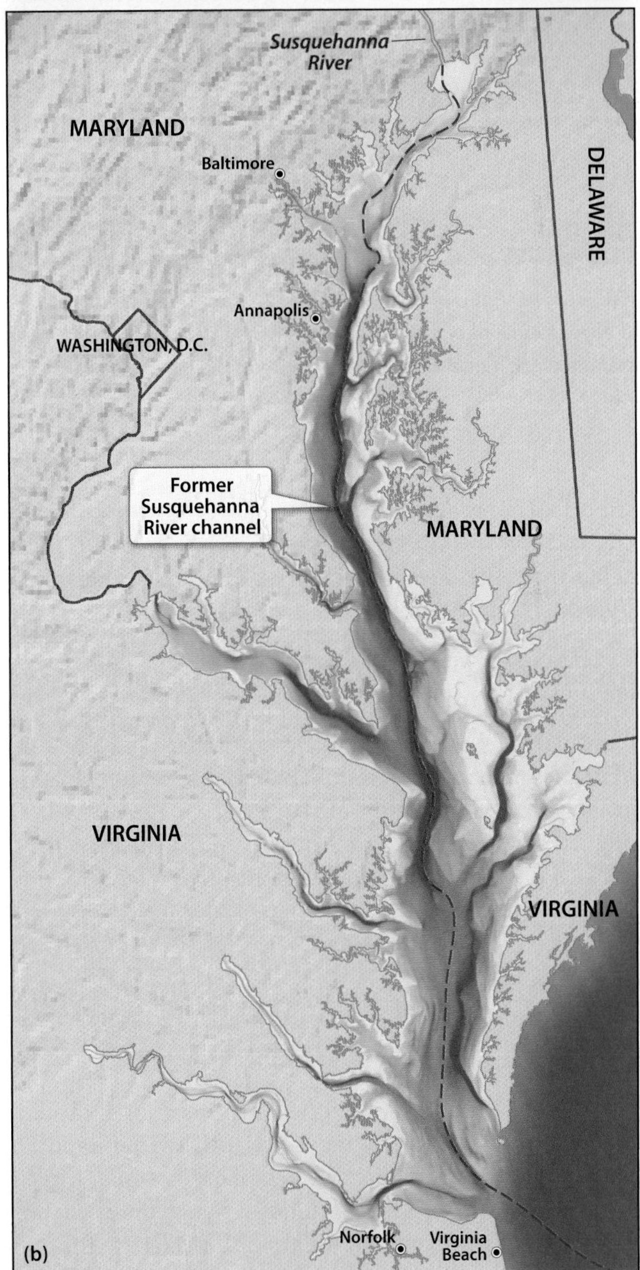

FIGURE 9-20 ▲ Chesapeake Bay: A Drowned River Valley
Twenty thousand years ago, sea level was 125 meters (38 ft) lower and the Susquehanna River flowed across the continental shelf east of Virginia **(a)**. Chesapeake Bay, the largest estuary in the United States, formed as rising sea level flooded the Susquehanna River valley **(b)**.

when sea level was slowly rising or fluctuating at levels not much different from those today. Sediment from the continental shelf may have been collected and moved toward the shore as sea level rose more rapidly prior to 7000 years ago. Since then, sediment has apparently resided in nearshore shallow water as barrier islands. This sediment can move in response to wave and tidal influences, but the total amount may not have changed much during this time.

Other important features of the Atlantic coast are former river valleys drowned by rising sea level. In places these evolve into semi-enclosed coastal bays, or estuaries, where freshwater from rivers mixes with seawater. The Chesapeake Bay, which borders Maryland and Virginia, is the largest and most productive estuary in the United States. As Figure 9-20 shows, it formed as the great continental ice sheets melted and rising

FIGURE 9-19 ▲ Barrier Islands
Assateague Island, Maryland, is just one of many well-developed barrier islands along the Atlantic and Gulf of Mexico coasts.

FIGURE 9-21 ▲ Estuary Pollution
People have found it easy to pollute estuaries with garbage, sewage, industrial waste, and surface runoff. This is the shoreline of the Mersey estuary in England.

seas flooded a broad river valley. | The Chesapeake Bay: geologic product of rising sea level | **Estuaries** are shallow, marshy, and semi-protected from the ocean. They are valuable habitat for a great variety of fish and other marine life.

Because estuaries commonly make wonderful harbors and ports, many people live and work along or near them. As a result, they can become heavily polluted. The pollution can be from industrial waste discharges, untreated sewage releases, runoff from urbanized areas or farmlands, or just from inappropriate garbage dumping (Figure 9-21). As we will see in Chapter 10, maintaining water quality and healthy ecosystems in estuaries is an ongoing challenge in many places.

What are some Earth systems interactions in estuaries?

FLORIDA'S LIVING COAST

Barrier islands are present along some sections of Florida's coast, too, but this southernmost part of the mainland United States also has regions where the biosphere is a major influence on the coastal features. The Florida Keys, a chain of islands that sweep southwestward around the tip of Florida, are the exposed tops of ancient coral reefs. Corals are marine creatures that use calcium carbonate from seawater to make solid structures to live in. When the corals die, the structures remain, become broken up by waves, and supply sediment for beaches. The white beaches of the Florida Keys, like that shown in Figure 9-22, are composed almost entirely of the remains of corals.

FIGURE 9-22 ▼ Coral Beach
The white sands of the Florida Keys are mostly calcium carbonate—the remains of the hard parts of organisms that lived in coral reefs.

In the shallow, sheltered waters of Florida's bays, mangrove trees grow along the shoreline (Figure 9-23). Mangroves thrive in slightly salty (*brackish*) water at the interface between the sea and the freshwater flowing from the land. They form dense thickets in swampy areas, anchored to the wet soil by thick mats of roots that trap sediment and gradually build out the shoreline. Like the sandy coasts of barrier islands, mangroves provide a buffer against coastal storms.

What are some Earth systems interactions associated with coral reefs and mangroves?

THE GULF OF MEXICO COAST

The coast of the Gulf of Mexico, 2660 kilometers (1650 mi) in length, is a passive continental margin that formed as seafloor spreading opened the sea between the Yucatán Peninsula in Mexico and the southern U.S. coast. Its low coastal plain merges with the shallow and broad continental shelf of the Gulf of Mexico. Barrier islands

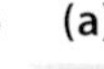

FIGURE 9-23 ▶ Coastal Mangrove Thickets
(a) The biosphere and hydrosphere are closely connected along the mangrove-lined shores of Florida. **(b)** The elaborate stilt-like root systems by which mangrove trees cling to the swampy soil help to trap sediment.

(a)

(b)

front about two-thirds of the coast (see Figure 9-19). Galveston Island, whose devastation in 1900 was described at the beginning of this chapter, is part of the barrier island coast of Texas. Stretches of mainland beach, coastal swamps, bogs, lagoons, and estuaries are interspersed among and inshore of the barrier islands.

The Gulf of Mexico is the location of the largest delta in the United States, the great Mississippi River delta in Louisiana (Figure 9-24), described in Section 7.1. Due to its size, extensive channel system, and large area of wetlands, the Mississippi River delta is a unique feature along U.S. coasts.

Hurricanes and other coastal storms present a major hazard to the Gulf's low-lying coastal communities. Flooding during storms can be especially devastating. The 1900 flooding of Galveston, Texas, is one example. The 2005 Hurricane Katrina flooding of New Orleans, Louisiana, is another.

(a)

FIGURE 9-24 ◀ The Mississippi Delta
The huge delta of the Mississippi River **(a)** is a myriad of streams and islands where the biosphere and hydrosphere intimately mix **(b)**.

(b)

THE PACIFIC COAST

The Pacific coast of California, Oregon, and Washington—2400 kilometers (1500 mi) long—is not a passive continental margin like the Atlantic and Gulf of Mexico coasts. In fact, it is a very active coast; both transform and convergent plate boundaries (discussed in Chapter 3) mark this edge of North America. Most of coastal California has been influenced by the transform San Andreas Fault system, while Coastal Oregon and Washington are strongly influenced by the convergent plate boundary of the Cascadia subduction zone (see Chapters 3 and 5).

The continental shelf along the Pacific coast is narrow and irregular compared to the Atlantic and Gulf of Mexico coasts. Its most common features are rocky cliffs and headlands separated by pocket beaches. The relentless pounding of the surf has carved distinctive and beautiful landforms from rocky headlands. Sea caves form where waves excavate a hollow area into the protruding headlands. If the cave cuts through to the other side of the headland, a sea arch is born, like the one in Figure 9-25. Sea stacks are towers or pinnacles of rock left stranded in the water as the cliffs erode and retreat landward.

FIGURE 9-25 ▲ The Sea as Sculptor
Erosion on the rocky Oregon coast creates sea arches, sea stacks, headlands, and pocket beaches.

Where sediment supply is abundant, large beaches—like the one shown in Figure 9-26—develop along the Pacific coast. Beaches are well developed in southern California in places like Malibu, Venice, and Huntington Beach. From northern California to Washington, longer beaches develop where rivers bring abundant sediment to the coast. These rivers cut steeply through the coastal mountains, scouring sediment from the bedrock and river bottoms. Several coastal towns have sheltered harbors where sand spits form across river mouths. Among the largest are Coos Bay, Grays Harbor, Tillamook Bay, and Willapa Bay. Rising sea level has flooded several large estuaries like the one shown in Figure 9-27, and bays on the northern U.S. Pacific coast add variety to the otherwise rugged coastline. These include San Francisco Bay, the mouth of the Columbia River, and Puget Sound.

THE ALASKA COAST

There is as much coastline in Alaska (10,700 kilometers, or 6640 mi) as in the entire rest of the United States. As you can see in Figure 9-28, it extends from the northeast Pacific Ocean west almost to Japan along the Aleutian Islands and north through the Bering Sea to the Arctic Ocean. Alaska's coastal features are more diverse than those of the other U.S. coasts. They include everything from glaciers cascading into the sea to isolated islands formed by stratovolcanoes.

Alaska's coastal diversity doesn't come from its tremendous length of coastline but from the complete range of tectonic settings developed along its continental margin. In southeast Alaska, a transform plate boundary (the Fairweather fault system) has led to a narrow continental shelf and rugged rocky coasts. Rising sea level has submerged this region, creating many islands and protected waterways among them. Glaciers extend from high mountains to the sea in places like Glacier Bay National Park.

A small crustal block (the Yakutat block) is colliding with Alaska along the eastern Gulf of Alaska coast. This causes the

FIGURE 9-26 ▼ California Dreaming
A plentiful supply of sediment has helped to create the long beaches of the Pacific coast.

FIGURE 9-27 ◀ **Pacific Estuary** Estuaries along the Pacific coast are river valleys drowned by rising sea level.

land to rise relative to sea level. A coastal plain with raised beach ridges, wave-cut terraces, and long sandy beaches characterizes the coast (see Figure 9-16). This is an emergent coast that is strikingly different from the submerged southeast Alaska coast.

The convergent plate boundary of southern Alaska has one of the longest subduction zones in the world (Chapters 3 and 5). Its westward extension into the Pacific Ocean creates the Aleutian Island arc (Chapter 6). Active plate tectonics along the convergent continental margin creates both emergent and submergent coastlines (Chapter 5).

The Bering Sea has been isolated from direct plate boundary influences for at least 40 or 50 million years. It has a broad continental shelf and a coastline with barrier islands and many very long beaches. Where mountains extend to the shoreline, headlands and sea cliffs have formed. Rivers, especially the mighty Yukon, bring abundant sediment to the Bering Sea

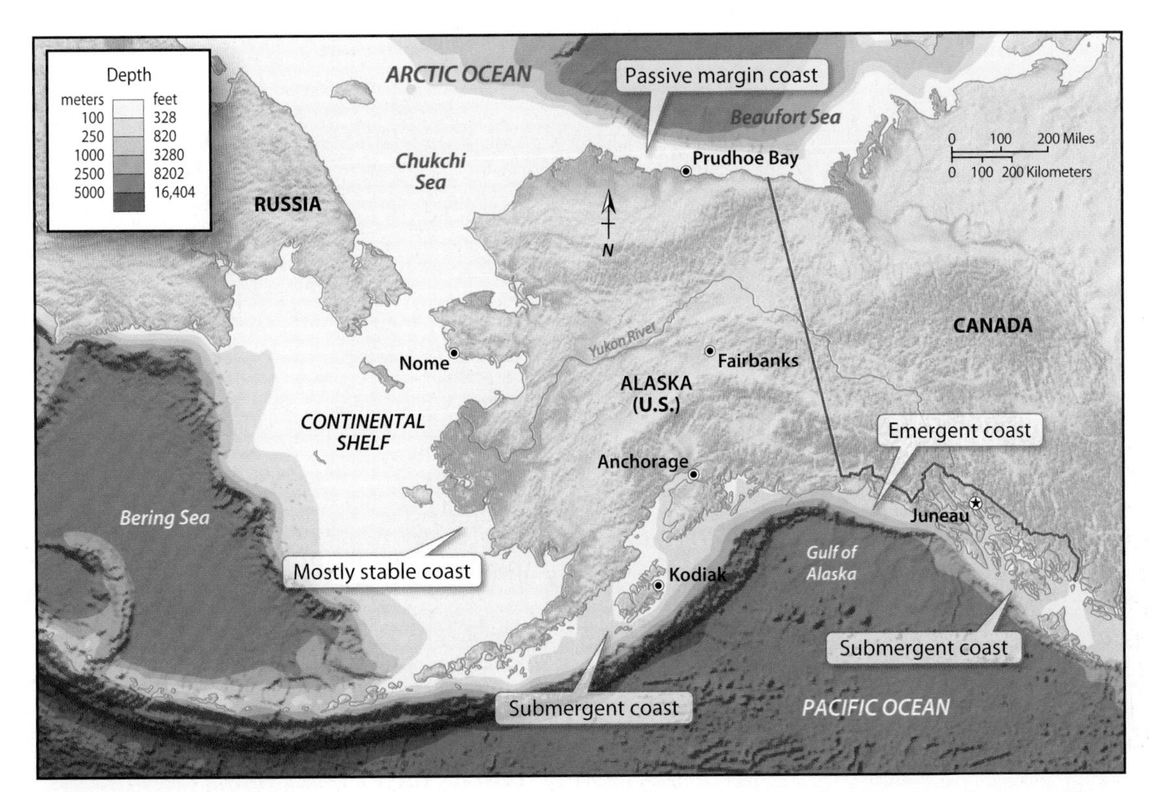

FIGURE 9-28 ◀ **The Alaska Coast** Features typical of the world's coasts are found in Alaska, a state with about as much coastline as all the other states combined. Large parts of the Bering Sea shelf are less than 100 meters (330 ft) deep. The connection between the Bering Sea and the Arctic Ocean was created when sea level rose between 20,000 and 7000 years ago.

coast. The Bering Sea coast merges with the coasts of the Arctic Ocean in northern Alaska. The Arctic coastline is on a classic passive continental margin that formed when seafloor spreading opened the Arctic Ocean. It has a wide coastal plain, barrier islands, and deltas at the mouths of rivers.

A CLOSER LOOK AT BEACHES

No matter where you are, from Maine to Alaska, beaches are the most distinctive and common feature developed along coasts. Formally, **beaches** can be defined as accumulations of loose water-borne material (generally sand and pebbles, often accompanied by mud, rocks, and shell fragments) deposited on the edge of a body of water—typically a gently sloping shore washed by waves or tides. Beaches are also where most people experience the coast. Imagine you are standing on the wet portion of a sandy beach on a calm summer day. Waves roll gently ashore and wash up the slope of the beach. When the sheet of frothy water retreats back to sea, you feel a gentle tug. If you stand still for 15 minutes or more, nothing obvious seems to change beyond the coming and going of the waves. To the untrained eye, the beach may seem as relaxed and unmoving as the people lounging lazily on the sand behind you.

However, appearances can be deceiving. The influences of atmosphere (wind), hydrosphere (waves), and geosphere (shoreline) come together at the beach. As a result, the only things really at rest on a beach are commonly the sunbathers. During winter storms, entire beaches can disappear in a matter of hours. Even on the calmest summer day, the sand is in motion. Propelled by the wind, grains skitter along the surface from the seaward edge of the dry beach toward inland dunes; on the wet beach, wave action moves sand down the shore. Indeed, beaches can be thought of as "rivers of sand" because longshore drift continually transports beach sediment along the coast.

Beach anatomy

Most people think of "the beach" exclusively as the dry platform of sand where people can sunbathe. However, to coastal scientists this is only one part of a broader system. The main components of such a system are diagrammed in Figure 9-29. The beach begins at an abrupt change in topography such as a sea cliff, bluff, or dune. Seaward of this point is a flat area of beach sediment composed of sand or gravel that is covered by water only during very high tides or large storms. This is called the **berm**. There may be one berm or several, stepped down toward the shore.

Continuing seaward, the next part of the beach is the flat ramp of sand leading down to the water's edge. This ramp, washed by waves each day, is known as the **beach face**. The low-tide line marks its seaward limit. Wading into the water, we enter the *nearshore zone*, where longshore and rip currents are active. In the *surf zone*, the water is cloudy with sand and other sediment. If you don a mask and snorkel and swim offshore, you may see a *longshore bar* right about where the waves start to break. A series of longshore bars may extend throughout the nearshore zone. The bars are part of the underwater realm of the beach that extends out to a depth of 10 to 20 meters (33 to 66 ft), where waves can still exert a moving force on sediment. The nearshore environment ends just beyond the point at which waves start to break. Beyond this point, longshore and rip currents are no longer active, although waves may still sweep sediment inshore from deeper sources, particularly during large storms.

Collectively, the berm, beach face, and nearshore zone represent the portion of the coastline within which sediment is

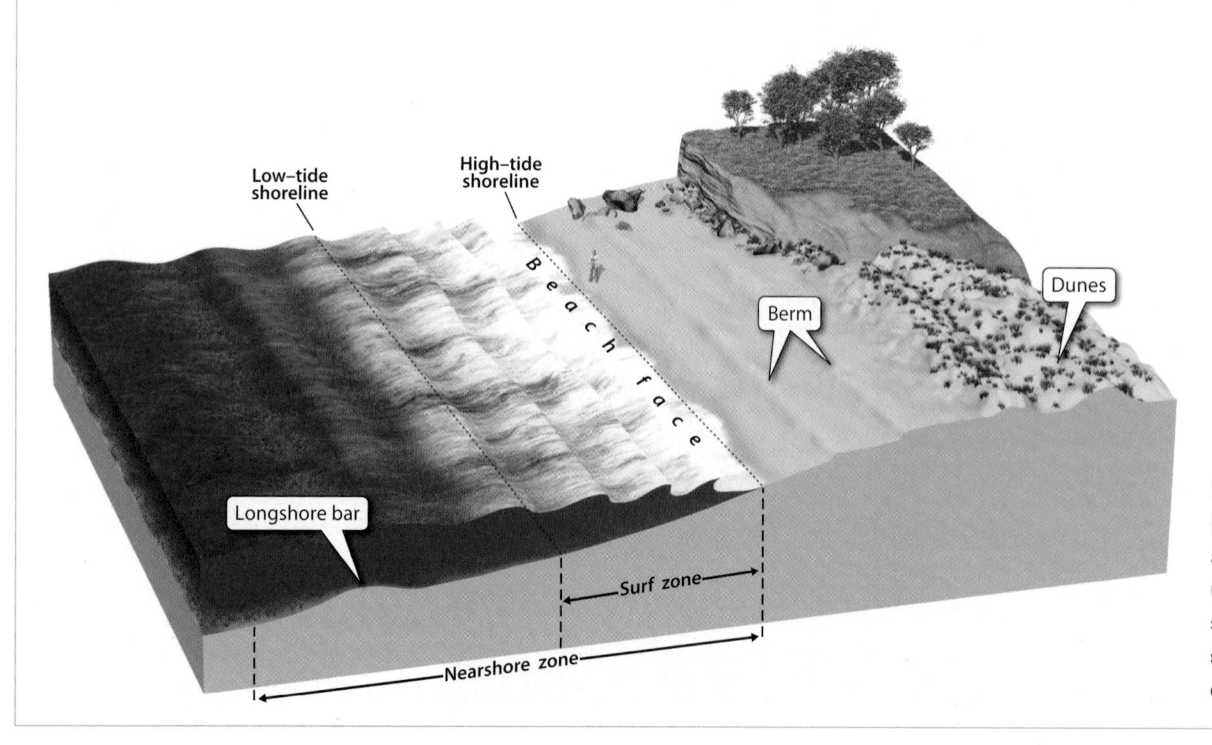

FIGURE 9-29 ◀ Beach Anatomy Distinctive parts develop on beaches as waves and winds come ashore under different conditions. The stronger waves and winds during storms are particularly effective at developing berms and dunes.

FIGURE 9-30 ▲ Hawaii's Black Beaches
Fragments of the volcanic rock basalt (Chapter 4) make beaches dark gray or black.

in motion, both on dry land and under the water. In this area winds from offshore blow sand off the berm and pile it into sand dunes; beach drift and longshore currents transport sediment along the shore; and waves transport sediment on and off the beach.

Beach materials

Close your eyes and visualize the perfect beach: It probably looks like a wide carpet of soft, light gray or tan sand sloping gently to the sea. The color of this classic beach image comes from minerals. Light gray or tan beach sand is rich in light-colored mineral grains such as quartz and feldspar (Chapter 4). Quartz, a relatively hard mineral resistant to weathering and erosion, is especially abundant in beach sand.

In tropical and subtropical climates, seashells or coral are commonly the primary or sole source of beach-building material. On islands like Hawaii, beaches are composed of eroded and weathered volcanic rock. This is the origin of the famous "black sand" beaches of Hawaii, shown in Figure 9-30.

In places, wave attack liberates a great deal of sediment from coastal cliffs. Locally derived beach material from sea cliffs can be gravelly, so much so that you would find it difficult to walk on. The continental shelf also contains great stores of sediment, some of which is the remnant of ancient beaches and dunes that stood above water when sea level was lower. Wave action may slowly sweep these materials ashore. However, most beach material is delivered to coasts by rivers.

Sediment supply

Beaches respond to the amount of sediment available by either growing wider or eroding. In the 1960s, scientists in California saw that coastlines could be segmented into a series of *littoral cells*. Each cell is a self-contained system where sediment enters or exits in various ways. Living with coastal change – coastal basics – littoral cells As indicated in Figure 9-31, longshore drift, rivers, and eroding cliffs bring sediment into the cell. Wind can carry sand out of the system by blowing it onto inland dunes. Sand may also leave the system if currents carry it far enough offshore that it can't get back to the beach. Each littoral cell therefore has a *sediment budget* consisting of credits (sand entering the cell) and debits (sand leaving the cell). When the debits exceed the credits, erosion results. Alternatively, if the credits outweigh the debits, the beach will grow.

FIGURE 9-31 ▼ A Littoral Cell
The Santa Monica littoral cell is just one of many on the California coast. Sediment that enters this littoral cell is carried along the shoreline and eventually lost to deep water through submarine canyons.

On the California coast, headlands and submarine canyons commonly mark the boundaries of littoral cells. Headlands interrupt the longshore currents that might otherwise bring sand in from other cells. On California's narrow continental shelf, deep submarine canyons lie relatively close to shore. If sand flows down into one, it moves to deeper water beyond the reach of currents or waves that might otherwise carry it back to shore.

The beach in action

You can watch a beach in action during a storm. That's when beach processes kick into high gear, and from one day to the next you can observe major changes in the beach. The larger storm waves strike the beach with more energy and may be carried up over the beach berm to the dunes or cliffs beyond. (The force of such waves can toss boulders about, so don't observe these processes from up close!) This removes sediment from the beach face, berm, and dunes—particularly the most fine-grained sand—and moves it inland. At the same time, stronger backwash and rip currents during storms carry much sediment from the beach back into the nearshore zone.

Sediments removed from the surface during a storm leave the beach with a higher proportion of coarse sand, pebbles, and/or cobbles. With the return of fair-weather conditions, much of the sediment removed to the nearshore zone may return to the beach, gradually rebuilding its berms and dunes and generally steepening its profile. Depending on the extent of the storm damage, it may take weeks to years to overcome the effects of a storm.

9.3 Coastal Erosion and Sedimentation

Living on the coast can be wonderful except for a couple of things. Erosion can remove parts of the coast that people use. And if there is erosion in one place, then there is sedimentation somewhere else—perhaps in places that people don't want it, such as shipping channels.

COASTAL EROSION

Coastal erosion is considered a costly natural hazard. An overview of coastal land loss: with emphasis on the southeastern United States However, defining erosion as a natural hazard is a matter of perspective. To a person who spends the summer in a Cape Cod bungalow, the sight of waves breaking closer and closer to the back door screams "erosion" and "hazard." Geologists see the same process, but use different terms to describe it, such as *shoreline recession, beach retreat*, and *barrier migration*.

But for the growing occupation of the coast, people would not perceive shoreline retreat as a serious problem. The beaches and cliffs would happily march landward and nobody would complain. Today, coasts are densely populated and valuable real estate. Recreational beaches, beachside hotels, and flocks of barrier-island bungalows are the lifeline of many communities. Not surprisingly, beachside towns do whatever they can to prevent erosion. Their hope is for the shoreline to stay put. But in many places this is at odds with nature. People find themselves caught between the land and the encroaching sea.

All over the world, coastlines are eroding. By one estimate, at least 70% of all beaches are eroding away. On the coasts of the United States, erosion affects 80 to 90% of all sandy beaches. The average rate of erosion of all the country's shores is about 30 centimeters (12 in) per year. Locally, rates of erosion vary considerably, with some coasts holding their ground, some growing slightly, and many receding.

Beach and sea cliff erosion consumes houses, commercial buildings, and the land itself. According to one estimate, 1500 homes and the property beneath them are lost to coastal erosion each year in the United States—at a cost of $530 million.

Beach erosion

In the United States, beach erosion is especially significant on coasts with barrier islands. Here, the loose sands facing the sea are vulnerable to wave attack. People's developments along these coasts have a long history, punctuated by both success and failure. Lighthouses offer particularly interesting examples.

The Cape Hatteras lighthouse in North Carolina is the tallest brick lighthouse in the country. When constructed in 1870, it was about 490 meters (1600 ft) inland from the shore (Figure 9-32a). Over the years, however, the shore crept closer. Barrier walls and sandbags couldn't stop the advancing sea as relentless waves washed the beach away. By 1987 the lighthouse was almost 40 meters (120 ft) from the waves and the National Park Service (NPS) had to do something if the lighthouse was to be saved. After much study, it was decided that the lighthouse had to be moved. In 1999, the Cape Hatteras lighthouse and its historic keepers' buildings were relocated 490 meters (1600 ft) inland from the shore (Figure 9-32b). This major engineering feat earned the NPS and its relocation team the Outstanding Civil Engineering Achievement Award from the American Society of Engineers.

Even though difficult, moving a lighthouse that's on land is a lot easier than moving one that has become surrounded by the sea. That's what has happened to the Morris Island lighthouse near Charleston, South Carolina. Morris Island has been the site of navigation aids since 1673. Morris Island Lighthouse, South Carolina at lighthousefriends.com The present lighthouse was 365 meters (1200 ft) from the shore when it was constructed in 1876, but beach erosion brought waves to its base by 1938. By 1962, the government gave up and decommissioned the lighthouse. Citizens' groups prevented its demolition and helped purchase it as an historical site for the State of South Carolina. But they couldn't stop the advancing sea. Now the Morris Island lighthouse is the only thing left above sea level, and it's not clear what its fate will be.

(a)

(b)

FIGURE 9-32 ◀ The Cape Hatteras Lighthouse
Erosion removed the beach separating the Cape Hatteras lighthouse from the sea. Attempts to prevent erosion, such as building a barrier extending out from shore to capture migrating sand—a groin, visible in **(a)**—were not successful. It was finally decided to move the lighthouse inland in 1999 **(b)**.

Sea cliff erosion

On coasts that have beaches backed by sea cliffs made of sedimentary rock, unconsolidated sand and gravel, and other relatively weak materials, the beach and the cliffs are inextricably linked (see Figure 9-29). If the beach is wide enough, it can act as a buffer against attack by storm waves. If the beach is narrow or absent, waves may be hurled against the cliff, undercutting it and ensuring its collapse into the sea. An estimated 72% of the California coastline is made up of eroding sea cliffs. Many of these cliffs are underlain by weak materials, such as sedimentary rocks, that are vulnerable to wave attack. Over the long term, sea cliffs recede about 30 centimeters (12 in) per year in most of California.

Most sea cliff erosion occurs during storms. The larger waves during storms are not only stronger than normal waves, but they can attack sea cliffs more directly. In addition to scouring and undercutting, waves erode sea cliffs by driving air into cracks and voids in the cliff face. This creates pockets of high-pressure air that split the bedrock like a jackhammer. Because of their great mass, large storm waves exert tremendous force on sea cliffs, breaking them down by a "wave hammer" effect. Storm waves have been known to throw boulders hundreds of meters (yards) into the air.

People's developments on coastal cliffs can increase erosion. All the things that make slopes unstable (Chapter 8) can play a role where people build on sea cliffs. Misdirected surface runoff, the weight of people's structures, or increased amounts of groundwater percolating through the subsurface soil and rock can decrease the resisting forces, increase the driving force, and help cause sea cliff collapse.

COASTAL SEDIMENTATION

Sediment eroded from one place doesn't disappear—it's just moved elsewhere. It will keep moving until it comes to a place where waves and currents cannot move it any more.

Where sediment is deposited

Sediment moving along a coast is deposited where the direct influence of waves and currents is diminished. The most common places are offshore in deeper water and inshore protected waters where currents weaken—for example, in estuaries. Sediment also tends to accumulate wherever the sediment supply is more than waves and currents can remove. Deltas at the mouths of rivers are classic examples of such accumulation.

If sediment is directed into deeper offshore waters, it can be lost indefinitely to the coastal system. This can happen where submarine canyons on the continental shelf are relatively close to shore, for example. Sediment directed into these canyons is no longer available to be moved or deposited along the coast (see Figure 9-31).

People can also cause seaward movement of sediment—for instance, by constructing barriers called **jetties** that extend out from the shore, like that shown in Figure 9-33 on p. 280). By deflecting longshore currents seaward, jetties can cause sediment to be carried away from the shore into deeper water. The main problem with this type of sediment deposition is that it creates a negative sediment budget on the shore and depletes the sediment supply needed to maintain beaches along a coast.

Inshore areas where sediment is deposited are commonly parts of bays, estuaries, and other areas protected from waves

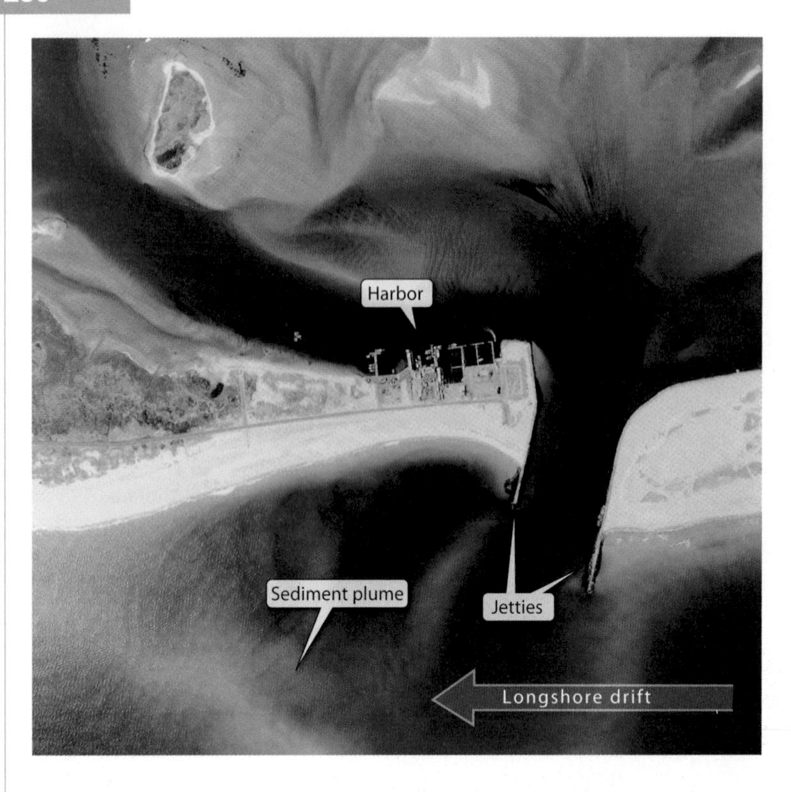

FIGURE 9-33 ◀ Jetties
Jetties—hard structures that extend perpendicular to shore to protect channels and inlets—affect longshore drift. They trap sediment on their updrift side; guide sediment seaward; and deplete sediment on their downdrift side, where erosion can then occur.

and currents. The shallow water *shoals* behind barrier beaches and along the margins of bays are examples. These areas are excellent places for salt marsh and coastal wetland development; if sediment doesn't continue to maintain these shoals, their valuable habitat can be lost.

Not surprisingly, places where sediment supply is high are often places where sedimentation is pronounced. A remarkable example is Turnagain Arm, an estuary at the head of Cook Inlet in southern Alaska (Figure 9-34a). Turnagain Arm is a fiord carved by glaciers descending from the nearby mountains. After the glaciers melted, the Susitna River brought tremendous amounts of sediment to upper Cook Inlet. Cook Inlet tidal currents, fueled by a tidal range of about 8 meters (26 feet), washed Susitna River silt and fine sand into Turnagain Arm. Even though tidal currents reach 24 kilometers per hour (15 mph), the Susitna River sediment filled up Turnagain Arm. There was just too much sediment with no other place to go. Now Turnagain Arm looks like a deep (but muddy) fjord at high tide (Figure 9-34b) and a very large mud flat at low tide (Figure 9-34c).

The Mississippi River delta

The Mississippi River delivers more sediment to America's coast than any other source. In fact, it is one of the biggest sources of sediment to the oceans in the world. This isn't too surprising given the tremendous size of its watershed (Chapter 7). Sediment becomes deposited as the river's currents diminish. Local thicker (and higher) deposits split the river's flow into the many distributary channels that characterize the delta (see Figure 9-24). Over the last several thousand years, the dis-

FIGURE 9-34 ▶ Turnagain Arm, Alaska
Tidal currents have filled Turnagain Arm, a fjord at the head of Cook Inlet in southern Alaska, with sediment from the Susitna River **(a)**. At high tide, Turnagain Arm looks like many other water-filled fjords **(b)**, but at low tide it turns into an expanse of mud and silt flats **(c)**.

tributary channels have shifted back and forth across the mouth of the river, creating different lobes of sediment and increasing the delta's overall area (Chapter 7). This process produced the largest coastal wetlands in the country—the winter home to 15 million water birds, 400,000 geese, and 5 million ducks (one-fifth of all the ducks in the United States). The Mississippi delta is an ecological resource unequaled along America's coasts.

Now the Mississippi delta is shrinking. Some natural processes—compaction and subsidence—are always at work lowering the delta, but for thousands of years these processes were kept in check by the amount of sediment added to the delta by the river. In fact, sedimentation was great enough to enable the delta to grow, not shrink. Today, the amount of sediment being deposited at the river's mouth has been greatly reduced.

Reduced sedimentation on the Mississippi delta has two principal causes. First, management of the river's flow in upstream parts of the watershed, especially the construction of numerous dams, has diminished the sediment load carried to the river's mouth (Chapter 7). One estimate is that there is almost 70% less sediment now than before upstream river management expanded. Second, channels constructed in the delta region itself direct much of the sediment that does arrive there on through. A lot winds up in deeper water rather than being deposited in the delta.

The net effect of natural subsidence and reduced sedimentation caused by people is the loss of over 900,000 acres of wetlands since the 1930s. The wetland loss rate is now 16,000 acres per year. At this rate, an additional 320,000 acres will be lost by 2050.

In the News

Sustaining the Mississippi Delta

A shrinking Mississippi delta has tremendous ecological and economic consequences. The habitat loss, combined with the impact on coastal fisheries and other resources, has alarm bells ringing in many places. Ecologists, engineers, government agencies, and local citizens are all recognizing that a shrinking Mississippi delta is not a good thing. But what can be done to reverse this trend?

Several government proposals to save Louisiana's coasts have been approved. The 1990 Coastal Wetland Planning, Protection, and Restoration Act is providing about $50 million a year for wetland enhancement in Louisiana. In 1998, state and federal agencies developed "Coast 2050: Toward a Sustainable Coastal Louisiana" proposals that defined some 500 projects with total estimated costs of $14 billion. Coast 2050: feasibility report This is the scale of effort needed, but attempts to develop this level of funding are still under way. In the meantime, the Mississippi delta keeps shrinking. Stay tuned—you're likely to encounter news reports of the delta's continuing changes and people's efforts to deal with them.

Geotimes – August 2007 - restoring the river

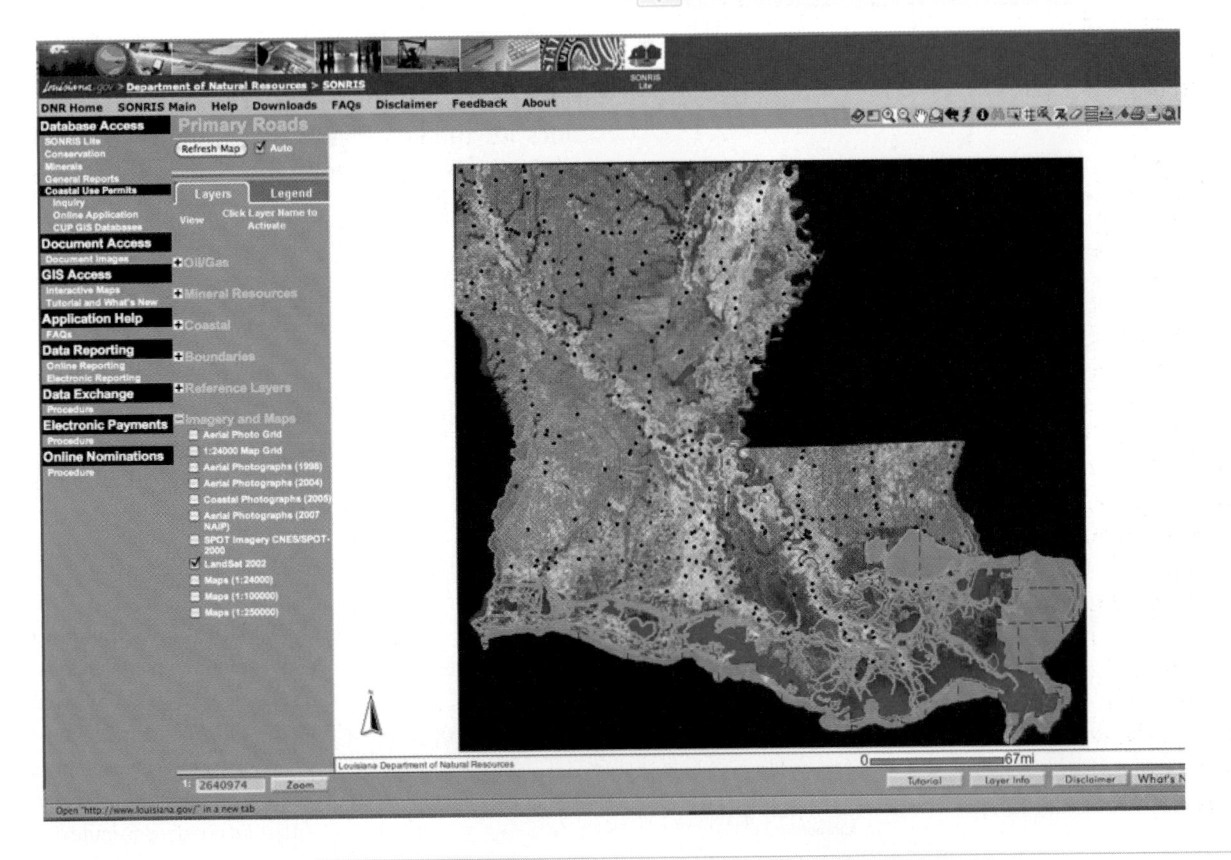

9.4 Coasts and Storms

Hurricane Katrina came ashore on the central Gulf of Mexico coast on August 29, 2005. Before Katrina died down on August 31, it had killed around 2000 people and caused more damage than any other hurricane in U.S. history. Most of the approximately $200 billion in damage was along coastal Louisiana, Mississippi, and Florida.

Katrina, and other hurricanes such as the four that struck Florida in 2004, are remarkable, devastating examples of Earth systems interactions. The devastation caused by hurricanes and other storms is possible because of the tremendous amounts of energy transferred among the atmosphere, the oceans of the hydrosphere, and ultimately the geosphere and biosphere where the storms come ashore. Because of the rates of these interactions (winds can blow ashore at over 240 kilometers per hour, or 150 mph) and the amounts of energy involved, coastal storms are dramatic if potentially deadly examples of Earth systems in action. It's during storms that flooding, erosion, and sedimentation processes become very obvious. It happens right before people's eyes.

People can anticipate the changes coastal storms will bring. This is because certain types of storms follow the seasons and varying inputs of radiation from the Sun. Even though different types of storms bring different levels of concern, they all bring some level of coastal flooding, erosion, sedimentation, and damage to property and habitat. In the United States, the principal storms of concern are hurricanes that affect the Atlantic and Gulf of Mexico coasts and winter storms that affect both the Atlantic and Pacific coasts.

HURRICANES

Hurricanes are vast rotating storms, some 500 kilometers (310 mi) or more across, which intensify by drawing energy and moisture from the warm surface water of the tropics. Atlantic hurricanes start as thunderstorms, many over west Africa. As these drift westward over the tropical Atlantic Ocean, the warm surface water heats the overlying atmosphere, and the storms strengthen as the hot air rises. The rising air causes the atmospheric pressure in the center of the disturbance to fall. Winds begin revolving counterclockwise around the center of low pressure. Storms with winds less than 62 kilometers per hour (39 mph) are called *tropical depressions*. When winds exceed 62 kilometers per hour, they become *tropical storms* and are assigned a name. When sustained winds in a storm reach 118 kilometers per hour (74 mph), it is known as a **hurricane** (or tropical cyclone). In a mature hurricane, diagrammed in Figure 9-35, thunderstorms surround the low-pressure center of the storm, forming the *eye wall*. The area of warm, calm air at the center of the hurricane is called the *eye*. Warm, moist air spirals inward toward the eye wall and then rises. Moisture in the rising and cooling air condenses into rain.

The potential for wind damage and flooding from a hurricane is ranked from 1 to 5 on the Saffir-Simpson Hurricane Scale, shown in Table 9-1. | Hurricane intensity scale The scale is based on the central pressure within the eye of the storm (the lower the pressure, the stronger the storm), the maximum sustained winds, and the potential for a dangerously high mass of water known as the **storm surge**. Major hurricanes—those of category 3, 4, or 5 storms—have the greatest potential for damage.

The Atlantic hurricane season is summer, when the Sun's radiation most effectively heats the tropical Atlantic. It officially begins June 1 and lasts until November 30, with activity usually peaking in September. In an average year, the Atlantic develops two major hurricanes. The frequency of hurricanes rises and falls cyclically, with periods of above or below normal activity lasting 25 to 40 years. For example, only four major hurricanes hit the U.S. East Coast from 1970 to 1994. During the previous 25 years, however, nearly four times this number made landfall.

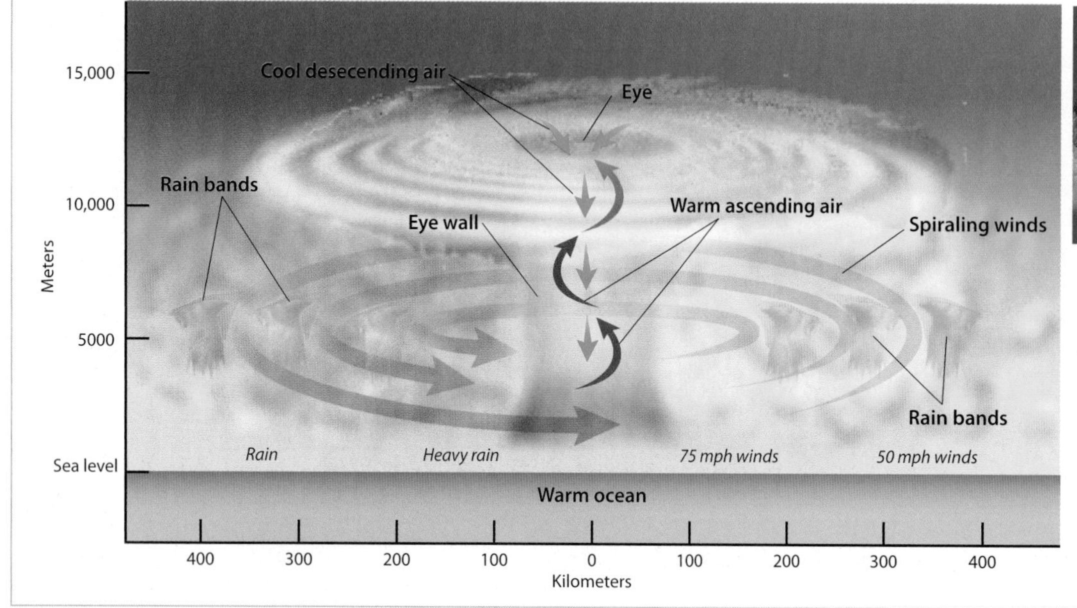

FIGURE 9-35 ◀ Anatomy of a Hurricane
The center of a hurricane is a low-pressure area of warm, calm air, called the eye. It is surrounded by a band of thunderstorms that form the eye wall. As warm, moist air spirals inward toward the eye wall and rises, it begins to cool and its moisture condenses into rain.

TABLE 9-1 THE SAFFIR-SIMPSON HURRICANE SCALE

Category	Winds	Effects
One	74–95 mph (119–153 kph)	No real damage to building structures. Damage primarily to unanchored mobile homes, shrubbery, and trees. Also, some coastal road flooding and minor pier damage.
Two	96–110 mph (154–177 kph)	Some roofing material, door, and window damage to buildings. Considerable damage to vegetation, mobile homes, and piers. Coastal and low-lying escape routes flood 2–4 hours before arrival of center. Small craft in unprotected anchorages break moorings.
Three	111–130 mph (178–209 kph)	Some structural damage to small residences and utility buildings with a minor amount of curtain wall failures. Mobile homes are destroyed. Flooding near the coast destroys smaller structures with larger structures damaged by floating debris. Terrain continuously lower than 5 feet above sea level may be flooded inland 8 miles or more.
Four	131–155 mph (210–249 kph)	More extensive curtain-wall failures with some complete roof structure failure on small residences. Major erosion of beach. Major damage to lower floors of structures near the shore. Terrain continuously lower than 10 feet above sea level may be flooded requiring massive evacuation of residential areas inland as far as 6 miles.
Five	greater than 155 mph	Complete roof failure on many residences and industrial buildings. Some complete building failures with small utility buildings blown over or away. Major damage to lower floors of all structures located less than 15 feet above sea level and within 500 yards of the shoreline. Massive evacuation of residential areas on low ground within 5 to 10 miles of the shoreline may be required.

From *http://www.aoml.noaa.gov/general/lib/laescae.html*

WINTER STORMS

Strong storms also develop in winter and cause a significant amount of damage and loss of life on both the Atlantic and Pacific coasts. These storms have different causes and go by a variety of names. For example, winter storms that form on the Atlantic coast near Cape Hatteras, North Carolina, are known as northeasters (or nor'easters). The warm, moist air that develops over the Gulf Stream ocean current rises, creating low pressure regions with strong northeasterly winds along the Atlantic coast. Cold air masses from Canada move toward the coastal low-pressure regions and collide with the warm air, causing intense precipitation. As they move up the coast, nor'easters can develop hurricane-force winds and can be deadly. Off New England, these storms have historically claimed many fishing vessels and their crews. The Perfect Storm October 1991

Winter storms pummel the Pacific coast, too. You learned about one type, the Pineapple Express, in Section 7.2. This funnel of moisture-laden, warm air originates in the western Pacific Ocean, tracks east and northeast, commonly passing over the Hawaiian Islands (from which it gets its name), and smashes into land at various places along the U.S. and Canadian west coast. The Pineapple Express is best known for the large amounts of precipitation (and resulting floods) it brings to the land, but all storms bring winds and waves as well. Although the winds themselves may not be extremely damaging, the long fetch and consistent direction they take across the Pacific Ocean creates waves that significantly affect the coast.

As if the Pineapple Express wasn't enough, other winter storms originate in the northern Pacific and swoop down on the Pacific coast. At times the winds can reach hurricane scale (118 kph, or 74 mph) but again, it's the waves smashing against the coast that have the most impact. Because the Pacific coast has so many headlands and sea cliffs, storm wave collisions with the coast can be spectacular—the scene in Figure 9-36 is typical.

FIGURE 9-36 ▼ Storm Waves on the Pacific Coast

Storms bring large waves and tremendous energy ashore during Pacific Northwest storms.

In some places, they have become tourist attractions. Local lodges advertise special rates for weekend "storm watching."

COASTAL STORM HAZARDS

The amount of energy expended against the coast during storms—by waves, storm surges, and currents fueled by high winds—is unequaled at any other time. The hazards that storms bring include high winds, inland flooding from torrential rain (Chapter 7), landslides (Chapter 8), high waves crashing on the shore (and anything that happens to be built on it), and storm surge. | Coastal change hazards: hurricanes and extreme storms | Wind tears off roofs and transforms debris into lethal projectiles. Most of Hurricane Andrew's estimated $15 billion damage in 1992 was due to wind. But the most direct effects of storms on coasts come from storm surge and from how storms affect erosion and sedimentation.

Storm surges are a major cause of coastal flooding. To an observer on the coast, the storm surge appears like an extremely high tide with storm waves superimposed on it. In addition to direct flooding of coastal land, the surge brings the surf within reach of structures that it normally would not touch. The two main causes of the surge are the low pressure within the storm and its strong winds. As a storm approaches the coast, the low atmospheric pressure allows the sea surface to rise upward, forming a surface bulge of water. More important, as strong winds approach the coast, they push the sea ahead of them. When the water reaches shallower depths, it has nowhere to go but up.

Coastlines with shallow, gently sloping continental shelves tend to develop the highest storm surges. (You can view an online animation of this process at the NOAA website.) | Storm surge | On the Atlantic and Gulf coasts, the surge in typical large storms is 4 to 5 meters (13 to 16 ft). The single worst loss of life in a coastal storm occurred in 1970 on the coast of Bangladesh, when a storm surge flooded the land and killed an estimated 300,000 people.

It is no coincidence that barrier islands are evacuated first when a hurricane targets the Atlantic and Gulf coasts. Low elevation and high population density on barrier spits and islands add up to the potential for major loss of life and property. The storm surge can completely inundate a barrier island, leaving people nowhere to run. This is what wiped out Galveston, Texas in 1900. New Orleans has still not recovered from the storm surge that came ashore with Hurricane Katrina in 2005.

(a)

FIGURE 9-37 ▼ New Orleans: 2005

(a) Levees could not hold back the storm surge of Hurricane Katrina, and 80% of the city flooded. **(b)** Human suffering in New Orleans was immediate and acute, but has also continued long after the storm was over.

(b)

HURRICANE KATRINA AND NEW ORLEANS

Hurricane Katrina originated in the Atlantic Ocean over the southeastern Bahamas on August 23, 2005. It moved northeast, crossing parts of Florida as a category 1 hurricane (see Table 9-1) on August 25. Its path then veered westerly into the Gulf of Mexico, where it gained strength and turned north toward New Orleans. | NCDC: climate of 2005: Hurricane Katrina | A hurricane watch was ordered for the Louisiana coast, including New Orleans, on the morning of August 27. Hurricane force winds, very heavy rain, and a building storm surge began impacting the Louisiana and Mississippi coasts on August 28. On the morning of August 29, Hurricane Katrina was a category 3 storm as its eye came ashore over the Mississippi River delta.

Katrina's devastation extended up to 160 kilometers (100 mi) inland along southeastern Louisiana, the entire Mississippi coast, and parts of Alabama. The storm's center passed to the east of New Orleans but it might as well have been a direct hit. The storm surge, estimated at 3 to over 4 meters (10 to 14 ft) in southeastern Louisiana, combined with rain falling at an inch per hour to quickly fill canals, lakes, and rivers. The levee system protecting New Orleans from flooding could not hold back the waters. Over 50 levees failed and 80% of New Orleans was flooded (Figure 9-37a).

Hundreds of thousands became homeless and over 700 city residents died. Human suffering continued for months as damaged or destroyed facilities and displaced or inadequate human resources prevented effective public services and timely disaster recovery (Figure 9-37b). The extent of the catastrophe is unprecedented in U.S. experience, and government's response to Katrina's damage continues to be a concern for the citizens of New Orleans.

The 2005 flooding and destruction of New Orleans is an example of the dilemma people face when they choose to live in coastal areas where flooding from storm surges is likely. Did you know that many of the hurricane's impacts on New Orleans were predicted? The destruction, the health and water quality problems, the homelessness and evacuation needs, even the time it would take to get rid of the floodwaters? You see, New Orleans knew it was at risk for flooding.

The initial settlement of New Orleans was on "high" ground. High ground here is several meters (tens of feet at the most) above sea level, and there isn't much of it. As the city's population grew, it expanded into lower and lower areas. Canals were built and pumps were installed to drain the low areas and create space for people. Now large parts of New Orleans are below sea level.

People have done more than just spread New Orleans into low areas. Over the years, shipping channels have been dredged through the nearby Mississippi River's delta. These channels have funneled the river's discharge to the Gulf of Mexico and not allowed it to continue building the delta and sustaining its wetlands. As a result, the land has subsided, making New Orleans even more vulnerable to river flooding and hurricane storm surges. The combined effect of canals, pumps, and channeling of the river has made New Orleans a disaster waiting to happen. And it will likely happen again.

You Make the Call

What Do We Do with New Orleans?

Reconstruction of New Orleans is under way. The city is a major commercial port and has a rich cultural heritage that many feel should be preserved. But the costs are tremendous and every American taxpayer helps pay them.

- Do you think flooding will be repeated in New Orleans?
- What would you do if you continually faced the tremendous flood hazards that the Mississippi and hurricanes bring its way?
- Do you think we can engineer New Orleans to safety?
- What do you think we should do with New Orleans?

In the aftermath of Katrina, parts of New Orleans have still not recovered. Many families may never return.

9.5 Living with Changing Coasts

In 1906, T. B. Potter, a real estate broker from Kansas City, saw a sandy spit across the mouth of Tillamook Bay west of Portland, Oregon, and imagined a resort without peer. The spit was about 6.4 kilometers (4 mi) long, with towering sand dunes backing a beautiful sandy ocean beach. He bought the land, subdivided it into lots, and began to build a town from scratch. The town would have a large hotel, a bowling alley, and an indoor swimming pool. Paved roads, streetlights, a freshwater system, and a telephone network further established the town, along with a narrow-gauge railway to convey the flocks of expected tourists. Potter even had his own newspaper—*The Surf*. By 1914, 600 lots had been sold. Potter christened the new resort town "Bayocean." But, as Figure 9-38 shows, erosion soon turned this oceanfront paradise into an Atlantis.

FIGURE 9-38 ▲ Bayocean, Oregon
This home, once in downtown BayOcean, was about to be washed away in 1947. There was plenty of sediment on the beaches and dunes of the Bayocean spit when the new community of Bayocean was constructed. Within 20 years, though, Bayocean began to lose its footing as erosion removed the beach and began eroding the dunes behind it. In 40 years, Bayocean was mostly history.

Tillamook Bay and the narrow inlet between the spit and the mainland also had great potential as a commercial port. However, tides continually clogged the inlet with sandy shoals, making navigation for larger ships difficult. The Army Corps of Engineers, with the enthusiastic support of the local community, built a stone jetty on the north side of the inlet to intercept sand flowing in the longshore drift southward and into the inlet. Completed in 1917, the jetty worked—but at a great cost to Bayocean. Starved of sand by the updrift jetty, the spit began to disappear. By the 1930s, the waves were consuming city sidewalks. The swimming pool fell victim in 1936 and collapsed. In 1948, a storm carved a new inlet through the spit. The last house on the spit was washed away in 1960 and the community ceased to exist, erased by erosion.

As the unfortunate residents of Bayocean discovered, living on coasts commonly puts people at odds with the processes that maintain beaches. Indeed, at times the relationship between humans and coastal erosion has been one of open warfare, complete with massive and expensive "coastal protection" and "coastal armoring" programs designed to keep the waves at bay. For as long as there have been populated coasts, people have erected structures of stone, wood, and—in modern times—steel in an attempt to protect their property from storms and erosion, to make nearshore waters more navigable, to create calm and safe harbors, or to enjoy an ocean view in their homes.

Modifying the natural coastal system has consequences, not all of them positive. Some structures meant to prevent erosion may actually make it worse. For example, the seawall at Galveston, Texas, that protects the shore from flooding during storms can lead to loss of the beach in front of it. Structures meant to protect a beach in one location may lead to increased erosion elsewhere.

Facing coastal erosion and storm threats, communities can choose from several broad strategies. They can build a wall against the sea or structures designed—not always successfully—to trap sand and bulk up the beach. They can mine sand offshore or from inlets and bays and add it to eroding beaches. Or they can simply let nature take its course and retreat landward along with the beach. Each approach has its costs and benefits, and the approaches are not mutually exclusive. It's possible to armor the coast, nourish beaches, and retreat all at the same time.

HARD STABILIZATION

Early coastal engineers, such as the Romans, strived above all to make the coast useful. Navigable waterways and protected ports were as essential to the empire as the aqueducts that supplied freshwater to cities and towns. Later, as coasts became more and more densely populated, erosion and storm hazards became more pressing matters. Thus was born the art of coastal armoring: the construction of stone, wood, and concrete "hardware" to hold back the sea.

A common term for engineered structures to stop, prevent, or reverse coastal erosion and protect coasts from storm damage is *hard stabilization*. Seawalls, breakwaters, groins, and jetties are the most widely used types of hard stabilization. For example, 9%, or a total of about 165 kilometers (103 mi), of the California coast is armored in some fashion, protecting coastal lowlands, dunes, and sea cliffs or bluffs.

Seawalls, bulkheads, and revetments

Seawalls, bulkheads, and revetments are all types of barriers built parallel to the coast. Depending on their intended function, they may stand below, at, or above the high-tide line. In general, **seawalls** are massive structures designed to withstand the full force of storm waves. The curved concrete seawall on the south shore of Galveston, Texas, is a good example of this kind of coastal armor (see Figure 9-2).

Bulkheads are usually smaller than seawalls and aren't designed for high-wave-energy environments. They are vertical walls typically constructed of timber, concrete, or steel. On beaches, they are often placed at the start of the berm and backfilled to create a flat surface. This type of bulkhead remains dry under calm conditions, but might have to withstand wave attack during storms. In ports, bulkheads are commonly used to line the inner harbor area. In this case, the base of the bulkhead is often located

FIGURE 9-39 ▲ Bulkheads and Revetments
(a) A bulkhead at work: Boston, Massachusetts. **(b)** A revetment at work: Toronto, Canada.

at or slightly seaward of the low-tide mark, and is in contact with water most of the time (Figure 9-39a).

Revetments are generally lighter-duty structures than seawalls and bulkheads. They provide a protective covering on embankments or beaches to resist erosion by near-shore currents or light wave activity. Many revetments are constructed from large blocks of broken rock, poured concrete slabs, or interlocking concrete or stone blocks (Figure 9-39b).

Particularly in the case of large seawalls, one of the primary disadvantages of hard stabilization is the cost, both for the initial construction and periodic maintenance. Continued wave attack may scour the base of the seawall, potentially causing an expensive failure. To prevent this, the foundations of the wall are set well below the sand and anchored with piles. Broken stone is commonly placed at the base of the wall to absorb wave energy. Seawalls may also be considered unsightly compared to a natural dune or beach. In some cases, they may completely block the view of the water from shore.

The most controversial aspect of seawalls concerns their potentially harmful effects on beaches. When waves strike a seawall, part of the energy is reflected back. This may increase erosion of the beach fronting the seawall, often causing its permanent loss. If erosion steepens the shore in front of a seawall, waves will break closer to shore, potentially increasing their effects.

Breakwaters

Breakwaters are barriers built parallel to the coast, like seawalls, but in the water, as you can see in Figure 9-40. The barrier causes waves to break offshore instead of onshore, reducing the energy expended on the shore. Most breakwaters rise above the water surface, although coastal engineers have experimented with submerged breakwaters. Breakwaters are commonly constructed to provide a calm area behind them for an artificial harbor or port.

In high-wave-energy environments, the most common type of offshore breakwater consists of a sloping wall of massive stone blocks. In places where the wave conditions permit, breakwaters may be made of less-massive materials. This includes sand-filled metal or concrete tubes, called cells, or sheet steel fastened to piles driven deep into the sea bottom.

Breakwaters diminish waves behind them in the surf zone, causing them to drop their sediment load. This tends to fill in the harbor, which may then require periodic dredging to remain useful.

Breakwaters can be used to protect eroding beaches. In this case, sediment falling out of the surf behind the breakwaters accretes onto the beach and locally develops a positive sediment budget. However, this type of breakwater may work too well, and peninsulas of sand may grow outward from the shore to the point that they contact the breakwater. This trapping of sediment disrupts longshore drift and commonly creates a negative sediment budget somewhere else along the shore where erosion can occur as a result.

FIGURE 9-40 ▼ A Breakwater at Work
Many breakwaters, like this one in the United Kingdom, are used to protect anchorages or harbors.

Groins and jetties

Groins (Figure 9-41a) are barriers built perpendicular to the shore. The intention is to capture sand from the longshore drift to thwart beach erosion or rebuild a beach that has eroded away—to create a positive sediment budget. Groins are usually built in groups, called *groin fields* (see Figure 9-41a), from stone block, concrete, sheet steel, or timber. Jetties too are built perpendicular to the shore, but they are used to stabilize inlets (see Figure 9-33). The jetties prevent the inlets from moving, as is their natural tendency, or filling in with sediment. A single jetty may be constructed on the updrift side of the inlet, or jetties may be constructed on both the updrift and downdrift sides.

The trouble with groins and jetties is that sediment accumulates on the updrift side of the barrier (where a positive sediment budget develops) but tends to erode from the downdrift side (where a negative sediment budget develops). This results in a distinctive pattern of sedimentation, with the contours of updrift accumulation and downdrift erosion forming a sawtooth pattern within the groin field or on both sides of the jetties. The downdrift effects of groin fields and jetties may extend a considerable distance down the coast. Sand that would have reached beaches downdrift remains within the groin field or on the updrift side of the jetties. This amounts to a redistribution of sediment that can cause serious downdrift erosion.

There are many examples in the record of erosion problems caused by groins and jetties. The history of the inlet and jetties at Ocean Beach, Maryland, is frequently cited. In 1933, a hurricane breached the barrier coast, creating the Ocean City Inlet. Jetties were constructed to stabilize the inlet, which provided welcome harbor access. Sand began to accumulate on the updrift jetty. By 1955, the accreting beach had nearly covered a recreational pier that extended out into the surf. On the south side of the inlet, the jetties caused severe erosion on Assateague Island. By 1962, erosion had severed the south jetty's connection to the island. It was repaired with dredged fill material.

To prevent erosion associated with jetties, harbors may be equipped with a sand bypassing system that dredges sand updrift of the jetty or jetties and pumps it to the downdrift side. This is expensive, however, and does not always counteract downdrift erosion problems.

It should be noted however, that seawalls and similar barriers are intended to protect property, not the beach. The irony of this type of coastal armoring is that a seawall, bulkhead, or

(a)

(b)

FIGURE 9-41 ◀ Groins
Groins, commonly constructed in sets called groin fields **(a)**, capture sand from longshore drift on their updrift side **(b)**.

revetment may succeed in its purpose—which is to prevent erosion or storm damage to valuable property and structures—while causing the loss or degradation of the very qualities of the coast that drew development there in the first place.

SOFT STABILIZATION

As people have recognized the financial costs and environmental consequences of hard stabilization, many beach communities have taken a different approach to addressing beach erosion known as *soft stabilization*. One form of soft stabilization now commonly used on many eroding coasts is **beach nourishment**. Sand is brought in and used to create a new beach as well as bulk up the underwater slope of the beach face. This develops at least a temporary positive sediment budget. Another soft stabilization tool is **dune restoration**. Using fencing, vegetation planting, and other methods, dunes are expanded and stabilized to provide a physical barrier to storm waves.

Since 1921, the United States has spent over $2.4 billion (in 1996 dollars) on beach nourishment. The aim is to create a natural-looking beach that restores the recreational use of the shore and protects beachfront property from erosion and storm damage.

Beach nourishment

Building a long-lasting beach that looks and feels natural is not a trivial task. Recall that the dynamics of the beach is a delicate dance between sea level, beach sediment supply, and wave conditions. Experience has shown that using the right kind of sediment is key. If the sediment is too fine, waves may simply scour it away and cause the water to become cloudy. Water clouded with fine sediment can harm habitat such as coral reefs.

One challenge in nourishing a beach is finding a suitable site for obtaining sediment. Dredging material from inlets and bays may quickly deplete it and can harm the local ecosystem, so most commonly beach nourishment projects seek their sediment from offshore sites. Finding the right quality of sediment, dredging it up, and transporting it to the beach is quite expensive. After the initial restoration, the beach must be renourished periodically to keep pace with the rate of erosion or periodic storm damage. Beach nourishment only temporarily reverses the negative sediment budget that characterizes eroding beaches.

Saving Miami Beach

Beach nourishment is common in Florida, since the beaches are so vital to the local economy. Miami Beach is a major tourist destination, having some of the most valuable real estate of its kind in the country. Huge hotels line this shore.

As you can see in Figure 9-42a, by the 1970s the natural beach separating the hotels from the sea had all but disappeared, replaced by a seawall. People would much prefer gazing at a beach than a seawall, so the city decided to rebuild the beach on an unprecedented scale. From 1976 to 1981, 11 million cubic meters (14 million yd^3) of sand were pumped onto the shore (Figure 9-42b). An additional 240,000 cubic meters (300,000 yd^3) were added in 1987. A protective dune was built at the top of the beach to provide protection from storms. The Miami Beach project nourished more than 16 kilometers (10 mi) of beach at a cost of $65 million.

The Miami Beach nourishment is widely perceived as a success of soft stabilization. Nourished beaches do not always perform as expected, however. For example, Ocean City, New Jersey, completed a $5 million nourishment in 1982 only to see most of the visible portion of the new beach disappear into the surf within a few months. This highlights the uncertainties that remain about the practice. On one hand, when beach nourishment lasts a long time—as in the case of Miami Beach—it can

(a)

(b)

FIGURE 9-42 ▲ Beach Nourishment in Miami

Miami Beach was just about gone in the 1970s **(a)**, but a major beach nourishment project brought sand back in the 1980s **(b)**.

offer significant gains to a community. On the other hand, it may be an unending chore.

Dune restoration

Restoring and preserving dunes is another method of countering coastal erosion without building hard structures. Dunes form when blowing sand encounters an obstruction, such as a clump of beach grass. This diminishes the wind speed, and the wind drops its sand load.

Dunes play a critical role in natural beach protection during storms. Waves surging over the berm may cut into the base of the dunes and drag sand seaward, preserving the beach. Dunes also form a physical barrier against high water levels during storms, much like a levee along a flooding river.

Dune-building programs seek to either restore dunes that have been damaged (as in Figure 9-43) or build new or larger dunes to protect property, usually on barrier islands. People usually create artificial dunes by installing sand fencing or planting beach grass. Another way is to just bulldoze sand into a pile and plant vegetation to help stabilize it.

In the 1930s, the U.S. government launched a massive dune-building project on the Outer Banks of North Carolina. The banks are part of that state's extensive barrier island coastline. The beaches were eroding, so the Depression-era government unleashed the Civilian Conservation Corps on the problem. They planted trees, shrubs, and beach grass to catch blowing sand over more than 48 kilometers (30 mi) of the coast, creating a virtually continuous wall of dunes.

FIGURE 9-43 ▼ Dune Restoration
Planting of native dune plants **(a)** is a common approach to restoring and stabilizing dunes **(b)**.

(a)

(b)

MANAGING SEDIMENTS

Sediment is not always deposited where it is wanted. This is especially the case in areas used by shipping. Sediment in unwanted places commonly leads to seemingly continuous efforts by people to control or remove it. Managing sediments is an ongoing effort—with mixed results—along many coasts.

Dredging

People remove unwanted sediment by **dredging**, either by scooping sediment off the bottom or pumping thick slurries away with suction dredges (Figure 9-44). Man-made channels are commonly places where unwanted sediment accumulates. At first, you might think that channels wouldn't influence sedimentation that much, except where they carry sediment farther out to sea. But in bays and estuaries, where waves and tidal currents rather than river currents dominate, shifting sediment can get trapped in channels. This is because the increased cross-sectional area of a channel decreases the velocities of currents that flow either along or

FIGURE 9-44 ▲ Dredging
This dredging ship is removing sediment from a channel and depositing it where it can protect low-land areas of the Netherlands.

FIGURE 9-45 ▲ Outbound over the Columbia River Bar
A sandbar produced by sediment from the Columbia River makes entering or leaving the river ports a hazardous undertaking.

across it. (Remember from Section 7.1 how changing channel geometry affects river flow?) The decreased current velocities lead to sedimentation. In these situations, dredging has to be done over and over again to keep navigational channels open.

An example comes from the mouth of the Columbia River at the Oregon–Washington border. Ports along the lower Columbia handle more than $15 billion of cargo a year. More wheat is exported from Columbia River ports than from any other ports in the country. But to get into the Columbia River, ships must pass over a sandbar developed by sedimentation where the river's currents encounter the waves and tides of the coast. As Figure 9-45 suggests, it's a notoriously dangerous passage, especially when storms come ashore from the Pacific Ocean. A channel has been dredged across the mouth to help ships pass over the bar. For years, this channel was about 180 meters (590 ft) wide and 12 meters (39 ft) deep. In 2004, Congress passed legislation that funded deepening the channel to 13 meters (43 ft). This doesn't sound like much, but it's enough to enable a ship to carry 300 more cargo containers or an additional 6000 tons over the bar.

Keeping the channel open over the Columbia River bar is an ongoing job. Sediments collect in the channel and dredging is almost continuously needed to remove them. With every dredging activity comes the question, where do we put the dredged sediments? People ask this question at all dredging sites. In the case of the Columbia River, the sediments have been dumped farther offshore, a common fate for dredged sediments along the nation's coasts. But the Columbia River's dredged sediments didn't completely disappear out into the ocean. They developed a mound on the seafloor that became shallow enough to affect surface waves and currents. The resulting sea conditions made shipping more difficult. Determining how best to manage sediments at the mouth of the Columbia River is a continuing challenge.

Sediment contamination

Careful consideration of sediment disposal choices is important wherever dredging takes place. It's especially important if the sediments are contaminated with pollutants. Because bays and estuaries are where people have built large ports and cities, many used for hundreds of years or more, sediments in them are commonly contaminated by industrial, shipping, or municipal waste disposal activities. The Environmental Protection Agency (EPA) estimates that 10% of the bottom sediment in U.S. waters is contaminated at levels that pose risks to ecosystems and people. From 5 to 10% of the 230 million cubic meters (300 million yd^3) of sediment that the Army Corps of Engineers removes through dredging each year has this level of contamination.

Dealing with contaminated sediment is very challenging. The sediment is difficult to carefully recover and difficult to remediate. It can't be used directly for sediment enhancement projects, and its disposal shouldn't pose new risks to the environment or people. There is enough of this material around to be a significant concern. The final report of the U.S. Commission on Ocean Policy in 2004 recommended that the EPA better understand how contaminated sediment is created and how to effectively prevent and treat it. Dealing with contaminated sediment still seems to involve more discussion and study than action.

MITIGATING COASTAL STORM HAZARDS

With improvements in storm forecasting and tracking, loss of life during major storms has fallen. However, hurricanes are still extremely dangerous and preparedness on all levels is critical. Hurricane-prone areas have disaster plans in place to notify people when it's time to evacuate the coast. In the United States, coastal states like Florida have extensive systems of storm shelters and evacuation routes.

To reduce structural damage, hurricane-prone coastal communities have adopted rigorous building codes. As with building for earthquake resistance, "storm proofing" a home really means preventing it from completely collapsing on the occupants. Special types of nails or reinforcing metal straps can be used to fasten the roof to the walls, preventing catastrophic "peel offs" in high winds. They are able to hold together during the 240 kilometer per hour (150 mph) winds of a category 4 hurricane.

COASTAL ZONE MANAGEMENT

U.S. coastal states and territories have all adopted some form of a comprehensive approach to living on coasts called coastal zone management. The 1972 Coastal Zone Management Act laid out some general goals for protecting the nation's coastlines and mitigating coastal hazards. States can get money from the Coastal Zone Management Program by building these overarching national goals into how they manage their coastlines.

You Make the Call

Dealing with Falmouth's Changing Coast

Falmouth, Massachusetts, is on the southern tip of Cape Cod. Initially settled in 1660, Falmouth was the site of much coastal commerce for many years. After the railroad arrived in 1872, people built summer homes on Falmouth's shores. In 1907, a channel from the shore to Deacon's Pond was dug, creating Falmouth's inner harbor. Since then, about everything people can do to a coast has been done at Falmouth. Homes crowd the shoreline, where armoring, jetties along the harbor channel, and groins have been constructed. The end result is a shoreline losing sediment. The beaches and dunes have narrowed or disappeared and waves wash against armored bluffs.

Citizens became concerned about the Falmouth shore. They organized the Falmouth Coastal Resources Working Group to study the coast and develop recommendations for dealing with the unwanted changes. Falmouth Coastal Resources Working Group In 2003 the group concluded that if nothing was done, the next 100 years would see their community facing "increased property damage from storms because protective beaches and dunes have shrunk or disappeared, increased need for armoring, reduced public access, loss of habitat, loss of attractive coastal vista, and rising costs to maintain a coast that cannot sustain itself." To avert this future, the group recommended that the community "acquire coastal open space, particularly in key sediment supply areas; encourage landowners to provide coastal buffers; undertake beach nourishment and removal of armoring beginning with town-owned parcels; encourage a public coastal pathway and promote coastal eco-tourism; develop town incentives for naturalizing the coast; develop a Flood Hazard Mitigation Plan; improve regulations to better protect the coastline; and provide public outreach concerning coastal processes."

In many areas along the densely developed and heavily armored coast of Falmouth, Massachusetts, beaches and dunes have been washed away.

- If you owned a home along Falmouth's shore, what would be your response to these recommendations?
- Would you want a coastal pathway and ecotourism developed?
- What if decreased armoring led to bluff erosion near your home?
- How much should the community spend on accomplishing these recommendations?
- Where should this money come from?

The recommendations for Falmouth seem sound and worthwhile, but implementing such changes in how people interact with coastal processes is tough to accomplish. Check in on Falmouth—is the community making progress?

SUMMARY

Lots of people live on or near coasts. People have benefited from the sea's resources and the global transportation that oceans allow for thousands of years. Global commerce is funneled through coastal ports around the world. But the coast is a place where Earth systems interactions are especially pronounced. As a result, this boundary zone for the atmosphere, hydrosphere, geosphere, and biosphere is very dynamic and changing. The biosphere, including people, adjusts in one way or another to these changes. In many cases, people find the natural processes along coasts challenging to live with. Scientists and engineers have developed ways to mitigate coastal hazards, but people's adjustments to living along coasts have consequences that may be unwanted. In places, people's actions have changed coasts in ways that increase risks from coastal hazards and jeopardize natural habitat. Living more sustainably with changing coasts is an ongoing challenge for people.

In This Chapter You Have Learned:

- Winds move surface and near-surface water to create waves and swells that transfer energy through the hydrosphere. Waves induce circular to elliptical motions of the water as they pass through it.
- Waves vary in size and are measured and described by their height (vertical distance between crest and trough), wavelength (distance between successive crests), and period (time between the passing of successive crests). A wave's velocity is its wavelength divided by its period.

- As waves enter shallow water, they begin to interact with the bottom at the point where the water depth is one-half of their wavelength. Friction with the bottom causes a wave's height to increase and its wavelength to decrease. As different parts of a wave encounter shallow water at different times, the wave bends (refracts) and becomes more parallel to the shoreline.
- Longshore drift develops where waves are oblique to the shoreline and move water and sediment along it. Water flowing back to the ocean from the shore can create local rip currents that are dangerous for swimmers.
- Tides, the very predictable twice-daily fluctuations of sea level, are caused by gravitational interactions of the Earth, Moon, and Sun. The hydrosphere, atmosphere, geosphere, and biosphere are constantly interacting in the area between high and low tide, the intertidal zone. Tidal currents are developed where water movement is constricted in inlets or other narrow passages.
- Global climate fluctuations change the amount of water stored in high latitude ice accumulations, and sea level changes as a result. During warm periods sea level is higher and during cold (glacial) periods sea level is lower. These sea level changes can be as much as 100 meters (328 ft) or more within several thousand years.
- Movements of the geosphere affect many coastlines. Gradual depressions or uplifts of the geosphere can result from the accumulation or melting of large amounts of overlying ice. Tectonic movements along plate boundaries have caused both uplift and subsidence on long lengths of coastline.
- The physical features along coasts vary depending on tectonic setting, local geology and geography, and sediment supply. Distinctive coastal features include headlands, bays, estuaries, barrier islands, coral reefs, deltas, and beaches of all sizes.
- Beaches have distinctive parts, including the berm, which is only underwater at very high tides and during storms; the sloping beach face, where waves wash back and forth; and the underwater nearshore zone, where longshore and rip currents flow. These three parts are the littoral zone where sediment is in motion along shores.
- Erosion, common along shores, occurs where there is a negative sediment budget—more sediment leaves than is supplied to the littoral zone. Shorelines directly attacked by waves, such as headlands and sea cliffs, are constantly being eroded.
- Sediment is deposited where waves and currents cannot move it. It can be deposited on beaches and deltas where there is a positive sediment budget and where wave and current energy is diminished, as in bays, estuaries, and behind barriers constructed by people.
- Coastal storms, including hurricanes, have major affects on coasts. The atmosphere, hydrosphere, and geosphere interactions that characterize coasts are amplified by storms. The biosphere, caught in the middle of these interactions, struggles to survive during storms, but over time has adapted to the dynamic changes always under way along coasts.
- People make many changes to enhance the use of the coastal zone. These changes can affect erosion and sedimentation in both positive and negative ways.
- The dynamic nature of the coastal zone makes it a challenging place for people to live and work. Physical features people construct to control coastal processes, such as groins, jetties, breakwaters, and seawalls, have had mixed results—as has dredging to mitigate sedimentation. Soft stabilization techniques such as beach nourishment and dune enhancement can help, but over the long term, coastal processes are not very controllable.
- People will continue to live along coasts. Understanding coastal processes and how they change the coast enables people to anticipate where their homes and businesses are in most jeopardy. Respecting coastal processes and the value of natural coastal resources, such as habitat and ecosystems, will be part of what people do to achieve a sustainable future.

KEY TERMS

barrier island (p. 270)
beach (p. 276)
beach drift (p. 263)
beach face (p. 276)
beach nourishment (p. 289)
berm (p. 276)
breakwater (p. 287)
bulkhead (p. 286)
dredging (p. 290)
dune restoration (p. 289)
estuary (p. 272)
groin (p. 288)
hurricane (p. 282)
jetty (p. 279)
longshore current (p. 263)
longshore drift (p. 263)
revetment (p. 287)
rip current (p. 263)
seawall (p. 286)
storm surge (p. 282)
surf zone (p. 262)
tide (p. 264)

QUESTIONS FOR REVIEW

Check Your Understanding

1. The coastline is a place where all Earth systems interact. Describe a few of the interactions of Earth systems at coasts.
2. Waves are characterized by height, wavelength, and velocity. Describe these fundamental properties of a wave.
3. What is fetch? Why are the waves of the Pacific Ocean shore generally much bigger than those found on other oceans worldwide?
4. Surfers typically prefer plunging waves with a tube-like hollow space underneath the breaking wave. What does this suggest about the characteristics of the seafloor at popular surfing spots?
5. The text illustrates how rip currents return water back to the ocean from the beach through the surf zone. Despite this seemingly innocuous description, they remain very dangerous to swimmers. What is the driving force behind rip currents?
6. Why does sea level rise during warmer interglacial periods? Illustrate your answer with examples of the two major processes that contribute to this rise.
7. What are barrier islands? Are these generally made of solid material suitable for construction of long-term structures? Why or why not?
8. Many of the large bays and inlets of the U.S. Atlantic coast are now understood to be submerged river valleys. Explain how these would form in a warmer, interglacial climate.
9. Florida is one of the few places in the mainland United States where land is built primarily by the action of the biosphere. Illustrate and explain how land is made up of biologically derived material in this region.
10. Unlike so many of the rivers on the Atlantic coast of the United States, the Mississippi River has not produced a submerged canyon and accompanying bay/inlet. What has it produced instead? Why?
11. The Pacific coast is characterized by cliffs and rocky shorelines interspersed with sandy beaches. Why is this coastline so much more rugged and higher in elevation than its eastern cousin on the Atlantic coast?
12. The features of many beach and nearshore environments include longshore bars. These are located offshore about where waves start to break. Why do they form here?
13. The text states that 80 to 90% of *all* sandy beaches in the Unites States are currently eroding, and all of the country's shores are retreating at the rate of about 30 centimeters (12 in) per year. What are the causes of this erosion?
14. The Mississippi delta is known to be shrinking, decreasing habitat and human living area as well as increasing hurricane risk to inland communities such as New Orleans. Why is this region shrinking, and what influences have combined to reduce sediment input to this region?
15. Coastlines with shallow, gently sloping continental shelves usually experience the highest storm surges from hurricanes and winter storms. Why is this? Explain your answer in terms of storm surge generation and the style of breaking waves generated by seafloor of this kind.
16. There are many kinds of hard stabilization strategies for coastlines. Describe the benefits and drawbacks of three of them.

Critical Thinking/Discussion

17. Soft stabilization strategies for coastlines often include beach nourishment or replenishment by pumped-in sand. Miami Beach in Florida has been successful (at a cost of $64 million), whereas efforts at Ocean City, New Jersey (which has spent $5 million on the project) have not been. What does this suggest about the volume of sand required to successfully protect eroding beaches?
18. Deepwater waves approaching shore typically start to increase in height and break when the rising seafloor reduces the water depth to half the wavelength of the wave itself, leaving the water (and its energy) with nowhere to go but up. What does this imply about the wavelengths of tsunami waves generated by earthquakes—which can rise high enough to flood vast areas of coastlines—compared to normal deepwater waves?
19. Tidal range is known to be a function of local factors such as the shape of the coast—especially how shallow and restricted a given bay or inlet may be. Islands far out to sea typically do not see major variations in tidal range because of this fact. Given that tidal flats provide a unique habitat for many interesting species that have adapted to a periodically submerged life, where are the most extensive habitats for creatures like this likely to be found, and why?
20. Geoscientists who study coastal systems have identified *littoral cells*, or systems of beach sand circulation localized around a particular area. Each area or cell has a sediment input and outflow that makes up a sediment budget. The outflow of sand is typically a submarine canyon or similar feature. What is a common source of sediment supply to a littoral cell, and what changes do people make to this source of sediment that threaten beaches?
21. What factors do you think should determine whether a natural coastal erosion process constitutes a coastal erosion hazard, and who should make this determination? Consider how the answers to these questions might bear on the level of intervention people choose to make in coastal processes.
22. Why is hard stabilization of coastlines by seawalls known to be problematic? Outline the benefits and drawbacks of this approach in terms of wave energy and sand budget for neighboring beaches.

ANSWERS TO IN-CHAPTER INSIGHT QUESTIONS

What are some Earth systems interactions associated with waves along a coast?

P. 263

- Waves build and break as they interact with the submerged geosphere.
- Waves and nearshore currents they create move geosphere sediment on coasts.
- Waves erode the geosphere.
- The surf zone along a coast is unique biosphere habitat.

What are some Earth systems interactions associated with tides?

P. 266

- Currents created by tides move geosphere sediment.
- The intertidal zone is unique biosphere habitat.

What are some Earth systems interactions associated with sea level changes?

P. 269

- Climate fluctuations affect the amount of water in hydrosphere reservoirs, particularly ice sheets and oceans.
- Sea level changes either submerge or expose coastal parts of the geosphere.
- Sea level changes influence coastal biosphere habitat.
- Geosphere movements along coasts can change local sea level.

What are some Earth systems interactions in estuaries?

P. 272

- Mixing of freshwater and ocean water in estuaries creates unique biosphere habitat.
- Sediment transferred to the coast by the hydrosphere can accumulate in estuaries.
- Estuary habitat, such as wetlands, can influence water quality.
- People's wastes can contaminate geosphere sediment, degrade water quality, and damage biosphere habitat in estuaries.

What are some Earth systems interactions associated with coral reefs and mangroves?

P. 272

- Biosphere organisms transfer material from the hydrosphere to the geosphere at coral reefs.
- Mangrove plants help stabilize and trap fine sediment and protect shores from erosion during storms.

FIGURE 10-1 ▶ The River That Burned
Pollution of the Cuyahoga River in Ohio enabled fires to start on its surface several times. This photograph is of a fire in 1952.

CHAPTER

10

WATER RESOURCES

Did you know that rivers can burn? The Cuyahoga River, which flows through northeastern Ohio before emptying into Lake Erie, has caught fire at least 10 times during the last 150 years (Figure 10-1).

The Cuyahoga flows past many sources of pollution, especially in the industrial and urban center of Cleveland. For years, rivers like the Cuyahoga were treated as convenient waste disposal systems. Raw sewage, industrial wastes, oil processing residues, paint, and other pollutants were dumped in the river. By the middle of the 20th century, the Cuyahoga had become filthy. Nothing could live in it, and flammable waste floating on its surface repeatedly caught fire and burned.

The most recent fire on the Cuyahoga was in 1969. It wasn't the biggest ever on the river, but it became the best known. Because it was highly publicized, the nation came to realize that the Cuyahoga and other rivers like it were being seriously mistreated. Just three years later, in 1972, Congress passed the landmark legislation that became the Clean Water Act. Since then, rivers cannot be treated as all-purpose dump sites; water that discharges to rivers or other surface waters must be permitted by regulatory agencies. Bodies of water like the Cuyahoga are closely monitored to ensure that new pollution is not occurring, and clean-up efforts are bringing life back to once-dead rivers. The Cuyahoga still has some problems, but it's definitely not going to burn again.

Protecting rivers like the Cuyahoga is part of a much bigger challenge. Remember the water cycle from Chapter 2? What happens in one part of the hydrosphere almost always affects other parts. Rivers in particular are key links between the land in watersheds, surface waters such as lakes, and underground water. As life depends on water, its quality and distribution are critical to sustaining a healthy biosphere—which includes us. However, many areas and populations suffer from lack of abundant, clean water, and the demands for water are increasing. People need and depend on water in many ways, not all of which are obvious.

Earth is unique in the solar system—the only planet with liquid water on its surface. The hydrosphere sculpts the landscape, helps regulate global climate, and sustains life. Our bodies are about 60% water, and people need some 2.9 liters (more than 3 quarts) of water each day from food and liquids to maintain good health. Deprived of water for a week, you will die. Where clean water is scarce, poor public health is usually the result. The availability of water controls how much food a country can provide to its people, determines whether and where it can develop industry, and strongly influences where people live.

Earth's water resources are part of a closed system: Water is not physically destroyed when it is used. This is why water is considered a renewable resource. But there is a catch. People can do things to water that change its availability and usefulness in ways that hundreds or even thousands of years can't repair. For this reason, being careful stewards of Earth's water resources is very important. Although sustaining these resources is possible, it is not always being accomplished.

IN THIS CHAPTER YOU WILL LEARN:

- The characteristics of various water resources, including surface water and groundwater, and the connections between them
- How Earth's water resources are distributed and how much water is available for people's use
- How people use water, both in familiar ways and in ways with which you may be less familiar, such as electric power generation
- The environmental impact of developing water resources by means of dams and groundwater pumping
- How natural water quality varies, what water pollution is, what common water pollutants are, and where pollutants come from
- How people can sustain water resources through careful use, prevention of pollution, and conservation

10.1 Water Resources

The historian A. Trevor Hodge once remarked that the most representative symbol of the Roman civilization is not the massive Colosseum (site of the notorious gladiator contests) but the aqueduct. "The aqueducts went everywhere Rome went, an outward symbol of all that Rome stood for and all that Rome had to offer," Hodge wrote in *Roman Aqueducts & Water Supply.* Essentially, the aqueducts were artificial rivers that conveyed spring or river water across the land to where it was needed, whether a village, a city, or a farm (Figure 10-2). They were a reminder to all of Roman power, ingenuity, and dominion over the empire. With a constant flow of clean water came amenities such as sanitary sewers, public baths and toilets, and (for an affluent minority of citizens) indoor water supplies. Aqueducts also delivered water to drive flour mills and to irrigate crops. Then, as now, the challenge to communities was, commonly, transporting water to where it was needed.

FIGURE 10-2 ▲ Water for the Empire
Two thousand years ago, aqueducts like this one brought freshwater to far-flung regions of the Roman Empire. Built to last, many of these structures still survive as monuments to Roman planning and engineering.

Do you think getting water is easy? For many of us all it takes is turning on the faucet. But providing clean, fresh (non-salty) water everywhere people need it isn't really that

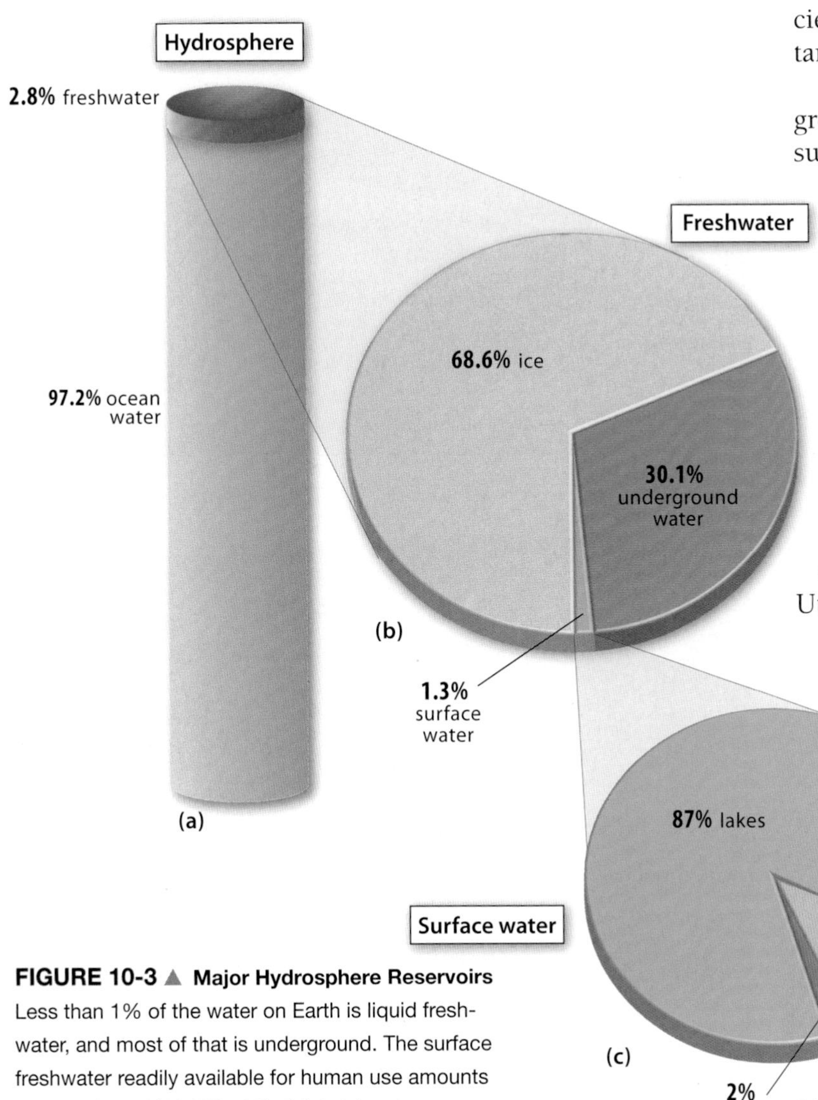

FIGURE 10-3 ▲ Major Hydrosphere Reservoirs
Less than 1% of the water on Earth is liquid freshwater, and most of that is underground. The surface freshwater readily available for human use amounts to only about 3/10,000 of Earth's total water.

simple. Over one billion people—a sixth of Earth's population—do not have access to adequate supplies of clean water. A main reason is that there really isn't that much freshwater available.

The hydrosphere does contain a tremendous amount of water—about 1360 million cubic kilometers (326 million mi^3). That is enough water to cover the entire United States to a depth of 145 kilometers (90 mi). Put another way, it is about 171 million gallons of water per person on Earth. So there is a lot of water. The problem is that, as Figure 10-3a indicates, 97.2% is ocean water, which is too salty to drink or grow crops with. Just 2.8% is freshwater.

The situation is even more challenging, as over two-thirds of Earth's freshwater is trapped in glaciers and ice sheets—mostly in the Arctic and Antarctic, far from large populations. The result is that only about 0.9% of all of Earth's water is liquid freshwater that people can potentially use. And as you can see in Figure 10-3b, almost all of this nonfrozen freshwater is underground. People mostly use surface water from rivers and lakes (Figure 10-3c), which contain less than 0.04% of Earth's water. This relative scarcity makes clean surface water valuable and explains why people around the world today, like the ancient Romans, capture surface water and transport it great distances to where it is needed.

Another important water resource for human use is groundwater. People are already relying on groundwater where surface-water supplies are insufficient. In many places, the use of surface water affects groundwater and vice versa. Sustaining water resources requires understanding surface water, groundwater, and how they are connected.

SURFACE-WATER RESOURCES

Surface waters are bodies of water—the oceans, springs, creeks, rivers, ponds, and lakes scattered across the landscape—that you can reach out and touch, perhaps even swim in or fish in. Most contain freshwater, but in some places, like the enclosed surface basins of the arid west, evaporation has turned lakes salty. The Great Salt Lake in Utah (Figure 10-4) is a well-known example.

Many societies, just like the Romans, have altered the landscape to exploit surface-water resources. People have dug canals and constructed aqueducts to carry the water; they have built artificial reservoirs to store spring rainwater and snowmelt for use in drier seasons. Cities located on or near large lakes or rivers draw water directly into water treatment facilities and purify it for drinking. All these things have happened along the Colorado River.

The Colorado River

The Colorado River flows 2300 kilometers (1430 mi) from its headwaters in the Wind River Mountains of Wyoming to a muddy delta on the Gulf of California. We discussed the challenges of managing Colorado River water in Chapter 2. To use this water people have constructed 19 dams and hundreds of kilometers (miles) of

FIGURE 10-4 ▲ The Great Salt Lake
Because of its high evaporation rate, Utah's Great Salt Lake is typically 3 to 5 times saltier than Earth's oceans. The increased density of the salty water makes floating easy.

FIGURE 10-5 ▶ Parceling Out the Colorado

(a) People have built dozens of dams, reservoirs, diversions, canals, and aqueducts to control flooding on the Colorado River and distribute its waters to thousands of municipalities and farms. This is the Palo Verde Dam, which diverts about 50 cubic meters (1800 ft^3) of water per second to irrigate 490 square kilometers (120,000 acres) of land on the west side of the Colorado River in California. **(b)** Because so much water is diverted for human use, the river that finally reaches Mexico and the sea is shallow and salty.

(a)

(b)

diversions and aqueducts (Figure 10-5a). So many parties tap this river for so much water that in some years, barely a trickle makes it to the sea (Figure 10-5b).

The competition for Colorado River water is tremendous, and it remains a great challenge to manage these water resources sustainably. Through seemingly ceaseless litigation, every drop of Colorado River water has been allocated. In fact, more water has been allocated than the river carries on average—a feat possible only because some water withdrawn for irrigation eventually finds its way back to the river.

A portion of the irrigation water spread on fields migrates underground or runs off the surface and flows back to the river. As this process occurs, evaporation increases the water's salt content, called its **salinity**. The salinity is only 50 **parts per million (ppm)** upstream in Wyoming but in lower parts of the river it reaches 1500 ppm or more. [One part per million is a measure of concentration that is equivalent to about one drop in 13 gallons (50 L) of water.]

Water with salt concentration greater than 550 ppm is not healthy for people, and water with salinities of 750 ppm and higher is unsuitable for either human consumption or irrigation. High salinity created a problem when water from the Colorado River was found to be damaging Mexican crops. Mexico reached a deal with the United States and in the early 1990s the United States constructed a large and expensive desalination plant to clean up Colorado River water before Mexico received it (Figure 10-6). The great cost of the plant's operations, combined with a few years of high-flow conditions on the river, caused the facility to be shut down and mothballed. However, overuse of Colorado River water has continued and the plant appears to have a future in helping manage the region's water resources. Desalination plan for desert

The Colorado River is just one example of how people have come to manage water resources. Dams, canals, aqueducts, and reservoirs move surface water to where it is needed for drinking, agriculture, and industrial use around the world. In the United States and other industrialized countries, the 20th century saw enormous water development projects, far beyond the scale of anything even the Romans could imagine. As a result, major river systems have been fully developed as water supplies.

Along with the development of water infrastructure in the American West came a system of water rights laws. Water was converted into a form of property. The result is that today, for every major water supply in the west, individuals or companies literally own a certain amount of the river's annual flow. In a

FIGURE 10-6 ▼ The Colorado River Desalination Plant

This plant was constructed to treat river water that had become salty through irrigation practices.

river system fully developed for water supply, like the Colorado, there is no more water to be allocated—all that is possible is trading and purchasing established water rights. The era of giant water development projects and massive dam building is effectively over in the United States. Or is it? Consider the case of the Great Lakes.

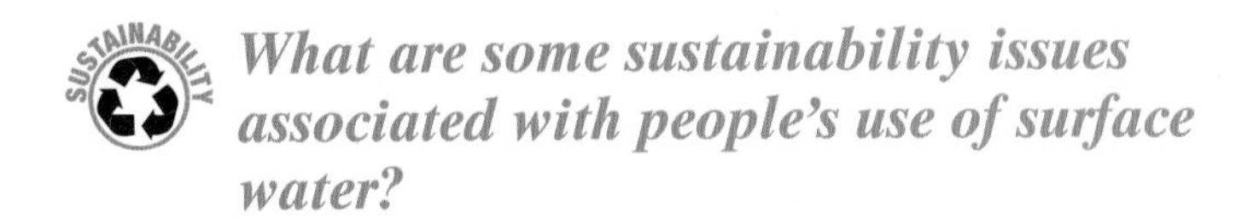

What are some sustainability issues associated with people's use of surface water?

In the News

Great Lakes Water Wars

Dozens of states have predicted they will have water shortages within the next 10 years. And the world as a whole needs water. The World Bank says that 4 billion people will have serious water needs by 2025. It's not surprising that once in a while people needing water focus their attention on the Great Lakes. A nation's growing thirst threatens a Great Lakes water war

The Great Lakes, bordered by eight states and two Canadian provinces, make up the largest group of freshwater lakes on Earth. They hold about 20% of Earth's fresh surface water, enough to cover the entire continental United States to a depth of 2.9 meters (9.5 ft). These five lakes have had a checkered environmental past as pollution, invasive species, and over-fishing have taken their toll, but they are a remarkable water resource.

But diverting Great Lakes water is a very sensitive subject. Previous attempts to divert these waters to areas outside the Great Lakes basin have been opposed or found to be much too expensive to justify. Now the states and provincial governments surrounding the lakes are trying to work together to prevent misuse of the water. However, developing agreements also requires the support of the Canadian and U.S. federal governments. The plans and regulations governing the Great Lakes water are a work in progress. Great Lakes – St. Lawrence River Basin Water Resources Compact Implementation It's good that planning is under way, as many people are expected to compete for this valuable resource in the future. Time will tell who ultimately controls this water and how it is used.

The Great Lakes, with a total shoreline of nearly 160,000 kilometers (10,000 mi), contain about 20% of the Earth's fresh surface water.

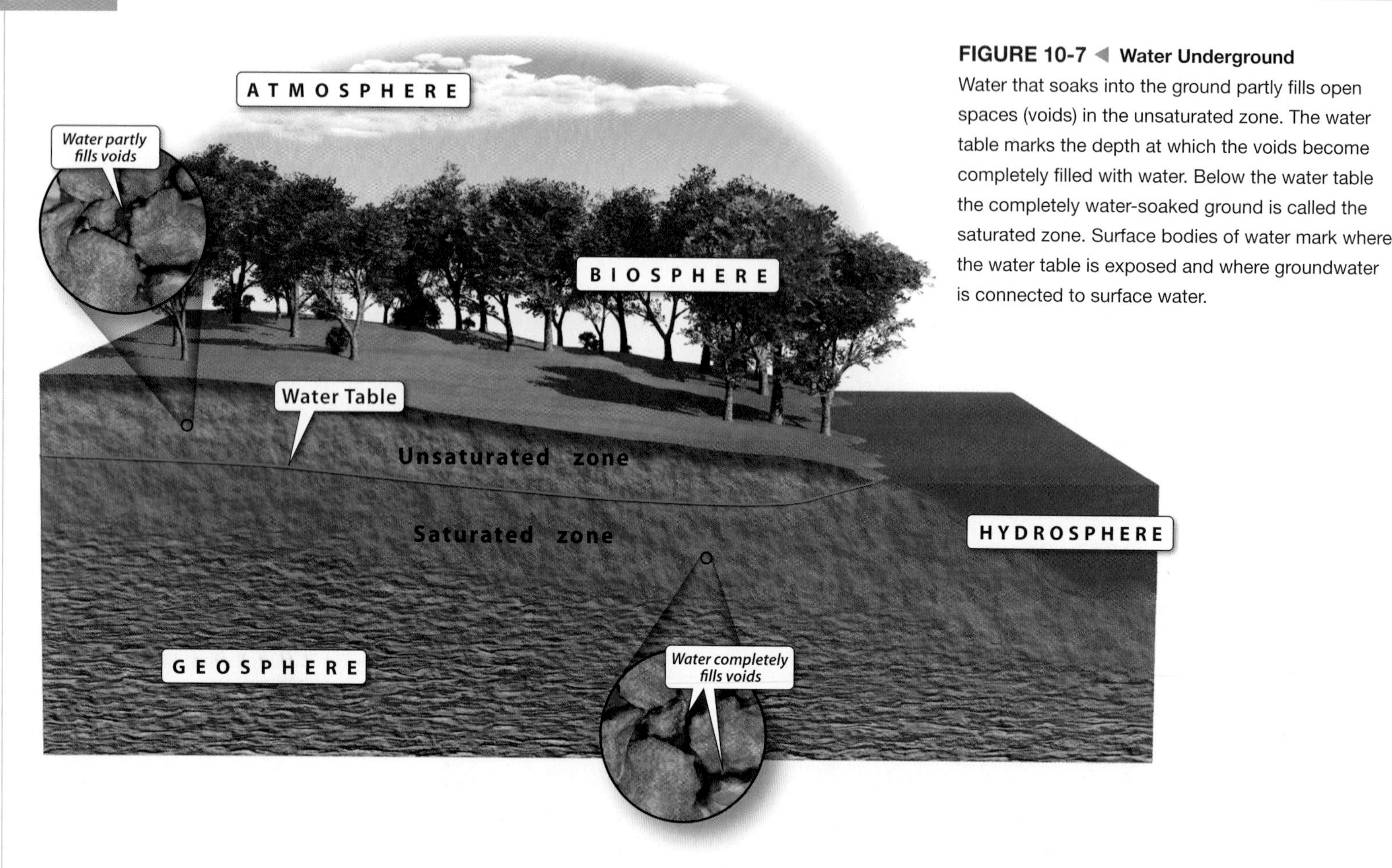

FIGURE 10-7 ◀ **Water Underground**
Water that soaks into the ground partly fills open spaces (voids) in the unsaturated zone. The water table marks the depth at which the voids become completely filled with water. Below the water table the completely water-soaked ground is called the saturated zone. Surface bodies of water mark where the water table is exposed and where groundwater is connected to surface water.

GROUNDWATER RESOURCES

Water that infiltrates the ground moves downward until it reaches a depth at which it completely fills all open spaces: voids as big as caverns, fractures in rocks, even tiny cavities between mineral grains. When all these spaces are completely filled, the ground is *saturated* with water. This water is termed **groundwater**.

The top of the *saturated zone* is called the **water table** (Figure 10-7). From the water table to the surface, the open spaces in the ground are only partly filled with water. This is the *unsaturated zone*, where some air is also present in the voids. The depth of the water table changes as the amount of available water changes. Drought conditions or water withdrawals from wells can lower the water table. On the other hand, extended periods of heavy rains may raise the water table. The water level in a well is at the water table, and the surface of streams and lakes are a visible part of the water table.

Earth materials differ in the amount of water they can store and the ease with which the water can be withdrawn, but many types of earth materials can function as underground reservoirs for water. If the reservoir can produce water at a quantity and rate useful to people, it is considered an **aquifer**. Whether or not an earth material saturated with groundwater can function as a useful water supply is determined by its *porosity* and *permeability*.

Porosity is the percentage of the earth material consisting of void spaces. The voids may take the form of the small openings (pores) between particles of sand, silt, clay, or gravel, or as fractures and hollows within a body of solid rock. If half of the material is void space, then the porosity is 50%. Fifty percent porosity is very high—porosities in the 20% to 30% range are considered excellent for aquifers.

High porosity alone does not mean that a groundwater reservoir is suitable as an aquifer, however. If the void spaces are not well connected, it will be difficult to recover useful amounts of water. To be an aquifer, the material must be both porous and permeable. **Permeability** is a way of measuring how easily water passes through an earth material. The more connections there are between the void spaces, the greater the material's permeability and the easier it is for water to flow through the reservoir.

Among the best aquifers are thick beds of sand and gravel. In this type of earth material, the void spaces are well connected, allowing groundwater to flow easily. Rock can also be sufficiently porous and permeable to serve as an aquifer. Some of the most productive aquifers on Earth are limestones in karst terrain (Chapter 8), where dissolution has created many connected underground openings, channels, and caves. Much of Florida depends on limestone aquifers, for example. Any rock that is fractured can also be an aquifer.

Aquifers can be classified on the basis of their relationship to the geological environment, as shown in Figure 10-8. An **unconfined aquifer** is overlain by permeable earth material. If a low-permeability formation overlies an aquifer, it is called a

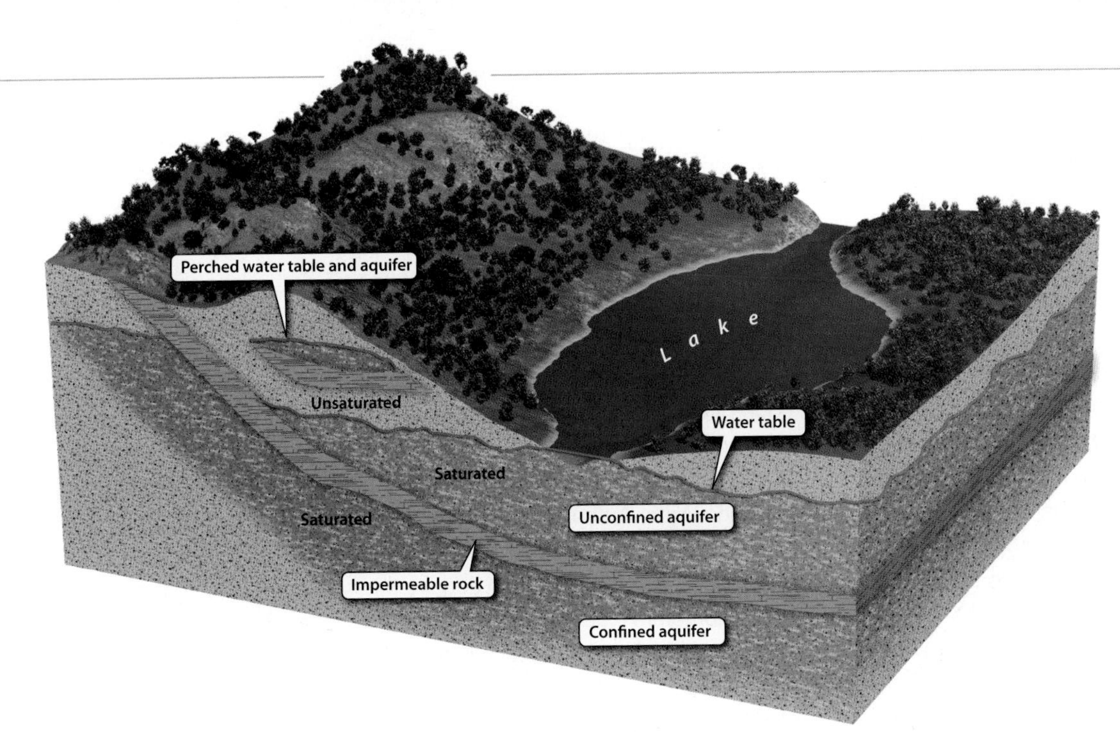

FIGURE 10-8 ◀ Aquifers
An unconfined aquifer can be easily recharged by water migrating freely down to it from the surface. A confined aquifer contains water below an impermeable layer. Because it is largely overlain by impermeable material, a confined aquifer can fill with water only through limited areas that are exposed at the surface–sometimes at a considerable distance from the main body of water. The water in such an aquifer is typically under pressure. A perched aquifer develops in an unsaturated zone where downward movement of water is stopped by an impermeable layer before it can reach the water table.

confined aquifer. In some places, a local impermeable barrier (such as a clay-rich layer) can inhibit downward movement of water in the unsaturated zone, creating a *perched aquifer* above the water table.

Whether an aquifer is confined or unconfined helps us understand how water gets into the aquifer in the first place. In unconfined aquifers, water can infiltrate downward directly from the surface. Precipitation that soaks into the ground migrates through the unsaturated zone to the water table and fills or "charges" the aquifer. By contrast, only parts of confined aquifers are located where water can directly migrate into them from the surface and charge the aquifer. This is why areas where water migrates into an aquifer are so important—it is here that the aquifer can be recharged if it has lost water. Aquifer recharge areas need special protection from damage or pollution, as what happens in a recharge area can come to affect the whole aquifer.

Aquifers can extend for hundreds of kilometers (miles) and be hundreds of meters (yards) beneath the surface. An excellent example is the High Plains aquifer, which underlies 450,000 square kilometers (174,000 mi^2) from Nebraska to Texas.

The High Plains aquifer

The High Plains aquifer underlies portions of eight states; it is critical to maintaining crop yields in one of the most productive agricultural regions in the world. The natural climate of the region is arid or semi-arid, so irrigation is essential to high crop yields. Farmers started to withdraw significant amounts of water from the aquifer in the 1930s. By 2002, the water was used to irrigate 5.14 million hectares (12.7 million acres) of alfalfa, corn, cotton, sorghum, soybeans, and wheat.

The aquifer is in mostly unconsolidated sand and gravel layers of the Ogallala Formation, eroded from the Rocky Mountains long ago. It is an unconfined aquifer, underlain by impermeable bedrock. As vast as the aquifer is, water tables began to decline soon after large-scale pumping started. As you can see in Figure 10-9, by 2005 the water table had dropped an

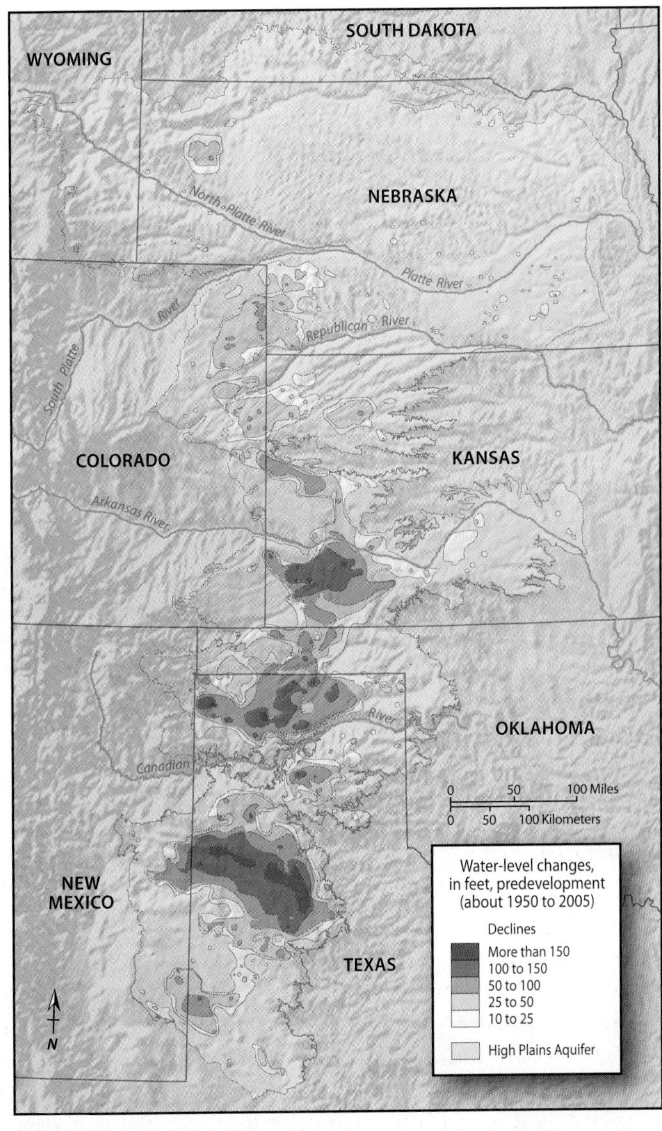

FIGURE 10-9 ▲ The High Plains Aquifer
Groundwater withdrawal from this unconfined aquifer has lowered the water table over 46 meters (150 ft) in places.

average of 4 meters (13 ft), and more than 30 meters (100 ft) in portions of Texas, Oklahoma, and Kansas. | USGS SIR 2006-5324: water-level changes in the High Plains aquifer | In these areas, recharge couldn't keep up with withdrawals and the groundwater was essentially mined. This is not sustainable management of an aquifer.

You Make the Call

Who Wins in Las Vegas?

Surface water is very scarce in the arid desert of southern Nevada, but groundwater aquifers are present in many different watersheds around the state. In the heart of the southern Nevada desert sits the city of Las Vegas, a virtual oasis. It's best known for its large hotels and casinos that thousands upon thousands of people visit each year. Las Vegas tries hard to lure visitors; it advertises its many attractions on national television.

The attractions include the world's most lavish water "features"—outdoor water decorations that flow, spray, and gurgle water 24 hours a day, 7 days a week. The most spectacular water feature is at the Bellagio, where an eight-acre lake contains over 1000 water "expressions" like the one shown in the photo—water jetted into the air to create a multitude of changing forms, all choreographed to music. | WET Design: water fountain and water features development for the Bellagio Hotel | Owners say that most of the water they use is for people inside the hotels and casinos, not to keep the water features going. But water features like the one at the Bellagio are fairly conspicuous in a city that's running out of water.

Las Vegas is also attractive to new residents; the 5000 or so people who move there each month make it the fastest growing city in the United States. In 2003, the USGS warned Las Vegas that it was going to run short of water by 2025 unless new supplies and conservation measures were developed. Some of the city leaders don't think they have a problem. The mayor has been quoted as saying they can "buy" their solution.

Las Vegas's water supplies are the responsibility of the Southern Nevada Water Authority. This agency has been actively buying up land in central Nevada valleys where groundwater resources are available. The authority intends to acquire this water and send it through 400 kilometers (250 mi) of pipelines to Las Vegas. The farmers and ranchers who live and work in the valleys targeted by Las Vegas are concerned. They fear that withdrawing the local groundwater will upset the balance between discharge and recharge in the aquifers and have negative consequences for habitat and farming.

Legally, groundwater in Nevada is a state resource. The ultimate decision about how much groundwater is withdrawn and sent to Las Vegas lies with Nevada's state engineer. Hearings allow some input from concerned citizens, but the state engineer makes the decision. This is a challenging decision that is heavily influenced by economic and political factors. | Desalination gets a serious look—*Las Vegas Sun* |

- If you were considering moving to Las Vegas, how would water availability influence your decision?
- Should communities manage water use within their jurisdictions? If so, how?
- How do you think groundwater resources should be allocated in southern Nevada or elsewhere around the country?

Displays like this water feature at the Bellagio Hotel in Las Vegas, Nevada raise questions of priorities and allocation of water resources in a very dry region.

SURFACE AND GROUNDWATER CONNECTIONS

So far we have discussed surface water and groundwater separately. In nature, however, surface-water and groundwater systems are closely connected. Indeed, they are part of the same system of freshwater storage and movement through the hydrosphere.

Commonly, the water table marks the point of contact between groundwater and surface water. Where the water table is close to the surface, it may intersect the bed of a stream, allowing water to move between the surface and the subsurface. In a **gaining stream**, groundwater discharges into the streambed. In some areas of the United States, such as a large portion of the Midwest stretching from Nebraska to Michigan, groundwater discharge is the main water supply for streams. Alternatively, the streambed may be a recharge point for the local groundwater system. Such a stream is known as a **losing stream**, because it loses water into the ground.

An excellent example of a groundwater reservoir that is recharged by losing streams is the Edwards aquifer in Texas. As you can see in Figure 10-10, this aquifer lies in karst

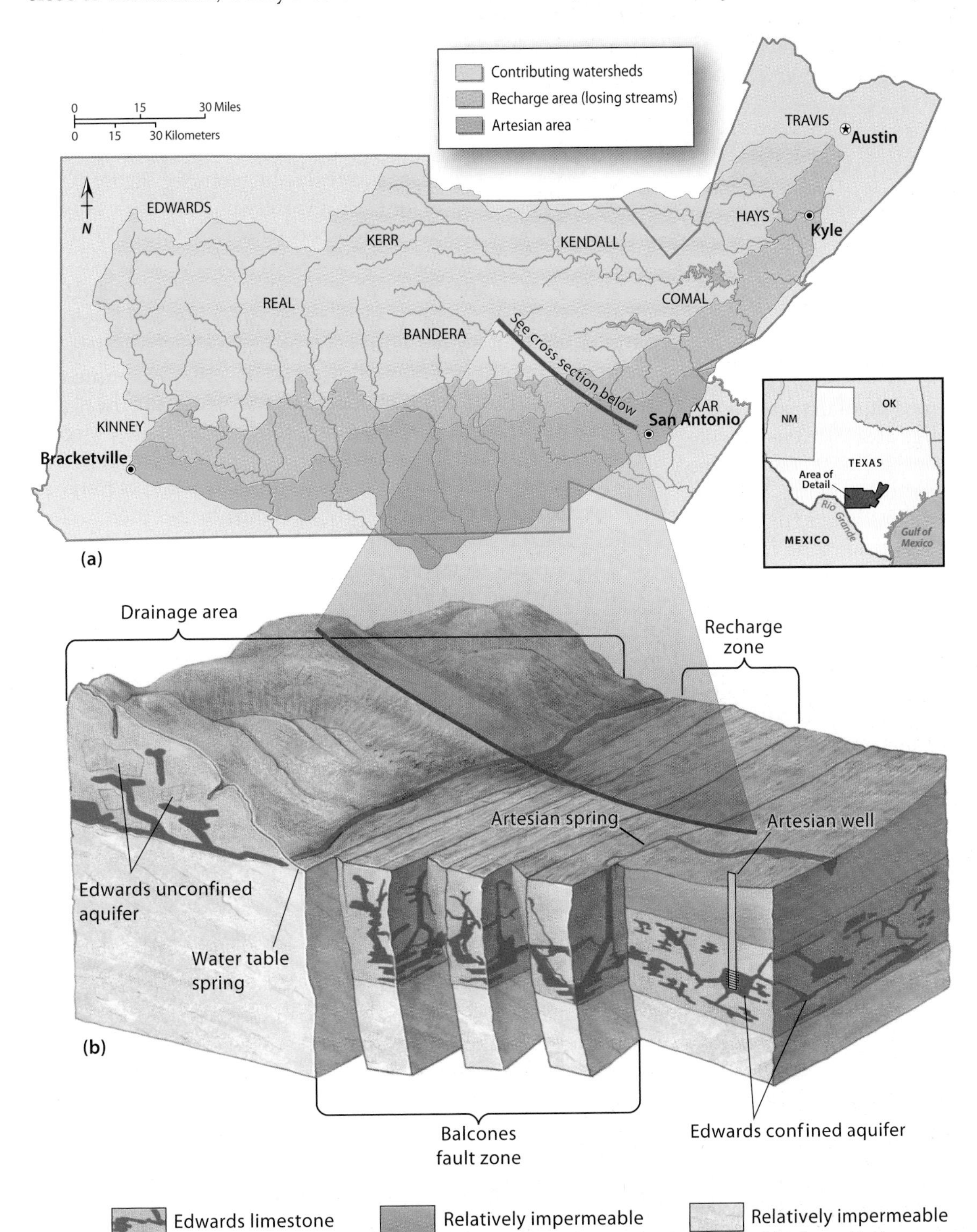

FIGURE 10-10 ◀ The Edwards Aquifer
This confined aquifer system extends across seven Texas counties **(a)**. It provides the water for the famous River Walk in San Antonio. Surface water in the recharge area comes from a northern part of the aquifer (called drainage area) that is separated from the southern part by uplift along a series of faults **(b)**.

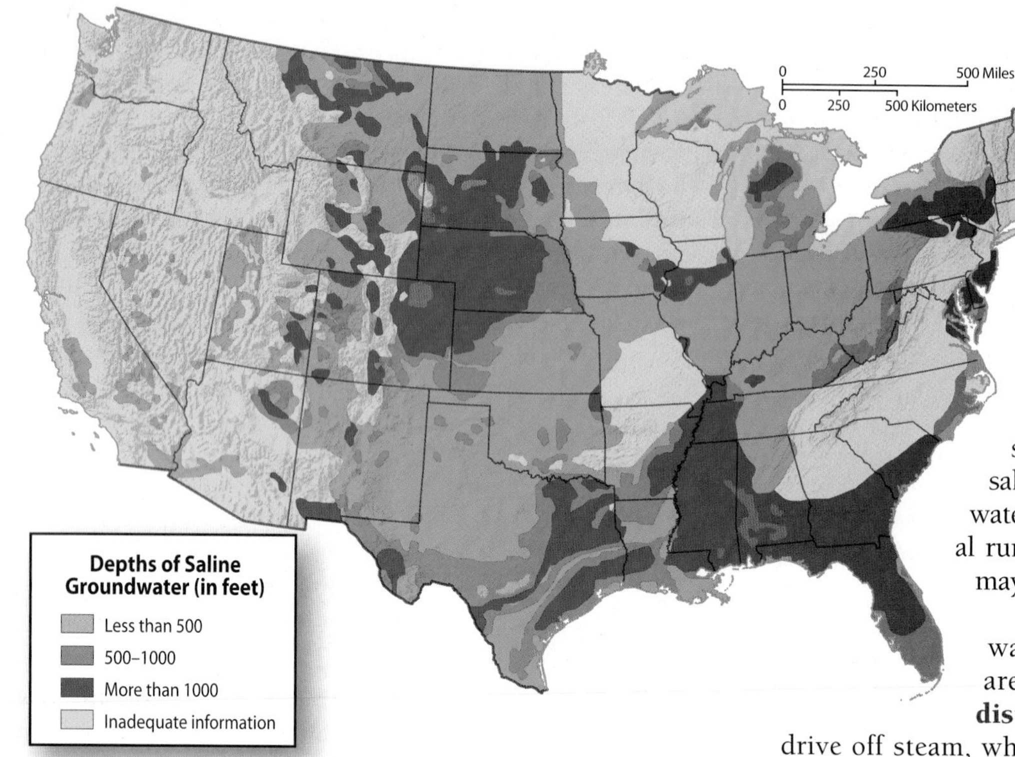

FIGURE 10-11 ◀ Saline Groundwater in the United States
Saline groundwater, a potential water resource, can be found at moderate depths through large parts of America.

terrain near Austin and San Antonio in west-central Texas. The Edwards aquifer website It is a confined aquifer that extends south in the subsurface from its recharge area. Streams flowing across the aquifer's limestone outcrops in its recharge zone (green in Figure 10-10) lose water to the aquifer. In some places natural springs form south of the recharge zone where the aquifer comes back to the surface (the tan zone in Figure 10-10). These springs are examples of **artesian flow**—flow from a confined aquifer in response to pressure developed by the weight of the water in higher parts of the aquifer. The pressure in the aquifer forces water up to the surface at springs or up well bores drilled into the aquifer (see Figure 10-10). If the pressure is great enough, water will flow to the surface in *artesian wells* drilled into a confined aquifer's artesian area.

The shore of a lake, wetland, or estuary intersects the local water table. Indeed, the surface of the lake or other body of standing water is essentially the water table exposed to the atmosphere. Clearly, then, if the water table drops, so does the lake level, and vice versa. Wetlands are particularly vulnerable to changes in the local water table, since many wetlands depend primarily on groundwater seepage for their water.

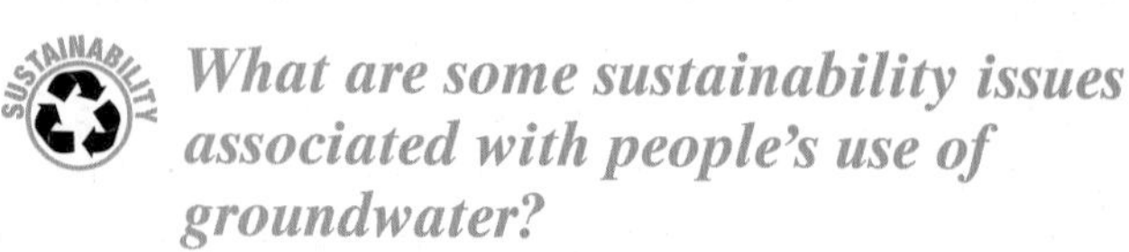

MAKING FRESHWATER—DESALINATION

Not all freshwater has to come from streams, lakes, or underground. Freshwater can be made from salty water by desalination processes that remove dissolved mineral salts. Most people are familiar with the desalination of seawater to produce drinking water. But reclaimed wastewater and agricultural runoff, like that in the lower Colorado River, may also require desalination (see Figure 10-5).

There are a variety of ways to desalinate water. The two most commonly used are distillation and reverse osmosis. In **distillation**, the salty water is heated to drive off steam, which is pure freshwater; the steam is then condensed back into a liquid. What remains is concentrated brine, containing sodium chloride (salt) and other minerals, heavy metals, and any chemicals used to pretreat the water before desalination (for example, to remove marine organisms from seawater).

In the **reverse osmosis** process, the salty water is pumped through special filters. The tiny openings in the filters allow water molecules to pass, but not salt and other minerals. As in distillation, the water is pretreated to remove large particles, such as bacteria and suspended solids, to prevent the filters from becoming clogged.

The amount of freshwater produced by desalination depends on the process used, ranging from 15 to 50 gallons of drinkable water per every 100 gallons processed. Under certain conditions, desalination is an economically viable way to supplement or supply freshwater. There are over 7500 operating desalination plants around the world. Not surprisingly, over 60% are located in the Middle East, where water needs are high and energy costs are low. Seawater desalination chapter 1

In the United States, the number of desalination plants is increasing. In 2004, the California Coastal Commission was reviewing proposals for 21 new desalination plants with a total capacity of 240 million gallons of water per day. These would more than double the number of existing seawater desalination plants on the California coast, which have a total capacity of about 4 million gallons per day. However, not all salty water comes from the ocean. A tremendous amount of groundwater is salty—not commonly as salty as ocean water, but containing enough salt and other dissolved solids to be unusable. The salty groundwater that is widespread at moderate depths in the central United States (Figure 10-11) is a potential water resource if desalination technology continues to advance and becomes more economical. | USGS Colorado water resources publication – desalination of groundwater: Earth Science perspectives |

The great potential value of desalination is that it could supply a virtually unlimited amount of freshwater. The main drawback is cost. Building the plant is a one-time expense, but most desalination methods also consume significant amounts of energy, which accounts for up to 50% of the cost of desalination. However, the overall cost of desalination is falling. A plant proposed for the Orange County (California) Municipal Water District would produce water at a maximum cost of about $1000 per acre-foot. (An acre-foot, the amount of water that would cover an acre to a depth of one foot, is equal to 325,821 gallons or 1233 cubic meters.) A decade ago, the costs were as high as $4000 per acre-foot. As a result, the difference in cost between desalinated water and imported groundwater has narrowed. In 1993, desalinated water in California was 300 times more expensive than imported groundwater. By 2004, it was only about twice as expensive.

Use of alternative power sources also has the potential to make desalination available to more people. In 2003, Abu Dhabi brought online a solar-powered desalination plant that produced 22,500 gallons of freshwater per day. This is enough to supply a village or town but not a large city. Meanwhile, there are serious discussions and planning in the works to harness nuclear power plants to drive desalination. It looks like desalination will be a bigger source of freshwater in the future.

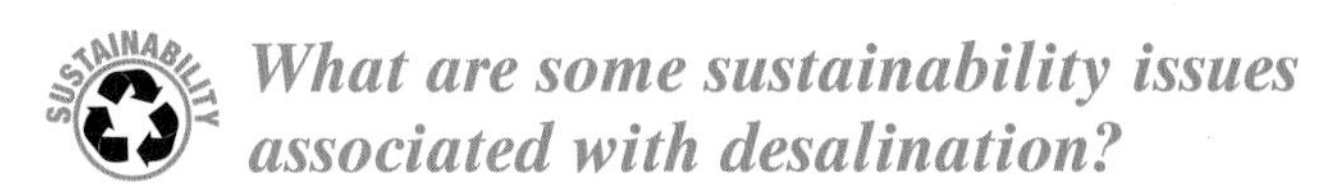

What are some sustainability issues associated with desalination?

10.2 How People Use Water

Think of the various things you do every day. In one way or another, virtually every one of them involves water. Flip the bathroom light on: It requires 21 gallons of water to produce 1 kilowatt-hour of electricity at thermoelectric power plants. Flush the toilet: Several gallons go down the sewer pipe. Take a nice long shower: Up to 50 gallons disappear down the drain. Then drive to school or work in a car: It took about 40,000 gallons of water to manufacture your car, including the tires.

Vast amounts of water are withdrawn from surface and groundwater reservoirs for human use, but most of the water is not destined for households. Growing crops and electrical power generation account for most withdrawals of freshwater. The water needed to cook, wash, drink, shower, garden, and wash cars accounts for a relatively small fraction of the total amount of freshwater withdrawn from the environment. However, personal water use adds up to about 80 gallons per day—or 200 gallons per day if outside uses like car washing are included. Did you know that you personally use this much water?

WATER USE IN THE UNITED STATES

Reliable, accurate statistics on water use in the United States are available thanks to the U.S. Geological Survey (USGS), which has tracked domestic water use since 1950. | Estimated use of water in the United States in 2005 | The sheer volume of water needed to run a country with the area and population of the United States is extremely large. In 2005—the most recent year for which there are statistics—the United States, Puerto Rico, and the Virgin Islands used over 400 billion gallons (1.5 billion m^3) of water each day. This adds up to about 1400 gallons per person per day. The leading water users are the electric power industry (48% of the total withdrawals) and crop irrigation (34%).

Water use by the power industry is an example of **nonconsumptive use**. After being used for steam generation and power plant cooling, the water can be used for other purposes or just discharged back to its source after cooling. Crop irrigation, on the other hand, is an example of **consumptive use**—the water is spread on fields, soaks into the ground, evaporates, or gets taken up by plants. Consumptive use disperses water to other system reservoirs. Your personal water use in your home, a consumptive use of about 200 gallons per day, is just 14% of the total water used to provide the goods and services you depend on each day.

What are some Earth systems interactions associated with consumptive water use?

FRESHWATER USE

Saltwater makes up only about 15% of the total water withdrawals in the United States; the other 85% is freshwater from surface or groundwater sources. In 2000, the United States and its territories withdrew an average of 320 billion gallons (1.3 billion m^3) of freshwater per day (Figure 10-12, p. 308). Of this amount, about 76% was from surface sources and the remainder from groundwater.

Public water supplies

Most of the water individuals use daily comes from municipal water utilities and other public sources—what many people refer to informally as city or town water. Utilities withdraw water from lakes and rivers and process it to filter out contaminants or kill dangerous microorganisms (Figure 10-13a, p. 308). This is what makes publicly supplied tap water safe to drink.

More than 140 million people in the United States, or about half the U.S. population, rely on groundwater for their drinking supply. This includes almost all the people living in rural areas, where water from wells is the main source of freshwater. In 2005, some 43 million Americans used self-supplied water, 98% of which came from household wells.

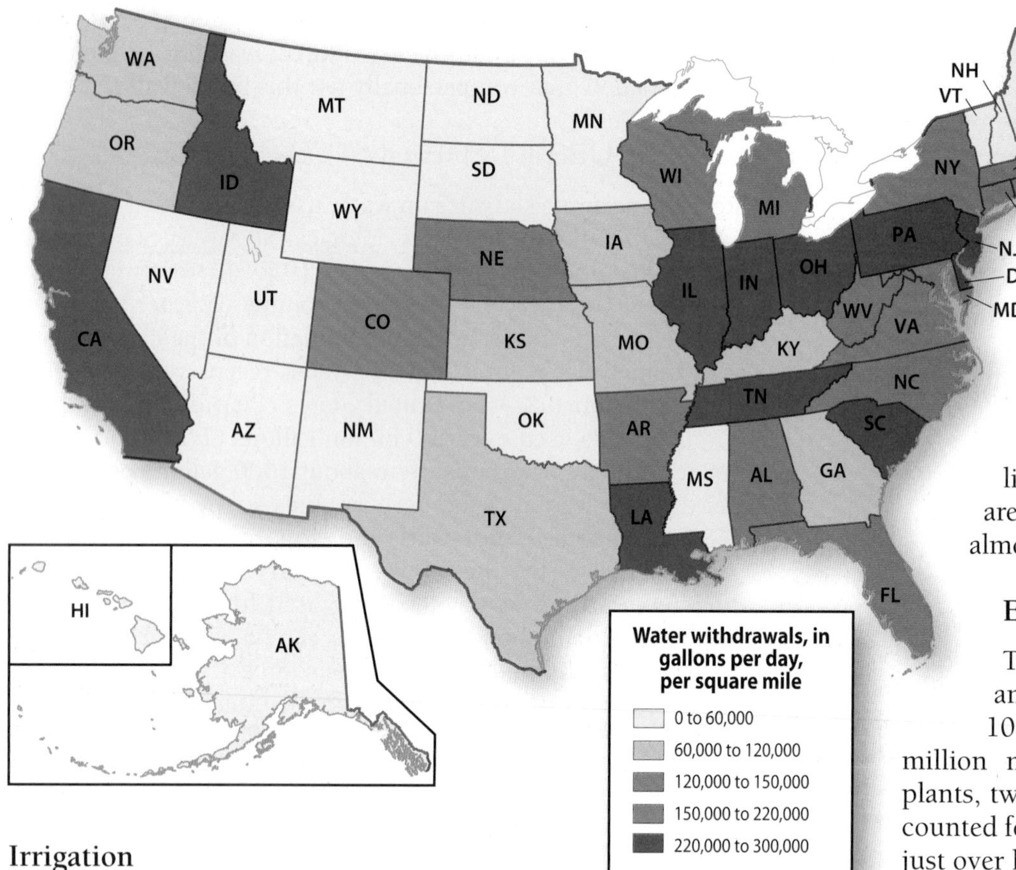

FIGURE 10-12 ◀ A Thirsty Nation Freshwater withdrawals in the United States, by state. Note that the withdrawals given in the key are per square mile. Altogether, the total daily withdrawal is about 350 billion gallons (1.3 trillion L).

Irrigation

Obtaining safe drinking water is a priority, but in rural America obtaining water for irrigation is also very important. Family farmers typically obtain their drinking water from a self-supplied source, commonly a well. Their livelihoods depend on large amounts of affordable water for irrigation. For every liter or gallon of water withdrawn to supply public water systems, 3.5 liters or gallons are taken for irrigation.

Irrigation accounts for 31%, or 128 billion gallons (485 million m^3) per day, of the freshwater withdrawn annually in the United States. This figure includes water used to water public parks, cemeteries, plant nurseries, and golf courses. The water may be sprayed or dripped directly on crops or spread across the land from canals and irrigation ditches (Figure 10-13b). As in the case of drinking water, most irrigation water comes from rivers and other surface sources. About 25 million hectares (62 million acres), or 16% of the total U.S. cropland, are irrigated. These irrigated lands account for almost half the value of annual crop production.

Electric power generation

The electric power industry taps enormous amounts of water to cool its facilities (Figure 10-13c). In 2005, 201 billion gallons (760 million m^3) were withdrawn for electric power plants, two-thirds of which was freshwater. This accounted for 40% of total freshwater use annually and just over half of the surface water that is withdrawn.

Other water uses

Public water supplies, self-supplied water, irrigation, and electric power generation account for about 93% of all freshwater used annually. The rest is used for watering livestock, operating fish farms and other forms of aquaculture, mining, and miscellaneous industries that operate their own water supplies, either from surface or ground sources. Meat processing, for example, requires significant amounts of freshwater, as does the manufacture of paper and chemicals. In 2005, industry used 18.2 billion gallons (69 million m^3) per day, which is 4% of the total freshwater withdrawn per year in the United States.

FIGURE 10-13 ▼ Uses of Freshwater
The water people use in their homes, commonly drawn from surface water reservoirs **(a)**, is a small part of their total water use. Irrigation **(b)** and power generation **(c)** are two other large uses of freshwater.

(a)

(b)

(c)

10.3 Water Withdrawal and the Environment

The process of withdrawing and transferring water to where it is needed has environmental consequences. These are independent of people's effects on water quality, such as their impact on the Cuyahoga River, which you learned about at the beginning of this chapter. Dams and related developments such as aqueducts change stream-flow characteristics. Groundwater pumping can deplete an aquifer, damage an aquifer, and cause land subsidence.

DAMS

Dams are about the most obvious thing people construct to manage water (see Figure 10-14a). They have several purposes: controlling floods (Chapter 7), generating hydroelectric power, providing recreation opportunities, and, of course, supplying water. A dam is commonly just the first step in controlling surface water, as canals and aqueducts take water hundreds of kilometers (miles) from dams to where it is needed (Figure 10-14b).

Dams were considered very important to developing America. The federal government passed the Reclamation Act of 1902 to establish surface-water management systems throughout the west. During the next 50 years, the United States was the world leader in dam construction—355 dams and 25,700 kilometers (16,000 mi) of canals were built. This included 49 dams constructed in the watershed of the Tennessee River (Figure 10-15, p. 310) to help this region recover from the economic hardships of the Great Depression. | TVA: from the New Deal to a new century | Today, hydroelectric power generated at dams supplies 15% of the nation's electricity. The electricity generated at dams is a clean supply of energy—it doesn't pollute the hydrosphere or atmosphere. But dams are not without their negative environmental consequences.

Dams change streams' natural flow and sediment-transport characteristics (Chapter 7). Sediment that gets trapped in the reservoirs behind dams can gradually fill them up. Lake Mead, Hoover Dam's reservoir on the Colorado River, lost 3% of its storage capacity in 14 years due to sedimentation. And the sediment-depleted flows below dams change river dynamics and habitat. Lack of sediment can cause river channels to become scoured and riverbanks to lose sand and gravel deposits. In the Colorado River, four species of native fish are now considered endangered and the biological productivity of its delta is only 5% of what it was before 19 dams were built to develop its waters.

(a)

(b)

FIGURE 10-14 ▲ Dams and Diversions
(a) Large dams, like the Grand Coulee on the Columbia River in Washington, are used for recreation, power generation, flood control, and water supplies. **(b)** Canals and aqueducts, like the Los Angeles aqueduct that transfers water from eastern California to the Los Angeles area, take surface water to where it is needed.

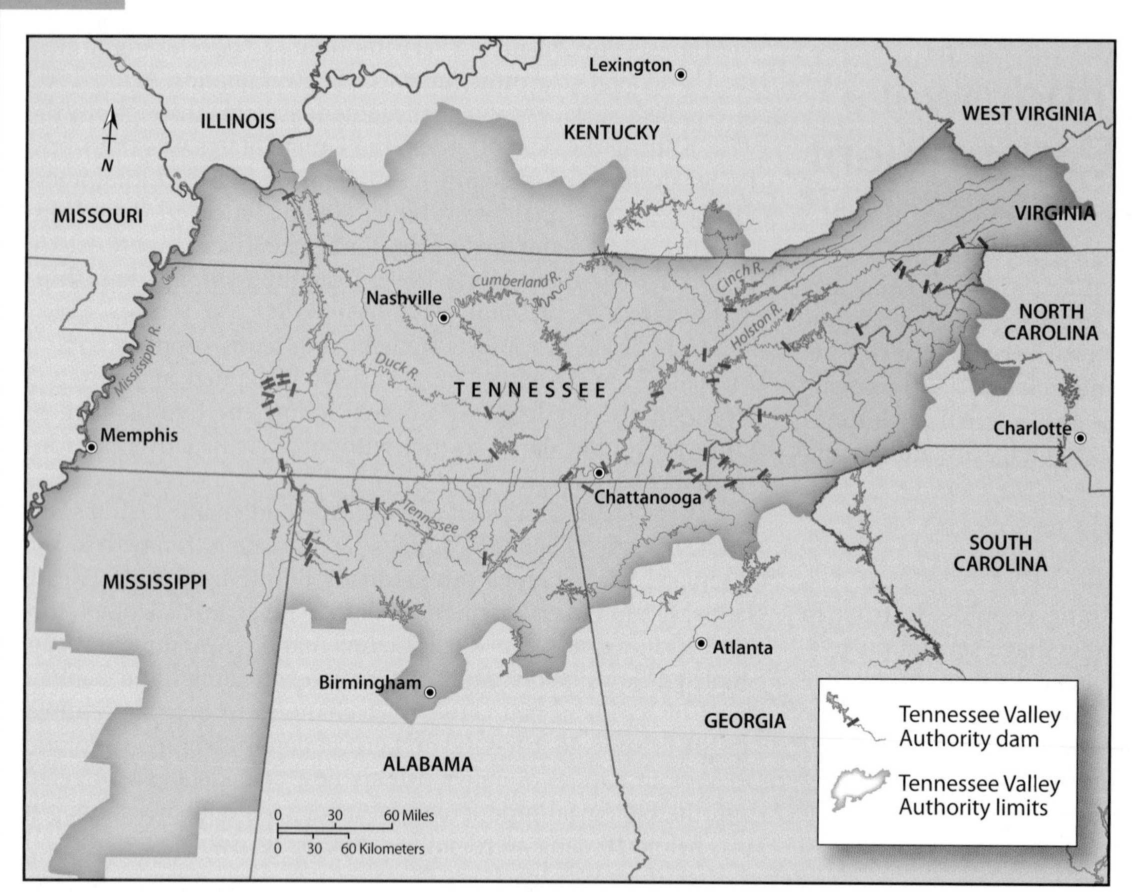

FIGURE 10-15 ◀ Dams in the Tennessee Valley Region
The Tennessee Valley Authority manages 49 dams in the watershed of the Tennessee River.

Dam construction has essentially stopped in the United States. The surface-water reservoir capacity has been about the same since 1980 (Figure 10-16). In some places, people are even considering tearing down dams. But most dams are here to stay, and the rivers they block are fundamentally changed.

What are some sustainability issues associated with dams and diversions of surface water?

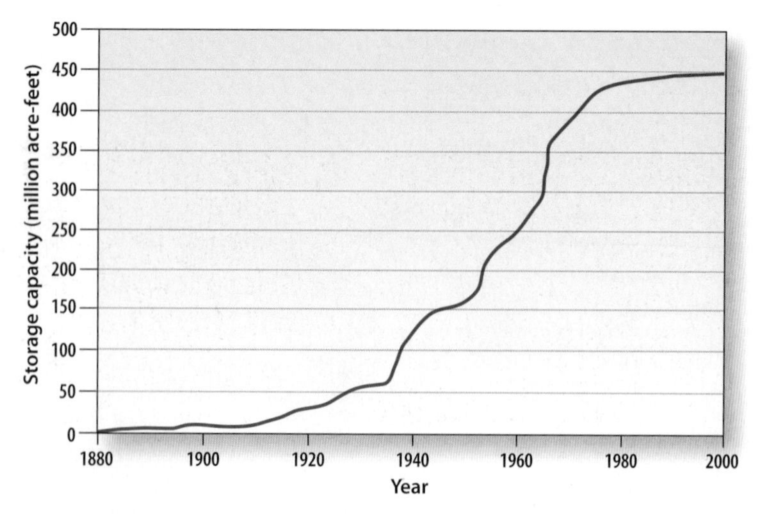

FIGURE 10-16 ▲ United States Reservoir Capacity
Reservoir capacity increased dramatically during the great period of dam construction from 1935 to 1975.

GROUNDWATER MINING

If the withdrawal from an aquifer is greater than the natural recharge, the resource is gradually depleted. If it is an unconfined aquifer that is being overdrawn, the depletion is evidenced by a drop in the water table. This drop is greatest at the point of the withdrawal, meaning the bottom of the well. The drop becomes gradually less with distance from the well, creating a **cone of depression** in the water table, as shown in Figure 10-17. The combined effect of large and widespread well withdrawals can lead to a significant decline in the depth of the water table over a large region (as in the High Plains aquifer, Figure 10-9). On human time scales, recharge may not occur soon enough to stave off water shortages. Pumping out more water than natural recharge can replace is called **groundwater mining**. Groundwater mining is essentially a one-time use of the water resource—it is not sustainable resource management.

LAND SUBSIDENCE

You learned about how withdrawing groundwater can cause land subsidence in Chapter 8. In the United States, groundwater withdrawals account for more than 80% of subsidence, affecting about 44,000 km^2 (17,000 mi^2) spread out over 45 states. Subsidence occurs where aquifers are most vulnerable to compaction, and once compacted, an aquifer may never regain its water-holding capacity.

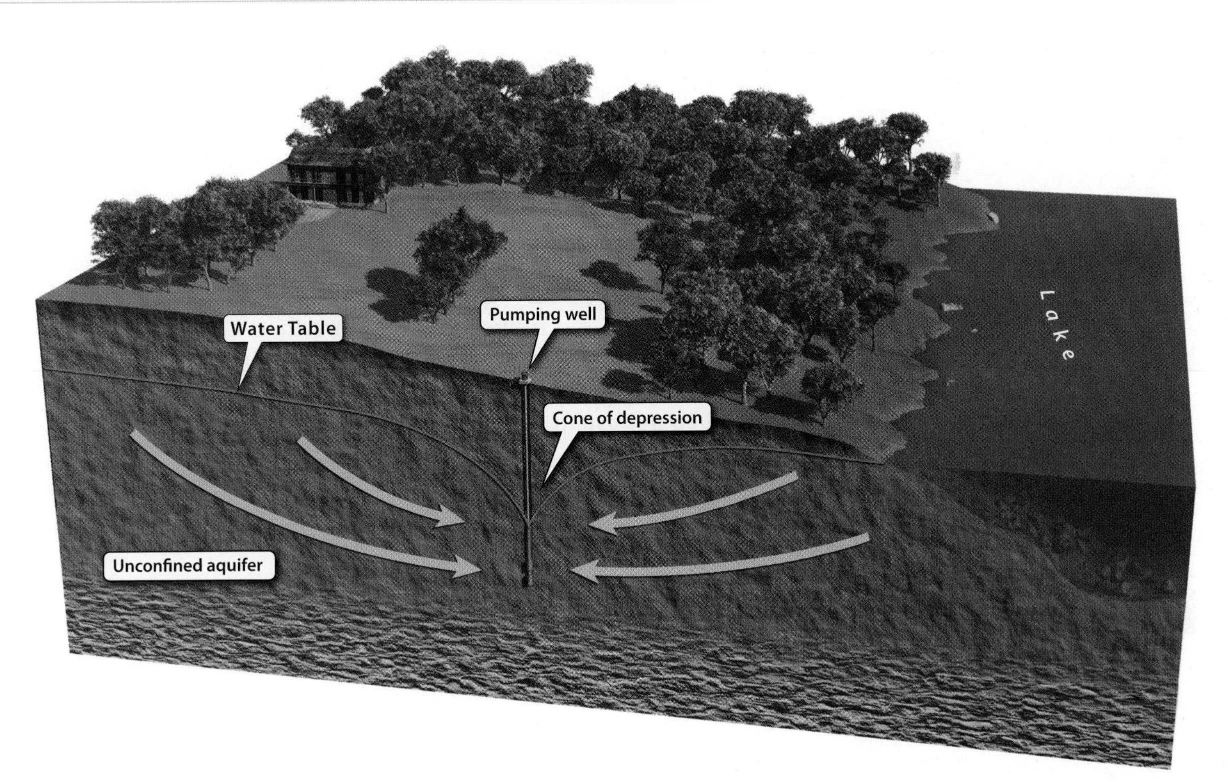

FIGURE 10-17 ◀
Aquifer Depletion
Pumping more water than natural recharge can replace lowers the water table. The drop is greatest where the water is being withdrawn, creating a cone of depression in the water table around the well site.

GROUNDWATER PUMPING AND SURFACE WATER

Heavy groundwater pumping can cause lakes, streams, and wetlands to lose water. | Effects of groundwater development on groundwater flow to and from surface water bodies | The most dramatic effects of groundwater pumping on surface-water bodies occur where the water table is shallow.

Imagine a municipal water well operating a short distance from a stream fed by groundwater. The cone of depression of the well intercepts water normally flowing into the streambed. Continued pumping may eventually draw water through the streambed and into the aquifer, a local reversal of the natural flow pattern. As the stream loses water, the habitat in and along it changes, as you can see in Figure 10-18. Nearby plants may die, fish and other aquatic life may become stressed, and wildlife living along the riverbank may lose its home.

SALTWATER INTRUSION

In coastal regions, aquifers containing fresh and salty water exist side by side. Differences in density prevent the two types of groundwater from mixing much, and the pressure of the column of freshwater (created by the weight of the water between its recharge area and the boundary with the salt water) keeps the fresh-saltwater contact stable. However, pumping water from a coastal aquifer faster than it can be replaced decreases

(a)

(b)

FIGURE 10-18 ▲ Effects of Groundwater Pumping
(a) A 1942 photograph of a reach of the Santa Cruz River south of Tucson, Arizona. **(b)** A 1989 photograph of the same site shows that the riverside trees have largely disappeared and the habitat has significantly changed. Groundwater pumping lowered the water table and created a losing stream in this area.

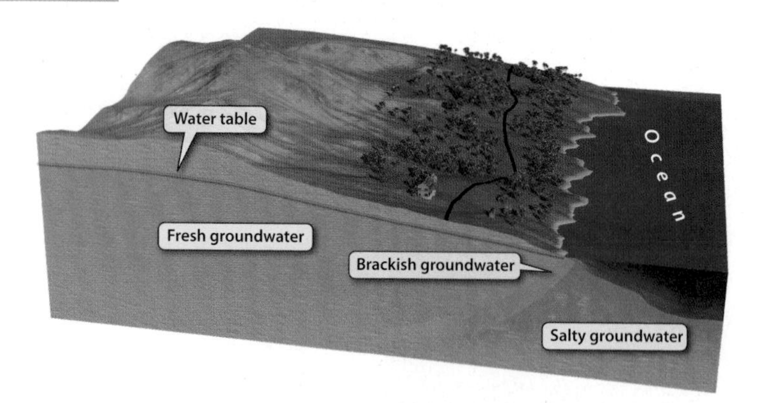

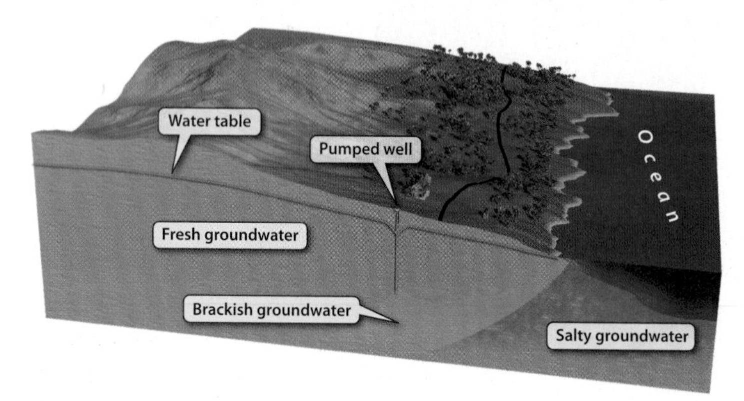

FIGURE 10-19 ▲ Saltwater Intrusion
Groundwater pumping can cause saltwater to move into a freshwater aquifer.

the pressure of the freshwater column and allows salt water to move into the freshwater aquifer, a process called **saltwater intrusion**, diagrammed in Figure 10-19. Once the freshwater portion of the aquifer is contaminated with salt, it is difficult to reverse the situation and flush the salt water out.

FIGURE 10-20 ▼ On the shore at Cape May, New Jersey
Salt water intrusion caused by groundwater pumping led to construction of a desalination plant at Cape May.

Saltwater intrusion is a problem along many parts of the U.S. coast. For example, Cape May, New Jersey, has struggled with saltwater intrusion since the late nineteenth century. This resort community lies on a peninsula between the Atlantic Ocean and Delaware Bay. Communities on the Cape draw from multiple confined aquifers, which extend to a depth of about 244 meters (800 ft) and are inclined slightly toward the ocean. Since 1940, saltwater intrusion has closed more than 20 wells used by towns and industries and more than 100 wells used by individual homeowners.

As municipal wells became contaminated, the city of Cape May enacted a plan to arrest saltwater intrusion into its aquifers. It reduced pumping rates at some wells and made up for the shortfall by constructing a special filtering plant to purify brackish water pumped from the Atlantic City aquifer, 244 meters deep (Figure 10-20).

The examples in this section show that the environmental impacts of capturing surface water and pumping groundwater can be significant. But as you learned at the beginning of this chapter, people can also degrade water's quality and make it unusable for many purposes. Polluting water resources is now against the law—but it happens.

10.4 Water Quality and Pollution

You could be surrounded by water and still die of thirst. Remember the classic image of the forlorn shipwrecked sailor on a desert island? He is surrounded by water but there is not a drop he can drink. It's not enough to have water available. To sustain people, water needs to be fresh (non-salty) and clean (without harmful constituents). It has to have what we call good quality.

What's your favorite image of clean water? Is it just clear water coming out of your kitchen faucet? Or perhaps a clear babbling brook in the mountains? You have an understanding of what good water quality is—it's healthful to drink and doesn't taste or smell bad. You can personally evaluate water's taste and smell, but what about determining how healthful it is? You probably cannot do this. Fortunately, doctors and scientists have learned how to identify what is healthful and unhealthful water. This wasn't always the case, however.

DR. JOHN SNOW AND WATER-BORNE DISEASE

The insight that dirty water could be a source of disease had its origin in the pioneering work of Dr. John Snow in London, England, more than

FIGURE 10-21 ▼ Tracing the Source of a Water-Borne Disease
(a) During an 1854 cholera epidemic in London, Dr. John Snow carefully recorded the addresses of the victims and where they got their water. The map he created (shown here in a slightly modernized version) helped to demonstrate that the disease was associated with contaminated water from the Broad Street well. When the pump handle was removed, the outbreak subsided. **(b)** This replica of the original Broad Street pump was installed in 1992 to honor Snow's pioneering scientific investigations.

(a)

(b)

150 years ago. Snow wanted to understand why an 1854 cholera epidemic that killed hundreds of people was concentrated in a particular part of London. He mapped where the deaths occurred and personally interviewed people who had and had not contracted the disease.

Snow discovered that the outbreak was centered on a well in Broad Street. Moreover, while the incidence of cholera was high among the people who used the well, others in the area who didn't depend on it for drinking water (such as those working in a local brewery) did not get the disease. The case was sealed when Snow determined that a local cesspool was leaking and that its effluent was contaminating the Broad Street well water.

Despite Snow's carefully marshaled evidence, getting the Broad Street well shut down was not easy. Scientists were just beginning to learn about germs and how they could live and be transported in water. But Snow persevered, and eventually use of the well was stopped—after which the outbreak subsided. His remarkable observational skills, careful documentation and mapping, and tireless efforts to integrate data into a comprehensive explanation of how a disease develops and spreads within an entire group of people has led to Snow being called the "Father of Epidemiology"—the study of disease in populations (Figure 10-21b). John Snow – a historical giant in epidemiology

NATURAL WATER QUALITY

An important characteristic of water is that its natural quality varies tremendously—it doesn't have to be polluted by people to be unhealthful. In fact, in nature there is no such thing as pure water. This is because water has some special chemical characteristics.

When two atoms of hydrogen (H) and one atom of oxygen (O) combine, a molecule of water (H_2O) is formed. The hydrogen atoms, which are small and carry a partial positive charge, share electrons with the slightly larger oxygen atom, which bears a partial negative charge. However, the hydrogen atoms are not symmetrically positioned around the oxygen atom. Instead, as you can see in Figure 10-22, the hydrogen atoms are both closer to one side of

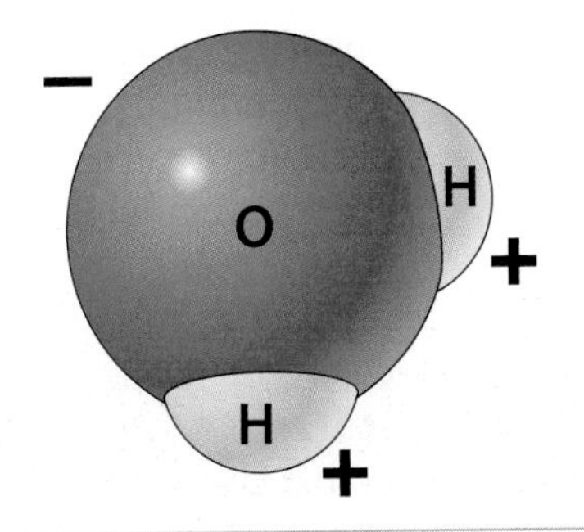

FIGURE 10-22 ◀ Water Molecules Have a Dipolar Electrical Character
The hydrogen atoms carry a partial positive charge, while the oxygen atom has a partial negative charge. The resulting asymmetrical charge distribution allows water molecules to interact with many different substances and to serve as an excellent solvent.

the oxygen atom than to the other. Consequently, one side of a water molecule has a positive charge (where the hydrogen atoms are located) and the opposite side has a negative charge. Such a molecule is called a *dipole*.

Water's dipolar character explains why it can interact with both positively and negatively charged ions as well as other dipolar molecules, separating them from one another and from the solid matrix of which they are a part. This ability makes water an excellent solvent (it's commonly called the "universal solvent"), and explains why it commonly contains dissolved substances in nature, even if only in small amounts.

As water passes through the atmosphere, geosphere, and biosphere, it dissolves or suspends materials that determine its quality. Some dissolved constituents may be present in minute or trace amounts, but they can still be potentially harmful to people. This is why "clear" water is not always clean water. Here are some examples of how natural processes affect water quality.

Pure rainwater?

Most people think rainwater is pretty pure water. They are right—rainwater is commonly cleaner than other water sources. However, even rainwater is not pure water (the only water that is commonly pure, just H_2O and nothing but H_2O, is water people distill). Water in the atmosphere comes in contact with suspended particles and gases. One of the natural processes that takes place in the atmosphere is the chemical reaction between water and carbon dioxide (CO_2). When water and carbon dioxide react, they form carbonic acid. Carbonic acid is not a strong acid, but it gives rainwater a slightly acidic composition. This is why rainwater that infiltrates the ground can dissolve limestone and certain other rocks to form karst terrain, discussed in Chapter 8. (The slightly acidic rainwater caused by reaction with carbon dioxide is not the "acid rain" caused by air pollutants that you will learn about in Chapter 13.)

Arsenic in natural water

Everyone knows that arsenic can be deadly. It is a poison that has caused many deaths in the real world (as well as countless more in novels, movies, and plays). For arsenic to be lethal quickly requires that people ingest significant amounts in their food or water over a relatively brief period of time. We understand such acute arsenic exposure, its consequences, and how to avoid it. It's chronic arsenic exposure—the kind of exposure that occurs when people eat or drink very small amounts over extended periods—that can sneak up and surprise us.

One of the major consequences of such exposure is increased risk of cancer. Though the effects may take years to develop, people who are chronically exposed to arsenic have higher rates of kidney, liver, lung, bladder, and skin cancer, as well as circulatory problems. | Chronic arsenic poisoning: history, study, and remediation

Chronic exposure to arsenic most commonly occurs through drinking water. There are many places, such as Bangladesh, where the arsenic content of drinking water is unhealthy, but the most studied example is in Taiwan. In fact, the many years of epidemiological studies in areas of Taiwan helped the U.S. National Research Council conclude that even arsenic levels below 50 micrograms per liter (μg/L) in drinking water pose an unacceptably great danger of bladder, lung, and skin cancer (Figure 10-23).

FIGURE 10-23 ▲ Chronic Arsenic Exposure Is Harmful
Many people in Taiwan, Bangladesh, and other regions use well water that is naturally contaminated with arsenic. Arsenic poisoning can lead to skin lesions like that on the thumb of this Bangladeshi woman. The effects of long-term ingestion of even small amounts of arsenic can include skin cancer.

How much is a microgram per liter? This is the standard unit used to measure trace amounts of substances in water. One μg/L is also one part per billion (ppb)—a very small amount. It's about one drop in 50,000 liters (220 bathtubs) of water.

In 2001, the EPA lowered the maximum contaminant level (MCL) for arsenic in drinking water from 50 μg/L to 10 μg/L. However, there are lots places in the United States where drinking-water arsenic concentrations are still a cause of concern. The results of drinking water monitoring by the USGS indicates that about one in 10 small community water systems exceeds the new 10 μg/L arsenic MCL. | Arsenic in groundwater of the United States

Where does this arsenic come from? It comes from continental crust rocks, which on average have an arsenic concentration of 1.8 parts per million. This is 1800 parts per billion, or 180 times the arsenic MCL of 10 μg/L (10 ppb) now allowed in our community drinking water. Some rocks contain as much as several percent arsenic and some have essentially none, but on average, traces of arsenic are widespread. Most of it occurs in arsenopyrite, a mineral containing arsenic, iron, and sulfur. When arsenopyrite reacts with oxygen in air or water, the arsenic it contains can be released and become dissolved in groundwater. This is the principal way groundwater can come to have unhealthy arsenic concentrations.

What are some Earth systems interactions associated with higher arsenic concentrations in natural waters?

Parasites in natural water

Can cold, clear mountain brooks have water quality problems? You bet. A common example is water that contains the microbe *Giardia lamblia*, shown in Figure 10-24. *Giardia lamblia* is a one-celled parasite that likes to live in airless places, such as the intestines of animals and people. The most common intestinal parasite in the United States, it causes people to have diarrhea, and feel generally awful, for up to two weeks or more after infection. Division of Parasitic Diseases – Giardiasis fact sheet

The form of *Giardia* that lives in intestines has flagella resembling tiny whips that help it move about some. It reproduces by division, producing countless copies of itself. As the parasites move through the intestines they change into cysts—tough little capsules that pass through the intestines along with feces and become part of the environment. Many animals that have *Giardia,* such as beavers and muskrat, actually live in water, and it's easy for *Giardia* to spread through streams and lakes. Migrating water birds, like Canada geese, may also help spread *Giardia.*

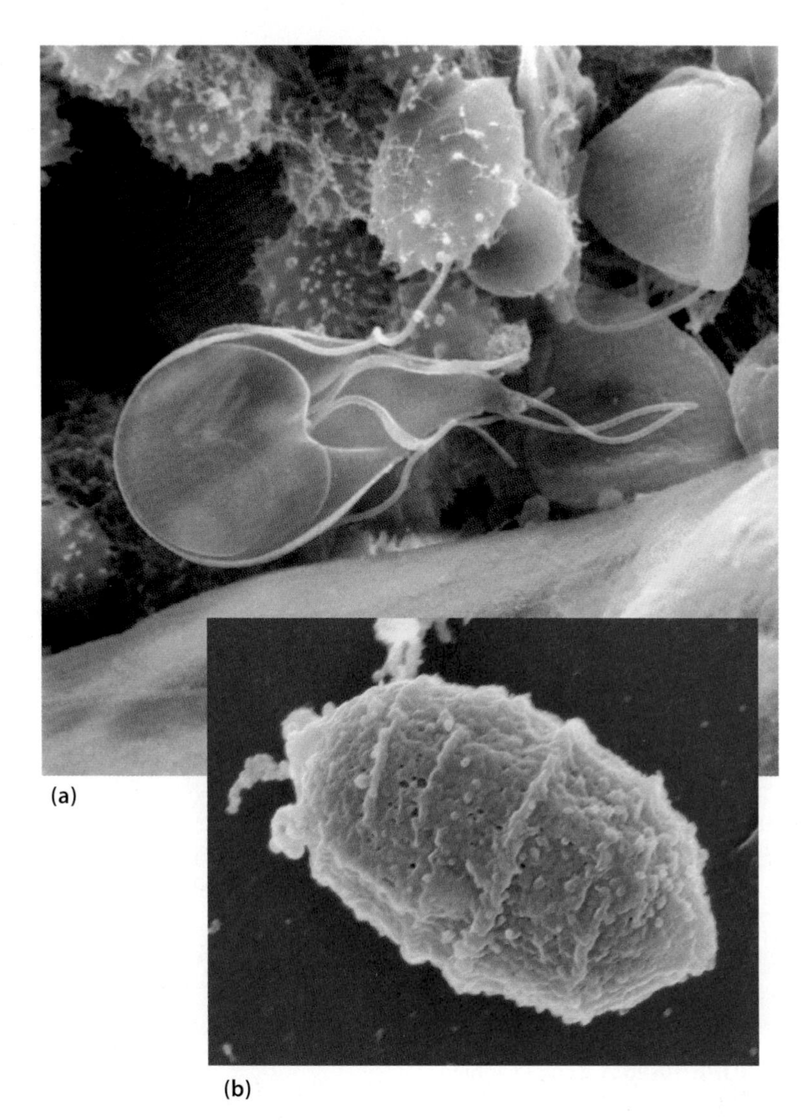

FIGURE 10-24 ▲ *Giardia lamblia*
Active form **(a)** and cyst form **(b)**.

Animals, though, weren't the original source of *Giardia* in North America. *Giardia* appears to have gotten its North American start in Aspen, Colorado, back in 1970. It seems that people visiting from Russia brought *Giardia* with them. Once it was loose it quickly spread; it even became widespread in remote Alaskan waters by the 1980s. Today, drinking water directly from clear mountain brooks, even in remote Alaska, can have serious consequences. Back-country travelers should boil or filter their water to be safe.

POLLUTION AND WATER QUALITY

Water pollution is the degrading of water quality by people, however they do it—from releasing toxic chemicals such as pesticides to discharging heated water (creating *thermal pollution* that can make surface water bodies too warm to be hospitable to many aquatic organisms) to clogging a river like the Cuyahoga with waste and debris. Pollution can make water resources unusable by people and uninhabitable by aquatic organisms.

With passage of the Clean Water Act in 1972 (which regulated discharge of pollutants to the nation's waters) and the Safe Drinking Water Act that followed in 1974 (which regulated the amounts of harmful substances allowed in drinking water), it became very important to understand what harmful pollutants are and how much of them it takes to be hazardous to people. The responsibility for gaining and applying this understanding fell to the EPA (Chapter 15).

The EPA has completed much research to identify potentially harmful water contaminants and the amounts of these contaminants that present unacceptable risks to people. The EPA now sets drinking water standards for about 90 contaminants. Most of these contaminants fall into three general categories: (1) microbes, (2) inorganic contaminants, and (3) human-made chemicals.

Microbes

You have already learned about *Giardia lamblia,* the most common intestinal parasite in the country. It gets into water supplies through animal waste and sewage. This is where other microbes of concern originate, too. Coliform bacteria (those that resemble *E. coli*, a common and usually harmless occupant of the human digestive tract) are a sign that water supplies are contaminated with fecal wastes—and thus perhaps with less benign strains of *E. coli* or with *Cryptosporidium,* a parasite that causes gastrointestinal distress. Municipal water treatment plants and septic systems (Figure 10-25, p. 316) are designed to remove such harmful organisms from water after people use it.

Inorganic contaminants

The inorganic contaminants are mostly metals and nitrogen compounds. They include the arsenic you learned about above. Some other metals of concern are chromium, copper, and lead. Metal concentrations can increase naturally or through people's activities. For example, water can dissolve lead that is often present in the pipes and plumbing fixtures

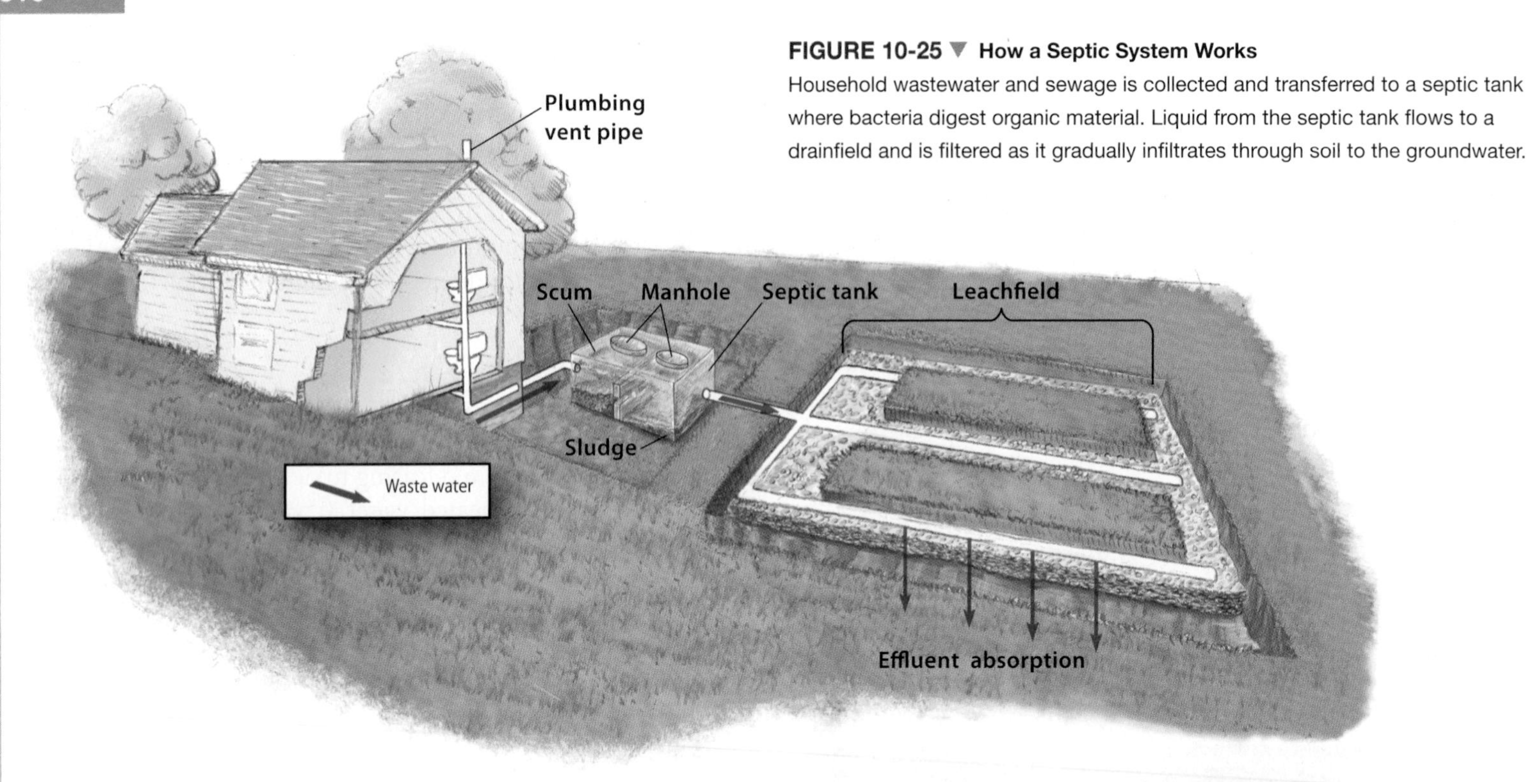

FIGURE 10-25 ▼ How a Septic System Works
Household wastewater and sewage is collected and transferred to a septic tank where bacteria digest organic material. Liquid from the septic tank flows to a drainfield and is filtered as it gradually infiltrates through soil to the groundwater.

formerly used in constructing homes around the country (Figure 10-26). Arsenic can come from pesticides and chemicals used to prevent wood from rotting. Mercury emitted from coal-fired power plants (discussed in Chapter 13) is an example of a metal pollutant that eventually finds its way up through aquatic food chains to fish that people eat.

Human-made chemicals

There are dozens of human-made chemicals that are of concern if they get into water supplies. These include pesticides and herbicides used around the home or on agricultural fields. You have probably heard of dioxin, PCBs (polychlorinated biphenyls), benzene, and MTBE (a gasoline additive). Many of these contaminants have been linked to increased risk of cancer if people are harmfully exposed to them. Manufactured materials such as solvents, paints, and other chemical products are commonly the source of these contaminants.

Sediment

Increased sediment loads are a form of pollution. Suspended sediment obviously makes water more dirty or muddy, and its deposition on the bottom of surface water bodies can also affect aquatic life and habitat. Sediment pollution commonly derives from plowed agricultural fields and timber harvest areas where removal of the surface vegetation has caused increased erosion (Chapter 7). Construction sites can be a source of increased sediment in streams, too. Have you driven by a road construction area and noticed hay bales in the ditches? They are there to decrease the sediment load of surface runoff from the construction area. Such runoff is commonly regulated to ensure that sediment pollution is minimized.

FIGURE 10-26 ▲ Lead Plumbing
Lead pipes were once used to bring water to homes. They can contaminate water with lead and should be replaced, as is being done for this older home.

What You Can Do

Investigate Your Water Quality

Are you a person who just turns on the faucet and hopes for the best? Do you want to learn more about where your water comes from and how clean it is? You easily can, if your water comes from a public water system (if you pay for your water, your system is public). The EPA regulates and monitors drinking water quality. Its Local Drinking Water Information site provides access to information about all the public water systems in the country. Here you can find out if your local water system has had any water quality violations. Local drinking water information The EPA requires public water systems to provide an annual water quality report. A few of these reports are online, but most are not. The EPA site will help you determine how to contact your local water supplier and request a copy of its annual water quality report.

If you get your water from your own well, there are ways to check your water quality, too. The EPA suggests you test your water quality if its taste or smell isn't quite right, if it stains plumbing fixtures, or if people drinking the water get ill. You will have to collect a sample and pay for a laboratory analysis, but this isn't too expensive in most cases. The EPA site also provides links to state agencies that check on and approve water quality labs in your area. Determining water quality is very standardized, and there are many laboratories that can help you.

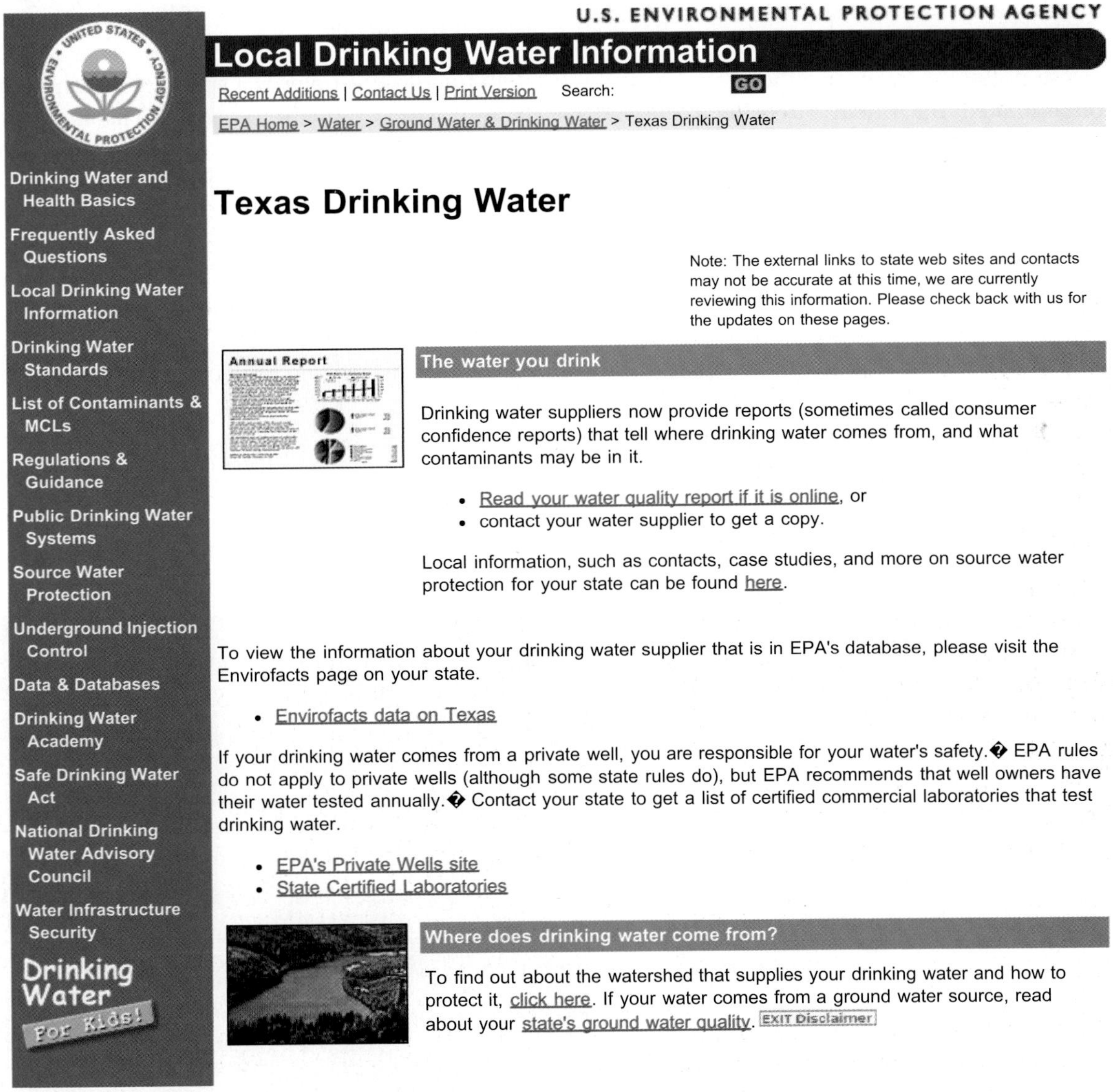

U.S. ENVIRONMENTAL PROTECTION AGENCY
Local Drinking Water Information
Recent Additions | Contact Us | Print Version Search: GO
EPA Home > Water > Ground Water & Drinking Water > Texas Drinking Water
Drinking Water and Health Basics
Frequently Asked Questions
Local Drinking Water Information
Drinking Water Standards
List of Contaminants & MCLs
Regulations & Guidance
Public Drinking Water Systems
Source Water Protection
Underground Injection Control
Data & Databases
Drinking Water Academy
Safe Drinking Water Act
National Drinking Water Advisory Council
Water Infrastructure Security
Drinking Water For Kids!
Texas Drinking Water
Note: The external links to state web sites and contacts may not be accurate at this time, we are currently reviewing this information. Please check back with us for the updates on these pages.
Annual Report
The water you drink
Drinking water suppliers now provide reports (sometimes called consumer confidence reports) that tell where drinking water comes from, and what contaminants may be in it.
• Read your water quality report if it is online, or
• contact your water supplier to get a copy.
Local information, such as contacts, case studies, and more on source water protection for your state can be found here.
To view the information about your drinking water supplier that is in EPA's database, please visit the Envirofacts page on your state.
• Envirofacts data on Texas
If your drinking water comes from a private well, you are responsible for your water's safety.� EPA rules do not apply to private wells (although some state rules do), but EPA recommends that well owners have their water tested annually.� Contact your state to get a list of certified commercial laboratories that test drinking water.
• EPA's Private Wells site
• State Certified Laboratories
Where does drinking water come from?
To find out about the watershed that supplies your drinking water and how to protect it, click here. If your water comes from a ground water source, read about your state's ground water quality. EXIT Disclaimer

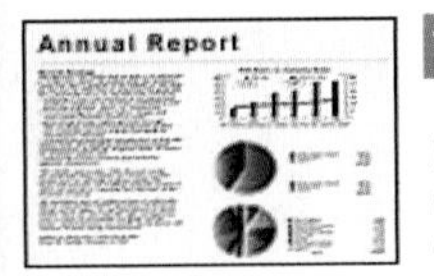

SOURCES OF POLLUTANTS

Pollutants can be released to the environment from just about every activity people engage in. The releases can be chronic, dragging on for years, or seemingly catastrophic, as when spills of hazardous materials take place. They can affect small or large areas and both surface water and groundwater. Several important sources of water pollution are related to people's development and use of natural resources including soils, minerals, and energy resources. These examples are discussed more fully in Chapters 11–13.

People trying to understand water pollution have found it helpful to categorize sources of contaminants as either point sources or nonpoint sources.

Point sources

Point sources release pollutants from specific sites. The leaky cesspool that contaminated the Broad Street well in 1850s London was a point source for the germs that cause cholera. Other examples of point sources are wastewater discharge pipes from an industrial facility, a leaking storage tank, or an animal feedlot (Figure 10-27).

Point sources are identified by sampling waters and mapping the concentration of the pollutant of concern. High pollutant concentrations will be close to the source because the farther a pollutant travels, the more diluted it becomes.

Nonpoint sources

Nonpoint sources release pollutants to the environment over a large area; they are not as easy to identify or clean up as point sources. Many of the classic examples come from agricultural areas where fertilizers and pesticides are widely spread. Urban areas are also increasingly important nonpoint sources of water pollution. Urban development covers the landscape with roads and parking lots. Storm water that flows off these impermeable areas carries with it the dirt, grime, and miscellaneous garbage that accumulate on them (Figure 10-28). A lot of the petroleum products that find their way into surface and groundwater get there from storm water runoff in urban areas (Chapter 13).

GROUNDWATER POLLUTION—A SPECIAL PROBLEM

Polluting groundwater is a special concern because once it happens, it is very difficult to reverse. This is why careful control of storm water runoff and other sources of pollution in karst terrain (Chapter 8) is so important. Surface water gets into groundwater very easily in karst areas. The Edwards aquifer is an example (see Figure 10-10). The residence time for water (and pollutants if they are present) in the Edwards aquifer is about 200 years. Groundwater aquifers with long residence times are very challenging to clean up. Natural recharge areas for aquifers need to be carefully managed and protected from pollution.

(a)

FIGURE 10-27 ▼ Point Sources of Pollution

(a) Untreated wastewater discharge can be a point source of pollution. **(b)** Although many feedlots are large, they have well-defined boundaries and can still be considered a point source of organic contaminants to surface and groundwater.

(b)

10.5 Sustaining Water Resources

Water resources are renewable and sustainable. Protecting water quality by preventing pollution and cleaning up water bodies such as the Cuyahoga River is one part of sustaining water resources. Other steps to sustain water resources include recycling, conservation, and wise resource management.

WATER TREATMENT

Cleaning up surface bodies of water like the Cuyahoga River is done mostly by preventing the continued release of pollutants and waste into them. The waters then gradually clean themselves as pollution becomes dispersed and diluted. The treatment of industrial and municipal wastewater plays a key role in preventing release of pollutants to surface water bodies. Cleaning up municipal wastewater is a good example.

All sizable towns and cities must have some way of treating municipal wastewater from surface runoff and household use including sewage. This wastewater is collected in underground sewer systems and delivered to treatment plants like the one shown in Figure 10-29. Treatment commonly takes place in three steps.

- Primary treatment removes solid materials such as sand; insoluble fats, oils, and greases (organic material that settles easily); and

(a)

(b)

FIGURE 10-28 ▲ Nonpoint Sources of Pollution
(a) Surface runoff that sweeps up scattered garbage and other water contaminants is a nonpoint source of pollution. **(b)** The widespread distribution of pesticides, as in this aerial spraying of a field, is another nonpoint source of pollution.

FIGURE 10-29 ▲ Municipal Water Treatment Plant
Communities collect sewage and other wastewater and treat it in plants like this one before releasing it back to surface water.

floating material. These materials are mechanically separated from the wastewater.

- Secondary treatment degrades the organic contaminants in the wastewater that have passed through primary treatment. The treatment is biological. Bacteria and protozoa that eat the organic contaminants are provided oxygen and surfaces to live on. They metabolize the organic waste, and after settling or filtering, the water that passes through secondary treatment contains only minor amounts of organic or suspended material.
- Tertiary treatment is used if the water needs to be disinfected or otherwise treated before it is released. The treatment is commonly by the addition of chemicals such as chlorine.

Solids separated during water treatment are disposed several ways. Solids like sand and garbage that makes its way to a water treatment plant may be taken to a landfill. The organic solids that are collected, called sludge, can be used as a fertilizer to improve soil quality. The treated water can be released to surface waters such as a river or allowed to infiltrate groundwater. There are rigorous guidelines and standards (Chapter 15) for all releases of treated solids and water from water treatment plants.

Primer for municipal wastewater treatment systems

RECYCLING WASTEWATER

Wastewater treatment commonly produces an effluent with a level of purity that allows it to be discharged to the environment—perhaps a nearby stream. However, treated wastewater can also be reused rather than released to the environment, and recycling of treated wastewater is becoming more and more common.

Processing wastewater to the point where it can be used by people, even for drinking water, is possible, but obtaining this level of purity is generally expensive and involved. Some cities have chosen a compromise option: Process the wastewater so that it is pure enough to reuse as irrigation water for lawns, gardens, and crops, as well as in certain industrial applications.

In the 1970s, St. Petersburg, Florida, found itself running short on drinking water as its population grew. The city is in west-central Florida, on Tampa Bay. It had no local water supply, forcing it to import freshwater from outlying areas. In addition, the city was facing increasingly stringent limits on the discharge of treated wastewater into Tampa Bay. The city chose to develop a "dual use" water system. The existing network of pipes would continue to deliver drinking water to homes and business. The city then built a parallel network of pipes to distribute reclaimed wastewater for watering gardens, lawns, and golf courses. Four treatment plants process the water to the point where it contains low levels of solid contaminants and very low levels of bacteria, viruses, and other disease-causing pathogens. The reclaimed water is also suitable for certain industrial and commercial uses, such as a cooling medium in building air-conditioning systems. Residents pay an initial charge to hook up to the system, which they can draw on afterward without metering. By 2001, reclaimed water met 42% of St. Petersburg's needs.

Water treatment technology is capable of making even sewage wastewater fit to drink. If communities can accept using this type of recycled wastewater, it can be a big help in meeting their needs. Orange County, California decided that it could help drought-proof its water supply with a toilet-to-tap treatment and recycle system. The half-billion-dollar plant, commissioned in January 2008, supplies 72,000 acre-feet (89 million m^3) of pure, drinkable water per year. However, the water doesn't go directly from the treatment plant to home faucets. It's used to replenish the county's groundwater aquifer. Groundwater replenishment system The treated wastewater is transferred to a surface reservoir in the recharge area of the county's aquifer. It infiltrates the aquifer and is eventually pumped back to the surface and distributed for home use. The Orange County approach to recycling wastewater could be a good example for other communities facing water shortages, high costs of importing water, and aquifer maintenance challenges.

Recycling water is becoming a standard practice in many industries. For example, it decreases the amount of water used in the mining industry (Chapter 12), in refining petroleum products (Chapter 13), and in electric power generation. Major recycling of water in the electric power industry started in the 1970s with passage of the Federal Water Pollution Control Act. This legislation focused on pollution prevention. In the case of the electric power industry, a goal was to decrease the release of heated waters to the environment (thermal pollution). The industry met this requirement by recycling, which had the additional benefit of conserving water.

CONSERVATION—USING LESS WATER

Conservation is an attempt to deal with increasing demand and limited supply by finding ways of consuming less water to meet our needs. Conservation can take place at all levels, starting in the home.

Household water conservation

On a household level, individuals can take many actions to reduce the use of water. These include:

- **Using Water-Saving Fixtures and Appliances:** Installing low-flow showerheads and toilets means using less water (low-flow toilets, for example, use less than 2 gallons per flush). Aerating faucet attachments mix air into the flow of water to the sink. Front-loading clothes washing machines use less water than top-loading appliances. Repairing dripping faucets can also save a lot of water. The 1992 Energy Policy Act required all federal facilities to use low-flow plumbing fixtures. This type of fixture is finding its way into more and more homes, particularly in places like California where water can be scarce.
- **Avoiding Water-Wasteful Gardening:** Daily watering of lawns and plantings is usually unnecessary. It is best not to water during the heat of the day, when water tends to evaporate rapidly. It is much more efficient to water in the morning or early evening. Using mulch helps to retain more moisture in the soil. Developing yards and landscapes that need smaller amounts of water is called

FIGURE 10-30 ▲ Xeriscaping Conserves Water
A key to successful xeriscaping is the use of vegetation that does not require much water to grow and remain healthy. Although common in arid environments, xeriscaping can be applied anywhere to conserve water resources. This partly xeriscaped yard is in the mountains of Colorado.

"xeriscaping." Xeriscaping, derived from the Greek words *xeros* ("dry") and *scape* ("scene"), develops outdoor areas using drought-resistant plants, soil additives that help hold water, and irrigation techniques that conserve water (Figure 10-30).

Xeriscaping does more than conserve water. It commonly results in less need for fertilizers, pesticides, and general maintenance. Las Vegas, Nevada, now requires new subdivisions to use xeriscaping in place of lawns.

Community water conservation

Communities often respond to water shortages with public education. Citizens are urged to take steps in the home to reduce water use. In times of drought, however, the local government may be forced to switch from the velvet glove of public education to the iron fist of the law. Denver's response to prolonged drought conditions is an example.

Denver institutes its "drought response plan" when reservoir levels get too low. One key aspect of the plan is mandatory watering restrictions. Watering is limited to two days per week for single-family residences. Watering grass is limited to 15 minutes. No watering is allowed between 10:00 a.m. and 6:00 p.m. and none at all from October 1 to May 1. No washing of sidewalks, driveways, or automobiles is allowed. City "water cops" can issue tickets for rule breakers, and Denver residents can call a special number to turn in their neighbors if they see them breaking the rules.

Water-saving agriculture

As agricultural consumption accounts for 31% of all the water used in the United States each year, becoming more efficient at irrigating fields is a very helpful way to conserve water. Wasteful structures and techniques include unlined and uncovered irrigation ditches and canals, unpressurized irrigation that allows water to flow over the land surface, and sprinklers that launch a lot of water into the air. These cause water to be lost into the ground or through evaporation. The alternative is targeted, pressurized irrigation through pipes, drip systems, or sprinklers that deliver only what is needed where it is needed (Figure 10-31). The whole irrigation system can be guided by instruments that monitor the moisture in the soil.

FIGURE 10-31 ▲ Irrigation Technology Can Conserve Water
This underground drip system will replace overhead sprinklers (one is visible in the distance) and conserve water in Nebraska.

Results of water conservation

Conservation and recycling are working. You can see in Figure 10-32 on p. 322 that since 1985, U.S. water consumption has leveled out at about 410 billion gallons per day. The USGS keeps tabs on national water use and notes that projections in the 1960s and 1970s for skyrocketing water demand have not been borne out. Instead, use has remained virtually unchanged for the last 20 years. This outcome, even in the face of increasing population, is attributed to water recycling in industry (especially the electric power industry) and conservation

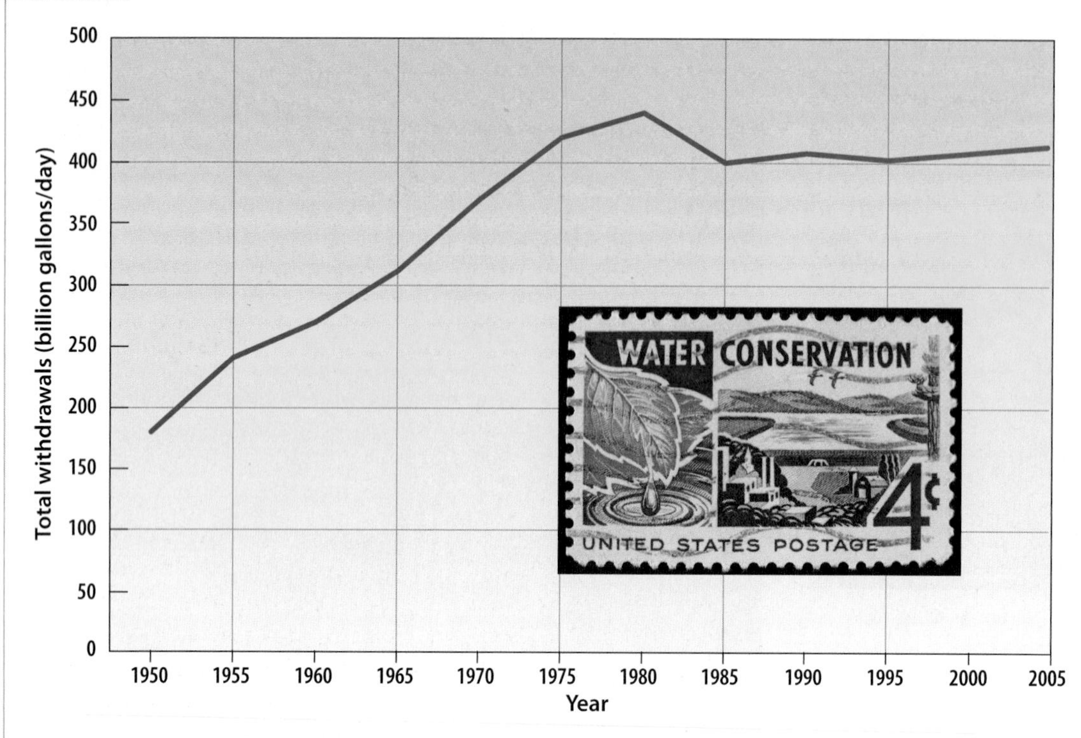

FIGURE 10-32 ◀ Annual Water Use in the United States
Water conservation is working. Although the amount of publicly supplied water has gradually increased, irrigation and electric power generation use has leveled off. Total water withdrawals have been at about 1.5 trillion liters (400–410 billion gal) per day since 1985.

efforts such as low-flow plumbing and low-pressure directed irrigation. At a national scale, this is good news, but it doesn't mean there are no water supply problems. There are still parts of the country that face very significant water shortages, especially in the western states.

RESOURCE MANAGEMENT—MAKING BETTER USE OF WATER RESOURCES

Living with natural limits on water supplies and using water resources sustainably require planning. Planning how water is obtained, distributed, and used is termed *water management.* Science and technology both play roles in water management.

Groundwater management

Sustainably managing groundwater resources is a challenge. As noted earlier, groundwater is a renewable resource if the amount of water withdrawn is the same as the amount of recharge. When withdrawals of groundwater exceed the natural rate of recharge, the aquifer will be depleted. To prevent this, communities have adopted a number of innovative practices and technologies collectively known as *groundwater management.*

Sound management must take into account the connected nature of ground and surface-water resources. An example is a city that draws on a river as well as a shallow aquifer for its drinking water. In the wettest months, it relies mostly on river water; in dry months, it switches to groundwater. The city closely monitors the water table and stream flows to guide decisions about tapping ground or surface water.

Suppose the city's population continues to grow, and eventually withdrawals during the dry months begin to exceed natural recharge rates. Like a person drawing too quickly on a savings account, the total water "capital" available to the community will gradually dwindle and cause water shortages. One thing the city could do is develop an artificial groundwater recharge program. This means using injection wells or recharge basins on the surface to allow either treated wastewater or excess stream flows to more effectively recharge the local aquifer (as Orange County, California is doing). Such a program allows the community to sustain higher withdrawals of groundwater without depleting the resource—an example of sustainable use.

Another tool available to groundwater managers is groundwater banking, which can help communities that depend chiefly on surface water create a hedge against drought. During wet years, when precipitation or spring snowmelt is heavy, the community stores the excess in a local aquifer. During dry years, it taps the stored water as needed.

People farming near Bakersfield, California, use groundwater banking to even out their erratic irrigation water supply, which can fluctuate greatly depending on rainfall from season to season. Farmers receive their water from the Arvin-Edison Water Storage District. It comprises about 53,500 hectares (132,000 acres), most of which is planted with vineyards, potatoes, cotton, and fruit orchards. Part of the district's allotment of water from the California Aqueduct is guaranteed; the rest varies from year to year depending on how much is available. The district banks as much as it can in the local aquifer. Two complexes of recharge basins do the work (Figure 10-33). From 1966 to 2000, the district banked 1.5 million acre-feet (1850 million m^3) of water and withdrew about 0.7 million acre-feet (860 million m^3). The difference, or 0.8 million acre-feet (990 million m^3), is the district's water savings account.

FIGURE 10-33 ▲ Recharge Basins
Surface water delivered to farmers in the Bakersfield, California area is routed to these reservoirs, where it percolates downward and recharges the local aquifer. This recharge system enables water to be stored, or *banked*, when more is available than needed.

During drought years, it can draw on this water to offset lower allotments from the California Aqueduct.

Comprehensive water management

Good financial planning addresses more than just your current assets and income; it looks to the future to make sure you'll have what you need in the long term. *Comprehensive water management* operates much the same way. Arizona, faced with dwindling groundwater supplies, enacted an innovative law—the Arizona Groundwater Management Code. It combines regulation of current groundwater reserves, mandatory conservation by water users, and planning for the future. In exchange for adopting the code, Arizona was assured federal funding for the Central Arizona Project, an aqueduct that diverts Colorado River water to the southern part of the state.

The code established two types of management zones in Arizona. Areas where pumping had begun to deplete aquifers were designated as Active Management Areas, or AMAs. There are five AMAs presently, largely coinciding with the most densely populated areas of the state. In these areas, groundwater use is closely regulated and various measures to conserve water are mandatory. For example, the state code requires towns to landscape road medians only with drought-tolerant plants from a pre-approved list, and water utilities must reduce the amount of water used per capita (that is, per person) over time. One way to reduce per capita consumption is raising the cost charged to customers per gallon as their total water use rises. Neither putting new agricultural land into production nor irrigating it with groundwater is allowed in AMAs.

The other type of water-management zone was established for agricultural areas. These are called Irrigation Non-Expansion Areas (INAs). In Arizona's three INAs, existing irrigation is regulated closely and no new irrigation is permitted.

Under Arizona's code, developers have to demonstrate that new land uses, such as a housing subdivision, will have a guaranteed 100-year supply of water. People who use groundwater must pay a fee to support the management program and the development of water supplies. In this way, the burden of preventing groundwater depletion becomes everyone's problem—not a looming crisis that a future generation will have to solve when the water finally runs out. Fundamentally, the code treats groundwater as a resource shared by all citizens, not as private property.

The ultimate goal of Arizona's code is to achieve a *safe yield* from aquifers by 2025 in the three AMAs that coincide with the cities of Phoenix, Prescott, and Tucson. *Safe yield* is defined as a long-term balance between "the annual amount of groundwater withdrawn in the AMA and the annual amount of natural and artificial recharge." In other words, safe yield means the sustainable use of the state's groundwater resources.

SUMMARY

Earth's water is a renewable and sustainable resource. However, almost everything people do with water has the potential to waste or degrade this resource. In some cases, misuse of water resources has effectively made them nonrenewable; aquifers have been damaged or depleted, for example. However, understanding and appreciation of water resources can guide their sustainable use.

In This Chapter You Have Learned:

- Water is abundant on Earth, but the water people most need, freshwater, isn't. Only 0.9% of Earth's water is fresh, liquid water, and almost all of this is underground.
- People rely largely on surface water, but groundwater is becoming increasingly important. Surface and groundwater are connected and use of one can affect the other.
- People in the United States use a little over 400 billion gallons (1.5 billion m^3) of water each day. Electric power generation (40%) and crop irrigation (31%) are the two biggest users. Personal water use is about 80 gallons (0.3 m^3) per day.
- Withdrawing surface and groundwater can have environmental consequences. Dams and other surface-water diversions can alter stream flow and habitat. Groundwater pumping can damage aquifers, deplete resources, cause land subsidence, and lead to saltwater intrusion.
- Natural water quality varies tremendously and some natural water contains unhealthful contaminants, including dissolved metals and disease-causing microbes.
- Pollution is water quality degradation caused by people. Any human activity in or around water can cause pollution.

Historically, water quality has been seriously degraded where people's wastes have been released to surface waters.

- Legislation to protect the nation's water quality is administered by the EPA. The EPA identifies potentially harmful pollutants and determines the concentrations allowed in wastewater discharges and in drinking water.
- Wastewater treatment, recycling, conservation, and sound water-resource management can make significant contributions to sustaining water resources.
- Sound water management recognizes the connections between surface and groundwater, the importance of aquifer recharge areas, and the need to maintain balance between aquifer recharge and discharge.

KEY TERMS

aquifer (p. 302)
artesian flow (p. 306)
cone of depression (p. 310)
confined aquifer (p. 303)
consumptive use (p. 307)
distillation (p. 306)
gaining stream (p. 305)
groundwater (p. 302)
groundwater mining (p. 310)
losing stream (p. 305)
nonconsumptive use (p. 307)
nonpoint sources (of pollution) (p. 317)
parts per million (ppm) (p. 300)
permeability (p. 302)
point sources (of pollution) (p. 317)
porosity (p. 302)
reverse osmosis (p. 306)
salinity (p. 300)
saltwater intrusion (p. 310)
unconfined aquifer (p. 303)
water table (p. 302)

QUESTIONS FOR REVIEW

Check Your Understanding

1. Water exists in all three physical states on Earth: solid, liquid, and gas. Why is this important to life on Earth?
2. The water cycle is considered to be a closed system on Earth. What does this mean, and why is it important in terms of water resources?
3. What are surface waters? Are all of these good water sources for human consumption? Explain why or why not, and give examples.
4. What are the ways in which the salinity of a body of water can increase?
5. What is the role of pores between grains of rock, fractures in rocks, and other underground openings with respect to groundwater?
6. How do the properties of porosity and permeability affect the usefulness of an aquifer as a water resource?
7. Why would aquifers developed in limestone be some of the most productive on Earth? Formulate your answer in terms of porosity and permeability.
8. What is the difference between a gaining stream and a losing stream? Where is the water table with respect to the stream bed in each case?
9. Why is the surface of a lake or wetland generally the same as the water table surface? Illustrate the interaction with groundwater that makes this so.
10. How do reverse osmosis and distillation remove salt from water in the desalination process?
11. Water is used in a tremendous number of processes that provide materials, food, and energy to people every day. What is the difference between nonconsumptive use and consumptive use in these processes? Illustrate your answer with examples.
12. Plants that produce electric power typically use large amounts of water, but it is usually quickly returned to the environment with no added chemicals or materials. What are the power plants using the water for and what kind of pollution do these plants release into the environment via their water use?
13. The reservoir behind Hoover Dam lost 3% of its storage capacity in 14 years due just to sediment capture and infill behind the dam. If this rate is constant, roughly how long will it take for the reservoir to be completely filled with sediment?
14. What is groundwater mining? Use the example of the High Plains aquifer to illustrate the potential negative effects of this practice.
15. Why does over-pumping of groundwater in communities near the ocean shore risk the long-term health and usefulness of the local aquifer?
16. The text states that there is no such thing as "pure water" in nature. What other kinds of materials or chemicals are normally associated with natural freshwater?
17. What are the common pollutant types that contaminate freshwater? What are the sources of these pollutants called? Give an example of each.
18. Why is storm runoff from urban areas, particular from streets, harmful to surface and groundwater?
19. What is groundwater banking, and how does it work?

Critical Thinking/Discussion

20. The text states that only 0.9% of Earth's 171 million gallons per person is liquid freshwater. This works out to roughly 154,000 gallons per person at any given moment. Yet water shortages are very real, so what causes them? What are some reasons why high quality water may be scarce for people in any location on Earth?

21. The Great Lakes are discussed as a truly remarkable freshwater resource for North America, holding about 20% of the world's supply of readily accessible fresh surface water. These lakes are bordered by many states in the United States and two provinces of Canada. Diversion of water from the Great Lakes out of this basin has been an unpopular idea for most residents of the region, but what could either country do if the other decided to divert water to more distant areas for its own purposes, from its own shoreline? Could the U.S. stop Canada from doing this? Could Canada stop the U.S. from doing something similar? How would such a situation likely be handled, legally and diplomatically? Discuss what kinds of scenarios and solutions might emerge.
22. Desalination is often seen as the best future solution for water supplies in arid and semi-arid regions such as the Middle East or the U.S. Southwest. However, it is not without problems. What are the economic and environmental tradeoffs and challenges associated with the shift to a desalinated water source in these regions?
24. In the United States, its territories, and its possessions, people use roughly 1360 gallons of fresh water a day to support themselves and their lifestyle. Based on the consumption figures provided in the text, outline three strategies for directly reducing the amount of water used per person, focusing on water use in energy production, food production, and personal use. Discuss in general terms the costs and difficulties of implementing your solutions and strategies.
25. Aquifer recharge zones are among the most critical to protect from pollution, as any water that goes into the aquifer necessarily affects the quality of the water withdrawn later from the aquifer. Point-source pollution can be easy to identify and mitigate. Why is protection of aquifers from nonpoint-source pollution so important, why is it so challenging, and what strategies can be used by lab managers to minimize the effects of nonpoint-source pollution on groundwater quality?
26. The text outlines the three levels of water treatment and purification, as well as recycling schemes that have been put in place by many communities. What is the underlying logic and benefit of installing recycling systems in municipal or local water districts? What are the economic, social, psychological, and practical limitations and challenges in implementing this approach to water conservation, especially compared with the other approaches outlined in the text?

ANSWERS TO IN-CHAPTER INSIGHT QUESTIONS

P. 301

What are some sustainability issues associated with people's use of surface water?

- Surface-water dams and diversions change biosphere habitat and biodiversity.
- Use of surface water can change its quality and availability.
- Surface-water allocation priorities can be challenging to determine and implement.
- Surface-water resources can be insufficient for meeting all needs.

P. 306

What are some sustainability issues associated with people's use of groundwater?

- Groundwater withdrawal can deplete aquifers faster than they recharge.
- Depletion of groundwater resources can lead to permanent damage of aquifers through compaction.
- Groundwater withdrawal that lowers the water table can affect surface-water levels and habitat.

P. 307

What are some sustainability issues associated with desalination?

- Separation of non-salty water by desalination processes leaves behind a very salty brine as a waste product. Disposing of this brine in an environmentally sound way is needed for desalination to be sustainable.
- Desalination uses financial resources that could be applied to other ways such as water conservation.
- Desalination could replace withdrawals from surface and groundwater and help protect or enhance natural ecosystems and habitat.

P. 307

What are some Earth systems interactions associated with consumptive water use?

- Water is transferred from hydrosphere reservoirs such as surface water bodies to other Earth systems reservoirs such as plants and the atmosphere.
- Water may be removed from Earth systems interactions if it is used as a component of manufactured material such as cement.

P. 310

What are some sustainability issues associated with dams and diversions of surface water?

- Dams and diversions affect biodiversity by changing habitat along and within rivers.
- Sediment accumulation in dam reservoirs can reduce their size and capabilities over time.
- Dams and diversions may not satisfactorily address shifting water resource needs in the future.

P. 314

What are some Earth systems interactions associated with higher arsenic concentrations in natural waters?

- Geosphere processes mobilize and concentrate arsenic in minerals like arsenopyrite.
- Reactions of arsenopyrite with oxygen in air or water releases arsenic to the environment.
- The biosphere is exposed to arsenic through ingestion of arsenic-bearing water.

FIGURE 11-1 ▶ From China, with Haze

In April 2001, a huge dust cloud from China (at left) blew across Japan (center) and the Pacific Ocean all the way to western North America. Haze obscured the view in the Grand Canyon (facing page) as well as many other places throughout the American West. The dust consisted largely of eroded soil.

CHAPTER

11

SOIL RESOURCES

In April 2001, people in downtown Denver had trouble seeing the nearby Rocky Mountains. Visibility was reduced to 16 kilometers (10 mi) by a whitish haze. In Aspen, Colorado, the concentration of particulate matter in the air—dust—was four times greater than usual. The culprit was a dust cloud 6 kilometers (4 mi) thick that blanketed western North America from the U.S.–Mexico border to Calgary, Canada—some 2000 kilometers (1200 mi) to the north.

The remarkable thing about this dust cloud was that it came from halfway around the world! It originated in northwest China—a dry region where increased farming and grazing are making the land more susceptible to wind erosion. Windstorms pick up soil from the land surface and blow it across China, across neighboring countries like Korea and Japan, and, within 4 to 10 days, all the way across the Pacific Ocean to the United States (Figure 11-1).

In China, the dust storms are so thick that they shut down airports; crews need to repeatedly clean municipal buildings, and residents caulk their windows during the dust storm season (Figure 11-2, p. 328). Dust storms cause pall over north Such concentrations of dust are a serious health hazard for older people and people with respiratory problems. But dust storms also pose serious long-term challenges for Chinese agriculture. The dust is soil eroded from farmlands. China may be losing the equivalent of 2300 square kilometers (900 mi^2) of farmland to the wind each year. This vast country is literally scattering its valuable soil resources to the wind.

China isn't the only place where soil erosion, by wind and flowing water, is significant. In fact, erosion is the principal way soil is degraded or lost all around the world. Other factors affect soil, too—and as you might suspect, people's activities are a major influence on the quality and amount of soil resources.

Soil is a mix of interacting Earth systems. Geosphere starting materials are weathered by interactions with the atmosphere and hydrosphere to begin the process of soil formation. The resulting loose mix of weathered geosphere materials and open voids filled with air or water provides habitat for an enormous number of very small but incredibly diverse organisms. These components interact continuously to develop mature, fertile soils—the soils people depend on for most of their food. But as you will see, soils are sustainable resources that do more than provide food for people—they also play significant roles in the transfer of matter and energy among Earth systems.

FIGURE 11-2 ▲ Earth Systems Interactions: Geosphere Meets Atmosphere Meets Biosphere in China
Dust storms continue to be a fact of life in China.

IN THIS CHAPTER YOU WILL LEARN:

- What soil resources are
- The key functions of soil
- How soils form
- How soils are characterized and how they vary
- The physical, compositional, and biological properties of soils
- How soils are degraded or lost
- How soil resources can be sustained

11.1 What Soil Resources Are

Soil is an essential resource for life. Like water, soil resources are used by people for basic sustenance. Most of the food people eat comes from soil—either directly (in the form of plants such as wheat and rice) or indirectly (in the form of meat and dairy products derived from animals that eat plants). Also like water, soil resources seem to be poorly appreciated; people frequently degrade and destroy them. Appreciating soil resources starts with understanding what soil is and how it functions.

SOIL DEFINITIONS

Everyone has an idea of what soil is and how it is used. Perhaps the most common perception of soil is that it's "dirt"—dirt in which people grow things, like flowers in a window box or corn in a field. It's the stuff that gets our hands dirty—soiled—digging in the backyard or playing football in the local park. But what really *is* soil?

To soil scientists, soil is pretty much the same thing most people think it is. It is the dirt that plants can grow in. This definition emphasizes one key function of soil, as a medium for growing plants.

To geologists, soil is any loose surface material produced by the physical and chemical breakup (weathering) of rocks or sediment. By this definition, plants do not need to grow in it

FIGURE 11-3 ▲ Footprints in Moon Soil
Moon soil consists of small rock fragments and dust (regolith), pulverized by countless meteorite impacts over millions of years.

for it to be soil. Vegetation-free deserts can have soil. From this perspective, the Moon's surface has soil, too (Figure 11-3).

To engineers, soil is a physical medium—surface material that can be broken up and moved without blasting. This is a useful definition for engineers because they build things like roads, bridges, and buildings on soil. Engineers need to understand the physical properties of the materials underlying the structures they build, and in most places this material is soil.

From the perspective of environmental geology, soil is all these things—a medium for growing plants, the loose surface material developed by weathering, and a foundation material for structures. Most soil is a mixture of mineral particles of varying size and organic matter of varying composition. Together, these components make up about 50% of a soil's volume. The remainder is air and/or water that fills voids (pores) in the soil. Soil components interact to perform many valuable functions.

SOIL FUNCTIONS

Soil plays many roles in the environment. Among the most important are:

- Supporting the plants that provide us with food and fiber
- Cleaning and storing water
- Recycling waste
- Providing habitat for diverse forms of life
- Transferring matter among Earth systems

All of these soil functions are useful to people.

Food and fiber

Green plants capture the Sun's energy and use it to make chemical compounds that store this energy. This ability makes plants the principal sources of food and fiber for people and animals. Some of these plants live in water, but most are rooted in soil.

It is estimated that cereal grains provide one-half or more of the protein and calories that people worldwide consume each year (Figure 11-4a). Over one-third of the grain

FIGURE 11-4 ▼ Food and Fiber Come from Soil
Food crops such as wheat **(a)** and trees, the principal source of fiber **(b)**, are not possible without soil.

(a)

(b)

produced feeds livestock that people raise for food. Fiber comes from crops like cotton and from the wool of animals like sheep (which feed on plants). Such fiber is still the raw material for much of our clothing. Most of the fiber we use, however, comes from wood (Figure 11-4b). Globally, people consume an estimated 3.4 billion cubic meters (120 billion ft^3) of wood each year, largely for building materials, fuel, and paper. One-fifth of the wood harvested each year is used to make paper.

Because of the productivity of plants, civilization itself can be thought of as rooted in soil. The ancient civilizations of Egypt, Mesopotamia, India, and China developed on river floodplains where rich soils and water were available for farming. The foundation of vibrant civilizations is their ability to produce food surpluses. This was possible along rivers because annual floods helped replenish the soil and kept cropland productive for hundreds of generations. With abundant food, people were free to pursue other endeavors—they could trade food for other resources, build cities, and investigate the world around them. They could build civilizations.

Water storage and cleaning

Soil both stores and cleans water. Because soil is a loose material, it is full of small voids called pores. Under most conditions pores are filled with both air and water, but during periods of heavy precipitation water can displace the air, completely filling the pores and saturating the soil. Soil's ability to soak up and store water helps diminish flooding and supports plants. Plants always need large amounts of water to grow—about 90 to 400 kilograms (200 to 900 lbs) of water for every 0.5 kilograms (1 lb) of new solid growth. A soil's ability to store and provide this water, even in dry periods, is essential to keeping vegetation alive and healthy.

Soils that are commonly saturated with water can become wetlands—mushy and swampy areas with thick mats of decaying organic matter (Figure 11-5). These areas are excellent water filters. Nutrients, sediments, and even chemicals and organic waste are removed from water as it slowly passes through wetlands.

Waste recycling

You learned in Chapter 10 how septic systems are used to treat domestic wastewater and sewage before it is released to the environment. The wastewater released from septic tanks percolates through drain fields to soil. Soil filters out small particles in the water and supports biological activity that further decays organic matter. As the water slowly makes its way to the water table, it

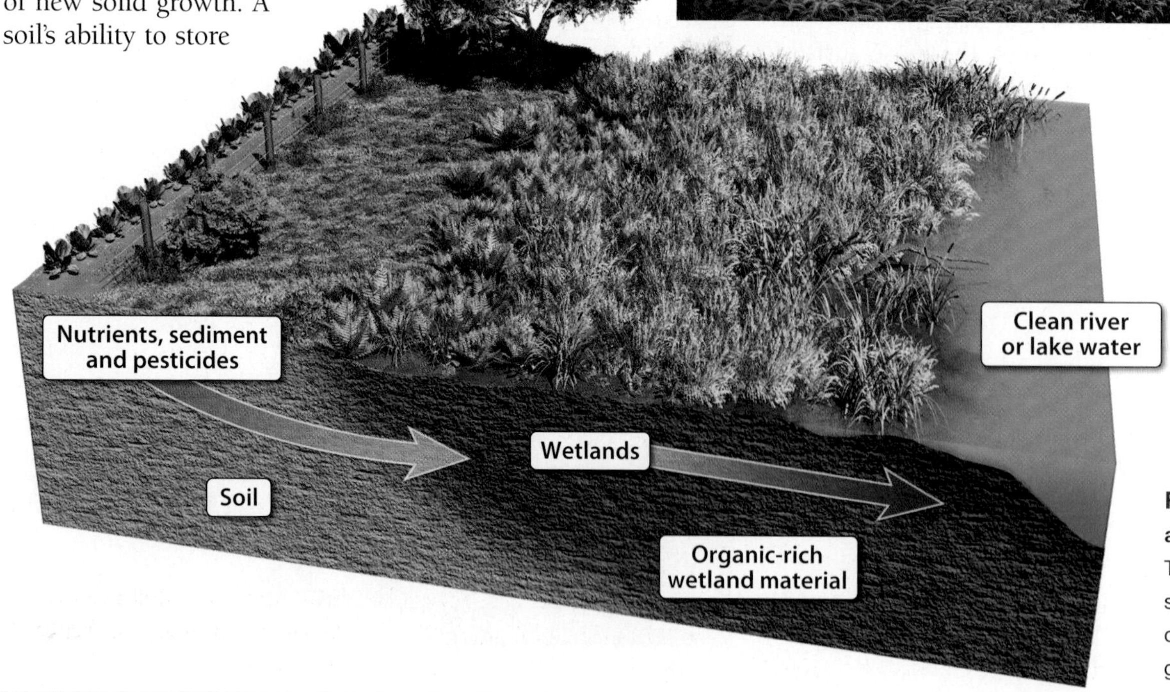

FIGURE 11-5 ◀ **Wetlands Store and Clean Water**
The thick organic material in wetland soils has high porosity and filters contaminants from water that migrates through (blue arrows).

becomes cleaner as the soil traps organic matter, including bacteria (Figure 11-6). Sewage sludge from community treatment facilities (discussed in Section 10.5) can be used as a fertilizer. About half of the nation's treated sewage sludge (called **biosolids**) is recycled in this way. | Biosolids frequently asked questions Other wastes can be recycled in soil, too, including drywall and the dust collected in the exhaust systems of electric power plants.

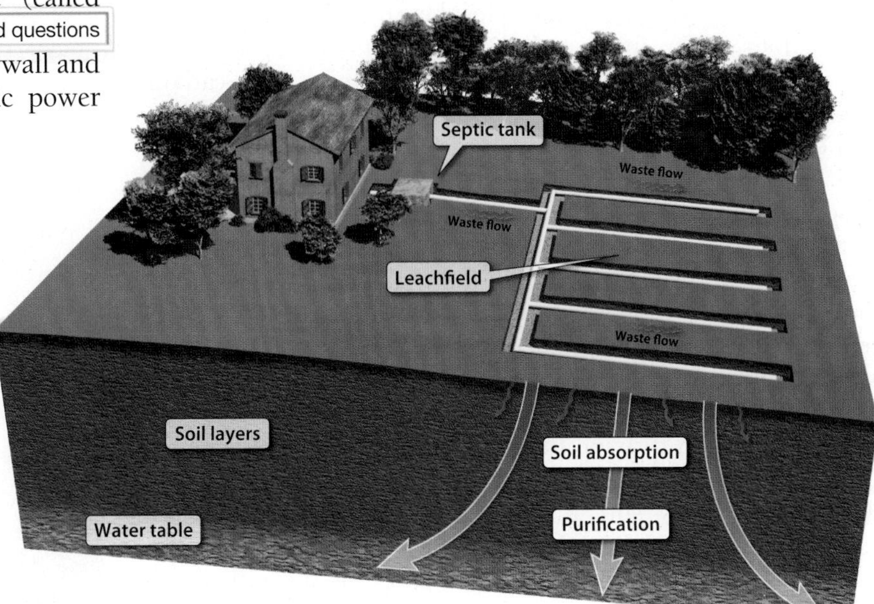

FIGURE 11-6 ▼ Soil Cleans Water
Water passing through soil is cleaned by a combination of processes: absorption, filtration, and the biological consumption of organic matter.

Soil as habitat

Soil is crammed full of life (Figure 11-7). It can contain more species and biodiversity than aboveground habitats, including places like rainforests. The living things in soil include bacteria, fungi, protozoa, arthropods, and worms of various types. Plants—converters of energy from the Sun, and the foundations of the biosphere's food webs—are literally rooted in soil. It is around plant roots that much soil life thrives.

FIGURE 11-7 ◀ Soil as Habitat
A complex food web in soil supports abundant life.

Earth systems interactions

Soil is a membrane that connects Earth systems. It intimately mixes components from all four Earth systems (geosphere, atmosphere, hydrosphere, and biosphere) and facilitates their interactions. As an interface between Earth systems, soil plays key roles in the transfer of matter among them. The carbon cycle, discussed in Chapter 1, is an example (see Figure 1-10). Soil contains about twice as much carbon (1580 billion tonnes, or 1740 billion tons) as does the atmosphere. But carbon resides in soil for only nine years on average because it is continually in flux to the atmosphere, biosphere, hydrosphere, or geosphere.

Photosynthesis by plants is the main process that provides carbon to soils. When plants (and animals that have fed on plants) die, their remains add carbon compounds to soils (Figure 11-8). This material is food for many organisms that live in soil. Decomposition of organic material in soil produces carbon dioxide that can escape to the atmosphere (a process called soil respiration) or be dissolved in groundwater. The ways soil influences carbon distribution in Earth systems give it a significant role in determining the atmosphere's content of greenhouse gases, as we will see in Chapter 14.

Perhaps one of the best understood transfers of matter in soils is between the geosphere and biosphere. All except four (oxygen, hydrogen, nitrogen, and carbon) of the 16 essential nutrients (chemical elements) that plants need to grow, come from the geosphere through soil. Soil makes essential nutrients like phosphorus, calcium, magnesium, potassium, sulfur, and iron available to plants. In natural settings, plants recycle nutrients to soil when they die, and thus help sustain the soil's ability to grow new plants. People's interruptions of this self-sustaining cycle—by harvesting plants, for example—can deplete soil of nutrients.

Because of soil's role in Earth systems interactions and in supporting life, soil resources need to be sustained. However, sustaining soils can be challenging. If they become degraded or destroyed, it can take a very long time for natural processes to form them again.

FIGURE 11-8 ▲ Soil and the Carbon Cycle
The global soil reservoir contains about 1580 Gt (billion tonnes) of carbon. Ongoing carbon exchanges between soil and other reservoirs (shown in Gt) are a part of the carbon cycle that people influence by their land use practices.

11.2 How Soils Form

Soils are dynamic natural materials, the end product of the weathering of the geosphere. You learned about physical and chemical weathering in Chapter 4. Making mature soils takes a combination of physical, chemical, and biological weathering. Over hundreds to thousands of years, these soil-forming processes combine to produce mature soils. Because soil-forming processes are affected by climate, the underlying geosphere materials, surface topography, and how long weathering processes have been at work, soils vary in character from place to place.

SOIL-FORMING PROCESSES

In Chapter 4 you learned that physical processes such as frost wedging break rocks into smaller and smaller pieces. The result varies, as Figure 11-9 shows, from a pile of rock rubble to a mixture of sand-sized (or smaller) mineral grains and rock fragments. Soils can develop directly from these broken bedrock materials or from sediments derived from them. These are the parent materials for soils.

FIGURE 11-9 ▲ Parent Materials of Soil
Soil formation begins when bedrock is broken down into smaller and smaller pieces.

Chemical processes further break down and change the parent materials. The key processes are those introduced in Section 4.2: hydrolysis, oxidation, and dissolution. In soils, these processes commonly produce material containing much clay, quartz, and oxide minerals. They also enable elements to be dissolved from minerals and removed or deposited in another part of the soil. Because warmer, wetter conditions facilitate chemical reactions, the changes brought about by chemical weathering occur faster and tend to be more complete in tropical climates.

Soil life aids weathering. Organisms churn soil, degrade and metabolize organic material, add their waste (and, when they die, their bodies), and emit carbon dioxide that reacts to form weak acids. Plant nutrients, including phosphorus and nitrogen, come from decomposing organic matter, and much of a soil's physical character—including porosity, permeability, and water-holding capacity—is influenced by its organic content.

A product of biological activity in soil is **humus**: organic matter that accumulates in shallow parts of soil, giving it a dark color. Humus facilitates many chemical and physical soil processes, especially where it is acidic.

The end result of effective soil-forming processes is a well-developed, mature soil that contains abundant open spaces (pores), organisms, decaying organic debris, clays, and grains of various minerals, including oxides and quartz.

THE SOIL PROFILE

Mature soils are not homogeneous. Soil-forming processes typically stratify soils into layers known as **horizons**. Climate, topography, and the nature of the underlying parent materials can strongly influence the rates and end results of soil development. Nevertheless, soil development tends to produce a typical layered sequence of horizons called a **soil profile**, like the one shown in Figure 11-10.

A soil's profile is defined by the horizons developed within it. Typical mature soil profiles have three principal or *master* horizons, designated A, B, and C.

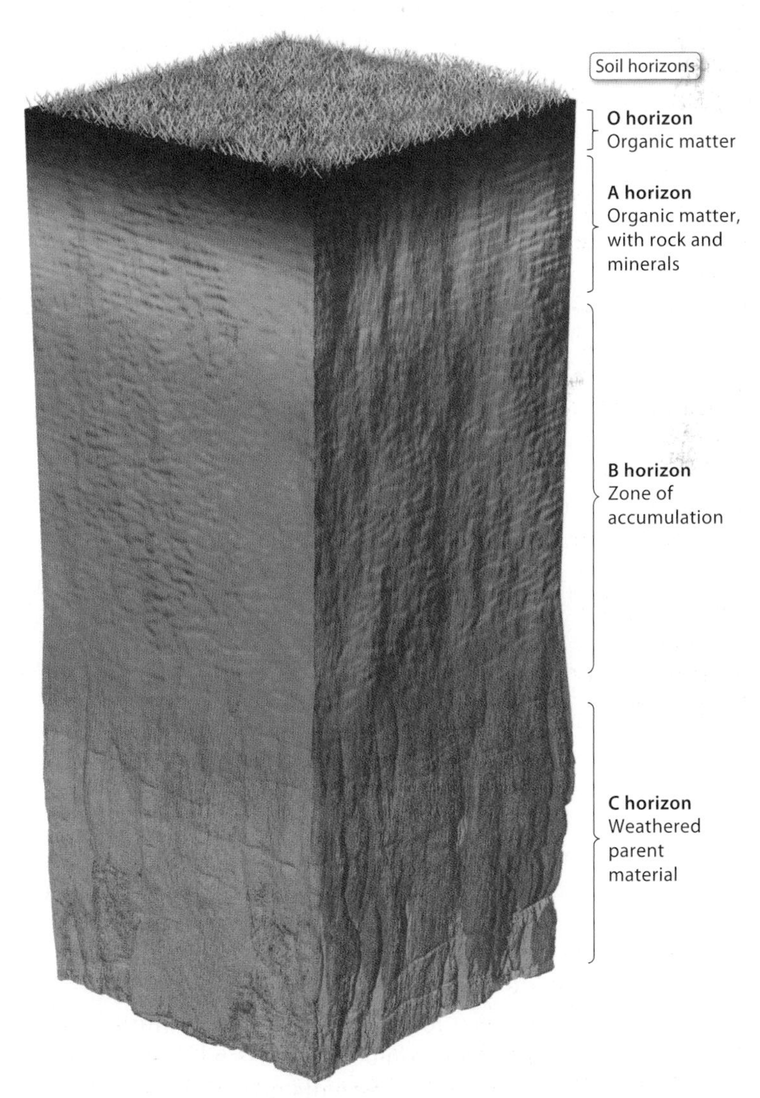

FIGURE 11-10 ▲ Typical Soil Profile
Soil components are leached or otherwise removed from the organic-bearing A horizon and accumulate in the B horizon. The C horizon contains remnants of the parent material. The O horizon is recognized where abundant organic material accumulates on the surface.

The A horizon (**topsoil**) is the upper layer in a soil (see Figure 11-10). This is where water seeping into the ground dissolves material and either carries it away or deposits it in lower layers. Water moving through the A horizon can also physically carry small mineral particles (especially clay) and decaying organic matter (mostly from vegetation). The downward transfer of material from this horizon commonly makes it light in color, but accumulated organic matter tends to make it dark near the surface. The material in this horizon is nutrient-rich and a fertile medium for growing plants.

The A horizon gradually changes (grades) downward into the B horizon. In the B horizon, much of the material removed from the A horizon accumulates. This is why clay minerals are common in the B horizon. Iron-oxide minerals commonly give the B horizon splotchy yellow, orange, or reddish colors.

The B horizon grades downward into the C horizon. The distinctive characteristic of the C horizon is that it contains remnants of the underlying parent material. Weathering processes are under way, but they have incompletely changed the parent materials into soil components like those in the A and B horizons. Unchanged mineral grains and rock fragments from the underlying bedrock can be found in the C horizon.

Master horizons provide a beginning framework for describing soil profiles. Soil scientists recognize many sub-horizons and other internal characteristics that enable them to more completely describe a soil and understand how it formed. For example, the "O" horizon is commonly recognized in soil profiles with a very organic-rich upper layer (see Figure 11-10).

What are some Earth systems interactions involved in forming soil?

SOIL VARIATIONS

The nature of soils varies locally and regionally across land surfaces. In fact, every state in America has a specific soil identified as its "state" or representative soil. State soils For well-developed, mature soils typical of large regions, the key factors that influence soil character are parent materials and climate. Topography is a factor that leads to more local soil variations.

Soils can develop on any type of bedrock or unconsolidated surface material. For example, parent materials vary from mafic to felsic igneous rocks and from mud-rich, sandy, or gravelly sediments to limestone. The composition of the parent material directly influences the abundance of nutrients and other compositional soil characteristics, such as the amount of calcite or quartz.

Climate directly influences the physical and chemical weathering processes that form soils. Warm, wet climates facilitate hydrolysis and dissolution of minerals and can cause soils to quickly become leached of nutrients needed by plants. Arid climates, on the other hand, may lead to soils with little organic matter, and with carbonate minerals that have been deposited in them rather than being dissolved and removed.

Topography can lead to soil variations where slopes are present. Both the steepness of slopes and their orientation with respect to the Sun, called **aspect**, influence soil development. Slope soil materials are susceptible to erosion, and water movement through them is commonly in a downslope direction. Over time, these factors combine to transfer soil materials from slopes to lowlands. Aspect will affect soil temperature, and thus influence the rates of physical, chemical, and biological weathering processes. Erosion, water movement, and variations in soil moisture and temperature commonly produce soils on slopes that are thin and poorly developed compared to those formed on level areas like valley bottoms.

Soil scientists carefully describe soils and use a systematic classification system to categorize and compare them. This enables them to define many soil variations called **soil orders**. There are 12 major soil orders, but four examples—*spodosols, aridisols, mollisols,* and *oxisols*—show how soils vary regionally. The twelve soil orders – soil taxonomy

Spodosols develop in cool, moist coniferous forest regions such as the Pacific Northwest, the Great Lakes region, and the Northeast states. They are acidic and characteristically have a subsurface accumulation of humus that is combined with aluminum and iron oxides or hydroxides. Their oxidized character commonly gives them splotchy brown to reddish colors (Figure 11-11). Spodosols – NRCS soils Their

FIGURE 11-11 ▲ A Spodosol Profile
Spodosols are acidic forest soils with strongly leached A horizons and B horizons with aluminum and iron oxides or hydroxides. The leached A horizon in this example is light gray in color.

lichen, and their byproducts that holds soil particles together (see Figure 11-12).

Mollisols develop in the grasslands (prairies) of the world. They are widespread in the central and western states. As you can see in Figure 11-13, the A horizon in mollisols is thick and dark-colored from the accumulation of organic material from plant roots. This makes mollisols very fertile and excellent for agricultural purposes. They are the soils that produce most of the wheat, corn, soybeans, and other crops people eat. Regions with well-developed mollisols produce so much food that they are commonly called "breadbaskets."

Oxisols—also called **laterites**—develop in warm, wet tropical forests. In the United States, they are found in parts of coastal Louisiana and southern Florida, Hawaii, Puerto Rico, and the U.S. Virgin Islands. Oxisols are deeply weathered soils that have been leached of much of their original mineral content. The result is an aluminum- and iron-oxide rich soil without abundant plant nutrients. Their oxide content gives

FIGURE 11-12 ▲ An Aridisol Profile
Aridisols form in arid climates and lack abundant organic matter. Soluble minerals such as calcite commonly accumulate in the B horizon (the variably thick, lighter colored material between 2 and 4 on the scale). Scale is in 10 cm (3.3 in) intervals.

FIGURE 11-13 ▲ A Mollisol Profile
Mollisols are characteristic of temperate climate grasslands. They have a nutrient-rich A horizon (the dark brown layer from the surface down to about 20 cm depth).

acidic nature means that people need to add neutralizing material, such as lime, to help agricultural plants, and even lawn grass, grow.

Aridisols develop in arid regions such as the Southwest states. Under these dry conditions, calcite and other minerals easily dissolved by water remain in the soil or are deposited in it (Figure 11-12). The soluble minerals accumulate in the B horizon through dissolution and redeposition by infiltrating waters from the surface. In addition, the attraction of water to mineral grain surfaces (its *adhesion*) promotes the upward movement of water in aridisols as they dry through surface evaporation. This wicking process, called *capillary action,* also helps deposit soluble minerals in the upper levels. Organic matter is not abundant in aridisols. In fact, many seem to lack vegetation entirely. But if you look closely, these soils commonly have a *biological soil crust* composed of cyanobacteria, mosses,

FIGURE 11-14 ▲ An Oxisol Profile
Oxisols develop in tropical climates and characteristically lack well-developed horizons. They are essentially thick, highly leached A horizons with abundant iron and aluminum oxide or hydroxide minerals. The scale is 1 meter (3.3 ft) long.

them the light to reddish colors seen in Figure 11-14. In some places, aluminum or nickel is concentrated enough that the soil can be profitably mined for its metal content (Figure 11-15).

These four examples illustrate how variable soils can be. The variation in soil character leads to large differences in their agricultural usefulness, the challenges they bring to engineering projects, and their susceptibility to degradation and loss. By understanding the properties of individual soils, people can determine how their actions will affect soils and whether they will help sustain soil resources.

11.3 Soil Properties

Soil properties determine how well soils can serve useful functions. Physical, compositional, and biological properties combine to give soils their individual character and to define their quality.

FIGURE 11-15 ▲ Mining an Oxisol
Oxisols are sources of iron, nickel, and aluminum. Bauxite, seen here being mined in Australia, is an oxisol rich in aluminum.

PHYSICAL PROPERTIES

Physical properties of soils include texture, structure, density, shear strength, and compressibility. A soil's texture is determined by grain size—the proportion of clay-sized, silt-sized, and sand-sized particles within it (Figure 11-16). Clay feels smooth and cohesive; wet, clay-rich soil is sticky and can be molded. Silt-sized particles feel gritty if they are rubbed against your teeth, and sand-sized particles are visible to the eye and feel gritty if they are rubbed between your fingers.

Soil structure reflects the way that soil particles aggregate together, which determines how the soil breaks apart when it is disturbed:

- Granular soils break into small pieces, approximately the same size—up to about 10 millimeters (0.4 in)—in each dimension;
- Blocky soils break into larger pieces—up to 50 millimeters (2 in) in each dimension;
- Platy soils break into small platy pieces;
- Prismatic soils break into elongate blocks.

These types of soil structure are shown in Figure 11-17. A soil's structure is a significant factor in determining its susceptibility to erosion and how air or water migrate through it. Granular soils are the best at allowing air and water to pass through.

Density is the mass of a soil per unit volume. Density depends on the type of solid material and on the proportion of voids (porosity) in the soil—the internal open spaces that can be filled by air or water. In most cases, the less dense a soil is the more porosity it has and the more air or water it can contain.

To understand its suitability as a foundation material for construction, engineers determine a soil's shear strength and compressibility. **Shear strength** is a measure of how well the soil resists forces before fracturing internally. **Compressibility** measures how a soil compacts under applied forces, such as

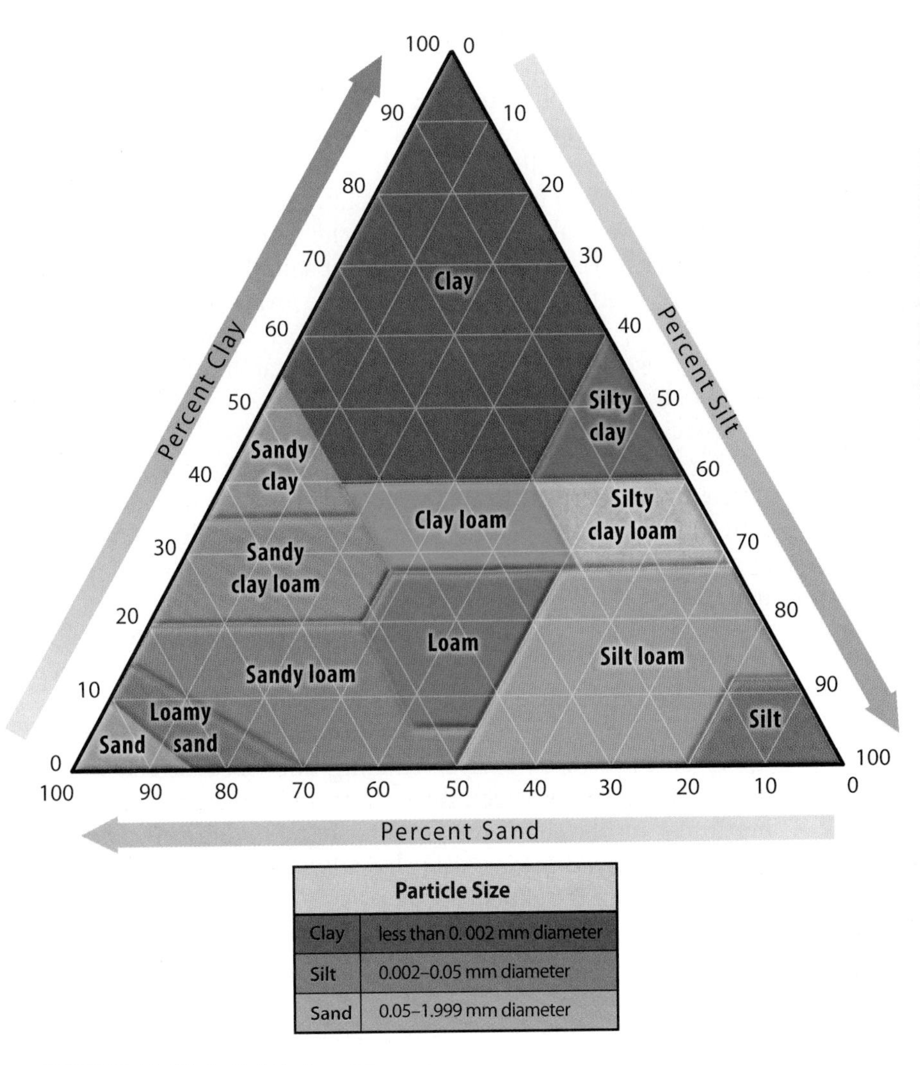

Particle Size	
Clay	less than 0. 002 mm diameter
Silt	0.002–0.05 mm diameter
Sand	0.05–1.999 mm diameter

FIGURE 11-16 ▲ Soil Texture Diagram
The proportions of clay, silt, and sand define soil textures.

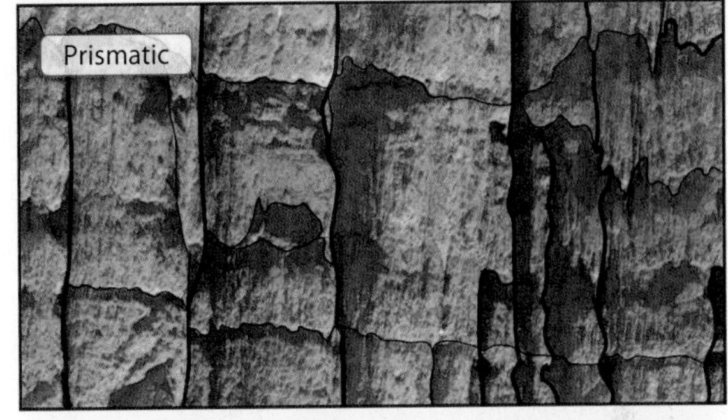

FIGURE 11-17 ▲ Soil Structures
Soil structure determines how the soil breaks apart when it is disturbed.

when a building is constructed upon it. Engineers take samples of soil and test them to determine these physical properties. In general, the higher the density of a soil, the greater its shear strength and the more suitable it is as foundation material.

The importance of soil strength is especially evident in surface excavations. The steep sides of soil excavations, as in trenches dug to install underground pipelines, can make soils unstable and lead to their collapse (Figure 11-18, p. 338). Because soil is heavy—a cubic meter of soil commonly weighs 1900 kilograms (3200 lb) per yd^3—people buried in soil don't have much of a chance. Collapsing soil excavations kill about 70 people each year in the United States. The following story, from the *Los Angeles Times* of June 24, 1993, is typical. The headline read, "Laguna Beach Man Killed in Trench Cave-In." Here is an excerpt from the account:

> The survivor said, "All day he had been asking me, 'If this caves in, where are you gonna go?' I asked him this morning, let's get some boards to shore this thing up and he said, 'We're almost done.' In five more minutes we would have been sitting at the table eating lunch." It took firefighters an hour to reach the man's wrist and determine he was dead. It took another five hours to pull his body from the trench.

COMPOSITIONAL PROPERTIES

We can determine a soil's physical properties, but they give an incomplete picture of a soil's character. Compositional properties combine with physical properties to help make soils distinct. Three key compositional properties of soil are the proportion of

FIGURE 11-18 ▲ Excavations in Soil Can Be Dangerous
This worker is in a dangerous position. About 70 people die each year in trench collapses and about ten times this many are injured.

air and water in its pores, its pH, and the amounts of nutrients available to plants.

The amount of water in a soil is its *moisture content*. The soil's pores can vary from completely filled with water (saturated) to completely filled with air.

A soil's **pH** is a measure of its acidity. On a scale of 1 to 14, a pH of 7 is neutral, less than 7 is *acidic*, and greater than 7 is *alkaline*, or basic (Figure 11-19). Most soil pH falls in the range of 4.5 to 9; wetter regions generally have lower-pH soils and arid regions have higher-pH soils. A soil's pH strongly influences chemical weathering reactions and the stabilities of minerals. Carbonate minerals, for example, will be dissolved in acidic soils and precipitated in alkaline ones.

The *nutrient content* of soil can vary significantly. Most nutrients are derived from the geosphere, but they can be removed from soil by plants, and can also be dissolved (and carried away) as water passes through. Plant nutrients If plant material is added back to the soil when plants die, the soil's nutrient content can be maintained. Mollisols (see Figure 11-13) are examples of soils that have maintained nutrient levels over long periods of time.

Oxisols, on the other hand, commonly have low nutrient levels because of extensive leaching and because the nutrients from decaying vegetation are quickly used by the growing forest rather than being added back to the soil. Oxisols need nutrient additions—fertilizer—in order to be productive farmlands.

An example of the limitations of oxisols comes from Brazil, where oxisols are well developed. These soils are characteristically acidic (pH = 5.5) and deficient in nitrogen, phosphorus, potassium, calcium, magnesium, sulfur, and other nutrients. Upland (non-irrigated) rice is generally the first crop planted when Brazilian forests are converted to farmland. But rice production varies tremendously depending on rainfall and the application of appropriate fertilizers. Under favorable rainfall conditions, fertilizer application can triple rice production. As Figure 11-20 indicates, adding nitrogen, phosphorus, and potassium fertilizers have positive effects. Without fertilizer, the oxisols of Brazil would be only marginally productive.

FIGURE 11-19 ▼ Soil pH
Soils range in pH from 3 (as acidic as vinegar) to 11 (as alkaline as ammonia).

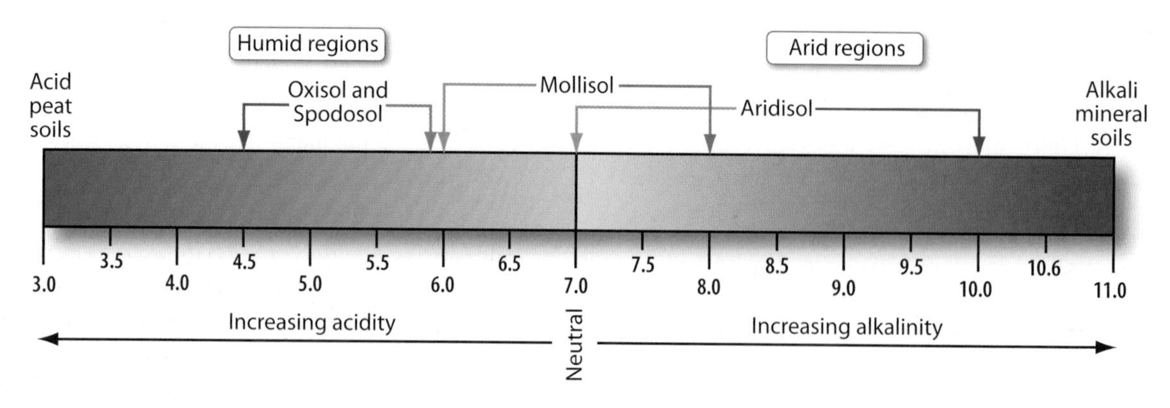

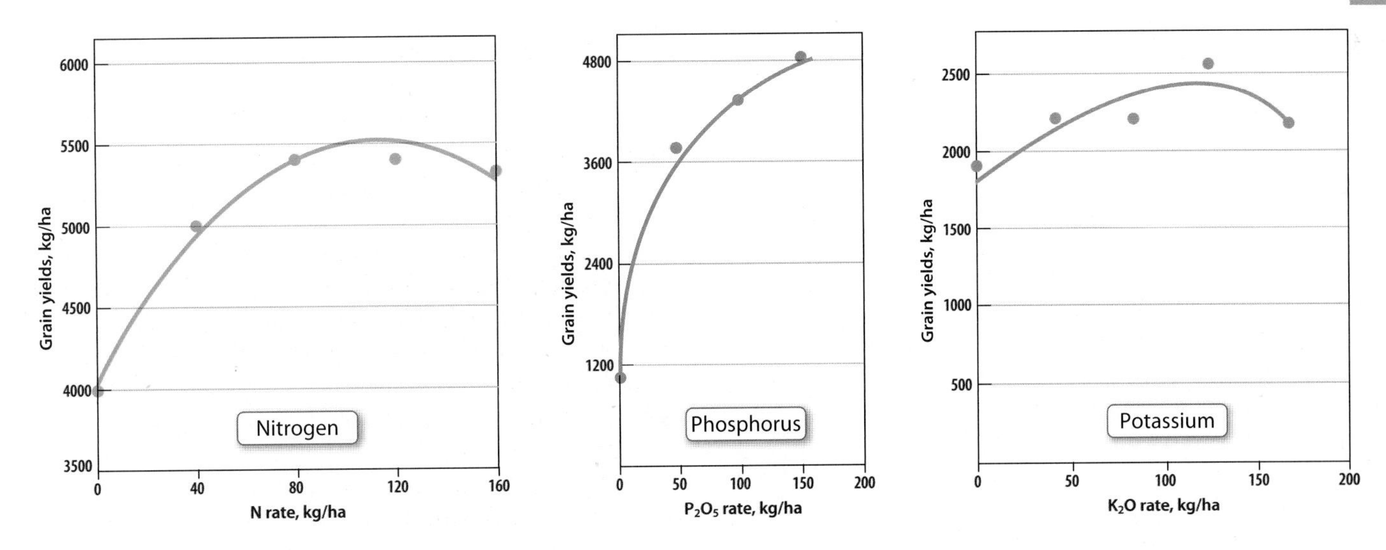

FIGURE 11-20 ▲ Effect of Fertilizer on Oxisol Rice Yields in Brazil

The addition of nitrogen, phosphorus, and potassium significantly increased rice yields as long as rainfall (or irrigation) was adequate to sustain healthy crops. The studies also showed that over-application of some nutrients can actually decrease yields. This is a common outcome in applying fertilizers, and is used to guide their efficient use.

BIOLOGICAL PROPERTIES

Soils are an ecosystem that commonly has more biological diversity than ecosystems on the surface. | NRCS soil biology As already noted, there can be more species in a shovelful of rich soil than in the entire Amazon rain forest. In many ways, soils are alive and their biological components are critical to the functions that soils perform. Life in soil includes bacteria, fungi, protozoa (single-celled organisms), worms, and arthropods (insects and similar creatures) (Figure 11-21). | Soil biological communities

A spoonful of soil can contain millions of bacteria. Some bacteria are primary producers of food by means of photosynthesis (Section 2.2), some fix nitrogen (take it from the air and convert it into forms plants can use), and some decompose organic matter.

TYPICAL NUMBERS OF SOIL ORGANISMS IN HEALTHY ECOSYSTEMS

		Agricultural Soils	Natural Mollisols	Natural Spodosols
Bacteria	Per teaspoon of soil (one gram dry)	100 million to 1 billion	100 million to 1 billion	100 million to 1 billion
Fungi	Per teaspoon of soil (one gram dry)	Several yards.	Tens to hundreds of yards.	Several hundred yards in deciduous forests.
Protozoa	Per teaspoon of soil (one gram dry)	Several thousand flagellates and amoebae, one hundred to several hundred ciliates.	Several thousand flagellates and amoebae, one hundred to several hundred ciliates.	Several hundred thousand amoebae, fewer flagellates.
Nematodes	Per teaspoon of soil (one gram dry)	Ten to twenty bacterial-feeders. A few fungal-feeders. Few predatory nematodes	Ten to several hundred.	Several hundred bacterial and fungal-feeders. Many predatory nematodes
Arthropods	Per square foot	Up to one hundred	Five hundred to two thousand.	Ten to twenty-five thousand. Many more species than in agricultural soils.
Earthworms	Per square foot	Five to thirty. More in soils with high organic matter.	Ten to thirty. Arid or semi-arid areas may have none.	Ten to fifty in deciduous woodlands. Very few in coniferous forests.

FIGURE 11-21 ◀ Life Is Abundant in Soil

Healthy soils are ecosystems with tremendous numbers and diversity of soil organisms. Agricultural practices tend to decrease the abundance of soil organisms, especially fungi, nematodes, arthropods, and earthworms.

Protozoa are mobile organisms in soil. They are much bigger than bacteria, which they eat, but a spoonful of soil can still contain thousands of them. The waste from protozoa is a source of plant nutrients such as nitrogen.

Fungi are stationary organisms that absorb food from their surroundings. Organisms like mushrooms and yeast are fungi, but there are over 100,000 known kinds (and probably several hundreds of thousands more that have not yet been identified), most of which are found in soil. Many fungi live around plant roots and help extend the plant's contact with soil nutrients (Figure 11-22). Enzymes secreted by fungi help decompose organic matter and make nutrients available to plants.

Worms—tiny roundworms, known as nematodes, and common earthworms—help recycle nutrients in soil. The earthworm They move through the soil eating just about anything they can get into their front end that is marginally nutritious. Out the other end comes waste (castings) that is essentially fertilizer. Earthworms produce castings equal to their own weight each day. In a year's time, they will produce over five tons of castings in an acre of fertile soil. Earthworms also churn the soil as they eat their way along. This promotes internal soil drainage, aeration, and the mixing of nutrients.

FIGURE 11-22 ▼ Fungi and Plants
Some fungi harm plants, but many more are necessary for healthy plant growth. The beneficial fungi live among or even within plant roots, where they make nutrients more readily available. **(a)** The potato plant on the right has benefited from having fungus living in its root system. **(b)** A scanning electron micrograph image of a fungus (white threads) on the roots of a vascular plant.

FIGURE 11-23 ▲ The Antlion—A Soil Predator
The antlion that lives in soil is the larval stage that eats ants and other insects. (Adult antlions are flying insects that feed on pollen and nectar.)

Arthropods (Figure 11-23) are widespread in soil. They churn soil as they move through it, helping to improve soil structure and making nutrients more available to plants. They thrive on decaying organic matter and also help control protozoa populations.

Most life in soil is not visible to people, but its abundance and species diversity is astounding. These biological components contribute to soil weathering processes, support vegetation growth, and cycle matter and energy among Earth systems. Without their biological components, soils would just be a mix of rocks and minerals. With them, they are a critical ecosystem, a living membrane at the boundaries of all Earth systems.

What are some Earth systems interactions that take place in soil?

SOIL QUALITY

The physical, compositional, and biological properties of soil combine to determine a soil's *quality:* its capacity to sustain plant growth and animal productivity, maintain or enhance water and air quality, and support human health and habitation. High-quality soils have a diverse biology, control water flow and water character, facilitate interactions among Earth systems, and store matter involved in Earth system cycles. High-quality soils are an exceptional resource that can be sustained if they are properly used and maintained. Too often, however, they are not, as we will see in the following section.

11.4 Soil Degradation and Loss

China's tremendous soil loss, evidenced by the 2001 dust cloud that obscured the Rocky Mountains, is just part of a global problem. Poor agricultural practices have degraded about one-third of the 1.5 billion hectares (3.7 billion acres) of cropland around the world (Figure 11-24). Wind and water erosion are the major causes of this soil degradation and loss, even in the United States. Erosion has removed one-third of our nation's topsoil (generally the upper organic-rich part of the A horizon). Some 16 million hectares (40 million acres) are losing 11 tonnes per hectare (5 tons per acre) each year to water erosion alone. Did you ever wonder what makes rivers like the Mississippi so "dirty" (Figure 11-25)? It is largely suspended sediment from soil eroded from the land that is being carried to the sea.

History has provided insight into the long-term effects of continuing soil loss. Just as the availability of rich soils fostered developments of the world's first great civilizations, the loss of these soils through erosion and degradation led to their demise. One of our nation's great soil scientists, W. C. Lowdermilk, investigated relationships between the ancient civilizations and their use of the land. Loudermilk found overwhelming evidence that soil loss (related to erosion as a result of agricultural practices) led to the collapse of civilizations. The lesson from history is clear—if soils are not sustained, civilization is not sustained. Loudermilk's report, *Conquest of the Land Through 7,000 Years*, was published in 1953 and remains a classic. NRCS conquest of the land through 7000 years It is an excellent overview of the relationship of people to soil.

FIGURE 11-25 ▲ Eroded Soil Makes Rivers Muddy
This photograph of the Mississippi River was taken near Memphis, Tennessee. It clearly shows where a lot of the soil eroded from farmlands in the river's watershed ends up—in the muddy river and headed for the Gulf of Mexico.

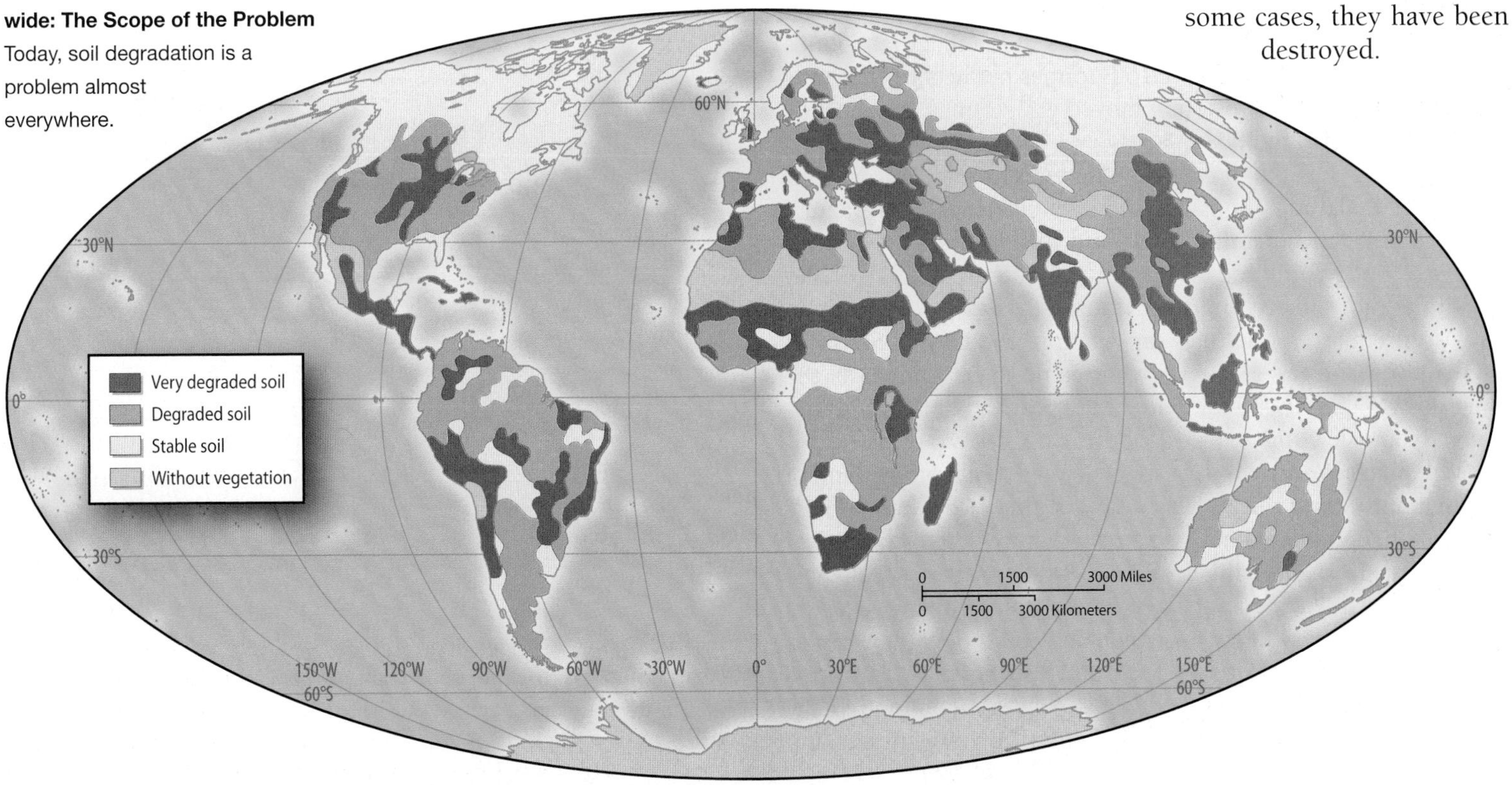

As important as erosion is and has been, it isn't the only cause of soil degradation and loss. People contaminate soil, deplete its biodiversity and nutrients, and cover it up or otherwise change it by their land uses. Some of people's effects on soil are reversible and some are not. In general, soils are a renewable resource but in some cases, they have been destroyed.

FIGURE 11-24 ▼ Soil Degradation Worldwide: The Scope of the Problem
Today, soil degradation is a problem almost everywhere.

Very degraded soil
Degraded soil
Stable soil
Without vegetation

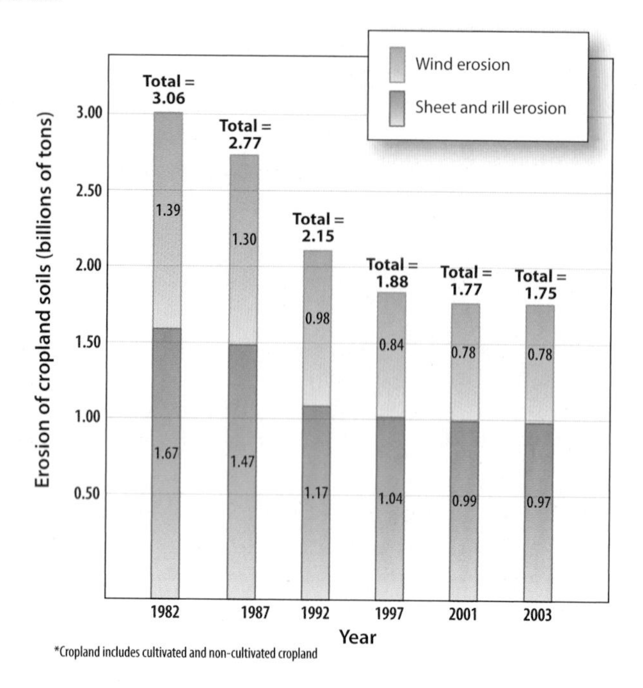

FIGURE 11-26 ▲ Wind and Water Erosion of Cropland Soils in the United States

Although soil conservation has been a national goal since the 1930s, erosion continues to be a significant problem in many places.

EROSION

As you can see from Figure 11-26, wind and water erosion combine to remove almost 1.8 billion tonnes (2 billion tons) of soil from America's cropland each year. Most of this erosion is in the watersheds of the large rivers draining the central part of the country where cropland is abundant. Although the amount of soil erosion has decreased in recent years, it is still a tremendous amount. Water erosion removed 5.8 tonnes per hectare (2.6 tons per acre) and wind erosion removed 4.7 tonnes per hectare (2.1 tons per acre) in 2003. Soil erosion – 2003 annual NRI – NRCS

Soil is susceptible to erosion wherever its vegetation cover is removed. Tilling, construction-site clearing, overgrazing, deforestation, and creating roads and hiking trails are known causes of both wind and water erosion (Figure 11-27). Wind erosion is more likely in arid regions. The type of devastation now occurring in China has happened in the United States. It occurred during the 1930s in what came to be known as the Dust Bowl—parts of Colorado, New Mexico, Texas, Oklahoma, Nebraska, and Kansas (Figure 11-28).

Following World War I, farmers flocked to the prairie land of this region to grow wheat. With tractors, they were able to plow up the native grassland and, if rains fell, grow large amounts of wheat. But 1931 brought drought conditions that would last until 1939. The lack of rain kept wheat from growing, and the bare soil was an easy target for winds that commonly blow across the plains. "Black blizzards"—dense and fast-moving dust storms like the one shown in Figure 11-29a—became common. They disrupted travel, damaged buildings, shut down businesses and

(a)

(b)

FIGURE 11-27 ▲ Soil Erosion

When vegetation is removed, soil becomes vulnerable to erosion by wind and water, as these examples from Madagascar **(a)** and the Chiapas highlands of Mexico **(b)** clearly show.

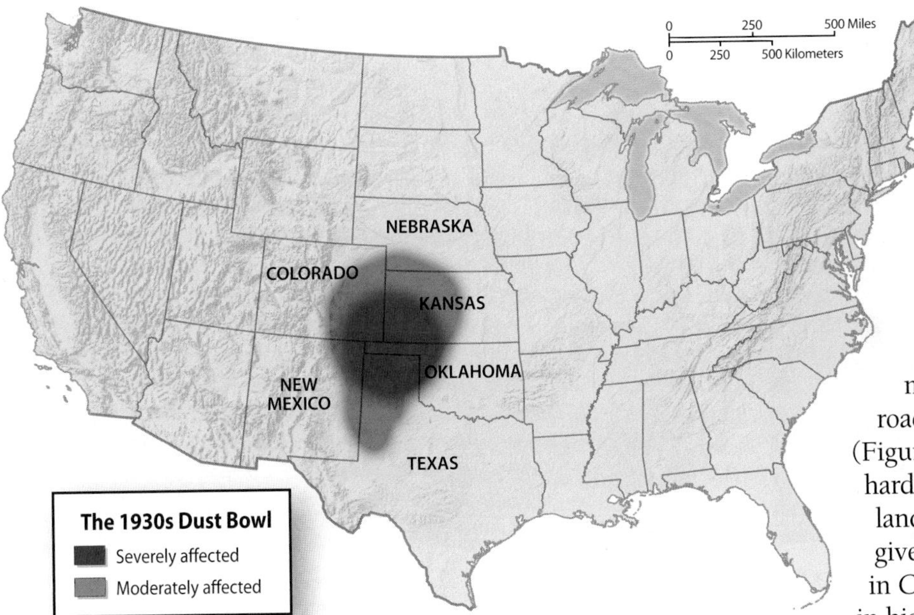

FIGURE 11-28 ◀ The 1930s Dust Bowl Extensive plowing of prairie followed by an extended period of drought led to devastating wind erosion of soil during the 1930s over large parts of Kansas, Colorado, New Mexico, and Oklahoma—the American Dust Bowl.

public institutions, made breathing difficult, and often produced longer-term respiratory harm in the most vulnerable. The windblown soil piled up along roads and fences and buried buildings and machinery (Figure 11-29b). Farming was impossible, and economic hardship accompanied the physical devastation of the land. Dust Bowl photographs Many people were forced to give up their farms and leave the region in search of work in California and elsewhere. As John Steinbeck described in his 1939 novel, *The Grapes of Wrath:*

> And then the dispossessed were drawn west—from Kansas, Oklahoma, Texas, New Mexico, from Nevada and Arkansas, families, tribes, dusted out, tractored out. Car-loads, caravans, homeless and hungry; twenty thousand and fifty thousand and a hundred thousand and two hundred thousand. They streamed over the mountains, hungry and restless—restless as ants, scurrying to find work to do—to lift, to push, to pull, to pick, to cut—anything, any burden to bear, for food. The kids are hungry. We got no place to live. Like ants scurrying for work, for food, and most of all for land.

The Dust Bowl experience brought widespread awareness of the destructive consequences of poor farming practices on lands susceptible to erosion. It was this episode that sent W. C. Loudermilk on his 1938 and 1939 investigations to learn the lessons of history about soil use, and sparked the creation of soil conservation organizations by both farmers and the government. Lessons learned in the Dust Bowl were translated into improved practices to prevent erosion and protect soils. But as Figure 11-26 shows, wind erosion still happens on American farmlands. It's responsible for about 45% of the 1.6 billion tonnes (1.8 billion tons) of eroded soil each year. Such losses increase the amount of fertilizer needed to keep soil fertile, make it more difficult for seedlings to survive, and damage crops. The costs of soil erosion are still high—tens of billions of dollars per year.

Erosion also occurs under wet conditions, when soil is exposed to rain and surface runoff of water. Water flowing across bare soil always becomes muddy. Small streamlets, or **rills**, and

(a)

(b)

FIGURE 11-29 ◀ Dust Bowl Scenes **(a)** This dust storm—a typical "black blizzard"—was photographed as it approached Stratford, Texas, on April 18, 1935. **(b)** Wind-blown soil buried abandoned buildings and machinery in the Dust Bowl.

(a)

(b)

FIGURE 11-30 ▲ Rill and Sheet Erosion
(a) Extensive rill erosion has washed away young plants in this Iowa field. **(b)** Sheet erosion commonly accompanies rill erosion in fields.

thin nonchannelized overland flow called **sheet flow** are responsible for most water erosion on cropland (Figure 11-30), but in some places erosion is concentrated in gullies that cut deeper into the soil. Farming practices like row cropping and plowing leave bare soils that can be battered by rain and easily eroded by runoff of surface water. Grazing that destroys the vegetation cover on soil makes it susceptible, too.

In places, deforestation has had the same effect. This process has essentially wiped out the magnificent cedar groves that once grew in the eastern Mediterranean region—the Cedars of Lebanon. Along with them went much valuable soil. Only about 300 trees remain (Figure 11-31) from a vast forest that covered the mountains of Lebanon. Deforestation and the resulting soil erosion, worsened by continual overgrazing, have made the Lebanon mountains a mostly barren landscape.

When soil erosion reaches high levels in arid regions, the land can become a desert. Whether the erosion is primarily due to natural changes that accompany drought conditions or due to misuses of the land, such as overgrazing, the lack of adequate vegetation cover leads to serious soil erosion. If dry, arid conditions continue, most of the soil can be lost and desertification can result.

FIGURE 11-31 ▼ Deforestation
After thousands of years of exploitation, few of the biblical Cedars of Lebanon now remain on mountains that were once heavily forested.

SOIL CONTAMINATION

Soil can be degraded by contaminating it with chemicals people use. Some contamination is deliberate, as when salt is applied to roads and carried onto soils by runoff. Some is accidental, as when petroleum or chemicals spill during transport. Salt, fertilizers, and pesticides have all contaminated croplands.

Salination

Salty soils decrease or prevent plant growth. The most widespread salt contamination of soil—**salination**—can occur naturally in arid regions. In areas of low rainfall, the water in soil contains dissolved mineral salts, including ordinary table salt (sodium chloride). The water is drawn by capillary action to the surface, where its evaporation leaves behind deposits of the mineral salts in the soil and on the soil's surface. Over time, this process naturally degrades soils and makes them toxic to vegetation.

FIGURE 11-32 ▲ Salination on California Cropland
The white area in this photograph is where irrigation has led to the accumulation of soluble salts and severe soil degradation.

Irrigation can lead to salination, especially in dry regions. Evaporation of irrigation water creates salt deposits on the fields (Figure 11-32). Irrigation that uses somewhat salty water to start with (like that of the lower Colorado River) makes salination more likely, but the continual recycling even of irrigation water that was originally low in salt can ultimately produce widespread degrading effects on farmland.

Salination caused by irrigation affects large areas in the arid southwestern states. This problem affected Mesopotamia 2000 years ago, when canal irrigation was first practiced. The rich farmlands of the Nile River valley began to suffer from salination after the Aswan Dam was constructed in the 1960s. The Nile's annual floods that covered the farmlands and helped flush away salts were prevented by the dam. Salt became concentrated in the soils as a result.

Fertilizers

Fertilizers can be a source both of pathogens and of toxic elements in soil. The use of natural organic fertilizers—fresh manure—is a well-known potential source of pathogens, including bacteria like *Salmonella* and *E. coli,* which can be deadly or at least make people miserable. Because the hazards of using manure as a fertilizer are well known, people have developed methods to treat it and remove pathogens before they apply it to fields. However, *Salmonella* and *E. coli* outbreaks from eating vegetables are becoming more common. Some croplands are being contaminated with pathogens, perhaps inadvertently from nearby animal production facilities or even from wild animals.

In the News

Pathogens in Your Produce

In late August and September 2006, at least 199 persons from 26 states were infected with a toxic strain of *E. coli* bacteria after eating packaged spinach. Three confirmed deaths included two older women and a two-year-old child. At about the same time (September 2006), 183 people in 21 states were sickened by *Salmonella* after eating tomatoes. Both *E. coli* and *Salmonella* are bacteria that live in the intestinal tracts of animals and humans. Most people are exposed to these bacteria through contact with animal feces.

How do bacteria from animal feces get on the vegetables people eat? It's not always clear. The bacteria can be in water used for irrigation, in manure used for fertilizer, in water that flows from livestock feeding facilities onto nearby fields, or in feces left by wild animals that pass through fields. The spinach contamination in 2006 was apparently caused by wild pigs migrating into fields through animal feedlots.

The number of food poisoning outbreaks involving produce contaminated with *E. coli* and *Salmonella* is significant. There were at least 60 and as many as 80 per year in the United States during the 1999 to 2004 period. In all, about 20,000 people were affected, some of whom died. You can see from the graph that the produce causing these outbreaks varied from lettuce to berries. Overall, produce-related infections affect more people than those traced to other sources. Until we better understand what causes produce-related *Salmonella* and *E. coli* outbreaks, they are likely to continue occurring—and be in the news.

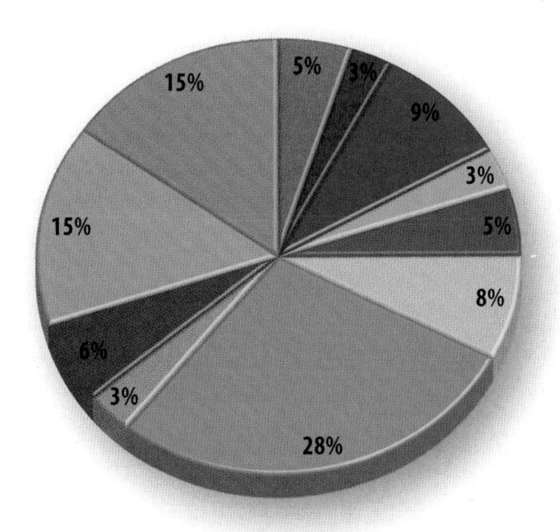

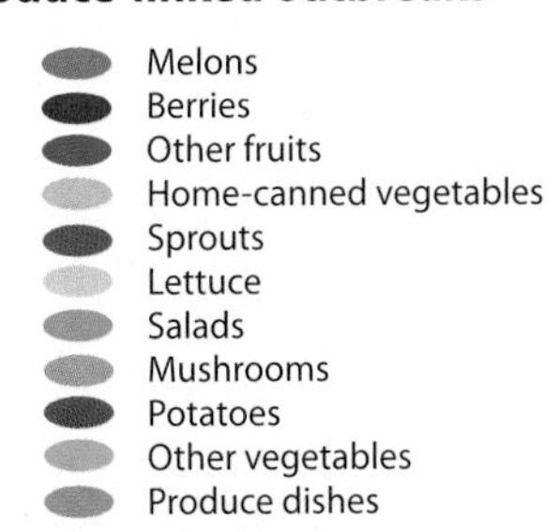

The sources of produce-related outbreaks, 1990–2004

FIGURE 11-33 ▲ Applying Pesticides to Crops
Pesticides protect crops from organisms in soil that cause plant disease and from insects that eat crops.

Manufactured fertilizers can also be a source of toxic elements in soils. Fertilizers that use waste materials from industrial processes as a source of plant nutrients (such as iron and zinc) can also have high concentrations of other elements such as arsenic, cadmium, and chromium. At low concentrations these metals are essential nutrients for plants, but at higher concentrations they become toxic to both plants and animals.

Pesticides

Pesticides are widely used around the world to kill insects that eat plants and microbes that spread plant disease (Figure 11-33). Pesticides that kill unwanted plants—weeds—are called *herbicides*. Pesticides are most commonly manufactured chemicals. Many are now designed to degrade after use, but some persist, and pesticide residues can be present in soils for decades. For example, elevated levels of arsenic in soils have been traced to former pesticide use in places like New Jersey and Texas.

In many cases, pesticide use is a two-edged sword. Consider the case of DDT. DDT is an organic chemical compound that is cheap and easy to make. In 1939, Dr. Paul Muller discovered that DDT is very good at killing insects.

During World War II, Allied military forces used DDT to protect troops from malaria and typhus spread by mosquitoes and lice. Its effectiveness at preventing disease led to its widespread use around the world after the war, and it was credited with saving millions of people's lives. Dr. Muller received the Nobel Prize in 1948 for his discovery of how to control insect populations with DDT.

The more people used DDT, however, the more they began to recognize its ecological impact. Many types of fish accumulate DDT in their bodies and pass it on to animals that eat them. As it is passed up the food chain from fish, its concentration in individual organisms increases, (a process called **biomagnification**). During the 1960s, fish-eating birds, such as pelicans, eagles, and osprey, were found to be producing eggs with abnormally thin, brittle shells—an effect attributed to the DDT in their prey species. With their ability to successfully hatch their young impaired, populations of these birds were declining sharply. On December 31, 1972, the EPA banned the general use of DDT in the United States.

Only 24 years passed between Dr. Muller's Nobel Prize and the banning of DDT's general use in the United States. DDT provides an excellent example of a new scientific discovery helping overcome a dreadful human malady. It is also an example of how understanding continues to evolve and gives us insight into the complexities of pesticide use and dispersal in the environment.

Today, evidence of pesticide use can show up on your dinner table. U.S. Department of Agriculture testing of fruits and vegetables found detectable traces of pesticides on 70% of samples; 50% of orange juice samples contained detectable levels of pesticides. Carefully washing pesticide-containing produce doesn't always guarantee its removal, but trace amounts of pesticides may not be harmful to people. A bigger concern about pesticide use is its effect on soil biodiversity.

BIODIVERSITY DEPLETION

Maintaining abundant and diverse life in soil is critical to soil quality. But constantly churning and tilling soil, growing only one type of crop (monoculture farming), and applying pesticides decrease a soil's biodiversity. Pesticide use is especially significant in this regard; pesticides can prevent many things from growing in soil.

NUTRIENT DEPLETION

Sixteen chemical elements—nutrients—are essential for plant growth and reproduction. Under natural conditions, nutrients taken up by plants—phosphorus, potassium, calcium, magnesium, and sulfur, for example—return to soil when plants die and decay. Nutrients are primarily lost from natural settings by leaching as water moves through soil, dissolves minerals and other materials, and carries nutrients away. Leaching is why oxisols, characteristic of wet and warm tropical regions, have lower nutrient contents than many other soils.

Farming practices lead to nutrient loss, too. The removal of crops prevents nutrients taken up by the harvested plants from being cycled back into the soil. Harvesting makes it a one-way trip for nutrients—from the soil to a dinner table or perhaps an animal feedlot.

Monoculture farming (tobacco, for example) is particularly notorious for causing nutrient loss. This problem seems always to have confronted farmers. In a letter to William Pearce dated December 18, 1793, George Washington wrote: "*My object is to recover the fields from the exhausted state into which they have fallen, by oppressive crops and to restore them (if possible by any means in my power) to health and vigor.*"

Nutrient loss can also be caused by heavy tilling of soil. Churning up soil helps crops by increasing porosity, mixing nutrients, and increasing oxygen levels. But at the same time, higher oxygen content accelerates chemical reactions that release nutrients, causes organic matter to decompose more rapidly, and decreases humus levels. In these situations, nitrogen is usually the first nutrient to be depleted to the point that it affects plant growth, and nitrogen-bearing fertilizers are soon needed on heavily tilled farmlands.

Nutrient loss also occurs when soil erodes and when increased acidity—from nitrogen-bearing fertilizers, for example—increases nutrient leaching. Maintaining healthy nutrient levels on cropland is challenging. Consider the following example.

You Make the Call

Is Ethanol a Sustainable Energy Resource?

The increasing price of gasoline and our country's dependence on foreign sources of petroleum have made alternative transportation fuels, such as ethanol, attractive to many people. The fact that ethanol is a renewable energy resource makes it appear even more promising. One result of ethanol's appeal was the 2005 Energy Bill, which mandated that use of biofuels (mostly ethanol) reach 7.5 billion gallons by 2012. In order to meet this goal, the gasoline you use in your car will have to contain 2% to 10% ethanol, and in some cases more.

Ethanol is an alcohol distilled from plants, mostly corn in the United States. (It is the alcohol in alcoholic beverages.) This is why it's called a renewable energy resource. All people need to do to produce more fuel is grow a new crop. But is growing corn, especially at the scale needed to supply significant amounts of transportation fuel, really sustainable?

Several aspects of large-scale corn farming for ethanol production raise concerns about sustainability. Large amounts of water are needed to grow and process corn, and more farmland would need to be dedicated to corn production. Moreover, extensive use of pesticides and herbicides accompany corn farming (corn is grown on about one-fourth of U.S. croplands, but requires two-thirds of the total herbicide use). Corn is a row crop, and soil erosion rates are higher for corn fields than for several other types of crops. Growing corn in the United States has also required large amounts of fertilizer to maintain crop yields.

From the standpoint of sustainability, the effect of corn farming on soil nutrient levels and quality are significant concerns.

- What do you think is happening to nutrient levels and biodiversity in corn fields?
- Do you think soil quality is being sustained in these fields?
- Over the long term, do you think ethanol is a sustainable energy resource?

FIGURE 11-34 ▶ Urbanization of Cropland
The Santa Clara Valley, now more commonly known as Silicon Valley, was once mostly orchards but now grows business parks and housing subdivisions seemingly overnight. Urbanization today covers many thousands of acres of the valley's rich soil.

URBANIZATION AND SOIL

Converting any natural land to human use affects soil. Converting native prairie grassland to plowed fields was a major cause of the 1930s Dust Bowl. Forests have been cut down in ways that contributed to widespread soil loss. A land-use choice that is just about irreversible when it comes to soil is urbanization. Once fields and orchards are replaced by roads and buildings, it's hard to go back.

Going back has happened. Hundreds to thousands of years after ancient cities collapsed, nature has reclaimed their lands. Forests now largely cover the ancient Mayan fields and cities of Central America, for example. Still, it's hard to imagine Silicon Valley in California being converted back to the orchards that once covered large parts of this region (Figure 11-34).

Urban land made up less than 3% of total U.S. land area in 2000. This may not seem like much, but the amount is increasing—between 1990 and 2000 it rose about 15%, to a total of 51 million acres. As the population grows, urban centers grow. Taking both urban and rural development together, about 2 million acres of farm or other land is converted each year for people to live on. About one-fifth of this total is prime cropland.

What are some influences people have on the sustainability of soil quality?

11.5 Sustaining Soil Resources

If soils are completely lost from the land or severely degraded, it will take hundreds to thousands of years—if it's possible at all—for nature to replenish them. From this perspective, people can in effect destroy soil resources, making them unavailable for so long that they become nonrenewable resources.

This doesn't have to be the case. People have learned how their activities have negative consequences for soil and have developed ways to mitigate or prevent soil degradation and loss. How well this knowledge is applied, at a global scale, is the key to sustaining soil resources.

SOIL CONSERVATION

Major soil conservation efforts evolved out of the Dust Bowl experience in the United States. In 1935, Congress declared soil erosion a "national menace" and created the Soil Conservation Service (now the Natural Resource Conservation Service) in the Department of Agriculture. This federal agency developed methods and provided education and training to protect soil and prevent its irreparable harm. In 1936, the agency established local conservation districts that engaged farmers directly in changing their practices to protect soil resources. Soil conservation districts are still going strong.

Soil conservation studies have identified a wide range of farming practices that help protect soil resources. These practices include the following measures:

- Contour farming—tilling perpendicular to surface slopes—creates plowed furrows that catch soil and water rather than let them run off freely (Figure 11-35).
- Terracing—converting steeper slopes into a series of flat terraces—has been used to reduce surface runoff and erosion for thousands of years. (Figure 11-36).
- Tilling fields when storms are not likely can decrease erosion.
- Using machinery that can aerate, plant, and weed fields without greatly churning the soil helps maintain its stability.
- Cultivating crops in parallel strips that can be harvested and tilled at different times—strip farming—ensures that some areas will always be covered with vegetation (Figure 11-37).
- Planting barriers that protect fields from wind—tree rows, for example—can help control erosion.

FIGURE 11-35 ▲ Contour Farming
Plowing fields parallel to slopes decreases erosion. This Wisconsin farm has plowed and planted alfalfa hay (green) and corn (yellow) in contoured strips.

- Rotating different crops in a field through the years can help prevent selective nutrient depletion. Corn and soybeans are a good example. Corn decreases soil nitrogen levels, but soybeans harbor nitrogen-fixing bacteria on their roots. If soybeans and corn are planted alternately over time in the same field, the soil's nitrogen levels can be maintained.
- Reducing tillage and adding organic wastes helps maintain or increase the organic content of soils.

FIGURE 11-36 ▲ Terracing
Terraces, horizontal benches along slopes, significantly decrease erosion and have sustained soils on steep slopes for thousands of years. These terraced rice farms on the lower slopes of Mount Batukaru in Bali are classic examples.

FIGURE 11-37 ◀ Strip Farming
Alternating crops in parallel strips (corn, wheat, and soybeans, for example) is a type of crop rotation that helps maintain soil nutrients. It is commonly combined with contour farming (see Figure 11-35) to decrease erosion. This farm is in Pennsylvania.

What You Can Do

Use Your Local Soil Conservation District

Soil conservation districts exist in all the states and some territories. In many states, there are districts for each individual county. You can explore what is happening in your own soil conservation district by going to the National Association of Conservation Districts website. National Association of Conservation Districts Using the "Districts on the Web" link you can go to district lists for each state. Find the district in your area and investigate its activities and the services it provides.

It's likely that your district will provide you information about such things as the type of soils in your area, how stream and wetland habitats can be protected, how to manage livestock areas and manure, and the native plants that grow where you live. It's common for soil conservation districts to provide information on how to have a soil's quality evaluated. Tests are available to determine a soil's pH, nutrient content, and the amount of organic matter. You can expect guidance on how to collect a soil sample, how to package and store the sample, and where to submit it for testing. Many districts do the testing themselves, at a cost that is generally less than $20 per sample. You can use soil test data to guide your garden and yard maintenance. The difference between struggling and healthy plants may be as simple as adding lime to your soil to decrease its acidity.

SOIL REMEDIATION

If a soil becomes contaminated with pollutants, people can clean it up or remediate it. A very wide range of technology has been developed to do this. Two approaches that can be used directly on soil (*in situ*) are *bioremediation* and *phytoremediation*.

Bioremediation

Do you remember that there can be thousands of bacterial species in a spoonful of soil? Somewhere in this mass of tiny life are species that will eat toxic substances that people have applied to or spilled on soil. Isolating the bacteria that will eat a specific substance requires considerable laboratory work, but it has been accomplished for several potential soil contaminants. In many cases, the key bacteria are already living in the contaminated soil. They just need to be assisted some, perhaps by adding oxygen, to increase their populations. Using these bacteria to eat the contaminants and metabolize them into nontoxic substances is called **bioremediation**.

A good example of bioremediation is using bacteria to eat petroleum or its products—oil, gasoline, diesel, and jet fuel, for example—that have been spilled on the land. It may take awhile, and soil conditions have to be right (temperature and oxygen content may have to be within a certain range, for example), but some bacteria thrive on these contaminants. Mixing the bacteria into soil and controlling soil conditions will eventually change the contaminants into nontoxic substances—mostly carbon dioxide and water in the case of petroleum and its products.

Phytoremediation

Using plants to clean up soil is called **phytoremediation**. Certain plants selectively take up and concentrate toxic substances, which they then degrade or release to the atmosphere in modified form. Some plant species are excellent accumulators of specific metals—arsenic, cadmium, and lead, for example. Growing these plants and then harvesting them can help reduce metal concentrations in the soil. Of course, the harvested plants must be properly disposed of. Metal-bearing plants are usually burned and the ash deposited in landfills. Over time, this approach can effectively clean soil contaminated with toxic metals.

Some hopeful entrepreneurs have proposed using specific plant varieties for mining applications, a technique called phytomining. | Phytomining One candidate for such use is *Streptanthus polygaloides* (milkwort jewelflower, shown in Figure 11-38), a hyperaccumulator of nickel. Hopeful miners would grow plants of this species in nickel-bearing soils, harvest them, convert them to ash, and sell the ash for its nickel content.

Desalination of soil

Remediating salty soils is very challenging. Salination has degraded soils over large areas, and there is more understanding of how to prevent this process than of how to reverse its effects. Irrigating with less salty water, using minimum amounts of irrigation water, and reducing water evaporation rates have all helped decrease salination in specific situations. After salination has occurred, flooding fields and draining the water away can help flush salts from soil. In the worst cases, however, salination due to irrigation has led people to abandon badly degraded soils. It takes many years of natural flushing for such soils to recover—and natural flushing may not be very effective in arid regions. It's better to prevent salination in the first place if that is at all possible.

SOIL PROTECTION

Other actions that help sustain soil quality include reforestation and judicious land-use choices. Reforestation restores natural vegetation cover, soil biodiversity, water quality and retention, and nutrient cycling abilities. Reforestation can take up to 200 years; it is being practiced over large parts of North America and other developed countries.

FIGURE 11-38 ▲ Milkwort Jewelflower
This pretty plant is a hyperaccumulator of nickel.

Controlling land use can be a very effective way of protecting and sustaining soils. For example, the Federal Agriculture Improvement and Reform Act of 1996 continued a federal government program that discourages agricultural use of lands at risk for soil loss. The government establishes long-term contracts with farmers who convert farmlands susceptible to extensive erosion to protective vegetation cover. Farmers are essentially paid to stop harmful farming and protect the soil. The program has reduced annual topsoil loss in the United States by 635 million tonnes (700 million tons), or 43 tonnes per hectare (19 tons per acre) under contract. However, when the prices of farm commodities rise (as they have since biofuel additions to gasoline have been required), there is a tendency for farmers to opt out of their contracts and place their lands back in production.

Just as farming practices and land-use choices can help sustain soil, so can grazing practices. Controlling the types of grazing animals, their numbers, and the times they are allowed to graze a specific area can protect soil. Rotational grazing—shifting grazing from one area to another—enables vegetation to recover and keeps soil stabilized.

What are some things people can do to sustain soil resources?

SUMMARY

Soils are an essential resource that people depend on. They facilitate material and energy transfers among Earth systems and in effect function as a membrane between them. The roots of all terrestrial life are literally in soil. Soil is a valuable renewable resource that is often misused and needs to be protected.

In This Chapter You Have Learned:

- Soil is the loose material on land that develops through weathering. Soil is a mix of air, water, minerals, organic material, and life that forms a membrane between Earth systems. Plants grow in soil and engineers build structures on it.
- The functions of soil include providing food and fiber, storing and cleaning water, providing habitat for incredibly diverse life, and helping transfer matter among Earth systems.
- Soils are formed by physical, chemical, and biological processes that affect geosphere materials on Earth's surface. It takes hundreds to thousands of years to make a mature and fertile soil.
- The soil-forming processes combine to develop stratified layers called horizons that define a soil's profile. The upper "A" horizon is where material is leached and removed (although organic material may accumulate here); the middle "B" horizon is where material that has moved downward from the A horizon accumulates; and the lower "C" horizon is where remnants of the weathered geosphere parent materials are present.
- The nature of soil varies from place to place depending largely on climate and the character of the parent materials. Four examples of the 12 major variations recognized by soil scientists—called orders—are spodosols (forest soils), aridisols (desert soils), mollisols (prairie soils), and oxisols (tropical soils).
- The physical properties of soils include texture, structure, density, shear strength, and compressibility.
- The compositional properties of soils include the proportion of water and air in pores, the pH (acidity), and the amount of nutrients available to plants.
- The biological properties of soil include its biodiversity and the amount of decaying organic matter that accumulates and changes within it. A small sample of mature, rich soil contains more species than an entire rain forest.
- The physical, compositional, and biological properties of soil combine to determine its quality. A high-quality soil is populated by diverse organisms, controls water flow and character, facilitates interactions among Earth systems, and stores matter involved in Earth system cycles.
- Soils are degraded or lost through erosion, contamination, biodiversity depletion, and nutrient depletion. Erosion is the largest cause of soil degradation and loss. Farming practices from tilling to irrigation, livestock grazing, and deforestation have all contributed to soil degradation and loss. Converting land to urban use effectively involves an irreversible loss of soil.
- Soil is a sustainable resource. Farming practices such as contour tilling, terracing, crop rotation, and no-till farming help conserve soil. Some contaminated soils can be remediated to remove or change components toxic to plants and animals. Both bioremediation and phytoremediation techniques have been used to clean soils. Soils can be protected by reestablishing appropriate vegetation cover and discontinuing harmful practices. Preserving wetland soils helps store and clean water.
- Scientists understand how to sustain soils, but fully applying this knowledge is a continuing challenge for people.

KEY TERMS

aridisol (p. 335)
aspect (p. 334)
biomagnification (p. 346)
bioremediation (p. 351)
biosolids (p. 331)
compressibility (p. 336)
horizon (p. 333)
humus (p. 333)
laterite (p. 335)
mollisol (p. 335)
oxisol (p. 335)
pH (p. 338)
phytoremediation (p. 351)
rill (p. 343)
salination (p. 344)
shear strength (p. 336)
sheet flow (p. 344)
soil orders (p. 334)
soil profile (p. 333)
spodosol (p. 334)
topsoil (p. 334)

QUESTIONS FOR REVIEW

Test Your Understanding

1. Give two examples of how Earth systems interact to form soil and describe the products that result.
2. How long does it typically take to develop a mature and fertile soil? What processes are involved in this development?
3. What basic benefits does soil provide to people?
4. Soil provides habitat for a wide range of species. Name two of them, describe how they live in the soil, and explain how they benefit soil development.
5. How is soil a part of the carbon cycle? Describe two processes that involve soil in the carbon cycle.
6. The chapter states that soils are "dynamic natural materials." What is meant by this, and in what sense are soils dynamic, i.e. changing through time?
7. What is humus, and how is it related to the fertility of soils?
8. How does the A horizon of soils typically differ from the underlying B horizon? What contrasting processes result in these very noticeable differences?

9. If you wanted to understand the parent material that a soil developed on, which of the three main layers of the soil horizon would provide the best information, and why?
10. How is water important in the formation of oxisols in tropical regions? How does this affect the nutrient content of these tropical soils?
11. Why is soil structure important in terms of air and water flow?
12. Why are soil shear strength and compressibility important factors for engineers and builders to consider?
13. Why are bacteria in soil critical to soil health and fertility?
14. How is soil erosion connected to the demise of many large civilizations? Is this still happening today? If so, provide two examples of ongoing soil erosion that should be of some concern.
15. Why is soil erosion a particular concern in arid and semi-arid regions? Discuss your answer in terms of the rate of soil formation vs. soil erosion.
16. How can irrigation of crops in arid regions lead to increases in the salt content of the soil?
17. How can pesticide and herbicide application lower the biodiversity in a soil?
18. Certain farming practices lead to greater soil conservation. Name and discuss how two of them work in preserving soil resources.

Critical Thinking/Discussion

19. Why is human civilization so dependent on soil? Discuss how the growth of civilizations has been affected by soil resources, and how, ironically, civilizations can also endanger that critical resource base through their own success and growth. What is the solution to this potential paradox?
20. Soil development depends on many interacting geographic factors. If you were mapping soil variations across a hilly region underlain by a single type of bedrock and with a single general climate, what types of variations might you observe and why?
21. For oxisols to be developed into good farmland, they have to be heavily fertilized. This is true in many tropical forests in Central and South America that have been cleared for use as farmland. The text describes the benefits of such fertilizer addition, but what drawbacks might there be to amending the soil with lime, nitrogen, phosphorus, and potassium fertilizers in terms of other ecological systems?
22. Why is it important to consider all the organisms present in a soil before applying fertilizer or chemical herbicides or fungicides to a crop? Discuss potential ecological systems effects that a farmer would have to consider carefully.
23. The case of DDT illustrates how a simple application of chemicals to soil can result in unintended consequences through biomagnification and other mechanisms. If you were overseeing development of new crop treatments or other chemicals for widespread application, what issues would you bear in mind as you tested new products for safety, and what kinds of oversight to look for unforeseen effects would you recommend once a product went into widespread use?
24. Urbanization is one source of soil loss. Why is this process hard or impossible to reverse? What steps can be taken in urban planning to make sure too much of available soil resources aren't lost through urban growth?

ANSWERS TO IN-CHAPTER INSIGHT QUESTIONS

P. 334

What are some Earth systems interactions involved in forming soil?

- Freezing and thawing of water helps break up geosphere parent materials to begin soil formation.
- Reactions with oxygen from the atmosphere change geosphere materials.
- Biosphere organisms mix and change soil components.
- Movement of water moves soil components.

P. 340

What are some Earth systems interactions that take place in soil?

- Nutrients such as carbon and nitrogen are cycled among the atmosphere, biosphere, hydrosphere, and geosphere in soils.
- Energy from the Sun captured through plant photosynthesis is transferred to soil organisms.
- Water from the atmosphere is changed (in pH, for example) as it migrates through soil to groundwater.

P. 348

What are some influences people have on the sustainability of soil quality?

- Many agricultural practices decrease soil quality by decreasing soil biodiversity.
- Erosion caused by people's land use commonly decreases soil quality by removing nutrient-rich topsoil.
- Many agricultural practices decrease soil nutrients and therefore soil quality.
- Addition of natural and man-made fertilizers can increase or maintain soil quality.

P. 351

What are some things people can do to sustain soil resources?

- Prevent erosion.
- Maintain biodiversity.
- Replenish nutrients.
- Preserve high-quality soils by guiding infrastructure development to other areas.

FIGURE 12-1 ▶ A Death in the Alps
In this mountain pass, hikers discovered the mummified body of a man (circle) who had died 5300 years earlier.

CHAPTER

12

MINERAL RESOURCES

Helmut and Erika Simon were two of the many visitors attracted to the hiking trails of the high Alps each summer. They wanted to experience these rugged mountains before fall weather set in, and on September 19, 1991 they found themselves crossing a pass at an altitude of 3200 meters (10,500 ft). The pass is commonly covered with glacier ice and snow, but summer melting had uncovered rock outcrops here and there (Figure 12-1). Wandering off the main trail, the Simons discovered more than exposed bedrock. Thawing from the ice along the bottom of a depression was the body of a man (Figure 12-2, p. 356). Discovering a dead person while hiking is surprising enough, but even more surprising was how long ago the person had died. The mummified remains were determined to be 5300 years old.

The deceased soon acquired a name: "Ötzi," or just "The Iceman." The circumstances of his death led to the remarkable preservation of his body, which quickly became the focus of extensive scientific study. The Iceman: discovery and imaging – Murphy et al. Preserved along with Ötzi's body were his tools, weapons, clothes, and shoes. Even remnants of his last meals could be investigated. As his tools show (Figure 12-3, p. 356), Ötzi was dependent upon mineral resources. His ax was made of copper and his knife was made of flint, a variety of quartz having very tiny crystals.

Ötzi's reliance on such materials is not surprising. In fact, human technological progress is largely defined by how people have come to use mineral resources. Human origins, origins that date to about 2 million years ago in the rift valleys of Africa, are marked by the use of stone. The *Stone Age,* from the first use of stone tools to the beginnings of agriculture and metal use some 6000 to 8000 years ago, saw increasing technological advances. Tools for cutting and scraping, exquisite points for spears, arrows, and other weapons, and techniques for the construction of shelters were widely developed. Eventually, almost everyone seems to have used a stone hand ax like the one shown in Figure 12-4, as they are among the most common artifacts recovered at Stone Age archaeological sites. Ötzi lived during the last phase of the Stone Age, when copper metal began replacing stone for tools.

As the Stone Age evolved into the *Bronze Age,* metals began to play a dominant role. There are places where copper occurs on the land surface as native copper metal—solid, malleable copper (Figure 12-5a). Somewhere, somehow copper-bearing materials were melted, perhaps first in pottery kilns, where the minerals may have been used for glazes. In several cultures, melting and processing copper into ornaments, utensils, and tools was the beginning of metalworking (Figure 12-5b). However, combining molten copper with molten tin was the significant technological breakthrough. The resulting mixture of metals, or **alloy**—bronze—is much stronger and more durable than either pure copper or pure tin. When people discovered how to make bronze, it changed their world.

The Bronze Age started at different times in different places, but it was generally between 6000 and 2500 years ago. People used bronze for strong tools, weapons, and building materials (Figure 12-5c). And because tin is not as widely distributed as copper, trading tin to make bronze

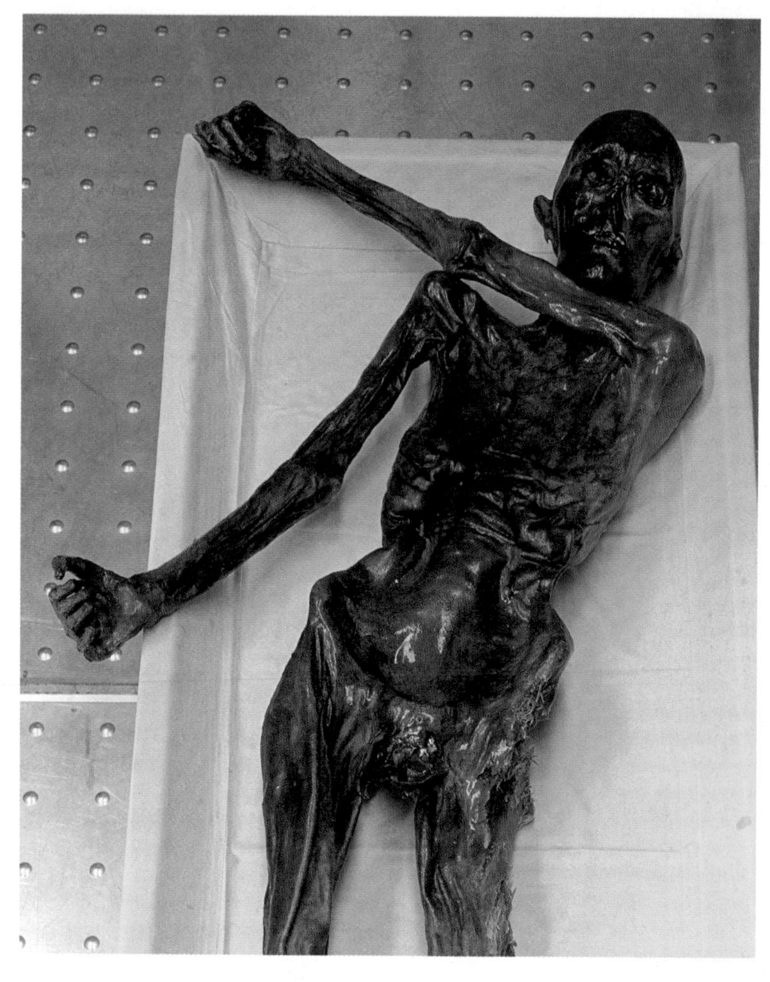

FIGURE 12-2 ▲ The Iceman
The body of the Iceman, or Ötzi, as he came to be known, as it was discovered.

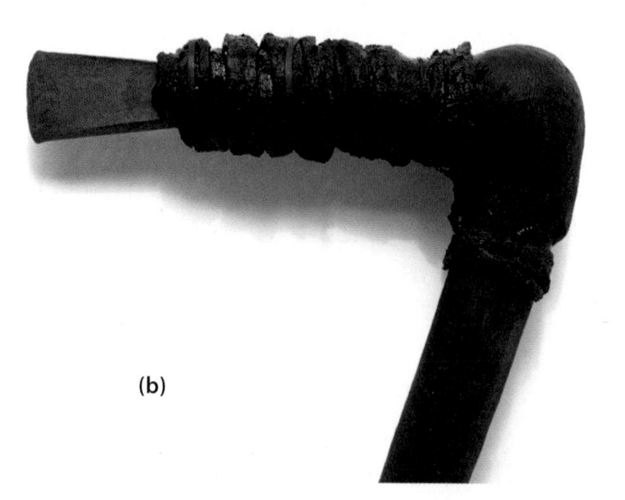

FIGURE 12-3 ▲ Ötzi's Tools
Ötzi lived at the end of the Stone Age, when metal implements had just begun to make their appearance. The knife (with scabbard) (**a**) is made of flint, but the ax **(b)** is made of copper.

FIGURE 12-4 ▶ A Stone Age Hand Ax
Based on the findings at many archaeological sites, these tools seem to have become almost as common as pocket knives today.

made the world smaller. A principal source of tin during the Bronze Age was Great Britain. Tin from Britain supplied bronze-makers in Europe and throughout the Mediterranean region.

Although bronze is strong and durable, iron is a more abundant metal than copper or tin. About 2500 years ago, iron use began to displace bronze use. Perhaps tin supplies grew difficult or costly to acquire. Whatever the reason, the *Iron Age* began. During this period, nearly everyone benefited from the use of metal tools and equipment (Figure 12-5d). The widespread availability of iron products meant that

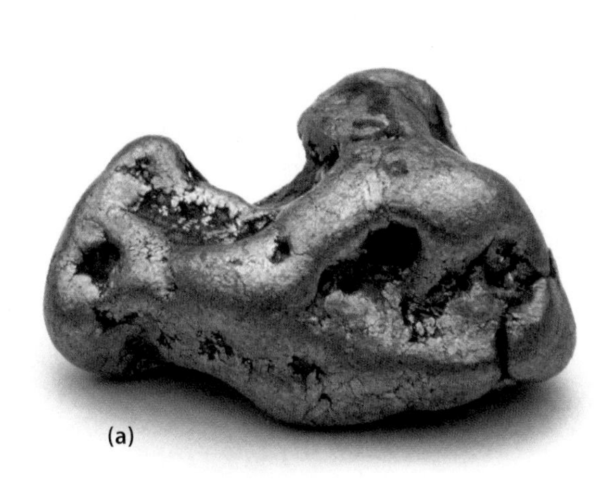

(a)

(b)

(c)

(d)

FIGURE 12-5 ▲ Early Metal Use
(a) A native copper nugget. This easily worked and easily melted metal can be found exposed on Earth's surface in places. **(b)** Copper tools and utensils were among the earliest products of metalworking. **(c)** Copper and tin can be melted together to make the alloy bronze, which is harder and stronger than pure copper. **(d)** Eventually, iron replaced bronze in the manufacture of tools and weapons.

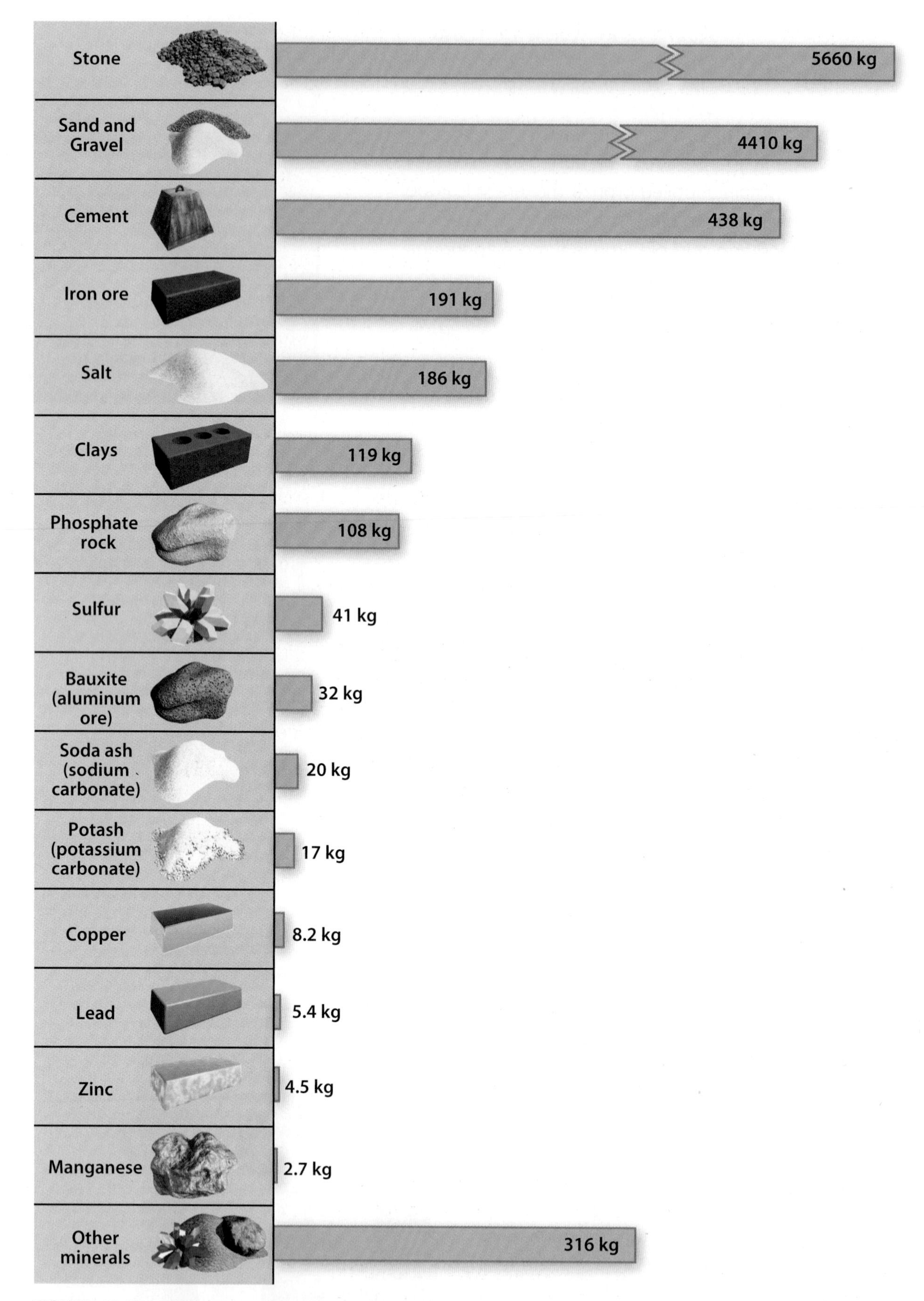

FIGURE 12-6 ▲ Mineral Use in the United States

Each year, mineral consumption in the United States amounts to nearly 21,700 kilograms (47,800 lb) per person.

they could be applied to help people in most of their daily endeavors. Further advances saw the making of hard, durable iron alloys—steel.

By the time of the Industrial Revolution 250 years ago, metal processing technology was very advanced. The long list of useful metals included copper, tin, iron, lead, zinc, mercury, and, of course, gold and silver. Today, people continue to expand their use of mineral resources. The U.S. Geological Survey (USGS) keeps track of the use, production, consumption, price, and sources of over 80 commodities (materials that are bought and sold) derived from mineral resources. These include metals like copper and iron that have been used for thousands of years, rare elements like europium used in color TVs and computer screens, minerals like kyanite (aluminum silicate) used in the steel industry, and common cement, sand, and gravel. In fact, minerals are everywhere in your life. Take computers, for example. Computers and their monitors contain over 30 components that come from minerals—lead, zinc, copper, aluminum, iron, tin, manganese, nickel, silver, and gold, for example.

Citizens of the United States consume more mineral resources per person than do people anywhere else in the world (Figure 12-6). We use these resources in almost everything—buildings, roads, power lines, vehicles, and electric devices of all sorts, from washing machines to cell phones. The more we strive to improve our lifestyle—with better transportation and communication systems for example—the more we depend on mineral resources. But mineral resources are nonrenewable. We need to understand everything we can about mineral resources in order to use them wisely.

You probably already understand that mineral resources come from the geosphere, and that the principal way they are recovered is by mining. Mining is just the first step in making mineral resources available for your use, however. Most mineral resources require one or more processing steps to make them tradable commodities. Every step in recovering and processing mineral resources has potential environmental consequences that can be a concern. Because these environmental consequences are well represented in the mining and processing of metal-bearing geosphere materials, this chapter emphasizes metal-bearing mineral deposits.

IN THIS CHAPTER YOU WILL LEARN:

- What makes minerals a resource for people
- The roles played by exploration, mining, and processing in providing people with mineral resources
- The environmental concerns associated with mineral resource production
- How negative environmental impacts can be avoided or mitigated
- The future challenges of mineral resource use

12.1 What Mineral Resources Are

Mineral resources are naturally occurring solid materials in or on Earth's crust from which we can currently or potentially extract a useful commodity. Economics plays a key role in defining mineral resources. Take lead, for example.

Lead is the metal in your car battery that makes it so heavy. Lead is widely distributed in small amounts—its average concentration in Earth's crust is 12.5 parts per million (ppm), or 0.00125%. A cube of average crust 100 meters (328 ft) on a side weighs almost 3 million tonnes (6.6 billion lb). If it has an average lead content of 0.00125%, such a cube contains over 3.7 tonnes (8200 lb) of lead. At the 2008 average price of $1.31 per pound, lead recovered from this cube could be sold for a total of over $10,000. This is equivalent to receiving about 0.3 cents for every ton of crust processed to recover the lead.

It's impossible to recover lead from crust at costs this low, however. The small amounts of lead scattered through the crust are not a mineral resource—they cannot be economically recovered, even if the price paid is many times higher. Lead concentrations thousands of times greater than the crustal average are needed if it is to be recovered economically and become a mineral resource.

What You Can Do

Investigate Mineral Resource Economics

You learned about the USGS's Mineral Commodity Summaries in Chapter 4. Revisit these summaries online. USGS minerals information: mineral commodity summaries Mineral Commodity Summaries 2008 has some especially interesting information about economic conditions and international markets, including the outlook for some mineral commodity prices. Because the summaries include some key data for five-year intervals, you can investigate things like consumption and price trends for specific commodities. For example, the price of silver increased from $6.69 to $15.85 per ounce between 2004 and 2008. Why do you think this is the case?

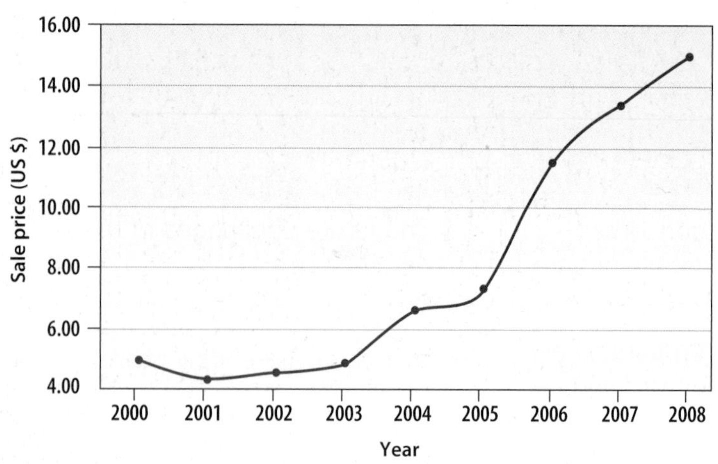

The price of silver has risen dramatically in recent years.

MAKING MINERAL DEPOSITS

Mineral deposits are concentrations of useful minerals in Earth's crust. If mineral deposits have high enough concentrations of valuable elements to allow profitable mining and recovery of a saleable commodity, they are called ore deposits. The part of a mineral deposit that can be mined and recovered profitably at current and projected commodity prices is **ore**.

Ore deposits form in many ways. They can involve crystallization processes in igneous rocks, surface processes that concentrate heavy metal-bearing minerals in sediments such as beach sands, and chemical precipitation on the seafloor. Many minerals form from the interaction of water-rich fluids and crustal rocks. Here are some examples.

- Hot, newly formed oceanic crust can heat ocean water that percolates downward into it. This hot water will dissolve metals—copper, lead and zinc, for example—distributed through the oceanic crust. Some of the metal-bearing hot water will migrate back to the ocean floor to form springs (the "black smokers" discussed in Section 3.3). Here the dissolved metals react with sulfur to form sulfide minerals that sink downward and accumulate on the seafloor. The ore deposits in the ancient copper mines on the island of Cyprus (Figure 12-7), a major center of bronze manufacturing and trading during the Bronze Age, were produced in this way.
- Some magmas that solidify in the shallow crust release hot fluids rich in water and dissolved metals. These fluids precipitate sulfide minerals containing metals such as copper. The deposits that form, like the Bingham Canyon copper deposit in Utah (Figure 12-8), can be huge, and much of the world's copper comes from deposits that form in this way.

FIGURE 12-7 ▼ A Copper Mine on Cyprus

FIGURE 12-8 ▲ The Bingham Canyon Mine in Utah

- When sediments are deposited on the ocean floor, they commonly contain seawater in the pores between sediment grains. The salty water becomes part of the sedimentary sequence and will come to contain dissolved metals, especially lead and zinc. This metal-bearing water may escape from the enclosing sediments along permeable pathways such as faults and emerge on the seafloor. Here the springs form precipitate sulfide minerals that create mineral deposits on the seafloor. The large lead and zinc deposits at the Red Dog mine in Alaska are examples of this type of ore deposit (Figure 12-9).

- You learned in Chapter 4 how rocks dehydrate and release water-rich fluids when they metamorphose. It is fairly common for these waters to contain some gold and other metals. If the metal-bearing waters can escape to shallower crustal levels and become localized along permeable structures they will precipitate minerals, commonly quartz with small amounts of dispersed gold (an ounce of gold for each ton of rock would be a high gold concentration). The gold-bearing veins of the California Mother Lode region are examples of this type of ore deposit (Figure 12-10, p. 362).

These examples illustrate some of the common processes that form mineral deposits in Earth's crust. Some areas may be relatively well endowed with mineral resources and some not. And mineral deposits, although rare geologic features, are found in all physiographic settings, from the frozen Arctic to the sweltering rainforests of the equatorial regions. This is why the environmental concerns associated with mineral resource production can require different responses, depending on a mineral deposit's location. Mining in a desert, for example, can have very different implications for surface and groundwater than mining in a rainforest.

What are some Earth systems interactions associated with the formation of mineral deposits?

FIGURE 12-9 ▼ The Red Dog Mine in Alaska

(a)

(b)

FIGURE 12-10 ▲ A Gold-Bearing Vein in the Mother Lode Region of California
(a) This gold-bearing white quartz vein is exposed underground in the Lincoln Mine. **(b)** Gold in quartz.

THE SHAPES OF MINERAL DEPOSITS

As Figure 12-11 suggests, ore deposits come in all shapes and sizes. Their physical shape and location relative to the land's surface help determine how they are mined. Some ore deposits, called **veins**, are long (hundreds to thousands of meters or yards), narrow (one to fifty meters or yards), sheet-like structures. These veins may extend vertically deep into the crust, for 2 kilometers (1 mi) or more, and some may not be exposed at the surface at all. Other ore deposits may be discontinuous, somewhat tube-shaped *pods* or disk-shaped *lenses*.

The ore deposits that extend deep into Earth's crust need to be mined by underground methods. Because these methods are expensive, the ore deposits need to be of a relatively higher **grade**—that is, have a higher concentration of valuable minerals—to be profitable. Overall, ore deposits mined underground tend to be small; many are in the 1 to 10 million ton range.

If the ore deposit is exposed or close to the surface, at least part of it can be mined by surface excavations called **open pits**. Inclined veins may quickly extend to depths that require underground mining, but if the exposed ore deposit is of somewhat equal extent in all three dimensions, larger proportions of it can be mined by open-pit methods. The large copper deposit at the top of the intrusion in Figure 12-11 is of this type. Such ore deposits have lateral and vertical dimensions that range from hundreds to thousands of meters

FIGURE 12-11 ▼ Ore Deposits That Can Form in Stratovolcanoes
Deposits containing copper, lead, zinc, gold, and silver commonly form within or near stratovolcanoes. The gold and silver deposits tend to be veins, whereas the lead, zinc, and copper deposits in carbonate rock tend to be lens-like or pod-like in shape. The largest deposits are copper-rich and form in the interior of the volcano near the top of an intrusion.

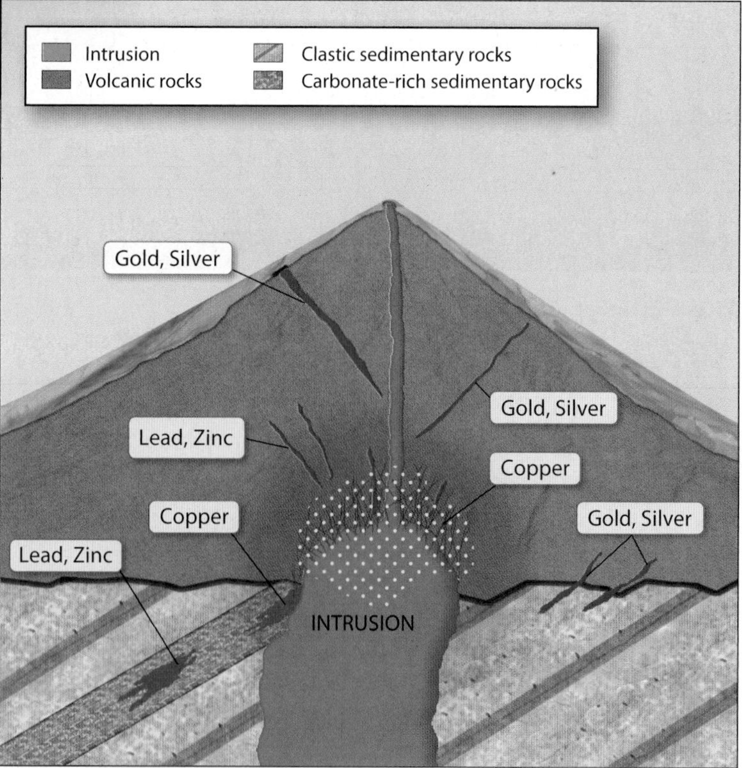

(yards). Because surface excavations are less costly than underground excavations, the grade of ore deposits mined this way can be lower. And many are huge—they can contain several hundred million tons or more of ore (see Figure 12-8).

12.2 Finding, Mining, and Processing Mineral Resources

Because the geologic conditions that lead to formation of mineral deposits are widely scattered and not common, the first step in making mineral deposits available for people's use is finding them through exploration. However, once a mineral deposit is discovered, it takes much more exploration to determine whether it contains an ore deposit. If an ore deposit is judged to exist, permission to mine must be obtained from one or more government regulatory agencies. If the agencies decide that the deposit can be mined in an environmentally responsible manner, permits will be approved and mining can start.

Mining commonly recovers material that needs to be processed to separate minerals containing useful elements. This step is called *beneficiation*. The separated minerals are then further processed in another step, called *metallurgy*, to separate useful elements from their mineral hosts.

FINDING MINERAL RESOURCES

The search for ore deposits—exploration—is an exciting and challenging endeavor. Geologists start from scratch by examining geologic maps and satellite imagery to identify areas that may be favorable for mineral deposits. They will then examine these areas more closely (commonly on foot), make more detailed geologic maps, and extensively sample surface materials such as rocks, soils, stream sediments, and vegetation (Figure 12-12). Surface samples are analyzed by *geochemical techniques* for evidence of telltale elements in the vicinity, and subsurface relations are investigated with *geophysical techniques* such as magnetic surveys. Surface observation, geochemical techniques, and geophysical techniques do not significantly alter or disturb the local area, so this stage of exploration raises little environmental concern.

FIGURE 12-12 ▼ Mineral Exploration
Collecting and analyzing rock samples can provide geologists with information that helps them to identify valuable mineral deposits.

If a mineral deposit is found by initial exploration efforts it is called a **prospect** until it is shown to contain ore that can be profitably mined. Exploration commonly finds hundreds of prospects before one containing ore is identified—no more than a handful of new ore deposits per year are discovered anywhere in the world. And to determine whether a mineral deposit contains ore is generally a very long process that involves a combination of trenching and drilling to determine the character of the prospect. Significant infrastructure, such as camps, roads, and airstrips, is generally needed to support this stage of exploration.

What are some Earth systems interactions that make geochemical techniques useful exploration tools?

Trenching

Surface materials commonly obscure the bedrock containing a mineral deposit. This material can be removed by using backhoes or similar equipment to dig trenches (like the one shown in Figure 12-13, p. 364) that expose the bedrock for detailed observation and sampling. Trenches vary in length—some are hundreds of meters (yards) long—but are generally shallow, less than a few meters deep. Geologists carefully observe, map, and sample the rocks exposed in trenches to help characterize the mineral deposit.

Drilling

The subsurface character of mineral deposits is determined by drilling. Drilling recovers subsurface samples in the form of rock chips or tubes of rock called **cores**. The drill rigs that obtain these samples are mobile machines that can be moved in pieces to very remote places, commonly using helicopters if roads are not present or are inadequate. Powered by gasoline or diesel engines and using air or water as their chief lubricant, they can cut into the subsurface rock to depths as great as 1000 meters (3300 ft). A drill rig, like that shown in Figure 12-14 on p. 364, collects rock chips from underground and may be truck-mounted or set up on a pad about 10 square meters (108 ft^2) or less in area. Geologists examine, describe, and take samples of the recovered rocks. They use the results to make maps and diagrams that help them understand the underground character of the mineral deposit.

Infrastructure

Onsite exploration of a mineral deposit generally takes place over several years and requires a variety of infrastructure to

(a)

(b)

FIGURE 12-13 ▲ An Exploration Trench
This trench **(a)**, created to help explore gold deposit in Alaska, was restored **(b)** after geologists had studied the subsurface materials.

FIGURE 12-14 ▼ A Mineral Exploration Drill Rig
This drill rig cuts and collects rock chips at a gold prospect in Australia. The white bags are full of rock chips recovered from specific intervals underground. These are the samples that are analyzed in a laboratory to determine the amount of gold that is present.

support drilling and other studies. Because most mineral deposits are eventually determined not to contain ore and do not become mines, exploration infrastructure is purposely temporary. Nevertheless, it can be significant—camps like that shown in (Figure 12-15) typically include an airstrip, trails or roads for equipment to move about on, housing for people, wastewater treatment facilities, and structures for maintaining equipment and handling samples. This is one reason that merely determining whether a given deposit is worth develop-

FIGURE 12-15 ▲ An Exploration Camp in Alaska
This is the exploration camp at the Donlin Creek gold project. About 100 people live and work here year-round.

ing may require the investment of millions of dollars. It is common for exploration projects to remove infrastructure when the project ends.

In the News

Donlin Creek Gold Deposit

The Donlin Creek gold deposit is in southwestern Alaska. It is remote—native villages of the Kuskokwim River region are its nearest neighbors—and it is huge. It has been extensively explored by drilling for over ten years, and is now known to contain over 850,000 kilograms (over 29 million ounces) of gold. But exploration continues to determine if a part can be mined profitably. The mining company in charge of this project is nearing the feasibility study stage that will determine whether an ore body is present. If it is, the company will apply for permits that will enable it to begin mine construction.

Permitting new mine developments is a complicated and involved process that includes evaluating the economic, social, and environmental consequences for the entire region, not just the mine site. This project faces all the challenges common to new mine developments in the United States, and you can expect it to be in the news for several years. Just search online to keep up-to-date. Donlin Creek gold, Alaska

MINING MINERAL RESOURCES

If exploration discovers a new ore deposit and appropriate permits are obtained, then mining can begin. Modern mines recover the ore from bedrock using a variety of technically complex and highly mechanized operations, and large facilities are needed to support them. These facilities are constructed to last the life of the mine and include roads, buildings, power systems, and water systems. There is a significant difference in size between open-pit and underground mine operations. Open-pit mines commonly excavate and process much more rock than underground mines.

Open-pit mining

Open-pit mines can be the lowest-cost mining operations in the world. They are also the largest. Mines like the one shown in Figure 12-8, which is 4 kilometers (2.5 mi) across and 1.2 kilometers (0.75 mi) deep, produce up to 136,000 tonnes (150,000 tons) of ore daily. As Figure 12-16 shows, open-pit mining blasts ore to break it up and uses large trucks to transport the broken ore to a crusher.

FIGURE 12-16 ▲ Open-Pit Mining
Blasting **(a)**, hauling **(b)**, and crushing **(c)** in an open-pit mine. The crushed ore is commonly transferred to a mill by conveyor.

FIGURE 12-17 ▲ Open-Pit Mines Produce Huge Quantities of Waste Rock
The hill in the background is all waste rock from the Bingham Canyon mine in Utah. (The area in the foreground has already been reclaimed.)

However, open-pit mines also need to remove large amounts of rock that surround the ore deposit. As the pit gets deeper and deeper, more and more of this kind of rock, called **waste rock**, must be removed to expose the ore. It's common in these mines to remove 2 or 3 tonnes (tons) of waste rock for every tonne of ore that is mined.

Because open-pit mines produce so much waste rock, properly disposing of it can be a challenge. It is generally stacked in huge piles, called *dumps*, someplace near the open pit. Waste rock dumps like those at the Bingham Canyon mine (see Figure 12-8) can be some 200 meters (660 ft) high and thousands of hectares (acres) in area (Figure 12-17). The EPA classifies waste rock as toxic waste. This is why mines are commonly recorded as exceptionally large generators of toxic waste. Environmentally sound disposal of waste rock is required of modern mining operations.

Underground mining

Ore deposits that extend deep into the subsurface are mined by underground methods. The ore deposits can be accessed in a variety of ways, including vertical shafts or openings that are driven into the side of slopes (**adits**) or inclined from the surface downward (*declines*). As Figure 12-18 shows, the underground openings where ore is removed can be large, but underground mines are generally much smaller operations than open-pit mines. They commonly produce only a few million tonnes (tons) of ore over their entire life and the amount of waste rock they generate, which is about the same as (or less than) the amount of ore they yield, may be disposed of underground. If some is brought to the surface the resulting waste

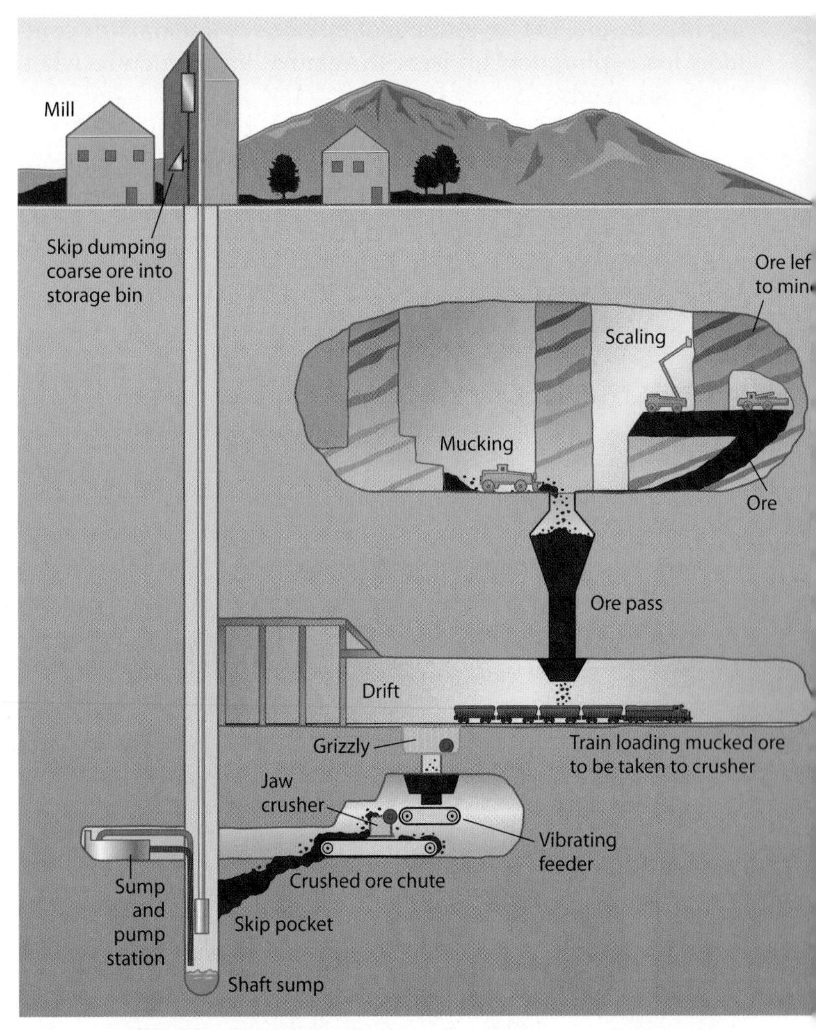

(a)

(b)

FIGURE 12-18 ▲ Underground Mining
(a) This diagram of a zinc mine in Tennessee shows the basic components of an underground mine. **(b)** Underground mine operations can use large equipment like this scaler, which removes loose materials left on the walls and roof after blasting.

FIGURE 12-19 ▲ **Abandoned Mine Land**
The waste rock dump at this abandoned mine in Colorado is the gray pile of loose rock at upper left.

rock dumps are relatively small (though still in some cases quite conspicuous).

Figure 12-19 shows a waste rock dump from an old mine in Colorado. Before 1900, mine operations were mostly underground and waste rock dumps like that in Figure 12-19 are characteristic of old mines and prospects.

The West's legacy—abandoned mine lands

From the time gold was discovered in California in 1848, mining helped develop the West. The Mining Law passed by Congress in 1872 encouraged people to explore and mine government lands throughout the country. For example, this law (which is still in effect) allowed prospectors to obtain the right to exploit the mineral resources of an area by staking a claim, which they could do simply by physically marking the corners of the area. How to stake a mining claim

At first, mineral exploration relied mainly on observation. To be discovered, mineral deposits had to be exposed at the surface, or in shallow excavations. People seem to have been digging for mineral deposits everywhere in the west. Physical disturbances such as prospect pits, trenches, exploration shafts or adits, and small waste rock dumps are common (see Figure 12-19).

People tried mining in many places, but the key to their projects being profitable was the discovery of high-grade ore deposits. Such deposits are generally not large, and most require underground mining. Moreover, they do not take long to exhaust ("mine out"), and early miners would quickly abandon them and move on to search for others. They left behind the disturbed mining areas and in places entire towns—ghost towns, like the one shown in Figure 12-20.

Abandoned mining areas are widespread through the West. Their disturbed areas may be conspicuous features on the landscape but many are small and of little environmental consequence. In some cases however, the environmental consequences are significant—especially where water quality is affected by erosion of waste rock or by release of polluted waters from mine openings. Identification and remediation of environmental damage from early mineral exploration and mining is an ongoing effort by federal and state agencies throughout the country. These sites are the legacy of the rapid and effective expansion into the West at a time when environmental concerns were of little interest.

PROCESSING MINERAL RESOURCES

Mining recovers rocks that are a mixture of valuable and nonvaluable minerals. For example, a typical copper ore contains the valuable copper-bearing sulfide mineral chalcopyrite along with other minerals such as quartz, mica, and the iron-sulfide mineral pyrite. **Beneficiation** separates and concentrates the valuable minerals such as chalcopyrite. The key steps in this process, milling and flotation, produce a waste material called *tailings*.

Milling

Milling grinds the ore into particles the size of silt or fine sand. The objective is to break the ore down into separate mineral grains. The mills that do this are cylinders (like the ones

FIGURE 12-20 ▼ **A Western Ghost Town**
This is Bodie, California, an 1800s mining town that is now a state park. Note the small waste rock dumps on the hill beyond the town.

FIGURE 12-21 ◀ Milling
The large cylinders (ball mills) hold steel balls that grind the ore as the mill rotates.

shown in Figure 12-21) containing steel balls that grind the ore as the cylinder rotates. Water is mixed with the ore in the mill, and when grinding is complete, a *slurry* containing the ground-up ore particles suspended in water is removed for further processing.

Flotation

The next step in the processing of sulfide mineral ores is commonly **flotation**. Flotation concentrates the valuable sulfide minerals and enables them to be separated from nonvaluable minerals such as quartz or pyrite (Chapter 4). Flotation is accomplished in vats, like those in Figure 12-22a, that contain the slurry from the mill and special bubble-forming reagents. The discovery of these reagents and their application to the processing of sulfide minerals in the early 1900s was a major technological breakthrough. The bubbles they make in the agitated vats are much like soap bubbles. Specific sulfide minerals such as chalcopyrite will selectively adhere to a bubble and float to the surface, where they can be collected as shown in Figure 12-22a. After the mineral-bearing bubbles are collected they can be dewatered, commonly by filtering, to leave behind a concentrate of the valuable minerals.

Left behind in the slurry after flotation are minerals of no value, such as quartz and pyrite. Once the valuable minerals have been removed, the slurry is termed **tailings**. The tailings are transferred to large ponds, known as *tailings ponds*, for storage and ultimately disposal (Figure 12-23).

FIGURE 12-22 ▼ Flotation
The watery slurry from the mill is pumped to flotation cells **(a)** where specific metal-bearing minerals attach to bubbles and float off the surface **(b)**. Filtering the bubbly mixture concentrates the metal-bearing minerals.

(a)

(b)

(a)

(b)

FIGURE 12-23 ▲ Tailings Disposal
The waste after milling and flotation, called tailings, is pumped to a disposal area **(a)**. The disposal area can be very large and collect pools of water on its surface where the tailings are less permeable **(b)**.

Tailings

Tailings are the principal focus of the environmental concerns associated with mineral beneficiation. This is because they commonly contain large amounts of pyrite—as much as 40%. As we will see, acidic conditions produced by the oxidation of pyrite can degrade both soil and water quality.

Tailings ponds are often located in areas lower than the mill facilities so that gravity can help as the tailings are pumped to the ponds. The ponds may be near the facilities or many kilometers (miles) away. For large mines like that at Bingham Canyon, Utah, the associated tailings ponds can be huge, over tens of square kilometers (thousands of acres) in area and about 100 meters (330 ft) deep. It is important that ponds like these be carefully designed and constructed to make them safe from failure, and properly located so that they cannot release material into environmentally sensitive areas such as streams.

Leaching

Some metals—mostly from certain types of copper- and gold-bearing ores—are removed from rocks and concentrated by **leaching**. These ores are piled in large *heaps* and specially formulated chemical solutions are allowed to percolate through them to dissolve the metals they contain. The metal-bearing solutions are collected from the base of the heap and processed to recover the metals. To prevent release of the solutions, which are acidic in copper leaching operations and cyanide-bearing in gold leaching operations, the base of the heaps are lined with impermeable synthetic or natural clay barriers as shown in Figure 12-24. Leaching does not require milling or generation of tailings, but the large piles of leached ore remain when operations cease.

RECOVERING METALS FROM ORE CONCENTRATE

Metallurgy removes desired elements from the valuable minerals that are concentrated by beneficiation. Melting sulfide minerals in a process called **smelting** is the most common metallurgical technique for recovering the metals they contain.

Smelting facilities, or *smelters*, have large furnaces in which the ore is placed and heated to its melting point. The molten metals sink to the bottom and are removed, while the impurities rise to the top. These impurities, mostly iron and silicon oxide, cool so quickly that they form a glassy solid called **slag**, the principal solid waste generated by smelting. Large dark piles of slag like the one shown in Figure 12-25 on p. 370—sometimes up to a square kilometer (247 acres) in area and 30 meters (100 ft) high—commonly mark the location of smelters.

FIGURE 12-24 ▼ Heap Leach Operation for the Recovery of Gold
Gold ore is being placed in a pile (heap) for leaching at this mine in Ghana (note the truck for scale). Solutions will be sprinkled on the surface that dissolve gold as they percolate through the heap. The solutions are trapped above an impermeable liner (the black sheet) at the base of the heap, and collected for processing to remove the dissolved gold.

FIGURE 12-25 ▲ Slag Is the Principal Solid Waste Produced by Smelting This pile of slag is in Leadville, Colorado. The inset shows the smooth, somewhat rounded surfaces characteristic of this glassy material.

Because it is glassy, slag is not highly reactive in the environment, but some does contain metals such as lead and arsenic that make proper disposal necessary. Sometimes, though, slag can be used for other purposes—as an abrasive, as base material for railroads, and even in traps on golf courses.

Smelters also release gases and particulate matter to the atmosphere. These emissions commonly contain sulfur dioxide and in some places troublesome elements such as lead and arsenic. As you will see, such emissions can be harmful to the environment and people if they are not adequately controlled.

12.3 Environmental Concerns

There are several environmental concerns associated with finding and producing mineral resources. The most obvious are the physical disturbances that accompany mineral production—excavations and waste rock piles, for example. But soil, water, and air quality can be affected, too. Fortunately, the environmental consequences of mining and mineral production are well understood. They can be mitigated or prevented if appropriately addressed.

PHYSICAL DISTURBANCES

Every step of finding, mining, and processing minerals physically disturbs the landscape. The physical disturbances from exploration, such as trenches and roads, are relatively minor compared to those of an open-pit mine with deep excavations and large waste rock dumps. And waste materials produced by beneficiation (tailings) or smelting (slag) create additional large piles on the land surface. The physical disturbances at underground mines are less dramatic than at many other mining operations, but all mines are industrial facilities that affect the land and nearby surroundings—often in highly visible ways. Mitigation of physical disturbances is required for all exploration, mining, and mineral-processing operations.

Mitigating physical disturbances can take place at various scales, from individual drill pads to complete mine sites, but it commonly involves three general steps: reshaping of the land surface to resist erosion, covering it with soil, and planting new vegetation to help stabilize the land surface. This reworking of the disturbed land, called **reclamation**, is commonly required by the permits for exploration and mining operations that government regulators issue.

Many permits also require bonding to guarantee that reclamation is carried out when operations are completed. Bonding involves posting or guaranteeing an amount of money that will be surrendered to the permitting agency to pay for site remediation if the mining company fails to do an adequate job.

At mine closure, it is common practice to demolish, salvage, or otherwise remove all mine support facilities. Reclamation is also carried out after mining and processing have been completed—to close waste rock dumps, tailings ponds, and slag piles, for example. Effective reclamation techniques vary depending on the nature of the waste material, because in many cases it's not just physical disturbances that need mitigation. As you will learn, improper disposal of mining and mineral-processing wastes can degrade soil, water, and air quality.

Reclamation is capable of changing large piles of bare rocks, tailings, or slag into stable, vegetated landscapes. However, it does not restore the land to its pre-mining conditions. And reclamation can modify and stabilize slopes in open pits, but these excavations are commonly not refilled at mine closure. Most open pits remain visible parts of the landscape, despite the fact that closed rock quarries may find new uses (Chapter 4). A sharp eye will also be able to identify reclaimed waste disposal sites, even when they provide new habitat for native plants and wildlife.

What are some Earth systems interactions influenced by reclamation?

SURFACE WATER QUALITY

Protecting water quality is a high-priority environmental challenge at mining and mineral-processing sites. Surface water, streams, lakes, and even ocean waters in some cases, can be degraded by accidental spills of toxic chemicals, erosion of waste materials, or discharge of contaminated water from mines or related facilities.

FIGURE 12-26 ▲ The Baia Mare Spill
These fish were killed by the accidental release of cyanide into Romania's Tiza River following the collapse of the Baia Mare tailings dam.

Spills

Accidental spills of toxic chemicals from storage or processing facilities are potential concerns wherever they are used. Where mineral processing employs toxic chemicals, such as cyanide, modern facilities are surrounded by structures (commonly *berms*) to contain potential spills. Sound operating procedures call for regular maintenance and monitoring of such facilities to check for potential structural failures that could cause a spill. However, accidents have happened. The collapse of the tailings dam at Baia Mare, Romania, caused a major cyanide spill into a river system.

Baia Mare, Romania, was the site of sulfide ore mining for centuries. The old tailings from this mining contained some gold—enough that a company decided to process them for their gold content in the 1990s. They constructed a cyanide-leach facility and began operations. Unfortunately, the facility was not properly designed.

After a period of heavy rains, the dam at the tailings impoundment failed on January 30, 2000. About 100,000 cubic meters (3.5 million gal) of tailings material, containing some 91,000 kilograms (100 tons) of cyanide, was released to the Tiza River drainage. Within a few days the contamination extended as far downstream as the Danube River in Yugoslavia, 400 kilometers (250 mi) away. The cyanide poisoned aquatic life, and pictures of dead fish were on the front pages of the world's newspapers (Figure 12-26).

The Baia Mare cyanide release is evidence that strong adherence to best practices is needed to insure safe use and handling of cyanide in gold leaching operations around the world. In fact, this experience led to a clarification of what constitutes "best practices" in the mining industry's design and operation of cyanide leach facilities—the *International Cyanide Management Code*. Individual facilities can now be audited to determine how well they meet these best practices. However, the list of gold mines fully certified as using cyanide according to industry standards remains incomplete. | International Cyanide Management Code

Industry best practices include engineering designs that use strong, impermeable barriers at the base of the leach pads and effective collection systems for the leach solutions that percolate through the ore. Operators closely monitor well-designed heap leach pads to prevent leaks from developing and solutions from being released to the environment. Because some leaching chemicals may be left behind after the metals have been removed, mines commonly employ a combination of rinsing, physical isolation, and detoxification of heap leach pads before they are reclaimed. Fortunately, cyanide degrades naturally in the presence of sunlight, atmospheric oxygen, and rainfall. It will not accumulate in the environment or maintain its toxicity for very long.

What are some Earth systems interactions that accompany the release of cyanide to the environment?

Erosion

Erosion of waste materials, especially waste rock and tailings, can affect surface water quality. Waste rock disposal areas are located as close to mines as possible to minimize haulage costs, but if they are not properly sited, metal-bearing materials can be eroded into streams or other bodies of water. When this occurs, the materials may react with water and oxygen to release metals and other potentially undesirable elements into the environment. Metals that become dissolved in surface water in this way are more **bioavailable** to organisms. For example, even trace amounts of dissolved copper can be harmful to freshwater fish. This copper affects the nervous system of fish and can lead to lesions in their gills. Severe exposure can damage the gills enough to cause death.

Regulations do not allow current mining operations to dispose of waste rocks or tailings where they can be eroded into surface bodies of water. As we saw previously, though, this was not the case during the early days of mining when the West was being explored and developed. Protection of surface water from improperly disposed waste rock and tailings is still needed in many old or abandoned mine sites.

Proper disposal requires placement outside active floodplains, as well as reclamation that stabilizes the waste rock or tailings. Figure 12-27 on p. 372 illustrates a simple but effective design for reclamation of waste rock. As you will see, many of these wastes can oxidize and generate acidic soils or waters. In these cases, people add lime or other materials that can neutralize acidity and then cover the rock with topsoil to promote vegetation growth. Revegetation and surface contouring to control runoff are additional reclamation steps that will inhibit water infiltration, stabilize slopes, and prevent erosion.

Discharge of acid rock drainage

Many excavations at mines are in effect a type of well—they extend down to the water table and will fill up with water if they are not continually pumped out. Water that collects in

FIGURE 12-27 ◀ A Typical Reclamation Cover Design This design was used in the flower-covered area, which consists of reclaimed waste rocks from a lead, zinc, and silver mine near the Bingham Canyon copper mine in Utah (see Figure 12-8).

mines or drains through them can become acidic and contaminated with toxic metals. Where this water is discharged to the surface, it can degrade nearby surface water quality. This is especially true in mines where the ore deposit is rich in sulfide minerals. The sulfide mineral that has the greatest effect on water quality is pyrite (iron sulfide). Pyrite is generally not valuable (although some has been mined to make sulfuric acid), but it very commonly accompanies other sulfide minerals in ore deposits.

When pyrite is exposed to air and water, it oxidizes. Oxidation of pyrite, which is enhanced by bacterial action, results in the formation of iron oxides and sulfuric acid (Figure 12-28). Where water encounters oxidizing pyrite it becomes acidic and dissolves metals such as copper, zinc, and silver. Acidic water produced by the oxidation of pyrite (and other sulfide minerals) is **acid rock drainage** (ARD). If production of acid rock drainage is not prevented or is left uncontrolled, the resulting acidic and metal-bearing water may drain into and contaminate streams and other surface water bodies. Some open pits have filled with acid rock drainage (Figure 12-29).

Acid rock drainage can be the most significant environmental impact of a mine. ARD that is removed during operations, commonly by pumping, must be properly treated and disposed of to ensure that surface water quality is not degraded. However, after mining ceases, the mine workings will fill with water and some may be discharged to the surface through mine openings. Because these waters migrate before being discharged through rocks that commonly contain pyrite, they can become acid rock drainage (Figure 12-30). If left unmanaged, significant volumes of acid rock drainage can form.

FIGURE 12-28 ▼ Generation of Acid Rock Drainage
Reaction of oxygen with pyrite, catalyzed by certain bacteria, produces sulfuric acid that can mix with and contaminate surface and groundwater.

Because acid rock drainage can have such a significant environmental impact, extensive efforts to prevent or control it are required. Approaches to the prevention of ARD all involve preventing oxidation of sulfide minerals, especially pyrite. Disposal of pyrite-bearing wastes in appropriate places, with impermeable materials at their base, and sound reclamation methods that inhibit water infiltration greatly reduce the generation of ARD and its release to the environment. Other prevention techniques include flooding underground mine openings, encapsulating pyrite-bearing rock in mines with impermeable coatings, and filling unused mine openings with material that neutralizes acid.

Whenever ARD cannot be prevented, it has to be treated. If the amount of ARD is small, the placement of neutralizing material such as limestone at ARD discharge points has helped protect surface and groundwater. But active treatment that captures ARD and neutralizes it, commonly with lime, is needed in many places. Such conventional ARD treatment facilities, like the one shown in Figure 12-31, require operation and maintenance that can go on indefinitely. They also generate a large amount of their own waste, a metal-bearing sludge that must also be properly disposed of. Generally, it's best to prevent ARD in the first place.

GROUNDWATER QUALITY

Mining operations and ARD can also affect groundwater. In the case of mine operations, withdrawing water from mine workings can lower the water table and change the way groundwater migrates in the subsurface. Dewatered mines can essentially create large cones of depression in the water table (Section 10.3) that focus groundwater movement toward them.

Because waste rock and tailings commonly contain pyrite, water that infiltrates them and migrates into the subsurface can contaminate groundwater supplies. Tailings produced from milling sulfide ores—primarily copper, lead, zinc, and nickel ores—are especially prone to generating ARD, as they may contain up to several tens of percent pyrite. Also, because mineral grains in tailings have been broken into small particles (the size of fine sand and smaller) they have large surface areas

FIGURE 12-29 ▲ The Berkeley Open Pit at Butte, Montana
The pit filled with acidic and metal-bearing water after mining (and water pumping) stopped.

FIGURE 12-31 ▲ A Water Treatment Plant for ARD
The addition of lime neutralizes ARD at this treatment plant on Silver Bow Creek, near Butte, Montana.

that make them more reactive with air and water. They are also saturated with water, and because they contain clay-sized particles called *slimes* their surfaces are relatively impermeable and collect pools of surface water (see Figure 12-23). This can help keep them water-saturated, and if surface water is not prevented from collecting on them, such tailings may indefinitely seep ARD.

As Figure 12-32 shows, reclaiming tailings is similar to reclaiming waste rock. However, it is commonly necessary to

FIGURE 12-30 ▼ ARD Being Released from a Mine Adit
This ARD, flowing from a small underground mine in Colorado, contains high levels of dissolved metals including copper, iron, aluminum, zinc, and arsenic.

FIGURE 12-32 ▼ Reclaimed Tailings
This old tailings pond near Rico, Colorado **(a)** was stabilized and reclaimed in the 1990s **(b)**.

place an impermeable barrier at the base of the tailings pond to prevent ARD seepage. Older tailings ponds that were constructed without impermeable bases have generated ARD that has caused surface and groundwater contamination.

ARD that migrates into groundwater will react with rocks and minerals and eventually lose its acidity, but significant levels of dissolved constituents, such as sulfate, remain. Although this degraded water is not highly toxic, it cannot be used for drinking or irrigation.

What are some Earth systems interactions that accompany the generation and release of ARD?

SOIL QUALITY

The bare, orange- or rust-colored slopes on the side of a mountain (see Figure 4-20) or on the surface of a waste rock dump are evidence of oxidizing pyrite and acidic soil conditions. Acidic soils prevent plant growth and leave the surface vulnerable to erosion from surface water runoff.

It is not just acidity that degrades soils. Elevated metal content can also be harmful to organisms, including people. Such metal concentrations can develop where waste rocks or tailings have been dispersed rather than contained in reclaimed disposal sites. If such soils are identified to be potentially harmful, they are commonly removed and placed in repositories, like that shown in Figure 12-33, to prevent their exposure to the biosphere.

Table 12-1 gives examples of metal concentrations that may be hazardous to people in unconsolidated surface materials (commonly called "soils" by environmental scientists and regulators even if they are materials like waste rocks). But understanding soil contamination can be complicated. Take the case of soil lead levels, for example.

TABLE 12-1 USEPA SCREENING LEVELS FOR POTENTIALLY HAZARDOUS CONCENTRATIONS OF METALS IN RESIDENTIAL SOILS (PPM)

Arsenic	Cadmium	Lead	Silver	Zinc
0.4	39	400	390	23,000

The U.S. Environmental Protection Agency's generic soil screening level for lead in residential soils is 400 parts per million (ppm). This is the concentration that the EPA considers elevated enough to make the area a high-priority regulatory concern. The EPA has required removal of soils at specific sites if they contain as little as 500 ppm lead. Mining-related residential communities have some of the highest soil lead levels in the nation. Many of these soils, with lead concentrations up to tens of thousands of ppm, have been targeted for soil removal cleanups. (An example is shown in Figure 12-34.) However, studies in many of these same communities have failed to find any health effect related to the lead levels. As Figure 12-35 indicates, such studies have shown that blood lead levels, the guidelines for identifying potential lead-related health concerns, in these communities are consistently below the national average and are not correlated with soil lead levels. With respect to lead exposure, mining communities appear to have been better places to live than some other areas of our country.

The reasons blood lead levels in mining communities are lower than in many other places are still being investigated. It appears that the mineral source of the lead, commonly galena (lead sulfide), plays a role. Lead in galena or other minerals in mining communities may be less bioavailable than the type of lead people are exposed to elsewhere. The use of lead as a gasoline additive or in paint has historically exposed many people to this metal. The declining blood lead levels for the nation (see Figure 12-35) probably reflect diminished exposure to these sources of lead after such use was discontinued as a result of EPA regulations.

Soil remediation methods (discussed in Section 11.5) can be applied to some soils that contain mining wastes. One approach is to add chemicals to the soil that react with the elements of concern, such as metals, and make them less mobile and less bioavailable. These reactions form new minerals in the soil and keep the elements of concern from being dissolved in passing water or otherwise released to the environment.

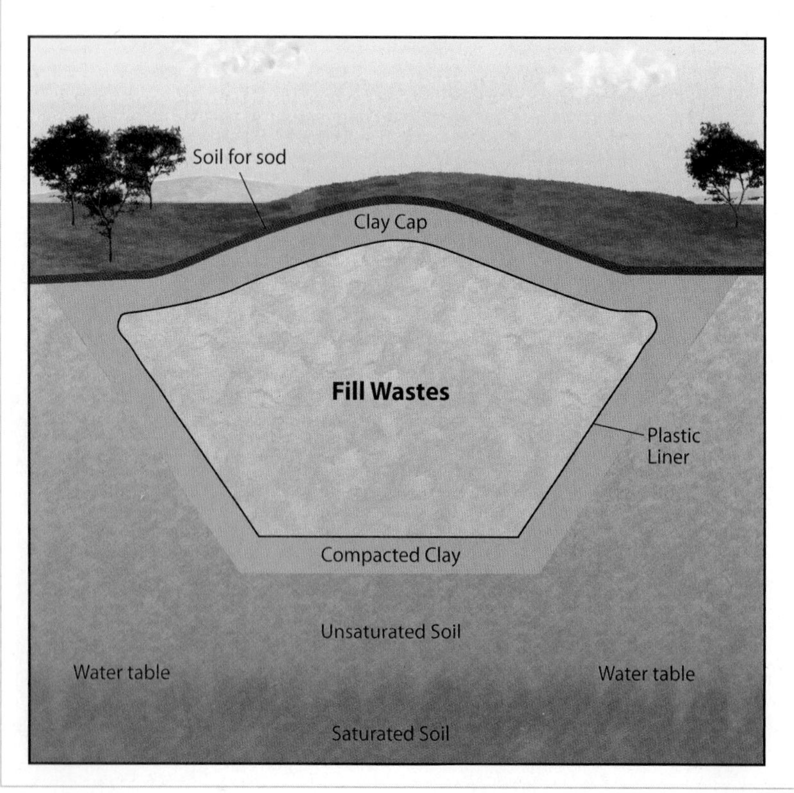

FIGURE 12-33 ◀ A Repository Design for Metal-Bearing Soil
Repositories physically isolate metal-bearing soil from interacting with the environment.

FIGURE 12-34 ◀ Removing Lead-Bearing Residential Soil
This soil removal operation is in a community near Salt Lake City, Utah.

Another soil remediation method is phytoremediation. This approach involves growing plants that take up elements of concern from the soil. Cultivating and harvesting these plants gradually decreases the toxic element content of the soil.

AIR QUALITY

Mining and mineral-processing operations can affect air quality. The principal concerns are dust generated at mine sites or blown off tailings ponds and the emissions from smelter operations.

Dust

Every step in recovering ore at mines is dusty. Drilling, blasting, hauling, and crushing rocks all create dust that is potentially harmful to anyone breathing it. This is why permits for mine operations all contain air quality standards and mine sites are routinely monitored to ensure the standards are met. Fortunately, procedures for controlling dust at mines are well developed. Water spray systems and vacuums are routinely used to diminish dust to acceptable levels.

If tailings pond surfaces dry up, the fine slimes in them can be a source of wind-blown dust, too. Satisfactory reclamation of tailings ponds controls their potential dust generation in addition to preventing erosion and ARD.

Smelter emissions

Controlling emissions is the biggest environmental challenge at smelters. Old smelters were not designed to mitigate these emissions and, as you can see in Figure 12-36 on p. 376, the environment was very seriously affected by them. Sulfur dioxide has been the key culprit in the emissions, as it reacts with water to form sulfuric acid that contaminates soils and water and kills vegetation. Some smelters also emitted concentrations of metals, such as lead, that were harmful to people nearby.

The hazards of smelter emissions are well known and major steps have been taken to control them. Smelters today,

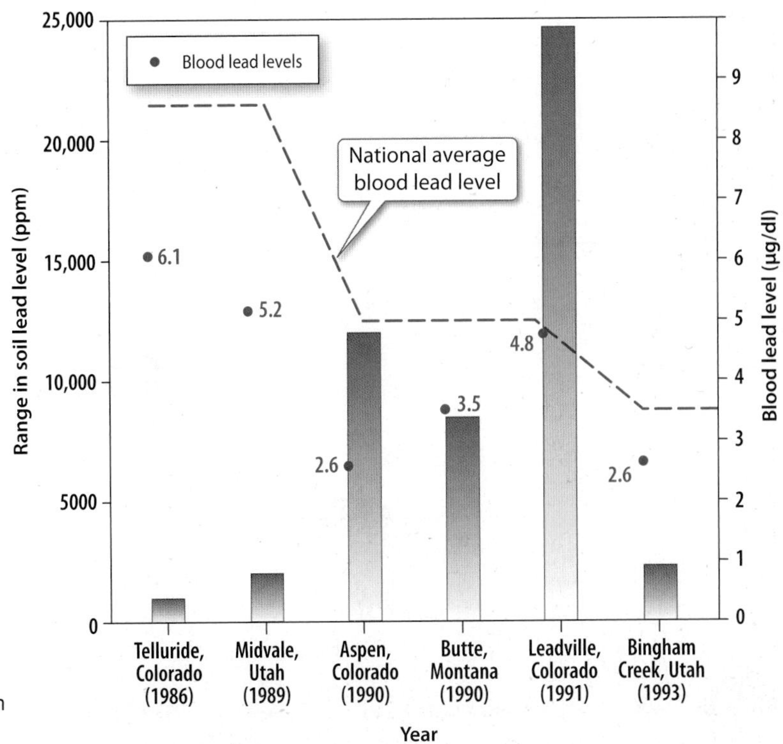

FIGURE 12-35 ▶ Blood and Soil Lead Levels in Mining Communities
People's blood lead levels in mining communities with high soil lead levels have been low.

FIGURE 12-36 ◄ The Bad Old Days
The first smelters, like this early 1900s operation in Idaho, did nothing to control their emissions.

like that shown in Figure 12-37, capture almost all (99.9%) of the sulfur dioxide they generate. Still, smelters process large volumes of material, and even small amounts of emissions can add up to levels that may be of environmental concern.

What are some Earth systems interactions that accompany sulfur dioxide emissions from smelters?

12.4 Mineral Resources in the Future

Providing the mineral resources people need involves several challenges. You have learned about the environmental challenges in the previous section, key ones like proper waste rock disposal, sound tailings pond construction, prevention of ARD, and control of smelter emissions. Other issues associated with mineral resource production and use include the rapidly increasing demand for mineral resources, the role of recycling in meeting mineral commodity needs, and the application of sustainability concepts to the use of minerals.

FIGURE 12-37 ◄ The Kennecott Smelter Near Salt Lake City, Utah
Modern smelters like this one, which operate under close regulatory oversight, have decreased their emission of sulfur dioxide and particulates to below permitted levels.

FUTURE MINERAL RESOURCE NEEDS

The demand for mineral commodities is increasing. Societies need mineral resources to sustain their economic well-being, and populous societies with expanding economies need even more. You learned in Chapter 1 that the human population is expected to rise to over 9 billion people by 2050. Population growth is a key long-term driver of mineral resource consumption. Also, the economies of several very populous countries are expanding at exceptional rates. The result has been rapidly increasing per capita consumption of mineral resources in such countries as Brazil, India, and China.

Expanding economies need more natural resources. When it comes to mineral resources, the increased demand in China alone has been staggering. For example, from 2000 to 2006, China's demand for iron ore increased as steel production more than tripled. During the same period, its consumption of copper increased by more than 14% per year; by 2020, China is expected to consume 23% of the world's copper production.

China's mineral commodity consumption has helped create supply shortages and driven up prices. For example, as Figure 12-38 shows, the price of copper quadrupled between 2003 and 2007. High mineral commodity prices of recent years plummeted in late 2008 as the global economic recession developed. Lower metal prices will curtail aggressive mineral exploration but companies are still exploring around the world, finding new ore deposits, and planning new mine developments. You can expect mining operations to increase in many places as the global economy rebounds and populous countries increase their metal consumption in the years to come.

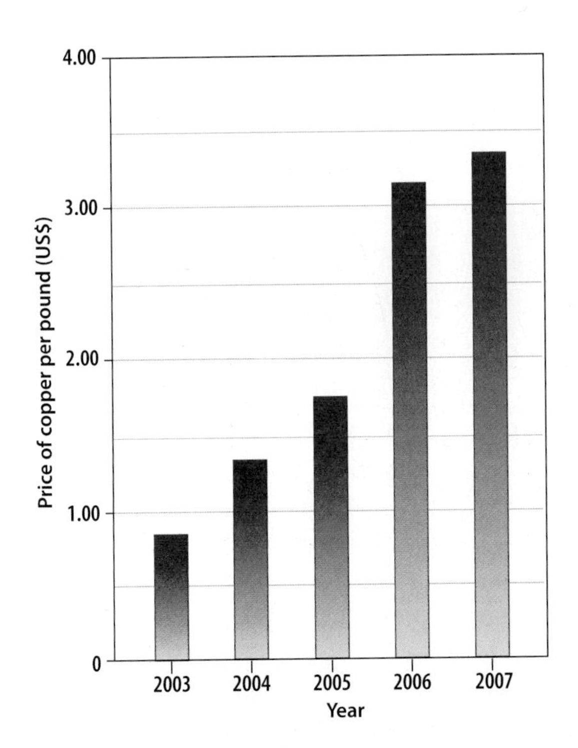

FIGURE 12-38 ▲ The Recent Price of Copper
The price nearly quadrupled in just four years.

RECYCLING

Recycling can not only extend the use of a finite resource, but also diminish the environmental consequences of mineral resource development and production. For any industry or nation on a mineral-resource budget, recycling and conservation are always prudent measures to consider.

Recycling is a source of many mineral commodities. At least 26 metal commodities are currently recycled to a significant extent in the United States. Flow cycles for recycling metals in the United States Overall, recycling meets over half of our nation's metal needs each year (Table 12-2). Can recycling provide more of our needed resources in the future? Yes and no. Consider, for example, the case of lead and zinc.

Lead, a malleable, blue-gray metal with a low melting point, has been used for many purposes for thousands of years—as a construction material, as part of solder alloys, in ceramic glazes, and as a constituent of crystal glass. For many years, lead was incorporated in paint and gasoline. These were, **dissipative** uses—ones from which we cannot recover lead. Such uses have been discontinued to decrease people's exposure to toxic forms of lead.

In 2007, 1.45 million of the 1.63 million tons of lead consumed in the United States, or 89%, was in lead-acid batteries. Recycling of lead from batteries is very effective—by 1997 almost all (97%) of lead-acid batteries were recycled. Lead recycling in general now meets about 70% of our needs and could reach 80% if dissipative uses were decreased. Recycling 70 to 80% of a mineral commodity is a very high rate of recovery.

We cannot do nearly this well with zinc. In 2007, we consumed 1.18 million tons—most of it (55%) for galvanizing steel (coating it with zinc to protect it from corrosion). Many uses of zinc, however, are inherently dissipative—uses in dies, pigments, fertilizers, animal feeds, fungicides, and pharmaceuticals, for example. Zinc recycling—primarily of zinc-bearing metal alloys (such as brass) or coatings in cars, metal roofing materials, and the like—provides about 35% of our annual zinc consumption. Given the nature of zinc's uses, a significant increase in this figure seems unlikely.

The lead and zinc recycling examples show us that we can get very good at recycling some metals but we won't be able to provide all of the metals we need in this way. For some metals, such as zinc, we may not even be able to recover half of the

TABLE 12-2 TOP RECYCLED METALS IN THE UNITED STATES, 2006

Commodity	Value	Percent of total annual consumption
Iron (and steel)	$36,100,000,000	55%
Aluminum	$21,800,000,000	40%
Copper	$20,800,000,000	31%
Nickel	$ 6,100,000,000	35%
Lead	$ 2,680,000,000	70%

metal we use through recycling. Also, recycling brings its own set of environmental concerns. Lead recycling, for example, has contaminated soil and caused elevated blood lead levels in people. In general, recycling has proved more valuable for helping to sustain our mineral resources than for reducing the environmental concerns related to mineral use.

SUSTAINABILITY AND MINERAL RESOURCE USE

Some mineral commodities, such as phosphate rock used for fertilizer, are completely consumed when they are used. Others, like metals, are not completely consumed, but dissipative uses prevent their complete recovery by recycling. Either way, it takes hundreds of thousands to a few million years for geologic processes to form new mineral deposits to replace what we consume. For these reasons, mineral resources are considered nonrenewable. We mine an ore deposit until it is depleted.

Because mineral resources are nonrenewable, their consumption may eventually lead to shortages of some mineral commodities. Industrial societies have faced mineral commodity shortages in the past. These situations were created by global events, such as world wars, that limited international trade and access to ore deposits. During such times, people learned how to substitute materials for needed minerals, use less mineral-intensive technologies, or more carefully use and recycle (conserve) less available mineral commodities. If mineral commodity shortages due to consumption develop in the future, it is likely that a combination of these responses, together with advances in mining technology, will enable people to meet their material needs. However, there is one step in the production of mineral commodities that is truly irreversible. Once an ore deposit is mined, it is gone and its valuable minerals sent off to whatever fate society has decreed for them. How can sustainability concepts be applied to nonrenewable resources like a mined ore deposit?

The exploitation of natural resources can produce great wealth and opportunity. Where mines are developed, they create jobs, stimulate the growth of related industries, and serve as a source of tax revenue. They help fund community developments such as schools, hospitals, roads, and electric power systems.

In these ways, the financial capital derived from mineral resources is converted into other forms of economic capital, social capital (enhanced community institutions) and intellectual capital (the educational advancement of new generations). Such conversion is a major contribution that mineral resource development can make to a sustainable future for society as a whole. The mineral resource capital, an ore deposit in this case, is converted to another form of capital that can provide sustainable benefits to society. There are many contemporary examples of this relationship. An instructive one is provided by the Inuit people of Nunavik Territory, who have allowed the Raglan Mine to be developed on their land.

The Raglan mine is in Quebec, Canada (Figure 12-39), some 1800 kilometers (1100 mi) north of Montreal. In this remote region there are no trees; the ground is frozen all year round, and the temperature averages –10° C (14° F).

FIGURE 12-39 ◀ The Raglan Nickel Mine in Nunavik Territory, Quebec

FIGURE 12-40 ▲ An Inuit Family from Nunavik, Canada, and Their Catch

| Raglan mine site infomine | The nickel- and copper-bearing deposits here have been known since the 1930s, but it wasn't until the 1990s that enough ore was found and the price of nickel and copper high enough for a mine to be developed. The new Raglan Mine was the first in Quebec to develop an environmental impact statement (Chapter 15) and to develop a mine and operations plan agreed to by the local indigenous people.

The 8000 Inuit people of Nunavik Territory (Figure 12-40) subsist off the land. They did not want the new mine to harm the fish and animals of the region on which they depend, nor disrupt their traditional ways. To achieve these aims the mining company and the Inuit developed the "Raglan Agreement." | The Raglan Agreement | This agreement established employment priorities for local people, contracting priorities for local businesses, and numerous environmental safeguards to protect water and wildlife resources. In addition, the inhabitants of the region received money for the use of their land and a share in the mining proceeds. The Raglan Agreement has many provisions for helping the local people and mining company work together.

But what about sustainability? How will the Inuit people of the Ungava Peninsula invest their new financial, social, and intellectual capital so that their future is what they want it to be after the ore deposit is depleted and the Raglan mine closes? The mine will operate only for a generation or two—then it will be necessary to replace mining with another industry capable of carrying on and contributing to the economy. The Inuit people will have to decide how the financial capital derived from the Raglan mine can be used to sustain them in the future.

You Make the Call

Where Should Mining Occur?

As we have seen, our consumption of mineral resources is increasing, and most people recognize that the mining industry provides goods that we want and need. So as a whole, U.S. citizens seem very accepting of mining—if it occurs somewhere else! But people don't seem to want mining in the United States. As the industry struggles to operate and develop new mines, it is often thwarted by community opposition and environmental litigation. As a result, exploration and new mine development have gone elsewhere and our domestic mining industry is shriveling. This has not been the case in environmentally sensitive countries such as Finland and Sweden, which accept and foster the mining industry as part of the nation's industrial capability. | A new look at mining and the environment: finding common ground, Geotimes, April 2006 |

An example from Wisconsin provides some insight into the challenges facing new mine development in America. The Crandon ore deposit is located in an area of northern Wisconsin with forests, rivers, lakes—and people very concerned about mining's effects on the environment.

The discovery of the Crandon deposit, in the mid-1970s, was a successful application of geophysical techniques and drilling, as the ore is buried beneath 15 to 107 meters (50 to 350 feet) of glacial deposits. It contains over 50 million tonnes (tons) of zinc- or copper-rich ore with lesser amounts of lead, gold, and silver. The Crandon deposit is pyrite-bearing, so prevention of ARD and surface or groundwater contamination is a key challenge to planning any mine development.

Mine developments in Wisconsin must be approved and permitted by the Wisconsin Department of Natural Resources. | Proposed Crandon mine information | Over the years, a series of mining companies developed plans for underground mining of the Crandon ore deposit. These plans included such steps as mixing pyrite removed from tailings with cement and pumping it back into the underground workings, injecting a grout blanket over the deposit as a barrier to groundwater migration into the mine, and on-site disposal of treated wastewater through groundwater seepage cells rather than discharge to surface water drainages.

Many people in northern Wisconsin, however, didn't trust either their state regulators or the mining companies. They were convinced that mining of the Crandon ore deposit would cause irreparable harm to the environment and vigorously opposed its development from the very first proposed mining plan. | No Crandon mine | The opposition declared success when the Forest County Potawatomi Community and the Sokaogon Chippewa Mole Lake Band completed purchase of the proposed mine site and the Crandon ore deposit in November 2006. These parties do not believe that current mining technology could allow the mine to operate safely. | Crandon mine sold to Forest County Indian tribes |

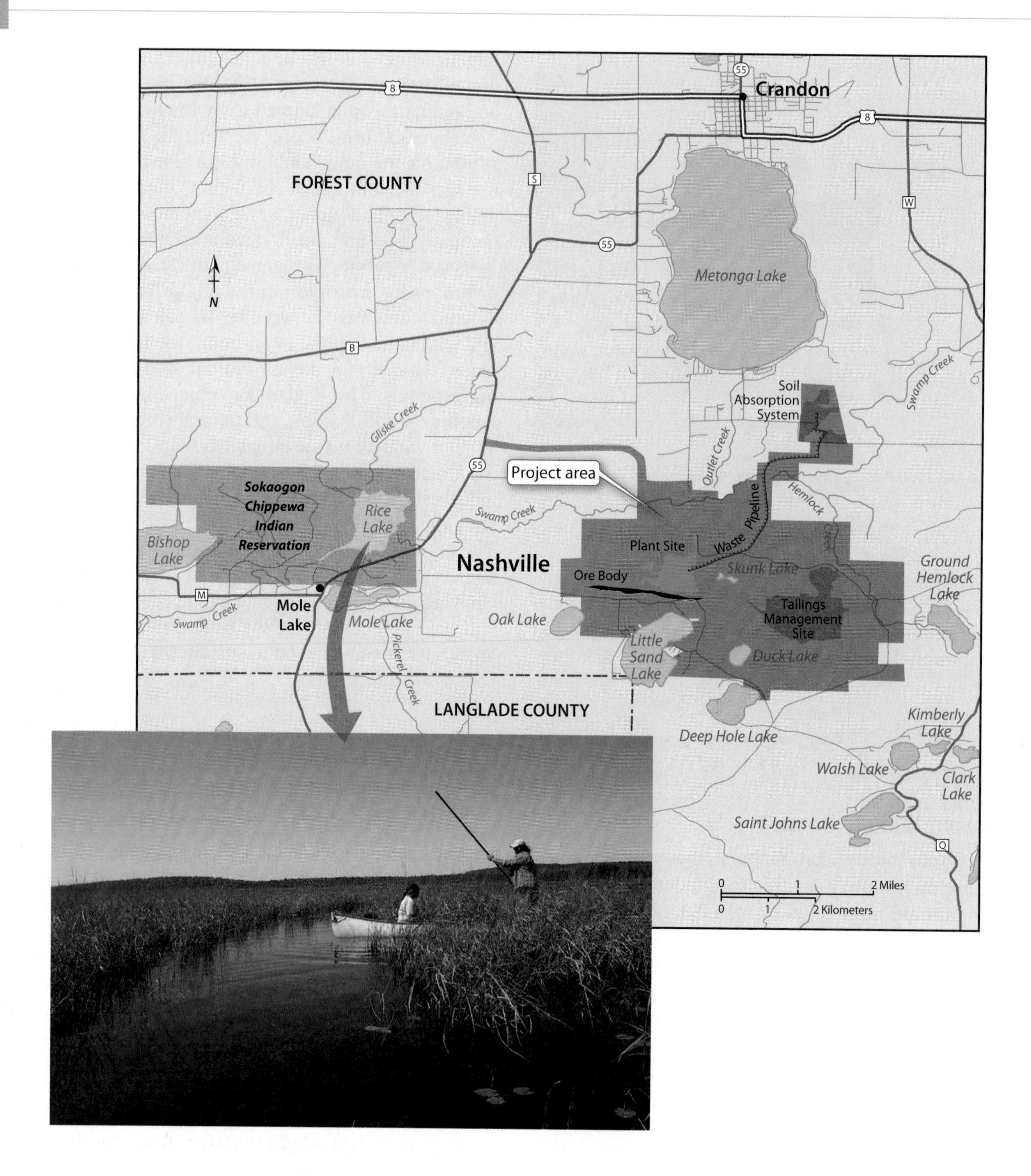

The long struggle over development of the Crandon ore deposit is an example of the challenges facing new mine developments in many parts of America. People rightly treasure the natural settings and habitat that they have grown up with and used for recreation and subsistence. And a mine is an industrial development that does bring changes to an area.

As you study the information available online about Crandon, consider the following questions.

- What is your reaction to the information, oversight, and decision making of the Wisconsin Department of Natural Resources?
- What is your reaction to mining plans brought forward by a mining company?
- Where do you think you could go to get information that you trusted about mining?
- Do you think you could take a side—for or against—mining the Crandon ore deposit?

SUMMARY

Exploration, mining, and mineral processing will supply needed mineral resources to the world for generations to come. Each step in providing mineral resources for people's use has some environmental consequences that change the landscape and potentially affect water, soil, and air quality. In the past, these environmental consequences left a legacy of disturbed land and degraded soil and water. Modern mineral resource production need not have the same environmental consequences. Sustaining mineral resources is partially achieved through recycling. The conversion of financial capital derived from mineral commodity production to other forms of capital can help develop and sustain other foundations of society.

In This Chapter You Have Learned:

- Mineral resources are naturally occurring solid materials from which economic extraction of a commodity is currently or potentially feasible. Over 80 mineral resource commodities play a significant role in the U.S. economy, and the supply of over half of these is dominated by imports from other countries.
- Concentrations of valuable minerals—mineral deposits—form in many ways. Many form where valuable minerals precipitate from metal-bearing waters in the shallow crust or on the seafloor. Ore is the part of a mineral deposit that at current or foreseeable commodity prices can be mined at a profit.
- Ore deposits are uncommon geologic features. Exploration to find them can be expensive and take long periods of time. Exploration involves surface observation and sampling combined with subsurface evaluation through trenching and drilling.
- Mining removes ore by underground or open-pit methods. Mining also removes waste rock surrounding the ore. The amount of waste rock is commonly two or three times the amount of ore in open-pit mines, but it can be less than the amount of ore in underground mines.
- Ore is processed by milling and flotation to separate and concentrate valuable minerals. The leftover nonvaluable minerals are a waste product called tailings.
- Leaching removes valuable metals, particularly copper and gold, from some types of ores. Leach solutions (commonly containing sulfuric acid or cyanide) soak through the ore, dissolve the metal, and carry it to processing facilities where the metal is removed.
- The process of smelting recovers metals from the minerals concentrated by milling and flotation. The solid waste from smelting is commonly an iron- and silicon-rich material called slag. Smelters are a source of gas (particularly sulfur dioxide) and particulate emissions to the atmosphere.
- Exploration, mining, and mineral processes lead to physical disturbances of the landscape. Physical disturbances such as industrial facilities, trenches, roads, waste rock dumps, tailings ponds, and slag piles can be reclaimed. Reclamation recontours the land, reestablishes vegetation cover, and protects waste materials from being eroded.
- Mining and mineral processing can affect water quality. Protection from spills of toxic chemicals is provided by containment structures and careful oversight and maintenance of facilities. Proper reclamation prevents erosion of wastes, such as tailings, that could contaminate surface water bodies.
- Pyrite-bearing materials in rock exposures in mines, waste rock dumps, and tailings ponds potentially affect surface and groundwater quality through generation of acid rock drainage (ARD). ARD can be prevented by inhibiting water infiltration, blocking migration pathways, and creating barriers between pyrite and oxygen sources. It can be treated by the addition of neutralizing materials such as lime.
- Dust can be generated during mining operations and from the fine materials (slimes) deposited in tailings ponds. Dust controls, such as water spraying, are used during mining operations, and sound reclamation prevents dust generation from tailings ponds.
- Sulfur dioxide emissions from smelters have historically been a major cause of environmental degradation. Sulfur dioxide reacts with water to form acid rain, which acidifies surface water (lakes and streams) and ultimately the soil itself. Modern smelters remove almost all the sulfur dioxide and particulates from their emissions.
- People will continue to need mining in the future. Population growth and expanding economies, particularly in countries like Brazil, India, and China, will significantly increase demand for mineral resources in years to come.
- Recycling can help sustain mineral resources, but it will not be efficient or complete enough to replace mining or new mineral resource production altogether.
- Sustainability can come partly from mineral resource production if the financial capital derived from production is satisfactorily converted to other forms of capital that can sustain society after the mineral resources are depleted.

KEY TERMS

acid rock drainage (ARD) (p. 372)
adit (p. 366)
alloy (p. 356)
beneficiation (p. 367)
bioavailable (p. 371)
core (p. 363)
dissipative (p. 377)
flotation (p. 368)
grade (p. 362)
leaching (p. 369)
metallurgy (p. 369)
milling (p. 367)
mineral deposit (p. 360)
open pit (p. 362)
ore (p. 360)
prospect (p. 363)
reclamation (p. 370)
slag (p. 369)
smelting (p. 369)
tailings (p. 368)
vein (p. 362)
waste rock (p. 366)

QUESTIONS FOR REVIEW

Check Your Understanding

1. What was the crucial transition in human use of natural resources that differentiated the Bronze Age from the earlier Stone Age?
2. Bronze is an alloy of copper and tin. What are alloys and why are they important?
3. What is meant by the term "mineral resource"?
4. Mineral resources are considered nonrenewable. What are the implications of this fact for future mineral use?
5. What general processes form mineral deposits?
6. Ore deposits are unusual geologic features and unevenly distributed around the world. How do the physiographic settings of ore deposits vary, and how do such variations affect their potential environmental impacts?
7. What is ore grade, and why is it relevant to the economics of mining?
8. How is a prospect different from a mine?
9. What are the general characteristics of mineral deposits that are mined by open-pit methods?
10. What are the general characteristics of mineral deposits that are mined by underground methods?
11. What is beneficiation, and why is it an important part of the mining process? What is the principal waste generated by beneficiation, and what is its environmental significance?
12. What is leaching, and why does this present environmental challenges in ore processing? Where are these challenges most acute?
13. Describe what smelting does in metal ore processing and how it works. What are the wastes generated by smelting?
14. Does reclamation of a mine site mean that the former appearance of the site will be restored? Why or why not?
15. What are the key aspects of reclamation and how does it help protect the environment?
16. What does it mean for an element to be "bioavailable" in mine waste?
17. What is acid rock drainage and why is it a problem with sulfide mineral deposits?
18. How is acid rock drainage prevented or mitigated?
19. In what ways is air quality affected by mining and mineral processing? How are air quality impacts managed at mining, beneficiation, and smelting sites?
20. Based on the discussion of resource sustainability in the text, why is it important to recycle metals whenever possible?
21. Why are human and ecological factors as important as geologic factors in the decision to open a new mining operation?

Critical Thinking/Discussion

22. The text outlines how the manufacturing of bronze "made the world smaller." In what sense was the growth of global trade tied to this and other alloys and materials?
23. The United States is a net importer of mineral resources and one of the biggest consumers of mineral commodities on Earth. What are the implications of these facts for the future of the United States economy, especially in light of the growth of other developing and industrialized nations elsewhere?
24. Mineral resources of all kinds exist almost everywhere on the planet at some level of concentration. However, only a very few of these are economically recoverable. Why? Describe the interaction of geologic and economic conditions that must exist for a mineral resource to be economically recoverable.
25. In almost all mineral formation processes outlined in the text, fluids are involved. What is the consistent role of fluids, especially warm to hot fluids, in creating a metal ore deposit that is of economic value?
26. Tailings ponds are a necessary but problematic part of the ore processing of many metals. Why is it so critical to build these well, and what are the consequences of poorly built tailings ponds?
27. Compliance with best practices in mining by many independent operators in many different countries can be difficult to enforce. Discuss strategies by which the industry can encourage widespread voluntary compliance with recognized best safety and environmental practices in mining.
28. In many cases, especially in large open-pit mines, reclamation is accompanied by other environmental protections,

including habitat or other improvements in regions near but not at the mine site. Why would permitting agencies and other stakeholders require this additional investment? Why is mine reclamation sometimes inadequate in the case of large earth-moving operations?

29. Why is acid rock drainage such a big problem for mines extracting sulfide mineral ores? Do natural exposures of these same materials create environmental impacts? How would mining-related exposures of sulfide minerals have different or more problematic environmental impacts than natural exposures? Explain your answer in terms of how sulfide minerals are exposed to the atmosphere and hydrosphere.

30. Working from the examples of lead and zinc cited in the text, describe another material for which dissipative versus concentrated use of a metal is likely to be a factor in recycling efficiency.

ANSWERS TO IN-CHAPTER INSIGHT QUESTIONS

P. 361

What are some Earth systems interactions associated with the formation of mineral deposits?

- Hydrosphere fluids react with the geosphere and dissolve elements such as metals. These fluids migrate through the geosphere to lower pressure and temperature regions and precipitate minerals containing valuable elements.
- Melting of the geosphere produces magmas that release water-rich, metal-bearing fluids when they crystallize.
- Recrystallization of the geosphere through metamorphism releases water-rich, metal-bearing fluids.
- Seawater trapped in sediments dissolves metals, migrates along permeable pathways in the geosphere to the seafloor, and precipitates metal-bearing minerals.

P. 363

What are some Earth systems interactions that make geochemical techniques useful exploration tools?

- The hydrosphere, atmosphere, and geosphere interactions involved in weathering and erosion processes release minerals, ions, and compounds that become dispersed in soils, waters, sediments, and plants.

P. 370

What are some Earth systems interactions associated with reclamation?

- Reclamation prevents erosion of geosphere materials by the hydrosphere or atmosphere.
- Reclamation controls infiltration of geosphere materials by hydrosphere precipitation or runoff.
- Reclamation fosters biosphere expansion and growth.
- Reclamation controls oxidation that accompanies interaction of the hydrosphere or atmosphere with geosphere materials.

P. 371

What are some Earth systems interactions that accompany the release of cyanide to the environment?

- Cyanide is dispersed through the hydrosphere.
- The respiratory functions of biosphere organisms, such as fish, are affected by exposure to cyanide in the hydrosphere.
- Reactions with oxygen in the hydrosphere or atmosphere changes cyanide to nontoxic compounds.

P. 374

What are some Earth systems interactions that accompany the generation and release of ARD?

- Reaction of oxygen in the atmosphere or hydrosphere with geosphere materials containing sulfide minerals (especially pyrite) generates ARD.
- Biosphere bacteria facilitate the chemical reactions that generate ARD.
- ARD released to the environment contaminates the hydrosphere and soils.
- ARD in soils and waters can negatively impact biosphere viability.

P. 376

What are some Earth systems interactions associated with emissions of sulfur dioxide from smelters?

- Roasting of geosphere materials in smelters releases sulfur dioxide gas to the atmosphere.
- Sulfur dioxide in the atmosphere reacts with water vapor to form sulfuric acid aerosols.
- Sulfuric acid in the atmosphere is transferred from the atmosphere to soils, water, and vegetation downwind of smelters.
- Biosphere vegetation may not be able to survive and grow in acidic air, water, and soil near smelters.

FIGURE 13-1 ▶ Going Up?
In the summer of 2008, the price of gas rose to over $4 per gallon in most parts of the United States. What determines this price, and what will the future bring?

CHAPTER

13

ENERGY RESOURCES

Do you remember the high gasoline prices in 2008, when the price at the pump soared to over $4.00 a gallon over large parts of the country? It may have cost you over $60 to fill up your car's gas tank (Figure 13-1). In late 2008 gasoline prices fell sharply, and within a few months they were less than half their earlier highs. What was causing these drastically fluctuating prices?

The Energy Information Administration informs us that, on average, 58% of the gasoline price you pay at the pump is due to the price of crude oil—when the price of crude oil goes up, so does the price of gasoline. | A primer on gasoline prices The other components of the price are taxes (15%), the costs of refining crude oil to make gasoline (17%), and the costs of distributing and marketing the gasoline (10%). These components lead to variations in gasoline prices across the country, but it's the price of oil that's the biggest influence on the price you pay at the pump. | U.S. national gas temperature map

Oil has been the most versatile and inexpensive source of energy used by people. As a result, everyone has come to depend on oil, most directly through the use of transportation fuels such as diesel and gasoline derived from oil. What causes the price of oil to change? Most fundamentally, it is the relationship between how much people want—the demand—and how much oil is available—the supply.

On the demand side, the world's need for crude oil significantly increased in the years leading up to 2008. Rapidly expanding economies in several populous countries drove crude oil consumption up to over 80 million barrels per day in 2008 (a barrel of oil is a unit of measure equal to 159 liters, or 42 gallons). Photos like Figure 13-2 (p. 386) dramatize how quickly populous countries such as India and China have increased their use of energy resources. China's demand for oil is now about 7 million barrels per day and has recently grown at rates of 5 to 10% per year. China is now the second largest consumer of oil, behind only the United States. Like the United States, China imports more oil than it produces and competes around the world for oil resources.

In mid-2008, demand for oil dropped as an economic recession developed around the world. Recessions are times when economies contract rather than grow—when people lose jobs and stop buying things other than necessities. Resource consumption, manufacturing, and trade decline as a result. But recessions don't last forever. When the populous countries around the world continue their economic expansion, the demand for oil can be expected to increase. This demand is expected to reach 118 million barrels per day by 2030. Will the Earth be able to supply this needed oil?

FIGURE 13-2 ▲ Rush Hour in Xiamen, China
China's oil demand is rising rapidly. It is expected to increase from 2.5 billion barrels per year in 2006 to 4 billion barrels per year in 2020.

On the supply side, the world has a lot of oil—about 1300 billion barrels of proven reserves (oil that is economically recoverable today). The main sources are shown in Figure 13-3. World proved reserves of oil and natural gas The USGS has estimated that the world's undiscovered oil totals 649 billion barrels and that better recovery from existing fields will add another 612 billion barrels of reserves. USGS world petroleum assessment 2000 At a consumption rate of 100 million barrels per day, the proven reserves (1300 billion barrels), undiscovered resources (649 billion barrels), and oil added by better production technology (612 barrels) add up to about a 70-year supply. This could stretch farther into the future if more oil is discovered, recovery improves, or people use less than expected as the price continues to rise. Any way you look at it, however, oil supplies could get tight at least during your children's lives.

However, oil supplies could be getting tight already. You see, the big question is not how long it will take to use most of the world's oil, it's when production will no longer be able to keep up with demand. There will come a time when everyone's best efforts will be

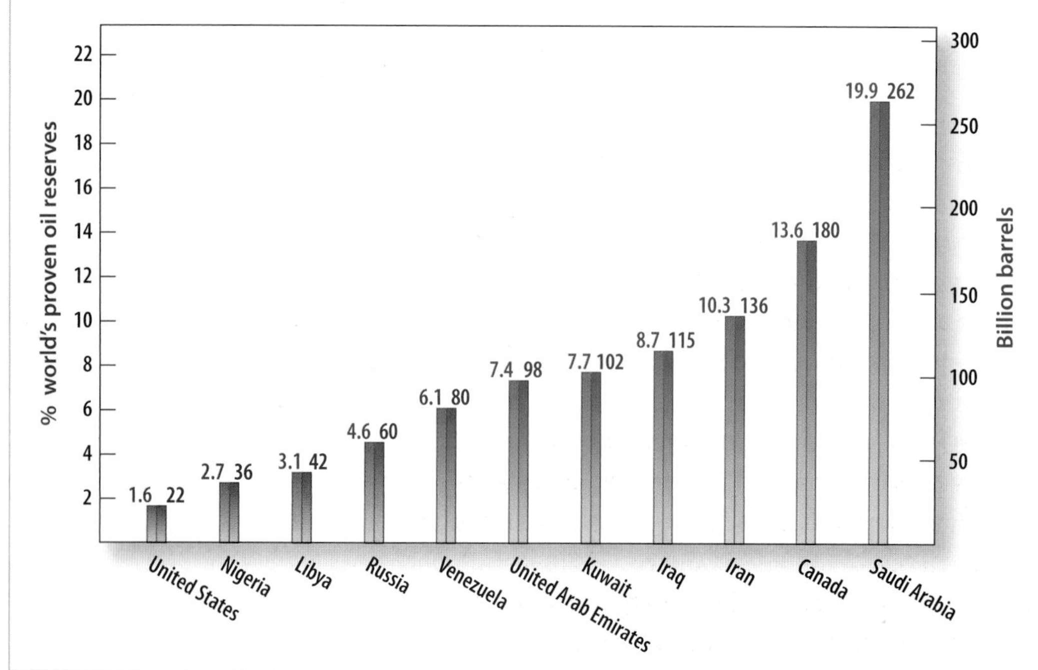

FIGURE 13-3 ◄ Countries with the Greatest Oil Reserves
As of January 2007, 11 countries accounted for about 86% of the world's proven oil reserves of 1317 billion barrels. (A barrel is equivalent to 42 gallons.)

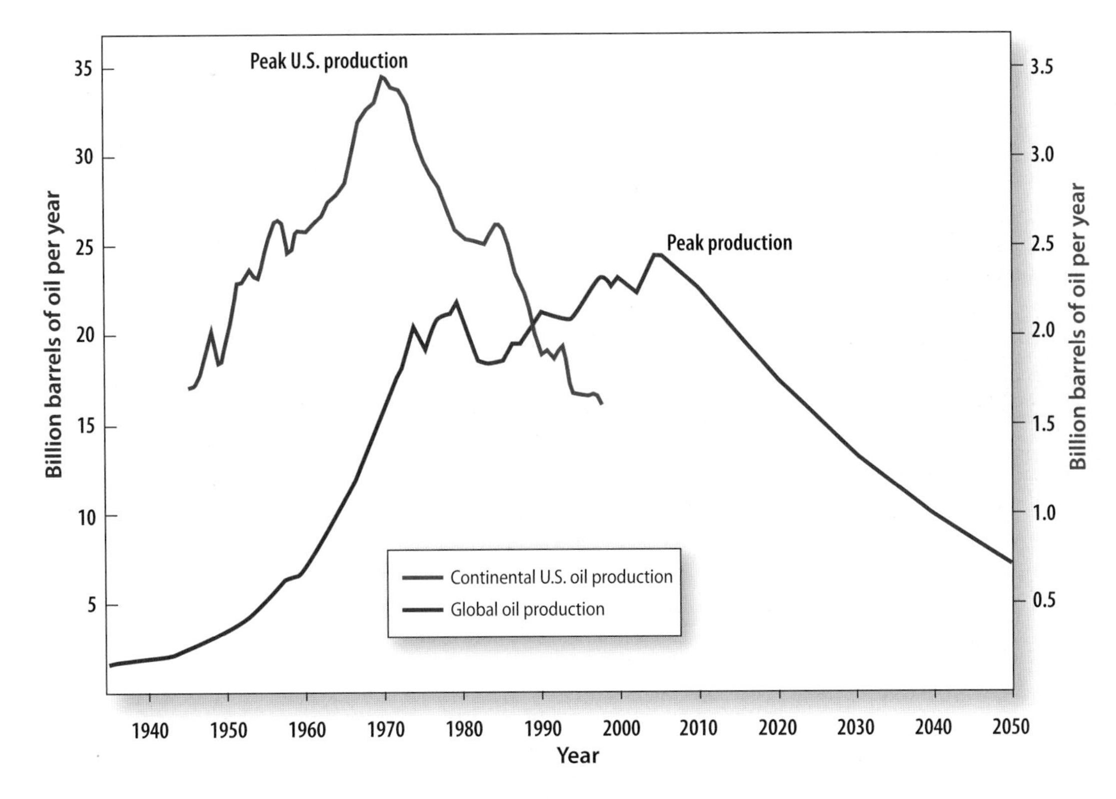

FIGURE 13-4 ◀ An Estimate of the Time of Global Peak Conventional Oil Production This estimate (blue curve) was made in 2006 by the Association for the Study of Peak Oil and Gas. |ASPO International| The shift from increasing or stable production levels to declining production has been called the "Big Rollover." Such a rollover in oil production occurred in the continental United States in 1970 (red curve).

unable to increase production—the time of **peak oil production**—and the amount of oil produced will begin to decline. The shift from increasing to decreasing production that you can track in Figure 13-4 has been called the "Big Rollover." |Les Magoon: Are we running out of oil?| This is when the competition for oil will be intense, prices will rise, and it will be a seller's market.

Some people think that oil production will peak soon, if it hasn't already. Estimates for the timing of peak oil production and the Big Rollover vary from 2005 (as in Figure 13-4) to about 2040. |Long-term world oil supply| Even if these estimates are wrong and the Big Rollover doesn't occur until later, it is probably coming in your lifetime. The combined consequences of increased oil demand, decreased production, and increasing oil prices will have profound effects on the world's economy and how our society will function.

The price of oil is a major influence on the value of all energy resources. This is because different energy resources are to some extent interchangeable, and so compete with one another in the marketplace, directly or indirectly. For example, natural gas can substitute for oil in many uses, even in fueling vehicles. Similarly, natural gas and coal compete as fuels for electricity generation. And when the cost of electricity generation using coal or natural gas goes up, then other sources of electricity—nuclear power and renewable energy sources such as biofuels, wind, and solar power—become more economically viable.

These relationships all build on the foundation of oil resources and their markets. Fundamentally, this is because oil has been the lowest cost and most widely used energy resource in the world. What happens to oil and its price affects all energy resources people can use. If the future supply of oil is insufficient to meet demand (if the Big Rollover is under way), then its price will rise, perhaps to levels that make other energy resources economically viable and needed by people as a replacement for high-priced oil.

The transition from an oil-based economy is already in progress. For example, President Obama's goals in his national energy plan include:

- Invest $150 billion to build a clean energy future.
- Put one million hybrid cars that get up to 150 miles per gallon on our roads by 2015.
- Obtain 10% of our electricity from renewable sources by 2012 and 25% by 2050.
- Reduce emissions of greenhouse gases by 80% by 2050.

These goals reflect our need for low-cost energy and our concerns for the environment. Barack Obama and Joe Biden new energy plan for America As you will see, it is not just the economics of supply and demand that will determine each energy source's future role. Every source of energy also has related environmental consequences that will help determine its cost and its role in meeting future demand. It's the interplay among supply, demand, and production costs—including the costs of addressing environmental concerns—that will drive changes in how people come to use energy resources.

This chapter focuses on helping you understand the environmental challenges associated with each energy resource that will play a role in your future.

IN THIS CHAPTER YOU WILL LEARN:

- Basic facts about energy, including the nature of energy resources, what Earth systems interactions produce such resources, the history of people's energy use, and the future of energy supply and demand
- The origin of fossil fuels (oil, natural gas, and coal) and the environmental concerns associated with their exploration, production, transportation, refining and processing, and consumption
- How nuclear energy works, how nuclear energy is used around the world, and the safety and environmental concerns associated with producing nuclear energy
- The sources of renewable energy resources (biomass, geothermal, hydropower, wind, and solar) and the environmental concerns associated with their use
- The future challenges of energy production and use, including the problem of emissions, the role of stewardship and sustainability, and the scope of decisions about energy resource development and use that people will face

13.1 Energy Basics

It is a lot easier to describe what energy does than to define what it is. We all have a sense of what energy does. We know when we have it (we are not tired) and we know when it does things for us—fuels our cars, heats our homes, or runs our appliances, for example. Scientifically, energy is a physical quantity that objects and systems can possess, and that can be transferred to other objects or systems in the form of *heat* or *work*. When such a transfer occurs—for instance, when the energy in propane is used to boil a kettle of water or the energy in gasoline is used to accelerate a car—the quantity of energy that is transferred can be measured.

When humans first started using energy resources by burning wood, it helped them survive. As they learned to use energy to do work (turn wheels, process metals, plow fields, and so forth), it helped them prosper. Today, the amount of energy people use is a good indicator of their standard of living. The United States is the classic example. The 300 million people in the United States (less than 5% of the world population) use over 20% of the total energy consumed in the world each year. Our high standard of living reflects this high level of energy use. This relationship is why economic expansion in other countries, like China, requires so much energy. Energy drives

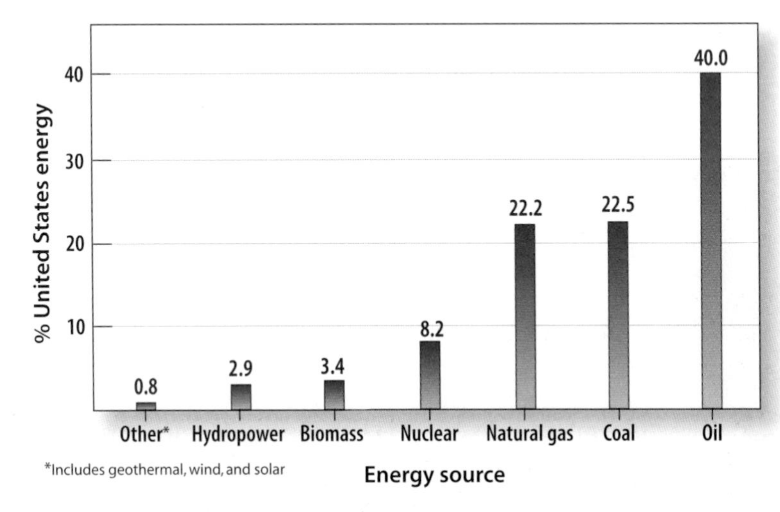

FIGURE 13-5 ▲ The Sources of Energy in the United States

everything needed to expand economies—to transport material and people, produce more food, manufacture goods, and support technology, for example.

The energy people use comes in many forms. Things that move have *kinetic energy* that can do work—falling water spinning an electrical generator, wind turning a windmill, or hot gases expanding in the cylinder of an engine are examples. Energy stored in molecular bonds (chemical energy), within atomic structures (nuclear energy), between objects attracted to each other (gravitational energy), and between positive and negative charges (electrical energy) are forms of *potential energy*. Energy in one form can be converted to another, but it is never destroyed when it is used.

In the really big picture, it's hard to overemphasize the role that the Sun's radiation has in supplying energy for people. Through photosynthesis, radiation from the Sun is converted into chemical energy stored in plants. The many food chains of the biosphere have their foundation in this energy conversion. Through various Earth systems interactions (described below), this energy conversion also leads to accumulations of stored chemical energy in coal, oil, and natural gas deposits. Because these deposits all derive from organisms of the biosphere that once lived on land or in the oceans, they are called **fossil fuels**. People have come to depend on fossil fuels for low-cost, plentiful energy (Figure 13-5).

People's harnessing of energy to do work began with burning wood. Even the first steam engines used wood for fuel. The superior heat production from burning coal enabled it to displace wood as a fuel. Some people still use coal to meet their daily energy needs, such as cooking and heating their homes, and coal is a major fuel for the generation of electricity around the world (over 50% of the United States' electricity is produced by burning coal). Coal, in turn, is readily displaced in many applications by oil and natural gas if they are cheaper. As we look ahead, the demands for energy will be so great that all sources of energy can be expected to show increased use, including nuclear energy and several renewable energy resources (biomass, geothermal, hydropower, wind, and solar). Because they are such dominant sources of energy, our discussion of the environmental consequences of producing and using energy starts with fossil fuels.

Fossil fuels all result from conversion in the biosphere of energy from the Sun into the chemical energy of carbon and hydrogen chemical bonds. This is why they are also called *hydrocarbon fuels* or just **hydrocarbons**. Because they have been plentiful, easily transported, and readily release their energy through oxidation (burning), they have been people's principal source of energy for the last 200 years.

But fossil fuels are completely consumed when we burn them. Because they take so long to form—at least millions of years to change organic remains into oil, natural gas, or coal—they are nonrenewable resources. It is impossible for nature to replace them at rates sufficient to keep up with the rates at which people use them.

There are environmental concerns associated with all steps in using fossil fuels—exploration, production, transportation, refining and processing, and consumption. Because all of these concerns have become priority issues in specific situations, all are introduced and discussed here. Overall, the environmental consequences of consuming fossil fuels present the biggest problems.

13.2 Oil, Natural Gas, and the Environment

Most oil and natural gas comes from the remains of organisms that lived in oceans, especially tiny floating plants (algae) and animals (zooplankton), like those shown in Figure 13-6. When these organisms die, their remains sink and accumulate on the ocean floor. The remains are a type of sediment that becomes buried by other sediments. Over time, as Figure 13-7 on p. 390 shows, the temperature and pressure that accompany burial by thick sedimentary layers change the organic remains into oil and natural gas.

FIGURE 13-6 ▲ Precursors of Oil and Gas
Microscopic algae like these diatoms are among the organisms whose remains give rise to oil and natural gas.

Photosynthetic organisms use energy from sunlight to make carbon and hydrogen-bearing compounds (organic compounds)

Oil and natural gas migrate to traps

Burial generates oil and natural gas from organic compounds in sedimentary rocks

Microorganism bodies create organic-rich sediments on the ocean floor

FIGURE 13-7 ▲ How Oil and Natural Gas Form

Burial of the remains of tiny marine life in ocean sediments changes them, over time, into oil and natural gas.

Initially, crude oil forms from the organic debris as it is heated at depth. Particular temperatures and pressures (and the length of time these physical conditions are maintained) are optimum for oil generation. If these conditions are exceeded—especially if temperature increases—the large, complex molecules in oil will break down into the simpler ones that make up *natural gas* (mostly methane but commonly including other gases such as ethane, carbon dioxide, and nitrogen). Under these conditions, natural gas can also be generated directly from organic material remaining in the source sedimentary rocks. This is why oil and natural gas commonly occur together and why natural gas is the dominant hydrocarbon at greater burial depths. Temperatures at which oil molecules break down and form natural gas commonly occur at depths of 7600 meters (25,000 ft) or more.

Both oil and natural gas are buoyant fluids that will migrate if the rocks containing them are permeable. They will keep migrating until they reach a place where impermeable barriers trap them in porous reservoirs, as shown in Figure 13-8. Such

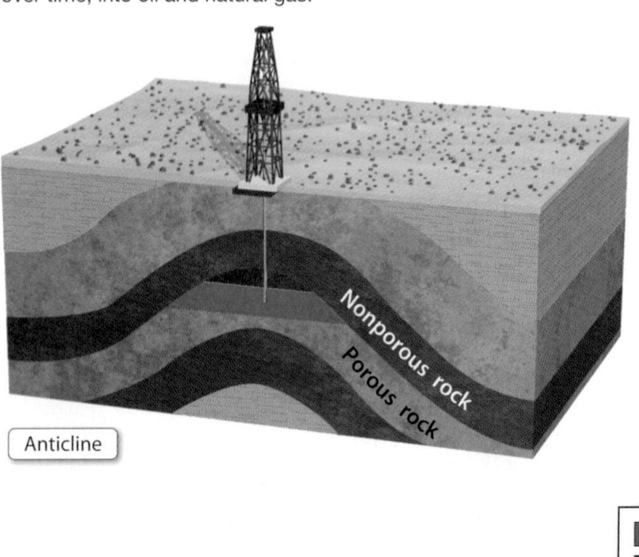

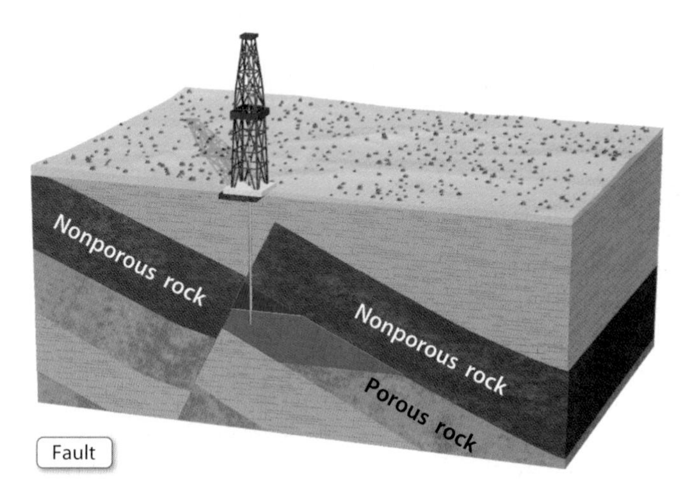

Oil
Gas

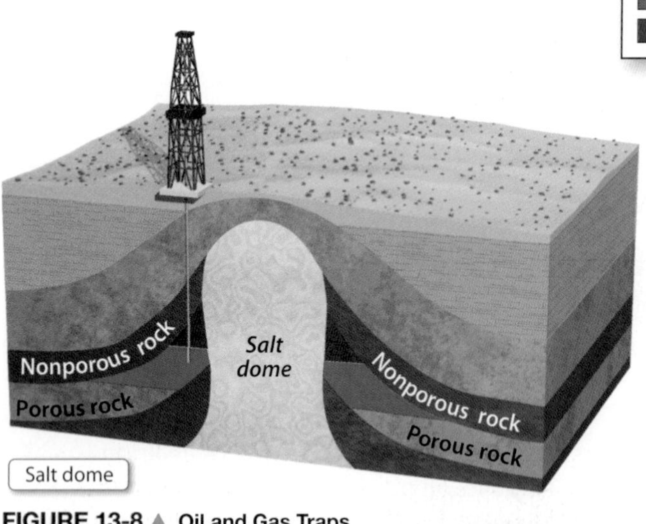

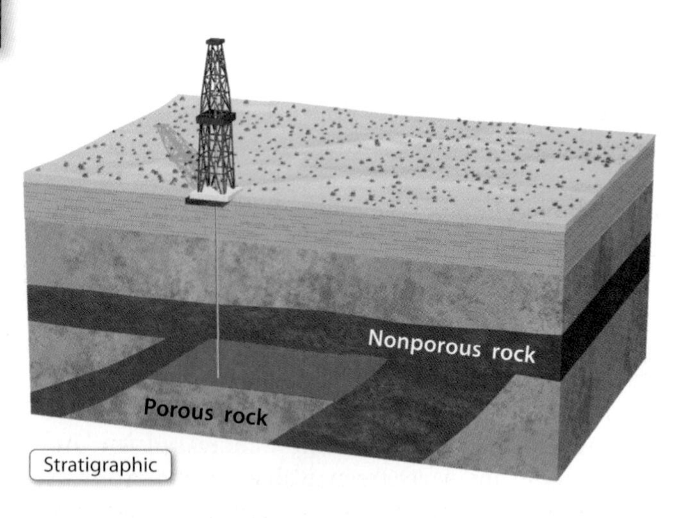

FIGURE 13-8 ▲ Oil and Gas Traps

Over time, oil and natural gas migrate through permeable pathways and into traps.

FIGURE 13-9 ▲ A Natural Oil Seep—Wind River Canyon, Wyoming

oil and natural gas **traps** are the target of exploration efforts to find new oil and natural gas fields. If traps are not present, oil and natural gas can migrate all the way to the surface and form natural seeps like the one in Figure 13-9.

Texas's Spindletop Dome is America's most famous oil trap, or at the least, the most historically significant. It is here that buoyant salt deposits pushed their way upward through overlying sedimentary rocks, deforming them into a broad dome 1.6 kilometers (1 mi) across. The dome created a small circular hill just 5 meters (16 ft) high on the otherwise flat Texas prairie. After the dome was formed, oil migrated into the area and became trapped below arched and impermeable sedimentary rocks. Pattillo Higgins, a local resident, recognized the hill and its associated natural gas and foul water seeps as an unusual feature, an anomaly in the landscape. He reasoned that rocks below the surface could contain oil and his persevering efforts lead to exploration drilling on Spindletop Dome.

As shown in the photograph in Figure 13-10, the exploration drilling encountered oil that rushed to the surface and spouted high above the drill rig on January 10, 1901. This spectacular *gusher* marked the discovery of America's first large oil field. Sixty million barrels have been recovered from the Spindletop Dome field. This discovery broke Standard Oil's monopoly and paved the way for an expanded oil industry in America. Spindletop was the first of literally thousands of oil fields cumulatively containing billions of barrels of oil scattered across onshore and offshore America. Oil went from being a source of lamp fuel and lubricants to the principal transportation fuel it is today after the Spindletop field was discovered.

What are some Earth systems interactions involved in the formation of oil and natural gas?

OIL AND NATURAL GAS EXPLORATION

As you can see in Figure 13-11 on p. 392, exploration for oil and natural gas takes place in the most remote onshore and offshore places in the world. The search for oil and natural gas is a very challenging endeavor that involves conducting **seismic surveys**—imaging the subsurface with seismic waves

FIGURE 13-10 ◀ The Spindletop Gusher
Blowouts were commonly called "gushers" in the early days of oil exploration.

FIGURE 13-11 ▲ Much Exploration for Oil and Gas Takes Place in Remote Parts of the World
This is a production platform for one of the oil fields discovered in the Cook Inlet region of southern Alaska.

(Chapter 5)—and drilling to determine if oil or natural gas is present.

Seismic surveys

Seismic surveys are done on land and on the ocean. As Figure 13-12 shows, they involve generating seismic waves that travel underground and reflect off boundaries between rock units (such as sedimentary layers) where physical properties change. The key environmental concern relates to the generation of the seismic waves. On land, seismic surveys use trucks that vibrate a pad against the ground or explosives placed in shallow drill holes (shot holes). The explosion produces about one-tenth as much vibration as a magnitude 1.5 earthquake—barely noticeable unless you are standing nearby. As a result, on-land seismic surveys don't raise much environmental concern.

Ocean seismic surveys, however, can be problematic. They use a device called an air gun that releases compressed air into the water. The high-pressure air quickly expands into a bubble that sends a shock wave through the water to the ocean bottom below. The key concern is the effect the air bubble and shock wave have on marine life, especially mammals with sensitive hearing such as whales, seals, and dolphins. If marine life is close to the air gun blast (a few meters and possibly up to a few hundreds of meters in some cases), physical damage can occur. Nearly everyone can relate to the discomfort of loud noises and it's not surprising that marine mammals have been observed to move 5 to 30 kilometers (3 to 18 mi) away to avoid offshore seismic survey areas. In addition, the long-range impact of seismic survey noise (low frequency vibrations that can travel across entire oceans) may handicap the communications of certain whale species. Researchers with very sensitive listening devices are being surprised by how noisy remote open sea regions can be.

Protecting marine animals from the effects of offshore seismic surveys is achieved primarily by minimizing their exposure to them. Investigators can also take care to increase the size of their air gun blasts gradually and pull the air guns through survey areas slowly, so that sensitive marine life can leave the vicinity before they are harmed. In some places regulations require that such surveys be conducted when sensitive marine life, such as migrating whales, are elsewhere.

There are places, like the Gulf of Mexico, where offshore seismic surveys have been conducted for decades. In general, fisheries have not been adversely affected by these activities. In U.S. offshore waters, the Department of Interior's Minerals Management Service oversees a permitting process that requires mitigation of the impacts of seismic surveys on marine life. However, because we do not fully understand marine life and its journeys, offshore seismic surveys can still be a cause for concern. They need to be planned carefully and evaluated on a case-by-case basis.

Drilling

Once seismic surveys delineate possible traps, drilling determines if oil or natural gas is actually present. The principal environmental concerns associated with exploration drilling involve infrastructure, disposal of drilling wastes, and accidental release of produced fluids (oil or natural gas) to the environment.

Oil and natural gas drilling on land requires access and space for equipment and operations, just as mineral exploration drilling does (Chapter 12). But oil and natural gas drilling is commonly 5 to 10 times deeper than mineral exploration drilling—the deepest onshore oil or natural gas well in the United States reached a depth of over 9500 meters (31,000 ft). Deeper drilling requires larger drill rigs. Most rigs are either hauled over roads (some are truck-mounted, while larger ones are hauled in pieces). The drill site is commonly about the size of a football field or larger, as shown in Figure 13-12. Because only one in ten (or more) exploration wells is successful, infrastructure for exploration drilling is commonly temporary. After drilling, the equipment is removed and the disturbed areas can be reclaimed if they have no further use.

If exploration drilling is offshore, it can be done from drill ships, floating rigs, or rigs with support legs that reach to the seafloor. Exploration drilling (and oil production) has taken place in waters nearly 6100 meters (20,000 ft) deep. A potential problem with offshore exploration drilling is the possible effects of such operations in areas where they may disturb endangered species or migrating species such as whales. Concern for migratory species can influence regulatory agencies when they consider issuing permits to allow offshore drilling.

Drilling requires fluids to lubricate the well bore and control the pressure within it. The fluid is a mixture of water, clay, the mineral barite ($BaSO_4$), and various chemicals; it is commonly just called drilling mud. As the well is drilled ahead, the drilling mud circulates down and up the well bore, carrying displaced rock chips with it. The rock chips—called cuttings—are separated at the surface and the mud is circulated back down the hole to collect more. The cuttings and mud, perhaps containing some oil, are waste that must be disposed of when drilling is complete. Today, this disposal is commonly done by grinding up the cuttings and injecting them, along with excess drilling fluids, into underground zones that are known to be isolated from interaction with groundwater.

Proper disposal of drilling wastes, guided by regulatory permits, is now common practice, but it wasn't always so. In offshore operations, drill cuttings and mud were often just dumped into the ocean. On land, the mud was mixed in pits, which then became repositories for waste cuttings and mud after drilling was completed. If the cuttings and mud contained traces of oil or other

FIGURE 13-12 ▲ Seismic Surveys
Surveys are done both onshore (as shown here) and offshore. Sound waves or vibrations travel through the geosphere and reflect or refract as they pass through materials with different physical properties. These interactions change the travel path and the time it takes for the waves to return to the surface, where geophones record their arrival. The variations in travel time are analyzed by a computer and displayed in ways that enable the subsurface geometry and character of rocks to be interpreted.

FIGURE 13-13 ▼ An Oil Exploration Drill Site
This exploration well is being drilled in the onshore Arctic region of Alaska. Exploration drilling here is done in the winter when ice pads can be constructed for the drill site. The trucks and other vehicles provide insight into the area needed for the drill site. Can you identify where the person is?

(a)

(b)

(c)

FIGURE 13-14 ▲ Mud Pit Remediation
(a) This quiet pond is actually an abandoned mud pit whose bottom is covered with drilling mud waste. **(b)** Remediation involves draining the pond and removing the old drilling mud and other waste that has accumulated on the bottom. **(c)** Remediation is complete when the pond is filled with clean soil and the surface restored to use.

toxic components, the unlined pits could become sources of soil and water pollution. Many old mud pits that have become environmental problems have subsequently been remediated and reclaimed (Figure 13-14).

Perhaps the biggest concern with exploration drilling, onshore or offshore, is the inadvertent release of oil or natural gas to the environment if a well becomes uncontrolled and "blows out"—like the Spindletop discovery well gusher (see Figure 13-10). Blowouts were most common when drilling encountered unexpected high-pressure accumulations of oil or natural gas, which escaped, uncontrolled, up the well to the surface. Blowouts like the one shown in Figure 13-15 have surprised drillers, created spectacular fires, and released large amounts of oil and natural gas to the environment. Intentional damage to producing oil wells by retreating Iraqi forces during the 1991 Gulf War caused blowouts and contributed to the largest oil spill in world history—at least 7.5 million barrels.

An unspectacular blowout offshore Santa Barbara, California, in January 1969 influenced many subsequent decisions about offshore oil and natural gas drilling—including its banning in California offshore areas. The Santa Barbara blowout wasn't a large one, as only 80 to 100 thousand barrels of oil were released, but it occurred in a populous and popular part of southern California and affected a very conspicuous 64 kilometers (40 mi) of heavily used seacoast.

The Santa Barbara blowout was largely caused by the lack of casing in the shallow part of the well. *Well casing* is a steel tube that prevents fluids from migrating from the well into adjacent rocks. It is commonly installed under standard permit guidelines, but government regulators had allowed it to be omitted in the blowout well. Oil and gas under pressure came up the well, and as operators attempted to block its escape it was forced into shallow surrounding rock formations, from which it migrated up to the seafloor.

Blowouts are now much less common than they once were because drillers can tell where oil and gas are under high pressure, and drill rigs have devices that automatically close wells if a blowout seems to be developing. Such protections are working. Oil released to oceans from oil and natural gas production activities accounts for less than 2% of all the oil that somehow makes its way into marine environments each year. And most production-related oil release is from discharge of oil-bearing water, not from blowouts.

FIGURE 13-15 ▲ The Ixtoc-1 Blowout and Oil Spill—Mexico
Ixtoc-1 was an offshore exploration well that blew out on June 3, 1979. Before the blowout was controlled on March 23, 1980, about 3.4 million barrels of oil were released to the environment. This was the second largest oil spill in history, exceeded only by the intentional release of 7.5 million barrels during the Gulf War in 1991.

Opposition to oil and natural gas exploration has led to several prospective areas being declared off-limits to seismic surveys and exploration drilling. Some of these areas are offshore western Florida, offshore California, and in the Arctic National Wildlife Refuge (ANWR) in Alaska. As oil becomes more and more expensive, you can expect renewed efforts to allow exploration in these areas. The decisions will largely be political, so you will have an opportunity to participate in the decision-making process.

OIL AND NATURAL GAS PRODUCTION

Oil and natural gas are fluids that flow to the surface through wells. Physical disturbances to the landscape and disposal of produced water are concerns that accompany their production.

Physical disturbances

Physical disturbances associated with oil and natural gas production include roads, well pads, pipelines, and local processing facilities that separate oil and natural gas from water that is produced along with them. With recent advances in drilling technology, the spatial footprint of oil and natural gas production facilities has been sharply reduced, as Figure 13-16 indicates. An example is the Alpine field on the North Slope of Alaska. You can see in Figure 13-17 on p. 396 how close wellhead spacing and horizontal drilling technology enabled this 57-square-kilometer (22 mi^2) field to be developed from only two drill sites and about 40 hectares (100 acres) of total footprint area, less than 1% of the total area of the oil field. Oil and natural gas fields eventually become depleted and their physical disturbances, even roads, can be reclaimed.

Produced water

Oil and natural gas wells commonly produce water, too. Initially, the amount of water is minimal, but as production continues it increases. Because most oil wells in the continental United States have been in operation for many years, today's production averages six barrels of water for every barrel of oil. Such water commonly contains enough salt and traces of oil that it cannot be directly used for drinking or irrigation, nor can it be safely discharged to surface drainages. It is in effect brine that must be properly disposed of once it is separated from the accompanying oil or natural gas.

Most of the 20 to 30 billion barrels of water produced each year are injected into oil or gas fields to maintain their reservoir pressure and increase oil recovery. However, about 35% of the nation's produced water cannot be recycled this way and requires disposal. As you would probably have guessed, disposal of produced water prior to 1970s environmental laws and regulations commonly did not consider its environmental impacts. Many old *salt scars* developed where the brines were simply discharged to the surface near the producing oil or natural gas well. As this water evaporated, salt was deposited that impregnated soil, killed vegetation, and potentially contaminated groundwater. Today these salt scars are being remediated by adding gypsum (a source of calcium) and organic fertilizers to soil, planting salt-tolerant vegetation, and irrigating the area to help flush and clean the soil (if groundwater contamination is not an issue).

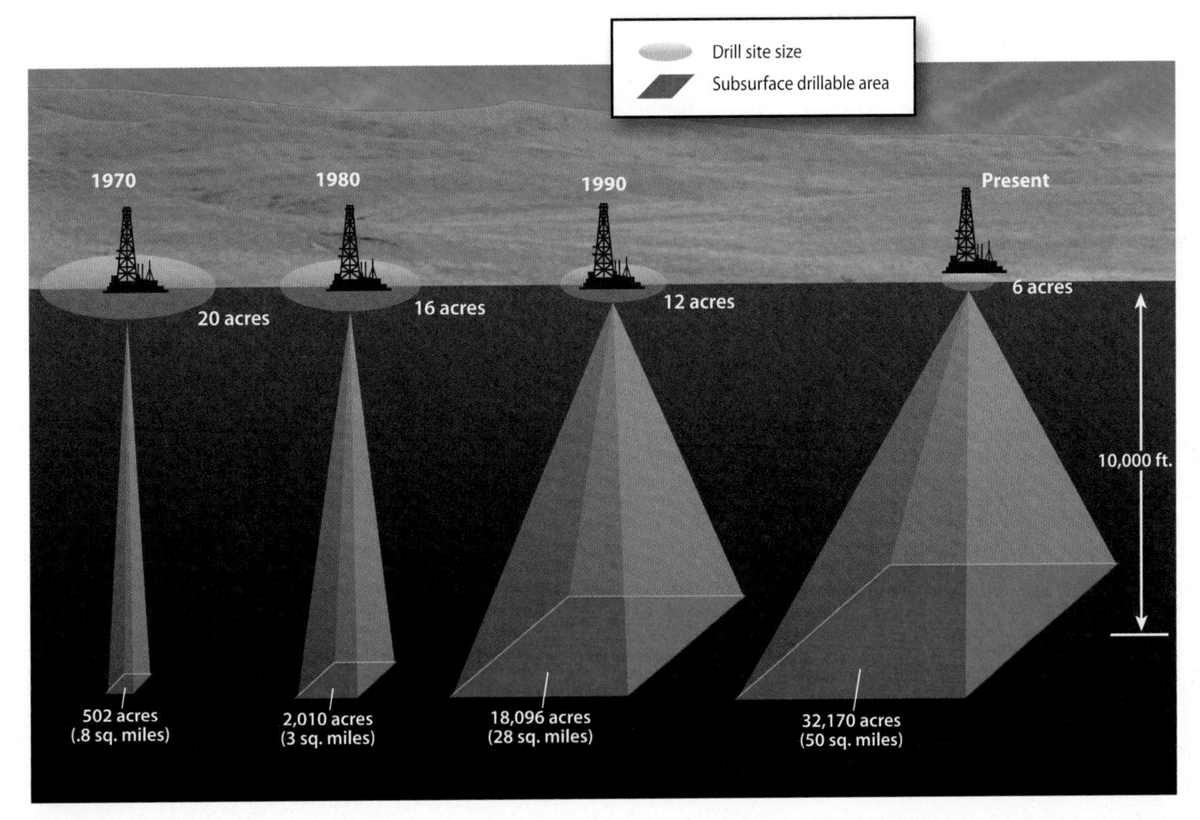

FIGURE 13-16 ◀ More from Less
Advances in drilling and production technology have decreased the surface area (footprint) needed to produce oil from larger and larger areas.

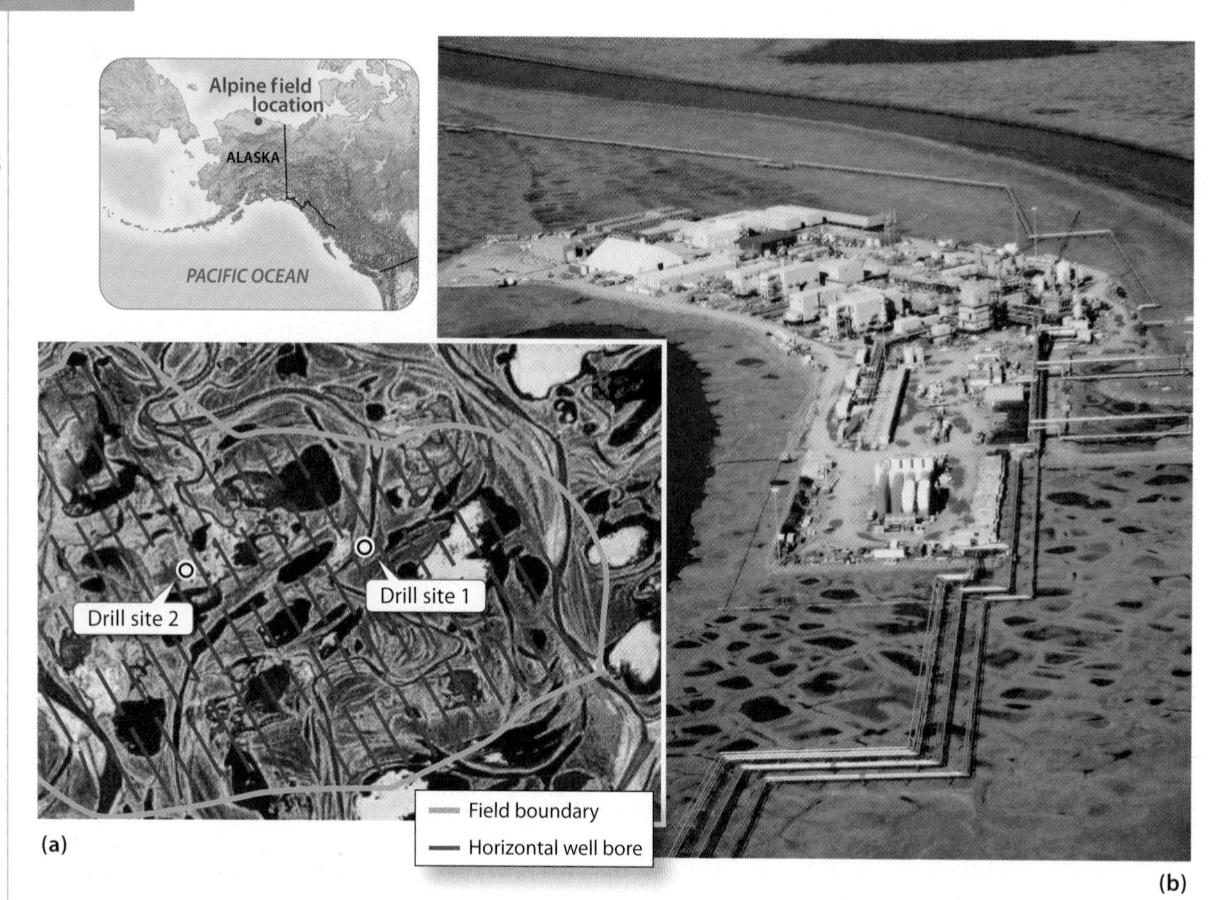

FIGURE 13-17 ▲
Horizontal Drilling
With the aid of horizontal drilling, a 57-square-kilometer (22-mi^2) Alpine oil field in Arctic Alaska **(a)** was developed from only two drill sites and a total footprint of no more than about 40 hectares (100 acres). The production facilities **(b)** are reached by plane in the summer and by ice road in the winter.

Produced water also comes from wells drilled into coal seams. Coal seams all have natural gas (methane) in them. This gas, **coal-bed methane (CBM)**, has historically caused unexpected underground explosions in coal mines that have killed many miners around the world. In some settings, where the coal seams are not too deep, wells can be drilled that produce CBM for use. The methane-bearing coal seams also contain much water, which is produced along with the natural gas. If the produced water is salty, it is like the water from conventional oil and gas wells and can't be safely discharged to the surface. But even if it is fresh, its production can prompt concerns about possible depletion of groundwater resources.

There are three ways to properly dispose of produced water from coal beds or oil and natural gas fields:

1. If the produced water is not too salty, it can be used for purposes such as irrigation. If it is, the salt can be removed before the water is used.
2. Produced water can be injected into deep porous rock formations where it can't contaminate fresh groundwater resources.
3. Produced water can be injected back into the rock formations from which it is obtained. This is a common disposal procedure, as it helps maintain pressures in the reservoir rocks and allows the recovery of more oil or natural gas.

OIL AND NATURAL GAS TRANSPORTATION

Oil and natural gas are transported in several ways, including through pipelines, on ships and barges, and in trucks. Pipelines can extend across continents, and very large tanker ships carry oil across oceans. On any given day, about 50 million barrels of oil are being transported in tanker ships. Most of the rest of the oil in transport, some 35 million barrels, goes by pipeline. Natural gas is also transported by ship and pipeline. Although protecting pipelines from accidental rupture and explosions is always necessary, oil spills from tanker ships are the biggest environmental concern associated with transporting fossil fuels.

Marine tanker transport

Oil spills from large tankers can have immediate and devastating effects on marine life, especially along coastlines. These spills are widely reported when they occur, and many people's understanding of the risks of oil transportation come from the news they receive about this type of accident. When spills occur, the birds, shore life, and marine mammals, such as seals and otters, that become covered by oil cannot survive. Oil-covered seabirds, like the one shown in Figure 13-18, have become the iconic image of these accidents.

Most of these accidents occur because of a human error or a mechanical problem—for example, when incorrect navigation or disabled engines cause a tanker to run aground. A collision between two ships caused the largest tanker spill in history. On July 19, 1979, the five-year-old *Atlantic Empress* collided with the *Aegean Captain*, another oil tanker, off Tobago Island in the Caribbean Sea. Twenty-six sailors were killed and the ships caught fire. The fire on the *Atlantic Empress* could not be controlled and on August 2 it sank under a cloud of billowing black smoke. A total of about 2 million barrels of

FIGURE 13-18 ▲ Oil Spill Casualty
Seabirds coated with oil have become the iconic symbol of the affect of marine oil spills on the environment.

FIGURE 13-19 ▲ Grounding of the *Exxon Valdez*
The *Exxon Valdez* strayed out of the shipping lane and struck Bligh Reef on March 24, 1989.

crude oil were released to the sea. Fortunately, this disaster occurred far out at sea and no oil came ashore.

The *Atlantic Empress* sinking is noteworthy because of the large amount of oil that was released into the environment. Another tanker spill with which you may be more familiar was also caused by navigation error. The *Exxon Valdez* ran aground 10 years later (March 24, 1989) in near-shore waters of Prince William Sound in the Gulf of Alaska (Figure 13-19). The *Exxon Valdez* released about 260,000 barrels of crude oil into the sea, about one-tenth the oil lost by the *Atlantic Empress* and only the 46th largest tanker spill in world history. However, in the case of the *Exxon Valdez*, a lot of the oil came ashore over a wide area, as Figure 13-20 shows.

Prince William Sound contains many sensitive ecosystems and significant commercial fishing industries. Moreover, liability for the oil spill very clearly

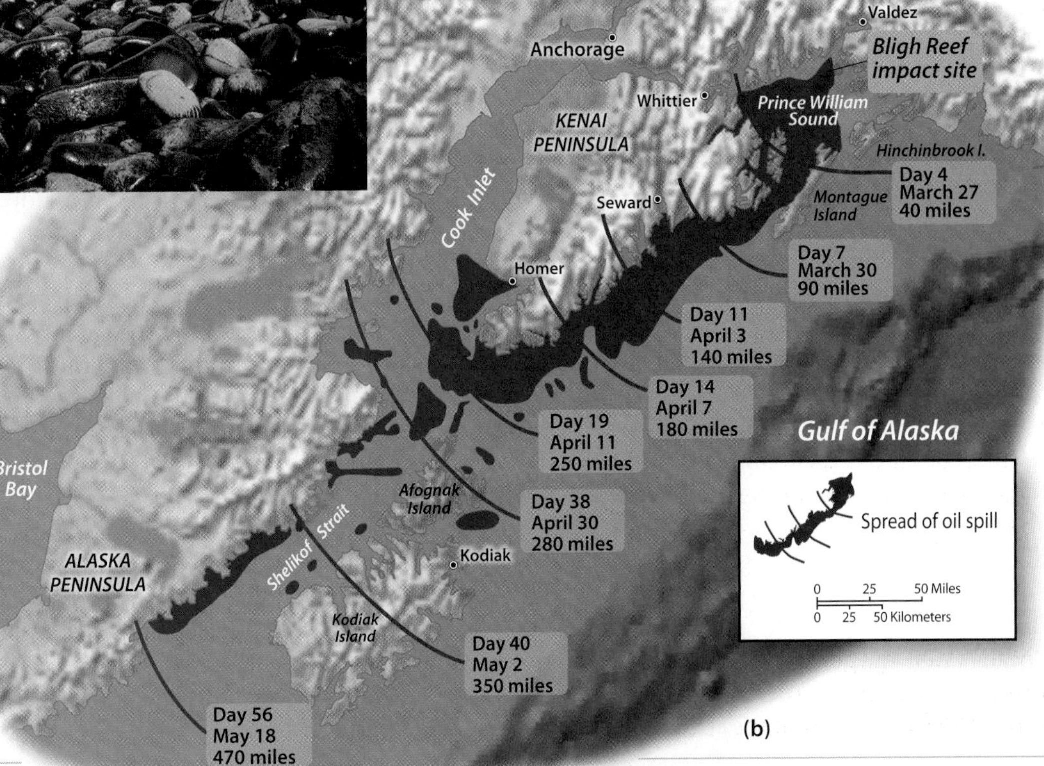

FIGURE 13-20 ▶ Spread of Oil from the *Exxon Valdez*
The *Exxon Valdez* oil spill mostly came ashore in Prince William Sound **(a)** but parts eventually spread southwest as far as Kodiak Island and Shelikof Strait **(b)**.

FIGURE 13-21 ▲ Cleaning Up
Intensive efforts to capture and remove the oil spilled by the *Exxon Valdez* included spraying high-pressure hot water on the oil-soaked shore. A lesson learned from this spill is that such cleanup efforts can extend the length of time it takes for shoreline habitat to recover.

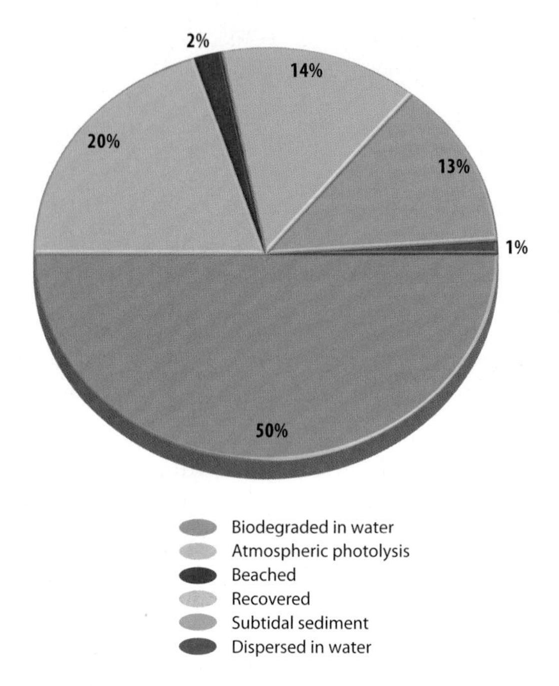

FIGURE 13-22 ▲ The Fate of Oil Spilled from the *Exxon Valdez*
Oil was transferred to the atmosphere by evaporation, emulsified and incorporated into seafloor sediments, and biodegraded or photolyzed. (Photolysis is the breakdown of chemical compounds by the absorption of sunlight.) Only 14% was recovered by the cleanup operations (which cost over $2 billion).

lay with Exxon (now ExxonMobil)—the largest independent oil company in the world. For these reasons, the *Exxon Valdez* oil spill became the most intensely studied, and perhaps litigated, in history. Over $2 billion was spent to clean up and otherwise deal with the disaster, and many lessons were and are being learned from this spill.

A large number of people and a great deal of equipment were mobilized to combat the spill (Figure 13-21). Steam and detergent systems were brought in to clean up the beaches. Skimming vessels were used to recover oil floating on the sea, and floating barriers were used to control the movement of oil slicks. The extent and degree of shoreline oiling resulted in the most intensive shoreline cleanup effort ever attempted, but monitoring studies have shown that the intensive treatment actually delayed the recovery of rocky shore intertidal communities. These studies demonstrate how an overaggressive cleanup, in some instances, can slow recovery of affected communities. NOAA Office of Response and Restoration—What lessons have we learned?

By 1993, many of the beaches that were covered with oil immediately after the *Exxon Valdez* spill had largely recovered, and very little residual oil was evident anywhere in the Sound area. The National Oceanic and Atmospheric Administration reported that green algae, rockweed, and barnacles repopulated many of the affected areas within two to three years, but some types of clams, limpets, and snails would take much longer. Residual oil can still be found buried by gravel on about 8 hectares (20 acres) of beaches. Still, as Figure 13-22 indicates, most of the oil is gone.

Natural processes do a good job of cleaning up after oil spills, but they may take awhile—from a few years to several decades. Oil in the sea may be volatized (converted to gaseous form), dissolved, emulsified (dispersed as tiny droplets in water), deposited in sediments, oxidized (chemically combined with oxygen), photolyzed (chemically broken down by sunlight), and biodegraded (chemically broken down by the action of organisms). Spilled oil in the sea Over time, these processes, in various combinations, can significantly alter the physical character and bioavailability of spilled oil.

What are some Earth systems interactions that accompany the natural remediation of marine tanker oil spills?

Oil in the sea and you

Do you think you have much to do with oil in the sea? Where do you think most of the oil in the sea comes from? It's not from tanker spills. In fact, as you can see from Figure 13-23, tanker spills and other releases associated with shipping oil (such as when valves break during tanker loading) account for only 3.6% of the almost 2 million barrels of oil released each year to North American marine environments.

The biggest source of oil in marine waters of North America is natural seeps—seeps that have flowed oil to the surface for thousands or even millions of years. Seeps release more than 1 million barrels (62.4%) of the oil entering North American marine environments each year. This is the equivalent of four *Exxon Valdez* spills each year. There were globs of oil on the beaches of Santa Barbara long before the 1969 offshore well blowout, and there will be globs of oil there indefinitely in the future. There are many offshore natural oil seeps in the Santa Barbara area like the one shown in Figure 13-24.

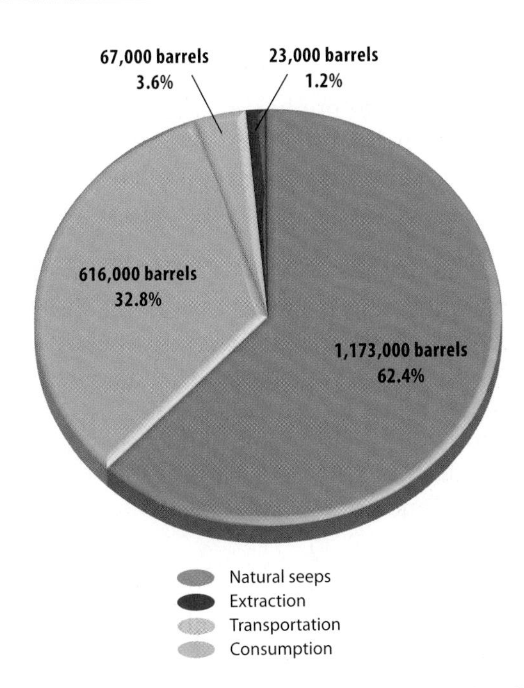

FIGURE 13-23 ▲ Sources and Amounts of Oil Released into North American Marine Waters

Nevertheless, sources of oil in the sea related to human activities are very significant. About 600,000 barrels per year, or 33% of the oil released to North American marine waters, can be traced to oil *consumers*. One of the biggest sources of consumption-related oil pollution in the marine environment is river runoff containing oil from streets and other onshore developments. Another is the transfer through the atmosphere of oil components—derived from **volatile organic compounds (VOCs)**, easily vaporized compounds of carbon and hydrogen—that are emitted by internal combustion engines, power plants, and industrial operations. Other consumption-related sources include two-stroke engines (outboard engines that require mixing oil with the gasoline for lubrication) and aircraft fuel dumping. There are many sources, but river runoff and two-stroke engines make the consumer the principal source of human-related oil pollution of the marine environment.

FIGURE 13-24 ▼ A Natural Oil Seep

This photo was taken on Carpenteria Beach, 10 kilometers (6 mi) east of Santa Barbara, California.

Are you finding it difficult to imagine how consumers, people like you, can contribute so much oil to the marine environment? Consider what happens to used motor oil. Over 600 million gallons of motor oil are used each year in the United States to lubricate vehicle engines and equipment. About half of this motor oil is used by individuals who change their own oil. Many people properly dispose of the used motor oil—for example, by placing it in recycling facilities at local landfills—but many don't. About 175 million gallons of used motor oil are thrown away with the household garbage, poured down storm drains, or just dumped on the ground. This is 16 times (each year!) the amount of oil spilled by the *Exxon Valdez*.

There's a lot of oil available on land to be flushed down streams and rivers to the ocean. Check your school parking lot. What percentage of the lot's surface is NOT covered by oil or grease spots? What do you think happens when heavy rains wash across the parking lot and all the other lots and streets in your neighborhood?

What You Can Do

Recycle Used Motor Oil

Most people would readily recycle used motor oil, especially if they knew where to do it. You can find out about where recycling centers accept used motor oil near you through the Earth911 website. Earth911 Enter "motor oil" and your zip code and the site will bring up a recycling center location, hours of operation, and other information you may need to become a regular recycler of used motor oil.

Even if nature is good at cleaning up oil spills, there is no getting around the very negative, immediate, and costly consequences spills can bring to marine life and coastlines when they occur. Accidents that send large amounts of oil into the environment in minutes to days need to be prevented. The *Exxon Valdez* oil spill had a profound effect on how oil is transported in North American seaways.

Congress passed the Oil Pollution Act of 1990 in response to the *Exxon Valdez* accident. This act requires that all oil tankers in U.S. waters have double hulls. Double hulls significantly reduce the chance of major marine oil spills. Other measures that help prevent marine oil spills include:

- Tanker personnel must be licensed, specially trained, and subject to random drug screening (like airline crews). The training includes simulations of real-world emergency situations.
- Navigation technology now allows constant determination of a tanker's location within 4.5 meters (15 ft). Alarms sound if the tanker is even slightly off course.
- Tankers continuously transmit identification and location information to help avoid collisions with other ships.
- Powerful tugboats are permanently located in areas where tankers may ground if they lose power. Tugboats also routinely guide tankers through difficult-to-navigate waters.

Marine tankers are transporting oil more and more safely. As Figure 13-25 indicates, the frequency of marine spills and the amount of oil spilled have both significantly decreased.

OIL REFINING

Processing petroleum to produce marketable products such as gasoline or diesel fuel is called **refining**. Petroleum refineries, like the one shown in Figure 13-26, are large industrial complexes that operate 24 hours a day. At night they are conspicuous, brightly lit, complex structures. By day they are still large complex structures, but their most conspicuous features are the billowing white clouds of steam rising above them. There are 149 operating refineries in the United States with the capacity to process 17 million barrels of crude oil each day. This processing requires handling and storing large amounts of volatile and potentially hazardous liquids and gases. The abundant pipes, connections, valves, and tanks in refineries are all possible points of mechanical failure.

Refining safety

Because refineries produce very combustible fuels like gasoline, it is not surprising that fires and explosions are a significant risk wherever they are located, and over the years there have been some big ones. In 1955, an explosion ripped apart a refinery in Whiting, Indiana, hurling massive chunks of metal thousands of feet. Some of it landed in a nearby residential neighborhood, where one person died. The resulting fire burned for eight days and consumed most of the refinery.

Explosions and fires like that at Whiting have occurred in other places where refineries are common—places such as Louisiana, Texas, and California. But the hazards are well recognized, and numerous steps have been taken to diminish the

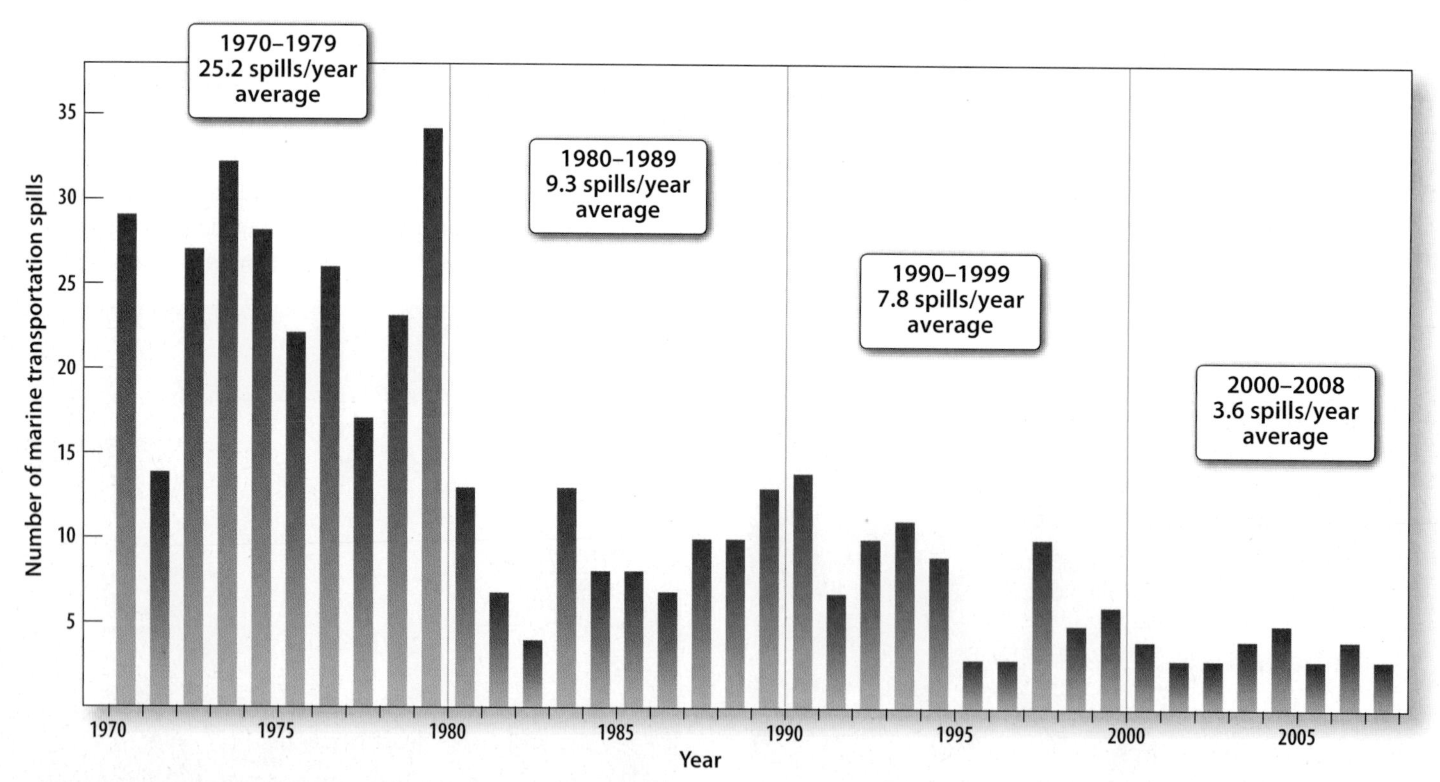

FIGURE 13-25 ▲ **The Frequency of Marine Transportation Spills and the Amount of Oil They Release into the Marine Environment Are Decreasing**

FIGURE 13-26 ▲ Oil Refineries Are Large and Complex Facilities

risks. These steps include making sure that refineries are not located close to residential areas, physically isolating storage tanks from one another, using fire- and corrosion-resistant construction materials, and employing elaborate monitoring and fire suppression procedures.

Safety procedures at refineries are working, as the incidence of fires and explosions is decreasing, but accidents still happen. Fifteen people died and 170 were injured in a Texas City, Texas refinery explosion on March 23, 2005 (Figure 13-27). There is still more progress to be made in achieving safe operations at refineries.

Refinery disturbances

Oil refineries are a conspicuous feature of the landscape, clearly a physical disturbance. The best way of diminishing the impact of oil refineries is to locate them where they do not conflict with other land-use choices, such as valuable wildlife habitat or where people want to live. Refineries are commonly best located where oil transportation facilities—ports or pipelines—are also available. If a new refinery is proposed for an area, it's good to make sure that other land-use choices have been fully considered. Once a refinery has been built, it can be around for a long time.

Refinery wastes

Oil refining removes impurities from oil and its products, such as gasoline. These impurities include sulfur and metals. In addition, the refining process uses caustic chemicals, microorganisms for treating used water, and various catalysts. These materials eventually become part of the waste stream along with thick oil-water emulsions that do not easily break down. Altogether, U.S. refineries generate about 3 million tons of waste each year. Refineries send 60% of these wastes back through the refining process and treat another 20% to make them inert. The less than 20% of refinery waste that remains to be disposed (about 600,000 tons each year) is placed in landfills that are designed to prevent their release into the environment.

Another waste commonly developed at refineries, and wherever oil is stored, is a thick sludge composed of tar-like material and sediment that sinks to the bottom of storage tanks. These "tank bottoms" are collected when the storage tanks are cleaned and either cycled back through the refinery or treated to remove the oil components. Treated tank bottoms are spread on nearby ground or placed in repositories.

Refineries emit particulate matter (a mix of tiny solid particles and liquid droplets), sulfur dioxide (SO_2), nitrogen oxides (NO_x), volatile organic compounds (VOCs), carbon monoxide (CO), and other contaminants such as benzene. These pollutants come directly from the processing and from burning hydrocarbon fuels, such as diesel fuel, in processing equipment. The VOCs can come directly from oil processing or be "fugitive VOC emissions" released by small leaks from valves, pipe connections, and storage facilities. Refineries produce a half pound of waste

FIGURE 13-27 ▼ Aftermath of the 2005 Texas City Refinery Explosion
Fifteen workers were killed and 170 people were injured in a series of explosions at this facility on March 23, 2005.

for every barrel of crude oil processed, and 60% of this waste is emitted into the atmosphere.

The Clean Air Act authorizes the EPA to establish emission standards and regulate refinery operations that emit substances into the atmosphere. Increasingly stringent standards have required refineries to reduce their emissions of atmospheric pollutants. Since the 1980s, many billions of dollars have been spent to repair and upgrade refineries. Some of the steps refineries have taken to reduce emissions include:

- Using processes that are more efficient and require less fuel.
- Installing emission-control devices, such as catalytic converters and SO_2 recovery systems.
- Using fewer and better valves and pipe connections to reduce fugitive VOC emissions.
- Cycling excess gases back to fuel streams.
- Installing vapor recovery systems on storage tanks.

In November 2007, the EPA cited British Petroleum for a number of Clean Air Act violations at its Whiting, Indiana oil refinery. |EPA reports violations at BP's Whiting refinery| The EPA's enforcement actions against BP are part of its national initiative to reduce air pollution from refineries. As upgrading older refineries to meet emissions standards is an ongoing process, further EPA enforcement actions are likely.

OIL AND NATURAL GAS CONSUMPTION

Of all the human activities involving oil and natural gas, it is their consumption—their use by people—that has the most significant consequences for the environment. All fossil fuels are burned (oxidized) to release the energy stored in their carbon–hydrogen bonds. This process produces pollutants that are emitted into the atmosphere. It also generates greenhouse gases—especially carbon dioxide—that influence global climate. (You will learn a lot more about greenhouse gases and climate change in Chapter 14.)

Burning gasoline and diesel fuel for transportation produces most of the air pollution in the United States. When these fuels burn, the engines of cars, trucks, and other machines emit exhaust gases to the atmosphere. As Figure 13-28 shows, the perfect burning of a pure hydrocarbon fuel produces carbon dioxide and water. However, internal combustion engines do not do a perfect job of burning fuel. As a result, the engine's exhaust gases contain small amounts of leftover hydrocarbons, nitrogen oxides (NO_x), carbon monoxide (CO), and other potentially harmful emissions. In addition, burning of some diesel fuel emits sulfur dioxide (SO_2), and VOCs leak and evaporate from engine fuel systems.

Hydrocarbons and nitrogen oxides are especially important atmospheric pollutants. In sunlight, these pollutants are involved in chemical reactions that produce ozone. Ozone is a form of oxygen—each ozone molecule contains three oxygen atoms. It occurs naturally in the upper atmosphere, where it helps protect people from the Sun's ultraviolet radiation (Chapter 2), but ozone

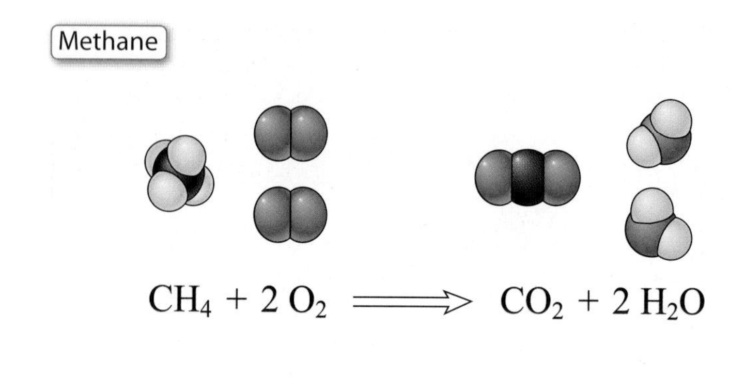

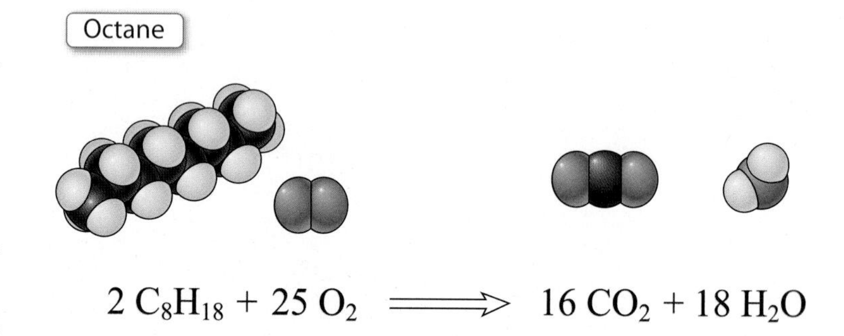

FIGURE 13-28 ▲ The Burning (Oxidation) of a Hydrocarbon Fuel
The products of complete combustion are carbon dioxide (CO_2) and water (H_2O).

in the lower atmosphere can be harmful. It causes respiratory problems, especially in people with asthma or respiratory disease.

Low-level ozone also combines with particulate matter to make **smog**—a dense, hazy air pollution that obscures visibility. Wherever vehicles are abundant, smog can develop, but some conditions combine to make it worse. Weather can be a big factor, especially when it is sunny and a *temperature inversion* develops—a condition in which the air at higher altitudes is warmer than at ground level, hindering convection and helping to stabilize the atmosphere. During temperature inversions, vehicle exhaust and ozone-bearing smog, rather than rising and dispersing into the atmosphere, build up near the ground to levels that can be dangerous. Increased death rates accompany serious smog events in major cities around the world.

The impact of smog is not confined to the regions where it forms. Smoggy Los Angeles air spreads throughout the southwest, making the sky hazy in places as far away as the Grand Canyon. However, regulations that control vehicle emissions have caused dramatic improvement, as you can see from Figure 13-29. The regulations have led to changes in both gasolines (which now contain less sulfur and burn more efficiently) and engines (for example, the addition of catalytic converters) that decrease smog-causing emissions.

Even though new cars are far less polluting than those of a few decades ago, more reductions are needed, and some are planned. For example, the EPA intends to make diesel fuel almost sulfur-free by 2010. Cleaner-burning transportation fuels reduce emissions of all pollutants except for carbon dioxide. Although carbon dioxide is not toxic or harmful at levels present in the atmosphere, its release is a concern because it is a greenhouse gas that influences global climate.

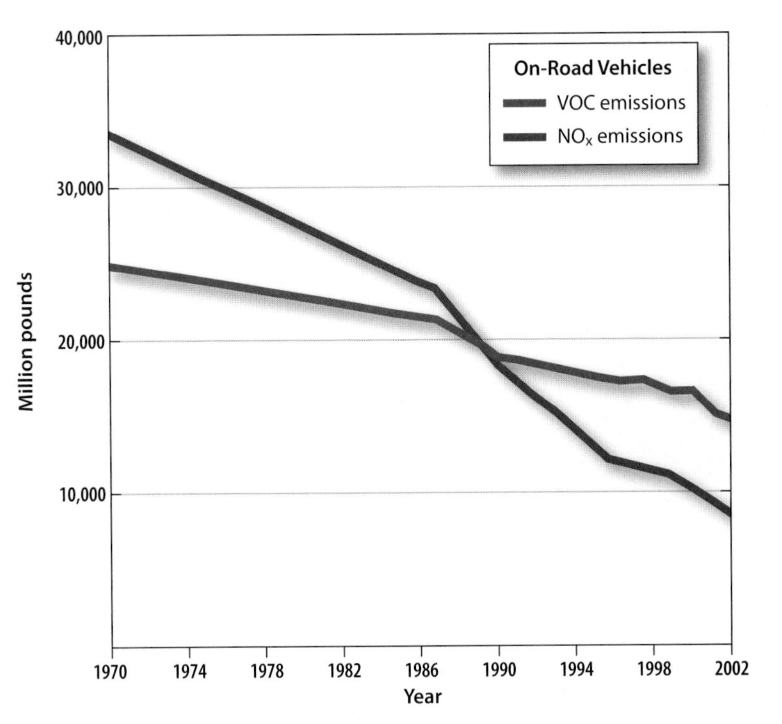

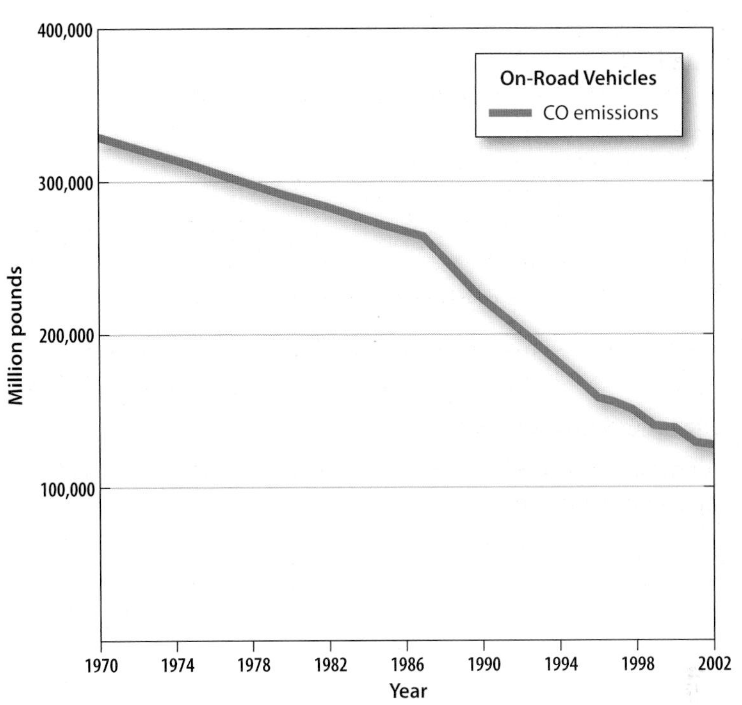

FIGURE 13-29 ▲ Vehicle Emissions of VOCs, NO_x, and CO
Emissions have significantly decreased even as the number of vehicles and miles driven annually have increased.

In the News

It Wasn't Just Dust from China

In Chapter 11 you learned about huge dust clouds that blow completely across the Pacific Ocean from China and obscure the Rocky Mountains. Unfortunately, these clouds contain more than eroded soil particles. As they blow across China, the dust clouds pick up other air pollutants produced by the combustion of fossil fuels, pollutants like carbon monoxide, low-level ozone, and aerosols that contain traces of arsenic and other metals. There are places on the U.S. West Coast where dust clouds from China raise air pollutant levels to 75% of federal standards. Don't be surprised to see air pollution—not just dust—from China again. | The Pacific dust express

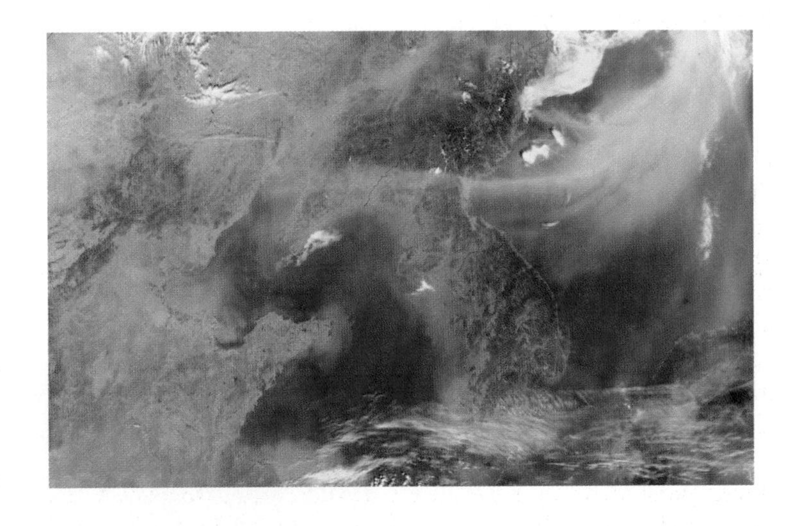

13.3 Coal and the Environment

Coal forms from the remains of vegetation that accumulates in swamps and other wetlands, usually along river floodplains or coastal lowlands. As plants continually grow and die, their remains form *peat deposits*—water-saturated, compacted, partly decomposed plant debris—that are eventually covered and buried by river sediments. Over time, thick layers of sediment bury the deposits. The increasing temperature and pressure that accompany burial to greater and greater depths change the plant remains into coal, as shown in Figure 13-30 on p. 404. The degree of change—coalification—varies from barely changed and soft (*lignite*), to moderately changed (*bituminous coal*), to extensively changed and hard (*anthracite coal*).

COAL PRODUCTION

As with mineral resources (Chapter 12), coal is mined from both underground and surface operations. The main methods are summarized in Figure 13-31 on p. 404.

Surface mines tend to be larger operations that affect more of the landscape. Some western surface mines are called *strip mines* because the highly mechanized operations continuously remove overburden (the material overlying shallow coal

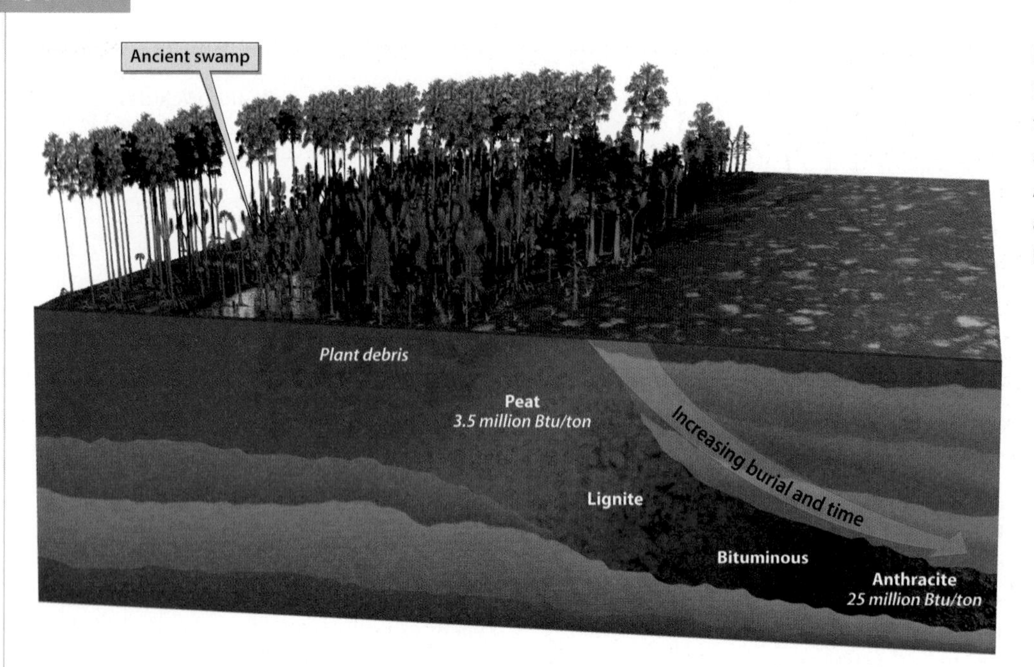

FIGURE 13-30 ◀ How Coal Forms
Partly decayed vegetation forms thick peat accumulations saturated with water. Burial by sediments heats and compresses the peat, gradually changing it to coal. Although it contains about one-seventh the energy of an equivalent weight of anthracite, dried peat has long been used as a fuel.

seams) and deposit it in adjacent strips of previously mined land, as shown in Figure 13-32.

Strip mines are not possible in the hilly and mountainous coal-producing areas of Appalachia. In this region, another type of surface mining has been developed called *mountaintop-removal mining,* or simply mountaintop mining. As Figure 13-33 shows, mountaintop mines can be large operations with very significant physical disturbances and it is not surprising that they are highly controversial. Mountaintop-removal mining – high resolution photos In some places the coal seams are everywhere within 300 meters (1000 ft) of the surface and the overburden is essentially the entire upper part of a mountain—the mountain's top. To expose the coal at the surface, the mountaintop material is removed and placed in nearby valleys. This placement of the overburden directly affects drainages, and as in disposal of waste rock at metal mines (Chapter 12), care must be taken to protect surface and groundwater.

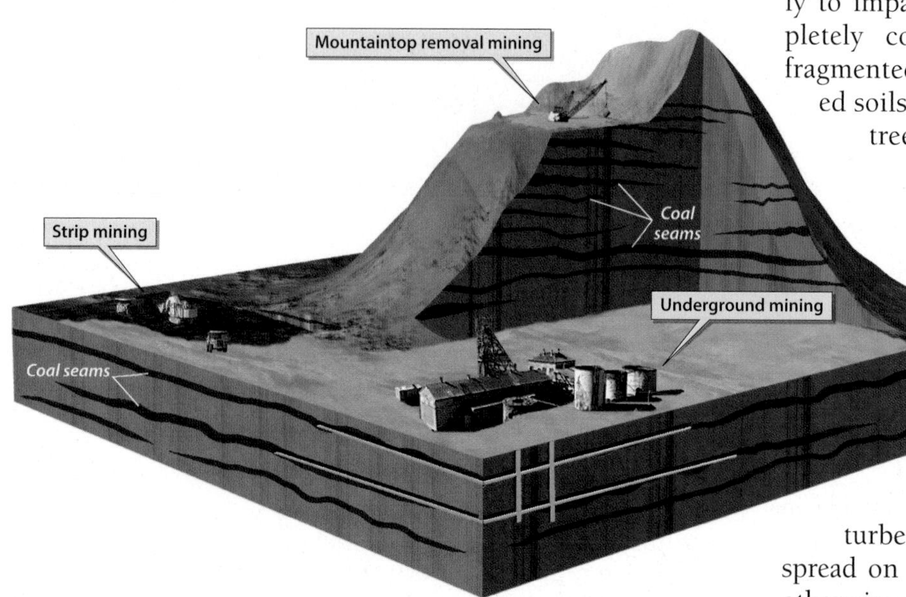

FIGURE 13-31 ▲ Three Methods of Mining Coal
Strip mining is common in western states such as Wyoming and Colorado, mountaintop mining occurs in the Appalachian region in places like West Virginia, and underground mining occurs in many places across the country.

The EPA has examined over 1200 stream segments impacted by mountaintop mining and valley fill and noted a number of potential environmental issues. Mid-Atlantic mountaintop mining—U.S. EPA The study found that zinc, sodium, selenium, and sulfate levels increased in some surface waters sufficiently to impact aquatic life. Some stream segments were completely covered by valley fill and forests were locally fragmented into disconnected sections. In addition, compacted soils in some reclaimed areas were found to hinder new tree growth and the reestablishment of forests. The EPA's main conclusion, that the cumulative environmental impacts of this technique have not yet been identified, suggests that continued assessment of the environmental impacts of mountaintop mining is needed.

The soil that is removed at surface coal mines is saved for reclamation. At large western surface mines, reclamation proceeds as mining progresses. The overburden is placed back in disturbed areas, the surface is graded and contoured, soil is spread on the graded overburden, and the area is seeded or otherwise revegetated (Figure 13-34, p. 406). Reclaiming surface disturbances at coal mines is required by regulatory permits that guide all contemporary mining operations.

The biggest environmental concern associated with coal mining is how it may affect surface and groundwater quality.

FIGURE 13-32 ◀ Coal Strip Mining in North Dakota
As mining advances (left to right in this photograph) overburden from the strip being mined fills the previously mined strip. Reclamation that includes leveling, replacing soil, and revegetation occurs as mining advances.

Improper disposal of overburden and waste rocks can be a major problem, just as it is in mining metal-bearing ores (Chapter 12). In coal mines the waste rocks are rocks adjacent to or interlayered with coal seams that are removed along with the coal. They can degrade water quality if they contain soluble components (mineral salts) or pyrite and are not safely disposed of.

The oxidation of pyrite in coal mine waste rocks can create the same problem that oxidizing pyrite in metal mines does: acid rock drainage, described in Chapter 12. Pyrite-bearing waste rocks must be physically isolated from interaction with oxygen in the atmosphere or hydrosphere to prevent acidic soil or water from developing. As in the case of metal mines, several reclamation methods are available to help prevent acid rock drainage from coal mine waste rocks. Impermeable barriers can be placed below and over them; runoff controls can be used to direct surface water away from them; alkaline material can be placed at selected locations to neutralize any generated acid; and they can be revegetated to limit water infiltration and surface erosion.

FIGURE 13-33 ▶ Mountaintop-Removal Mining
In this controversial method, commonly used in Appalachia, entire mountaintops are removed to get at the underlying coal seams and the waste material is deposited in nearby valleys.

(a)

(b)

FIGURE 13-34 ▲ Surface Coal Mine Reclamation
(a) A surface coal mine operation before closure and reclamation. **(b)** After reclamation has recontoured the surface, replaced the soil cover, and established vegetation, the previously mined area is usable for other purposes.

COAL PROCESSING

Newly mined coal may need to be separated from incorporated rock and fine material and cleaned before it can be used. The separated material is waste that must be disposed of. It probably won't surprise you that this waste is not necessarily environmentally benign. It is generally a slurry that contains fine coal, rock fragments, clay, and other things—perhaps pyrite (Chapter 12). Properly disposing of this waste involves ensuring that it is not eroded into streams and doesn't contaminate surface or groundwater, especially with acid rock drainage. The same safeguards for properly disposing of metal mine waste rock or tailings (discussed in Section 12.3) apply to properly disposing of coal-processing waste. Unfortunately, they have not always been employed.

On October 11, 2000, 250 million gallons of coal slurry poured out of an impoundment in southeast Kentucky near the border with West Virginia. The slurry flooded through abandoned underground coal mine workings that collapsed under the impoundment. It escaped through two mine portals into surface drainages. Before it was over, some 120 kilometers (75 mi) of streams in two watersheds were visibly contaminated and several hundred thousand fish died. Called by the EPA one of the worst environmental disasters ever in the southeastern United States, the spill prompted Congress to ask the National Research Council (NRC) to investigate coal slurry impoundment safety. The resulting 2002 NRC report identifies engineering design, impoundment monitoring, and more efficient coal-processing technologies that can reduce the potential for similar accidents in the future.

Coal slurry spill settlement reached

COAL COMBUSTION

Burning coal has been a source of air pollution for a long time. The use of coal to heat homes in London led to a lingering, sooty haze over this city in the mid-1900s. A mixture of fog and smog in early December 1952 shut down airports and highways for days, forced drivers to abandon their cars on roads, and led to the deaths of several thousand people (Figure 13-35). Although coal is no longer so widely used for home heating in

FIGURE 13-35 ▼ The Great Smog of 1952
This blanket of pollutants, produced mainly by coal fires, led to the deaths of over 4000 people in London.

cities, it does provide the energy needed to produce much of the world's electricity. This is why burning coal is still a significant source of air pollution.

The burning of coal in power plants releases SO_2 and NO_x to the atmosphere. By reacting with water vapor, these compounds form sulfuric acid and nitric acid aerosols—acid droplets scattered through the air. The acid aerosols make rain acidic enough to damage vegetation, as well as buildings made of limestone or marble. Where acid rain makes surface water acidic, aquatic life can be harmed, too. The principal cause of acid rain around the world is emissions from power plants that burn coal.

Pollutant emissions

In August 2004, poor visibility resulted in the collision of ships in Hong Kong Bay. The cause was smog, created by special weather conditions and emissions from coal-fired power plants on the Chinese mainland. The emissions are some of the same ones produced by burning gasoline and diesel fuel—SO_2, NO_x, and particulates—but because coal naturally contains more impurities and is not "refined" like petroleum, it also contains other pollutants such as mercury. The amount of coal burned by power plants is tremendous, and so is the amount of emitted pollutants.

The SO_2 and NO_x emissions from coal-fired power plants have been a cause of acid rain. These oxides react with water vapor in the atmosphere to form tiny droplets of sulfuric and nitric acid. The rain that includes these droplets is acidic and damages vegetation, increases chemical weathering of some building stone (limestone and marble), and forms acidic streams and lakes that are hard for fish to live in. Acid rain, discussed more fully in Chapter 14, can affect large regions downwind of coal-burning power plants.

U.S. coal-fired power plants have been reducing their emissions. Regulations have required burning lower-sulfur coal and more effective capture of SO_2 in exhaust gases. The exhaust (flue) gases are "scrubbed" by injecting calcium-bearing materials (limestone or lime) that react with SO_2 to precipitate solid gypsum (hydrated calcium sulfate). This gypsum dust is collected and used for making wallboard (Sheetrock) and other purposes.

Coal burning produces not only gases but also small particles. It is largely through the escape of particulate matter that burning coal releases other pollutants to the environment. Because coal is a natural rock formed from plant material, it contains small amounts of many different elements and compounds. The EPA classifies 187 substances as potentially hazardous air pollutants. Of these, 15 occur naturally in coal. However, only one—mercury—is released to the environment in amounts considered potentially dangerous to people by the EPA.

Mercury is a metal that is a easily volatilized, so it enters the atmosphere as a gas along with other emissions at coal-burning power plants. Over 36 tonnes (40 tons) of mercury escape into the atmosphere each year from burning coal in the United States (Figure 13-36a). This mercury becomes very widely distributed, but eventually makes its way to soils and surface waters. There, microbes change it to *methylmercury*, a chemical form that can be harmful to people.

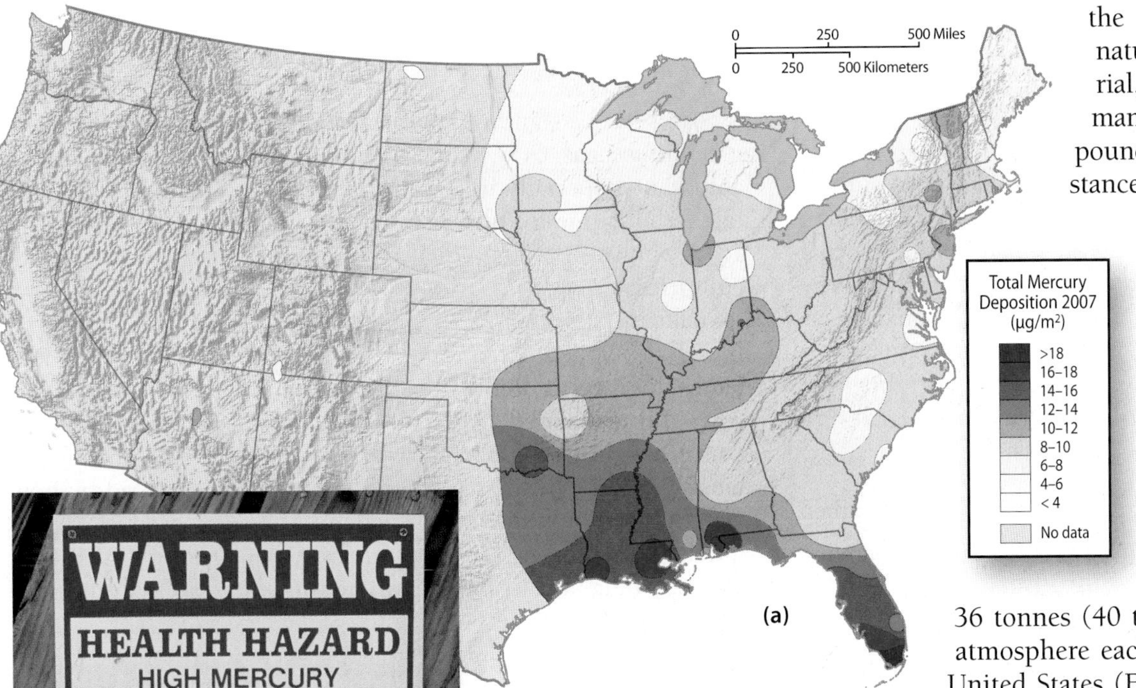

FIGURE 13-36 ▲ Mercury as a Pollutant

(a) The burning of coal to generate electric power, particularly in the eastern United States, releases mercury into the environment. **(b)** When this element is deposited in wet environments, microbes convert it into toxic methylmercury, which can accumulate in the food chain.

One reason that methylmercury is dangerous is that its concentration can increase as it passes up the food chain, a process known as *biomagnification*. The methylmercury consumed by small fish that feed on plants, for example, remains unmetabolized in their tissues. Larger fish that eat these fish may in turn accumulate enough mercury to make them unsafe for human consumption. The mercury levels in some fish—albacore tuna, for example—are high enough to affect brain development in fetuses if consumed by pregnant women. Some 8% of women of childbearing age, according to EPA estimates, already have sufficient mercury in their systems to pose a health risk. This is why the EPA issues warnings about eating too much of certain species of fish from time to time (Figure 13-36b), and is continuing its efforts to lower mercury emissions from coal-burning power plants.

What are some Earth systems interactions associated with the generation and dispersal of methylmercury?

Greenhouse gas emissions

As we noted previously, the greenhouse gas carbon dioxide (CO_2) is produced when all fossil fuels are burned. In 2007, the Supreme Court, by a 5 to 4 decision, ruled that the EPA has the express authority to regulate CO_2 emissions from vehicles as an air pollutant, not because it's hazardous, but because it's a major greenhouse gas that contributes to global climate change.

If we use more fossil fuels than anyone else, it won't surprise you to learn that we also emit more greenhouse gases—about 20% of the world's total. The largest single source is coal-burning power plants, with internal combustion engines (particularly in vehicles) also a major contributor. Carbon dioxide is by far the most abundant greenhouse gas in these emissions.

Our future use of fossil fuels will, at least in part, be determined by how we come to control the release of CO_2. Several possibilities exist for controlling CO_2 emissions:

- Use fuels such as natural gas (methane) that contain less carbon.
- Become more fuel efficient, especially in vehicle fuel consumption.
- Replace hydrocarbon fuels with hydrogen where possible. The byproduct of burning hydrogen is water.
- Prevent the release of CO_2 in emissions by capturing and storing (sequestering) it where it cannot escape to the atmosphere.
- Finding uses for CO_2, such as injecting it into subsurface reservoirs to increase oil recovery.

Carbon dioxide can dissolve in oil, lowering its viscosity and enabling it to flow to the surface more easily. In addition, CO_2 injection helps maintain oil reservoir pressure. This combination enhances recovery at 74 oil fields in America. All but four of these projects get their CO_2 from natural accumulations—CO_2 trapped in the subsurface just as natural gas (methane) is. This natural CO_2 flows to the surface in wells and is piped to the oil fields where it is needed. The human-produced CO_2 that is used in four oil fields is a byproduct of natural gas processing or fertilizer manufacture. The future may find more and more oil fields using CO_2 generated by people's activities, such as burning coal in power plants.

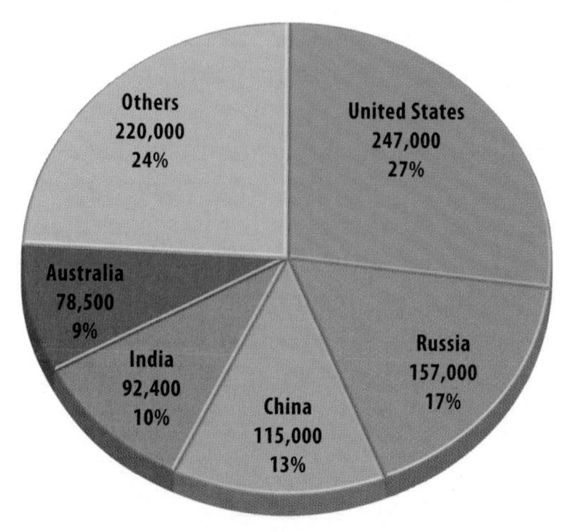

FIGURE 13-37 ▲ Countries with the Highest Proportion of World Coal Reserves

Five countries account for 76% of the world's proven coal reserves of nearly a trillion tonnes. (Numbers, in millions of tonnes, are for the end of 2006.)

COAL'S FUTURE

If coal is such a significant source of air pollutants, including the greenhouse gas CO_2, what should its future role be? Coal is abundant, especially in North America. The United States contains about one-fourth of the world's coal reserves; as Figure 13-37 shows, over three-quarters of these reserves are in just five countries. BP statistical review of world energy—June 2007 Technology to make marketable gas (gasification) and synthetic oil from coal is well developed. Expanding use of coal will enable the United States to meet a higher proportion of its energy needs, but with this expanded use will come increasing environmental concerns—from the effects of mining to those of atmospheric emissions. The future use of coal should be influenced by how well these environmental consequences are addressed. However, much future coal use is likely in rapidly developing countries such as India and China where the influence of environmental concerns is less than in the United States.

13.4 Nuclear Energy and the Environment

Currently, 103 nuclear reactors in 31 states generate 20% of the nation's electricity. But, as Figure 13-38 shows, we rank far below some countries in our reliance on this source. In France, for instance, nearly 80% of the electricity comes from nuclear reactors. The increasing cost of fossil fuels makes it more likely that expansion of U.S. nuclear power will happen in your lifetime. A new generation of nuclear power plants in the United States

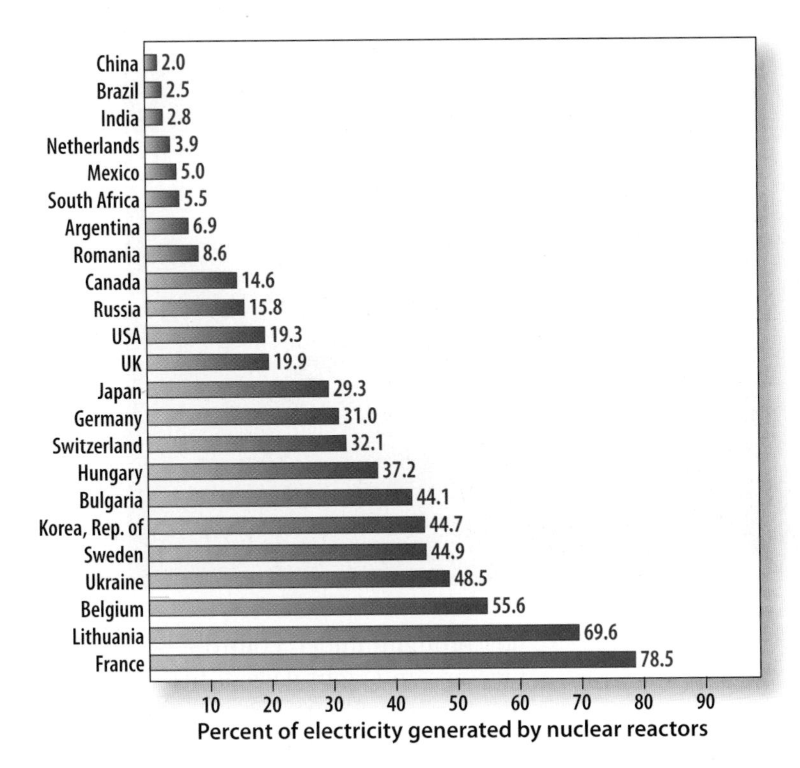

FIGURE 13-38 ▲ **Worldwide Use of Nuclear Energy**

NUCLEAR ENERGY

The nuclei of atoms—the source of nuclear energy—are made up of protons and neutrons and are surrounded by orbiting electrons. Some *isotopes* (forms of atoms that vary in their number of neutrons) have too few or too many neutrons in their nuclei to be stable. These unstable isotopes can spontaneously change into more stable atoms by emitting particles—often neutrons—in a process called *radioactive decay* (Chapter 2). Some unstable nuclei can even split apart, a phenomenon known as **nuclear fission**. Fission releases energy in the form of radiation and heat.

A uranium atom can be induced to undergo fission by bombarding it with a neutron. This nuclear reaction, diagrammed in Figure 13-39, produces two lighter element atoms, neutrons, and energy. Once fission starts in a uranium fuel it is self-sustaining because the released neutrons split other uranium atoms, creating a *chain reaction*. Small amounts of uranium produce huge amounts of energy, and the process does not release greenhouse gases. U-235, with 92 protons and 143 neutrons in its nucleus, is the isotope used for nuclear fuel because it easily undergoes fission. This isotope, however, makes up less than 1% of the uranium found in nature; thus, making uranium fuel requires a process of concentration, or enrichment.

When a fission chain reaction is uncontrolled, it can cause a tremendous explosion, as it does in an atomic bomb. In a nuclear reactor, however (Figure 13-40, p. 410), fission is controlled so that it produces energy at a usable rate. Fuel rods enriched in U-235 are immersed in water, and the rate of induced fission is regulated by means of neutron-absorbing material (control rods) that can be inserted between the fuel rods to slow the reaction. The heat from the fission process is used to make steam, which is then used to drive electrical generators.

Nuclear reactors produce electricity and radioactive wastes. There has been considerable concern about the safety of nuclear reactors, but the key environmental issue associated with using nuclear energy is the disposal of radioactive waste.

NUCLEAR REACTOR SAFETY

Radiation can be dangerous. Large doses of radiation from uncontrolled nuclear reactions—as in atomic bomb blasts—can kill or irreparably harm thousands upon thousands of people. On the other hand, controlled radiation is a very valuable tool. For example, in medicine radiation is used to create images of internal organs and tissues, as well as to treat many cancers.

In the United States, we haven't constructed a new nuclear power plant since 1979 because of high costs, regulatory hurdles, and public opposition. Opposition has come because of two nuclear power plant accidents, or meltdowns. A **meltdown** can occur if a plant's cooling system and related safety devices fail, allowing the reactor to overheat to the point where its fuel and containment structures melt. Steam explosions may also occur during a meltdown. All of these events can release radioactivity into the atmosphere.

The first meltdown accident happened on March 28, 1979 at Three Mile Island near Middletown, Pennsylvania (Figure 13-41, p. 410). Although some radioactivity was released, the meltdown was brought under control and no one was hurt. However, the outcome could have been much worse. Fact sheet on the Three Mile Island accident And in fact, seven years after the Three Mile Island accident a far more serious nuclear power plant accident occurred at Chernobyl in the former USSR (now Ukraine).

The Chernobyl meltdown, caused by design and procedural failures, occurred on April 26, 1986. It was

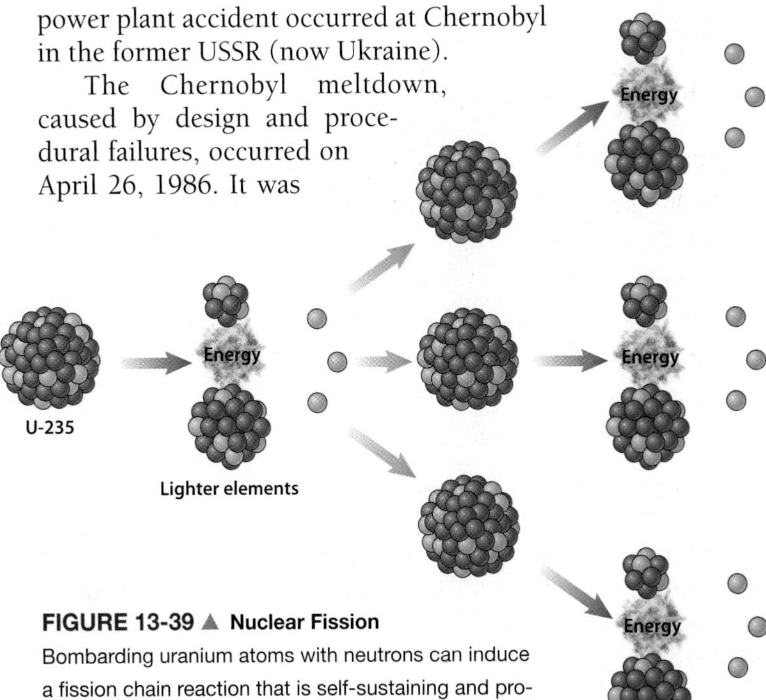

FIGURE 13-39 ▲ **Nuclear Fission**
Bombarding uranium atoms with neutrons can induce a fission chain reaction that is self-sustaining and produces large amounts of energy.

FIGURE 13-40 ▶ A Nuclear Reactor
Fission of U-235 in the fuel rods at the core of a nuclear reactor generates heat, which is used to make steam. The steam turns the turbine of an electrical generator. (The adjustable control rods absorb neutrons, allowing the rate of the chain reaction to be regulated.)

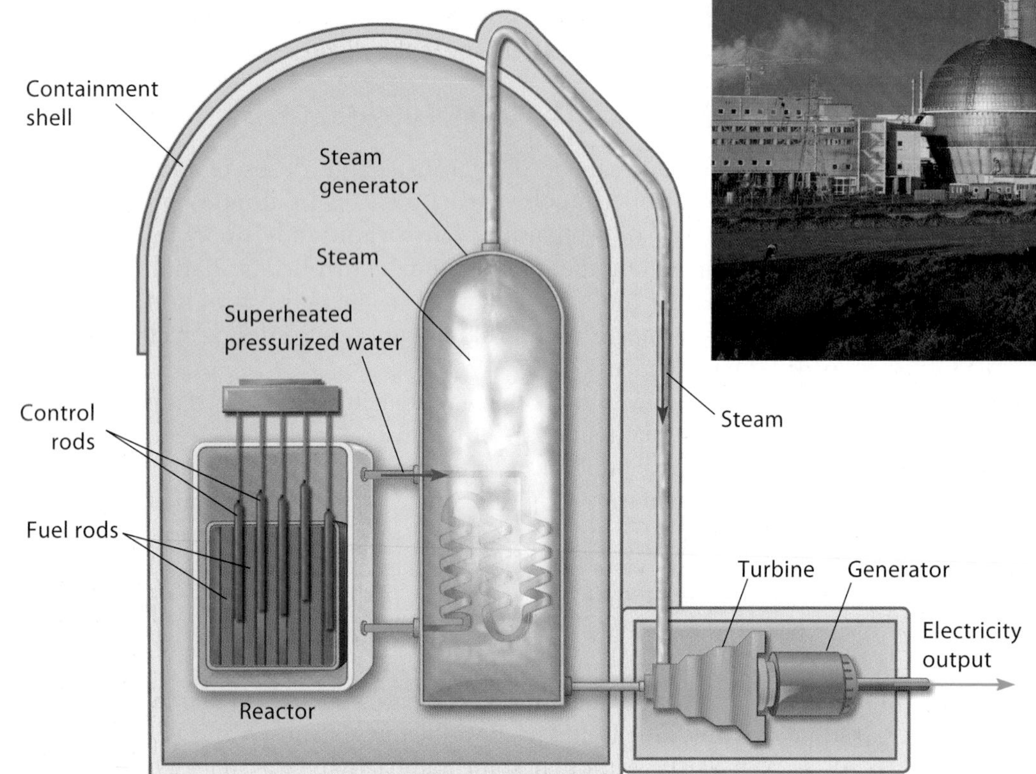

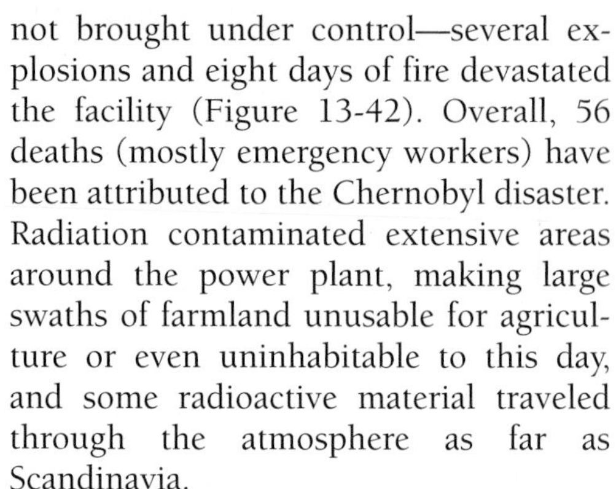

not brought under control—several explosions and eight days of fire devastated the facility (Figure 13-42). Overall, 56 deaths (mostly emergency workers) have been attributed to the Chernobyl disaster. Radiation contaminated extensive areas around the power plant, making large swaths of farmland unusable for agriculture or even uninhabitable to this day, and some radioactive material traveled through the atmosphere as far as Scandinavia.

The Three Mile Island and Chernobyl meltdowns are the worst nuclear power plant accidents that have occurred to date. Because its design was intrinsically less safe than that of U.S. reactors, Chernobyl has never been a good analog for U.S. nuclear power plants. In addition, the Three Mile Island experience has guided the U.S. nuclear power industry and regulators toward much safer designs and operational procedures.

In 2007 there were 28 new nuclear power plants under construction around the world. Sixty more were in the planning stages. By 2009, it's expected that the energy industry will request permits for 31 new nuclear power plants in the United States. Geotimes – November 2008 – trends and innovation – nuclear in a carbon-based world The new reactors use safer designs that have evolved in the years after the Three Mile Island accident as nuclear power plant development continued around the world. The key challenge facing the future of nuclear power generation is not the safety of nuclear plants but the safe disposal of the radioactive waste that they generate.

FIGURE 13-41 ◀ The Three Mile Island Nuclear Power Plant in Pennsylvania
On March 28, 1979, the plant almost experienced a meltdown.

FIGURE 13-42 ▲ The Chernobyl, Ukraine Nuclear Power Plant Disaster This poorly designed and carelessly operated facility was destroyed by a meltdown and resulting explosions on April 26, 1986. Overall, the deaths of 56 people were attributed to the disaster and even today large areas of farmland many miles from the plant remain unsafe for habitation or use.

RADIOACTIVE WASTE DISPOSAL

Over time, radioactive waste gradually emits less and less radiation as the radioactive elements in it decay. However, some waste—including that from nuclear power plants—can take thousands of years to decay to safe levels. What we do with such waste is the big question challenging the present and future use of nuclear power. The only strategy that seems feasible at present is physical isolation—that is, storing the waste permanently in a way that protects people and the environment from any exposure to radioactivity.

Radioactive nuclear power plant waste comes from fuel remnants (called *spent fuel*) and from contaminated liquids developed during reprocessing of spent fuel—**high-level radioactive waste**, or **HLW**. NRC: high-level waste Because nuclear power plants keep operating, they continue to generate more nuclear waste each year. This waste, mostly spent fuel rods, is temporarily stored in water-filled containers. But "temporary" is turning out to be a long time, because there is no operational long-term disposal site for this waste anywhere in the United States.

This is not for lack of trying—but attempts at storing radioactive waste anywhere have encountered tremendous design as well as political and environmental challenges. The design for any storage facility needs to ensure that it will be a stable repository for at least 10,000 years. This is necessary because the natural decay rates of the wastes are slow—some have half-lives of more than 10,000 years (Chapter 2). But even if a satisfactory design can be devised, no one seems to want a radioactive waste repository in their state, let alone near their community.

Nevertheless, a storage site has been selected. It is a place in Nevada called Yucca Mountain. OCRWM – Yucca Mountain repository Yucca Mountain will store radioactive waste underground in a geologic repository, diagrammed in Figure 13-43. This type of repository must be in a highly stable geologic setting and located well above the water table. However, Yucca Mountain is in a region with ongoing earthquake activity and small volcanic eruptions as recent as 80,000 years ago. Fracture systems in the mountain could allow movement of surface water down through the repository to the groundwater table below. Although extensive government studies have concluded that Yucca Mountain is geologically safe, this conclusion is controversial.

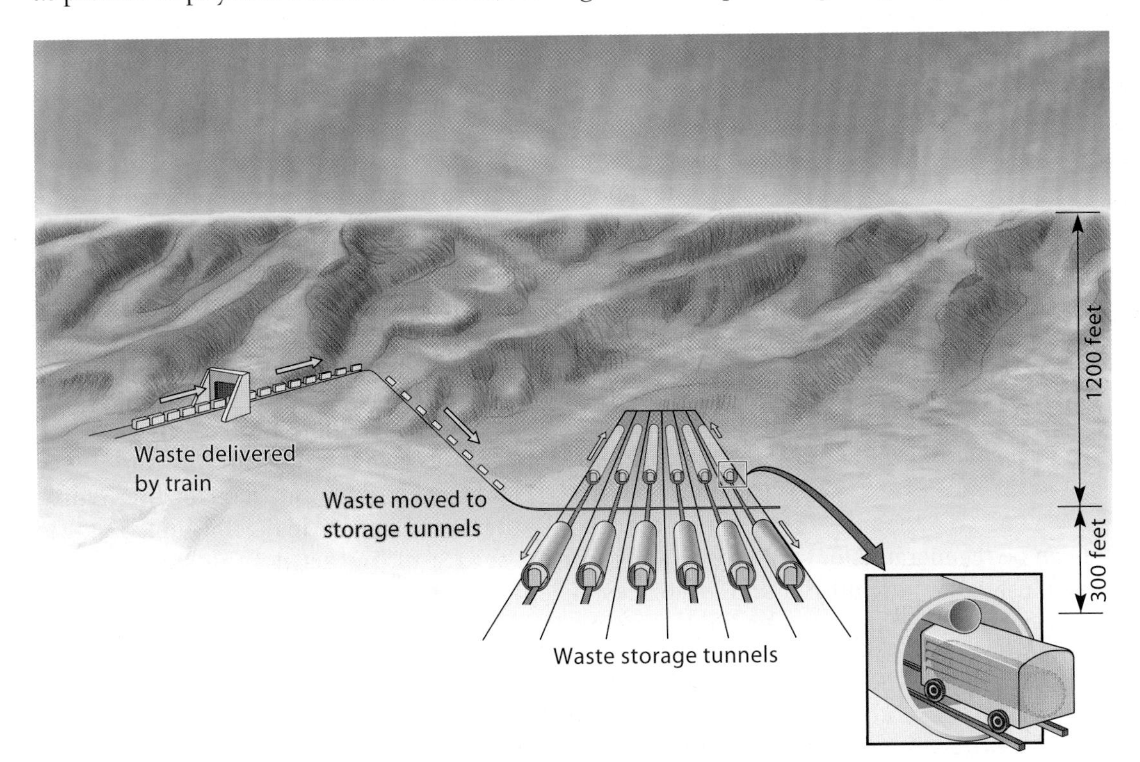

FIGURE 13-43 ◀ The Geologic Repository at Yucca Mountain, Nevada This site has been chosen by the Department of Energy as the nation's long-term underground repository for spent nuclear fuel and high-level radioactive waste.

You Make the Call

Where Do We Put Nuclear Waste?

The Yucca Mountain repository will be located some 300 meters (1000 ft) below the surface, in a layer of volcanic rock more than 1800 meters (nearly 6000 ft) thick. The water table will be another 300 meters below the storage tunnels. Extensive studies since 1979, costing billions of dollars, have prompted the Department of Energy to conclude that Yucca Mountain is a safe repository. |Yucca Mountain standards – radiation protection – USEPA| Others have concluded just the opposite, and many people living in the area (or near the railways and roads along which the waste would have to travel) are strongly against the project . |Yucca Mountain – transporting nuclear wastes| You can keep up with the controversy on the Web. |Eureka County, Nevada—home page for oversight of Yucca Mountain|

The decisions surrounding disposal of nuclear waste are complicated and unresolved.

- What would your reaction be to the location of a facility like Yucca Mountain in your state?
- Who should decide what we do with nuclear waste?
- Where do you think nuclear waste should be disposed?

13.5 Renewable Energy and the Environment

Renewable energy sources are those that can be regenerated, or at least not significantly depleted by their use. They include biomass, geothermal energy, hydropower, wind, and solar energy. Together, they currently contribute about 7% of the energy produced in the United States each year (see Figure 13-5). Renewable energy sources are commonly considered environmentally benign or "clean" compared to fossil fuels or nuclear energy, but each has presented environmental concerns to at least some people in specific cases.

BIOMASS

Biomass is essentially plant debris in all its various forms, including even household garbage and livestock manure. Mostly though, it is wood, field crops like corn or sugar cane, and alcohol (ethanol) distilled from plant material. In a sense, methane (natural gas) generated in landfills or from manure compost is a biomass fuel.

Biomass holds the record for the longest running energy source used by people. People have been using biomass since the first wood fire for protection, heat, and cooking. There are about 1000 wood-fired power plants in the United States today—most are in paper and pulp mills where wood waste products are readily available for fuel. Only about a third of these power plants sell electricity.

Household garbage is a form of biomass that can also be used as a fuel. It is burned to generate electricity in 90 plants across the country. Although it takes 2000 pounds of garbage to produce the same amount of energy as burning 500 pounds of coal, generating electricity from garbage is still an economic operation in some places, especially if communities save money by burning their garbage rather than placing it in a landfill.

When most people think of biomass they think of biofuel—especially ethanol distilled from crops like corn or sugar cane for transportation fuel. But in Chapter 11 you learned that growing crops like corn to produce ethanol has some drawbacks—increased demands on water resources, increased pesticide use, increased soil erosion, and degraded soil quality, for example. Renewable biomass energy sources are not as environmentally benign as they might first appear. Using crops like corn for biofuels also raises food prices, especially in developing countries that are net food importers. The World Bank estimates that a 1% rise in food prices reduces calorie intake by 0.5% by the poor.

The high oil prices of 2007–2008 rekindled interest in algae as a source of biofuel. Algae use energy from sunlight to metabolize CO_2 and make sugars and other hydrocarbon compounds. Some microscopic algae multiply very rapidly, produce large amounts of oily hydrocarbon compounds called *lipids*, and need only water (containing nutrients like nitrogen and phosphorus) to grow in. Research is under way to determine how best to grow algae and recover the hydrocarbon compounds they produce to make biofuel. If algae become a significant source of biofuel it could help sustain soil resources, keep food prices from rising, clean wastewater (by using it as a growth medium), and recycle CO_2 (rather than just adding it to the atmosphere by burning hydrocarbons).

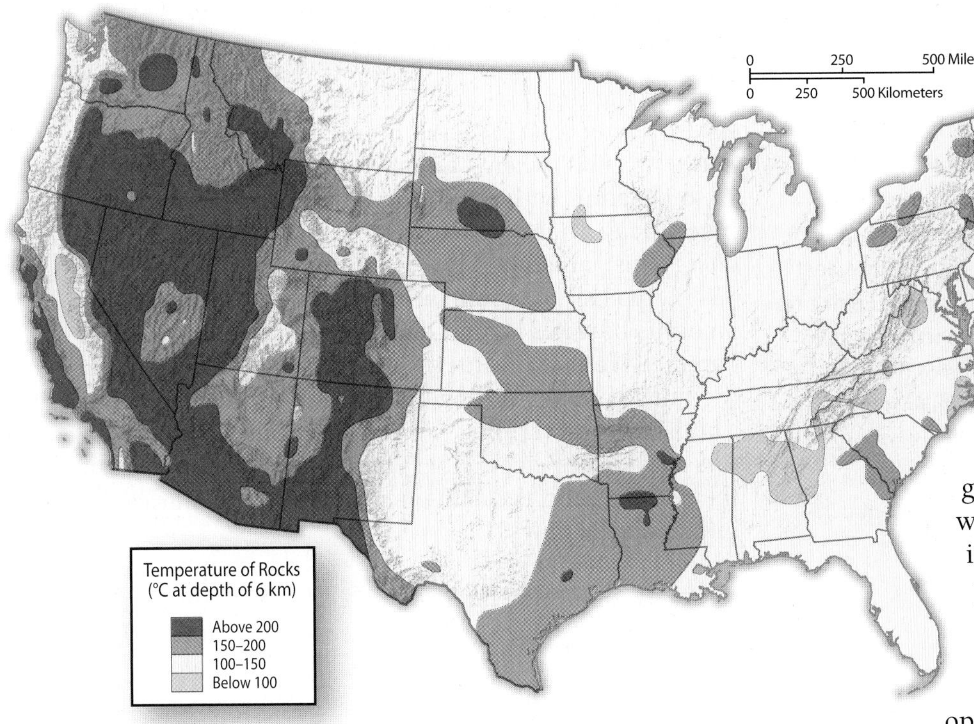

FIGURE 13-44 ◀ Regions of Exceptionally Hot Rock
This map shows the temperature of rocks at a depth of 6 kilometers (3.7 mi) in the continental United States.

GEOTHERMAL ENERGY

Heat stored in the Earth's crust, called **geothermal energy**, is another renewable energy source. |Geothermal energy—clean power from the Earth's heat| Earth is constantly producing heat at depth, primarily from the decay of natural radioactive elements. This heat (together with heat left over from Earth's early formation) gradually moves to Earth's surface, where it is lost to the atmosphere. In general, the deeper into the crust we probe, the hotter the rocks, but in some places, rocks are even hotter than expected, or *anomalously hot*. Volcanic activity (Chapter 6) is a major cause of anomalously hot rocks. In the United States, as you can see in Figure 13-44, especially hot rocks are located mainly in western states.

If rocks are hot enough they can turn groundwater to steam. Wells can tap the heated water or steam and bring it to the surface, where it is used to drive turbines and generators to produce electricity. California, Nevada, Utah, and Hawaii are among the places where this type of geothermal energy has been developed. The 70 geothermal power plants now in operation in those four states produce enough electricity for 1,500,000 homes.

Iceland is another place where geothermal energy has been widely developed. Along the Mid-Atlantic Ridge beneath the island, magma wells up from the spreading center where two tectonic plates are diverging (Chapter 3). Consequently, Iceland has abundant hot groundwater that has long been exploited for heating (it is used in 89% of all buildings) and a variety of other purposes (Figure 13-45). Overall, more than 20% of Iceland's electricity is provided by geothermal resources (hydropower supplies the rest). Iceland will probably be the first country not to need fossil fuels. |Iceland's energy lessons—*Newsweek* leadership and the environment|

Environmentally, geothermal energy is fairly benign. For example, the emissions are only some 1% of those from a similar-sized coal-burning facility. The proper disposal of produced water, which can be salty or contain other dissolved minerals, has also been a concern, although now almost all such water is pumped back into the ground to help maintain the heated groundwater system.

FIGURE 13-45 ◀ Geothermal Resources Are Used by Everyone in Iceland

What You Can Do

Use a Geothermal Heat Pump

The Earth's internal heat can be used everywhere people live. The geosphere characteristically stays at a fairly constant temperature below the surface—commonly in the range of 45° F (7° C) to 75° F (21° C) depending on latitude. By circulating certain fluid through underground pipes, as shown in the diagram, people can capture heat from the geosphere during winter and release heat during summer. These geothermal **heat pumps** significantly reduce energy costs to heat and cool buildings. They can be used in all sizes of buildings, from individual homes to large office buildings. Their initial costs can be high, but over time the energy they save can make them very economically viable. You can learn more on the Web about how these systems work (including utility-company or government economic incentives) and how you can install one in your home.

| *Geoexchange*

In summer, heat is collected from the building and transferred to the ground.

In winter, heat is collected from the ground and transferred to the building.

HYDROPOWER

The use of moving water to do work goes back thousands of years to the first paddle-wheel mills that helped grind grains. **Hydropower** was first used to produce electricity in the United States in 1880, when a Grand Rapids, Michigan, chair factory lit some lamps with electricity generated by a water-driven turbine. Today, 7% of the nation's electricity (and 2.9% of its total energy) is generated by water-driven turbines (see Figure 13-5). Typically, hydropower is produced by first trapping large amounts of water behind dams (Figure 13-46a). Release of the dammed water converts its gravitational potential energy into the kinetic energy of flowing and falling water (Section 13.1). The flowing water moves through turbines that drive generators to produce electricity (Figure 13-46b). As you can see in the photographs, the dams and their power plants can be huge, impressive facilities.

This type of energy generation is commonly thought to be relatively environmentally benign, but there are some environmental consequences. As you learned in Chapter 10, dams invariably change habitat for animals and fish along rivers, and in some places they change the availability and character of groundwater.

Although the construction of dams to produce electricity seems to be over in the continental United States, it is still happening in other places. The huge Three Gorges Dam in China (Figure 13-47) is the latest example. It has displaced 1 to 2 million people and dramatically changed river ecosystems. It is now the largest hydroelectric power generator in the world.

FIGURE 13-46 ▼ Hydroelectric Power

(a) To generate hydroelectricity, enormous volumes of water are first trapped by dams, such as the Kerr Dam on the Flathead River in Montana, shown here.
(b) Some of the water is released, and as it falls it spins the generators that produce electricity. These generators are at the Bonneville Dam on the Columbia River near Portland, Oregon.

(a)

(b)

FIGURE 13-47 ▲ The Three Gorges Dam in China
This dam is the largest hydropower electricity generator in the world.

Damming rivers is the best-known way of using the energy of moving water, but the movement of ocean water caused by waves, tides, and currents can be used, too. People have put wave motion to work generating electricity in Portugal, and a similar project was recently funded in Scotland. These "wave farms" employ a type of floating machine, shown in Figure 13-48, that uses the up-and-down and side-to-side motion of waves and swells to drive a hydraulic motor.

FIGURE 13-48 ▼ A Wave Machine Generating Electricity off the Coast of Portugal
The up-and-down motion of passing waves moves hydraulic rams within the machine that drive electricity generators. Each machine is 140 meters (460 ft) long and 3.5 meters (11 ft) in diameter and produces enough electricity for 500 homes. The first commercial field off the Portuguese coast went into operation in September 2008.

Ocean power delivery ltd

The environmental concerns associated with using waves, tides, and ocean currents depend on local conditions and whether the facilities will affect fish populations and sediment movement. In the open ocean, such as offshore Portugal and Scotland, safe operations during storms and around shipping are probably bigger concerns than possible effects on the environment.

WIND POWER

The uneven heating of Earth's atmosphere by energy from the Sun produces variations in temperature and pressure that set air in motion. Moving air—wind—has provided power for people for thousands of years. Sailing ships and windmills are the classic examples of early wind use. Only now, however, are advances in technology making electricity generation by wind power a viable option. We still don't produce much of our electricity this way—only about 0.1%—but the amount is growing rapidly.

Figure 13-49 shows an example of a wind farm in California. The wind-driven turbines in California produce roughly the same amount of power as a single nuclear reactor—enough to supply about 1% of the state's electricity needs. That number may grow to 5% by 2015—still considerably less than in

FIGURE 13-49 ▼ The Altamont Pass Wind Farm in California
Altamont, one of the first large wind farms, was developed after the energy crisis in the 1970s. It uses over 4900 small, relatively closely spaced, and now obsolete turbines that have killed a few thousand birds each year. More efficient, larger, and slower-turning turbines are gradually replacing the old ones.

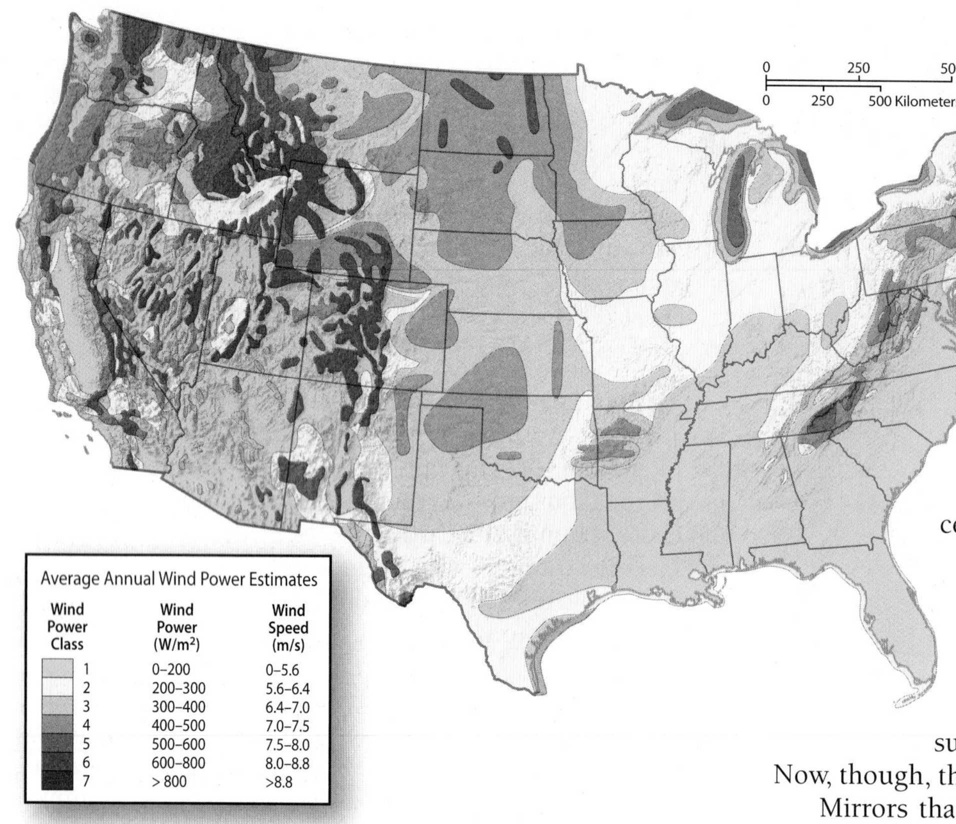

FIGURE 13-50 ◄ Some Places Are Windier Than Others
This map shows the average annual wind power estimates for the continental United States.

countries such as Denmark, for example, which gets more than one-fifth of its total electric power from wind.

Wind energy is environmentally clean, but some people nonetheless have concerns about this energy source. As Figure 13-50 indicates, some places are windier than others, and wind farms are best located in open areas or mountain passes. Wind-driven turbines sit atop towers that are sometimes more than 120 meters (390 ft) high. Such tall features are very visible on the landscape. A new wind farm proposed for the Flint Hills in Kansas met much local opposition because of the aesthetic impact it would have on the surroundings. In many places, this is the primary objection local residents have to wind farms.

Some people are also concerned about the risks wind turbines present to birds, and even bats in some places. |Bat fatalities at wind turbines: investigating the causes and consequences| A lot of the worry about bird safety seems to derive from experience at one of the early wind farms in California, Altamont Pass (also the site of an infamous Rolling Stones concert in 1969). At this site (shown in Figure 13-49), about 1000 birds, including red-tailed hawks, golden eagles, and burrowing owls, are killed each year by 5000 wind turbines. Altamont Pass is a migratory bird route and many golden eagles nest there. From the viewpoint of risk to birds, Altamont Pass wasn't an ideal place to build a wind farm. It also doesn't help that modern wind turbines are not in use at Altamont. New wind turbines have larger and slower-moving blades that birds find easier to avoid.

Bird fatality studies at other wind farms around the world indicate that they are not killing birds at high rates. Plate-glass windows, power transmission lines, communication towers, collisions with vehicles, and neighborhood cats take tremendous tolls, at least many hundreds of millions of birds each year. The fatality rate for wind turbines is one to two deaths per turbine per year. Careful siting and up-to-date wind turbine technology can combine to make the risks to birds from wind farms acceptably low.

SOLAR POWER

People can tap the energy of direct sunlight in various ways. The first use of solar energy was to heat water for homes. Rooftop tubs in sunny climates made this application fairly easy. Now, though, the Sun's energy is being used to make electricity.

Mirrors that focus the Sun's rays can concentrate enough heat to turn water into steam, which can then be used to generate electricity. However, **photovoltaic cells**, also just called solar cells, are the most common means by which sunlight is converted into electricity. Solar cells have layers of silicon between layers that conduct electricity. Electrons released by the silicon when sunlight strikes it travel through the conducting layers and generate usable electricity.

Solar cells are costly to make and install. Moreover, their efficiency is not great—no more than about 15% at best. As of now they do not produce much electricity for general use. In some cases, however, they can supply enough electricity to power a home (Figure 13-51). Solar power is also very helpful

FIGURE 13-51 ▼ Power from the Sun
A typical rooftop array of solar cells.

FIGURE 13-52 ▲ The New Solar Power Station Near Seville, Spain
Mirrors focus sunlight onto the tower, where it converts water into steam. The steam drives turbines and generators producing enough electricity for 6000 homes.

where other energy sources are not available or are very expensive. Many cruising sailboats recharge their batteries with solar power, and it is often used for devices such as handheld calculators and temporary roadside signs.

Solar power technology is advancing rapidly. One sign that it is gaining a place in the mix of energy resources is the debut of Europe's first commercial solar power station, shown in Figure 13-52, which went online in 2007. Inhabitat—Seville's solar power tower

The really good news is that Earth receives tremendous amounts of energy from the Sun. One minute of intercepted solar radiation has as much energy as all the fossil fuels consumed in a year. And of all Earth's energy sources, solar energy has the fewest environmental concerns associated with its production and use. It does take large areas to capture significant amounts of solar energy, and so far the technology is expensive compared to that used with other energy sources. Progress is needed to make solar energy less expensive and to overcome the challenges of cloudy weather and the darkness of night, but the costs are continuing to decrease. Solar energy will almost certainly be more important in the future.

13.6 The Energy Challenges Ahead

Energy resource issues are part of everyday life. The most obvious issues are economic, like the price of gasoline, fuel oil, or electricity. But other, less obvious issues are all around us: whether (and where) to construct new energy-related facilities like oil refineries; government efforts (using our tax dollars) to provide safe disposal of radioactive waste from nuclear power plants; and international relations among countries supplying and using fossil fuels. If you live in a city, you may experience firsthand the effects of using oil to fuel the millions of vehicles on our highways—poor air quality. Perhaps the scope of energy's involvement in your life hasn't been that obvious to you. Looking ahead, however, energy issues will be hard to avoid.

THE COST OF OIL

Many of the most pressing energy issues arise from changes in the cost of oil. At the beginning of this chapter, you learned how the relationship between the world's demand for oil and its ability to produce oil influences oil price. When global production is unable to keep up with demand, or when production actually begins to decline (the Big Rollover), the price of oil will increase. It will be a seller's market. It's possible that we have already reached this stage in the history of oil use.

Higher oil prices have more far-reaching consequences than just rising gasoline prices that we pay at the pump. As oil prices go up, so does the cost of many other products that are made from petroleum or that involve transportation (of raw materials or finished products) in their manufacturing or marketing. Under these conditions, other sources of energy become more competitive and economically viable to produce. Oil use will give way to increased use of other energy resources as the price of oil rises.

ENERGY TRANSITIONS

Historically, people have made several significant transitions in the use of energy resources. Daily energy consumption increased significantly about 600 years ago as wind, water, and animal power were adapted to farming, milling, and transportation uses. However, you can see in Figure 13-53 on p. 418 that even more significant energy transitions began about 150 years ago as the changes brought about by the industrial revolution took hold.

Biomass, in the form of wood, was the energy resource that kicked off the industrial revolution. Wood's role as a fuel is epitomized in the movies by numerous scenes of a steam locomotive's boiler being frantically fed wood in an effort to make more steam, increase speed, and outrace outlaws intent on robbing the train. In the United States, the use of wood biomass peaked about 1875.

As Figure 13-53 shows, the age of wood gave way to the age of coal as this resource became a cheaper and more efficient source of energy. Next came the expanded use of oil starting in the 1920s (and natural gas a few years later) as the availability and cost-effectiveness of these fuels made them the preferred energy resources for diverse uses. The tremendous increase in oil use makes the period from about 1950 to the present the age of oil. But as you can see in the figure, coal use did not decrease to low levels during this time and has actually increased since about 1960; the world contains very large amounts of coal resources. This is not the case for oil.

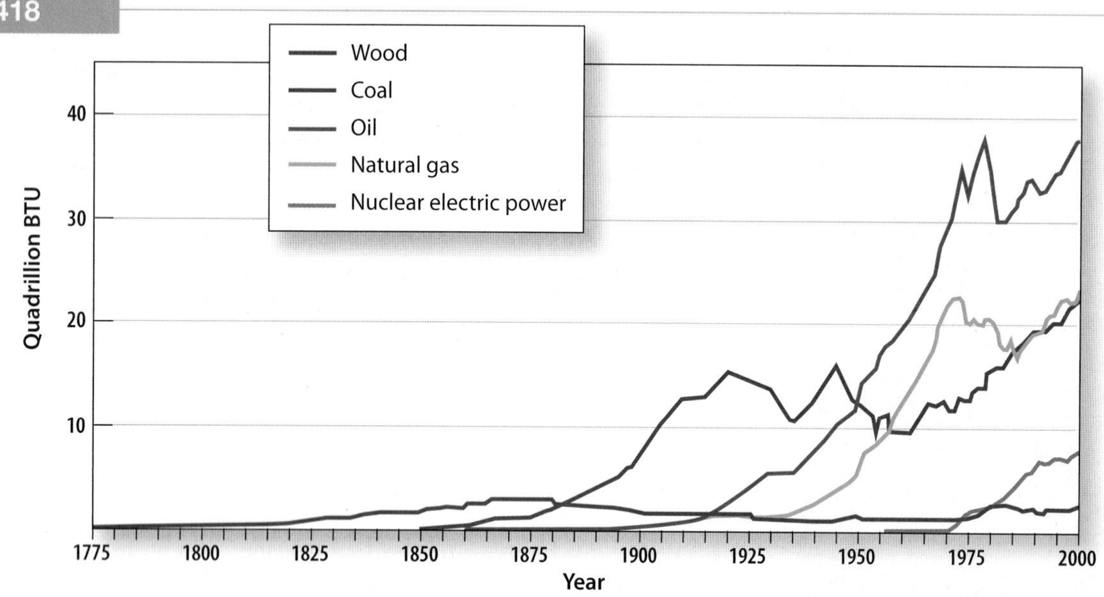

FIGURE 13-53 ◀ **Historical U.S. Energy Use**

As you learned earlier in this chapter, oil is a nonrenewable resource that seems to be nearing the limits of its availability. The age of oil appears to be coming to a close. The big question now is: How will people transition from the age of oil? All energy resources have a role in the future. However, even if we are successful in providing 25% of our electrical power generation from renewable resources as President Obama plans, we will still need to supply 75% of our energy from other sources. In the more distant future, energy from hydrogen or even nuclear fusion may be important. However, in the near term, say the next 20 years or so, the energy resources that seem especially important—because of the size of their potential contribution, and because of the environmental issues related to their use—are other sources of hydrocarbons, including natural gas, coal, and what are called unconventional resources. As you will see, we may soon be transitioning to the age of natural gas, but unconventional oil resources—*oil sands* and *oil shales*—are some of the first that are being affected by higher oil prices.

OIL SANDS AND OIL SHALES

In a few places, oil has migrated near to the surface along permeable pathways in the geosphere, commonly sandstone layers. Here its more volatile components evaporate and the oil is stripped of some components by passing water and degraded by biological activity. This near-surface degradation changes the oil into a viscous form called *bitumen*. Bitumen is a thick tar-like substance (tar is man-made by distilling oil) that must be heated or mixed with less viscous hydrocarbons before it will flow. Sandstone impregnated with bitumen is called **oil sand** or **tar sand**. | Tar sand basics Over 80% of the world's known oil sand is in the province of Alberta, Canada. When Alberta's oil sands became economical to exploit (as described in the following *In the News* feature), they increased Canada's oil reserves by at least 175 billion barrels, making this country second only to Saudi Arabia in proven oil reserves.

In the News

Mining Oil

Alberta's oil sands are a huge resource. They underlie an area larger than the state of Florida and contain 1.7 trillion barrels of oil, of which at least 315 billion barrels are estimated to be potentially recoverable. | Alberta energy—about oil sands About 10% of this resource is within 75 meters (250 ft) of the surface and can be mined by open-pit methods, just like those used to strip-mine coal. If the price of oil is above about $30 to $40 per barrel, mining Alberta oil sand is economically feasible. As the price of oil hasn't been below $30 per barrel since 2004, oil sand mining is going strong in Alberta. Fort McMurray, the city nearest the new oil mines, grew in population from 37,000 in 1996 to 80,000 in 2006 (not counting some 10,000 people living in nearby work camps). Oil sand production is now over 1 million barrels per day.

Producing oil from oil sands has environmental consequences. The physical disturbances that accompany large surface mines are obvious, but other impacts are also significant. The production process uses 2 to 4.5 barrels of water for every barrel of recovered oil. The mines commonly encounter the water table, and pumping to remove water affects groundwater levels and flow in the area. Large amounts of natural gas are needed to upgrade the bitumen to a synthetic crude oil and the process overall (including mining) emits two to three times more carbon dioxide per barrel than conventional oil production.

As you can see, mining oil is a mixed bag. On one hand it provides needed energy and jobs. On the other hand, it significantly changes the landscape, affects habitat and water resources, and emits increased amounts of carbon dioxide. It's not surprising that mining oil in Alberta is controversial. The rate at which oil mining expands will depend on oil prices, but it is likely to remain in the news—production from Alberta oil sands could quadruple by 2020.

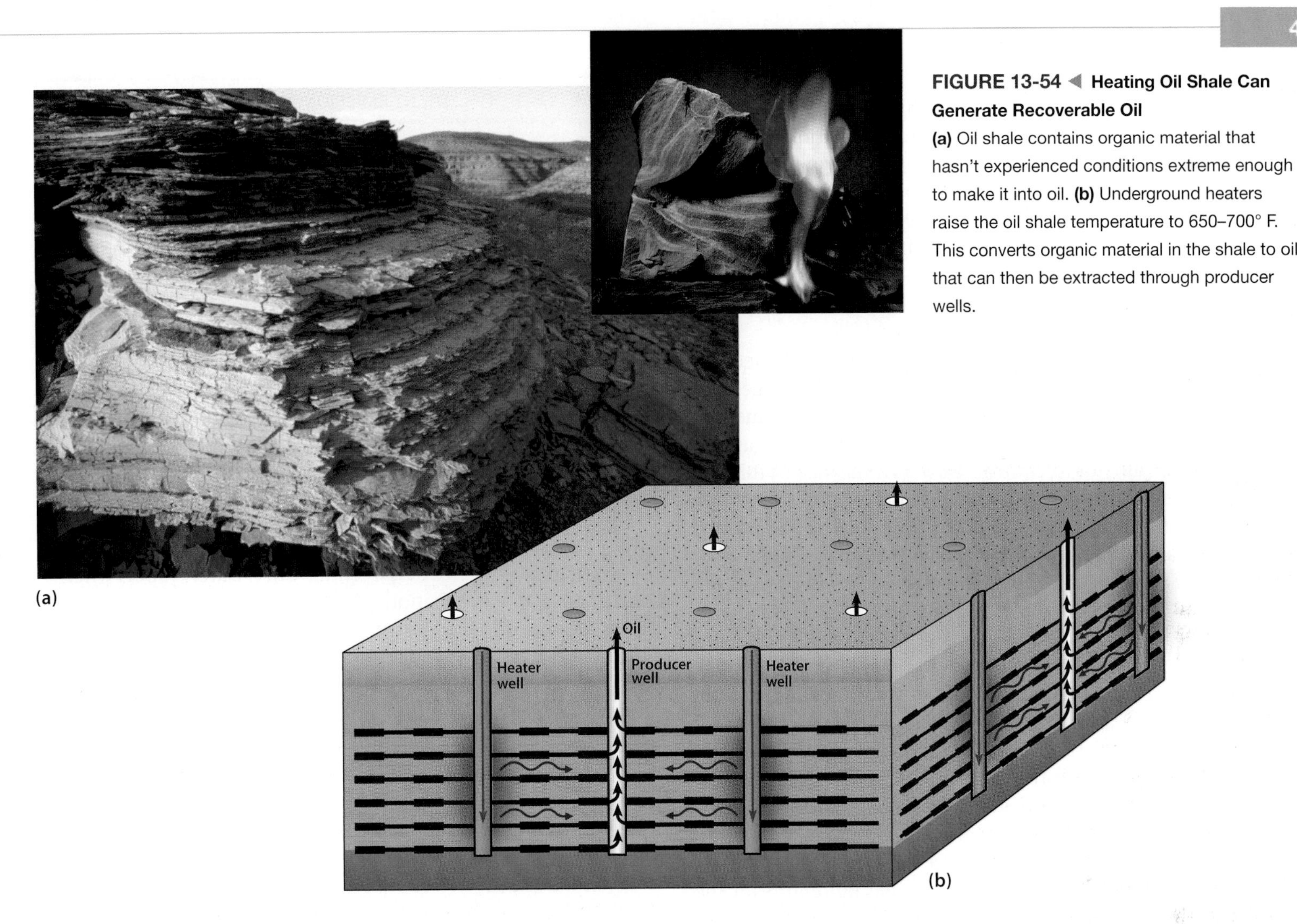

FIGURE 13-54 ◀ Heating Oil Shale Can Generate Recoverable Oil
(a) Oil shale contains organic material that hasn't experienced conditions extreme enough to make it into oil. **(b)** Underground heaters raise the oil shale temperature to 650–700° F. This converts organic material in the shale to oil that can then be extracted through producer wells.

Another unconventional source of oil is oil shale. **Oil shale** (Figure 13-54a) is an organic-rich sedimentary rock that is capable of generating oil (an oil *source rock*) but that hasn't experienced the temperatures and pressures needed to convert the organic material (called *kerogen*) to oil. By heating oil shale, people can complete this conversion and recover oil from oil shale. This is accomplished either by mining the oil shale and heating it in a *retort* (essentially an oven that allows gases to escape and condense into liquid) or by heating it in place, as shown in Figure 13-54b. In the United States, oil shale is primarily located in Utah, Colorado, and Wyoming, where it is estimated to contain 800 billion barrels of recoverable oil or over three times the current oil reserves of Saudi Arabia. About oil shale Commercial production from U.S. oil shale resources has not yet occurred, but research is under way by companies such as Shell to develop the technology needed for economical operations. It's too early to accurately predict the environmental consequences, but if oil shale production becomes economically viable, we can expect environmental issues such as physical disturbances, water supply and quality, and waste disposal to arise.

A NATURAL GAS AGE?

The quantity of potential natural gas resources on Earth is huge. Global conventional natural gas reserves, together with USGS mean estimates of undiscovered resources and potential reserve additions from better recovery, total over 15,000 tcf (trillion cubic feet). This is 150 years of supply at the 2007 global consumption level of about 100 tcf per year. (The United States consumes about 22 tcf each year.) However, conventional resources could be just the tip of the iceberg when it comes to natural gas. Consider the potential of tight gas and natural gas hydrate.

Tight gas is natural gas trapped in very low-permeability reservoirs. Tight gas will not readily flow to conventional wells drilled into the reservoir, so special drilling and production techniques, such as horizontal wells and induced fracturing, are required to recover the gas. Tight gas reservoirs are scattered through the Rocky Mountains south from Wyoming, east through Texas, Oklahoma, and Louisiana, and in Appalachia. These reservoirs contain 105 tcf, or over half of the nation's total natural gas reserves of 196 tcf.

Natural gas hydrate is ice with included methane. It's a solid in which the frozen water molecules form a cage-like structure that traps the methane molecules. Natural gas hydrate is widely distributed in seafloor sediment along continental margins at water depths of 500 meters (1600 ft) or more. It's also present in sediments that underlie Arctic onshore regions down to depths of about 2000 meters (6600 ft).

The amounts of natural gas hydrate are so large that they are challenging to comprehend. Gas hydrate contains more than twice the amount of carbon in all existing and consumed fossil fuels combined, including coal. | USGS fact sheet 021-01: natural gas hydrates: vast resource, uncertain future | The global resource estimate varies from 100,000 to 300,000,000 tcf of natural gas (by comparison, the global resource estimate for conventional natural gas, as noted above, is about 15,000 tcf). The USGS estimates the in-place natural gas hydrate resources in the United States alone to be 320,000 tcf.

However, technology does not yet exist to safely recover methane from natural gas hydrate. A particular challenge, and hazard, is the large volume expansion that accompanies melting of natural gas hydrate; the very large volumes of methane gas that are released are difficult to control. However, the size of this potential resource justifies efforts to solve the production technology challenges. If the challenges are overcome, natural gas hydrate could significantly change the global energy resource future.

Although abundant supply is important, the energy content of natural gas per mass of carbon is also a factor potentially influencing its future use. As Figure 13-55 shows, natural gas yields more energy per tonne of carbon than either oil or coal. This means that burning natural gas will produce less carbon dioxide than burning oil or coal to obtain the same amount of energy. That is why natural gas is sometimes called the energy resource bridge to the future. It can readily replace oil and coal in most uses and it decreases greenhouse gas emissions in the process. Also, natural gas is in general a cleaner hydrocarbon fuel than either oil or coal—you can see from Table 13-1 that its combustion releases smaller amounts of other pollutants for the same energy yield. Perhaps we will have a natural gas age.

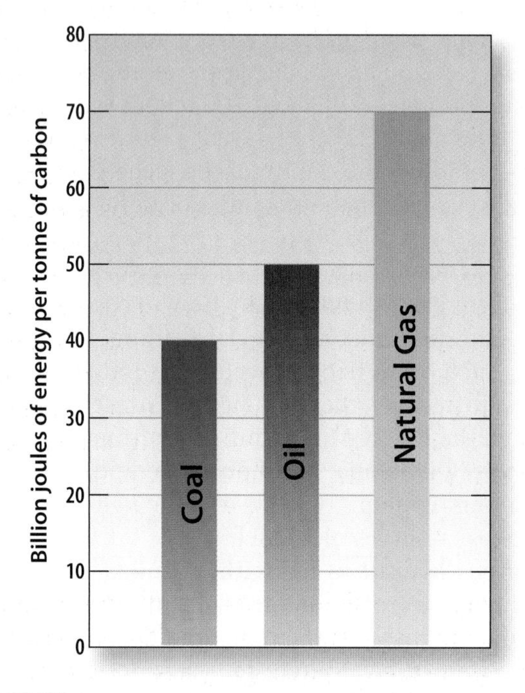

FIGURE 13-55 ▲ The Energy Yield of Fossil Fuels

Natural gas produces more energy for the amount of carbon it contains than any other fossil fuel. Figures are in billion joules of energy per tonne of carbon (one joule equals 0.239 calories).

TABLE 13-1 POLLUTANT EMISSIONS FROM FOSSIL FUELS

FOSSIL FUEL EMISSION LEVELS - KILOGRAMS PER BILLION BTU OF ENERGY INPUT			
Pollutant	**Natural Gas**	**Oil**	**Coal**
Carbon dioxide	53,000	74,000	94,000
Carbon monoxide	18	15	94
Nitrogen oxides	42	203	207
Sulfur dioxide	0.5	509	1175
Particulates	3	38	1244
Mercury	0.000	0.003	0.007

Source: EIA - Natural Gas Issues and Trends 1998

A HYDROGEN AGE?

Hydrogen is a clean energy resource that some have called the future foundation of the global economy. | Princeton hydrogen economy | It is a carbon-free fuel that emits only water vapor when it burns. In some applications, hydrogen isn't even burned; it is electrochemically combined with oxygen in **fuel cells** to produce electricity. | NREL—hydrogen basics—fuel cells | There are many types of fuel cells, but all basically use a catalyst to split hydrogen molecules into positive ions and electrons. The electrons flow and produce an electric current that can be used to power machines, even cars (Figure 13-56).

Hydrogen fuel cells work. Although technologic progress is needed, the economics get better and better the more expensive fossil fuels become. A bigger technological challenge than developing efficient fuel cells is developing a renewable source of hydrogen. The principal source of hydrogen is currently natural gas, so this hydrogen isn't really carbon-free.

Water could be a virtually unlimited hydrogen source but it takes energy to produce hydrogen from water. The challenge, therefore, is to develop renewable sources for the energy used to release hydrogen from water. Research is focusing on using solar or wind energy but the challenges are significant. | NREL—hydrogen basics—hydrogen production | However, under the right situations, they are not insurmountable. Iceland, with abundant and renewable geothermal energy and hydropower, is committed to a carbon-free hydrogen fueled economy. | YouTube—Iceland's hydrogen economy | Jules Verne may have been on target in 1870 when he wrote (in *L'île mystérieuse*):

> I believe that water will one day be employed as fuel, that hydrogen and oxygen which constitute it, used singly or together, will furnish an inexhaustible source of heat and light, of an intensity of which coal is not capable. I believe then that when the deposits of coal are exhausted, we shall heat and warm ourselves with water. Water will be the coal of the future.

AN ENERGY WILD CARD?

A wild card, by comparison with the resources we have just looked at, is nuclear fusion. **Nuclear fusion**, the merging of atomic nuclei of lighter elements, releases tremendous energy.

FIGURE 13-56 ▲ A Prototype Hydrogen Fuel Cell Car
Several major car manufacturers have adapted fuel cell technology to power automobiles.

Fusion reactions power the stars and the explosions produced by hydrogen bombs. They occur only at very high temperatures and are thus very challenging to control on Earth.

The expected fuel for nuclear fusion is deuterium, the isotope of hydrogen with one proton and one neutron in its nucleus (see Section 2-5). Deuterium can be readily extracted from seawater and is considered an essentially unlimited resource. The theoretical foundations of nuclear fusion are understood but the technology to control nuclear fusion reactions in commercial-scale reactors has not yet been developed. Several governments around the world have research under way to develop this technology, and experimental fusion reactors are in use. |Nuclear fusion power—the Encyclopedia of Earth |YouTube nuclear fusion It's possible that your children will use energy from this source.

So, energy transitions are under way as the age of oil gives way to the expanded use of other energy resources. Looking back from a hundred or so years in the future, the age of oil may seem like a very short-term blip in the history of energy use. Coal and natural gas will be part of the future energy mix but technological advances are likely to make other energy resources, from renewables to nuclear fusion, more and more important. It is not yet clear whether we will have an age of natural gas or hydrogen, but as the future unfolds, the most significant remaining question seems to be not whether energy resources will be available but rather how energy will be produced and distributed to people.

THE SIGNIFICANCE OF EMISSIONS

The environmental concerns associated with all forms of energy production and use will influence decisions about everything from specific proposed projects to national policies. You can expect continued adversarial encounters between those wanting to expand fossil fuel availability and nuclear power and those opposed to new energy developments. The debates will involve very entrenched and predictable positions. Some people see wilderness values as more important than new energy resources; risks to the environment from drilling, producing, and transporting oil too great to allow new petroleum developments; and coal mining so inherently damaging to the environment that it should not be allowed. Others see new energy developments as crucial to helping our country decrease its dependence on Middle East oil and creating needed jobs and government revenues. Many who hold this view believe that improvements in technology enable environmentally acceptable developments to proceed, even in remote and previously undisturbed areas. Depending on the specific project proposal, the debate will bring out different environmental concerns. Even renewable energy developments will be opposed by some who consider them aesthetically obtrusive and undesirable in their neighborhoods. A lot of these debates are essentially political. It is probably unrealistic to expect that these highly polarized positions will change much in the future.

What you can expect is that people's need for energy resources will continue to grow and that, one way or another, this demand will be met. Regardless of the entrenched perspectives, people will provide themselves with energy. The challenge is to meet people's energy needs in environmentally sound ways.

The technical challenges of providing environmentally sound energy production from fossil fuel, nuclear, and renewable energy resources are well understood and can be achieved if people consider doing so a priority. However, of all the environmental concerns related to energy production and use, from drilling to mining, the one that has not been satisfactorily met is control of atmospheric emissions, including greenhouse gases. People have identified technical options and have made advances, especially in controlling noxious vehicle emissions, but we have not controlled greenhouse gas emissions at significant scales. Satisfactorily dealing with atmospheric emissions is the main environmental challenge to future fossil fuel use and to the orderly transition from the age of oil to a new mix of energy resources that involves expanded roles for natural gas and coal. How we come to deal with atmospheric emissions is the biggest remaining environmental question related to future energy production and use. (We will examine this subject more closely in the next chapter.)

SUSTAINABILITY

You have learned a lot about sustainability in previous chapters. Sustainability concepts apply directly to resource use, and become particularly challenging when we deal with the use of nonrenewable resources and their depletion (see Chapter 12). With regard to energy resources, it's the depletion of global oil reserves that brings sustainability issues to the forefront.

The goal of sustainable energy production and use requires that renewable energy resources play a much expanded role in the future. The European Union wants 12% of its energy to come from renewable sources (wind, solar, and wave power) by 2010. These countries are leading the development of renewable energy resources in the world. But the use of renewable energy will have to expand more than this to attain sustainability. Sustainable energy production is likely to remain more of a goal than an accomplishment without national policies that recognize it as a priority. The United States has been far behind Europe in this regard but the energy policies supported by President Obama may reverse this.

Sustaining energy resources as much as possible is part of sound stewardship (responsible management of something you don't own), and conservation is the shortest path to better stewardship. Consider just two energy conserving examples: better fuel economy in our vehicles and the use of compact fluorescent lightbulbs.

The amount of oil that could be conserved by better vehicle fuel economy is staggering. This is because we have so many vehicles and we use them so extensively for our transportation. Two-thirds of the 20-plus million barrels of oil used each day is for transportation. Seventy-five percent of this use is on our highways. We have had government-mandated vehicle fuel economy standards since 1975.

Corporate Average Fuel Economy (CAFE) standards are defined by federal law and administered by the Department of Transportation. CAFE overview CAFE standards define what the average fuel economy (in miles per gallon, or mpg) should be for passenger cars and light trucks sold in the United States. Vehicle manufacturers are charged a penalty if they do not meet CAFE standards. The first CAFE standards set by the Energy Policy Conservation Act of 1975 doubled average passenger car fuel economy to 27.5 mpg by 1985. The CAFE standards were revised in 2007 to require an average fuel economy for both passage cars and light trucks of 35 mpg by 2020. President Obama would like to do better than this, achieving 35.5 mpg by 2016. Achieving a goal of 35 mpg will save 1.1 million barrels of oil and reduce carbon dioxide emissions by 473,000 tonnes (529,000 tons) every day.

Several approaches to increasing fuel economy are currently being explored. One is the use of diesel-powered vehicles. Diesel engines are more fuel efficient than gasoline-burning engines. We could save 1.4 million barrels of oil a day if one-third of our vehicles were diesel powered. Clean diesel technology is the key. By 2010, Honda is expected to produce a diesel-powered car that gets over 60 miles per gallon with emissions comparable to the cleanest gasoline-burning cars of today. Another avenue is the development of electric or hybrid gasoline-electric cars (cars that have a gasoline engine and battery power), which operate more efficiently than conventional vehicles and can achieve significantly higher fuel economies. The main point, though, is that a small amount of fuel economy goes a long way. Because we use so much oil for transportation purposes, it's easy for conservation to pay off quickly and handsomely.

FIGURE 13-57 ▶ Compact Fluorescent Lightbulbs Save Energy Bulbs like this one use less than a quarter of the energy of a conventional incandescent bulb to produce the same amount of light.

Now consider the case for using compact fluorescent lightbulbs (CFBs) in place of our regular incandescent lightbulbs. CFBs, like the one shown in Figure 13-57, are now available in regular lamp sizes, and you could replace every one of the incandescent lightbulbs in your home with them. They are a little more expensive initially, but one will last as long as about 10 incandescent bulbs.

The energy savings associated with CFBs are spectacular. CFBs use 75 to 80% less energy to emit the same amount of light as an incandescent lightbulb. These energy savings pay for the higher initial cost of CFBs in about five months of use. Lighting accounts for about 9% of the electricity use in a home, but consider this perspective: If every home (110 million in the United States) replaced just one regular 60-watt incandescent bulb with a comparable CFB, the amount of energy saved could power a city of 1.5 million people. That's a lot of energy savings. Guess who is trying to get at least one CFB into every American home? Wal-Mart! How many lightbulbs does it take to change the world? One. And you're looking at it

These are just two examples of how fairly small changes can have very significant cumulative effects as they propagate through the U.S. population. There are so many of us using so much energy that a bit of conservation really adds up. Looking ahead, we can do ourselves, our country, and the environment a big favor by becoming more energy efficient in our daily lives.

What You Can Do

Save Energy

There are many ways you can personally save energy. Your vehicle and electric light use are just two examples. Energy efficient appliances and homes are two other examples. Want to learn how to increase energy efficiency in your life? Go to the Energy Savers: Home website. There are many links to useful information on saving energy at this site. Energy savers: home

If you want to investigate the energy efficiency of your home go directly to the Home Energy Saver website. Here you can quickly compare how much energy-efficient homes save (in energy cost) over average homes in your community (just enter your zip code). You can also complete a detailed energy audit of your own home and compare it to an energy efficient home. You may be surprised how much energy efficiency can save. The home energy saver

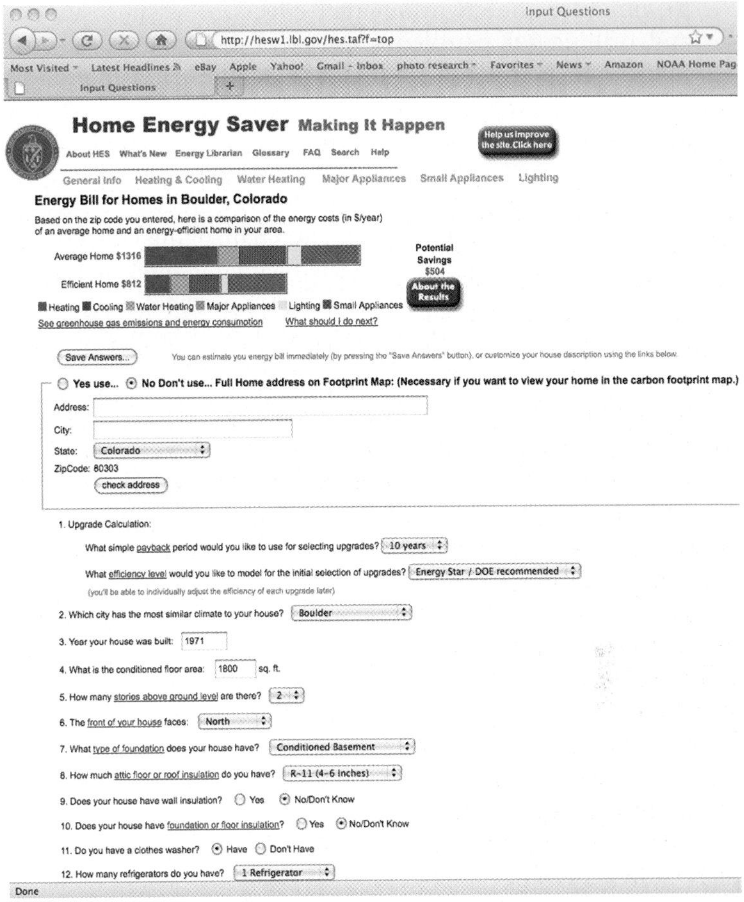

SUMMARY

The scope of the issues and decisions you will face concerning energy resources and the environment is very broad. You have personal choices to make about what kind of car to drive, home to build, or lightbulb to use. You are also likely to live in communities that will make choices about public transportation systems, utility expansions, and facility construction. Perhaps a new energy development will be proposed near you—a new coal mine, oil refinery, or nuclear power plant—about which you will have concerns. And of course U.S. energy policy is a work in progress that would benefit from your involvement.

Everyone depends on energy in daily life; our energy situation is changing as global demands increase, and we will face many choices as a result. The objective of this chapter is to help you be prepared to participate in making these choices.

In This Chapter You Have Learned:

- Energy resources include fossil fuels (oil, natural gas, and coal), nuclear energy, and renewable energy (including biomass, geothermal energy, hydropower, wind, and solar energy).
- Fossil fuels form from the remains of organisms in the biosphere. Plant remains accumulate in swamps and wetlands, and the remains of floating marine organisms accumulate on the seafloor. These accumulations are sediments that are buried by other sedimentary deposits. Increasing temperature and pressure changes plant material into coal and the remains of marine organisms into oil and natural gas.
- Energy resources do work for us and enable us to prosper. The more energy people use, the higher their standard of living. Expanding economies in populous countries like China and India are increasing the demand for energy resources. The result of this global economic expansion is increased competition for energy resources and higher oil prices.
- Expanding energy needs are going to challenge the world's ability to deliver adequate energy resources. The production of oil may not be able to meet demand sometime in the coming decades. As this Big Rollover occurs, oil prices will rise and other sources of energy will become more economically viable.
- Energy supply, demand, and price—including the cost of associated environmental concerns—will determine the mix of energy resources that people come to use as oil prices rise. This mix will include all the key energy resources, but it is likely to see expanded roles for natural gas, coal, nuclear energy, and renewable resources.
- There are environmental concerns associated with the exploration, production, transportation, refining and processing, and consumption of fossil fuels. There are several ways

to minimize the environmental effects of seismic surveys and drilling for petroleum, physical disturbances at coal mines and petroleum production facilities, transportation of petroleum in pipelines, barges, and tankers, and the disposal of wastes generated by drilling, petroleum refining, and coal processing. Emissions to the atmosphere from the consumption of fossil fuels, including greenhouse gases (Chapter 14), are a continuing and challenging concern.

- Nuclear energy generates electricity at large nuclear reactor facilities. Commercial nuclear reactors are being safely operated in many countries where they are key components of their energy mix. The principal environmental concern associated with nuclear energy is the disposal of radioactive waste. The U.S. Department of Energy has developed a geologic repository for such waste at Yucca Mountain, Nevada, but many people oppose the start-up of this facility.
- Renewable energy resources currently provide a small part of the energy used in the United States. Developing a sustainable energy future will require expansion of these sources of energy. In many cases, new renewable energy developments have been opposed by people concerned about their visual impact on the landscape. Solar and wind power facilities do need large areas for development. However, renewable energy has fewer environmental consequences, such as emissions to the atmosphere, than other energy sources.
- The significance of energy in your daily life is hard to overstate. Global changes with respect to energy resource supply and demand affect people in many ways—from choices in their homes to choices their national governments make about energy developments, energy conservation, and relationships with other countries.
- Sound stewardship and sustainability of energy resources are especially important as the transition from the age of oil to other energy resources unfolds. Science and technology are addressing the challenges of developing and expanding other energy resources in environmentally sound ways.

KEY TERMS

biomass (p. 412)
coal-bed methane (CBM) (p. 396)
fossil fuel (p. 389)
fuel cell (p. 420)
geothermal energy (p. 413)
heat pump (p. 414)
high-level radioactive waste (HLW) (p. 411)
hydrocarbon (p. 389)
hydropower (p. 414)
meltdown (p. 409)
nuclear fission (p. 409)
nuclear fusion (p. 420)
oil sand (tar sand) (p. 418)
oil shale (p. 419)
peak oil production (p. 387)
photovoltaic cell (p. 416)
refining (p. 400)
seismic survey (p. 391)
smog (p. 402)
trap (p. 391)
volatile organic compound (VOC) (p. 399)

QUESTIONS FOR REVIEW

Check Your Understanding

1. What is peak oil? State your answer in terms of production rate and demand.
2. Oil is a nonrenewable resource. How do availability and demand for this resource affect its price? How does the price of oil affect the use of other energy resources?
3. What is energy? Classify the different sources of energy discussed in this chapter as renewable or nonrenewable, and explain the basis for your classification.
4. Explain the difference between potential energy and kinetic energy, and give one example of how this is relevant in terms of oil production.
5. Why are temperature and pressure important for the formation of crude oil? How do these variables contribute to the formation of natural gas from oil?
6. Because oil and natural gas are less dense than the rock materials they are generated in, they naturally tend to migrate to the surface. Why then do we have any oil and gas trapped underground yet available to extract for energy use?
7. If oil and gas generation processes are natural and ongoing, why is it that these are not considered renewable resources relative to human energy needs?
8. How is it that oil and gas drilling footprints on the landscape have been reduced over recent years despite steady increases in production?
9. What are blowouts in oil wells, and why are they dangerous to drillers and the environment?
10. What is coal bed methane, and why is it both dangerous and a potential energy resource?
11. What is the role of water in fossil fuel recovery, especially in late-stage oil field development and coal mining? How can it be both a benefit and a nuisance to energy producers?
12. What is the biggest source of oil in North American oceans? How does this source compare in size to pollution from man-made disasters?
13. What is an oil refinery? What do refineries do and why are they important?
14. Why do we burn hydrocarbons for energy? Where does the energy ultimately come from?
15. How is peat involved in the formation of coal?
16. How is burning coal related to the problem of mercury pollution, and why is such pollution dangerous to human beings?
17. Coal is a huge natural resource in the United States, but from an atmospheric gas perspective, why is it a potential problem in terms of climate change?

18. What is nuclear fission, and what materials are used in this process to generate energy?
19. What is a meltdown and how might one occur? Why are meltdowns dangerous?
20. What provides the power in a geothermal energy plant?
21. The text discusses two sources of hydroelectric power. What are they and how do they work?
22. Why does the increasing price of oil drive further development of unconventional hydrocarbon resources?
23. What are natural gas hydrates? Why are they a focus of current energy research?
24. What is the difference between "stewardship" and "sustainability" in terms of energy?

Critical Thinking/Discussion

25. The text states that all fossil fuels and renewable energy resources ultimately come from the Sun. How is this so? Illustrate your answer with examples.
26. Mountaintop-removal coal mining has been very controversial in areas like West Virginia. Why is this so? Consider in your answer not only ecological consequences, but also human consequences for communities located in remote, rural valleys between coal-rich mountains and ridges.
27. Oil refining yields products from very light volatile organic compounds (VOCs), through gasoline and oil, to tarry residue. What does this suggest about the range of chemical composition of the components of crude oil? Is crude oil a homogeneous or heterogeneous source of hydrocarbons?
28. The text outlines a few cases in which environmental disasters have ironically resulted in major improvements to safe environmental handling and development of energy resources. Pick two examples of this outcome involving different energy sources, and discuss the changes made in light of the disasters presented.
29. If CO_2 (carbon dioxide) and H_2O (water) are the theoretical products of hydrocarbon burning, but many nitrogen- and sulfur-containing compounds, as well as other, lighter-weight hydrocarbons, are produced by burning in real settings. What does this fact imply about the purity and efficiency of most hydrocarbon burning? Explain in general terms where these non-ideal compounds come from.
30. Why is long-term nuclear waste disposal a problem? Why is it hard, in realistic repository scenarios, to ensure that such materials can be safely isolated for the period of time necessary for them to decay? Consider the potential problems related to long-term geologic repositories and storage.
31. Wind is an energy resource that is potentially useful in many parts of the United States. What are the relative advantages and disadvantages of this power source? Discuss in terms of infrastructure, public resistance, and relative power output.
32. What are the big challenges facing our society as the "age of oil" comes to a close, especially in terms of transportation fuels?
33. Why isn't hydrogen fuel truly carbon-free in most cases?

ANSWERS TO IN-CHAPTER INSIGHT QUESTIONS

What are some Earth systems interactions involved in the formation of oil and natural gas?

P. 391

- The biosphere converts energy from the Sun into the chemical energy of hydrogen and carbon bonds in hydrocarbon molecules.
- Sedimentary processes that involve interactions of the hydrosphere and biosphere accumulate hydrocarbon-rich organic material in sediments.
- Organic-rich sediments are buried and become solid materials within the geosphere.
- Burial in the geosphere converts organic material into oil or natural gas.
- Oil and natural gas are fluids that can move within the geosphere.

What are some Earth systems interactions that accompany the natural remediation of marine oil tanker spills?

P. 398

- The more volatile oil components are transferred to the atmosphere by evaporation.
- Some oil components are dispersed or dissolved in the hydrosphere.
- Less volatile oil components become physically dispersed in the hydrosphere and incorporated in ocean floor sediments.
- Biosphere bacteria consume and metabolize oil.

What are some Earth systems interactions associated with the generation and dispersal of methylmercury?

P. 408

- Trace amounts of mercury are in the vegetation that becomes coal in the geosphere.
- Burning coal vaporizes mercury and transfers it to the atmosphere.
- Mercury is dispersed through the atmosphere and transferred back to Earth's surface by precipitation.
- Biosphere microbes in soil and surface water convert inorganic mercury to methylmercury in soil and surface water.
- The biosphere transfers and biomagnifies methylmercury in food chains.

FIGURE 14-1 ▶ Free as Air? If you had to pay as much for the air you breathe as scuba divers do, it would cost you nearly $35 a day just to stay alive. Fortunately, the Earth's atmosphere is a free resource—but one that people can degrade.

CHAPTER

14

ATMOSPHERE RESOURCES AND CLIMATE CHANGE

Take a breath. Air goes in and air goes out. In the process, your lungs extract oxygen and expel carbon dioxide, the fundamental steps in respiration that keep you (and most of the rest of the biosphere) alive. How long can you go without a breath? A minute and a half is a long time for many people. Free divers have set the records—8 minutes and 6 seconds for men and 6 minutes and 6 seconds for women. For most of us, anything over two or three minutes would be a record.

Breathing air makes the atmosphere a resource we depend on. How much of the atmosphere do you think we use? One breath is about 0.5 liters (0.0177 ft^3) in volume, and you breathe 10 to 20 times a minute—let's say 15 times. This means that every day, 10,800 liters (381 ft^3) of air go into and out of your lungs. For the human population as a whole (6.8 billion people at the beginning of 2010), that adds up to about 73 trillion liters (2.6 trillion ft^3).

Suppose that we had to buy our daily supply of air. Can you imagine going to a store and buying nearly 400 cubic feet of air each day? Scuba divers (Figure 14-1) normally pay about $5 to fill up their dive tanks. But most scuba tanks hold only about 80 cubic feet of air—enough to last just a few hours. It's a good thing we don't have to pay for the air we breathe.

The atmosphere is essential for generating much of the energy that we use. In the process of combustion, oxygen from the atmosphere combines with fossil fuels in our machines, enabling us to travel, heat our homes, and power our industries. We also separate gases from the atmosphere to use in industrial processes—pure nitrogen, oxygen, and argon, for example. The atmosphere is our principal source of these useful elements.

Fortunately, the atmosphere is a renewable resource. Even the oxygen we use in every breath is restored by photosynthesis in plants and algae. The basic composition of the atmosphere is essentially in a steady state with respect to its main ingredients. But physically, the atmosphere is very fluid and definitely not stable. It's the most dynamic of all Earth's systems, churning, mixing, and blowing across the land. The movement of the atmosphere in response to variable energy inputs is remarkable. Winds can exceed 240 kilometers per hour (150 mph), clouds can rise several kilometers (miles) in an afternoon, and cold Arctic air masses can seemingly turn spring into winter overnight.

The atmosphere's dynamic character makes it a key connection among Earth systems in biogeochemical cycles that support life, such as the carbon cycle (Chapter 1) and the water cycle (Chapter 2). Transfers of energy and matter in the dynamic atmosphere are foundations of the Earth system interactions that shape the land and make life possible. The atmosphere is an indispensable resource for people and for most of the rest of the biosphere as well.

However, as you might expect, people both use and abuse the atmosphere. It's treated as a gigantic waste disposal system for emissions from our vehicles and industries. You learned about the biggest source of these emissions in Chapter 13: our consumption of fossil fuels. Products of fuel combustion pollute the atmosphere, causing smog, acid rain, and the global dispersal of toxic elements like mercury (Chapter 13). Emissions resulting from human activities also change the amount of minor components of the atmosphere, especially carbon dioxide (CO_2). Carbon dioxide is a greenhouse gas that is influencing global climate. Increased concentrations of CO_2 in the atmosphere, traceable to our consumption of fossil fuels, is leading to warmer climates around the world—global warming. The atmosphere plays a key role in global climate change and has done so throughout Earth history.

How people affect the atmosphere and influence global climate has become a major environmental concern of many around the world. This concern is leading to laws, regulations, and economic changes that affect people and industries in many ways. This chapter addresses environmental concerns, but it emphasizes an understanding of global climate change.

IN THIS CHAPTER YOU WILL LEARN:

- The nature of atmospheric resources, including the atmosphere's composition, physical character, and role in Earth systems interactions
- The nature and sources of atmosphere pollutants, including NO_x, SO_2, VOCs, low-level ozone, CO, and greenhouse gases, especially CO_2
- The effects of atmospheric pollution, including smog, acid rain, and changes in the natural stratospheric ozone layer
- How variations in Earth's orbit influence climate by changing the amount of solar radiation Earth receives
- How variations in the atmosphere's composition influence climate, what greenhouse gases are, where they come from, and how they work
- How scientists study climate change
- How Earth's climate has undergone major fluctuations between warm and cold periods, and how it has changed during the last few million years
- How human activities have changed the composition of the atmosphere, especially its CO_2 content, how these changes have altered climate and contributed to global warming, and what future challenges people will face as a result
- How science is helping people understand global warming and develop ways of dealing with it, including the capture and storage of CO_2

14.1 What Atmosphere Resources Are

You learned about the atmosphere in Chapter 2, including its basic composition, structure, evolution, and how it is involved in Earth systems interactions. Here we will revisit and expand on some of these characteristics to get a better understanding of the atmosphere as a resource people use.

COMPOSITIONAL CHARACTERISTICS

One aspect of a natural resource is that it provides useful commodities or materials. The atmosphere's composition is shown in Figure 14-2. People find several components of the atmosphere useful. The most obvious, of course, is the oxygen we breathe to stay alive, but the three elements that make up over 99.9% of the atmosphere—nitrogen (78.08%), oxygen (20.95%), and argon (0.93%)—have many uses.

Individual elements can be separated from air by **cryogenic distillation**. In this process, the temperature of the air is lowered until it becomes a liquid. The temperature is then allowed to rise to the point at which a specific element boils and becomes a gas. Nitrogen boils at –195.7° C (–320.42° F), argon at –185.86° C (–302.55° F), and oxygen at –182.96° C (–297.33° F). The gas that boils off and is collected at a specific temperature can be very pure.

Pure nitrogen (N_2) is commonly shipped and used in its very cold, liquid state. It can be used to cool furnace electrodes during glass manufacturing and to instantly freeze biological specimens (baseball star Ted Williams was preserved in liquid nitrogen!). It is nonreactive and can be used to purge vessels of reactive gases like oxygen, prevent food spoilage, and refrigerate perishables during shipping. Nitrogen prevents corrosion of metal surfaces during high-temperature processing such as annealing.

Pure oxygen (O_2) is used in iron and steel smelting and in cutting and welding torches to achieve higher temperatures. It's an oxidizing agent in processes that produce a variety of chemicals. Adding oxygen increases the efficiency of waste incinerators and water treatment facilities. And pure oxygen is used in medical applications to assist and sustain people's respiratory functions.

Argon (Ar) is a colorless, odorless, nontoxic, and nonreactive gas that forms no known chemical compounds. It is used to create inert environments for growing crystals used in semiconductors and for protecting materials against corrosion during different types of metal processing. It fills the "air" space in double-pane insulating windows and is the gas in incandescent and fluorescent lightbulbs.

The atmosphere's water vapor and other variable components, such as methane and carbon dioxide (see Figure 14-2), benefit people because they function as greenhouse gases and keep Earth's climate from being unbearably cold. Earth's surface would be 33° C (59° F) colder on average and completely frozen if there were no greenhouse gases in the atmosphere. You will learn more about greenhouse gases, their sources, and their role in climate change later in this chapter.

PHYSICAL CHARACTERISTICS

The general structure of the atmosphere is shown in Figure 14-3 on p. 430. Although the atmosphere extends about 1000 kilometers (620 mi) into space, most of the gas molecules are within the lower atmosphere—90% of the atmosphere's mass is in its lower 10 kilometers (6.2 mi) and 99.9997% is within the lower 100 kilometers (62 mi). There is no top to the atmosphere. The number of gas molecules just gradually decreases outward into space.

The lowest layer of the atmosphere, the troposphere, is incredibly dynamic. Its responsiveness to small changes of energy input from the Sun, land, and oceans keeps it constantly churning and moving. These movements facilitate Earth systems interactions critical to life. They make the biogeochemical and water cycles work. And the troposphere is where the day-to-day (even minute-by-minute in some cases) changes in the atmosphere—what we call weather—occur.

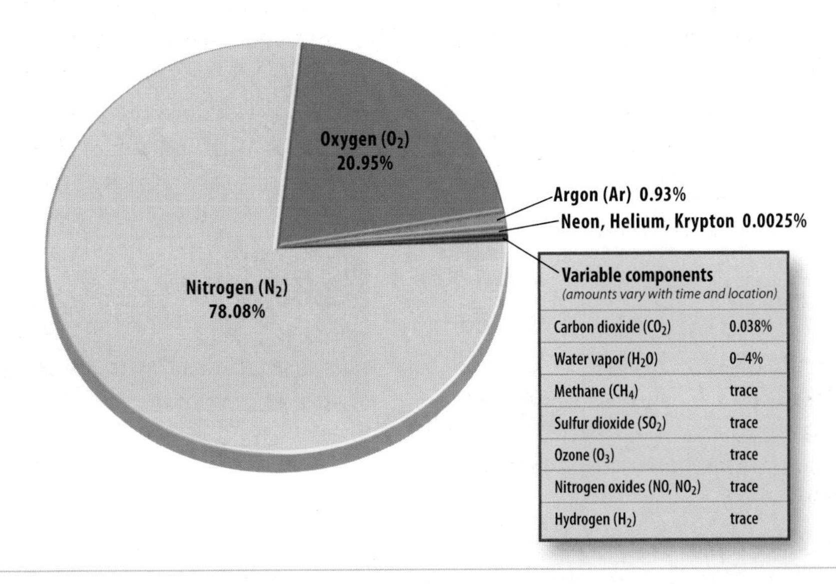

FIGURE 14-2 ◀ **Composition of the Atmosphere, Including Variable Components (by Volume)**

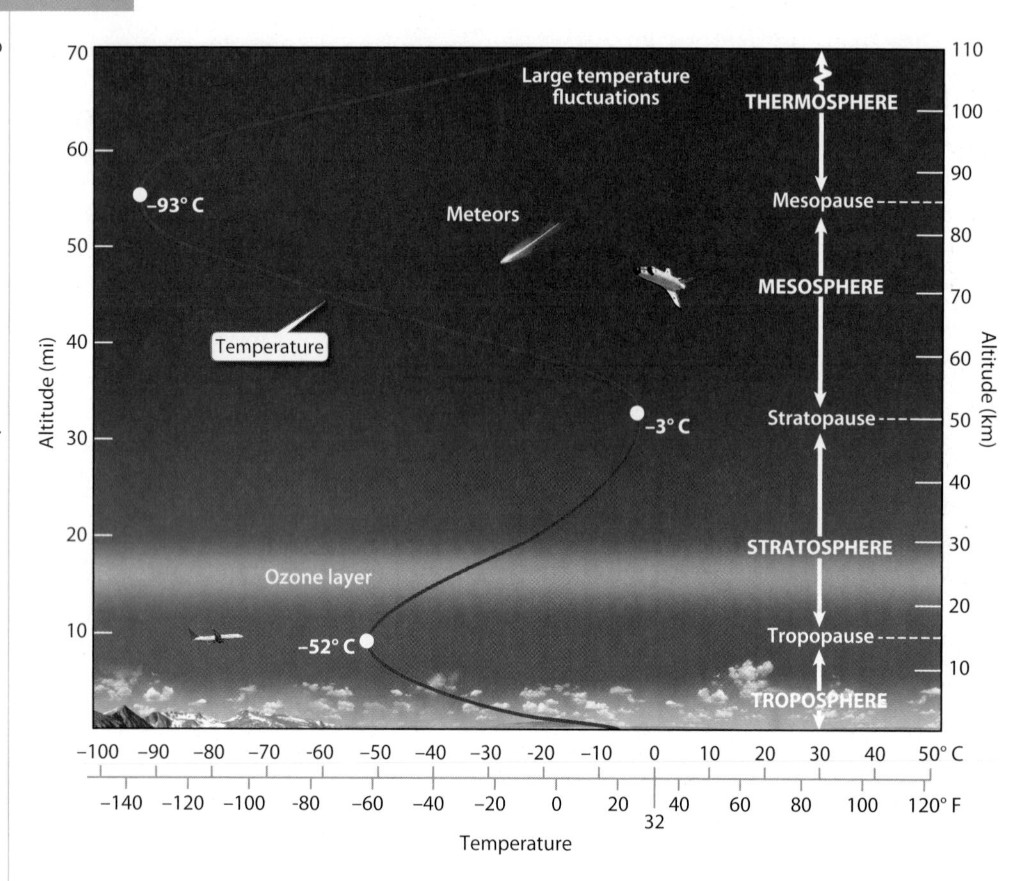

FIGURE 14-3 ◀ The Structure of the Atmosphere Temperature variations define the atmosphere's four principal layers: the troposphere, stratosphere, mesosphere, and exosphere. The ozone layer is within the stratosphere.

What You Can Do

Monitor the Weather

You can learn about current weather conditions across the United States—precipitation, temperature, humidity, and snow depth, for example—at the Weather Underground website. Weather underground The satellite view shows where the major cloud cover is. Current wind directions and velocities appear on a map of the United States. The animation command will show you how the weather parameter you have chosen has varied over the last few hours. Check in on the weather at home or where you will be traveling at Weather Underground.

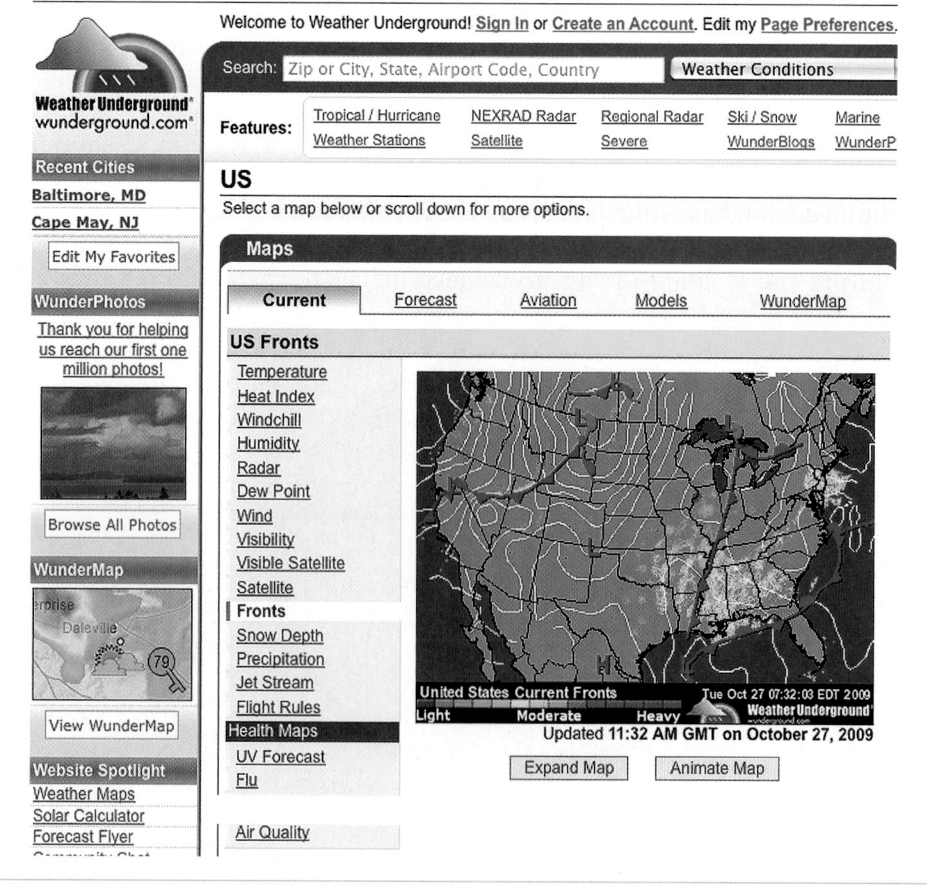

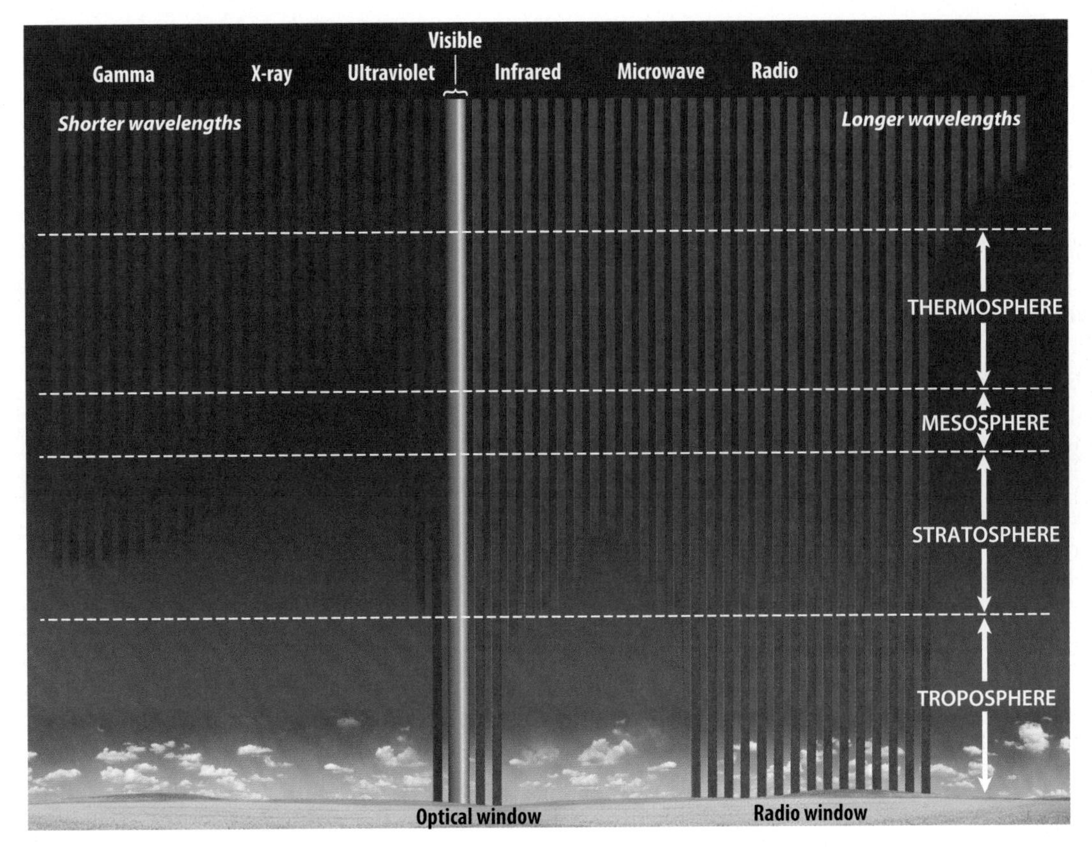

FIGURE 14-4 ◀ The Atmosphere Screens Earth from Harmful Solar Radiation Shorter wavelength gamma and X-ray radiation and large amounts of infrared radiation are completely absorbed by the atmosphere. The ozone layer absorbs the most harmful ultraviolet wavelengths. Only radio waves, visible light, and some ultraviolet radiation reach Earth's surface relatively unimpeded.

You can see in Figure 14-4 that the atmosphere as a whole acts as a giant sunscreen that absorbs different wavelengths of solar radiation. Much of this radiation could be harmful. The atmosphere completely absorbs the short wavelength gamma rays. This is a good thing—gamma rays are highly energetic and pass right through people. They are the main cause of the radiation damage and sickness that come from nuclear bomb explosions. X-rays have longer wavelengths and are less energetic than gamma rays, but too much of them isn't a good thing, either. The solar radiation that does reach Earth's surface includes ultraviolet, visible, infrared, and radio wavelengths. Of these, ultraviolet radiation can be harmful. Fortunately, a part of the atmosphere screens out a lot of the harmful ultraviolet radiation.

Within the stratosphere, some 15 kilometers (9 mi) up at higher latitudes and 35 kilometers (22 mi) up at low latitudes, is a layer that contains higher than average concentrations of ozone—the ozone layer. NAS ozone layer Ozone molecules contain three oxygen atoms (O_3) rather than the usual two (O_2). Ozone forms when ultraviolet radiation breaks apart some O_2 molecules and makes individual oxygen atoms available to combine with other O_2 to form O_3.

Ozone is not abundant in the atmosphere, but it is present at concentrations of a few parts per million in the ozone layer. The ozone layer is very effective at screening out the most harmful, shortest-wavelength ultraviolet radiation (see Figure 14-4). The slightly longer-wavelength ultraviolet radiation that causes sunburn and damage that can lead to skin cancer is almost entirely screened out, too—it's 350 million times weaker on Earth's surface than it is at the top of the atmosphere. Without the ozone layer, people would need to avoid exposure to direct sunlight.

14.2 Air Pollution

We all have a feeling for what clean air is; at least, we seem to know it when we breathe it. Everyone likes to go out for a "breath of fresh air," and most people can recall with pleasure the special quality of the air along the seashore, in a forest, or on a mountain. It just seems to smell and feel good to breathe—even though it's supposed to be odorless and tasteless!

A lot of the time we can also tell if air isn't clean. It smells bad, isn't clear, or is actually uncomfortable to breathe. Sometimes you can't really tell, though, because some of the gases that make air unsafe are colorless and odorless. Things that make air unsafe or dirty are **pollutants**. As you will see, anthropogenic pollutants include some of the same components that occur naturally in small but variable amounts in the atmosphere.

POLLUTANTS

The EPA considers the common air pollutants to be volatile organic compounds (VOCs), nitrogen oxides (NO and NO_2), sulfur dioxide (SO_2), and carbon monoxide (CO). In addition, a recent Supreme Court ruling requires the EPA to address carbon dioxide (CO_2), along with other greenhouse gases, as a pollutant because of its role in causing climate change. The common greenhouse gases are water vapor (H_2O), methane (CH_4), carbon dioxide, and nitrous oxide (N_2O).

Volatile organic compounds

As we saw in the previous chapter, *volatile organic compounds* (VOCs) are chemicals that contain carbon and hydrogen and vaporize easily. Chemicals like benzene, toluene, butane, and propane are examples. When you smell odors from paint thinners, solvents, plastics (like "new car smell"), and gasoline you are breathing air that contains VOCs. VOCs evaporate from crude oil when it is exposed to the atmosphere, and some escape from oil refining operations. Trees release VOCs, too.

VOCs are an air pollution concern because in the presence of sunlight they participate in chemical reactions that lead to smog. The concentration of VOCs is generally greater indoors than outdoors. High levels of exposure can be unhealthy. Because VOCs are so common in household products (paint, cleaning agents, aerosol sprays, and air fresheners, for example), careful use and disposal of these materials will help decrease VOC release to the atmosphere.

In the News

Why Do We Like "New Car Smell"?

Many people like the smell inside new cars. Why we find it pleasant isn't clear, but it's so desirable that people buy aerosol sprays that will make the insides of their cars smell "like new." | New car smell | The *Daily Telegraph,* an English newspaper, has called enjoying new car smell "akin to glue sniffing." The source of this smell are VOCs that volatize from vinyls, plastics, adhesives, and sealants built into the car. Measurements inside new cars have identified over 50 VOCs—including some that are potentially harmful—at high concentrations. The concentrations of VOCs decrease significantly within a few weeks, but they may rise again during hot weather. Japanese car manufacturers are now working to decrease VOC emissions in new cars, and other manufacturers are likely to follow. It will be in the news.

Nitrogen oxides

Nitrogen oxides are a family of nitrogen-oxygen compounds. Collectively they are commonly referred to as "NO_x" because the ratio of nitrogen to oxygen atoms varies from compound to compound. In air pollution, the most common are nitrogen oxide (NO) and nitrogen dioxide (NO_2). Another family member, nitrous oxide (N_2O), is "laughing gas." It's not toxic like NO and NO_2, but it is an effective greenhouse gas. Nitrogen oxides are produced during combustion of fossil fuels. At high combustion temperatures, as in your car engine, nitrogen and oxygen in the air/fuel mixture combine to form NO_x that is emitted as part of the exhaust gases. Nitrogen involved in this reaction comes from the air, and in the case of coal and oil, from the fuel as well. NO_x is a key ingredient in the formation of smog. It also contributes to developing acid rain (Section 13.3) and causes respiratory problems for people.

The EPA has regulated NO_x emissions since the 1970s. These regulations have led to vehicle engine designs and operating procedures at coal-fired power plants that have kept NO_x emissions below the national standard since the 1980s. As Figure 14-5 shows, national NO_x emissions are trending downward and the EPA recently established additional regulations to ensure this downward trend continues. On March 2, 2007, the EPA issued new rules that will limit NO_x emissions from diesel-burning trains and ships by 80%.

Sulfur dioxide

Sulfur is a common ingredient in oil and coal. During combustion of these fuels, the sulfur combines with oxygen to form sulfur dioxide (SO_2), which becomes part of exhaust emissions. In the atmosphere, SO_2 combines with water to form sulfuric acid (H_2SO_4), which can become a source of acid rain. Sulfur dioxide can also cause respiratory problems for people, and react with nitrogen in the atmosphere to form ammonium sulfate. The small particles of this chemical create atmosphere haze and decrease visibility across large regions.

People understand sulfur dioxide's role as an air pollutant. The EPA's regulations have led to lowered SO_2 emissions from industrial facilities, especially coal-fired power plants, and from the burning of sulfur-bearing vehicle fuels (see Figure 14-5). Historically, diesel fuel has contained sulfur and been a source of SO_2 emissions, but EPA regulations now mandate the use of low-sulfur diesel fuel for both cars and stationary diesel engines (such as power plants). Efforts to limit SO_2 release into the atmosphere have been very successful, but over 10 million tonnes (tons) are still emitted each year from various sources (Figure 14-6). That is why work continues to further decrease SO_2 emissions.

Volcanic eruptions also release SO_2 into the atmosphere. Volcanic SO_2 has a special effect—large amounts erupted into the stratosphere can cause global cooling (see Chapters 3 and 6). Sulfur dioxide erupted from volcanoes into the stratosphere forms sulfuric acid aerosols. By reflecting incoming solar radiation, such aerosols cause the temperature below to cool. It can take a few years for these aerosols to be removed from the atmosphere, so over this time, cooling of the troposphere can cause global cooling.

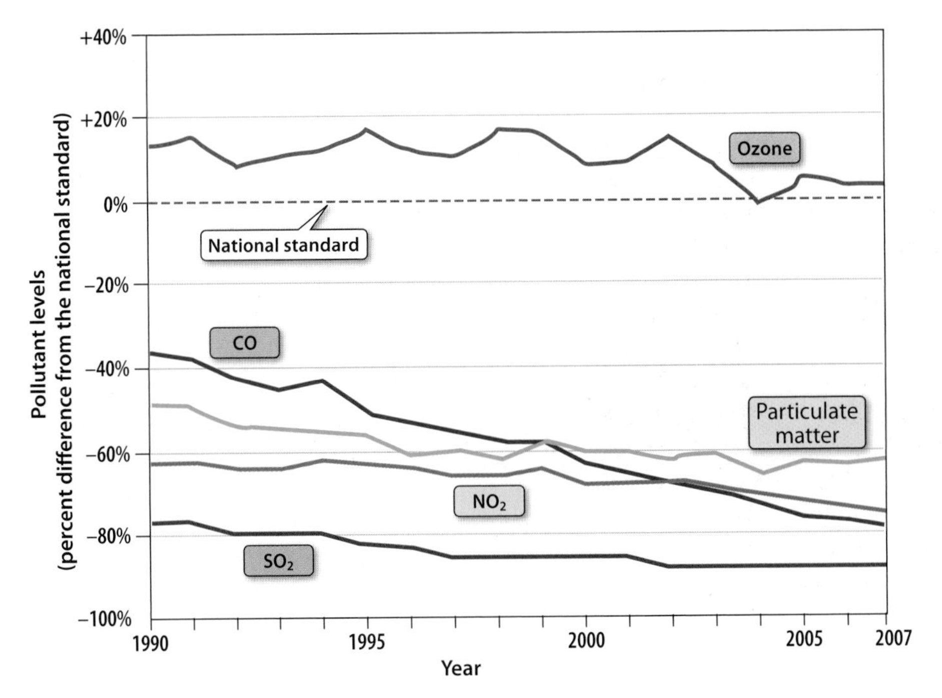

FIGURE 14-5 ▲ Emission Trends

Regulations developed by the EPA have had a major influence on pollutant emissions in the United States. A a result, air pollution levels are near or significantly below national standards and the trends continue downward.

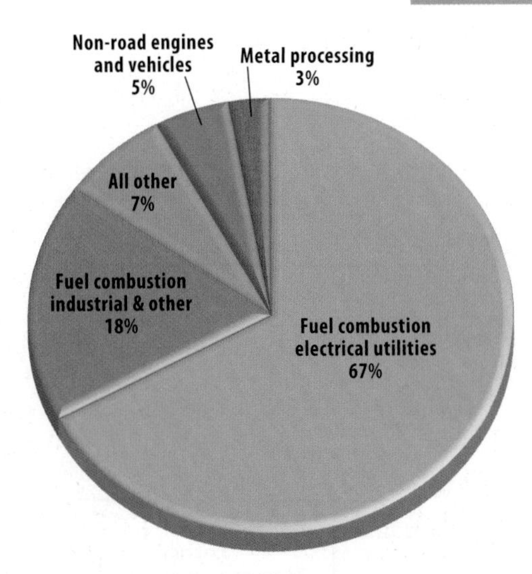

FIGURE 14-6 ▲ Sources of SO_2 Pollution in the Atmosphere

Although SO_2 emissions in the United States have significantly decreased, over 10 million tonnes (11 million tons) are released into the atmosphere each year.

Carbon monoxide

Carbon monoxide (CO) is a colorless, odorless, tasteless, and dangerous gas (Figure 14-7). It is produced by the incomplete combustion of fossil fuels. Charcoal burners, kerosene heaters, and gasoline-powered generators used within confined spaces can gradually increase the concentration of CO to dangerous levels. (Each year, many people die when they allow their cars to idle in closed garages.) Carbon monoxide asphyxiates people by inhibiting their ability to absorb oxygen from the air. Victims of CO poisoning don't see or smell it coming.

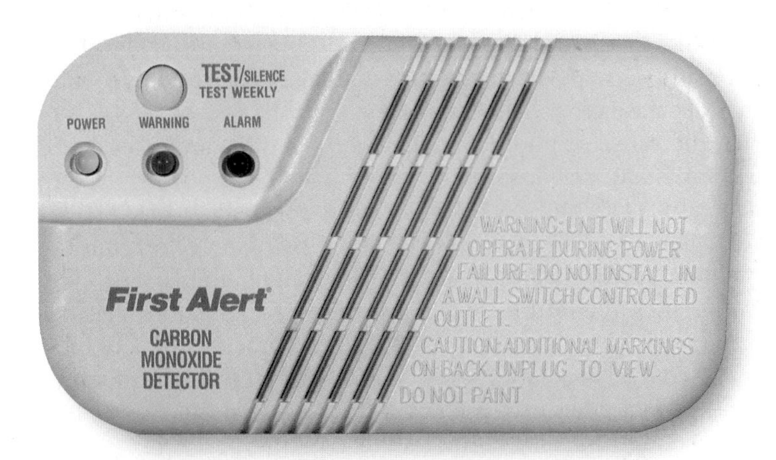

FIGURE 14-7 ▲ Carbon Monoxide Is Deadly

This gas is produced by the incomplete burning of fossil fuels. People can become exposed to dangerous levels in confined spaces such as homes, recreation vehicles, cars, and boats. | Carbon monoxide awareness campaign | Because it is colorless, tasteless, and odorless, electronic detectors must be used to warn people of its presence.

All internal combustion engines emit CO. Vehicles, industry, and burning forests and grasslands are the major sources of CO released to the atmosphere. Carbon monoxide is thus a good general indicator of air pollution; it can be detected and mapped by satellite-borne instruments (Figure 14-8 on p. 434). Because 75% of CO emissions come from internal combustion engines, the EPA's vehicle emissions regulations have been particularly effective at lowering national levels of CO in the atmosphere (see Figure 14-5).

Carbon dioxide

Our metabolism of food is the source of carbon dioxide (CO_2) that we exhale with every breath. Volcanic eruptions also release large amounts of CO_2 from time to time, but the main source of CO_2 emissions into the atmosphere is burning fossil fuel, which produces CO_2 as a major product. Government efforts to keep CO_2 from being considered a pollutant—because it isn't toxic, and classification as a pollutant would require that a lot of money be spent to control it—prompted litigation that was finally resolved by the Supreme Court on April 2, 2007. The Supreme Court concluded that greenhouse gases are atmospheric pollutants that the EPA can regulate under the Clean Air Act. The decision gives the EPA statutory authority to regulate CO_2 emitted by motor vehicles and other sources. This ruling signals changes ahead for CO_2 emitters everywhere, as climate change and global warming become priority environmental concerns. For example, EPA permits for new coal-fired power plants must now consider a "best available control technology" for CO_2 emissions just as it does for other air pollutants.

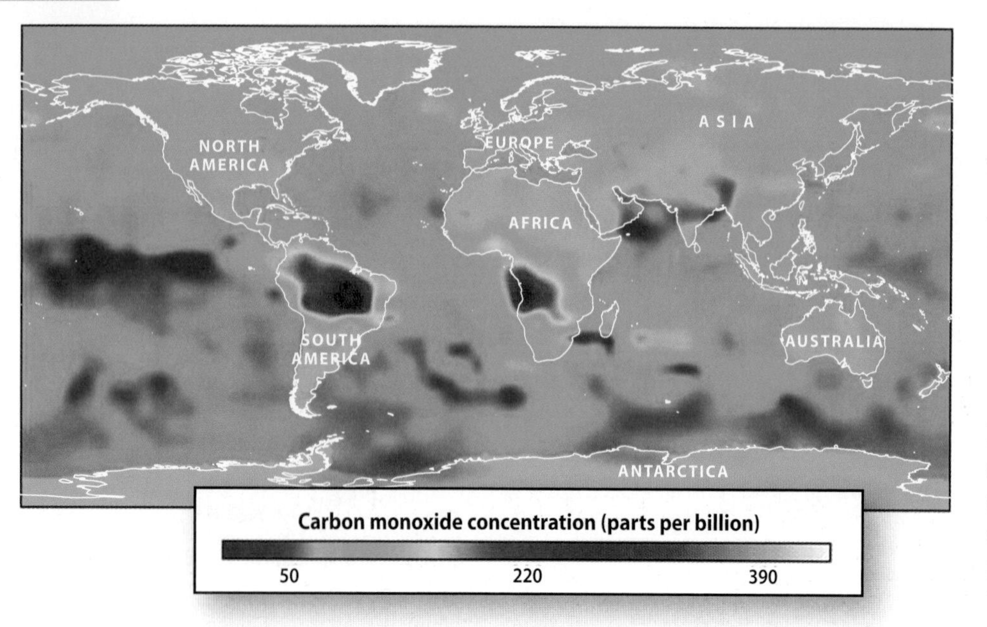

FIGURE 14-8 ◀ Satellites Can Now Map Carbon Monoxide and Other Air Pollutants at Global Scales
This false-color map shows atmospheric carbon monoxide concentrations on October 30, 2000. The two large and intense plumes of carbon monoxide centered over South America and southern Africa were caused by forest and grassland fires.

SMOG

Coal has been a key source of air pollution since it became a major fuel. In the 1800s, cities like London became known for severe episodes of air pollution (like the one described in Section 13.3 caused by burning coal for household heating). In 1905, a doctor concerned about the health effects of this air pollution drew attention to how smoke from coal burning combined with fog to create hazardous conditions. From this work came the new term **smog**, the combination of smoke and fog that characterized large cities of the time.

The burning of coal to heat homes decreased as its role in creating deadly smog conditions became better understood. Although coal-burning can still cause air pollution (as we saw in Chapter 13), it is no longer the key cause of smog in most cities. The smog that now plagues major cities around the world is a different type. Instead of emissions from coal burning being the culprit, CO, NO_x, and VOC emissions from vehicles cause most of today's smog problems.

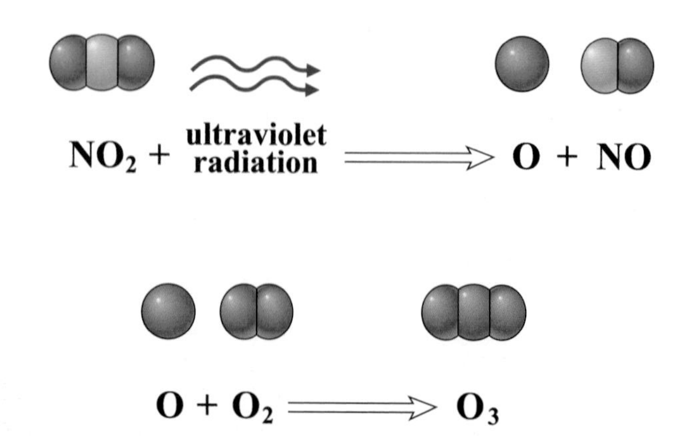

FIGURE 14-9 ▲ Two of the Reactions Involved in Forming Ozone (O_3) and Smog
There are several other reactions involving NO_x, VOCs, CO, and O (in the presence of sunlight) that also contribute to ozone and smog formation.

VOCs and NO_x are highly reactive in sunlight. They are involved in photochemical reactions (reactions that need energy from sunlight), like those shown in Figure 14-9, and other reactions involving CO that result in the formation of ozone. This is unhealthful "low-level ozone" (in contrast to the protective stratospheric ozone layer discussed previously), and it is the key ingredient in photochemical smog. *Photochemical smog* is a brownish haze that settles over cities, especially on warm sunny days. The brownish color, clearly visible in Figure 14-10, comes from NO_x. This smog has a smell, too. The smell comes from ozone (the odor at your local laser printer is caused by ozone).

Photochemical smog gets worse when air is trapped over cities. Places like Los Angeles and Mexico City, where hills and mountains create basins in which pollution can accumulate, are particularly susceptible. At one time, the smoggy haze became so bad in southern California that football fans sitting on one side of the Rose Bowl couldn't clearly see fans on the other side.

Authorities monitor pollution levels and issue "smog warnings" recommending that people stay inside and avoid strenuous activity—especially people with respiratory problems like asthma. The most dangerous component is low-level ozone—it irritates lungs, causing wheezing, coughing, breathing difficulties, and even permanent damage at high levels of exposure. Low-level ozone also damages vegetation in ways that decrease crop yields and forest growth.

The cause of photochemical smog is well understood and federal, state, and local regulations have been put in place to decrease VOC and NO_x emissions, especially from vehicles (Chapter 13). This has been accomplished by reformulating gasoline and by using devices like catalytic converters, electronic engine controls, and exhaust gas recirculators that enhance or complete the combustion process in engines. Smog-control regulations are working. Even though the number of vehicles has increased over the years, national ozone levels decreased from 1990 to 2007 (see Figure 14-5).

ACID RAIN

As shown in Figure 14-11, both NO_x and SO_2 can react with water in the atmosphere to form droplets of acid, but SO_2 emissions have been the principal cause of acid rain. Historically,

(a)

(b)

FIGURE 14-10 ▲ **Smog Is a Significant Form of Air Pollution in Many Large Cities**
This is Los Angeles on a clear day **(a)** and on a smoggy day **(b)**.

FIGURE 14-11 ▼ **Formation of Acid Rain**

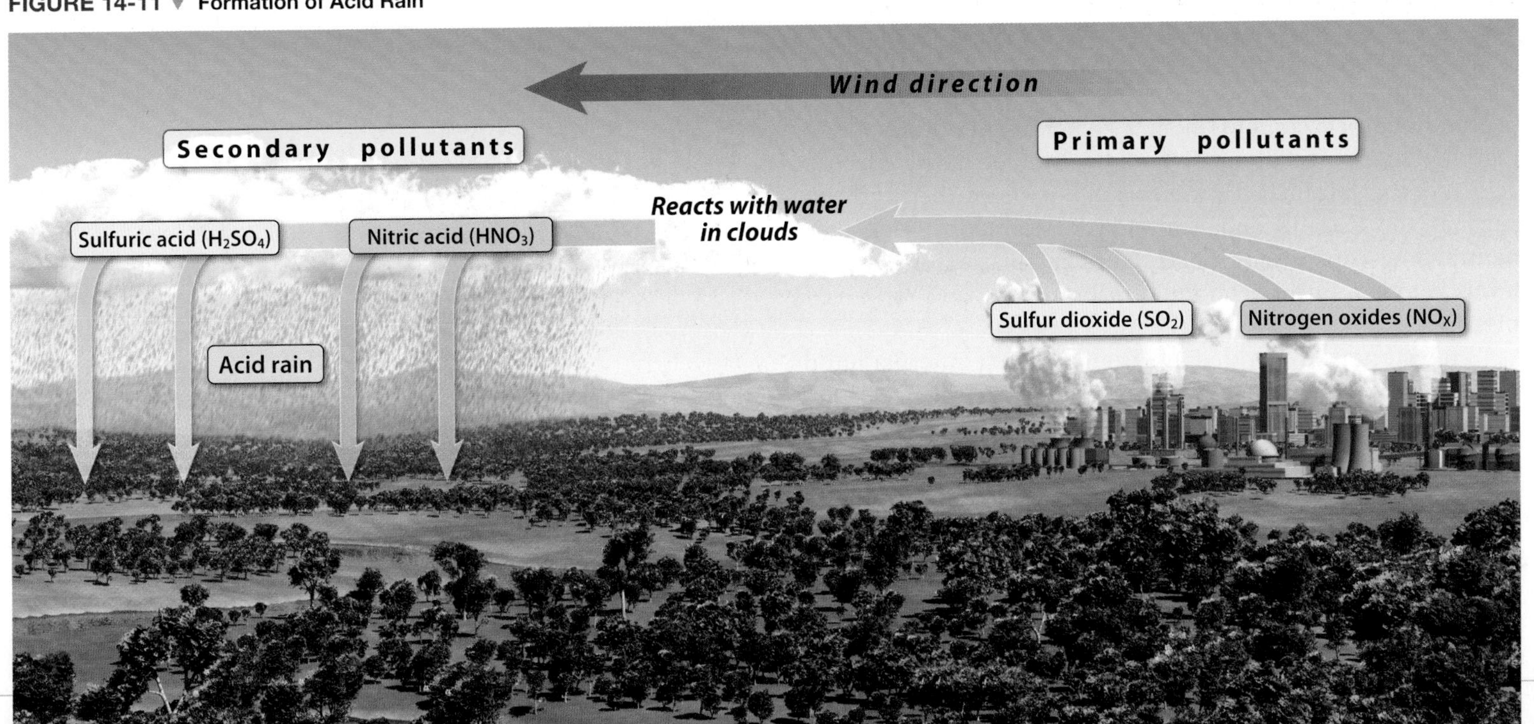

FIGURE 14-12 ▲ **Acid Rain Devastated the Vegetation and Soil of the Copper Basin in Tennessee and Georgia**
The extreme acid rain that caused this desolation resulted from emissions of sulfur dioxide by copper smelters from the 1800s to the early 1900s.

SO_2 emissions were not controlled. Some of the most extreme examples of acid rain and its effects can be seen at old sulfide ore smelters (Section 12.3). The Copper Basin, in Polk County, Tennessee and Fanin County, Georgia, is an example.

Copper-bearing sulfide deposits were discovered in the Copper Basin in 1843. Mining and smelting took place in the area into the early 1900s. The nearby forests supplied wood to fuel the early smelters. However, once uncontrolled SO_2 emissions covered the land with acidic soil, air, and rain, the vegetation was doomed over an area of 78 square kilometers (30 mi^2). The resulting bare soil was easily eroded and the area became the wasteland that you can see in Figure 14-12.

Uncontrolled SO_2 emissions and the resulting extreme acid rain are things of the past in the United States. However, as we saw in Section 13.3, SO_2 emissions still occur, primarily from coal-burning power plants. These plants release 70% of the over 20 million tons of SO_2 emitted to the atmosphere each year, and thereby contribute to the production of acid rain.

Rain is naturally slightly acidic due to the reaction of CO_2 with water to produce carbonic acid. Acid rain is still more acidic—by most definitions, it must have a pH less than 5.6. (Recall from Section 11.3 that a pH of 7 is neutral.) In the northeast United States, tiny droplets of sulfuric acid cause rain to have pH levels that average 3.6 (Figure 14-13a). Prevailing winds blow across the United States from west to east, so the northeast region is downwind of most of the nation's coal-burning power plants, as you can see in Figure 14-13b. Acid rain in the northeast has affected vegetation, aquatic systems, and buildings.

Acid rain will damage plant leaves and degrade soil by dissolving nutrients and harming soil life (Section 11.1). Where acid rain accumulates in surface waters, acidity can reach levels that are unhealthful for fish and other aquatic life. Where soil and bedrock contain alkaline materials like limestone or marble that can react with acid and neutralize it, the effects of acid rain are minimal. However, large parts of the northeast, such as the Catskill and Adirondack Mountains, lack these neutralizing capabilities. Some lakes in this region have pH levels below 5; Little Echo Pond in Franklin, New York, has a pH of 4.2.

Limestone and marble have been used to construct buildings in many cities. Acid rain that falls on them will react on the surface and along tiny cracks in the building stone. The reaction dissolves bits of the building stone and pits, etches, and discolors the building surfaces. In places, structural components are weakened and need to be replaced. In many cities of eastern Europe, buildings, bridges, and even statues are in need of ongoing repair due to acid rain (Figure 14-14 is typical).

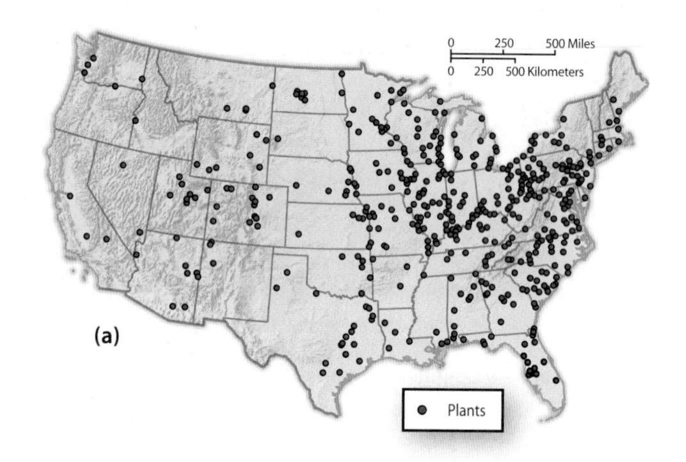

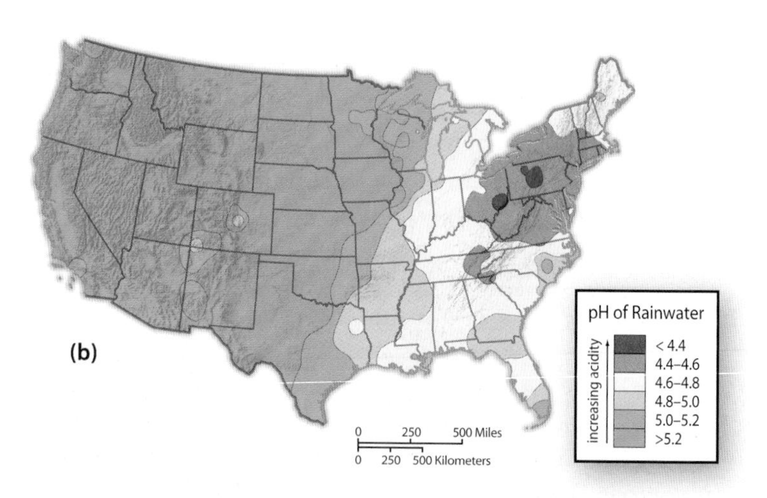

FIGURE 14-13 ▲ **Acid Rain and Power Plant Emissions**
(a) The location of coal-burning power plants. **(b)** The acidity of rainwater in different parts of the United States. The distribution of plants combined with prevailing winds (from west to east) make acid rain common in the east and northeast parts of the country.

FIGURE 14-14 ◀ Anything Made of Limestone or Marble Can Be Damaged by Acid Rain
This statue in front of a Cracow, Poland church was replaced because of acid rain damage.

What You Can Do

Investigate Acid Rain Damage in Washington, DC

Over 20 million people visit Washington, DC, each year. Perhaps you will be one of them someday. If you are, you can take a self-guided tour of buildings that have been damaged by acid rain.

Washington, DC, is famous for its many national monuments, museums, and government buildings. Many of these buildings were constructed with limestone or marble that is exposed to the elements on exterior surfaces. Some of these surfaces have withstood acid rain fairly well, but some have not. The USGS has provided a guide to the effects of acid rain in Washington, DC. Acid rain and our nation's capital This guide includes many photographs, descriptions, and directions that will lead you to places around the city where acid rain has attacked buildings. If you don't expect to visit Washington, DC, take a virtual tour through the USGS guide—you will learn how to recognize the effects of acid rain on building surfaces. Most communities have at least some stone buildings, and if you live in the east they may have been affected by acid rain. You can take your own local tour to discover them.

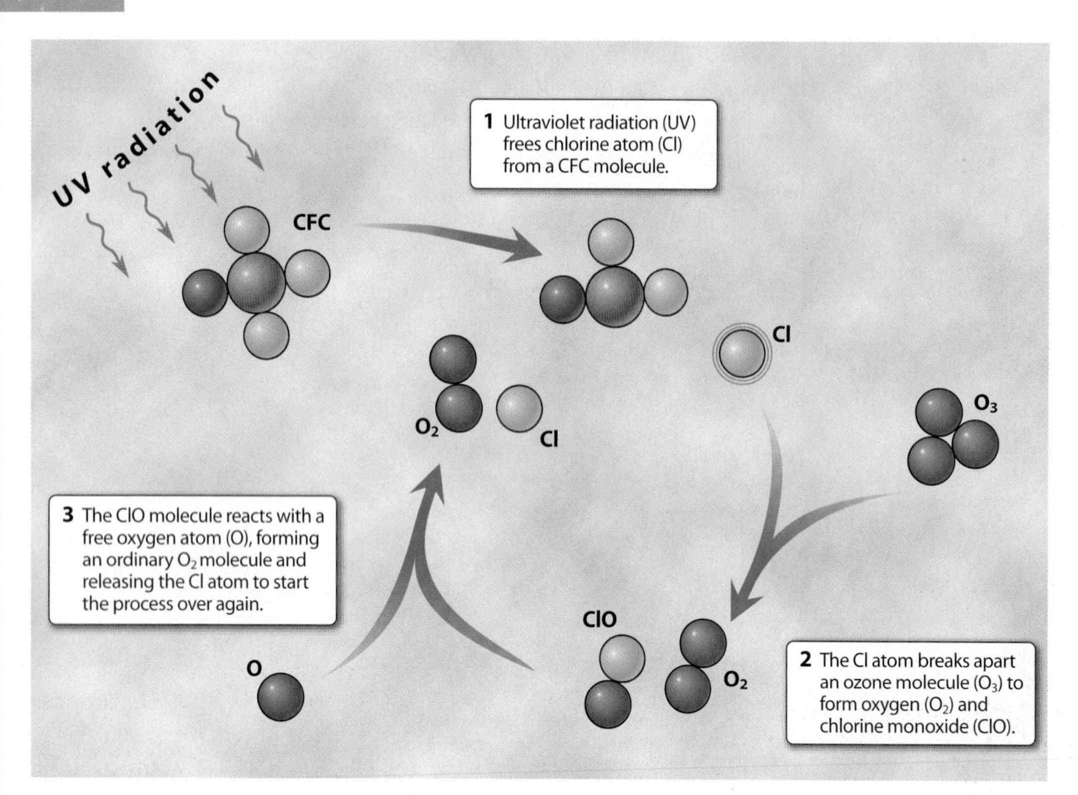

FIGURE 14-15 ◀ How Chlorine Destroys Ozone
CFCs have such a devastating effect on ozone because the reactions involved form a cycle. Once a single chlorine atom is dislodged from a CFC molecule, it can just keep reacting with one ozone molecule after another.

AIR POLLUTION AND THE OZONE LAYER

The ozone layer in the stratosphere absorbs shorter-wavelength ultraviolet radiation in sunlight that would otherwise be lethal to surface life (see Figure 14-4). The origin of the ozone layer goes back to the time when oxygen began to accumulate in the atmosphere, billions of years ago, as described in Section 2.2. Photosynthesis by algae in the world's oceans released oxygen, some of which migrated to high altitudes and reacted with ultraviolet radiation to form ozone. By 500 to 600 million years ago, the ozone layer was developed enough to effectively shield surface life. This allowed life, which previously could live only in the ocean, to evolve and move onto the land. The ozone layer has continued to shield surface life ever since.

Ozone was discovered in 1840, and its role in absorbing ultraviolet radiation in the ozone layer was identified by measurements made in 1879–1881. It wasn't until the 1920s, though, that G. M. B. Dobson developed an instrument that enabled routine, ground-based measurements of atmospheric ozone concentrations. People could then monitor the ozone layer. Early monitoring showed that ozone, which is preferentially formed over tropical regions (where solar radiation is most intense), is carried to higher latitudes by stratospheric winds. (These winds were in fact discovered by monitoring ozone distribution.) It was about this time that ozone's worst enemy was first recognized—chemicals called chlorofluorocarbons (CFCs).

CFCs are hydrocarbon compounds that contain fluorine and chlorine. (Closely related chemicals containing bromine have similar properties.) These nontoxic, nonflammable compounds found many roles in the home and industry because of their stability and safety. Starting in the 1930s, they became widely used as coolants in refrigerators and air conditioners, aerosol propellants, and cleaning solvents for electronic devices. When released to the atmosphere, CFCs can persist unchanged for a long time, some 50 to 200 years. They don't dissolve in rain and get thoroughly mixed in the troposphere. Some make it all the way to the stratosphere and create a problem: They destroy ozone.

In the 1970s, researchers identified the effect of CFCs (and some other chlorine- and bromine-containing chemicals, such as carbon tetrachloride and methyl chloroform) on ozone. When these compounds reach the stratosphere, ultraviolet radiation breaks them down and releases chlorine and bromine atoms that react with ozone, as shown in Figure 14-15. Chlorine and bromine atoms are very effective at destroying ozone—one chlorine atom can be responsible for the demise of 100,000 ozone molecules. After this threat to the ozone layer was recognized, the United States, Canada, Norway, and Sweden banned the use of CFCs in spray cans. But ozone-depleting substances weren't abandoned—their use actually increased in the 1980s. Consumption in 1988 is estimated to have been over a billion kilograms (2.2 billion pounds).

The discovery of a hole in the ozone layer over Antarctica changed everything. The British Antarctic Survey had been monitoring atmospheric ozone levels for decades, and in 1985 reported that October ozone levels were 35% lower than the average October levels during the 1960s. NASA confirmed these results with satellite data. The seasonal depletion of

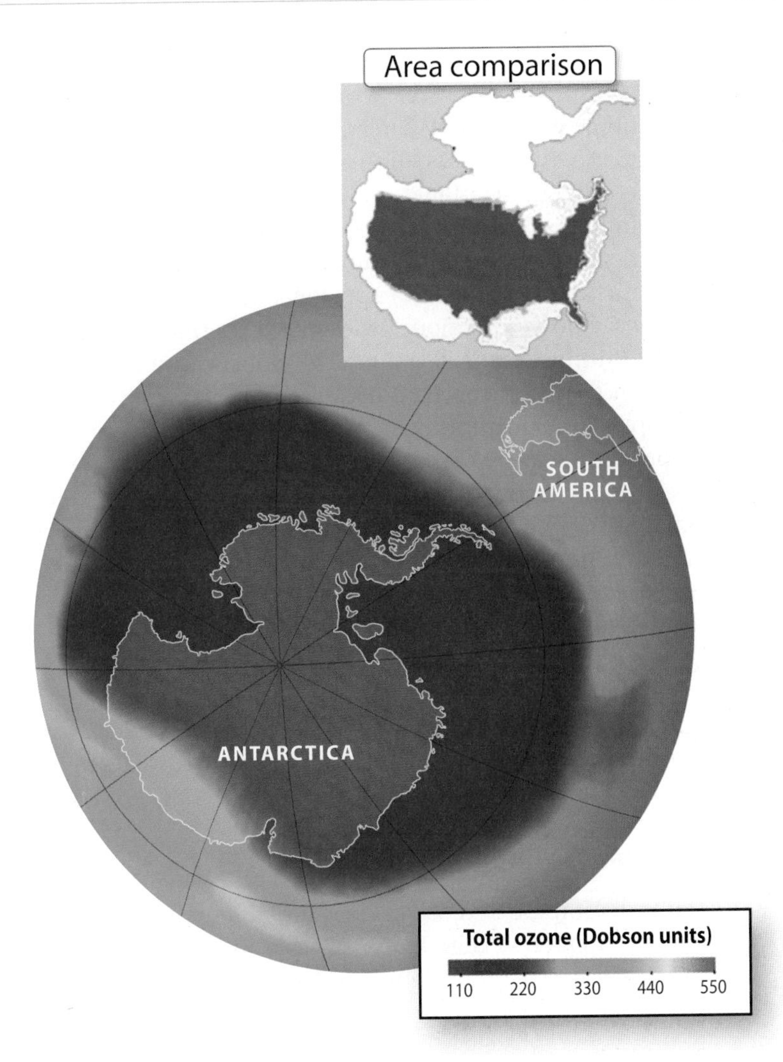

FIGURE 14-16 ▲ The Ozone Hole over Antarctica
The stable cold air over Antarctica and the presence of high stratospheric clouds in the Antarctic spring can lead to ozone depletion over an area about the size of the United States.

The ozone hole ozone over Antarctica is so dramatic—it can cover an area bigger than the United States—that it became known as the ozone hole (Figure 14-16).

This distinctive feature of the ozone layer has expanded since 1976. Airborne measurements confirmed that the ozone-depleting chemicals were compounds containing chlorine and bromine. They are especially effective in destroying ozone over Antarctica because of the very cold air mass that stabilizes over this icebound continent. The cold winter air produces thin, wispy clouds in the stratosphere, like those seen in Figure 14-17. As seasonal solar radiation starts to increase in August, ice crystals in the clouds facilitate reactions between ozone and chlorine (or bromine), and the ozone hole forms. The ozone hole is generally well developed between August and late November, the Antarctic spring.

The ozone hole was clear evidence of the destructive character of CFCs and other compounds on the ozone layer. If this destructive trend were allowed to continue, it could have severe consequences for people and the biosphere. In response, the international community, through the United Nations, adopted the *Montreal Protocol on Substances That Deplete the Ozone Layer* in 1987. The protocol has been revised several times in light of new scientific data, and in 1992 established a strict schedule for complete phaseout of CFC use by 2030. It was signed by over 100 countries representing 95% of global CFC consumption. Trade sanctions were imposed on nations not signing the protocol, and bans were extended to other chemicals, such as methyl chloride and carbon tetrachloride. The 1990 amendments to the U.S. Clean Air Act created measures to protect the ozone layer even stronger than those required by the Montreal Protocol.

FIGURE 14-17 ▲ Polar Stratospheric Clouds
Ice crystals in clouds like these facilitate reactions that destroy ozone.

Because of the long-term stability of CFCs, it will take several decades for seasonal development of the ozone hole to stop. The good news, though, is that the international community got serious about the pollution threats to the ozone layer. Scientists identified the problem and its causes, and decision-makers used scientific data and analysis to guide public policy decisions. The way in which threats to the ozone layer were dealt with is an excellent example of how science can serve people at a global scale.

What are some Earth systems interactions associated with air pollution?

14.3 The Atmosphere and Climate Change

Earth's climate—essentially, the average weather over long periods—is strongly influenced (what climate scientists call *forced*) by several factors. The atmosphere's composition and the amount of solar radiation Earth receives have been fundamental climate controls throughout Earth's history. Other factors, especially those that influence circulation in the oceans and atmosphere, can play significant roles, too.

ATMOSPHERE COMPOSITION AND CLIMATE; GREENHOUSE GASES

The components of the atmosphere that influence climate are water vapor; gases that have both natural and human sources, including carbon dioxide, methane, and nitrous oxide; and air pollutants, including CFCs. These are all greenhouse gases.

Greenhouse gases exert their effects through their special way of interacting with radiation, diagrammed in Figure 14-18. These gases let shorter-wavelength incoming sunlight through to Earth's surface, where it warms the ground, the ocean, and the lower regions of the atmosphere. These in turn reradiate some of their heat energy back at longer wavelengths (mostly infrared) that are absorbed by greenhouse gases, in effect trapping energy in the atmosphere. Part of this trapped energy is, in turn, reradiated back to Earth.

This trapping of heat in the atmosphere is often likened to the effect of glass in a greenhouse—hence the term "greenhouse effect." Greenhouse effect: background information However, a greenhouse is an imprecise analogy. In a greenhouse, the glass allows sunlight to warm the ground and the air above it. The glass keeps the air from escaping and mixing with cooler air, so the interior air warms up. Greenhouse gases in the atmosphere act somewhat differently. They too allow sunlight to warm Earth's surface, but they trap that warmth by absorbing some of the heat energy radiated from Earth's surface and sending it back toward Earth.

This greenhouse effect is very important to the hydrosphere and biosphere. It's what keeps Earth's surface temperature in the range where water can be a solid, a liquid, and a gas; it's the range in which plants, animals, and people can live. It's why Earth is a habitable planet.

The common greenhouse gases vary widely in their effectiveness at heating Earth and the atmosphere. Most greenhouse gases are much more capable of causing warming than carbon dioxide. Over a 20-year period, for example, methane will have 56 times and nitrous oxide 280 times the global warming potential of carbon dioxide. However, because these gases have such low atmospheric concentrations, the role of carbon dioxide is much more significant. Water vapor too is a very potent greenhouse gas, but the amount the atmosphere can hold is dependent on temperature. When people add water to the atmosphere (such as steam from power-generating plants) it is soon removed as rain (water additions, whether natural or man-made, have only one to two week residence times in the atmosphere). On the other hand, atmospheric carbon dioxide concentration, which has varied significantly many times in Earth's history, is now being influenced by people. As we shall see, Earth's climate can significantly change within decades or a few hundred years in response to changes in atmosphere greenhouse gas concentrations.

SOLAR RADIATION AND CLIMATE

The solar radiation Earth receives does much more to influence climate than interact with greenhouse gases in the atmosphere. Solar radiation drives atmosphere and hydrosphere circulation, the circulation that shifts air masses, and water, and energy

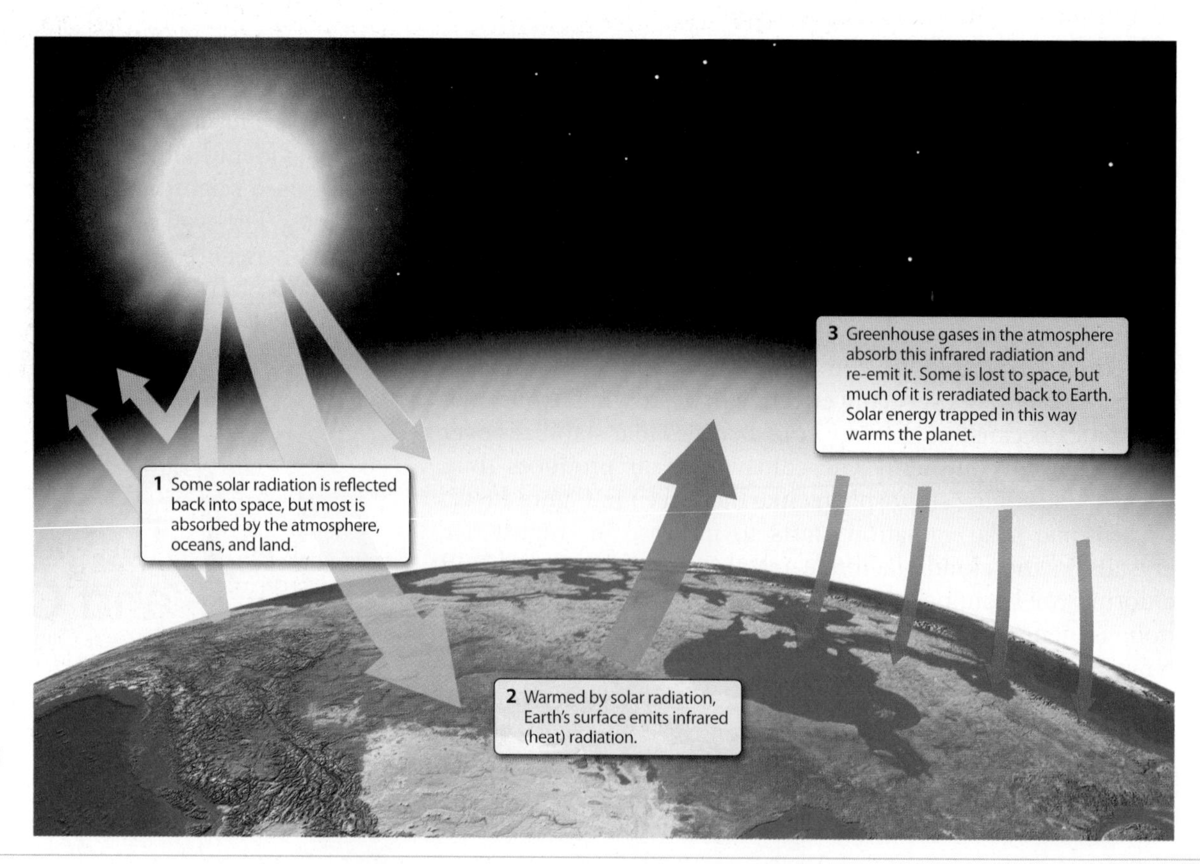

FIGURE 14-18 ▶ The Greenhouse Effect
Solar radiation reflected or reradiated by Earth's surface is absorbed by greenhouse gases and warms the atmosphere. The warmed atmosphere reemits radiation back to Earth, causing surface warming.

around the world. The amount of energy radiated by the Sun (its *luminosity*) has changed through Earth's long history, but in addition, the amount of solar radiation Earth receives depends on its changing orbital positions with respect to the Sun. Changes in the angle of tilt of Earth's axis, variations in the distance between Earth and Sun, and the way Earth wobbles as it orbits the Sun all change the amount of solar radiation Earth receives. These orbital changes occur in cycles ranging in length from tens to hundreds of thousands of years, and can be correlated with the regular climate fluctuations they cause.

Luminosity

When the Earth formed 4.5 billion years ago, the Sun was smaller and dimmer. Scientists estimate that the Sun was 70% less luminous then. The gradual increase in the energy radiated by the Sun has been important to climate evolution through Earth's long history. This gradual increase is continuing and could raise average global temperature by 10° C over the next two billion years, but such long-term change is not a concern today. However, there is a change in energy output from the Sun that does occur over periods of interest to people. The energy emitted by the Sun varies a small amount over 11-year periods called the *sunspot* (or *solar*) *cycle*.

Sunspots are areas of intense magnetic activity, visible on the Sun's surface as relatively dark areas. When sunspot activity increases, the amount of energy the Sun radiates also increases. The change is very small, only about 0.1% of the Sun's total energy output. These small variations were directly detectable only once satellite instruments were in place starting in the 1980s, and they are a continuing focus of research.

Axis tilt

The angle of tilt of Earth's axis is not constant. As Figure 14-19a shows, it gradually shifts back and forth between 22.1 and 24.5 degrees. This shift happens very regularly—once every 41,000 years. Today it is at an angle of 23.5 degrees.

The tilt of Earth's axis as it orbits the Sun determines the seasons. You can see in Figure 14-19b that sunlight striking Earth's surface at an angle must pass through more of the atmosphere and is spread out over a larger surface area. The steeper the angle, the weaker the solar radiation reaching the surface. Thus, the part of the planet tilted toward the Sun receives solar radiation more directly and experiences summer, while the part tilted away receives the Sun's rays more obliquely and experiences winter (Figure 14-19c and d). The amount

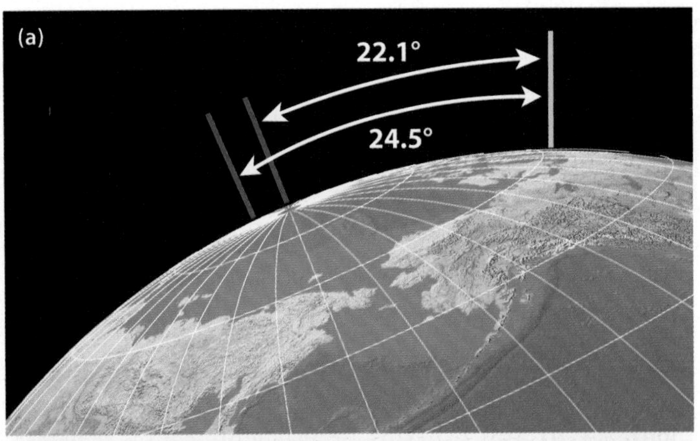

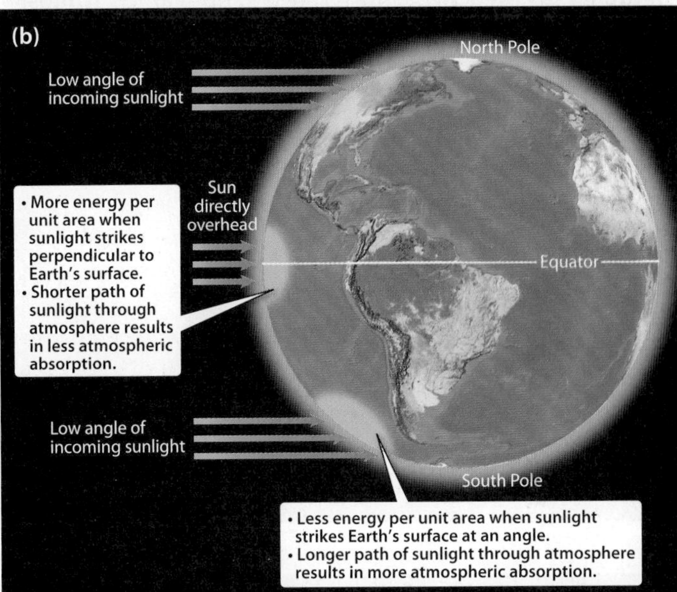

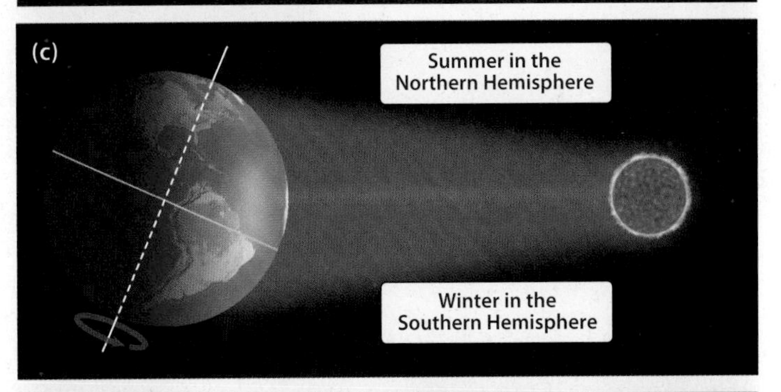

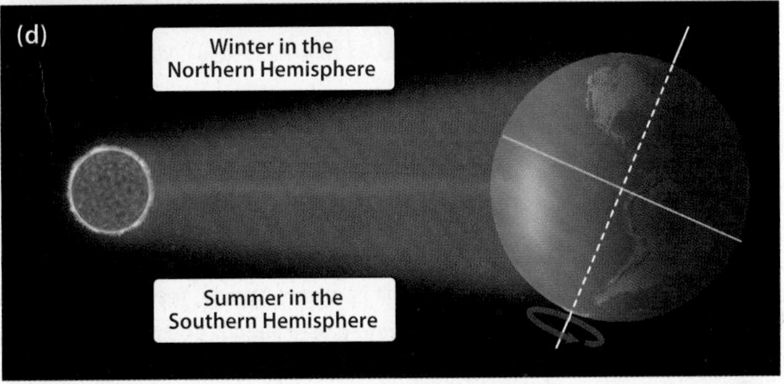

FIGURE 14-19 ▶ How Axis Tilt Affects the Intensity of Solar Radiation
The tilt of Earth's axis of rotation changes from 22.1 to 24.5 degrees and back in a 41,000-year cycle **(a)**. It is because this axis is tilted that we have seasons. Solar radiation reaching Earth's surface is strongest when the Sun is directly overhead and becomes weaker when Sun's rays strike at a more oblique angle **(b)**. In **(c)**, the northern hemisphere is tilted toward the Sun and experiences summer while the southern hemisphere experiences winter. Six months later, when Earth is on the opposite side of its orbit **(d)**, the northern hemisphere is tilted away from the Sun and experiences winter while the southern hemisphere experiences summer.

of sunlight received at low latitudes doesn't change that much from season to season and is not greatly affected by small changes in the tilt of the axis. At higher latitudes, however, even the shift of a few degrees can cause a significant change in the intensity of solar radiation. If the amount of tilt is greater, higher latitudes receive solar radiation more directly in summer than when axis tilt is less. When the angle of tilt is low, the amount of solar radiation reaching polar regions in summer is spread out over a larger area and may not be intense enough to melt the winter's accumulation of snow and ice. These are cold times capable of forming and growing permanent ice sheets. They become times of glacial expansion.

Orbital eccentricity

Earth's orbit varies over time from being perfectly circular to slightly elliptical, as shown (considerably exaggerated) in Figure 14-20. When Earth's orbit is elliptical (as it most often is), it is closer to the Sun at some times than at others. The degree to which the orbit varies from a perfect circle is called its **eccentricity**. For Earth's orbit, the gradual change from a circular to a maximum elliptical shape and back takes about 100,000 years. The principal effect of these cyclic changes is to contribute (along with axis tilt) to variations in the length and severity of the seasons.

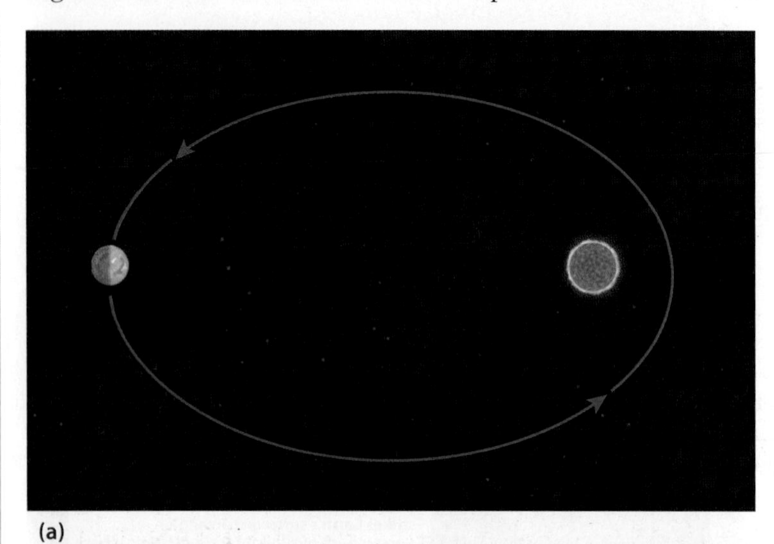

(a)

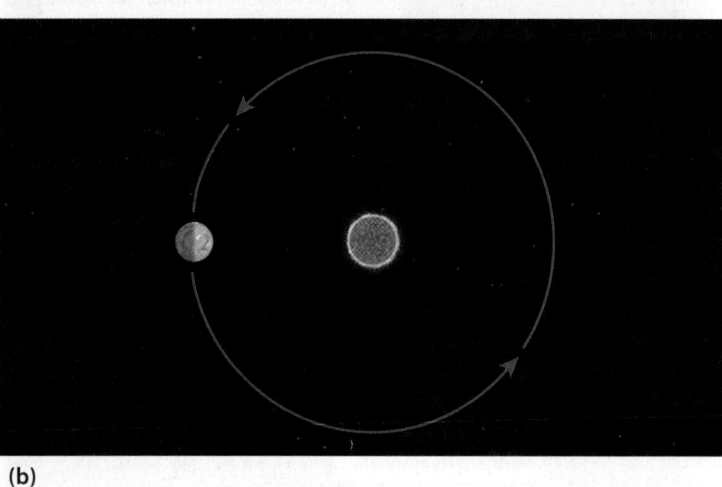

(b)

FIGURE 14-20 ▲ How Orbital Eccentricity Changes the Amount of Solar Radiation Earth Receives

Earth's orbit around the Sun changes from more elliptical **(a)** to less elliptical **(b)** and back about every 100,000 years. (The extent of the elongation in part (a) has been exaggerated.) When the orbit is more elliptical, the Earth moves both closer to and farther from the Sun during a year than it does when the orbit is less elliptical.

Earth's precession

Earth wobbles as it orbits the Sun, much as a top wobbles as it spins. This wobble, called **precession**, gradually rotates the poles in a circle. You can see in Figure 14-21 that precession changes the direction of Earth's tilt about its axis, but not the amount of tilt. Rotation of the poles shifts them slightly toward or away from the Sun. All of this happens slowly—a complete precession cycle takes 22,000 years.

Together, Earth's precession and orbital eccentricity vary Earth's distance from the Sun at different seasons, and therefore the total amount of solar radiation Earth receives. They help change the length and severity of the seasons, but it is axis tilt that causes the seasons.

Milankovitch cycles

The cyclic changes of axis tilt (complete in 41,000 years), orbital eccentricity (complete in about 100,000 years), and precession (complete in 22,000 years) combine to cause changes in the solar radiation received by Earth that strongly influence climate. Their possible relations to climate change were systematically evaluated by Milutin Milankovitch, a Serbian engineer and mathematician.

Beginning in 1911, Milankovitch carefully calculated *by hand* the amount of solar radiation Earth had received at different latitudes as these cycles varied through the past several hundred thousand years. His calculations showed that the amount of solar radiation received by Earth can vary as much as 10% from the average. He proposed that cyclic changes in axis tilt, orbital eccentricity, and precession cycles, now called Milankovitch cycles in his honor, led to cycles of colder climate (**glacial maximums**—periods when ice sheets reached their maximum extent) and warmer climate (*interglacial periods*) within the last ice age. Milankovitch died in 1958 but scientists have since demonstrated that Milankovitch cycles have influenced global climate as he first proposed.

TECTONIC PROCESSES AND CLIMATE

Tectonic processes determine the size and distribution of continents, mountain ranges, and volcanoes. These features of the geosphere in turn affect circulation patterns in the atmosphere and hydrosphere as well as the concentration of greenhouse gases in the atmosphere—all factors that have a powerful influence on climate.

Continent size and distribution

Continents and other landmasses cover about 30% of Earth's surface. During Earth's history, continents have been amalga-

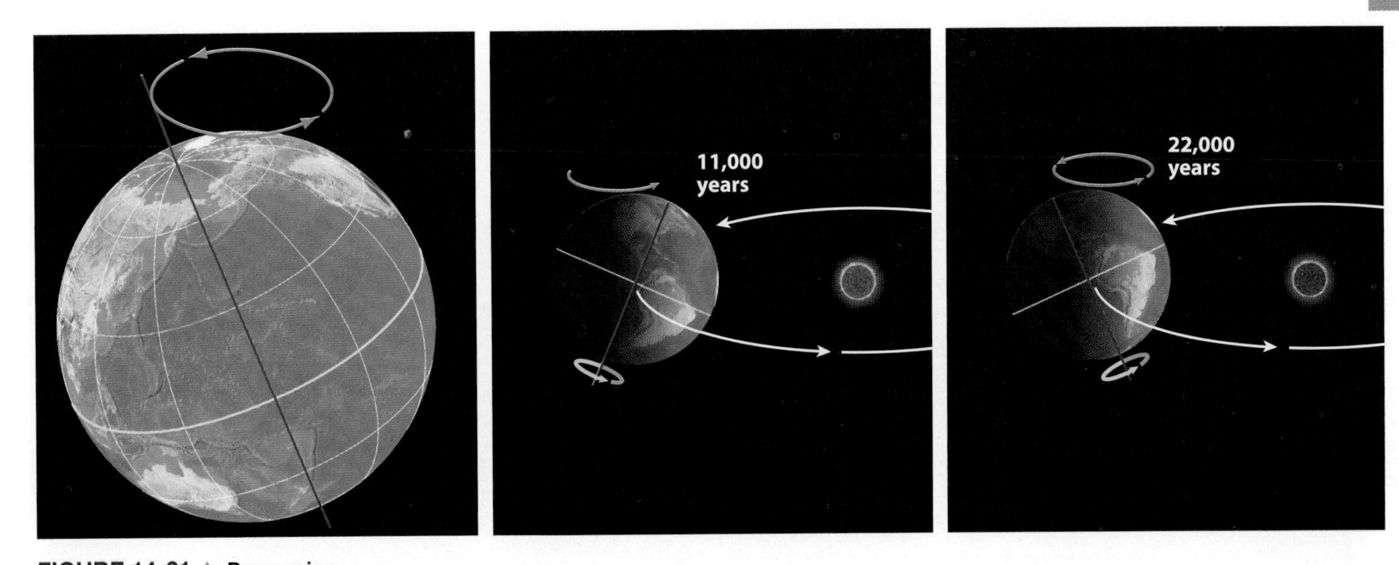

FIGURE 14-21 ▲ Precession
The Earth makes one complete wobble as it rotates about its axis every 22,000 years.

mated into supercontinents such as Pangaea (Section 3.1), and at other times (such as today) have split into several large and small landmasses. The size and distribution of these landmasses, and their intervening oceans, directly affect circulation patterns in the atmosphere and hydrosphere that influence climate. These factors can also affect the concentration of carbon dioxide in the atmosphere through their influence on the chemical weathering of silicate minerals in the geosphere.

Recall the discussion of the carbon cycle in Chapter 1 (see Figure 1-10). Geosphere volcanism transfers CO_2 to the atmosphere; the atmosphere transfers CO_2 to the oceans; and CO_2 accumulates in ocean sediments (in the form of calcium carbonate and organic compounds). However, the atmosphere transfers CO_2 to the geosphere as well. This transfer accompanies weathering processes, as described in Chapter 4. Rain naturally contains dissolved CO_2, which reacts with water to give carbonic acid (H_2CO_3). When this slightly acidic rain falls onto the land it becomes involved in chemical-weathering processes, which break down silicate minerals in the Earth's crust (Chapter 4). By consuming carbonic acid from the atmosphere, these reactions work to decrease atmospheric CO_2 concentrations.

The extent of silicate weathering is related to the distribution and character of continental crust. Silicate weathering is likely to be more extensive under the following conditions:

- If plate tectonics splits continents into smaller fragments surrounded by oceans, creating wetter conditions;
- If the continental pieces are preferentially located at lower latitudes, creating warmer conditions;
- If mountain belts form here and there, increasing crustal exposure and erosion. |Raymo – uplift hypothesis

Mountain ranges

Plate tectonics, especially convergent plate processes, commonly create large mountain ranges. You learned about these and their role in influencing atmosphere circulation and precipitation patterns in Section 3.4. Two key examples today are the Himalayas in Asia and the Andes Mountains in South America. These mountains are direct influences on atmosphere circulation and precipitation over large parts of their host continents. And, as described above, extensive mountains increase exposure of silicate minerals to chemical weathering that decreases CO_2 concentrations in the atmosphere.

Volcanoes

You learned about how volcanoes release gases to the atmosphere and affect global climate in Chapter 3 (Section 3.4). Volcanic gases such as sulfur dioxide that reach high altitudes and form aerosols can cool Earth for a few years. Some volcanic eruptions, though, if particularly voluminous and sustained, may release enough CO_2 to the atmosphere to affect global climate for long periods of time, as we noted in Section 6.3. These processes may have affected global climate at different times in Earth's history but their influence takes millions of years to be felt.

14.4 History of Climate Change

Earth's geologic record shows evidence of past climates both colder and warmer than today. These past climates, summarized in Figure 14-22 on p. 444, are related to complicated interactions involving geosphere processes, greenhouse gas cycles, the distribution and size of continents and oceans, and variations in the amount of solar radiation Earth receives. The role of solar radiation is not obvious over the longer-term periods (million of years) of ancient climate episodes, but it does influence the waxing and waning of cold and warm times within specific periods. As you will see, variation in the amount of solar radiation Earth receives is a significant control on the global climates people have experienced.

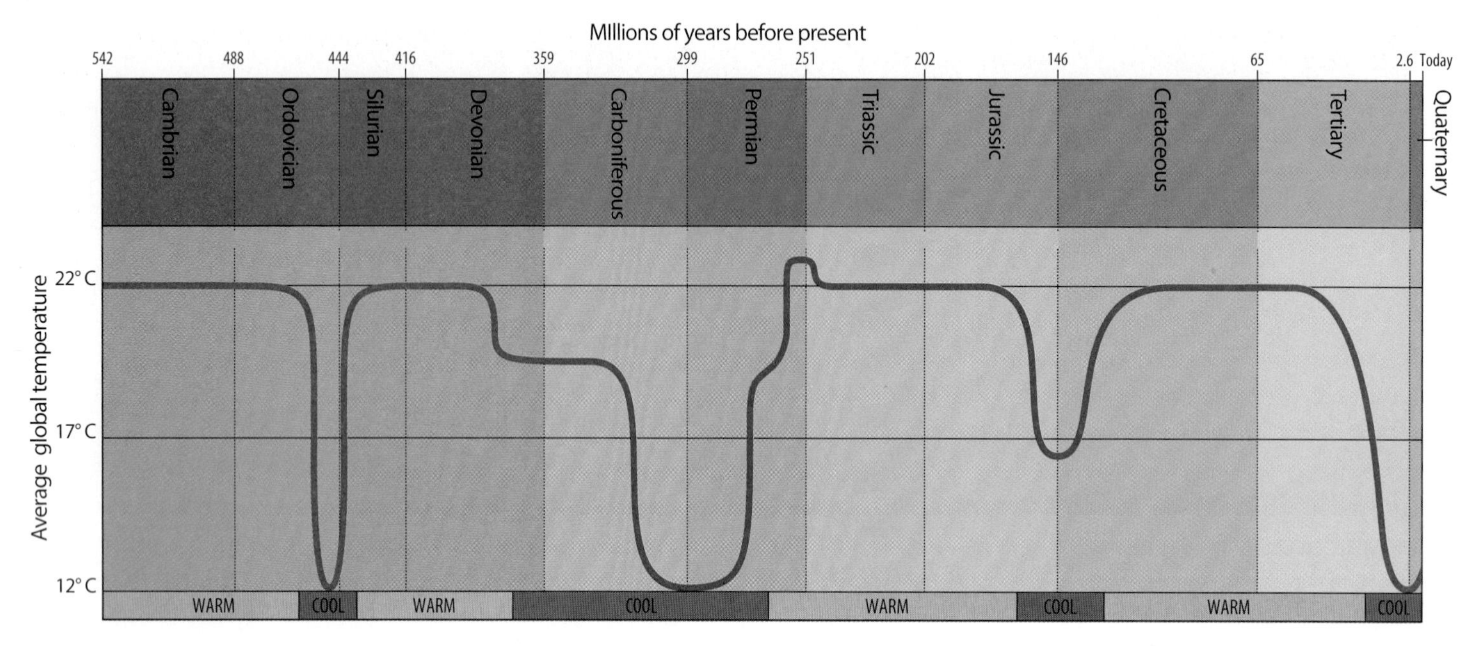

FIGURE 14-22 ▲ Average Global Temperature during the Phanerozoic
Over the last 540 million years, Earth was mostly warmer than it is now.

PALEOCLIMATOLOGY—STUDYING PAST CLIMATES

Paleoclimatologists—scientists who study past climates—use several approaches to decipher the history of climate change. Some of the key tools they use include the character of sediments and sedimentary rocks, especially those deposited in the oceans; the character and distribution of fossils; variation of oxygen isotope concentrations in seawater; the composition of atmospheric gas trapped in Antarctic and Greenland ice; and sea level history. Paleoclimatologists have identified many ancient climate changes but they have been particularly successful at identifying global climate change during the time humans have been on Earth, the last 2 to 3 million years.

Sedimentary records

Sediments record millions of years of Earth history. They can be exposed on land or collected from the seafloor by drilling and coring. The types of sedimentary rocks present can provide evidence of the general environmental settings in which they formed. Evaporites (salt-rich sediments), for example, indicate arid conditions; coal seams commonly indicate warm, moist conditions; and tillite indicates cold, glacial conditions.

Tillite is rock formed from an unstructured mix of mud and rock fragments, like those shown in Figure 14-23a. Tillites

FIGURE 14-23 ▼ Tillite Is Evidence of Ancient Glaciations
(a) Outcrop of Precambrian tillite in Nambia. **(b)** A polished and striated boulder in Precambrian tillite from Mauritania.

(a)

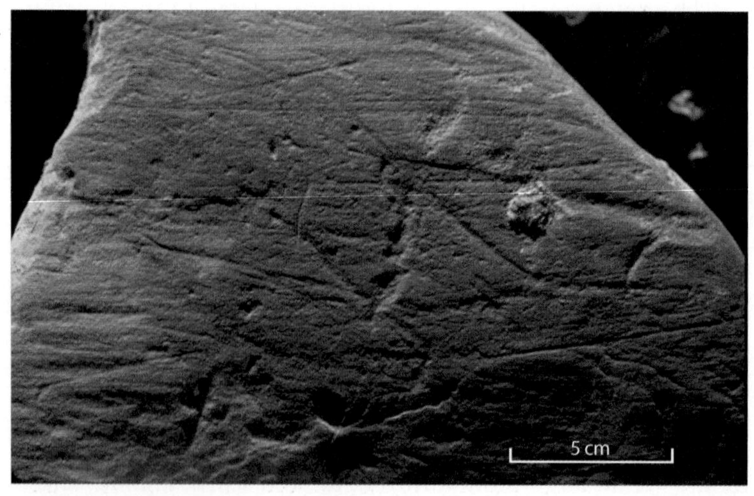

(b)

(a)

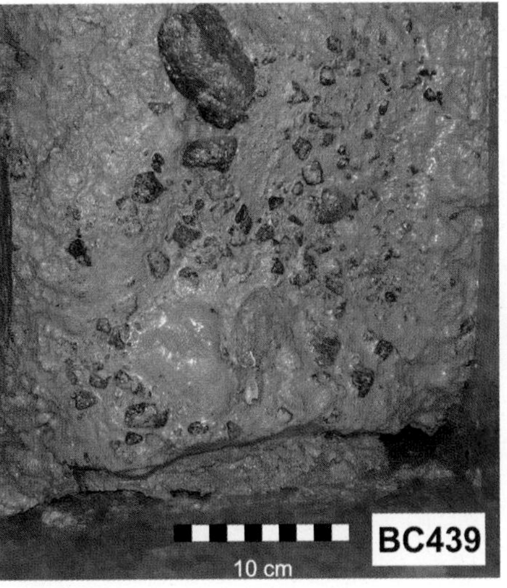

(b)

FIGURE 14-24 ◀
Icebergs and Dropstones
Rocks entrained in icebergs like this one are floated out to sea **(a)**. When the iceberg melts, the rocks sink to the seafloor and become included as dropstones in ocean sediments **(b)**.

are broken and ground up geosphere material that has been mixed, moved, and deposited by glacial ice. The rock fragments in tillites commonly have polished and striated surfaces caused by ice grinding them against each other and bedrock (Figure 14-23b).

Cores of deep-sea sediments are especially helpful in identifying when glaciers and ice sheets sent icebergs off to sea. Icebergs like the one shown in Figure 14-23a commonly contain rock debris that was incorporated as the ice ground its way across the land. As icebergs melt, they drop their included continental rock debris onto the seafloor. Because seafloor sediments are commonly mud-rich, even minor amounts of coarser, angular, and perhaps polished and striated continental rock debris can be very obvious, as shown in Figure 14-24b. Times when icebergs were abundant are times when ice sheets and glaciers were well developed.

Fossils

The remains of organisms—fossils—can help scientists determine environmental conditions, too. There are billions upon billions of floating animals and plants (zooplankton and algae) in the oceans. When they die, their skeletons sink to the ocean floor and become incorporated in sediments. Because different species thrive at different water temperatures, their presence and abundance can be indicators of warm or cold water conditions.

Plants are very adaptive to climate change and distinctive plant fossils, such as palms or conifers, can be used to indicate past climates in a region. Fossil plant spores and pollen have become particularly useful in studying ancient climate. Wind and surface waters carry these tough and distinctive parts of plants to places where they can be incorporated in sediments. Because plants closely adapt to climate conditions, pollen has been particularly good for distinguishing between warmer and colder or dryer and wetter environmental conditions. The distribution of distinctive, climate-adapted plants indicates how extensive particular climate conditions were. Because spores and pollen are easily transported before they are deposited, however, care must be taken in inferring plant distribution from them.

Oxygen isotopes

As we have seen, all atoms of an element don't necessarily have the same number of neutrons in the nucleus. Recall from Section 2.5 that atoms of the same element with different numbers of neutrons (and therefore different atomic masses) are called *isotopes*. An example that's important to paleoclimate studies is oxygen. Almost all oxygen has an atomic mass of 16—these are ^{16}O, or O-16 atoms. However, two other rare but stable oxygen isotopes have atomic masses of 17 and 18—they have one or two more neutrons in the nucleus than the most abundant oxygen atoms. Because O-18 is five times more abundant than O-17, it is used, along with O-16, for paleoclimate studies.

Because O-18 is slightly heavier than O-16, water containing this isotope does not evaporate as easily as water containing the more common O-16. Thus, when seawater evaporates, the escaping water vapor is slightly enriched in O-16 and the seawater left behind is slightly enriched in O-18, as Figure 14-25a on p. 446 shows. During cold periods, O-16-enriched water vapor is constantly being added to ice sheets through precipitation or direct condensation. This process is enhanced by the fact that as the water vapor moves (generally to higher latitudes), the O-18 it contains becomes incorporated into rain more easily than does O-16. This makes the remaining water vapor moving towards the poles even richer in O-16. By the time it gets to high latitudes where it can be added to ice accumulations, its concentration of O-16 is distinctly elevated. The continual loss of O-16-enriched water vapor to ice sheets gradually increases the O-18 to O-16 ratio in seawater, as diagrammed in Figure 14-25b.

Marine organisms incorporate this oxygen in their shells and skeletons, thus locking in a sample of the oxygen and its isotopic ratios from the time they grew. In this way, they

Warm Climate
Water vapor enriched in O-16
Ratio of O-18 to O-16 is unchanged
Higher sea level
(a)

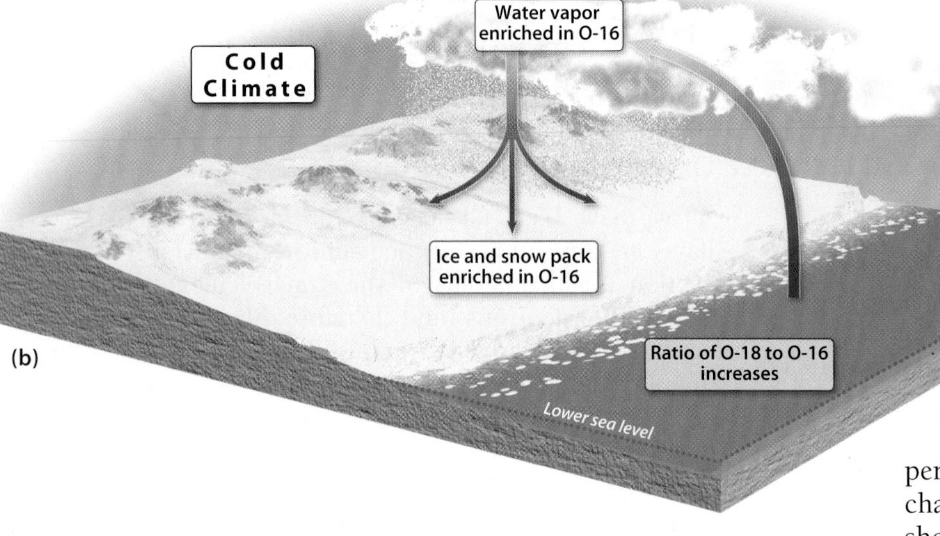

FIGURE 14-25 ◀ How Oxygen-16 Becomes More Concentrated in Glacial Ice
(a) Because O-16 is lighter than O-18, water vapor evaporated from the oceans is slightly enriched in O-16. Because most of this water eventually finds its way back to the oceans in rain and runoff, however, the ratio of the two isotopes in seawater is largely unaffected. **(b)** During times of glaciation, water vapor enriched in O-16 becomes trapped in ice accumulations and doesn't return to the oceans. As a result, the relative concentration of O-18 in seawater increases.

record times of expanding ice sheets (higher O-18 to O-16 ratios) and contracting ice sheets (decreasing O-18 to O-16 ratios)—global cold and warm periods. The history of O-18 to O-16 ratio changes in seawater for the last 540 million years, shown in Figure 14-26, has been determined from carbonate minerals deposited from seawater, including those in fossil shells and skeletons.

The water that is incorporated in ice sheets provides another sample of oxygen isotopes. As the ice sheets grow, they accumulate water evaporated from the oceans during cold periods—water enriched in O-16 compared to the seawater from which it came (see Figure 14-25). The amount of enrichment is less during warm periods, so the fluctuating O-18 to O-16 ratio in samples of ancient water from ice sheets records changes between warm and cold periods too.

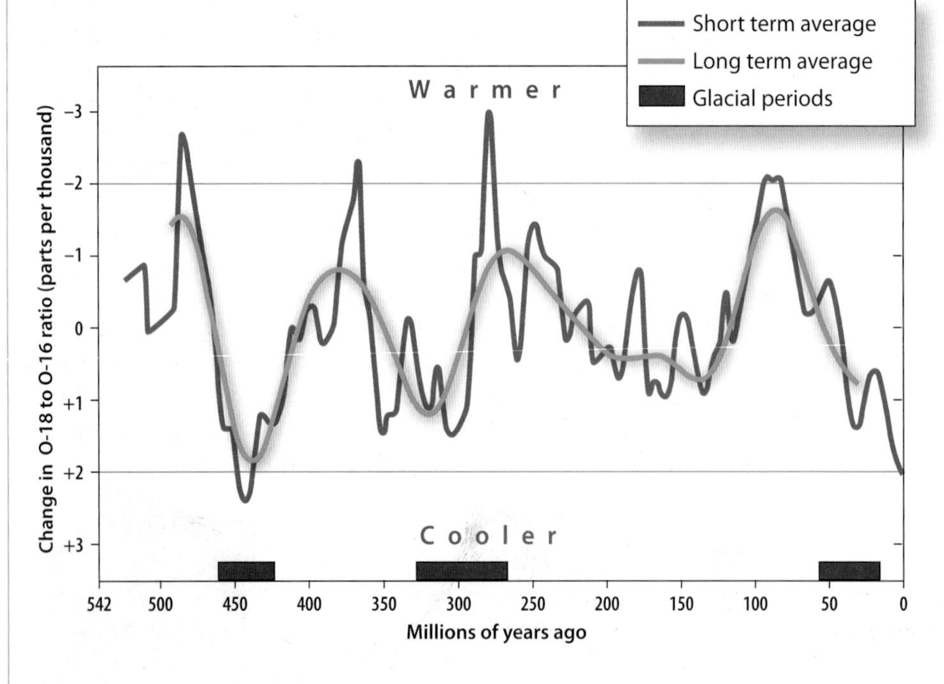

FIGURE 14-26 ◀ The Phanerozoic Oxygen Isotope Record of Seawater
The variation in the ratio of O-18 to O-16 is determined from measurements of oxygen isotopes in fossil marine shells and skeletons. A greater O-18 to O-16 ratio indicates colder periods (see Figure 14-25).

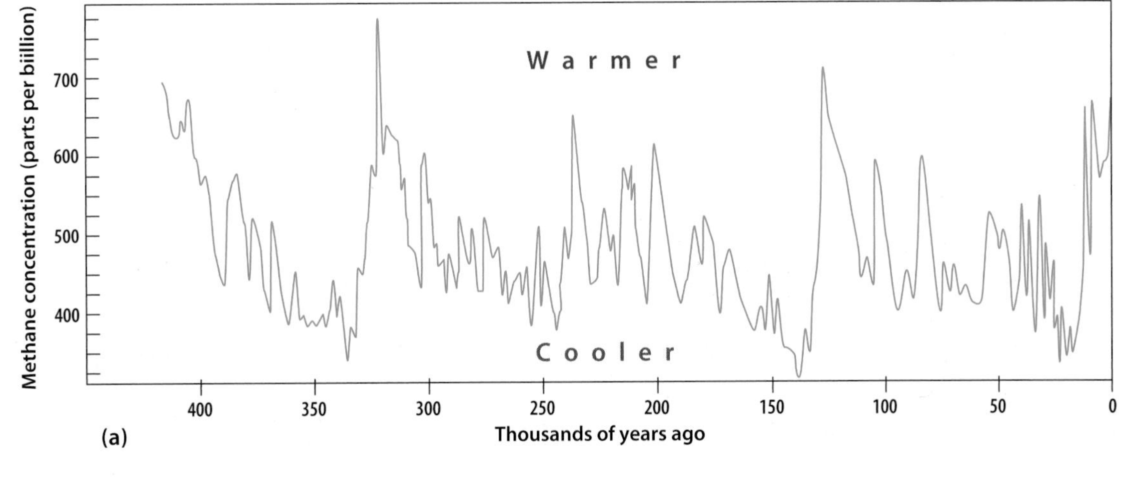

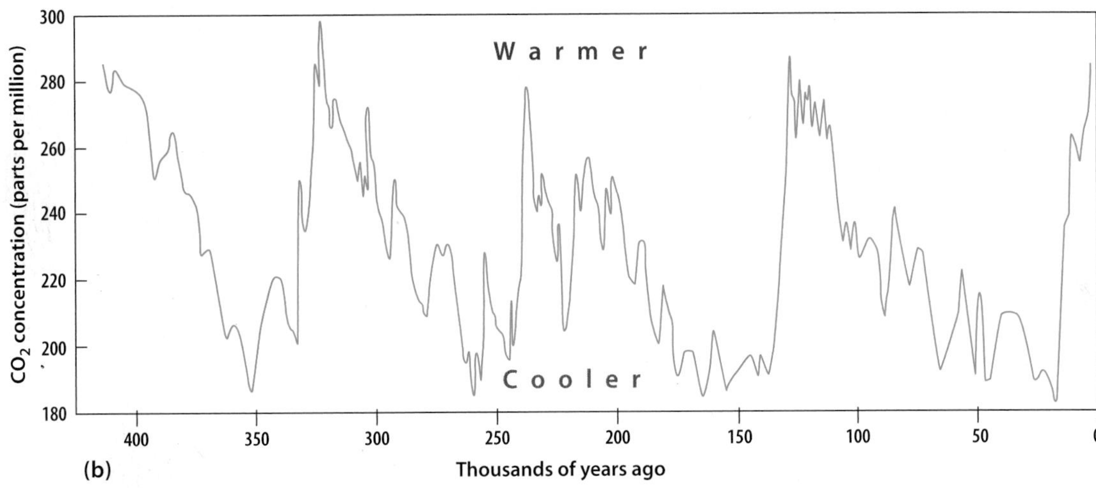

FIGURE 14-27 ◀ Atmospheric Gases Trapped in Ice at Vostok Station, Antarctica

Higher concentrations of methane **(a)** and carbon dioxide **(b)** in air bubbles trapped in ice cores indicate warmer interglacial times. Earth entered a warming interglacial period 20,000 years ago.

Atmosphere samples in ice cores

The ice sheets of Greenland and Antarctica preserve annual precipitation additions that go back hundreds of thousands of years. Air is included in the snow that accumulates on the ice sheets. As the snow becomes compressed under the weight of new snow, it turns into ice. The enclosed air becomes trapped in the ice as bubbles. Ice cores provide samples containing the bubbles of ancient air. Scientists can analyze these bubbles for their content of greenhouse gases and other constituents such as sulfate (ultimately derived from SO_2 emissions somewhere). This is the key way in which prehistoric atmospheric concentrations of methane (CH_4) and carbon dioxide (CO_2), like those recorded in Figure 14-27, have been determined. Ice-core analysis has been a very valuable tool for paleoclimate research.

Sea level history

Have you heard about how global warming could make sea level rise? It certainly has in the past. Many times, the shift from cold to warm climate marked a change from expanding permanent ice sheets to melting ice sheets. In Chapter 10 you learned about how much freshwater is stored in today's ice sheets. The volume is huge, but nothing compared to what it was just 20,000 years ago. Major cold periods make major continental-scale ice sheets, which trap tremendous volumes of water evaporated from the oceans. The net effect is a lowering of sea level. When the ice sheets were at their maximum 20,000 years ago, as shown in Figure 14-28a on p. 448, sea level was some 110 to 125 meters (360 to 410 ft) lower than it is today.

The opposite happens during warm periods. The great ice sheets melt and send tremendous amounts of water back to the oceans, and sea level rises as a result. Studies of continental margins and oceanic islands have identified many episodes of sea level rise and fall. Not all these sea level changes are related to shifts between warm and cold climates, but as our recent history shows (Figure 14-28b), when these types of climate changes occur, sea levels change significantly as well.

What are some Earth systems interactions that produce evidence of ancient glaciations?

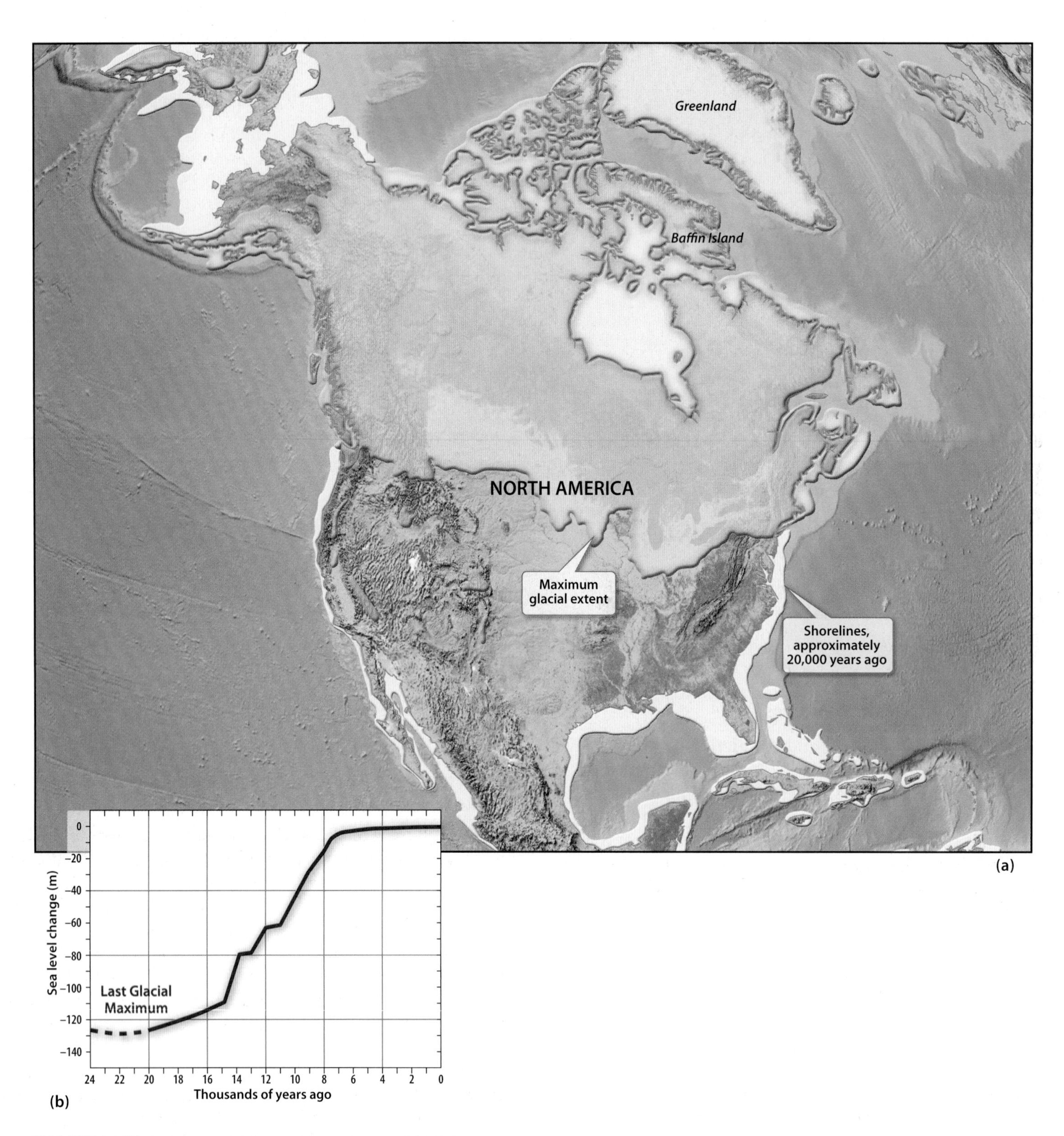

FIGURE 14-28 ▲ Sea Level Change During the Last 20,000 Years

(a) The maximum extent of ice and the location of sea level in North America 20,000 years ago. **(b)** Measurements from many locations around the world show how sea level changed from 20,000 years ago to today. Sea level rose rapidly from 20,000 years ago to about 7000 years ago.

PRECAMBRIAN ATMOSPHERE COMPOSITION AND CLIMATE

Do you remember from Chapter 2 how the composition of the atmosphere has evolved over the last 4.5 billion years? Hydrogen and helium were probably abundant in Earth's first atmosphere, but Earth's gravity couldn't keep these light elements from escaping into space. Earth's second atmosphere was probably rich in CO_2 and water vapor. These greenhouse gases, and perhaps others like methane, must have made Earth's climate very warm, even though the Sun was 20 to 30% less luminous then. Surface temperatures may have reached 70° C (158° F) and glaciers and ice sheets could not have formed.

Once oceans developed and photosynthetic organisms evolved about 3.8 billion years ago, the atmosphere's composition began changing. Carbon dioxide was removed from the atmosphere as it was consumed by photosynthesis, dissolved in the oceans, and incorporated into sediments. The period from 3.8 to 0.54 billion years ago was generally warm, but there must have been colder periods as well, as there is evidence in the geologic record for episodes of glaciation during this time.

Tillites and other glacially-related deposits identify major cold periods during the time of the second atmosphere, the Precambrian (Figure 14-29). The glacial deposits are widely scattered on the continents leading some geologists to conclude that the ancient ice sheets of these times extensively covered both land and ocean, even at low latitudes. Such extensive ice cover and the cold global climate it requires has been called "snowball Earth." Snowball Earth But how extensive the ice cover was during these times is still the subject of investigation.

If landmasses were small, widely distributed, and mountainous, silicate weathering could have been a cause of lowered atmospheric CO_2 concentrations at times of ancient Precambrian glaciations. Other factors, including a decrease in atmospheric methane concentrations—caused by reactions with the oxygen that was added to the atmosphere by early life—could also have played a role.

However, the farther back in Earth history we investigate, the more challenging it becomes to understand the forces affecting climate. Determining the cause and extent of Precambrian glaciations is a work in progress. Whatever the cause, the key point is that ancient shifts between warm and cold climates occurred even 2.4 billion years ago.

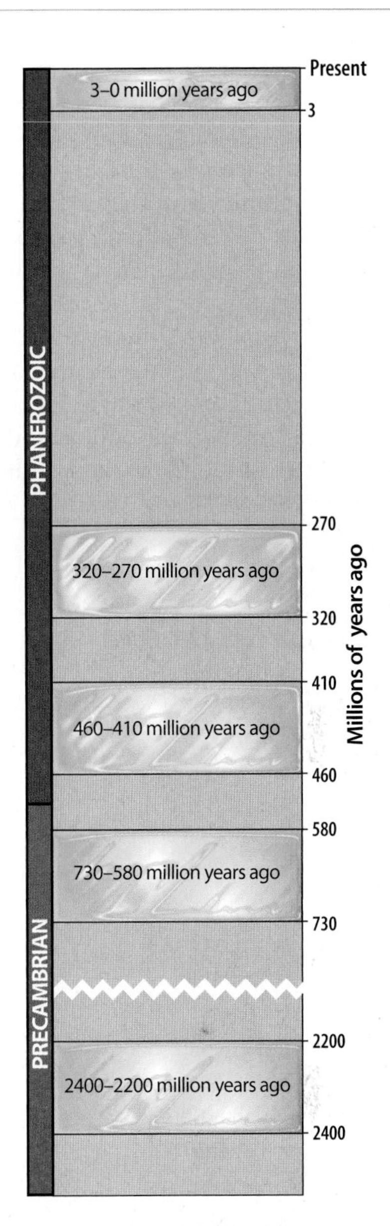

FIGURE 14-29 ▲ Major Ice Ages in Earth's History

PHANEROZOIC ATMOSPHERE AND CLIMATES

Factors such as silicate weathering, ultimately tied to plate tectonic processes, are long-term influences on global climate that exert their effects over millions of years. They not only influence the geosphere's role in the carbon cycle but can also change atmospheric and oceanic circulation patterns that help control climate. Such long-term factors are clearly involved in developing cold climates during the Phanerozoic Eon—the last 540 million years.

By 540 million years ago, oxygen levels were sufficient for the ozone layer to form and life to live on land. This was the time of Earth's third atmosphere (Chapter 2), whose composition was like today's except for variations in the concentration of minor components, including greenhouse gases. These fluctuated in ways sufficient to contribute to significant changes in climate. For example, this period saw atmospheric CO_2 vary from near 10,000 parts per million (ppm) at the beginning to as low as 180 ppm at times of major glaciations.

During the last 540 million years, climates have varied from extensively tropical, with average temperatures several degrees centigrade higher than today, to cold periods when large continental glaciations occurred (see Figure 14-29). The cold periods from 410 to 460 million years ago (Silurian and Ordovician) and 270 to 320 million years ago (Permian and Carboniferous) were times when supercontinents were located over the South

Pole. The extent of ice coverage from 270 to 320 million years ago fluctuated significantly and reached to within 30° of the equator. This period is considered the coldest during the last 540 million years, the time of the third atmosphere. Long-term factors tied to plate tectonics and the carbon cycle must have been major influences on these cold periods. They surely controlled global cooling and ice age development during the last 50 million years.

You can see from Figure 14-22 that the last 50 million years was a time of gradual global cooling, culminating in a true ice age that started about 3 million years ago. The movement of the continents during this time is well understood. Antarctica has been centered over the South Pole during this time and ice sheets developed there by 34 million years ago. Earth's other tectonic plates moved, changing ocean and atmosphere circulation patterns. Some of these changes included:

- Seaways that allowed equatorial flow and global distribution of warm waters were blocked by converging continents.
- Seaways were opened around Antarctica that led to circumpolar ocean currents and isolation of this continent from flows of warm equatorial water.
- Convergent plate boundaries created mountain belts—for example, the collision of India with Eurasia created the tremendous Himalaya mountains. Mountain formation led to the increased removal of atmospheric CO_2 by silicate weathering processes.

What You Can Do

Investigate Plate Tectonics and Climate Change

Earth's climate history during the last 540 million years is wonderfully summarized on global maps by the Paleomap Project. Climate history The global maps overlay climate information from the geologic record on the reconstructed location of the continents for specific geologic times. You can investigate any time of interest but click through those for the last 50 million years, the Early Eocene, Late Eocene, Oligocene, and Miocene epochs. This map series will help you visualize the closing of equatorial seaways, the opening of the seaway around the South Pole seaway, the collision of India with Eurasia, and how climate changed as a result. Don't forget to check out the many plate tectonic animations too!

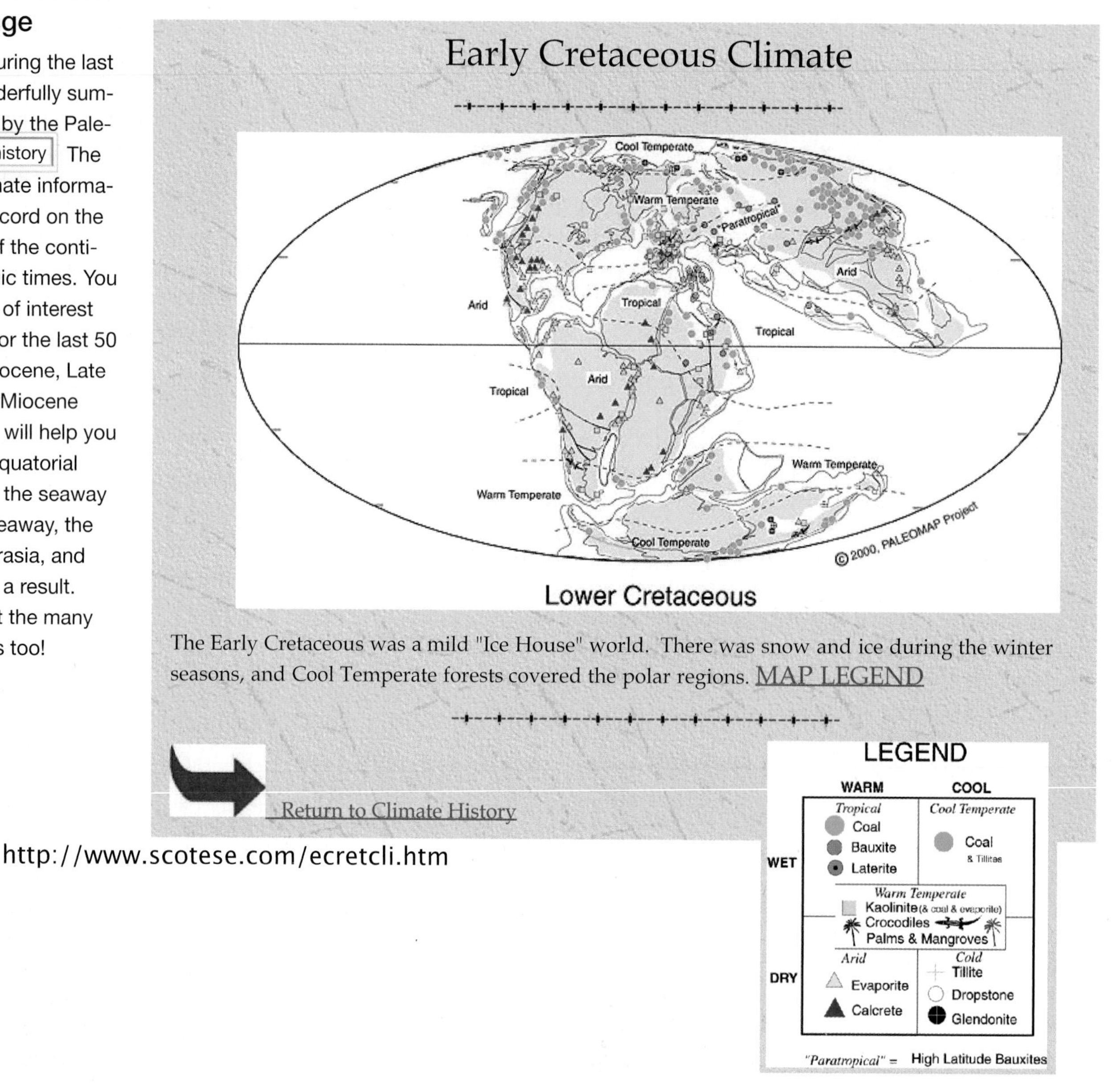

http://www.scotese.com/ecretcli.htm

THE LAST FEW MILLION YEARS

Our understanding of global climate gets better and better as we examine more recent changes. Our direct ancestors appeared about 2.5 million years ago, although as we saw in Chapter 2, fully modern humans didn't evolve until quite a bit later. Climate has changed very significantly through the past 2.5 million years.

Paleoclimatology has been very good at deciphering climate changes through this period of Earth history. The Antarctic ice sheet was extensive by 14 million years ago, and has continued as a permanent continental-scale ice accumulation ever since. However, because the north polar region is an ocean and not a continent, it wasn't until 2.75 million years ago that the climate became cold enough for large ice sheets to form at high latitudes of the northern hemisphere. These northern hemisphere ice sheets didn't become permanent, even though the climate continued to cool.

Deep-sea sediment cores have revealed that northern hemisphere ice sheets formed and melted some 40 to 50 times between 2.75 million years ago and the present. This remarkable and detailed record enables close comparison between the variations in solar radiation and the history of growing and melting northern hemisphere ice sheets. There is a clear tie between higher and lower amounts of solar radiation and northern hemisphere glacial fluctuations. As Figure 14-30 shows, from 2.7 to 0.9 million years ago, glacial cycles came and went every 41,000 years (as axis tilt varied) or sometimes within a 22,000-year interval (along with Earth's precession cycle). From 0.9 million years on, however, the glacial episodes came and went every 100,000 years or so. As climate gradually cooled during the last few million years, it became harder to completely melt northern ice sheets during cycles of higher solar radiation. Starting 0.9 million years ago, it took all the orbital influences working together—which could happen only every 100,000 years—to melt the northern hemisphere ice sheets.

As you can see in Figure 14-27, atmospheric CO_2 variations, as measured in ice cores, coincided with these cycles during the last 400,000 years. Higher CO_2 concentrations developed during the interglacial warm periods. In the fluctuations between warmer interglacial and colder glacial times, atmospheric CO_2 changes are thought to be a consequence of climate change, not a cause or *driver.* Variation in the solar radiation received by Earth was the driver.

It was during the last 3 million years of general cooling and cycling of glacial conditions that evolution brought modern humans into existence. Our ancestors adapted to some pretty harsh and extensive cold climates. Over the last 900,000 years, ice-free conditions like those of today were present only about 10% of the time. The last glacial maximum occurred about 20,000 years ago, when the great northern ice sheet extended across most of Canada and into what is now the United States (see Figure 14-28). Forests retreated southward, deserts expanded, and tundra developed across northern Europe. It was a cold, windy, and dusty time in the central part of the United States.

The great northern ice sheets didn't stay. A cycle of increasing solar radiation started melting the ice. Increasing solar radiation continued to about 10,000 years ago when the cycle reversed and the solar radiation received by Earth started decreasing. Sea level changes during the last 20,000 years (see Figure 14-28) closely follow this cycle, except for one thing. Solar radiation cycles, cycles that have controlled dozens of shifting warm and cold climates for over 2 million years, should have started a shift to a cold, glacial climate by 5000 years ago. Significant melting stopped about when it should have (7000 years ago), but sea level has not fallen and ice sheets have not expanded since then as predicted by solar radiation cycles. One possibility is that people and their greenhouse gas emissions started influencing global climate during this period, and a cold glacial period was at least postponed as a result.

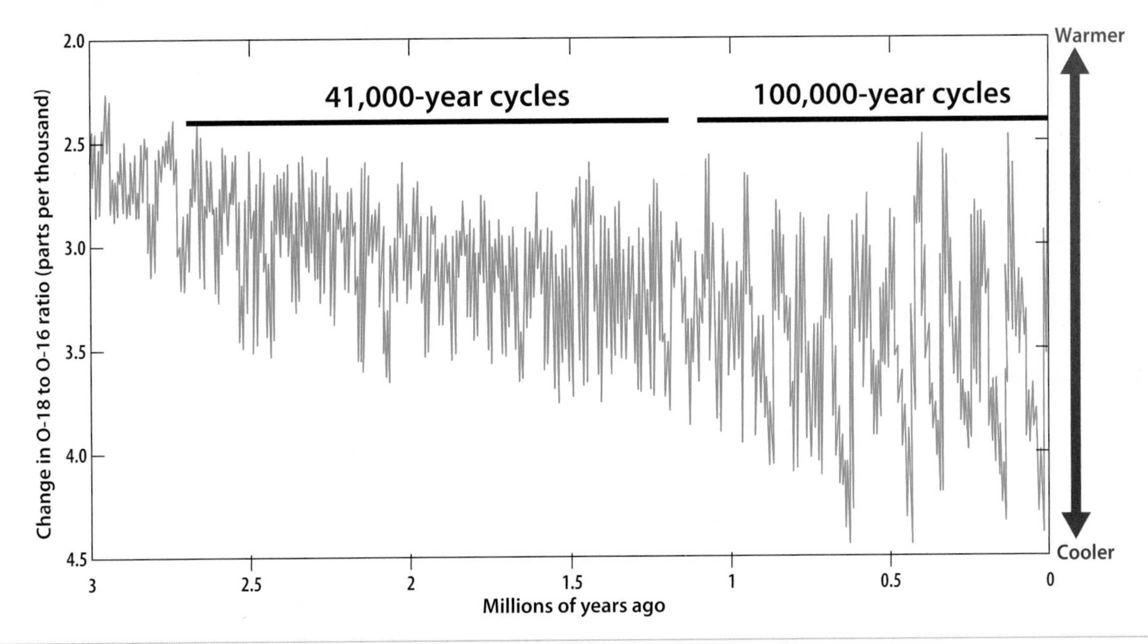

FIGURE 14-30 ◀ Shifts between Warm and Cold Periods during the Last 3 Million Years
The plots are based on deep-sea sediment oxygen isotope data. Up to 0.9 million years ago, shifts from warm to cold climates occurred every 41,000 years (the axis tilt cycle). Since then the shifts have been every 100,000 years.

14.5 People and Climate Change

People have become an influence on global climate. We are burning up so much fossil fuel that we are changing the concentration of carbon dioxide in the atmosphere. In the last 150 years, since the beginning of the industrial age, the concentration of atmospheric CO_2 has increased from 280 to 380 parts per million, enough to cause global warming. This recent change is an exceptional rate of increase, but people may have influenced atmospheric greenhouse gas concentrations for thousands of years.

PEOPLE AND GREENHOUSE GASES

Ice cores and the record of atmospheric greenhouse gas concentrations they contain help us better understand the climate of the last 10,000 years. The ice core record for the last several hundred thousand years shows that atmospheric CO_2 concentrations systematically fluctuated between high concentrations during warm periods and low concentrations during cold periods (see Figure 14-27). The record also shows the expected trend of decreasing CO_2, starting about 10,000 years ago, that accompanies a cooling global climate, as predicted by solar radiation cycles.

However, about 7000 years ago, as you can see in Figure 14-31, CO_2 concentrations started to rise instead of decline along the expected trend. It seems that something new was influencing the concentration of CO_2 in the atmosphere—something different from the factors that had controlled it during warm and cold cycles of the past few hundred thousands of years. The same type of concentration change was apparent in ice core methane data, but as you can see in Figure 14-31, in the case of this greenhouse gas, the shift from decreasing to increasing concentrations took place 5000 years ago. The new influence on greenhouse gas concentrations that changed the natural shifts appears to have been people.

To explain the unusual history of atmospheric CO_2 and methane change during the last 10,000 years, the paleoclimatologist William F. Ruddiman has proposed that humans began influencing the atmosphere's greenhouse gas concentrations as early as 8000 years ago, when they started farming. Ruddiman's hypotheses, set forth in his book *Plows, Plagues, and Petroleum,* tie several aspects of human history to climate and atmosphere change. These hypotheses are now being tested by other scientists and are not yet widely accepted, but they present a very interesting story.

Farming often involves the conversion of forested land to agricultural land. Typically, the forest is cut

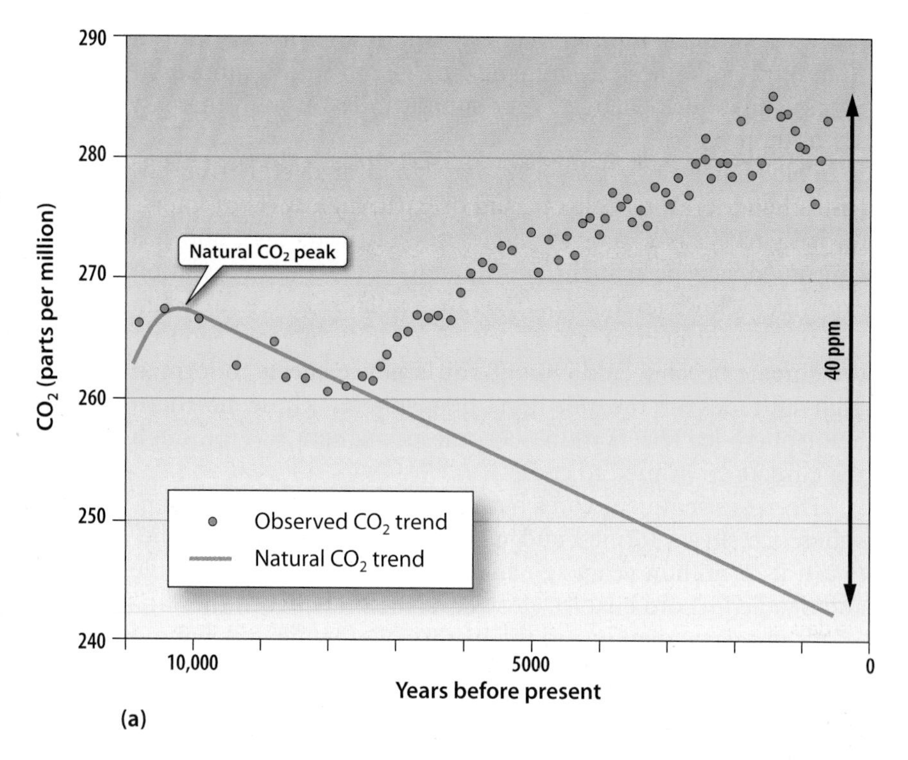

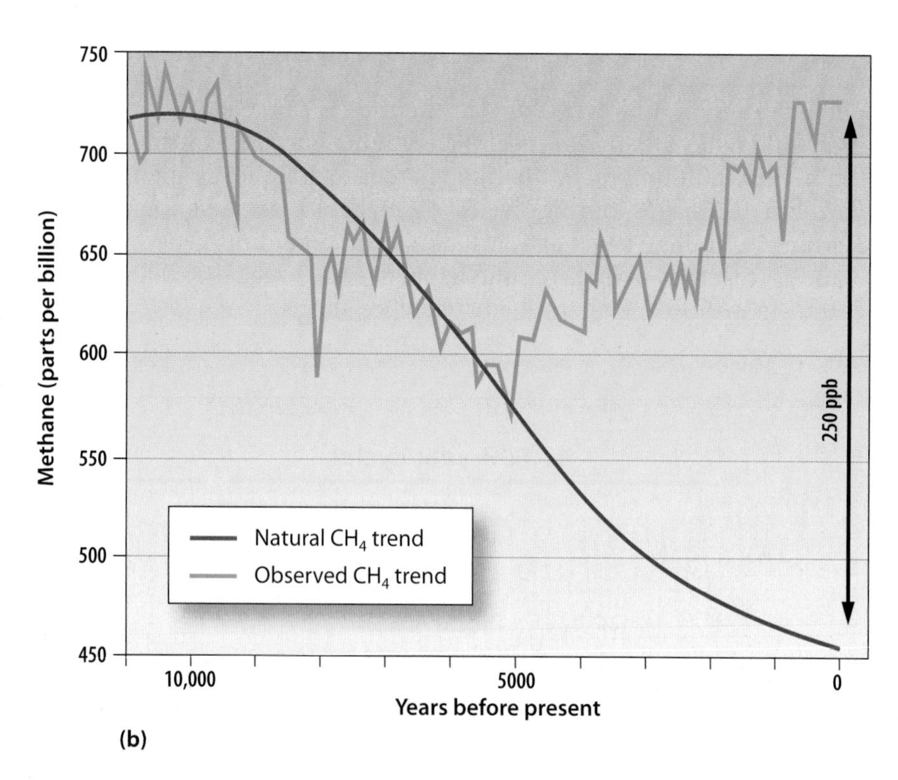

FIGURE 14-31 ▲ The Last 10,000 Years of Greenhouse Gas Concentrations The curves represent the concentrations of atmospheric CO_2 **(a)** and methane **(b)** as determined from ice cores. Note the shift from the longer-term natural trends at about 7000 years ago for CO_2 and at 5000 years ago for methane. If data for the last 200 years were included, the CO_2 concentration would go off the chart—CO_2 concentrations increased to 380 ppm during this time!

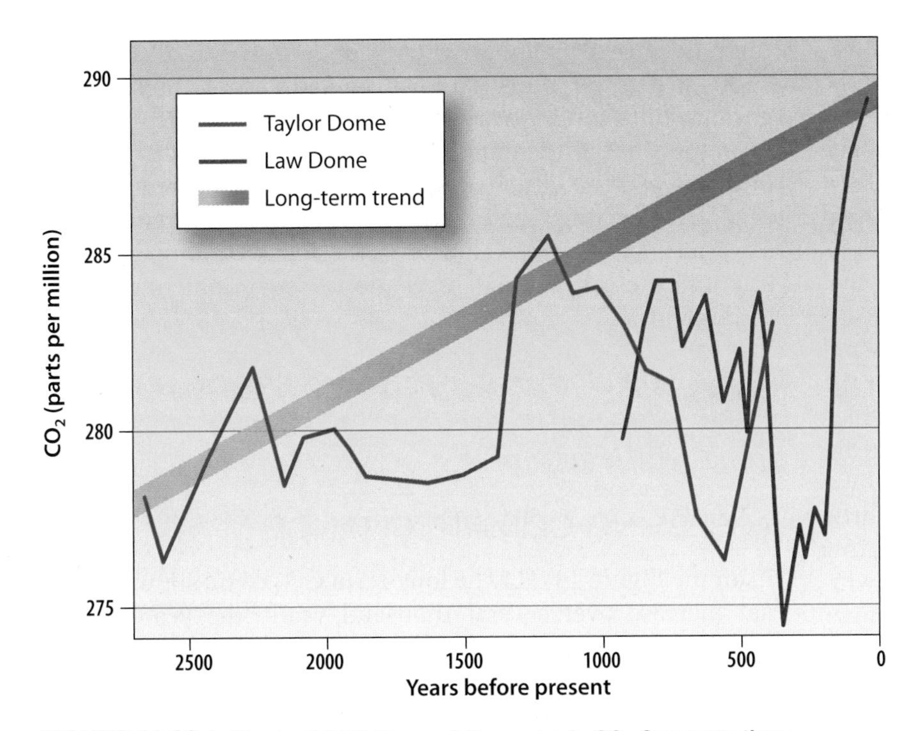

FIGURE 14-32 ▲ The Last 2500 Years of Atmospheric CO_2 Concentrations
The data are based on Antarctic ice cores from Taylor Dome and Law Dome. Note the lower CO_2 concentrations between about 900 and 200 years ago.

down and burned, releasing CO_2. The net effect of deforestation is a release of carbon to the atmosphere of 1000 to 3000 tons per square kilometer (0.4 mi^2).

Ruddiman suggests that significant deforestation began about 8000 years ago. Then, as people progressed from the Stone Age to the Iron Age, population and agriculture expanded. Historical records help us understand their impact. A survey completed in England upon William the Conqueror's takeover in 1089 A.D. revealed that 85% of the countryside was already deforested—700 years before the Industrial Revolution. Ruddiman suggests that these changes could have significantly contributed to increasing the atmosphere's carbon dioxide concentration from 260 ppm 7000 years ago to the preindustrial level of 280–285 ppm 200 years ago.

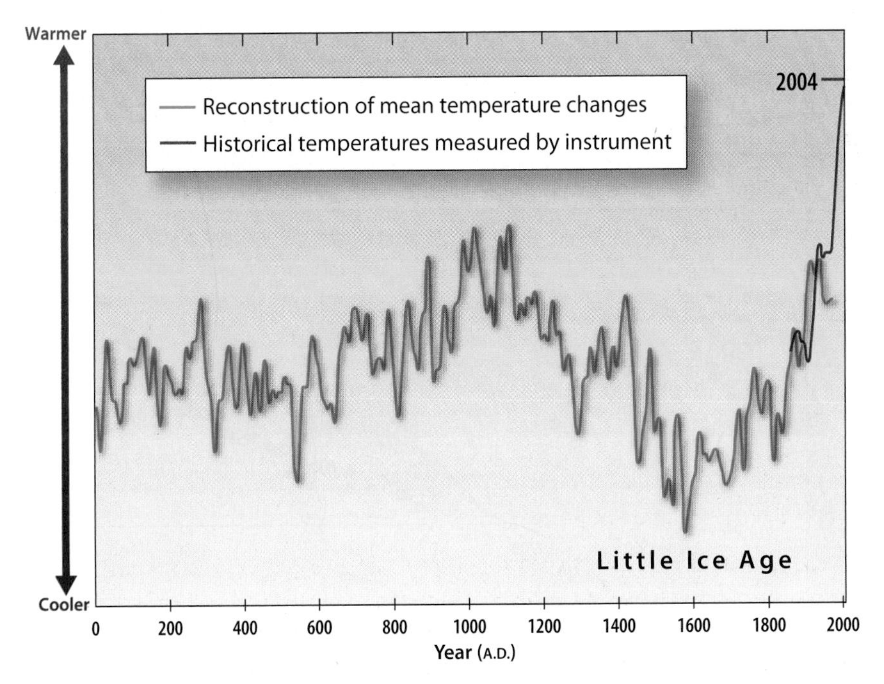

FIGURE 14-33 ▲ Estimated Mean Surface Temperature Variation during the Last 2000 Years

What about that other greenhouse gas, methane? Its concentration started increasing instead of naturally decreasing about 5000 years ago (see Figure 14-31). Ruddiman proposes that wet rice farming in Southeast Asia may have been the cause. Methane is generated by plant decay in wetlands. Irrigation essentially creates wetlands, and wet rice farms are a good example. Rice paddies are tremendous methane generators. People contribute methane to the atmosphere in other ways, too—from livestock they raise, biomass burning, and their own waste, for example. But the expansion of wet rice farming through Southeast Asia could be the most significant factor in increasing atmospheric methane concentration through the last 5000 years.

The preindustrial-era increases in atmospheric greenhouse gases that date from the start of farming may have prevented ice-age conditions from developing on Earth. We would expect such increases in CO_2 and methane concentrations to create warming of about 2° C, just enough to keep northern ice sheets from growing. If you look closely at the ice core CO_2 data shown in Figure 14-32, however, you can see that the shift to higher concentrations in preindustrial historical times has not been a smooth and steady process—there have been a number of reversals.

The most dramatic reversals in the general trend took place between 1100 and 1800 A.D. People had some hard times during this period. Bubonic plague, smallpox, typhus, and cholera killed millions. The pandemic caused by bubonic plague in the late 1300s may have wiped out as much as 33% of Europe's population. Smallpox killed 80 to 90% of the indigenous population of the Americas, perhaps as many as 50 million people. Ruddiman suspects that the population decrease that accompanied the spread of deadly disease influenced atmospheric CO_2 concentrations. As entire villages died out, the land was abandoned and forests reestablished themselves. This may have reversed the effects of converting forest land to farmland and led to decrease of atmospheric CO_2 concentrations.

Decreasing atmospheric CO_2 leads to global cooling, and that's just what happened. Earth's climate was teetering on the edge between warming and cooling. The lower CO_2 concentrations sent it on its way toward a glacial period. The time between about 1250 and 1850 A.D. has become known as the Little Ice Age (Figure 14-33). Alpine glaciers expanded down valleys, across fields, and through villages. Crops froze in late spring and grape cultivation had to shift 500 kilometers (310 mi) south in Europe. The Viking colonies of Greenland died out by 1400 A.D. Sea ice packed around Iceland and kept the fishing fleets home for increasing parts of the year. Around 1800 A.D., the fleet was icebound for about half the year, but today this condition hardly ever develops. The River Thames froze again and again between 1607 and 1814 A.D. New York harbor froze in 1780 A.D. Cold, long, and hard winters characterized the time. It's said that the cold conditions made trees grow more slowly, creating the dense wood that made Stradivarius's violins world famous.

The last remnants of the glaciers covering mainland Canada during the last glacial maximum are centered in the Baffin Island region (see Figure 14-28). During the Little Ice Age, the remnant ice caps stopped melting and began to grow. The expansion of Baffin Island ice is evidenced today by a surrounding area of dead lichen. Lichen are scruffy little combinations of fungi and photosynthetic bacteria that grow on tundra and rock surfaces. They can handle long, cold winters, but they need at least some short periods of summer sunlight to survive. As Figure 14-34 indicates, the lichen around the Baffin Island ice didn't survive. It must have been covered by snow and ice year-round for a long time during the Little Ice Age. The expanding Baffin Island ice during the Little Ice Age could have been the beginning of a new glacial period. But the expansion didn't continue. Ice in the Baffin Island area ice is now rapidly melting and retreating.

There is no consensus as to when the Little Ice Age started (dates between 1250 and 1650 A.D. have been suggested), but there is consensus about when it ended—it was in the mid-1800s, when the Industrial Revolution and fossil fuel burning got into full swing. A very rapid rise in atmospheric CO_2 concentration began then and has continued ever since. This not only stopped the Little Ice Age but started the current episode of global warming (see Figure 14-33). The very recent shift to a warmer and warmer climate is the focus of unprecedented scientific research and international environmental policy discussions.

FIGURE 14-34 ▲ The Baffin Island Area of Northern Canada
Snow accumulated around the now melting ice caps during the Little Ice Age, covering lichen and causing them to die.

What are some ways greenhouse gases have influenced Earth systems interactions?

ASSESSING CLIMATE CHANGE

Examine Figure 14-31. The long-term CO_2 trend shows a gradual increase over several thousand years, the years during which people expanded agriculture around the world. As you have learned, there were some reversals of this trend from time to time, apparently when people faced difficult times, population decreased, and forests grew back on agricultural lands. But the really big change in atmospheric CO_2, as you can see in Figure 14-35, was the rapid increase that started about 1850 A.D. Since then, atmospheric concentration of CO_2 has risen from 280 parts per million to 380 parts per million, an increase of 36% in just 150 years. This rapid change coincides with increasing use of fossil fuels as the world industrialized and moved into the age of oil (Chapter 13).

Increasing atmospheric CO_2 concentration has started surface and ocean temperature rising. As Figure 14-36 indicates,

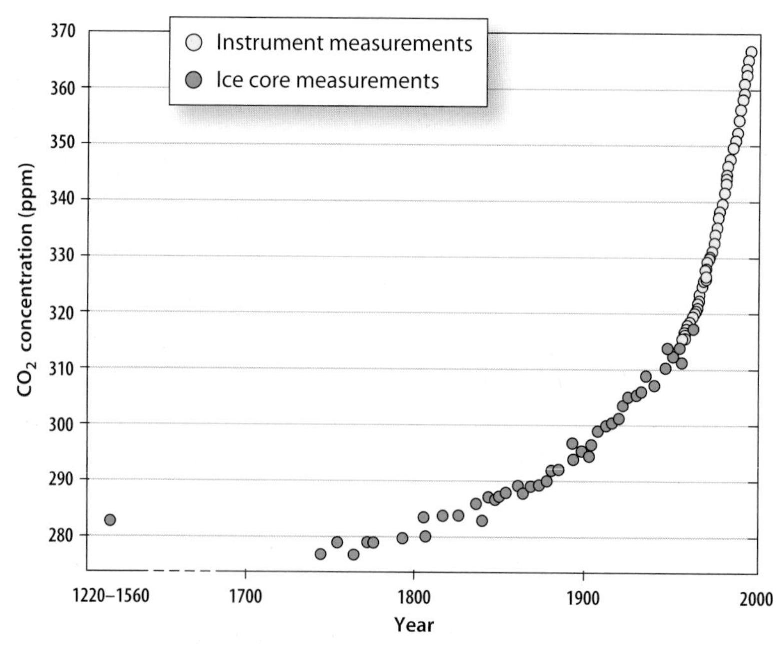

FIGURE 14-35 ▲ Atmospheric CO_2 Concentrations Increased Very Rapidly during the Last 200 Years
The change from about 280 ppm in 1800 to 380 ppm today reflects the impact of the Industrial Revolution and the increased use of fossil fuels.

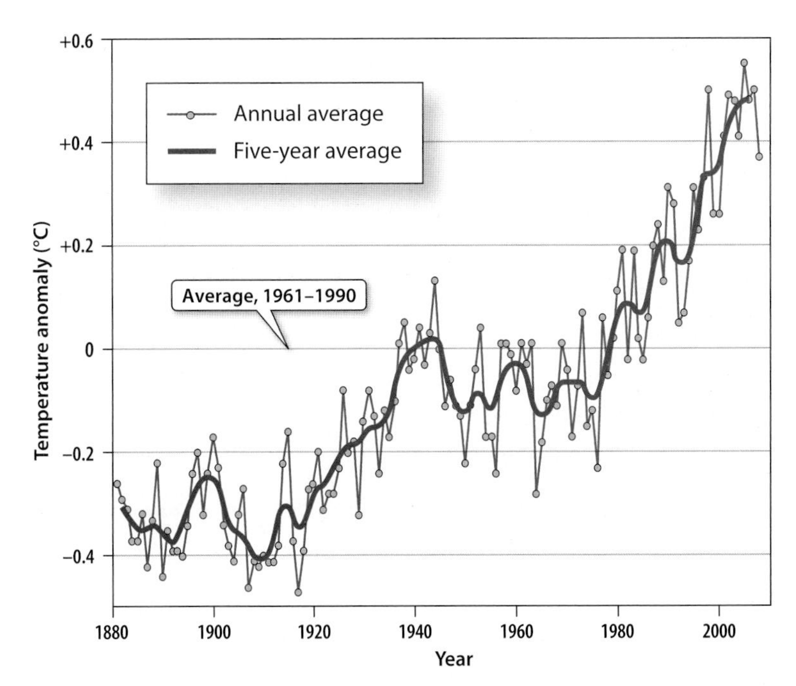

FIGURE 14-36 ▲ Average Global Temperature Rise over the Past Century Instrument-measured air temperatures are compared to the average from 1961 to 1990 (the zero line) to calculate the observed temperature anomaly.

the average temperature of air near Earth' surface has increased about 0.6° C (1.1° F) since the 1960s. Confirming the causes and predicting the consequences, including future climate change and its effects, is the subject of intense scientific investigations. An international effort, the Intergovernmental Panel on Climate Change (IPCC), has diligently assessed these issues since 1988.

The IPCC

The IPCC is a team of scientists established by the World Meteorological Organization and the United Nations Environment Program. Its task is to assess the results of scientific and socioeconomic investigations of human-induced climate change, predict possible outcomes, and communicate its conclusions to policymakers around the world. IPCC reports present a consensus of our current understanding of recent and future climate change and are a foundation of international policies included in the United Nations Framework Convention on Climate Change (1992) and the Kyoto Protocol (1997). The IPCC's fourth assessment report was released in 2007, with a fifth scheduled for 2010.

IPCC – Intergovernmental Panel on Climate Change

Climate models

Because the global climate system is very complicated, the IPCC's attempts to understand the causes and effects of climate change must rely heavily on models of how the system works. A scientific *model* is a physical or mathematical representation of system relationships and processes. Such models are a fundamental tool of scientific research. They can be fairly simple, like those used to evaluate the amount of solar radiation reflected by different amounts of ice cover, or very complex. The most complex are called general circulation models, or **global climate models (GCMs)**.

GCMs are quantitative simulations that represent the interactions of the major components of the climate system: solar radiation, atmosphere, oceans, land, and ice. They are constantly being refined and updated in an attempt to depict more realistically how the system responds to various factors, such as changes in atmospheric greenhouse gas concentrations. Although complex, GCMs share characteristics of all models, including:

- They are incomplete. Models represent what scientists consider to be the *most important* components; ocean temperature, solar radiation levels, atmosphere and ocean circulation patterns, and greenhouse gas concentrations, for example.
- They require assumptions. Scientists must assume many quantitative aspects of the climate system components and their relationships to one another—for example, the rate at which CO_2 is being added to the atmosphere each year. They base such assumptions on what they can reasonably project from past experience and on the best existing information.
- They incorporate variable parameters. The components of the climate system vary in their physical character through space and time; for example, ocean water temperatures vary around the world.

There are many variable parameters in climate models but what makes them even more complex is how the variables change as climate changes. Many global climate system components, such as the amount of water vapor in the atmosphere, will change if climate becomes warmer or colder. Sophisticated models take these changes into consideration as they simulate future conditions. The key is to recognize *feedback mechanisms*. **Feedback mechanisms** respond to a change in the climate system and either amplify or diminish the change. Those that amplify the change are called *positive feedbacks*. Those that diminish the change are *negative feedbacks*. The global climate system is full of feedback mechanisms. Feedback mechanisms in climate Here are some examples.

- Increasing ocean temperatures decrease the amount of CO_2 oceans can absorb. This in turn increases atmospheric CO_2 concentrations and their greenhouse gas effect causes additional ocean temperature warming. This is an example of a positive feedback.
- With climate warming, ice sheets melt and reflect less solar radiation. This in turn causes more surface warming, a positive feedback.
- Increasing air temperatures enable more water vapor to be stored in the atmosphere. As water vapor has a significant greenhouse gas effect, more water vapor increases surface and air temperatures, a positive feedback.
- If increased air temperatures lead to increased water vapor in the atmosphere, another outcome can be increased cloudiness. Clouds reflect incoming solar radiation and decrease surface and air temperatures, a negative feedback.

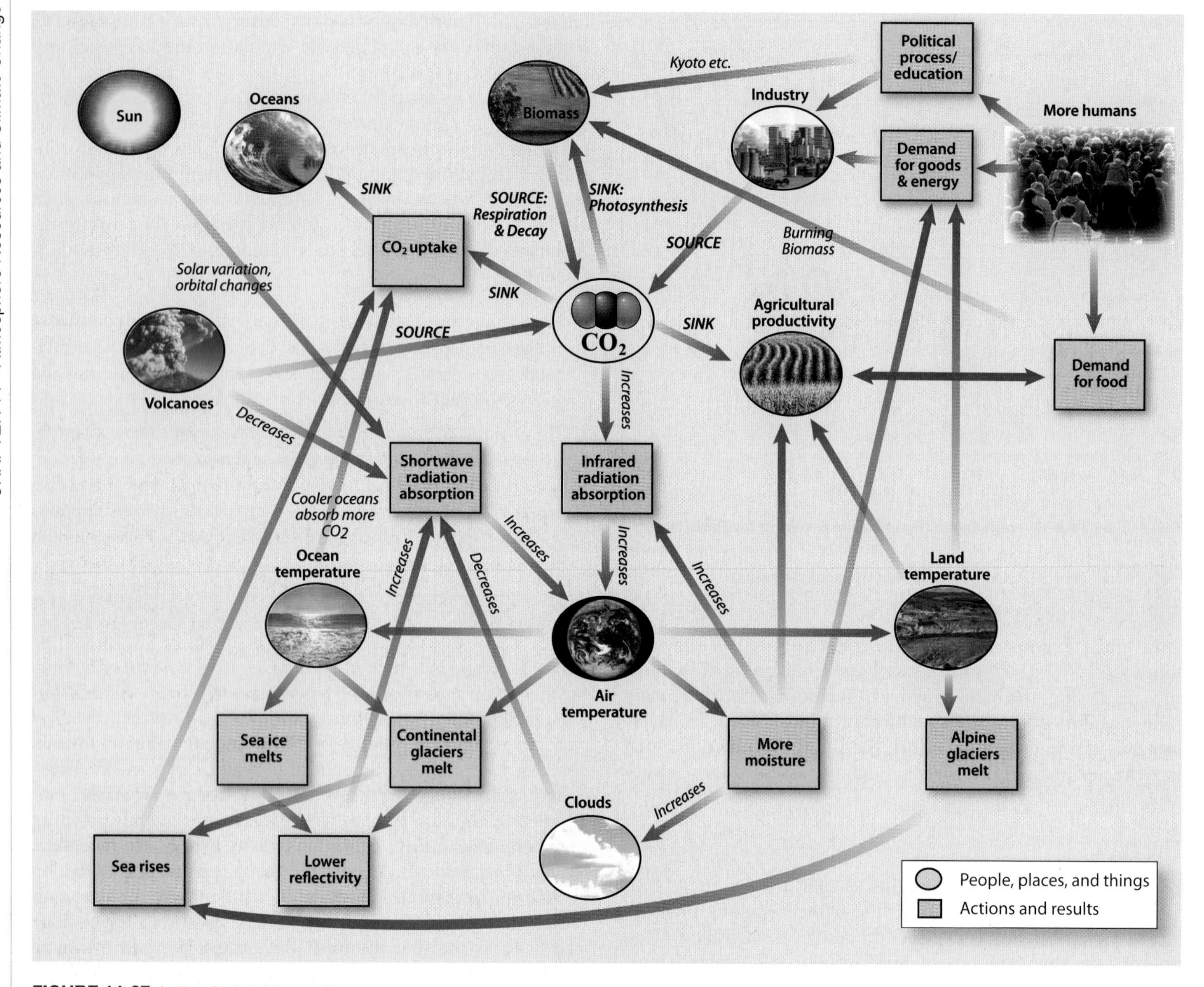

FIGURE 14-37 ▲ The Global Climate System Is Very Complex
This figure shows some of the many interlocking feedback mechanisms that affect global climate. (You don't need to worry about the details—the purpose of the figure is just to suggest why the system is so challengng to model.)

The global climate system, in other words, is *dynamic* and changes in response to its changes. The models that represent the global climate system are challenged to incorporate feedback mechanisms (such as those shown in Figure 14-37 and many others) as they simulate future conditions and attempt to predict the outcome of climate change. This is why climate models all have built-in uncertainties. The global climate system complexities are not fully understood and are difficult to represent quantitatively. Nonetheless, through investigation of very diverse data and a wide range of possible scenarios, climate studies and models provide a wealth of information to constrain conclusions about climate change. Most significantly, robust GCMs have been shown to accurately simulate past climates for which historical and geologic records are well developed. This is a key reason many scientists are confident that GCMs can be used to predict future climate.

IPCC story lines and scenarios

The IPCC has developed detailed and comprehensive methods for predicting future greenhouse gas emissions. These methods build on plausible paths that global drivers of greenhouse gas emissions (such as population growth and economic development) may take. The IPCC calls these *story lines* and has developed four of them to define the scope of possible changes the world may experience by the year 2100. The IPCC's story lines are:

- The A1 story line describes a future world of very rapid economic growth, low population growth, and the rapid introduction of new and more efficient technologies. Major underlying themes are convergence among regions, capacity building, and increased cultural and social interactions, with a substantial reduction in regional differences in per capita income.
- The A2 story line describes a very heterogeneous world. The underlying theme is self-reliance and preservation of local identities. Fertility patterns across regions converge very slowly, resulting in high population growth. Economic development is primarily regionally oriented, and per capita economic growth and technological change are more fragmented and slower than in other story lines.
- The B1 story line describes a convergent world with the same low population growth as in the A1 story line, but with rapid changes in economic structures toward a service and information economy, with reductions in material intensity, and the introduction of clean and resource-efficient technologies. The emphasis is on global solutions to economic, social, and environmental sustainability, including improved equity, but without additional climate initiatives.
- The B2 story line describes a world in which the emphasis is on local solutions to economic, social, and environmental sustainability. It is a world with moderate population growth, intermediate levels of economic development, and less rapid and more diverse technological change than in the B1 and A1 story lines.

These plausible global outcomes establish the framework for specific *scenarios* that quantitatively estimate key factors such as population, economic conditions, and technological development that influence greenhouse gas emissions. The IPCC has defined 40 scenarios in order to describe the range of possibilities within the story lines. Figure 14-38 shows the range of future CO_2 emissions predicted by the 40 scenarios.

Understanding 40 different projections into the future is difficult, so the IPCC has selected six scenarios to exemplify the range of possibilities. These *marker scenarios*, are typical representatives of the four scenario families (A1, A2, B1, and B2). To see how these IPCC projections work, let's examine the possible relationships between global warming and just one of its expected effects: sea level rise.

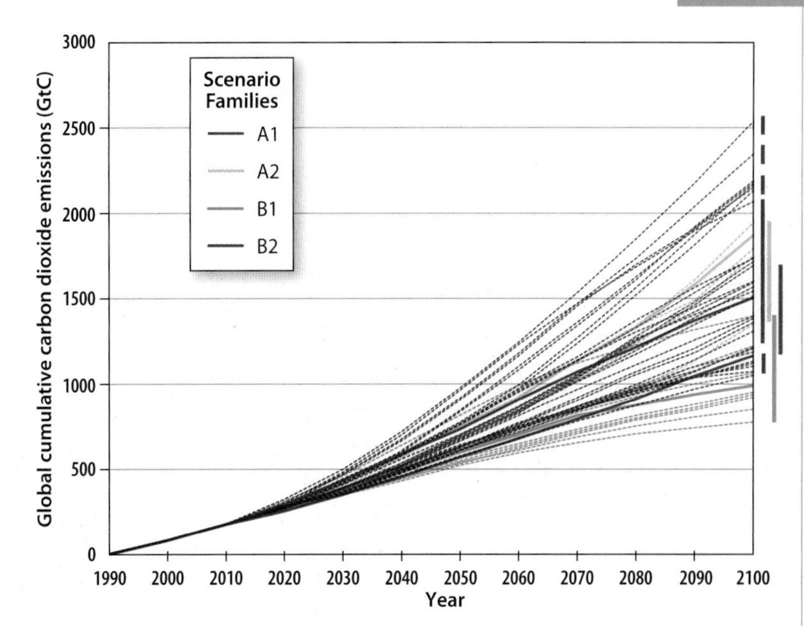

FIGURE 14-38 ▲ Intergovernmental Panel on Climate Change (IPCC) Carbon Dioxide Emissions Predictions
Each prediction is for a specific scenario within one of the IPCC's four story lines. The general grouping of the predictions for each story line is shown by the color bars on the right side of the diagram (A1 = red, A2 = gold, B1 = green, and B2 = blue).

Projecting sea level rise: an example

Sea level rise has received much attention; its effects on coasts and people are commonly included in discussions of possible climate change impacts. Global warming causes sea level to rise in two ways: the oceans expand as they warm and they receive water from melting ice sheets. As we saw in Chapter 9, specific coastlines may experience their own relative sea level changes due to geosphere movements such as subsidence or uplift. At a global scale, however, GCMs can be used to estimate average sea level changes in response to changes in greenhouse gas concentrations and global temperature rise. But predicting the exact amount of sea level rise is challenging and complicated. It requires making numerous assumptions about how future greenhouse gas concentrations will change, how Earth's temperature will respond to that change, and how the climate system will in turn react. The six estimates of global warming and sea level rise based on the IPCC marker scenarios are shown in Figure 14-39 on p. 458.

There are several things to note about the global sea level predictions in Figure 14-39. First, the range of a specific scenario prediction reflects the uncertainty in the estimates. For example, the A1F1 scenario (a fossil-fuel intensive future) predicts that there will be a globally averaged mean sea level rise of as little as 0.26 meters or as much as 0.59 meters (0.85 to 1.9 ft). Second, notice that many different scenarios lead to somewhat similar estimates of global sea level change. The overlap is particularly evident in the lower parts of the estimate ranges. This indicates that a rise of about 0.2 meters (0.66 ft) is essentially built into the climate system regardless of scenario (this rate is similar to the 1 to 2 mm rise per year Earth has experienced during the last 100 years).

The expected sea level rise by 2100 will lead to costly adaptations and displacements of people living along coasts. Storms that come ashore—likely to be stronger and more frequent due to the increased solar energy trapped on Earth—will be devastating, especially on low-lying coasts prone to extreme weather. This is the case along the northern Indian Ocean coasts, where millions of people will either be displaced or remain in harm's way.

The IPCC 2007 assessment

The most recent IPCC assessment was released in 2007. This assessment is based on the results of extensive scientific investigations of climate change ranging from direct measurements (such as atmospheric CO_2 concentration, surface temperature,

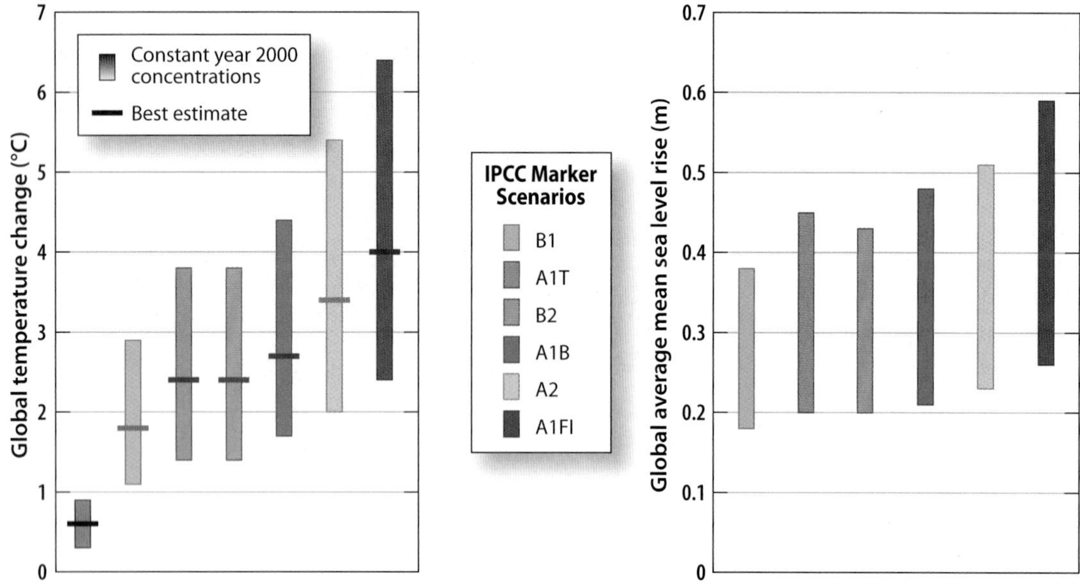

FIGURE 14-39 ◀ Six Possible Futures? Predicted average increases in **(a)** global temperature and **(b)** mean sea level rise for six IPCC marker scenarios to the year 2100.

and ocean water temperature) to the most complex GCM simulations. Here are some of their conclusions about the causes and possible effects of ongoing climate change.

- Warming of the climate system is unequivocal, as is now evident from observations of increases in global average air and ocean temperatures, widespread melting of snow and ice and rising global average sea level.
- Global GHG (greenhouse gas) emissions due to human activities have grown since preindustrial times, with an increase of 70% between 1970 and 2004.
- Most of the observed increase in global average temperatures since the mid-20th century is *very likely* due to the observed increase in anthropogenic GHG concentrations.
- It is *likely* that there has been significant warming caused by human activities over the past 50 years averaged over each continent (except Antarctica).
- There is *high agreement* and *much evidence* that with current climate change mitigation policies and related sustainable development practices, global GHG emissions will continue to grow over the next few decades.
- Continued GHG emissions at or above current rates would cause further warming and induce many changes in the global climate system during the 21st century that would *very likely* be larger than those observed during the 20th century.
- A wide array of adaptation options is available, but more extensive adaptation than is currently occurring is required to reduce vulnerability to climate change. There are barriers, limits, and costs, which are not fully understood.

These are selected conclusions from the IPCC's 2007 assessment report. You are encouraged to scroll through at least the summary document to see the scope of questions the IPCC has addressed. IPCC 2007 summary for policymakers As you do so, notice that the IPCC uses a set of likelihoods to characterize the confidence they have in their conclusions. These different confidence levels reflect the uncertainties inherent in assessing climate change and the range of model predictions that have been made.

Figure 14-40 shows what the IPCC now thinks some significant impacts of global warming could be by the year 2100. The possible effects of climate change on people and their land uses, including agriculture, are ongoing research topics, as are most aspects of climate change. As science continues to unravel the very complicated details, there will be no better place to learn about them and their likely costs (economic and otherwise) than the IPCC reports.

Climate change after 2100

As we have seen, the IPCC does not predict climate change beyond 2100. Can we look further into the future? A paleoclimatologist who has looked ahead a few hundred years is William Ruddiman, whose work was referred to earlier in this chapter. Using his assessment of the future helps us get a feeling for how the current episode of global climate change could play out.

Figure 14-41 is a very general estimate of how global temperatures could change if CO_2 concentrations reach two to four times their preindustrial levels. In Ruddiman's view, CO_2 concentrations may reach twice preindustrial levels if emission controls are moderately successful. If this occurs, the Greenland and Antarctic ice sheets would remain, because it takes them thousands of years to fully respond to climate changes. The Greenland ice sheet would melt some at its margins, but the Antarctic ice sheet would likely grow because of increased precipitation in this very cold region. Arctic sea ice could disappear as could most of the mountain glaciers, but sea level rise would be close to IPCC predictions. Frozen ground at high northern latitudes (*permafrost*) would not completely thaw, but forests would expand northward into tundra areas. Because movements in the atmosphere (weather) can respond very quickly to climate change, more intense storms (both wet and dry) are likely. The growing season would be longer at middle latitudes and episodes of cold, invading Arctic air masses would diminish in the winter.

What if CO_2 concentrations reach four times preindustrial levels? To reach four times preindustrial levels requires that people essentially do nothing to limit greenhouse gas emissions. Given the international awareness and the ongoing and developing efforts to understand and deal with global warm-

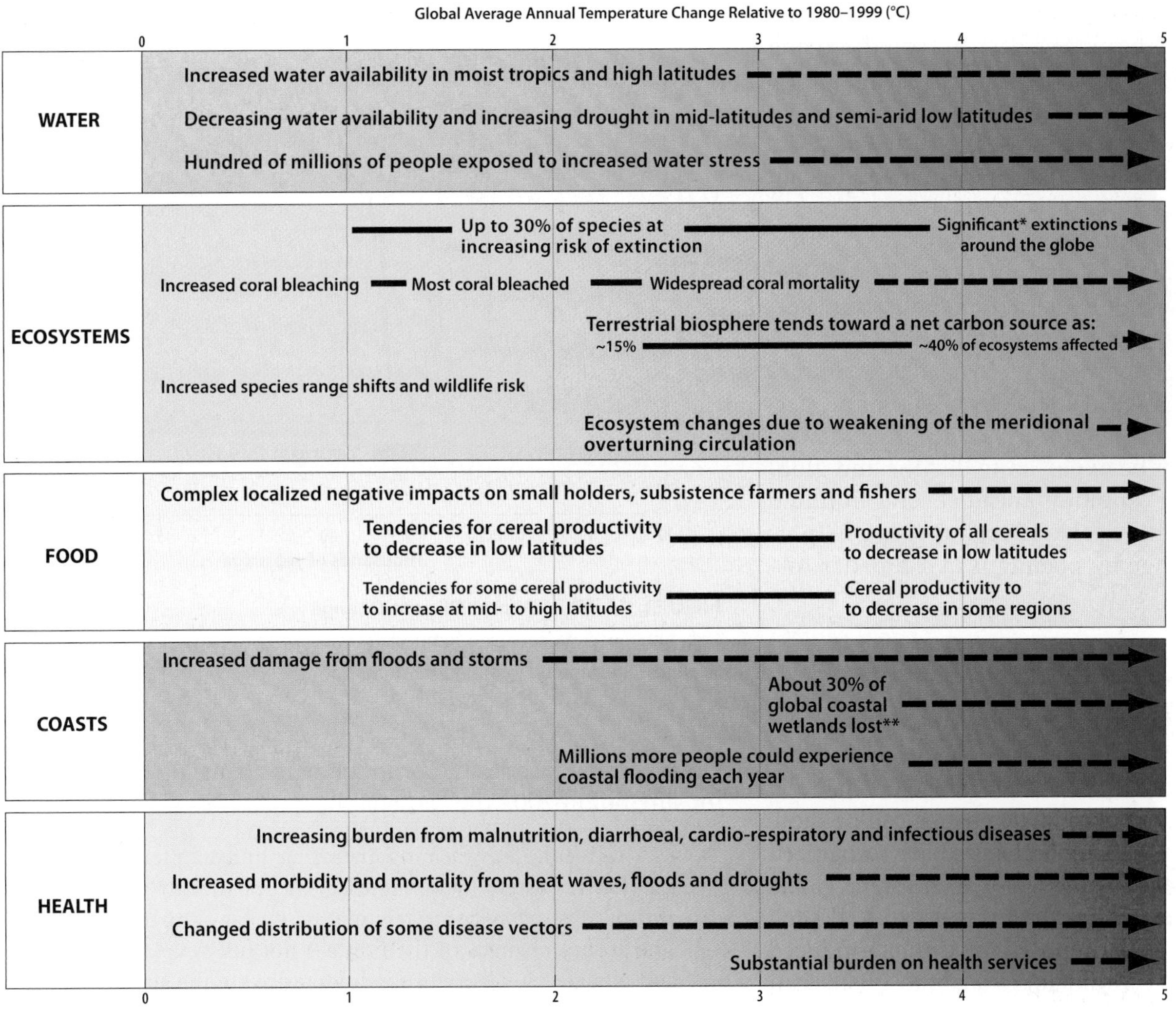

FIGURE 14-40 ◀ Examples of Impacts Associated with Different Amounts of Global Average Temperature Change

The impacts, as projected here by the IPCC, will vary depending on the extent of adaptation, the rate of temperature change, and socioeconomic factors.

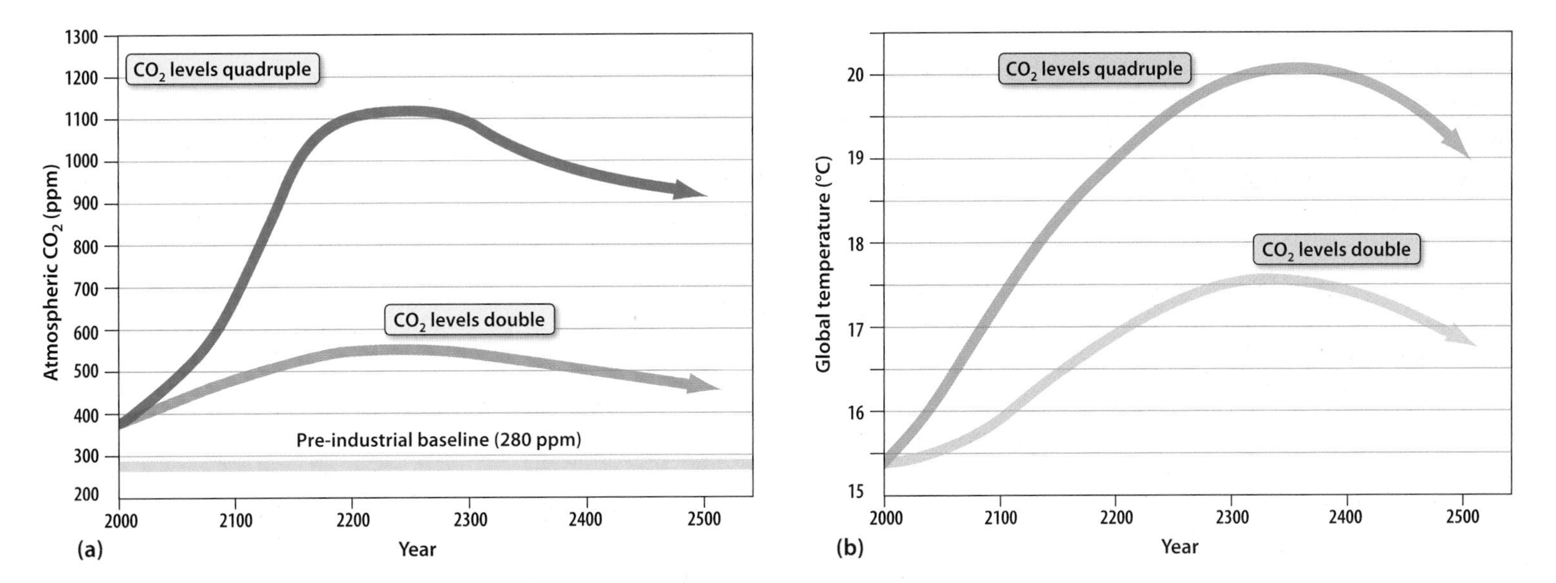

FIGURE 14-41 ▲ One General Estimate of Longer-Term Global Temperature Changes Depending on Future Atmospheric CO_2 Concentrations

(a) Future atmospheric CO_2 concentrations may reach two times preindustrial levels if greenhouse gas controls are in place, and four times preindustrial levels if they are not. **(b)** Temperatures will continue to increase until CO_2 levels begin to fall in response to eventual decreased use of fossil fuels.

ing, this seems unlikely. However, if it were to occur, mountain glaciers and sea ice would be gone. Forests would expand northward and there would be little tundra or permafrost. The Greenland and Antarctic ice sheets would shrink but not go away. Sea level would rise more, perhaps 1 to 2 meters (3 to 6 ft) according to Ruddiman—enough to cause significant flooding of low coastal areas.

Examine Figure 14-41 again. Note that within a few hundred years, CO_2 emissions will begin to decrease. Fossil fuel emissions will not influence global climate indefinitely. Perhaps people will control emission levels, but in any case fossil fuels will eventually be replaced by other energy sources (Chapter 13). As the use of fossil fuels diminishes, and perhaps as emission controls become more and more effective, atmospheric CO_2 levels can be expected to decline and global climate to begin cooling. Looking ahead several hundred years in this way is inherently speculative, but it can help us to understand climate change by placing it in a broader historical context.

ABRUPT CLIMATE CHANGE—A WILD CARD

Although our ancient ancestors adapted to fluctuations between warm and cold climates many times, a world with billions of people will find it difficult to adapt to climate change, especially if it occurs rapidly. As it turns out, climate change can be more than rapid, it can be abrupt. Paleoclimatologists have recognized that significant climate changes, some regional and some global in scope, can occur on a timescale as short as decades. If such sudden climate change is superimposed on human-induced global warming, it could be very difficult to adapt to. Historically, people have not experienced abrupt climate change.

Abrupt climate change occurs when the change is more rapid than the responsible forcing mechanism. |A paleo perspective on abrupt climate change| Well-studied examples occurred 12,800 and 11,600 years ago, at the beginning and end of a period called the Younger Dryas (after a cold-adapted flower, *Dryas octopetala,* that expanded its distribution through Europe during this colder time). As Figure 14-42 indicates, 12,800 years ago the average annual temperature in Greenland dropped 10° C (18° F) within decades, and the region fell back into ice-age cold conditions for 1200 years. At about 11,600 years ago the reverse happened—average Greenland temperatures rose 10° C in about 10 years!

The causes of the abrupt regional climate changes at the beginning and end of the Younger Dryas period are not clear. Some have proposed that changes in circulation of water through the Atlantic Ocean are a factor. Whatever the cause, the North Atlantic became a cold region again very rapidly. Twelve hundred years later, warm water returned to the North Atlantic (for reasons also not clear) and abruptly caused regional temperatures to rise.

Abrupt climate change would present very significant challenges for people. Those that have been identified in the past are related to natural processes, but researchers are now wondering if there will be connections between human-induced climate

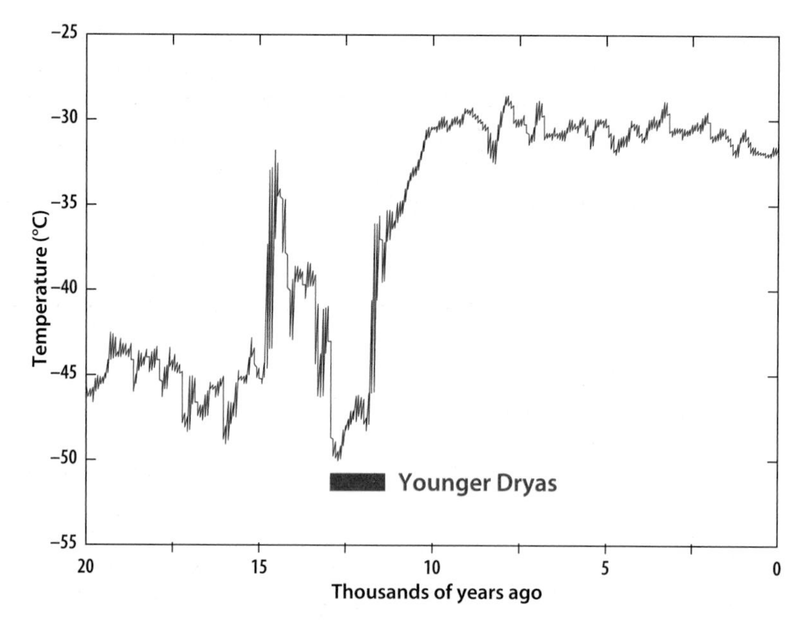

FIGURE 14-42 ▲ Abrupt Climate Change
Rapid changes in Greenland temperatures marked the beginning and end of the Younger Dryas period in the northern hemisphere—11,600 to 12,800 years ago.

changes and episodes of abrupt climate change. No one knows for sure but in 2002 the National Research Council concluded:

> . . . [G]reenhouse warming and other human alterations of the Earth system may increase the possibility of large, abrupt, and unwelcome regional or global climatic events. The abrupt changes of the past are not fully explained yet, and climate models typically underestimate the size, speed, and extent of those changes. Hence, future abrupt changes cannot be predicted with confidence, and climate surprises are to be expected. |Abrupt climate change: inevitable surprises|

14.6 Dealing with Climate Change

The future of global climate change depends greatly on how much CO_2 continues to be emitted. Fossil fuels will be a continuing source of energy. Oil use may start to decline some, but use of coal and natural gas is likely to increase as a result (see Chapter 13). Efforts to control and limit CO_2 emissions from fossil fuel burning are under way.

THE KYOTO PROTOCOL

Through the United Nations, the international community started trying to understand and respond to global warming, much as it had when pollution threats to the ozone layer became known. The United Nations Framework Convention on Climate Change, an international environmental treaty, was initiated at a meeting at Rio de Janeiro in 1992. A 1997 amendment to this treaty, known as the Kyoto Protocol, identifies CO_2 emission-reduction goals for all the countries that agree to participate. | United Nations Framework Convention on Climate Change | | Kyoto Protocol | It came into force on February 16, 2005, after countries responsible for 55% of global CO_2 emissions agreed to the treaty.

As of May 2008, 181 countries are parties to the agreement. The Kyoto Protocol imposes national caps on greenhouse gas emissions at a level 5% below their 1990 levels (to be reached by 2012). The caps are not the same for every country; European Union countries have an 8% reduction target, and Iceland can actually increase greenhouse gas emissions by 10%. Developing countries like China and India are not required to limit their emissions so that their economic growth can continue.

The problem is that a lot of countries with a lot of CO_2 emissions are not participating—the United States, for one. The United States did not ratify the treaty because of possible negative effects on its economy, the lack of participation of the biggest emitters among developing countries, and the billions of dollars in costs the treaty imposes on developed countries.

No one knows how effective the Kyoto Protocol will be. Ratification doesn't necessarily mean compliance. On the other hand, countries like the United States can start reducing greenhouse gas emissions even though they have not ratified the treaty. Now that the Supreme Court has required the EPA to treat CO_2 as an atmospheric pollutant, controlling CO_2 emissions is a recognized environmental concern. President Obama's goal for the United States is to reduce greenhouse gas emissions 80% by 2050. As a consequence, a lot of scientific and engineering research focuses on learning how to diminish, capture, or use CO_2 emissions rather than just release them into the atmosphere.

CARBON SEQUESTRATION: A KEY TECHNOLOGY

Remember from Chapter 13 that regardless of how much we come to use other energy resources, people will continue to burn large amounts of hydrocarbons in the coming decades. Capturing and storing CO_2 emitted by hydrocarbon burning is termed **carbon sequestration**. This is a key technology that could mitigate CO_2 emissions from the use of fossil fuels, especially large point sources like coal-burning power plants. Developing this technology is the focus of much current research.

There are many technologies for capturing CO_2 from large point sources. | NETL: CO_2 capture | These vary from physical membranes that separate CO_2 from exhaust streams to the use of sorbents that chemically react with emitted CO_2, as diagrammed in Figure 14-43. All these processes require energy and increase power generation costs, but some are very promising. A bigger challenge is storing CO_2 after it's captured.

Captured CO_2 can be stored for long periods in a few ways and in some cases it can be reused. Long term storage includes reacting CO_2 with metal ions such as calcium, magnesium, and iron to form stable carbonate minerals (Chapter 4). The technology for making carbonate minerals with captured CO_2 requires considerable energy.

A technology that is already viable in some places, and may become more important in the future, is storage in underground geologic reservoirs. Underground CO_2 storage, or **geo-sequestration**, injects captured CO_2 into permeable reservoirs such as depleted oil fields or saline water-filled formations. These reservoirs need to be overlain by impermeable layers that effectively trap the CO_2 and keep it from leaking upward. In some cases, the injected CO_2 is expected to react with geosphere materials and form carbonate minerals.

The challenge of geo-sequestration is identifying satisfactory reservoirs. Depleted oil fields may be limited in size but saline water-filled reservoirs are widespread and potentially huge. Geo-sequestration seems to be the direction that much CO_2 storage is going. | Department of Energy – DOE awards $126.6 million for two more large-scale carbon sequestration projects |

Reusing captured CO_2 is possible in some cases. CO_2 injection in declining oil fields is a proven technology for enhancing oil recovery. In the United States, most of the CO_2 used for this purpose is recovered from natural underground accumulations and piped hundreds of kilometers (miles) to oil fields for use. However, CO_2 captured at a coal-gasification plant in North Dakota is now being used to enhance recovery in an oil field in Canada.

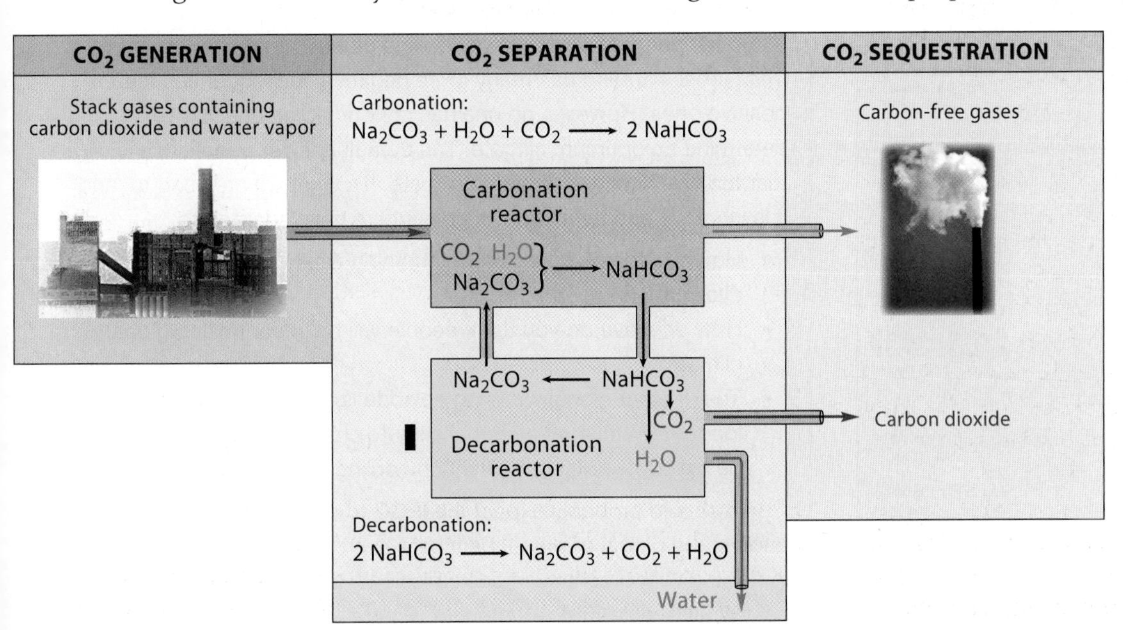

FIGURE 14-43 ◀ **An Example of Carbon Capture Technology**
CO_2 emissions from a large point source first combine with solid sodium carbonate in the carbonation reactor. The resulting sodium bicarbonate is heated in the decarbonation reactor. This releases concentrated CO_2 and regenerates sodium carbonate that is recycled back to the carbonation reactor.

Another potential use for recaptured CO_2 is in the making of methanol (wood alcohol, CH_3OH). Methanol has many uses including as a fuel. The reaction of CO_2 and hydrogen ($CO_2 + 3H_2 \rightarrow CH_3OH + H_2O$) requires energy (perhaps from renewable sources) but the many and diverse uses of methanol make it a valuable commodity. U.S. plants now produce 2.6 billion gallons of methanol each year.

Carbon sequestration, in one way or another, is the key to controlling future CO_2 emissions. The viable technologies identified through ongoing scientific and engineering research will balance cost and benefits, including their environmental consequences. The active research on carbon sequestration around the world is another example of how science is helping address significant environmental challenges, in this case global climate change.

What You Can Do

Calculate Your Contribution to Greenhouse Gases

Large point sources are not the only emitters of greenhouse gases. Do you want to get a feel for how much greenhouse gas you generate? Go to the EPA's Personal emissions calculator online. Fill in the short data blanks, and up comes the amount of greenhouse gas emissions that are released to support you and your family's home and transportation needs. This calculator also helps you determine how selected activities, such as using more energy-efficient lightbulbs, will reduce your contribution to greenhouse gas emissions.

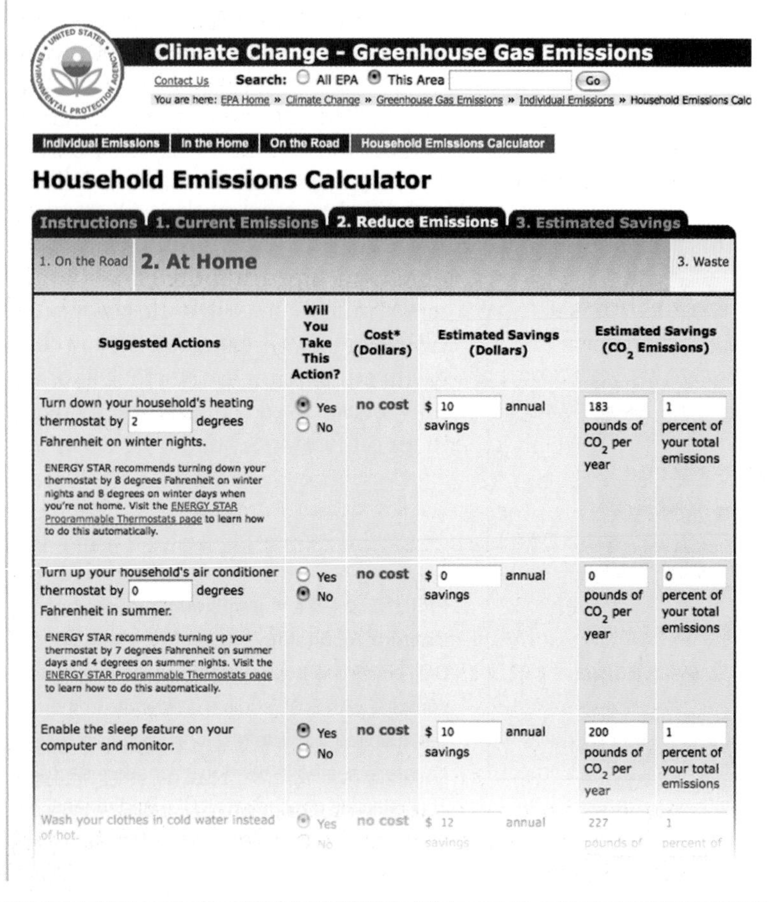

CONTROLLING CLIMATE

Having recognized their role in influencing climate change, people are now moving to control global climate. Concern about global warming has focused attention on the role of greenhouse gases, especially CO_2, as international agreements and other actions attempt to limit their release to the atmosphere. In addition, engineering approaches to control global warming are being discussed. These "geoengineering" proposals include fertilizing oceans to increase their CO_2 uptake, dispersing reflective chemicals or particles in the atmosphere to decrease the amount of solar radiation that reaches Earth's surface, or even building reflective objects that orbit Earth to block incoming solar radiation. Some research is also focused on recovering CO_2 directly from the atmosphere in hopes of developing technologies that could be used for a global thermostat. Human ingenuity, a remarkable trait, is now being directed at controlling global warming.

In general, people have a mixed track record of "managing" nature. (John McPhee's *The Control of Nature* is a wonderful book on the subject.) Watershed management, wildlife and forest management, and basic resource sustainability efforts, as people have historically carried them out, have not always been successful. Controlling climate seems like a stretch, but we are headed down the climate-control road and the pace is picking up.

You Make the Call

What's the Optimum Global Climate for Earth?

If you could choose a global climate, what would it be? This may seem like a fairly presumptuous question, but people are already into the global climate control business. We sure didn't plan it this way, yet it's happened. We are proactively working to control climate—to stop, or at least dampen, the global warming process that came with our use of fossil fuels.

So far, people's efforts at controlling global climate have assumed that global warming has many more negative consequences than positive ones. However, no one has conducted an analysis to determine an optimum climate. The default answer people use is that today's climate is the best climate. It's what we are used to, and it is what we and the rest of the biosphere have adapted to.

- How would you go about determining what the "optimum" global climate is?
- How effective do you think people will be in controlling global climate?
- The present global warming episode is essentially a blip in the longer-term history of global climate change. Will global cooling be welcomed? If so, how much cooling?

We should probably expect the IPCC to continue its efforts to understand global climate indefinitely, even if the world is successful in dealing with greenhouse gas emissions. It really isn't too soon to ask: What should Earth's optimum climate be?

SUMMARY

The challenges of dealing with global climate changes are likely to be in the news the rest of your life. Climate change will influence energy use and development decisions, and it will affect you personally in ways that are still not clear—perhaps it will even raise your taxes. The momentum behind dealing with global warming is unprecedented. Whatever people decide about controlling climate, it will involve dealing with the atmosphere's composition and processes. That's why this chapter has focused on helping you understand the atmosphere as a resource and as a key player in climate change.

In This Chapter You Have Learned:

- The atmosphere is a resource that people use. It supports life, supplies useful elements including oxygen, nitrogen, and argon, and serves as a critical link in Earth systems interactions.
- The atmosphere's composition makes it an effective shield against harmful solar radiation. The atmosphere absorbs gamma rays, X-rays, and the most harmful forms of ultraviolet radiation before they reach Earth's surface. The stratospheric ozone layer is especially effective at screening out harmful ultraviolet radiation.
- People's activities pollute the atmosphere. Atmospheric pollutants include nitrogen oxides (NO_x), SO_2, volatile organic compounds (VOCs), low-level ozone, CO, and greenhouse gases—especially CO_2. The principal source of NO_x, SO_2, CO, and CO_2 is the burning of fossil fuels in vehicles and industry. Low-level ozone forms by photochemical reactions involving VOCs and NO_x. VOCs are emitted by plastics, paint thinners, solvents, oil, and gasoline.
- Greenhouse gases include CO_2, methane, water vapor, and N_2O. They occur naturally and also as a result of human activities. They are minor constituents of the atmosphere, and people's emissions can significantly change their atmospheric concentrations.
- Smog forms from photochemical reactions involving VOCs and NO_x that produce low-level ozone. Combustion controls have significantly lowered NO_x emissions since 1980.
- Acid rain forms from reactions between NO_x and SO_2 and water vapor in the atmosphere. Limits on the emissions of NO_x and SO_2 have prevented severe acid rain, but some acid rain still forms from the SO_2 emissions of coal-burning power plants.
- Very stable chlorine- and bromine-bearing compounds, mostly chlorofluorocarbons or CFCs, are emitted to the atmosphere through their use in refrigerators, air conditioners, spray cans, foaming applications, and as cleaners in the electronics industry. CFCs stay in the atmosphere for 50 to 200 years. Because they destroy ozone in the ozone layer, CFCs have been banned by international treaty, but it will take decades for the remaining CFCs to be removed from the atmosphere.
- Greenhouse gases in Earth's atmosphere influence global climate. These gases absorb heat radiated by Earth and trap it in the atmosphere. Part of this heat energy is reradiated back to Earth, causing surface and ocean temperature increases. This so-called greenhouse effect keeps Earth from being frozen and uninhabitable.
- Variations in solar radiation caused by cyclic changing of the tilt of Earth's axis (a complete cycle every 41,000 years), Earth's distance from the Sun due to orbital eccentricity (a complete cycle every 100,000 years), and Earth's wobble as it orbits the Sun (a complete cycle every 22,000 years) are significant controls on global climate. Shifts between warm periods and cold, glacial periods have accompanied variations in incoming solar radiation.
- Paleoclimatology, the study of past climates, uses sea level history, the character of sediments and sedimentary rocks, the fossil record, oxygen isotope data, and samples of ancient air trapped in Greenland and Antarctic ice to identify ancient climate changes.
- Shifts between warm and cold climates have occurred for large parts of Earth's history. During the last 2 to 3 million years, when the evolution leading to modern humans was under way, there have been 40 to 50 cycles between warm climates and cold, glacial climates. Earth's orbital cycles controlled these climate changes.
- The last glacial maximum occurred 20,000 years ago, when northern hemisphere ice sheets extended across Canada and south into what is now the United States. As incoming solar radiation increased between 20,000 and 10,000 years ago, the ice sheets melted and sea level rose some 110 to 125 meters (360 to 410 ft). Solar radiation started decreasing about 10,000 years ago, but Greenland and Antarctic ice sheets have not expanded, nor has sea level fallen. The natural shift back into a cold, glacial climate was apparently averted by additions of greenhouse gases to the atmosphere. The CO_2 additions may have come from the effects of deforestation caused by people as they expanded agricultural lands starting about 7000 years ago.
- Addition of greenhouse gases, especially CO_2, from fossil fuel burning during the last 200 years has not only continued averting a glacial climate, it has led to global warming. CO_2 levels may reach two to four times their preindustrial levels before the use of fossil fuels declines in a few hundred years. The ice sheets of Greenland and Antarctica will shrink but not disappear under these conditions, but the Arctic Ocean may become ice-free, mountain glaciers may melt, and sea level may rise. More intense weather and longer growing seasons at middle latitudes can also be expected as a result of global warming.
- Controlling greenhouse gas emissions and mitigating global warming is now an international environmental priority. People's efforts to understand and control global climate will continue indefinitely and will affect your life in many yet-to-be-determined ways.

KEY TERMS

carbon sequestration (p. 461)
cryogenic distillation (p. 429)
eccentricity (p. 442)
feedback mechanism (p. 455)
geo-sequestration (p. 461)
glacial maximum (p. 442)
global climate model (GCM) (p. 455)
pollutant (p. 431)
precession (p. 442)
smog (p. 434)

QUESTIONS FOR REVIEW

Check Your Understanding

1. Why is the atmosphere so much more dynamic than the other Earth systems, such as the geosphere and the oceans?
2. Early in this chapter, the text states that the Earth's atmosphere is in essentially a steady state compositionally. With the evolution of humans on Earth and especially since the Industrial Revolution, why is this statement at best only approximately true?
3. What are the three gases that make up 99.9% of our atmosphere, and which of these are for the most part nonreactive?
4. Ozone is a component of air pollution at lower levels in the atmosphere, so why is it so helpful and necessary high in the stratosphere?
5. What are VOCs, and why do they present a pollution concern?
6. What does the "x" signify in NO_x? Where do these compounds come from in industrialized society?
7. Acid rain is caused by the formation of sulfuric acid in the atmosphere within rainwater. What are the main nonnatural sources of compounds that lead to this type of pollution?
8. How does SO_2 released into the atmosphere by volcanic eruptions play a role in global cooling events?
9. What is smog, and how does the photochemical version of this type of pollution form?
10. What are CFCs, and why are they so dangerous to stratospheric ozone?
11. What is the Montreal Protocol, and why is it important that it has been revised often in light of accumulating new scientific information?
12. How is climate different from weather? Do weekly changes in weather tell you anything directly about climate changes? Why or why not?
13. During the time of Earth's "second atmosphere" in the Precambrian, how is it possible that the surface may have had temperatures up to 70° C (158° F) while the Sun was actually 20–30% less bright than now?
14. Describe how enhanced weathering of the crust due to tectonic uplift can actually reduce atmospheric CO_2 and induce global cooling.
15. What is the basic climatological significance of Antarctica being an isolated continent covering the south polar region of the planet?
16. Describe three ways that the sedimentary rock record can be used to understand past climate changes.
17. How do the skeletons of old marine microorganisms tells us something about climate in the past?
18. How do oxygen isotopes in water from ice sheets provide information about past ocean temperatures?
19. What is a Milankovitch cycle? What are the three main parts of such cycles, and what is the characteristic time scale of each?
20. How may human activities over the last 5000 years have slowed the natural return of the Earth to ice-age conditions expected from orbital factors (Milankovitch cycles) alone? Why has that change been especially pronounced over the last 150 years?
21. What is a feedback mechanism in the climate system? Illustrate your answer with an example.
22. What is the Kyoto Protocol? Is it likely to be completely effective by itself in reducing the rate of climate change? Why or why not?
23. What is carbon sequestration, how does it work, and why is it important to climate change?

Critical Thinking/Discussion

24. The Sun directly heats the land and oceans, which transmit heat to the base of the atmosphere. How do these processes cause the dynamic behavior characteristic of the atmosphere, especially at its lower levels?
25. Why is it politically, economically, and environmentally significant that CO_2 is now regulated by the EPA in the United States? CO_2 is also produced naturally in large volumes, so how will the EPA need to focus its regulatory and scientific efforts to implement this regulation?
26. Why is it so important in places like Los Angeles and Mexico City to regulate vehicle emissions to control smog? Explain how reducing vehicle emissions helps, which type of emissions need to be reduced, and other factors related to the geography of these cities that is relevant to the formation of smog.
27. Fish in lakes and rivers in the American northeast were dying in large numbers from acid rain before the advent of regulatory controls on the sources of this pollution. Why was this region particularly impacted by acid rain, and what changed to start these ecosystems on the road to recovery?

28. Examining the history of the atmosphere and the history of life, one could argue that by 500 to 600 million years ago, new forms of life were able to emerge onto land as a result of sustained air pollution created previously by other species. In what sense is this true, and what were the processes and "pollution" that made it all possible?
29. The text outlines the three "atmospheres" the Earth has had in its history. What were the major driving forces in changing the atmosphere at each stage, especially the influence of life and tectonics?
30. Imagine a paleoclimatologist discovered warm-climate plant spores preserved in deep sediments currently located in a region with a cold climate. What would she conclude about change through time at that location? What other kinds of evidence should she look for to confirm her hypothesized climate change conclusions?
31. You are approached by a friend who is hoping to find and investigate 20,000–15,000-year-old archaeological sites along coastlines around the world. Given what you know from the chapter about ice sheets and sea level change, where should he go look for these sites and why?
32. The evolution of modern humans has mostly taken place during the last 2–4 million years, which has been a period of relatively low concentrations of CO_2 in the atmosphere and frequent ice ages. Many anthropologists and primate biologists speculate that relatively cool climates and ice ages spurred much of our development as a species. How might changing vegetation (especially the retreat of tropical forests and the growth of mid-latitude grasslands), as well as weather and rainfall patterns resulting from climate change, have contributed to our development from a tree-dwelling species to an upright-walking, ground-dwelling, tool-using species.
33. How has the rise of agriculture likely contributed to global warming and reversing of the cooling trend expected from analysis of Milankovitch cycles? Is this likely to become more or less pronounced as industrialization of agriculture increases and the human population grows? Why or why not?
34. Why is it crucial that the IPCC is an international body and is made up of both scientists and policy representatives from governments around the world? Explain why this arrangement leads to a body with significantly more influence than a scientific-only or one-nation body of this sort?

ANSWERS TO IN-CHAPTER INSIGHT QUESTIONS

P. 439

What are some Earth systems interactions associated with air pollution?

- Pollutants are transferred from the geosphere to the atmosphere by people's activities.
- Pollutants are moved in the atmosphere at regional to global scales by energy transfers between the atmosphere, hydrosphere, and geosphere that cause winds.
- Pollutants in the atmosphere react with geosphere materials.
- Pollutants in the atmosphere affect the biosphere.
- Pollutants in the atmosphere are transferred to the hydrosphere and geosphere by reactions and by precipitation of rain, snow, and ice.

P. 447

What are some Earth systems interactions that produce evidence of ancient glaciations?

- Ice in the hydrosphere breaks, move, mixes, and deposits geosphere materials in tillites.
- Ice in the hydrosphere transports geosphere material and transfers it to ocean sediments when icebergs melt.
- Atmosphere evaporation and transfer of water vapor leads to changes in the isotopic composition of ice sheets, oceans, and fossil shells and skeletons.

P. 454

What are some ways greenhouse gases have influenced Earth systems interactions?

- Increasing surface and air temperature increases hydrosphere evaporation and influences hydrosphere circulation.
- Increasing surface and air temperature influences atmosphere circulation.
- Increased surface temperatures influence biosphere range and habitat.
- Increased melting of ice sheets raises sea levels and changes Earth systems interactions along coasts.

FIGURE 15-1 ▶ Not in My Backyard
Long Island's garbage sailed the seas for two months on the barge *Mobro 4000* in search of a final resting place. After its 9700 km (6000 mi) journey it eventually arrived back in port in New York. Part was eventually incinerated and part was buried in a landfill.

CHAPTER

15

MANAGING PEOPLE'S ENVIRONMENTAL IMPACT

Over 7.5 million people live on Long Island, which stretches into the Atlantic Ocean east of New York City. Like people elsewhere in the United States, Long Islanders produce a lot of garbage—on average more than 2 kilograms (4.4 lbs) per person per day. Historically, most of this garbage found its way into landfills, where it was dumped, compacted, and eventually covered with soil. But Long Island landfills had a problem in the 1980s. Some were releasing pollutants to groundwater and soil and some were filling up. Long Islanders needed alternatives for disposing of their garbage.

An entrepreneur from Alabama proposed a solution. For a price, he would load the garbage on barges and haul it off to landfills in North Carolina. A deal was struck and the first barge, the *Mobro 4000,* was loaded with 2.8 million kilograms (3100 tons) of baled garbage and towed out to sea on March 22, 1987 (Figure 15-1). The *Mobro 4000* docked in Morehead City, North Carolina, to unload, but local citizens saw the garbage coming and protested to city officials. The Morehead City mayor got the governor of North Carolina involved and the *Mobro 4000* was not allowed to dock anywhere in the state, let alone unload. North Carolina would have nothing to do with Long Island's garbage.

The *Mobro 4000* put out to sea to find a home for the garbage. Landfills were sought in places like Louisiana, Texas, Belize, and Mexico, but no one would take it. Mexico's navy wouldn't even let the barge into Mexican waters. Before the *Mobro 4000* gave up and returned to New York in May, six states and three countries had refused to let Long Island's garbage off the barge. It was a fruitless two-month, 9600-kilometer (6000-mile) journey that was chronicled daily in the news and late-night talk shows. The fate of Long Island's garbage on the *Mobro 4000* became a national joke. It was a joke with a message, though: Getting rid of people's garbage isn't that easy.

Upon its return to New York, the *Mobro 4000*'s cargo was eventually incinerated and the ash buried in a Long Island landfill. Since then, Long Island has dealt with its large volume of garbage by incineration (some off-island), burning some to help generate power, and increasing recycling. The garbage disposal challenge faced by Long Islanders has been repeated in many places around the country. The experience has helped people better understand how much garbage they generate and how it is disposed of.

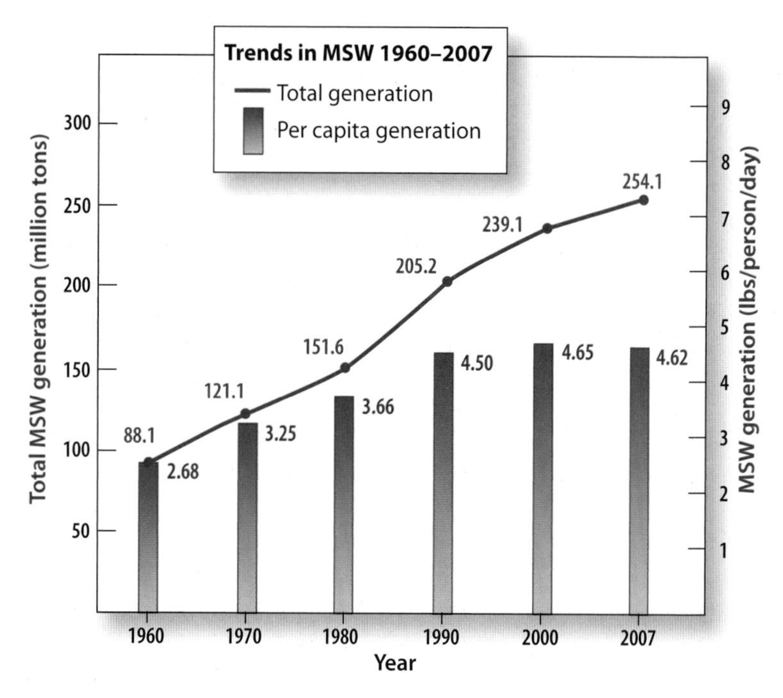

FIGURE 15-2 ▲ Municipal Solid Waste (MSW) Generation in the United States
The amount of municipal solid waste—garbage—that each of us generates has remained more or less the same since about 1990, but our increasing population makes the total amount of garbage continue to increase.

People in the United States generate 230 billion kilograms (254 million tons) of garbage—technically called *municipal solid waste*—each year. The amount generated per person has not changed much in recent years, but as Figure 15-2 shows, the total amount continues to increase as population increases. Most of this waste is sent to landfills. Landfills have gotten bigger and bigger, but their number has steadily declined, from 8000 in 1988 to 1654 in 2005. Many are running out of space. What are we going to do with all our garbage, and how can we protect the environment in the process?

A big part of managing people's effects on the environment is managing how they dispose of waste—everything from household garbage to waste rock at mines, emissions from vehicles, and spent nuclear fuel. All of these are wastes of one sort or another. It is also about their land-use choices and how they exploit and use natural resources—everything from animals and plants to soil, water, air, mineral, and energy resources. Essentially, it's about managing people's interactions with Earth systems.

Through a combination of experience and analysis, people have developed several approaches to managing their environmental impacts. As you will see, though, there is still room for improvement.

IN THIS CHAPTER YOU WILL LEARN:

- What environmental policies are and how they are established
- What environmental regulations are and how people implement and enforce them
- How people use litigation to oppose environmental regulatory decisions
- How business opportunities evolve from environmental regulation and public interest
- How economic incentives influence people's actions and the environment
- How adopting a collaborative rather than an adversarial approach can help people make better environmental decisions

15.1 Environmental Policy

Policy is a term that encompasses both specific plans of action and principles that guide future decisions. Any person or organization can have policies. People have policies guiding their financial decisions (I will not use a credit card.) or eating habits (I will not eat potato chips.), for example. Environmental policies are established to guide how people's actions affect the environment. An individual's environmental policy might be a commitment to recycle household garbage as much as possible. Companies can have polices to manage the environmental effects of their business activities and operations. Take the global investment bank, Goldman Sachs, for example.

Goldman Sachs employs 26,500 people around the world. Through its banking, trading, and financial services businesses it generates about $40 billion of revenue and $10 billion of profit each year. It is one of the most successful investment services businesses in the world. You may wonder why bankers in big downtown office buildings need an environmental policy. But Goldman Sachs's financial fingers reach into many businesses that raise environmental concerns.

As stated in its *Environmental Policy Framework*, the company recognizes that it affects the environment through the goods it purchases, the manufacturing and production it finances, and the investments it makes. | Goldman Sachs environmental policy framework | For example, Goldman Sachs's environmental policy explicitly prohibits "financing or investing in industrial activity in certain limited areas that are so environmentally sensitive that they must be preserved in their present condition." Companies like Goldman Sachs are corporate citizens whose environmental policies are as effective as their employees make them. Environmental policies in such companies can be so much window dressing unless all employees understand the policies and apply them in company decision making. These companies commonly review their annual environmental performance by comparing the past year's activities to their environmental policy guidelines.

What You Can Do

Investigate Corporate Environmental Policy

Is there a company in your hometown or in a particular line of business that you think should devise and follow an environmental policy? Are you curious to know if this company has even thought about the environmental consequences of its activities and operations? Many companies maintain websites that provide information about their business activities and performance. Try some Web searches for companies such as Ford or ExxonMobil and see what you come up with. | Ford Motor Company environmental policy | | ExxonMobil environmental policy | If you can't easily find a company's environmental policy, then it probably doesn't have one—or at least it doesn't consider the one it has to be very important.

Environmental policies are not very helpful if they are primarily established for public relations purposes. Continue your Web-based investigations by comparing a company's environmental policy to its annual reports on environmental performance (sometimes these are within other reports with titles like "Corporate Citizenship" or "Health and Safety"). The first such Goldman Sachs report is very impressive. | Goldman Sachs environmental policy: 2006 year-end report | This company does seem to be taking its environmental policy seriously.

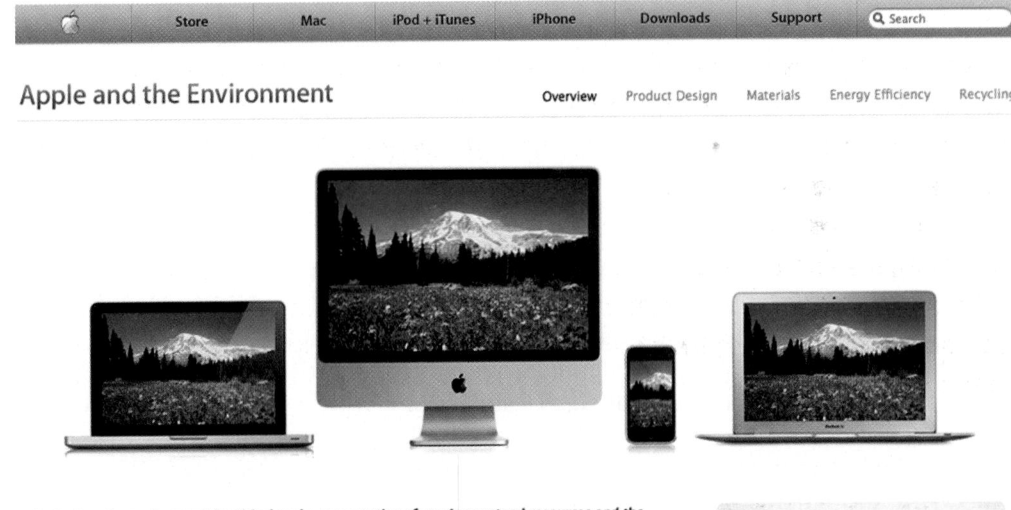

Personal and company environmental policies are as effective as individuals choose to make them. Another source of environmental policy carries much more authority. These are laws established by local, state, and federal governments that define how environmental issues will be addressed by all citizens, including companies and government organizations themselves. Laws establish authoritative polices because they can be legally enforced—compliance with this type of environmental policy is not voluntary. Environmental policies established at the national level—federal laws—are the most influential as they commonly guide environmental polices established at state and local levels.

NATIONAL ENVIRONMENTAL POLICY

National environmental policy is defined by federal laws or acts. Many acts establish some aspect of environmental policy, but those with the most significant and ongoing influence on environmental decisions in the United States are the National Environmental Policy Act (NEPA), Clean Water Act (CWA), Clean Air Act (CAA), Resource Conservation and Recovery Act (RCRA), and Comprehensive Environmental Response, Compensation, and Liability Act (CERCLA). As you can see, acronyms are an established part of the language in environmental management.

Major environmental laws

National Environmental Policy Act

The **National Environmental Policy Act (NEPA)** brought environmental responsibility to the federal government. It became law in 1970 "to declare a national policy which will encourage productive and enjoyable harmony between man and his environment; to promote efforts which will prevent or eliminate damage to the environment and biosphere and stimulate the health and welfare of man; to enrich the understanding of the ecological systems and natural resources important to the Nation." This act requires federal agencies to adhere to these guidelines in all their major or significant actions, which include the actions of others that the agencies authorize or regulate and includes major projects that federal agencies help fund. With the federal budget now approaching $3 trillion each year, NEPA can apply to a lot of projects—everything from new highway construction to offshore oil and gas leasing. Some federal actions are exempt from NEPA oversight. For example, the Department of Defense does not follow NEPA guidance where national security is the priority concern.

NEPA has become a central component of how environmental affairs are managed in the United States. Every significant federal undertaking requires a formal report evaluating its environmental consequences. Environmental Assessment reports may determine that the undertaking does not have a significant impact—but if it does, a formal **Environmental Impact Statement (EIS)** is required as a necessary step in planning and eventual approval of the undertaking. Projects that can significantly affect the environment cannot go forward without these statements (Figure 15-3).

FIGURE 15-3 ▲ A Typical Environmental Impact Statement
Environmental Impact Statements, like this one for a highway crossing of the Columbia River, are typically extremely long and detailed documents.

The application of NEPA to the federal government's business requires that the scope of concerns addressed in an EIS be very broad. Environmental Impact Statements don't just cover the obvious things like possible effects on water quality or habitat. They must also address how the undertaking will impact economic, social, health, and cultural resource factors. This aspect of NEPA explicitly acknowledges that significant projects affect many interrelated factors. Evaluation of alternatives, including the decision to take no action, is also required in an EIS.

NEPA also purposely involves the public and all other interested or affected parties—**stakeholders**—in the EIS process. Stakeholder input is received at public hearings that are required parts of developing an EIS. Written comments can also be submitted to the federal agency taking the lead in developing an EIS during a defined comment period. Stakeholder input is most commonly solicited at the beginning of the EIS process to help clarify the scope the EIS studies should take. This is the stage at which federal agencies define the scope of an EIS—what it will include and how it will be structured. The scope of an EIS can include new scientific or cultural studies to address issues and concerns that agency staff and stakeholders identify.

Stakeholder input is required by law and federal agencies can be held accountable if it is not satisfactorily provided. Environmental Impact Statements can be legally challenged if any part, including stakeholder input, is thought to be inadequate.

In the News

The Alaska Natural Gas Pipeline

Over 850 billion cubic meters (30 trillion ft^3) of natural gas resources have been identified in Arctic Alaska. Arctic Alaska, however, is a long way from where the gas is needed—the continental United States—and until recently it was not economical to produce. With higher energy prices, though, incurring the tremendous cost of constructing a pipeline to transport Arctic Alaska gas to market has become more feasible. A proposal to build a gas pipeline that connects Arctic Alaska to existing pipelines in Canada has been favorably received by Alaskan leaders. The Alaska Gasline Inducement Act The proposal is to build a pipeline 2750 kilometers (1710 mi) long, including 1200 kilometers (745 mi) and six compressor stations in Alaska, that would transport 130 million cubic meters (4.6 billion ft^3) per day. Constructing this pipeline is a huge project that will take years to permit and complete.

The construction of the pipeline will require an EIS. The EIS will need to cover a very wide range of issues, from designs to protect against natural hazards to the possible impacts on indigenous people and their subsistence resources. You can keep track of the project online and watch the EIS process unfold. There will even be ways for you to submit comments as part of the EIS process. Stay tuned, the permitting and construction of this pipeline will be in the news.

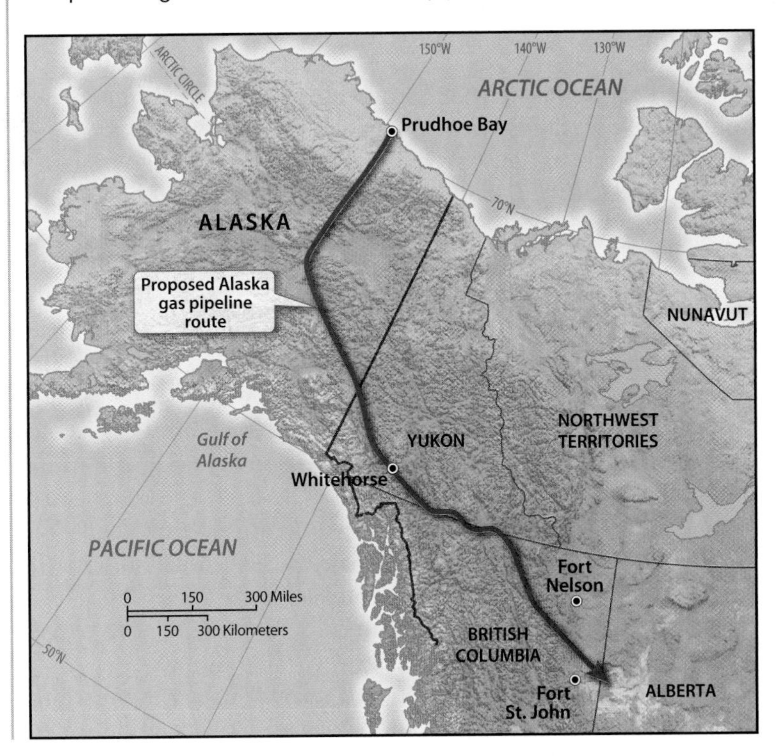

FIGURE 15-4 ▲ City Storm-Water Runoff Is Managed by NPDES Permits
NPDES permits are required because storm water that runs off streets, parking lots, and buildings is transferred to surface bodies of water such as lakes and rivers.

Clean Water Act

Controlling water pollution has origins in laws that go back to 1948. Increased public concerns about water pollution, especially after the Cuyahoga River in Ohio caught fire (see Chapter 10), led to reorganization and expansion of the Federal Water Pollution Control Act in 1972. This act, commonly just called the **Clean Water Act (CWA)**, is amended just about every year. Its primary goal is to restore and maintain the quality of surface and groundwater resources so that they are satisfactory for public water supplies, fish and aquatic life, and industrial purposes—in common terms "swimmable and fishable."

The CWA establishes a system for managing the discharge of pollutants to the nation's waterways and it provides mechanisms for enforcing its requirements. Under the CWA, all discharges of water pollutants must be formally permitted. The permitting system is called the **National Pollutant Discharge Elimination System (NPDES)** and it covers everything from point sources, such as industrial wastewater and marine sanitation devices, to nonpoint sources, such as storm-water runoff from urban areas (Figure 15-4).

As you might expect, laws are as effective as their enforcement and the CWA defines several mechanisms that ensure it is truly the law of the land when it comes to water pollution. Enforcement mechanisms provided by the CWA include financial penalties and both civil and criminal legal remedies.

The EPA has considerable discretion in how it applies its enforcement authority. An example comes from Massachusetts. Between June 2004 and February 2006, the Taunton

(Massachusetts) Municipal Lighting Plant did not satisfactorily monitor its wastewater discharge to the Taunton River. NPDES permits require water quality monitoring to verify that discharge requirements are being met. Scheduled sampling and water quality measurements need to be made and systematically recorded. If this is not done, the permit holder is in noncompliance and enforcement actions can result. The EPA found the Taunton Lighting Plant in noncompliance and, in addition to a fine, required it to contribute to an environmental management plan for a nearby lake. You will learn more about enforcement but, as this example shows, the Clean Water Act authority not only controls specific pollutant discharges, it can also become used to accomplish other environmental objectives the EPA may have.

Clean Air Act

The **Clean Air Act (CAA)** is another federal law that has evolved over the years. Its original purpose in 1963 was to fund the study and cleanup of air pollution, but amendments have expanded its scope and authority. It now regulates activities that have the potential to cause everything from acid rain to stratospheric ozone depletion. The CAA authorizes national air quality standards and an emissions permitting system much like that controlling water pollution under the CWA.

You have learned how the Clean Air Act has been effectively applied to limit pollutant emissions, especially those from vehicles, in Chapter 14 (see Figure 14-5). Implementation of the CAA by the EPA has been very successful in cleaning up air quality and it appears to be paying off. A study reported in 2009 concluded that air pollution reductions, especially of particulate matter, had helped increase life expectancy of people living in U.S. cities. The CAA will become an even bigger influence on national environmental policy as the Supreme Court's April 2, 2007 decision—that greenhouse gases are air pollutants that must be regulated under the Clean Air Act—is implemented by the EPA.

Resource Conservation and Recovery Act

The **Resource Conservation and Recovery Act (RCRA)**, commonly pronounced "rick-rah," establishes policies that cover the generation, storage, transport, and disposal of solid and hazardous waste. Solid waste includes things like construction debris and people's garbage; hazardous wastes include anything toxic, ignitable, corrosive, or reactive. Examples of hazardous wastes include oil, solvents, materials with heavy metals, and acids (Figure 15-5). RCRA's goals are to:

- Protect human health and the environment from the potential hazards of waste disposal.
- Conserve energy and natural resources.
- Reduce the amount of waste generated.
- Ensure the wastes are managed in an environmentally sound manner.

FIGURE 15-5 ▲ Regulations Guide How Hazardous Wastes Are Handled and Disposed Of

Hazardous wastes, defined as chemicals that are toxic, ignitable, corrosive, or reactive, include solvents, paints, acids, and organic chemicals.

In keeping with RCRA's overall goals, EPA guidance emphasizes integrated waste management systems that provide for reductions in the source of wastes and recycling as much as possible. However, solid waste is still generated in large amounts and must be disposed of. The fate of some, like that on the *Mobro 4000* (see Figure 15-1), is to be incinerated, but most is placed in landfills.

RCRA addresses current and future operations, not wastes left over from past operations. Facilities that deal with hazardous waste receive permits and must meet waste management and treatment standards. They must rigorously account for the amounts of hazardous materials generated, accumulated, transported, and disposed of—a "cradle to grave" accounting system. For example, mining operations that dispose of metal-bearing waste rocks (Chapter 12) must report the amounts annually. Even though mine waste rocks are natural materials, RCRA reporting commonly identifies mines as very large generators of "toxic" waste.

Landfills that receive solid and hazardous waste must meet specific EPA criteria for location, design, construction, operation, monitoring, and eventual closure. The typical engineered landfill design is like a "dry tomb" that physically isolates the waste from interaction with air, surface water, and groundwater.

Hazardous waste landfills (called Subtitle C landfills after the section in RCRA that pertains to them) are especially tight "dry tombs." They have at least a double layer of impermeable material (liner) at their base and an overlying impermeable cover that also controls surface runoff, as shown in Figure 15-6.

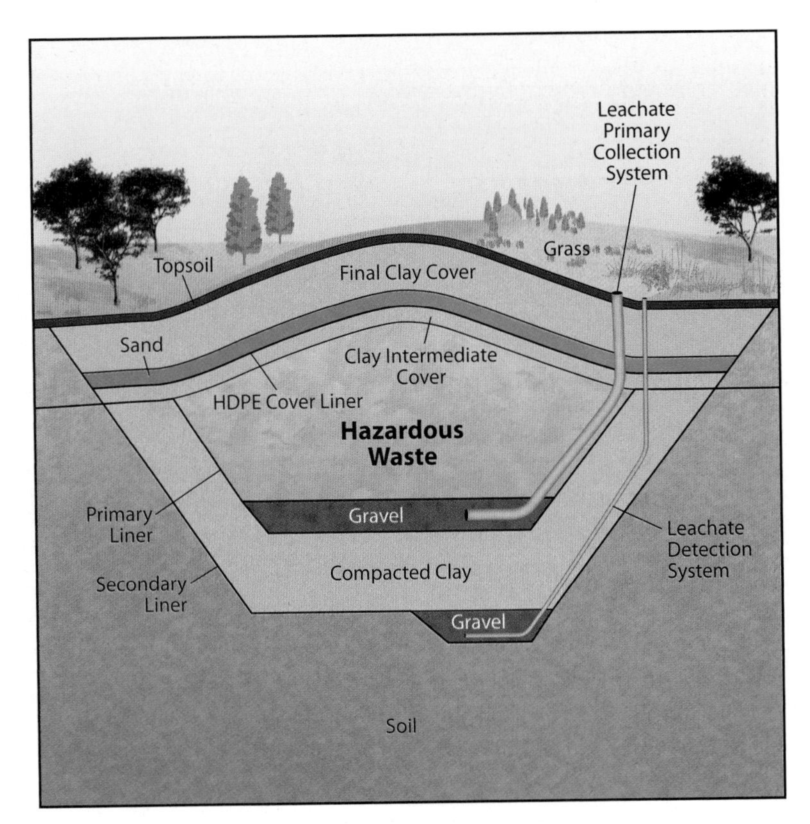

FIGURE 15-6 ▲ A Subtitle C Landfill Design
The leachate detection system checks for leaks that could contaminate groundwater.

The design allows for collection of liquid (called **leachate**) at the base of the landfill. Leachate is derived from the enclosed waste or from surface water that infiltrates the landfill. If leachate forms, it is removed and treated. Groundwater around landfills is monitored as a check on groundwater contamination. The EPA provides regulatory oversight of Subtitle C landfills unless it has specifically authorized oversight by state agencies in its place.

Landfills for nonhazardous municipal solid waste (household garbage, refuse, construction materials, and so forth) are called Subtitle D landfills. They are similar in design to Subtitle C landfills but commonly have a single impermeable liner and compacted clay at their base. They may also have collection systems for both leachate and methane generated from the decay of organic waste in the landfill. More and more Subtitle D landfills are becoming bioreactors that produce methane for use (Chapter 13). Such bioreactor landfills allow more interaction of air and water with the landfill waste. RCRA directs regulatory oversight of Subtitle D landfills to state and local governments.

We've learned from Long Island's experience in disposing of their municipal solid waste that landfills reach their limits. As RCRA emphasizes, source reduction and recycling will continue to be priorities when it comes to managing solid waste.

What are some sustainability issues associated with solid waste generation and disposal?

Comprehensive Environmental Response, Compensation, and Liability Act

Laws like NEPA, CWA, CAA, and RCRA are designed to regulate people's current actions that can affect the environment. What about the environmental impact of actions that occurred before these laws were in place? The environmental consequences of historical actions, such as mining and manufacturing that have gone on for hundreds of years, are addressed by the **Comprehensive Environmental Response, Compensation, and Liability Act (CERCLA)**. CERCLA, enacted in 1980, authorizes the federal government to respond to releases of hazardous substances from closed and abandoned sites, provides ways to allocate liability for these releases, and establishes a trust fund (with tax revenues from the chemical and petroleum industries) to clean them up. Because of this funding capability, CERCLA is commonly called **Superfund**.

The EPA evaluates uncontrolled waste sites with a Hazard Ranking System (HRS). This mechanism determines if a site has released (or has the potential to release) hazardous substances into the environment, considers the characteristics of the substance (such as its toxicity), and evaluates whether people or sensitive environments have been affected by the release. If the site is found to be sufficiently hazardous, it is added to the **National Priority List (NPL)** and recognized as a priority for continued study and potential cleanup. Superfund focuses on cleaning up NPL sites.

There are now over 1200 Superfund sites, but the trust fund is only large enough to clean up some of them. Therefore, the mechanisms for allocating liability that CERCLA provides have become important to funding cleanups of hazardous sites. Under CERCLA, any "**potentially responsible party (PRP)**" can be liable for cleanup costs. PRPs can be owners or operators, as well as those who arranged disposal or transport of a hazardous substance, pollutant, or contaminate to a site. Even an owner that obtained the site many years after the environmental problem was created can be held accountable for its cleanup.

Once a PRP is identified and ordered to clean up a site, noncompliance can lead to fines of $25,000 per day. Under CERCLA it just takes one PRP to get cleanup started. It is this lead PRP that becomes responsible for identifying other PRPs to share the cost. As a result, PRPs commonly wind up suing each other to force the sharing of cleanup costs under CERCLA. This has led to a tremendous amount of litigation, and in some cases legal costs have exceeded eventual cleanup costs.

A very large Superfund site is the area in and around Butte, Montana. Actually, if one includes all the related sites, it may be the largest area being regulated under Superfund in the country. The area of concern extends from Butte downstream along Silver Bow Creek, includes the old smelter town of Anaconda, and continues down the Deer Lodge Valley for a total distance of 193 kilometers (120 mi).

Butte is located on what was once called the "richest hill on Earth." The mines here produced almost 10 million tonnes (11 million tons) of copper between 1880 and 2005. In its early days, Butte was a rowdy and bustling city, the largest between

Minneapolis and Seattle. Like other western mining towns, business and politics ruled the day. It was a hotbed for labor movements and progressive politics, electing a Socialist mayor in 1914. What wasn't high on the list in early Butte was the environment.

Waste rocks were expeditiously dumped near mine operations, tailings from mills were discharged down Silver Bow Creek, and a huge smelter at Anaconda spewed gases containing arsenic and sulfur dioxide into the air (Figure 15-7). The mine workings extended deep below the water table, and today acid rock drainage is filling an open pit (see Figure 12-30). As the contaminated water accumulates in the pit, it threatens groundwater quality. Butte has become a poster child for the worst legacy environmental issues associated with mining.

The environmental cleanup of the Butte area is done under the EPA's Superfund regulatory authority. EPA Silver Bow Creek/Butte Superfund site An oil company that purchased the mines in 1977 became the lead PRP and must now clean up impacted soils, stream sediments, and water regardless of when or why it occurred. The cleanup will go on for years.

FIGURE 15-7 ▲ The Anaconda Smelter in Operation
This large copper smelting facility operated from 1918 to 1981.

What You Can Do

Investigate Superfund Sites Near You

Would you like to know if you live near a Superfund site? The EPA identifies the location and the cleanup history of these sites. Superfund sites where you live The sites identified as being on the National Priority List are those that are being managed under CERCLA authority. If a site is listed as active, cleanup is ongoing there. You can learn what the environmental problem is and what is being done about it.

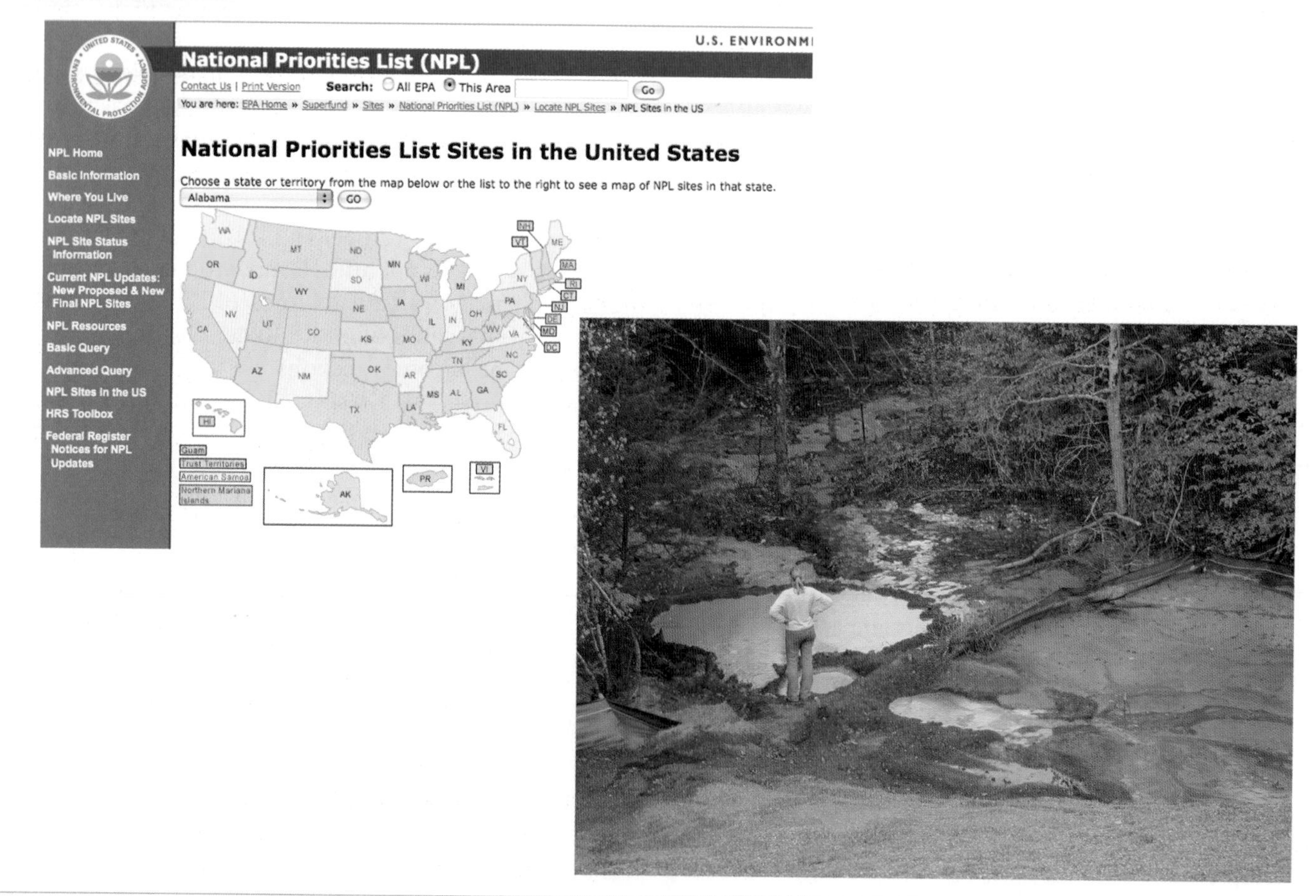

LAND-USE DESIGNATIONS

Land-use designations establish environmental policies through the activities they allow or disallow. These designations are made at local, state, and national levels (parks are an example). Federal lands are the largest areas affected by land use designations. The laws that guide these designations include the Antiquities Act, Wild and Scenic Rivers Act, Fish and Wildlife Act, and National Wildlife Refuge System Improvement Act. In addition to acts of Congress, presidential executive orders can withdraw public lands for various purposes such as national parks, preserves, and monuments. The National Park Service now controls over 32 million hectares (79 million acres)—two-thirds of them in Alaska—and there are more than 34 million hectares (84 million acres) in the National Wildlife Refuge System. These are areas set aside to protect habitat, wildlife, and recreation values of the land.

What are some ways in which sustainability is related to federal land use designations such as national parks?

STATE ENVIRONMENTAL POLICIES

Much of the authority developed by national environmental laws is delegated to states in a number of ways. States independently establish environmental polices, too. It's common for states (and local governments) to have many policies covering environmental concerns such as recycling, littering, and household waste disposal. For example, in the state of Washington, the **State Environmental Policy Act (SEPA)** is modeled after NEPA. It requires state and local agencies to consider the likely environmental consequences of a proposed action before approving or denying it. Colorado's Voluntary Cleanup and Redevelopment Act encourages landowners and other stakeholders to identify contaminated sites, propose remedial actions, and carry out cleanups. Colorado has state policies to diminish legal or other barriers to cleaning up contaminated sites.

SPECIAL INTEREST GROUP POLICIES

Special interest groups also develop and promote environmental policies. These groups work to mobilize public opinion and influence government decisions and actions—they are lobbyists. There are literally hundreds of political advocacy groups, and dozens exist to influence environmental policies. Two examples are the Sierra Club and the Wilderness Society. | Sierra Club | Wilderness Society The Sierra Club "seeks to engage a broad spectrum of citizens around the values of protecting wildlife, public lands, and special places, and blocking threats to these lands from commercial logging, mining, abusive recreation, and overgrazing." | America's wild legacy Organizations like the Sierra Club, commonly just called "environmental groups," firmly hold that wilderness values are more important than other values and aggressively work to limit natural resource development.

Organizations with very different environmental policies, generally advocates of multiple use of public lands, tend to be industry specific. The Alaska Miners Association, for example, works to keep lands available for mineral exploration and development—just the opposite of the Sierra Club's goal. | Alaska Miners Association When specific issues arise, organizations like the Sierra Club and the Alaska Miners Association can be expected to be on opposite sides. Organizations with long-standing advocacy policies are very predictable in their positions, and they are generally not sources from which other interested parties can obtain unbiased information about issues.

PUBLIC AWARENESS AND ENVIRONMENTAL POLICY

Environmental policies in the United States have been strongly influenced by public awareness of specific environmental issues or circumstances. There seems to be a national tendency for Americans to react to rather than anticipate environmental concerns, and there are several examples of how public awareness has come to influence environmental policy.

Public reporting of air pollution episodes that affected thousands of people, killing some, led to public support of the strengthened and expanded CAA. The pollution that enabled the Cuyahoga River to catch fire (see Figure 10-1) was widely reported and helped galvanize public support for a stronger CWA. Nationally reported health problems (and struggles with responsible businesses and government agencies) of people living near a buried toxic waste dump in Niagara Falls, New York—an area called Love Canal (Figure 15-8)—helped CERCLA become law.

Rachel Carson's famous book *Silent Spring* (published in 1962) alerted people and government leaders to the dangers of indiscriminate pesticide use. This very well-written book explains how unexpected and undesirable effects of pesticide

FIGURE 15-8 ▲ Moving Out at Love Canal
This family was forced to abandon their home after it became known that toxic chemicals had been buried in their Love Canal neighborhood.

use can propagate through the environment and even endanger human health. The connected nature of Earth systems that this book emphasizes became a foundation of the environmental movement in the United States.

Today, efforts to increase public awareness of and engagement with environmental issues are key components of environmental advocacy. It's owing to such efforts that the Arctic National Wildlife Refuge (ANWR) is now embedded in the national consciousness. It's also why Al Gore's film about global warming, *An Inconvenient Truth,* was made.

15.2 Environmental Regulation

In matters of environmental enforcement, regulations implement the authority conveyed by laws. They turn the general guidance embodied in laws into specific guidance. NEPA, for example, requires the federal government to *use all practicable means to create and maintain conditions under which man and nature can coexist in productive harmony.* Regulations define the specific ways this will be done.

Government agencies create regulations. Regulations set environmental standards (as for air and water quality), implement standards through permitting processes, and enforce standards and procedures. Like laws, regulations are legally enforceable; the agencies that create them enforce them.

ENVIRONMENTAL STANDARDS

Environmental standards are specific measures of environmental quality—they define the acceptable levels of pollution in air, water, and soil. These standards define acceptable pollution discharge levels and establish criteria for responses to contamination, such as cleanup actions and regulatory enforcement proceedings. The EPA is the principal definer of environmental standards in the United States. State agencies can also create environmental standards, but they must be at least as strict as EPA standards.

Air quality standards

Six pollutants are used to define and measure air quality: carbon monoxide, lead, nitrogen dioxide, particulate matter, ozone, and sulfur oxides (Table 15-1). There are many other air pollutants, such as volatile organic compounds (VOCs, discussed in Section 14.2), but these six "criteria" pollutants have been found satisfactory for measuring general air quality. The primary standards are set to protect human health and the secondary standards are those that are set to protect visibility, animals, vegetation, and infrastructure such as buildings (see Table 15-1).

There are also standards for air pollution discharges. For example, standards limit the amount of VOCs that industrial operations such as petroleum refineries can emit. Pollutant discharge standards are linked to operation permits. Operations monitor and report their emissions to regulatory agencies, and enforcement actions result if emissions exceed the permitted standards. Vehicle emission standards have been a major influence on cleaning up air quality (see Figure 14-5).

Water quality standards

Water quality standards to protect human health (drinking water standards) and aquatic life have been established for 126 priority toxic pollutants. Current national recommended water quality criteria The pollutants fall into several categories such as heavy metals (copper and silver, for example) and VOCs (benzene and chloroform, for example). The standards are based on the

TABLE 15-1 NATIONAL AMBIENT AIR QUALITY STANDARDS FOR SIX CRITERIA POLLUTANTS

Primary standards protect human health. Secondary standards protect visibility, animals, vegetation, and infrastructure such as buildings. Different standards may apply to different time periods (averaging times). Units of measure for the standards are parts per million (ppm) by volume, milligrams per cubic meter of air (mg/m^3), and micrograms per cubic meter of air (μg/m^3).

NATIONAL AMBIENT AIR QUALITY STANDARDS

Pollutant	Primary Standards	Averaging Times	Secondary Standards
Carbon monoxide	9 ppm (10 mg/m^3)	8-hour	None
	35 ppm (40 mg/m^3)	1-hour	None
Lead	1.5 μg/m^3	Quarterly average	Same as Primary
Nitrogen dioxide	0.053 ppm (100 μg/m^3)	Annual (arithmetic mean)	Same as Primary
Particulate matter ($PM_{2.5}$)	15.0 μg/m^3	Annual (arithmetic mean)	Same as Primary
	35 μg/m^3	24-hour	Same as Primary
Ozone	0.08 ppm	8-hour	Same as Primary
	0.12 ppm	1-hour (Applies only in limited areas)	Same as Primary
Sulfur oxides	0.03 ppm	Annual (arithmetic mean)	----
	0.14 ppm	24-hour	----
	----	3-hour	0.5 ppm (1300 μg/m^3)

expected use of the water—standards for drinking water are more stringent than those for irrigation water, for instance. States use EPA standards as guidelines for their own standards.

EPA water quality standards play a key role in determining how much pollutant discharge regulatory agencies will allow. As we noted in the previous section, the Clean Water Act (CWA) created the National Pollutant Discharge Elimination System (NPDES) to monitor and control pollutant releases to the nation's waters. Discharges are controlled so that pollutant levels do not exceed water quality standards in the receiving water bodies. The amount of pollutant that a water body can receive without exceeding its water quality standard is called its **total maximum daily load (TMDL)**. For drinking water, the highest level of a contaminant that is allowed is called the **maximum contaminant level (MCL)**.

The application of water quality standards is illustrated by those in place for treatment of the highly acidic and metal-bearing water removed from an open pit (Berkeley pit) in Butte, Montana (Figure 15-9). The acid rock drainage (Section 12.3) filling the pit must be treated before it can be released to Silver Bow Creek or be used for industrial purposes. A water treatment plant to accomplish this was completed in November 2003. The treated water is now used for the ongoing mining operations but over the long term discharges will be made to Silver Bow Creek. These discharges must meet standards listed in Table 15-2.

TABLE 15-2 WATER QUALITY STANDARDS FOR DISCHARGE FROM THE HORSESHOE BEND WATER TREATMENT PLANT, BERKELEY PIT, BUTTE, MONTANA

Metal concentrations are given in micrograms per liter (parts per billion).

Contaminant	Average Pit Concentration	Discharge Standard
Aluminum	270,000	87
Arsenic	700	10
Cadmium	2100	0.8
Copper	180,000	30.5
Iron	900,000	1000
Nickel	1200	100
Uranium	700	30
Zinc	620,000	388
pH	2.7	6.5–9.5

Soil quality standards

Environmental standards like those for air and water quality have not been established for soil quality. However, soil pollution can seriously degrade habitat and be hazardous to people. Linking human health and soil pollution requires careful analysis of the nature of the contamination and how people will be exposed to it. To help determine if this type of analysis should be done, the EPA has developed a soil screening guidance that uses levels of contaminants in soil to identify areas of potential concern—areas that need further analysis to determine the human health risks they present.

The EPA has defined **generic soil screening levels** for 110 chemicals, including metals like cadmium and lead (Table 15-3, p. 478). If the concentration of one of these chemicals in a soil exceeds the screening level, then further analysis of the risk to people is indicated. Depending on how people use the contaminated area, regulatory action may or may not be required. Generally, residential areas that contain soils with contaminant concentrations above generic soil screening levels have come to be judged hazardous by the EPA. Many cleanups of these areas have been enforced as a result.

An example of determining risks from exposure to contaminated soils and establishing site-specific soil cleanup standards, or **action levels**, comes from the Silver Bow Creek/Butte Superfund

FIGURE 15-9 ◀ Remediating Acid Rock Drainage
At the Horseshoe Bend water treatment plant, shown here, acidic water from the Berkeley pit is treated for reuse.

TABLE 15-3 EPA GENERIC SOIL SCREENING LEVELS FOR SELECTED METALS

These concentrations (in ppm) assume exposure through ingestion. If people are exposed to soils containing these metal concentrations, more detailed studies of the health risks they present are warranted.

Element	Concentration
Antimony	31
Arsenic	0.4
Barium	5500
Beryllium	0.1
Lead	400
Nickel	1600
Selenium	390
Silver	390
Vanadium	550
Zinc	23,000

site. As you can see in Figure 15-10, residential areas of Butte are adjacent to old mines and materials like waste rock have become scattered through residential areas. Metal-bearing soils in these areas expose people, especially children that play in yards, to high levels of lead in Butte. After extensive study, including measurements of blood lead levels in children, the EPA established cleanup levels for Butte community soils. The soil in yards that contain lead concentrations above the lead action level of 1200 ppm must be removed and placed in an appropriate disposal area.

What are some ways in which Earth systems are related to environmental standards?

FIGURE 15-10 ▼ Residential Areas of Butte, Montana near Past Mining Operations

The dark triangular structures are headframes that mark openings (vertical shafts) to underground mine workings. The Berkeley pit is in the distance. Some local soils were found to have high lead levels, prompting efforts to remove them.

PERMITTING

Regulatory agencies implement environmental standards through the permitting process. Permits are written permission from an agency to conduct an operation or action. Permits specifically define how the action will be undertaken, how it will be monitored and reported, and in many cases how it will be completed (for example, when a mine closes, as discussed in Section 12.3).

You now understand how environmental concerns are interconnected through Earth systems interactions—how effects on water, air, and soil quality can be related. People's actions, from driving cars to operating large industrial facilities, can impact all Earth systems. This is fundamentally why environmental permitting is an extensive process that involves many government agencies—all those that regulate air, water, and soil quality. Human activities that can have significant environmental consequences invariably lead to overlapping regulatory jurisdictions among government agencies.

Just determining what permits apply to a specific action can be challenging. A government handbook helps people understand how permits are applied in Washington state. Environmental Permit Handbook Permits fall into 13 different categories (Table 15-4) and there are commonly a dozen or more in each category—21 different permits cover water quality, for example. Permitting operations in or around wetlands are especially complicated and illustrate how overlapping jurisdictions come into play (Table 15-5).

Remember how difficult it was for Long Islanders to get rid of their garbage? Why didn't they just develop a new landfill or expand an old one? Do you think there is a business opportunity in developing a new landfill? The answer to this question isn't simple. It takes a lot of time, money, and effort to develop a landfill, and every step involves permitting. Consider the process in Kentucky. Landfill permitting overview

In Kentucky, the first step in developing a landfill is to determine if the proposed landfill is consistent with the local county plan. If it is, then the next step is to submit a Notice of

TABLE 15-4 THE ENVIRONMENTAL PERMIT CATEGORIES USED IN WASHINGTON STATE

- Air Quality Permits
- Aquatic Resource Permits
- Archaeology and Historic Preservation Permits
- Federal Requirements
- General Requirements
- Land Resource Permits
- Livestock Permits
- Local Permits
- Pesticide Permits
- Waste and Toxic Substance Permits
- Water Quality Permits
- Water Resource Permits
- Wetland Permits

TABLE 15-5 LIST OF GOVERNMENT AGENCIES INVOLVED IN WETLAND PERMITTING
• Local Jurisdiction (City or County Planning)
Floodplain Development Permit
Shoreline Substantial Development, Variance, or Conditional Use Permit
Growth Management Critical Areas Ordinance Requirements
• State Jurisdiction
Aquatic Use Authorization
Hydraulic Project Approval
Section 401 Water Quality Certification
Coastal Zone Consistency Determination
Noxious Aquatic and Emergent Weed Transport Permit
• Federal Jurisdiction
Section 404 Permit (Army Corps of Engineers)

Intent to the Division of Waste Management (and publish it in a local paper). At this stage, all relevant agencies and the applicant identify the permits that are required, everything from water discharge to transportation permits. Next, the proposed landfill site is characterized and the general plan for developing the landfill is submitted to the lead regulatory agency, the Kentucky Division of Waste Management. This information is made available to the public. Any interested party can submit comments about the plan during a 30-day comment period; stakeholders can request public hearings at this stage. Finally, the technical details for the proposed landfill are submitted. If these are accepted by the regulatory agency, it triggers another 30-day public comment period that can lead to additional public hearings if a stakeholder disagrees with the regulatory decision. If this happens, the new public hearings are called "adjudicatory." This means that the issues leading to the public hearing are adversarial and all sides must be represented by an attorney. The proposed landfill will not be approved (receive a construction permit) until the issues are resolved.

On average, the Kentucky landfill permitting process takes 18 to 24 months and $750,000 to $1,200,000 to complete. The costs include those to develop information in the permit applications (groundwater flow patterns, for example) as well as fees to cover the review costs of the permitting agencies. It is a long, complicated, and costly process. Even if construction is approved and completed, another permit is required to operate the landfill. Operation will not be approved until financial assurance bonds (money set aside for closure costs) are in hand. The end result of a successful permitting and development process is a landfill with many technical components that protect the environment (see Figure 15-6).

The Kentucky landfill permitting process is fairly typical of the permitting process for all substantial industrial facilities in the United States (you can see another example in Figure 15-11 on p. 480). The process is followed in order to protect the environment and to assure that concerns raised by stakeholders are addressed. Development projects such as new mines take even longer, are more complicated, and are more expensive to permit. Scrolling through the document summarizing Alaska's large mine permitting process will give you a feel for the number and scope of permits that guide development of new mines. | Large mine projects | It can take many years and many millions of dollars to gain regulatory approval of new mining operations in the United States.

ENFORCEMENT

Environmental standards and regulatory permitting implement specific guidance to protect the environment, but this guidance could be just so much paper if it were not effectively enforced. Regulatory enforcement starts with the permits themselves and includes penalties for noncompliance and civil and criminal legal remedies.

The first step in enforcing environmental standards is tied to the permitting process. By denying permit approval, regulatory agencies prohibit undertakings that do not, in their judgment, adequately protect the environment and stakeholder interests.

Permitted operations are required to monitor and report their performance. For example, discharges of water are routinely tested to determine the amount of pollutants they contain. If the discharges do not meet permitted guidelines, regulatory agencies can impose financial penalties. These administrative penalties can be substantial, up to thousands of dollars per day, but they have defined upper limits. Civil penalties can also be imposed but these require court approval. The civil penalty fines can be up to $27,500 per day and they can be assessed until the violation is corrected.

The most serious environmental violations can be enforced by criminal penalties. Individuals can be held accountable and penalties include jail time in addition to fines. Filing false statements and knowingly causing pollution that endangers people are examples of violations that can lead to criminal penalties. The prosecuted individuals can be those directly responsible for environmental compliance as well as their supervisors.

Criminal penalties can be significant. Alexander Salvagno and his father Raul Salvagno were convicted of conspiring to violate the Clean Air Act and other federal laws in 2006. They operated a business that removed asbestos-bearing materials from buildings in New York state. They used removal methods that endangered workers, falsified air sample data, and left asbestos in buildings they declared to be safe. The Salvagno's were required to forfeit $3,740,614 in assets and pay $45,915,182 in restitution to defrauded clients. Alexander Salvagno was sent to prison for 25 years and Raul Salvagno for 19 years. These are the longest prison terms for environmental crimes in U.S. history.

The 800-pound gorilla in the world of environmental enforcement is the EPA. The EPA may share its authority with other government agencies, commonly at the state level, but it never really gives up its own authority. You can expect the EPA to constantly review the application of environmental standards

PERMITS

The following permits are among the approximately 40 permits required for the construction of the Southern Delivery System Project. These permits would be obtained before construction begins.

Permit Title	Agency Issuing	Purpose / Description
Clean Water Act Section 404 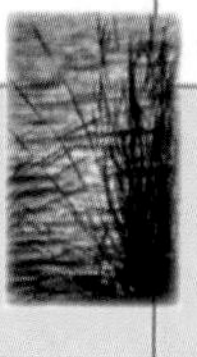	**U.S. Army Corps of Engineers**	**Regulates the discharge of dredged or fill material into waters of the U.S., including wetlands** The SDS pipeline would cross waters of the U.S., including Fountain Creek, Jimmy Camp Creek, and Williams Creek. Construction would occur near wetlands associated with these waterways. Either of these proposed reservoir locations would require an individual 404 Permit.
Clean Water Act Section 401	**Colorado Department of Public Health and Environment**	**Requires that any discharge into waters of the U.S. would not cause or contribute to a violation of state surface water quality standards.** A 401 Water Quality Certification is required of individual Section 404 permits. The state Water Quality Control Division issues the 401 certifications for individual Section 404 permits. An anti-degradation review may be required, which would look at 12 parameters associated with waters. The permittee would be notified if any special conditions may need to be resolved or applied. Individual Section 404 permits require a public notice reviews for anti-degradation and for certification. After the notice and review, the permittee would be notified of acceptance , with standard permit conditions.
Section 7 Consultation	**U.S. Fish & Wildlife Service** 	**Requires consultation with Fish & Wildlife related to impacts to any threatened or endangered species which may be in the project area** A survey will be completed of the Southern Delivery System project area for threatened, endangered and state species of concern. The project will be evaluated for effects on these species. The need for a survey depends on the extent of disturbance to potential habitat. As project limits are determined, there is a potential that construction may impact vegetative wetland or waterways.
Air Pollution Emission	**Colorado Department of Public Health and Environment**	**Ensures control of "fugitive dust," to protect air quality** State air permits for fugitive dust would be required for construction activities that disturb more than 25 acres or that take longer than six months, as is the case for the SDS project. The permit requires completion of, and adherence during construction to a Fugitive Dust Control Plan for: • Onsite, Unpaved Roads. • Disturbed Surface Areas • Transference of Mud and Dirt onto Paved Surfaces Control strategies may involve watering, applying chemical stabilizer, controlling vehicle speed, paving, graveling, minimizing disturbed areas, revegetation, furrow locations, compaction of disturbed soil, installation of windbreaks, synthetic or natural cover for steep slopes, prevention, and cleanup of paved areas.
Construction Stormwater General Permit 	**Colorado Department of Public Health and Environment (CDPHE)**	**Ensures quality of stormwater that may run into roadway drains and into waterways** The permit requires control of sand, dirt, trash, cement wash, vehicle fluids or other pollutants from entering into stormwater. Development and implementation of a Stormwater Management Plan (SWMP) is required. This permit regulates the discharge of stormwater runoff from construction sites that are greater than one acre. Because the SDS would disturb more than one acre, this permit would be required for construction of each component.
Dewatering	**Colorado Department of Public Health and Environment**	**Ensures quality of groundwater and stormwater that may run into roadway drains or waterways** This permit authorizes the discharge of groundwater and stormwater from excavation sites into State waters, and imposes effluent limitations and associated monitoring and reporting requirements.
Minimal Industrial Discharge	**Colorado Department of Public Health and Environment**	**Regulates wastewater discharges into State waters, and permits those with minimal impacts** This permit authorizes the discharge of waste-waters from various sources that are thought to have a minimal impact on water quality. For the SDS project, this permit would cover water discharge from the hydrostatic testing of the pipeline into state waters. The test is needed prior to the final tie-in to the existing pipeline.
Reservoir Plan & Dam Safety 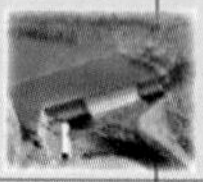	**Office of the State Engineer**	**Requires plans and steps to secure dam safety** The Colorado Department of Natural Resources regulates dam safety. It requires applications for review and approval of plans for the construction, alteration, modification, repair, enlargement, and removal of dams and reservoirs; quality assurance of construction; acceptance of construction, non-jurisdictional dams, safety inspections, owner responsibilities, emergency preparedness plans, and fees; and restriction of recreational facilities within reservoirs. The SDS would include the construction of Jimmy Camp Creek Dam and Reservoir and/or Williams Creek Dam and Reservoir
Special Use	**Cities & Counties in the Project Area**	**Regulates construction where equipment is fenced or in a building** Certain facilities may not be constructed until a special use permit has been issued. Facilities include water reservoirs, sewage lagoons, switching yards, pumping stations, and other component equipment installations on land owned or leased, where the equipment is fenced or placed in a building. The SDS project may involve these conditions.
Excavation Permits	**Cities & Counties in the Project Area (Road & Bridge Depts.)**	**Requires written permission to undertake construction excavation on Pueblo County Right-of-Way (ROW)** Prior to commencing any work on any Pueblo County ROW, the excavator shall obtain written permission to undertake such work. Permits issued by the County only allow work within the County-owned ROWs, and do not permit disturbance of existing installations owned by others.
Pueblo County 1041 	**Pueblo County**	**Requires coordination with Pueblo Board of County Commissioners to obtain a permit for the development of "areas of public interest."** The 1041 permit is required for the following (in Pueblo county): – Site Selection and construction, or major extensions of domestic water and sewage treatment systems – Site Selection and construction of major facilities of a public utility within the unincorporated areas of Pueblo County
Pipeline Encroachment License Agreements	**Railroads, such as Union Pacific Railroad (UPRR) & Burlington Northern Santa Fe (BNSF)**	**Ensures proper insurance is issued, ensures UPRR specifications are met and gives permission to install a pipeline on UPPR ROW.** Pipeline installations may be an encroachment, crossing, or both: • Encroachment – Pipeline enters the railroad ROW and does not leave the ROW or follows along the ROW for some distance. • Crossing - Pipeline crosses the railroad trackage from one side to the other of the ROW, in a straight line. A UPPR rail line is located near the I-25 corridor between Colorado Springs and Pueblo. If the SDS crosses or parallels this line within the UPPR ROW, a license agreement would be required.

FIGURE 15-11 ▲ The Permitting Process

A typical example of the array of permits needed for any environmentally sensitive project. This project is a proposed regional water delivery system designed to serve the water needs of several Colorado cities through 2046.

and regulatory compliance. It always reserves the right to independently assess compliance and initiate enforcement proceedings. EPA enforcement activities required polluters to spend $11.8 billion in 2008 on pollution controls, cleanup, and environmental projects—a record for this agency.

15.3 Third-Party Litigation

Although the processes of regulatory permitting and oversight are rigorous and comprehensive, some people may not agree with the decisions regulatory agencies reach. In these cases, individuals or organizations may initiate civil legal proceedings to reverse regulatory decisions. This reliance on litigation remedies is fairly common in the United States. It is especially common when people disagree with regulators about whether new resource development projects should be allowed. An example is Shell Oil Company's proposed 2007 exploration program in the Beaufort Sea offshore northern Alaska.

Shell successfully permitted a $200 million exploration program involving two drill ships during the summer open-water season in the Beaufort Sea. The permits were approved by several agencies, including the federal Minerals Management Service (MMS). A local government (the North Slope Borough, which is like a county in other states) and the Alaska Eskimo Whaling Commission filed a civil suit contending that Shell and the MMS had not fully evaluated the exploration program's potential affects on endangered bowhead whales. A federal court agreed to review the MMS studies over a period of time that would take longer than the summer exploration season, and the court prohibited Shell from conducting exploration while it was reviewing the case. This shut down the 2007 program and created uncertainty about its future. At the same time, environmental groups filed suits contending that Shell's permits had not satisfactorily addressed how noise and possible oil spills would affect marine life. At a minimum, litigation stalled Shell's exploration program and as of early 2009 it was still not under way.

This type of litigation by environmental groups is common. Many environmental groups believe that resource development on public lands should not occur, and they proactively use the legal system to oppose regulatory decisions that approve exploration or development projects. In a way, this approach is just another part of the American system of checks and balances. However, it is increasingly used to pursue specific environmental agendas, agendas that commonly want to preserve natural lands rather than develop them.

Third-party litigation is now largely expected whenever and wherever new resource development projects or large industrial facilities (even landfills) are proposed. At a minimum, they can cause delays and increase costs for these projects. In some cases they can change regulatory decisions and policies. The controversy surrounding the northern spotted owl is an example. Ethics and the spotted owl controversy

The small, brown spotted owl (Figure 15-12) thrives in old-growth forests of the northwest United States. As these forests became extensively logged, the population of spotted owls dwindled. Using authority provided by the federal Endangered Species Act, the U.S. Fish and Wildlife Service declared the spotted owl threatened. Logging restrictions were imposed to protect it. Environmental groups pursued litigation to help ensure that adequate steps were taken to protect the spotted owl. Over a period of several years, the controversy evolved many new forest management plans and practices. Of spotted owls, old growth, and new policies: a history since the interagency scientific committee report The spotted owl is still a threatened species, and efforts to protect it are still considered insufficient.

FIGURE 15-12 ◀ The Spotted Owl Is an Endangered Species in Northwest Forests Efforts to protect the spotted owl have included restrictions on logging.

In the News

Protecting the Spotted Owl

Actions to protect the spotted owl are still a work in progress. Litigation settlements require federal agencies to develop forest management plans that protect spotted owl habitat. However, recent independent reviews of the proposed plans have concluded that they will not adequately protect this species. Spotted owl plan fails peer review What happens next is not clear. It will take many years to resolve this issue.

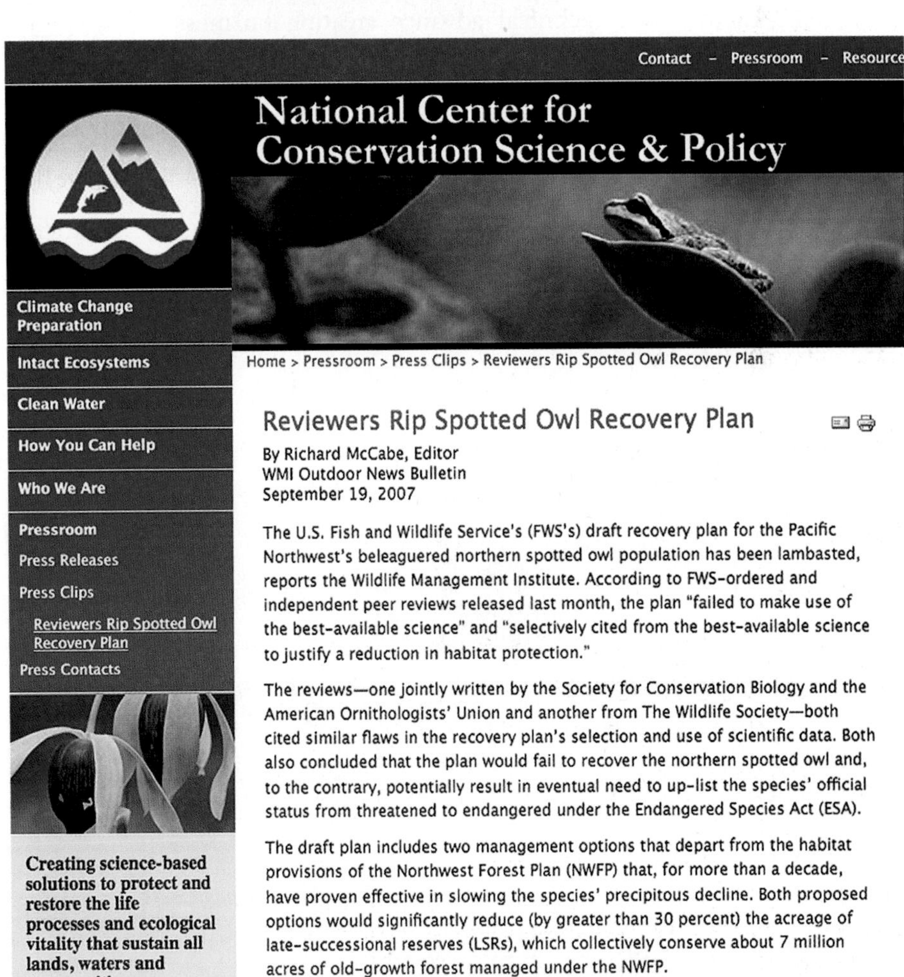

Reviewers Rip Spotted Owl Recovery Plan

By Richard McCabe, Editor
WMI Outdoor News Bulletin
September 19, 2007

The U.S. Fish and Wildlife Service's (FWS's) draft recovery plan for the Pacific Northwest's beleaguered northern spotted owl population has been lambasted, reports the Wildlife Management Institute. According to FWS-ordered and independent peer reviews released last month, the plan "failed to make use of the best-available science" and "selectively cited from the best-available science to justify a reduction in habitat protection."

The reviews—one jointly written by the Society for Conservation Biology and the American Ornithologists' Union and another from The Wildlife Society—both cited similar flaws in the recovery plan's selection and use of scientific data. Both also concluded that the plan would fail to recover the northern spotted owl and, to the contrary, potentially result in eventual need to up-list the species' official status from threatened to endangered under the Endangered Species Act (ESA).

The draft plan includes two management options that depart from the habitat provisions of the Northwest Forest Plan (NWFP) that, for more than a decade, have proven effective in slowing the species' precipitous decline. Both proposed options would significantly reduce (by greater than 30 percent) the acreage of late-successional reserves (LSRs), which collectively conserve about 7 million acres of old-growth forest managed under the NWFP.

Northern spotted owls are dependent on old-growth forest habitat. The species was listed as federally threatened in 1990 due to the destruction of more than two thirds of its native range by logging. Spotted owl pairs need large amounts of land for hunting and nesting and are generally intolerant of habitat disturbance.

In addition to reducing the amount of available spotted owl habitat, the plan further deviates from current management by placing primary importance on

15.4 Economics and Environmental Management

Economic incentives and disincentives play significant roles in obtaining desired environmental outcomes. Providing new ways for people to make money or save money can be powerful influences on their actions. Examples of such influences include creating business opportunities, making undesired actions more expensive, using markets to achieve environmental objectives, and directly funding needed or preferred actions.

BUSINESS OPPORTUNITIES

The development of businesses that contribute to achieving environmental objectives may be prompted by technical advances, environmental regulations, or public interest in the environment. An example of a technical advance creating business opportunities that help the environment is the in situ (in place) leaching of uranium.

In situ uranium leaching

Many uranium deposits form in permeable sandstone where groundwater with dissolved uranium encounters chemically reducing conditions—for example, where carbonaceous material from plant debris is present. Chemical reactions then cause uranium minerals to precipitate. Such deposits, which make up 18 percent of the world's known uranium reserves, have historically been mined by conventional underground or open-pit methods. In situ leaching technology uses the reverse of the chemistry that formed the deposit. It pumps oxidizing solutions (such as groundwater with added oxygen and carbon dioxide) through the deposit to dissolve the uranium. Production wells recover the uranium-bearing solution (Figure 15-13), and ion-exchange mechanisms separate the uranium from the leach solution.

In situ leaching technology now accounts for about 20 percent of the world's annual uranium production. It uses methods that control the underground flow of uranium-bearing solutions and it restores groundwater conditions when the uranium deposit is depleted. In situ uranium leaching replaces conventional mining that would create surface disturbances such as large open pits, as well as waste rock disposal areas like the one shown in Figure 15-14 that change the landscape and threaten soil and water quality if not properly designed. For deposits where groundwater flow can be well-managed, it is also very cost effective. Because of its lower costs, companies consider using this technology wherever it can be appropriately applied, and the environment benefits as a result.

FIGURE 15-13 ▼ How In Situ Leaching of Uranium Works
Solutions injected into the uranium-bearing layer dissolve uranium, are extracted through recovery wells, and transferred to plants where the dissolved uranium is recovered.

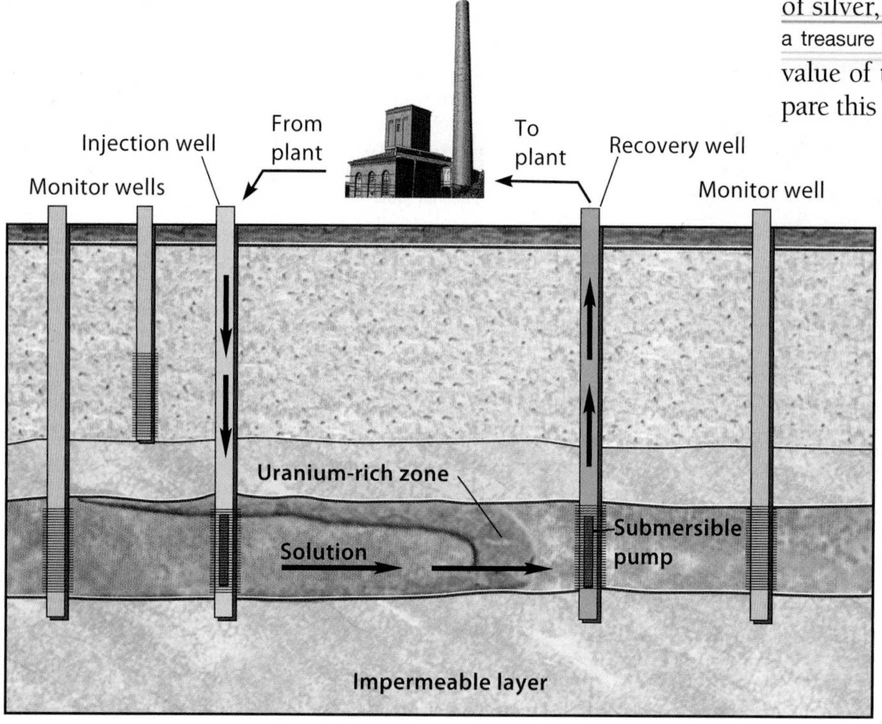

Recycling consumer electronics

Another business opportunity that has some parallels to mining is the recycling of consumer electronic products (Figure 15-15). Such products seem to become obsolete fast—especially things like cell phones and computers. Is there money to be made recycling these devices? Take cell phones, for example. A metric tonne (1000 kilograms, or 2205 lb) of cell phones contains 140 kilograms (309 lb) of copper, 3.1 kilograms (6.8 lb) of silver, and 310 grams (10.6 oz) of gold. | Recycled cell phones – a treasure trove of valuable metals | At recent metal prices, the gross value of this tonne of cell phones would be over $10,000. Compare this to the about $1000 (late 2009 prices) value of a tonne of very rich gold ore, which contains about 1 ounce of gold per tonne. It seems as if money could be made here, but recycling of electronic devices today usually happens only when someone pays to have it done.

There are many recyclers you can pay to take your old electronic devices. In many places, however, regulations have been needed to get this business going. Mandatory electronics recycling became the law throughout the European Union in 2005. In California, the Electronic Waste Recycling Act of 2003 required charging a recycling fee with every new purchase of electronic equipment. These fees pay recyclers to process electronic equipment, and all Californians can now recycle electronics rather than just throw them away. But don't be surprised if high metal prices soon lead recyclers to compete for your old electronic equipment, and perhaps even pay you for it.

FIGURE 15-14 ▼ A Conventional Open-Pit Uranium Mine
Impacts such as physical disturbances, waste disposal, and dust create long-term environmental consequences at this type of mine.

Landfill energy

People are making money from ordinary garbage. Landfills full of municipal solid waste generate significant amounts of methane. In the past, this methane, a greenhouse gas, was commonly just allowed to gradually escape, or was collected and burned. Today it can be a valuable energy resource (Chapter 13). The Yolo County Landfill in California is a 3.8 hectare (9.4 acre) site that uses the methane it generates to produce electricity—enough to power 3000 homes.

Nationwide, 400 landfills are producing energy from the methane they generate and about 560 more could. Using this methane reduces greenhouse gas emissions by the equivalent of 24 million tonnes (26 million tons) of CO_2 annually. This is the amount of CO_2 emitted by 17 million cars in a year.

Using landfill methane could be just the beginning of putting our garbage to better use. New technologies that can create energy, such as sewage-eating bacteria that generate electrical current and extremely high temperature (plasma) gasification, could make municipal waste water treatment self-sufficient and perhaps do away with the need for landfills.

| Waste not: energy from garbage and sewage; the prophet of garbage

Ecotourism

Business opportunities also come from people's interest in the environment. Ecotourism is an example. People pay for guided trips through Earth's great variety of natural habitats—from the frozen Arctic to tropical rainforests and harsh deserts. Whale watching, kayak rentals, glacier tours, and float trips on wild rivers are some of the related businesses that develop as a result. These types of businesses could help save Madagascar's rainforest.

Madagascar is a large island off the east coast of Africa. It has tremendous biodiversity, and 150,000 of its 200,000 species live nowhere else. It is particularly famous for the rare lemurs (like the one shown in Figure 15-16) and 1000 species of orchids that live in its rainforests. Between 2000 and 2005, logging cut these forests down at the rate of 37,000 hectares (91,000 acres) a year. The agriculture methods that have long been practiced on the island change forest into rice fields that deplete the soil of nutrients within a few years. The fires that clear the land for agriculture denude up to a third of the island each year, and extensive soil erosion soon follows.

Madagascar is a poor country and its forest products, such as ebony and rosewood, are valued around the world. Can the ecosystem values of its rainforests provide enough economic opportunity to replace this source of revenue? Some think so,

FIGURE 15-15 ▲ Recycling Consumer Electronics
Today you may have to pay to recycle your old computer, cell phone, or other electronic device—but perhaps the recyclers should be paying you.

FIGURE 15-16 ▲ A Madagascar Lemur
The physical isolation provided by the island of Madagascar has enabled lemurs—a group of primates, relatives of the monkeys and apes—to thrive and evolve into many species found nowhere else in the world.

FIGURE 15-17 ◄ Cruise Ship Visiting Sitka, Alaska
Sitka, a town of 9000 people, is in the Tongass National Forest, where logging and mill work once provided jobs for many people. Now tourism is a mainstay of the economy as 250,000 people visit each year on their cruises to Alaska.

and ecotourism could be the key. Nature guiding, marketing local crafts, and providing services for tourists are new businesses based on establishing environmentally sound access for people who want to experience Madagascar's unique wildlife and habitat. The people and their culture can also be valued parts of the experience that draws visitors to this island. Ecotourism has succeeded elsewhere—many people who logged the forests of southeast Alaska now make their living providing goods and services to the many tourists who visit each summer on cruise ships (Figure 15-17).

ECONOMIC INFLUENCES ON PEOPLE'S CHOICES

There is probably no more effective way to influence the choices large numbers of people make than by manipulating their costs. Lower-cost alternatives that meet people's needs or wants will be chosen over higher-cost options time after time. This well-understood relationship has been employed repeatedly by businesses and governments. Governments apply it through their tax policies. Tax breaks foster desired outcomes such as home ownership (deducting home mortgage interest from income taxes is seen as a way to vest people in the civic and economic well-being of the nation). Tax burdens, on the other hand, can be used to constrain less desirable activities.

Take gasoline taxes, for example. Gasoline is more expensive in Europe than anywhere else in the world. Filling up in Amsterdam can cost almost three times as much as in the United States, and about two-thirds of Amsterdam's cost is taxes. European nations keep their gasoline taxes high for several reasons. These taxes serve as a valuable source of general revenue. They also encourage use of public transportation systems, increase the fuel efficiency of private vehicles, and decrease reliance on foreign petroleum sources.

This high-tax policy seems to have worked. European cars on average get 32.1 miles per gallon, whereas U.S. cars average 21.6 miles per gallon. This is possible because 40% of the private vehicles in Europe use more efficient diesel engines (diesel fuel is commonly taxed less than gasoline, too). Europeans also use less gasoline and diesel than Americans—about 80% less per person. As gasoline prices go up, traffic goes down. Taxes are the biggest factor in high European gasoline prices. The environmental benefits of using less gasoline or diesel come directly from decreased vehicle emissions—less air pollution and less greenhouse gas emissions.

Gasoline price rises due to higher oil prices have affected U.S. drivers. How do you think this has influenced people's vehicle choices? The sale of fuel-efficient hybrid vehicles has increased each year since 2005, as Figure 15-18 shows. In June 2008, General Motors Corporation announced that it was closing four truck and SUV manufacturing plants and increasing production of smaller cars. More and more people buy fuel-efficient cars when the price of gasoline rises.

Perhaps increasing U.S. taxes on gasoline and diesel fuels would help the environment. The issue is complicated. Americans are scattered about more than Europeans and they treasure their ability to travel cheaply. If Americans paid as much as Europeans for their vehicle fuels, however, it's very likely that fuel efficiency of our nation's vehicles would improve.

You Make the Call

Raise Gasoline Prices?

- Do you think gasoline taxes should be higher in the United States?
- If gasoline taxes were higher, what do you think the effects would be?
- How would you use the increased revenues from higher gasoline taxes?

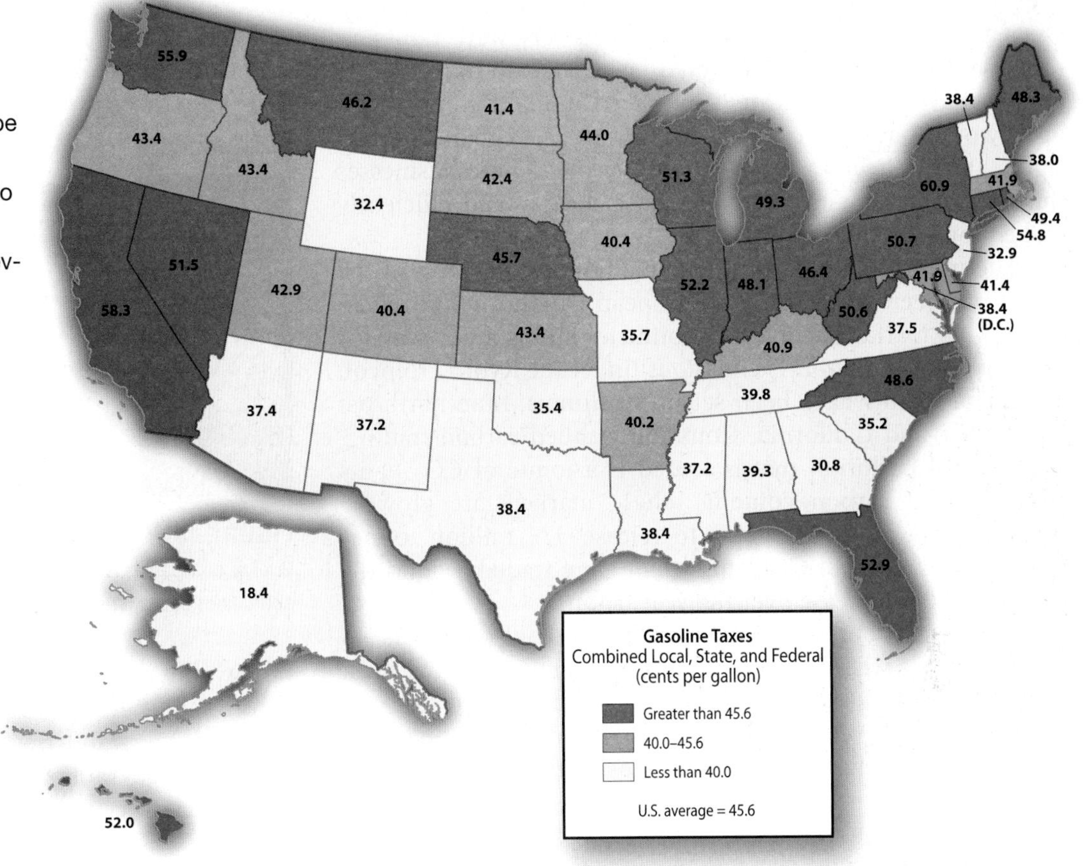

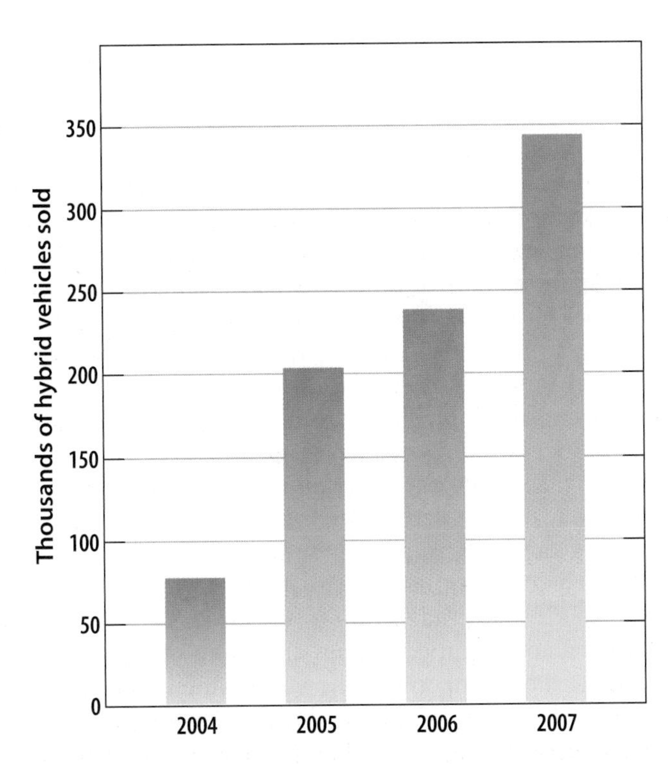

FIGURE 15-18 ▲ The Sale of Fuel-Efficient Hybrid Vehicles Is Increasing The economic downturn in 2008 lowered total vehicle sales by 18% but hybrid sales were only down 9.9%.

USING MARKETS TO ACHIEVE ENVIRONMENTAL OBJECTIVES

Some countries have imposed taxes on CO_2 emissions to help decrease the amount of this greenhouse gas that is released to the atmosphere. However, using markets to deal with carbon emissions is now the principal way countries address this and many other pollution issues.

A government authority creates a pollution market, also called a **cap and trade program**, by

- defining the amount of pollution that will be allowed (it sets a *cap* on the total amount of pollution),
- allocating the right to emit specified amounts of pollution (credits or allowances) to individual polluters (such as an industrial manufacturer or a power company), and
- making it possible for companies to buy and sell (*trade*) the rights to release pollutants.

If a company emits less carbon than its allocation allows, it can sell its rights to the unused allocation to another company. Conversely, if a company needs to emit more than it is allowed to, it can buy unused allocation from another company. Thus, the total emissions cannot exceed the cap.

Creating and using such markets helps achieve pollution limits at the lowest possible cost. The market for sulfur dioxide (SO_2) emissions in the United States is an example. The 1990

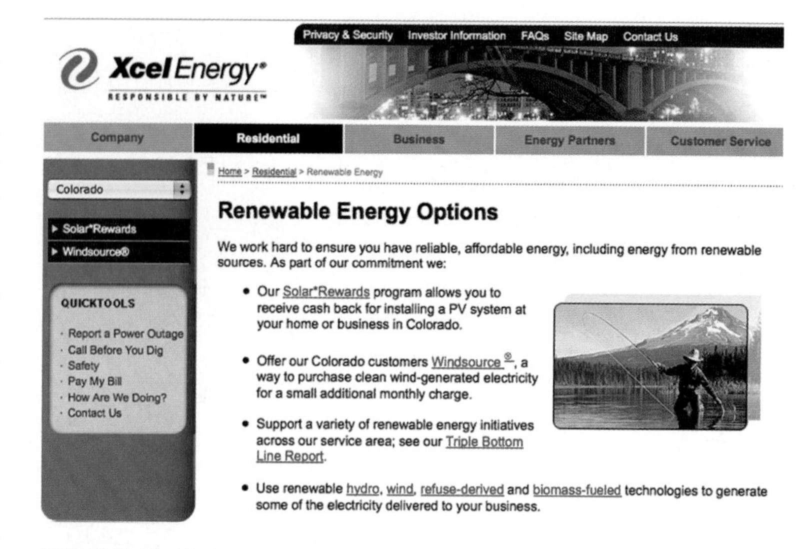

FIGURE 15-19 ▲ Green Energy Options

Energy companies are increasingly providing consumers with choices about the sources of energy they use.

Clean Air Act Amendments created an SO_2 emissions market. Over the years, this program has worked very well. Emission reductions have exceeded targeted levels and up to $1 billion in emission control costs are being saved annually compared to what would have been incurred under direct regulatory oversight. | Lessons learned from SO_2 allowance trading Those businesses that can decrease emissions the most cheaply and efficiently make money in this system.

The most ambitious and dynamic pollution market in the world has been set up by the European Union for CO_2 emissions. This market involves 27 countries and is a key way for them to control their CO_2 emissions under the Kyoto Protocol. Similar programs have been set up in Illinois, nine northeast U.S. states, and California. Commonly called carbon trading, (the unit of pollution that is traded is a tonne of CO_2 or its equivalent in carbon content), these markets are growing tremendously. In 2005, rights to release 374 million tonnes (412 million tons) of CO_2 equivalent were traded around the world, a 240% increase over the year before.

Markets to control pollution are not without detractors. In order to work well, initial allocation of pollution rights must reflect expected releases and methods are still needed to verify actual pollution levels. In some cases, it may also be necessary to decrease the overall pollution cap over time. This can be accomplished by removing (retiring) emission credits or allowances through government or third-party purchases. The program could also require that a proportion of all trades be retired. As the U.S. experience with SO_2 emission markets shows, using markets to control and manage pollution can be very effective.

FUNDING INCENTIVES

Funding, especially government funding, is used to achieve environmental goals through subsidies and direct investments, particularly in research and development activities. In Chapter 11 you learned how the federal government pays farmers—subsidizes them—not to cultivate certain areas, thereby protecting soil from erosion. Federal funding in Germany has helped renewable energy developments through direct subsidies, tax breaks, or low-interest loans. In the United States, wind energy development is financially supported by a production tax credit and rapid depreciation schedules (another tax advantage). In addition, some 20 states require commercial electric suppliers to obtain certain percentages of their power from renewable energy sources. In some cases, power companies can now offer their customers choices about the sources of the electricity they use (Figure 15-19).

A major commitment of federal funds to deal directly with environmental issues in one way or another is funding of the EPA. The EPA's 2008 budget is over $2 billion. | FY 2008 EPA budget in brief Over $750 million of this is for science and technology—research and development efforts to better understand and manage people's interactions with the environment.

However, every federal government agency has budget elements that deal with environmental issues. The Department of Defense cleans up environmental contamination at its old facilities, the Department of Energy funds research to identify technology for decreasing CO_2 emissions from coal-fired power plants, and the U.S. Geological Survey has a major effort under way to understand the effects of climate change. The list of ways your federal tax dollars are spent to address environmental issues is very, very long.

15.5 Decision Making

Having made your way through this book, you now understand how complicated environmental issues can be. It starts with the interrelated nature of Earth systems—what affects one system commonly affects the others as well. What we first think to be a small or isolated problem, perhaps emissions from a power plant, is part of something bigger that can degrade air quality, soil, and water over entire regions and potentially cause immediate or long-term harm to living things.

Another level of complication comes from concluding that sustaining natural resources, including biodiversity and habitat, is important for the future. Managing the environmental consequences of people's activities in ways that contribute to sustainability commonly requires tradeoffs, tradeoffs that have economic and social implications. Preserving biodiversity, for example, may require dedicating lands to natural habitat rather than letting them be used for new subdivisions, highways, or landfills. It is often quite difficult to reach decisions that involve these kinds of tradeoffs.

Another complication comes from the different perspectives people have concerning the environment and natural resource development. On one hand, people need natural resources, such as those that provide energy, minerals, food, and fiber. Some people work aggressively to prevent such resource development, even though many of them use the natural resources such developments provide. Preventing or

FIGURE 15-20 ▶ Environmental Protest Public protests on environmental issues are common in the United States, where the relationship between environmentalists and advocates of development is often adversarial.

mitigating negative environmental consequences of such developments is technically possible in many cases, and is required by environmental regulations in our country. As long as people need natural resources, developments such as mines and oil fields will take place somewhere. Overall, it's better for the environment for these developments to take place where environmental policies and regulations are strong.

Opposition to new industrial or resource development projects, even landfills, is common in the United States (Figure 15-20). Such opposition has helped create a complicated and costly regulatory system and provided much work for attorneys as litigation strategies are pursued to limit or prevent new developments. The overall U.S. system for managing people's effects on the environment is adversarial rather than cooperative.

The complicated nature of environmental issues suggests that an alternative approach, one that strives for consensus and relies on objective science to understand environmental consequences, would serve society and the environment better. This approach has worked at the Stillwater platinum and palladium mine in Montana.

THE STILLWATER MINE EXAMPLE

You may not know much about platinum (Pt) and palladium (Pd) other than that platinum is used for some jewelry. In fact, these closely associated metals play a variety of roles in medicine and industry, but the largest is at the heart of the catalytic converters that lower exhaust emissions from your car. As we increase our efforts to control air pollution, we need more and more platinum and palladium.

The only place where platinum and palladium are produced in the United States is the Stillwater mine in Montana (Figure 15-21). To meet the growing need for these metals,

FIGURE 15-21 ▼ The Stillwater Mine—Montana Plans to expand this facility led to an agreement that addressed the environmental concerns of local residents.

FIGURE 15-22 ◀ Worth Protecting The Stillwater mine is in a part of Montana with beautiful mountains, trout streams, and farmlands.

the Stillwater Mining Company wanted to expand its operations, in part by opening of a new mine nearby. Montana regulators watched closely over the planned expansion, but this wasn't enough oversight for many local residents. They still had concerns about more mining in their area—an area where fishing, hunting, camping, and agriculture were important (Figure 15-22).

Local groups first opposed the mine expansion through litigation, but eventually everyone involved came to realize there was a better way than being adversarial. All the concerned parties joined together, and after much discussion and effort developed an agreement that was legally binding—the "Good Neighbor Agreement." This agreement included provisions for eliminating discharge of wastewater, decreasing waste rock piles, expanding water quality monitoring, protecting thousands of acres in conservation areas, and decreasing vehicular traffic.

The Good Neighbor Agreement was the result of all interested parties—the stakeholders—coming together with good will and forthrightness. They tackled complicated issues but managed to balance the needs of everyone affected by expansion of the Stillwater Mining Company's operations. Their successful efforts set a good example for others trying to manage environmental impacts and the many complicated issues associated with new industrial development.

OUR SYSTEM'S STRENGTHS

In a sense, everyone is a stakeholder when it comes to achieving a healthy, sustainable environment. Given the complicated nature of environmental issues, how can people be confident that the best decisions are being made? Whom can you trust when it comes to information and decisions about the environment?

Think about the American people and culture you know from your own experiences. We are a country with:

- A strong environmental ethic—our national environmental laws are an excellent indicator.
- Strong professional ethics—scientists and engineers hold themselves to high ethical standards.
- Tremendous scientific and technical ability. When we recognize a problem we can focus excellent technical capabilities on understanding how to deal with it.

These are reasons the public should have confidence in the U.S. environmental permitting and regulatory system. In fact, this system could be streamlined and made more efficient with increased public confidence and support. The Stillwater mine stakeholders, including government agencies, gave us an example of how to better deal with complicated environmental issues. As such issues continue to be in the news (and perhaps in your neighborhood), keep in mind how the Stillwater mine stakeholders addressed their problems.

SUMMARY

The environmental concerns facing people are complicated. Now you know why. Most environmental issues involve the interaction of Earth systems, and people are an integral part of these interactions. People live where Earth systems can affect them—where earthquakes occur, volcanoes erupt, and coasts change. People rely on Earth resources for sustenance and materials essential to modern society. The ways in which people dispose of their wastes affect all Earth systems. Every interaction of people with Earth systems has environmental consequences. Understanding and dealing with these consequences are complicated. They require balancing different needs and making tradeoffs.

As people increase in number and continue their economic advancement, their effects on the environment are likely to gain in significance and require tougher and tougher decisions. These decisions will benefit from the understanding and technology that science and engineering bring to them, but in the end it will be up to people like you to make the decisions. Everyone is a stakeholder when it comes to balancing the needs of people and those of the environment. Achieving this balance is necessary to sustain human life, Earth's resources, and the environment. This is the challenge facing you and future generations.

In This Chapter You Have Learned:

- Environmental policies can be established by individuals, companies, and governments. The United States has established strong environmental policies through national, state, and local laws.
- People's activities that can affect the environment are guided by regulations that implement the authority created by laws. Regulations are applied through a permitting process and compliance is enforced by fines or civil and criminal penalties.
- Litigation can be used to challenge regulatory decisions. Third parties that employ litigation in this way are fairly common in the United States.
- Economic factors influence human activities that affect the environment. Businesses can use technological advances that have positive environmental consequences; respond to opportunities created by regulatory decisions; or take advantage of public interest in the environment.
- Economic incentives or disincentives influence people's choices that affect the environment. Governments use tax policies and subsidies to exert such influence.
- Government-created markets are increasingly used to control pollution. This effective approach leads to lower-cost control of pollution levels.
- Direct funding, especially of relevant science and technology, is used to understand, clean up, and protect the environment. This type of spending is a significant part of the U.S. federal budget.
- Managing people's activities that affect the environment is complicated due to the complex interactions of Earth systems, the need for sustainability, and the inherent tradeoffs that many choices require. Dealing with these complicated issues in the United States is commonly adversarial. Consensus-building, as demonstrated by the Stillwater mine example, is an alternative approach.

KEY TERMS

action levels (p. 477)
cap and trade program (p. 485)
Clean Air Act (CAA) (p. 472)
Clean Water Act (CWA) (p. 471)
Comprehensive Environmental Response, Compensation, and Liability Act (CERCLA) (p. 473)
Environmental Impact Statement (EIS) (p. 470)
generic soil screening level (p. 477)
leachate (p. 473)
maximum contaminant level (MCL) (p. 477)
National Environmental Policy Act (NEPA) (p. 470)
National Pollutant Discharge Elimination System (NPDES) (p. 471)
National Priority List (NPL) (p. 473)
policy (p. 469)
potentially responsible party (PRP) (p. 473)
Resource Conservation and Recovery Act (RCRA) (p. 472)
stakeholder (p. 470)
State Environmental Policy Act (SEPA) (p. 475)
Superfund (p. 473)
total maximum daily load (TMDL) (p. 477)

QUESTIONS FOR REVIEW

Check Your Understanding

1. Why is waste disposal best thought of as a land-use decision?
2. What is the difference between a government-issued environmental policy and an environmental regulation? Give an example.
3. What are stakeholders? How are they involved in the preparation of environmental impact statements?
4. What is the Clean Water Act and what does it do to advance the mission of the EPA?
5. What is RCRA, and why is it significant that this act governs only current and future operations?
6. What is leachate in a landfill, and why is it so important to contain?
7. The text states that RCRA emphasizes source reduction and recycling of solid wastes. Why is this important insofar as landfills are concerned?
8. What is CERCLA, and how is it related to issues and activities covered by RCRA?
9. Why is a federal land use designation relevant to environmental policies and land use on parcels of federal land?
10. What is a political advocacy group, and why are such groups relevant in the formulation of environmental policy?
11. What are environmental standards and how are they related to environmental regulations?
12. How does the expected used of water influence the water quality standards that regulate it?
13. What is the difference between the total maximum daily load and the maximum contaminant level for a reservoir of drinking water?
14. Soil quality also has environmental standards. How is soil quality linked to human health?
15. What is an environmental permit, and when is one required?
16. Is environmental regulation the same thing as environmental enforcement? Why or why not?
17. What is third-party litigation? Explain its importance for development projects with environmental implications.
18. How can a recycling fee influence business activity and environmental regulatory compliance?
19. How have high fuel taxes helped European countries achieve environmental and economic goals?

Critical Thinking/Discussion

20. Most environmental debates about setting a sustainable course to development and environmental health involve tradeoffs. What are some of these tradeoffs? Using examples, describe the complexity inherent in effective environmental regulation.
21. Why can internal corporate environmental policies sometimes be more effective than government-mandated environmental policies? What is dangerous about relying only on corporate environmental policies?
22. Why is it significant for the environment near and on military facilities that the Department of Defense is not bound to follow NEPA guidance? How might this create problems, especially overseas?
23. The U.S. Environmental Protection Agency (EPA) features prominently in environmental policy and regulation in this country. While it is not a cabinet-level agency, its administrator (the top job) is a presidential appointee and has been granted cabinet-level rank in each administration since that of Richard Nixon. Who proposed the agency in the late 1960s? From what you've learned in this chapter about the central importance and wide-ranging authority of this agency, why is it significant that its leader is a political appointee, and what effect, if any, do you suppose this has on the crafting and enforcement of environmental policy?
24. Under the guidelines established by CERCLA, the determination of the "potentially responsible party" is crucial. Why is this so under this act, and why is it all the more important for companies or individuals considering buying existing sites with possible environmental issues to be very diligent in their investigations prior to taking ownership of a property or facility?
25. The text states that, "There seems to be a national tendency for Americans to react to rather than anticipate environmental concerns . . ." The history of environmental regulation and policy certainly bears this out. Why do you think this is the case, and what influence does this national tendency have on our ability to deal with large-scale environmental concerns, such as climate change or crises in biodiversity?
26. In the permitting process for landfills, closure costs for the facility are factored in. Do you think this a good or a bad idea in letting a landfill project go forward? Why or why not?
27. Environmental regulations are commonly seen as a source of extra costs for businesses, developers, or any other entity seeking permits for operations. However, these are often a source of new revenues and business opportunities. How can businesses make money from imposed environmental regulations? Cite examples in your answer.
28. "Cap and trade" carbon emissions programs create a dynamic market for trading pollution allowances. Many companies actually would rather face a preset carbon emissions per-tonne tax rather than a dynamic, market- and government-driven cap and trade program. Why might this be so? Discuss the advantages and drawbacks of both systems from the perspective of governments and large companies.

ANSWERS TO IN-CHAPTER INSIGHT QUESTIONS

P. 473

What are some sustainability issues associated with solid waste generation and disposal?

- Material that becomes waste in landfills is not being used sustainably.
- Waste generation that is built into product design and use does not contribute to sustaining energy and materials; Styrofoam cups are an example.
- Landfills include materials, such as plastics, that do not degrade over long periods of time; these materials effectively make landfills a permanent use of the land.
- Landfill surface areas can become open space usable for selected purposes after closure; developing and sustaining wildlife habitat for example.

P. 475

What are some ways in which sustainability is related to federal land use designations such as national parks?

- Lands such as national parks are established to sustain selected resources such as wildlife habitat and biodiversity.
- Lands such as national parks preclude use of resources such as timber and minerals; these resources could be considered either removed from sustainability issues or preserved for future generations.
- People's use of lands such as national parks is restricted to selected activities that can be sustained indefinitely.

P. 478

What are some ways in which Earth systems are related to environmental standards?

- People's exposure to contaminated Earth systems (air, water, and soil) is a factor in determining environmental standards.
- Environmental standards anticipate interactions between the biosphere, such as aquatic life, and contaminated Earth systems.
- Earth systems interactions that control the distribution and concentration of contaminants (erosion, dissolution, and precipitation of contaminants for example) determine where environmental standards need to be applied.

GLOSSARY

A

Aa Aa is a Hawaiian term for a type of basaltic lava flow typified by a rough and jagged surface.

Absolute age Absolute age is the numerical age of any physical entity or event in years before the present.

Acid Rock Drainage (ARD) Acid rock drainage is acidic water produced by the oxidation of pyrite and other sulfide minerals.

Action level An action level is the concentration of a contaminant in soil that indicates a cleanup action is required.

Adit An adit is a horizontal opening into a hillside.

Aggregate Aggregate is a general term for rock material that is used for construction. It includes sand, gravel, and crushed or broken stone.

Alloy An alloy is a homogeneous mixture of two or more metals.

Alluvial fan An alluvial fan is a gently sloping deposit of sand or gravel (alluvium), shaped like an open fan, deposited by a stream where its gradient rapidly decreases as it approaches a local base level.

Amplitude The amplitude, or size, of a seismic wave is defined as half the vertical distance between its trough and peak as measured by a seismometer.

Angle of repose The angle of repose is the characteristic maximum slope (measured from a horizontal plane) at which loose material will come to rest on a pile of similar material.

Annual flood An annual flood is a stream's peak discharge in a given year.

Apparent polar wander curve An apparent polar wander curve is a line connecting the seemingly changing locations of the magnetic poles over time for a specific continent, as determined by paleomagnetism studies of rocks of different ages.

Aquifer An aquifer is a groundwater reservoir that can supply enough water to wells and springs to be useful to people.

Aridisol Aridisol is a soil order characteristic of arid regions; these soils have a low concentration of organic matter and commonly one or more subsurface soil horizons where minerals such as calcite, gypsum, or soluble salts have accumulated.

Artesian flow Artesian flow is flow from a confined aquifer in response to pressure developed by the weight of the water in higher parts of the aquifer.

Ash Ash is fine pyroclastic material composed of glassy solidified lava, rock, and crystal fragments.

Aspect Aspect is the orientation of a slope with respect to the rays of the Sun.

Asthenosphere The asthenosphere is part of the upper mantle. It is mostly solid, but pliable (like putty) and can flow under pressure. It is a weak part of the upper mantle where partial melting and slow movements of material occur.

Atmosphere The atmosphere is the envelope of gases that surrounds Earth. It is mostly nitrogen (78%) and oxygen (21%) but contains at least nine other gases.

Atom An atom is the smallest possible division of a chemical element that maintains all the characteristics of that element.

B

Bank-full stage The bank-full stage is the elevation of the water surface of a stream flowing at full channel capacity.

Barrier island A barrier island is a long, low-lying ridge of sand and gravel isolated from the mainland by shallow lagoons, bays, or marshes.

Basalt Basalt is a volcanic rock formed from partially melted mantle that migrates to Earth's surface. The most abundant volcanic rock on Earth, it is denser, on average, than rocks of the continental crust because it contains more iron and magnesium.

Base level Base level is the elevation at which a river cannot flow farther or erode deeper into its bed.

Beach A beach is an accumulation of loose water-borne material (generally sand and pebbles but also boulders and shell fragments in places) deposited on the edge of a body of water—typically a gently sloping shore washed by waves or tides.

Beach drift Beach drift is the gradual movement of sediment in a zigzag pattern along the shore, caused by waves that come ashore obliquely.

Beach face The beach face is the flat ramp of sand leading down to the water's edge at a beach.

Beach nourishment Beach nourishment is a soft stabilization method that brings in sand to create a new beach and bulk up the underwater slope of the beach face so as to develop at least a temporary positive sediment budget.

Bed A bed is an individual layer of sedimentary rock that can be distinguished from other layers.

Beneficiation Beneficiation is the process of crushing ore, separating minerals, and concentrating valuable minerals.

Berm A berm is a flat area of beach sediment composed of sand or gravel that is covered by water only during very high tides or large storms.

Bioavailable Bioavailable refers to the relationship between the amount of a substance an organism is exposed to and the amount the organism absorbs. The more of the substance retained by an organism as a result of a specific exposure, the more bioavailable the substance is.

Biodiversity Biodiversity refers to the full range of variability within the living world at all levels, including genomes, species, and ecosystems. The number of species in an area is commonly used as a measure of its biodiversity.

Biomagnification Biomagnification is the process whereby chemicals become more concentrated in individual organisms as they are passed up the food chain.

Biomass Biomass is essentially plant debris in all its various forms, including even household garbage and livestock manure.

Bioremediation Bioremediation is the technology that uses natural or enhanced microbial action in the soil to degrade contaminants.

Biosolids Biosolids are treated sewage sludge—the solid organic matter recovered from a sewage treatment process that can be used as fertilizer.

Biosphere The biosphere is made up of all the living things that inhabit Earth.

Body waves Body waves are seismic waves that travel through the interior of the Earth.

Braided channel A braided channel is a stream channel that frequently branches and rejoins around bars or islands of sediment.

Breakwater A breakwater is a barrier built in the water parallel to a coast to break the force of the waves and protect a harbor, anchorage, beach, or shore area.

Bulkhead A bulkhead is a vertical wall, typically constructed of timber, concrete, or steel, usually smaller than a seawall and not designed for high-wave-energy environments.

C

Caldera A caldera is a large circular-to-oblong depression that forms when magma chambers erupt their contents and the volcanic mountain above them collapses into the empty magma chamber.

Capacity A stream's capacity is its ability to transport sediment, the maximum bed load and suspended load it can carry.

Cap and trade program A cap and trade program creates a financial incentive for emission reductions by assigning a cost to polluting. The rights to emit defined amounts of pollutants are bought and sold; those who pollute less than their allowed amount can make money by selling their right to release pollutants to others.

Carbon sequestration Carbon sequestration is the capturing and storing of CO_2 emitted by hydrocarbon burning.

Carrying capacity Carrying capacity is the number of people that Earth can support sustainably at a specified level of economic and social well-being (standard of living).

Channel A channel is the bed where a natural body of surface water is usually present and the main current normally flows.

Cinder cone A cinder cone is a conical hill formed by the accumulation of cinders and other ejected volcanic material, normally of basaltic or andesitic composition.

Clastic sedimentary rock A clastic sedimentary rock such as shale, sandstone, or conglomerate is formed from clastic sediment (sediment formed from rock fragments and mineral grains).

Clay Clay is a general term for a group of water-bearing aluminum silicate minerals having a layered structure. They commonly form from the hydrolysis of other silicate minerals.

Clean Air Act (CAA) The CAA is a federal statute to protect air quality in the United States; it regulates activities that have the potential to cause everything from acid rain to stratospheric ozone depletion and CO_2 emissions. The CAA authorizes national air quality standards and an emissions permitting system.

Clean Water Act (CWA) The CWA is a federal statute to protect surface water quality in the United States; it defines a variety of ways to reduce direct pollutant discharges into waterways, finance municipal wastewater treatment facilities, and manage polluted runoff.

Cleavage A cleavage is a weak plane in a mineral's internal structure along which it will break.

Coal-Bed Methane (CBM) Coal-bed methane is methane that is generated and trapped in coal.

Cohesion Cohesion is the force created by attractions between grains of material that makes them stick together. Derived largely from electrical attractions between particles at the atomic level, it is affected by the amount of moisture present.

Comprehensive Environmental Response, Compensation, and Liability Act (CERCLA) CERCLA is a federal statute that authorizes the federal government to respond to releases of hazardous substances from closed and abandoned sites, provides ways to allocate liability for these releases, and establishes a trust fund (with tax revenues from the chemical and petroleum industries) to clean them up. Because of this funding capability, CERCLA is commonly called Superfund.

Compressibility Compressibility is a measure of how a soil compacts under applied forces, such as when a building is constructed upon it.

Cone of depression A cone of depression is a cone-shaped lowering of the upper surface of a body of groundwater around a well from which water is being withdrawn faster than it is replenished.

Confined aquifer A confined aquifer is an aquifer that is overlain by low-permeability material.

Consumptive use A consumptive use of water (such as evaporation, transpiration, incorporation into products or crops, or consumption by man or livestock) disperses water to other system reservoirs and causes depletion of the source.

Continental crust Continental crust is the outermost part of Earth's geosphere that underlies the continents—it makes up Earth's landmasses and their shallowly submerged edges. It is characterized by rocks that contain abundant silicon, oxygen, aluminum, potassium, and sodium.

Continental drift Continental drift is a general term for many aspects of a hypothesis, originally proposed by Wegener in 1912, that continents move relative to one another. Subsequent research established the theory of plate tectonics, which explains many of Wegener's observations.

Continental margin A continental margin is the part of a continent, especially the submerged part (the continental rise, slope, and shelf), adjacent to oceanic crust.

Convergent plate boundary A convergent plate boundary is the boundary along which two tectonic plates move toward each other.

Core A core is a cylinder of solid rock cut by a drill and brought to the surface for geologic examination and sampling.

Core The core is the iron- and nickel-rich, innermost part of the geosphere. It includes a solid inner core and an outer molten core.

Creep Creep, a very slow type of earthflow, is a more or less continuous downslope movement of mineral, rock, and soil particles under the influence of gravity.

Crust The crust is the thin, outermost part of the geosphere. Continental crust, with an average thickness of about 35 km (22 mi), underlies the continents; oceanic crust, with an average thickness of about 7 km (4 mi), underlies the oceans.

Cryogenic distillation Cryogenic distillation is distillation at the very low temperatures at which gases become liquids and boil; it is used to separate individual gases such as nitrogen from the atmosphere.

Crystal A crystal is a homogeneous solid structure (a mineral grain) that often exhibits a regular geometric form and planar surfaces reflecting its orderly internal atomic structure.

Cutbank A cutbank is a steep bare slope formed by lateral erosion of a stream.

D

Debris flow A debris flow is a water-rich moving mass of rock fragments, soil, and mud, in which more than half the particles are larger than sand size.

Delta A delta is a deposit of sediment at the mouth of a stream where it enters a body of water such as a lake or the ocean.

Differentiation Differentiation is the process by which the composition of magma changes through the physical separation of early-crystallized minerals (as when they sink in a magma chamber).

Discharge Discharge is the rate of flow of surface water at a given moment, expressed as volume per unit of time.

Dissipative A dissipative use of a substance is one that leaves it dispersed and unrecoverable.

Distillation Distillation purifies water by heating it to drive off steam (which is pure freshwater) and then condensing the steam back into a liquid.

Divergent plate boundary A divergent plate boundary is the boundary along which two tectonic plates move away from each other.

Divide A divide is an area of high ground, such as a ridge or summit, marking the boundary between two adjacent watersheds.

Dome A dome is a steep-sided, rounded accumulation of lava extruded from a volcano to form a dome-shaped or bulbous mass of solidified lava above and around the vent.

Dredging Dredging is the removal of unwanted sediment by scooping it off the bottom or pumping it away as a thick slurry using a suction dredge.

Driving force The driving force of slope failure is the force of gravity, which pulls downward on surface materials.

Dune restoration Dune restoration is a soft stabilization tool that uses fencing, vegetation planting, and other methods to expand and stabilize dunes.

E

Earth systems The Earth systems are the four basic components of our planet: atmosphere, hydrosphere, geosphere, and biosphere.

Earthflow An earthflow is a debris flow composed of fine-grained material such as soil, sand, or silt.

Earthquake An earthquake is a vibration in the Earth caused by the release of elastic strain on a fault.

Earthquake cycle The earthquake cycle is the repeated generation of earthquakes by the buildup and release of elastic strain on a fault.

Eccentricity Eccentricity is the degree to which an orbit, such as Earth's around the Sun, varies from a perfect circle.

Elastic rebound theory The elastic rebound theory states that movement along a fault is the result of an abrupt release of a progressively increasing elastic strain between the rock masses on either side of the fault.

Elastic strain Elastic strain is the change in shape or size of material caused by stress along a fault that can be reversed when stress is released.

Element An element is a pure chemical substance that cannot be broken down chemically into other substances. It is distinguished by its atomic number, the number of protons in the nucleus of its atoms.

Environment In environmental geology, the environment is very broadly defined to include all the physical and biological components involved in Earth systems interactions.

Environmental Impact Statement (EIS) An EIS is a document that must be publicly filed when the federal government takes a "major Federal action significantly affecting the quality of the human environment." An EIS comprehensively examines the environmental and human impacts the action may have.

Eon An eon is the major interval of geologic time on the geologic time scale. Geologic time is divided into four eons.

Epicenter The epicenter is the point on the Earth's surface that is directly above the focus of an earthquake.

Erosion Erosion is the transport of geosphere materials from one place to another by natural movements of water, wind, or ice (glaciers).

Estuary An estuary is a semi-enclosed coastal bay where freshwater from rivers mixes with seawater.

F

Fall A fall, the fastest mass movement, is characterized by the tumbling, rolling, or free fall of materials down a steep slope or cliff.

Fault A fault is a place where rock has broken and the blocks on opposite sides of the break have moved relative to each other.

Fault creep Fault creep is the slow and gradual movement along a fault that doesn't cause significant earthquakes.

Fault trace A fault trace is a linear feature that marks the intersection of a fault plane with the ground surface.

Feedback mechanism A feedback mechanism is a response to a change in the climate system that either amplifies the change (positive feedback) or diminishes the change (negative feedback).

Feldspars Feldspars are a group of silicate minerals that contain silica, aluminum, potassium, sodium, and calcium.

Felsic Felsic is a compositional name for igneous rocks with abundant quartz and feldspar.

Ferromagnesium minerals Ferromagnesium minerals are silicate minerals that contain abundant iron or magnesium in their internal atomic structure.

Fissure A fissure is a volcanic opening at the Earth's surface having the form of a long crack.

Flash flood A flash flood is a local and sudden flood of relatively great volume and short duration, generally resulting from brief but heavy rainfall over a relatively small area with steep slopes or the failure of an ice jam or constructed dam.

Flood basalt Flood basalt is a term applied to basaltic lavas that occur as vast accumulations of essentially horizontal flows erupted from fissures in rapid succession over large areas.

Flood crest Flood crest is the highest stage of a flood.

Flood frequency curve A flood frequency curve is a graphic illustration of annual flood discharges and their recurrence intervals.

Flooding Flooding is the covering of normally dry lands with water. It occurs along rivers when the discharge is so great that the water rises and overtops the river's natural or artificial banks.

Floodplain A floodplain is a flat or nearly flat lowland bordering a stream that may be covered by its waters at flood stages.

Flotation Flotation is a process that separates sulfide mineral grains from a slurry by attaching them to specific types of bubbles.

Flow Flow is a general term for mass movements in which the material moves like a liquid, with particles in motion independently of each other.

Flux Flux is the rate of transfer of matter among systems.

Focus The focus is the point underground where rock first ruptures to generate an earthquake.

Foliation A foliation is a two-dimensional sheeted structure in rocks.

Fossil A fossil is the remains or indications of former life, preserved in rocks.

Fossil assemblage A fossil assemblage is a group of fossil species found together in a specific sedimentary layer or sequence. The fossils are commonly inferred to have lived together.

Fossil fuel Fossil fuel is a general term for hydrocarbon deposits (principally oil, natural gas, and coal) ultimately derived from organisms that once lived on land or in the oceans.

Fossil succession Fossil succession refers to the sequence of fossil types in a distinctive order that represents changes in the biosphere through geologic time.

Frequency Frequency is the number of vibrations that a seismic wave completes in a given period of time; it is commonly measured in cycles per second (hertz, Hz).

Friction Friction is resistance to movement along a contact between two bodies such as blocks of rock or sand grains.

Fuel cell A fuel cell is a device that uses a catalyst to combine hydrogen and oxygen, producing an electric current that can power machines.

G

Gabbro Gabbro is the igneous rock that crystallizes from mafic (basaltic) magma in the crust.

Gaining stream A gaining stream is one that receives groundwater discharges because its channel lies below the water table.

Generic soil screening level A generic soil screening level is the concentration of a particular chemical in a soil that signals when further analysis of the risk to people is indicated.

Geologic time scale The geologic time scale is a chronologic arrangement of the periods and events in geologic time.

Geologic time Geologic time refers to periods of time long enough to give us a perspective on and understanding of Earth's history.

Geo-sequestration Geo-sequestration is the storage of carbon in underground reservoirs.

Geosphere The geosphere is the solid Earth and all its parts, including molten parts within Earth.

Geothermal energy Geothermal energy is heat stored in the Earth's crust.

Glacial maximum A glacial maximum is a period during which ice sheets reached their maximum extent.

Glacier A glacier is a large mass of ice, formed at least in part on land by the compaction and recrystallization of snow, moving slowly by creep under the influence of gravitational forces, and surviving from year to year.

Global Climate Model (GCM) A global climate model is a quantitative simulation that represents the interactions of the major components of the climate system: solar radiation, atmosphere, oceans, land, and ice.

Gondwanaland Gondwanaland is the name of a supercontinent that included the southern continents of South America, Africa, Antarctica and Australia, along with India. It is thought to have formed from the splitting apart of Pangaea about 200 million years ago.

Grade Grade is the concentration of a valuable mineral or element in a mineral deposit.

Gradient Gradient is the "slope" of a stream—the angle between the water surface or the channel floor and the horizontal, measured in the direction of flow.

Granite Granite is a type of rock, typical of continental crust, containing less-dense elements including calcium, sodium, potassium, aluminum, silicon, and oxygen.

Greenhouse gas A greenhouse gas is a minor constituent of Earth's atmosphere, such as carbon dioxide, that can absorb and reradiate energy, causing the atmosphere's temperature to rise.

Groin A groin is a low, narrow barrier built perpendicular to the shore to capture sand from the longshore drift, thwart beach erosion, or rebuild a beach that has eroded away.

Groundwater Groundwater is water that infiltrates the ground and completely fills all open spaces: voids as big as caverns, fractures in rocks, even tiny cavities between mineral grains. It is the part of the subsurface water that is in the saturated zone.

Groundwater mining Goundwater mining is the pumping out of groundwater faster than it can be replenished.

H

Half-life The half-life of a radioactive isotope is the time it takes for half of the atoms in a sample to decay.

Heat pump A heat pump is a device that circulates a fluid to warm or cool a building by transferring heat from a relatively low-temperature reservoir to one at a higher temperature.

High-Level radioactive Waste (HLW) High-level radioactive waste is waste from nuclear reactors including used (spent) fuel and material left over from reprocessing spent fuel.

Horizon A horizon is a stratified layer within a soil approximately parallel to the land surface and differing from adjacent layers in physical, chemical, and biological properties.

Hot spot Hot spots are places where voluminous mantle material rises and melts to form mafic magma.

Humus Humus is a general term for the organic matter that imparts a dark color to upper levels of soil.

Hurricane A hurricane is a tropical cyclone, especially in the North Atlantic and eastern North Pacific basins, with sustained near-surface wind speed of at least 64 knots (73 mph).

Hydrocarbon A hydrocarbon is any organic compound consisting solely of carbon and hydrogen.

Hydrograph A hydrograph is a plot of a river's discharge (or height in some cases) over time.

Hydropower Hydropower is power, mostly in the form of electricity, derived from water flowing under the influence of gravity.

Hydrosphere The hydrosphere is all the water in the oceans, lakes, rivers, underground, and in the permanent ice accumulations.

Hypothesis A hypothesis is a tentative explanation that is consistent with all we know about a situation or problem.

I

Ice sheet An ice sheet is an area where glaciers coalesce and cover more than 50,000 km^2 (31,000 mi^2).

Igneous rock An igneous rock is a rock formed from molten material (magma in the crust and lava on Earth's surface).

Intensity Intensity is a measure of the amount of shaking and other effects by an earthquake at a particular location. It is measured in terms of its effect on people and structures.

Interglacial period An interglacial period is the time interval between two successive glacial maximums.

Intraplate earthquake An intraplate earthquake is one that occurs within a tectonic plate rather than at a plate boundary.

Intraplate volcano An intraplate volcano is one that is far from any plate boundary and therefore considered unrelated to subduction or seafloor spreading processes.

Ion An ion is an atom or molecule that has an electric charge, either as a result of chemical reactions or from interactions with radiation.

Ionosphere The ionosphere is the upper part of the atmosphere, coinciding with the thermosphere, where solar radiation ionizes some of the constituent gas molecules.

Island arc An island arc is a chain of volcanoes in an ocean that forms above a subduction zone.

Isotope Isotopes are forms of an element with different atomic masses. All atoms of an element have the same number of protons in their nucleus, but the atoms of different isotopes differ from one another in having different numbers of neutrons.

J

Jetty A jetty is an engineering structure extending more or less perpendicularly from the shore into a body of water, designed to direct and confine the current or tide, to protect a harbor, or to prevent shoaling of a navigable inlet. Generally larger than groins, they are often built in pairs on either side of a harbor entrance, inlet, or the mouth of a river.

K

Karst terrain Karst terrain is a type of topography that is formed on limestone, gypsum, and other soluble rocks, where bedrock dissolution has created underground openings such as caves.

L

Lahar A lahar is a wet debris flow originating on the flanks of a volcano. Lahars are a slurry of ash, lava debris, and water, as well as any other material such as soil, rocks, and trees that the flow picks up on the way down the slope.

Landslide Landslide is a general term for mass movements that involve relatively coherent blocks of soil and rock material.

Laterite A laterite is an oxisol rich in oxides of iron, aluminum, or both that develops in a tropical or forested warm to temperate climate as a residual product of weathering.

Lava Lava is molten rock erupted onto Earth's surface.

Lava tube A lava tube is a roofed conduit that lava flows through from an eruptive vent to a depositional site.

Leachate Leachate is liquid formed by leaching, such as a solution containing contaminants picked up through the leaching of waste in a landfill.

Leaching Leaching is a mineral processing technique that removes metals from an ore by dissolving them in a percolating solution.

Levee A levee is an artificial bank constructed along a stream to increase its channel area and enable more discharge if needed.

Limestone Limestone is a sedimentary rock consisting chiefly of calcium carbonate, primarily in the form of the mineral calcite.

Liquefaction Liquefaction is the change of a water-saturated and unconsolidated surface material, such as soil, from a solid to a liquid state when internal strength is lost—typically caused by earthquake shaking.

Lithified Sediment is lithified when it changes into solid rock, commonly through processes related to compaction.

Lithosphere The lithosphere is the shallowest physical layer in the geosphere, made up of strong, rigid rocks that can break when they move. It includes rocks of the oceanic crust, the continental crust, and the top portion of the underlying upper mantle.

Load Load is the material that is moved or carried by a stream, including bed load, suspended load, and dissolved load.

Longitudinal profile A longitudinal profile shows the elevation of a stream channel along its length, from the source to the mouth.

Longshore current A longshore current is a current that flows along a shore as a result of waves approaching the coast at an angle. They are generally confined to the surf zone.

Longshore drift Longshore drift is sediment that is moved along a shore by longshore currents.

Losing stream A losing stream is one that loses water into the ground because its channel lies above the water table. It serves as a recharge point for the local groundwater system.

M

Mafic Mafic is a compositional name for igneous rocks with abundant ferromagnesium minerals. Mafic rocks are rich in iron, magnesium, and calcium, and are most characteristic of oceanic crust.

Magma Magma is molten rock within the geosphere.

Magma chamber A magma chamber is a reservoir of magma in the shallow part of the lithosphere from which volcanic materials are derived.

Magnetic reversal A magnetic reversal is a 180-degree shift in the orientation of Earth's magnetic field—the north magnetic pole becomes the south magnetic pole and vice versa when a magnetic reversal occurs.

Magnetic stripes Magnetic stripes are alternating regions of strong and weak magnetic strength that occur in oceanic crust parallel to spreading centers at mid-ocean ridges.

Magnitude Magnitude is a measure of the strength of an earthquake as determined by seismograph observations.

Mantle The mantle, the largest part of the geosphere, is the part of Earth's internal structure between the *crust* and the *core*. The mantle has parts that are dynamic and changing (asthenosphere) as well as an upper part, adjacent to the crust, that is rigid and part of the lithosphere.

Mass extinction A mass extinction is a relatively short period in geological time during which a significant percentage of Earth's species become extinct.

Mass movement A mass movement is an individual episode of mass wasting.

Mass wasting Mass wasting is a general term for the downslope movement of earth materials under the influence of gravity.

Maximum Contaminant Level (MCL) A MCL is the highest level of a contaminant that is allowed in drinking water.

Meander Meanders are sinuous curves, bends, loops, or turns in the course of a stream, formed as the stream shifts its course laterally as it flows across its floodplain.

Meltdown A meltdown is a severe overheating of a nuclear reactor core, resulting in melting of the core and escape of radiation.

Mesosphere The mesosphere is the part of the Earth's atmosphere above the stratosphere where temperatures decrease upward.

Metallurgy Metallurgy is a set of processes that separate metals from their host minerals.

Metamorphic rock A metamorphic rock is a rock that has changed in response to temperature and pressure.

Milling Milling breaks ore into individual mineral grains, tiny particles the consistency of silt, sand, and clay.

Mineral A mineral is a naturally occurring inorganic solid made up of an element or a combination of elements (a compound) that has an ordered arrangement of atoms and a characteristic chemical composition.

Mineral deposit A mineral deposit is a concentration of useful minerals in Earth's crust.

Modified Mercalli Scale The Modified Mercalli Scale ranks earthquake intensities on a 12-point scale, expressed as Roman numerals from I (not felt by people) to XII (damage nearly total).

Mohorovičić discontinuity (Moho) The Mohorovičić discontinuity is the place within the geosphere where seismic waves passing downward abruptly accelerate at the base of the crust. It marks the boundary between the crust and upper mantle.

Mollisol Mollisol is a soil order characteristic of grasslands (prairies); in these soils the A horizon is thick and dark-colored from the accumulation of organic material from plant roots.

Moment magnitude Moment magnitude (M_W) is a numerical scale of the amount of energy released by an earthquake. It is calculated from the total area of the fault that ruptures, how far the rocks move along the fault during the earthquake, and the strength of the rock that ruptures.

Moraine A moraine is a body of rock debris carried and moved by glaciers.

Mouth The mouth is the place where a stream discharges into a larger stream, a lake, or the ocean.

N

National Environmental Policy Act (NEPA) NEPA is a federal statute that requires all federal agencies to further a national policy "to promote efforts which will prevent or eliminate damage to the environment and biosphere" and "enrich the understanding of the ecological systems and natural resources important to the Nation." It also established a Council on Environmental Quality. The act can be applied to any project, federal, state or local, that involves federal funding or work performed by the federal government, and requires EISs for such projects.

National Pollutant Discharge Elimination System (NPDES) NPDES is a permit program, established by the CWA, that controls water pollution by regulating point sources that discharge pollutants into waters of the United States.

National Priority List (NPL) The NPL is the list of hazardous waste sites recognized as a priority for continued study and potential cleanup under CERCLA.

Natural selection Natural selection is the process in nature by which organisms better adapted to their environment tend to survive, outreproduce less well-adapted members of their population, and transmit their genetic characteristics in increasing numbers to succeeding generations.

Nonconsumptive use A nonconsumptive use of water is one that allows it to be used subsequently for other purposes or discharged back to its source.

Nonpoint sources (of pollution) Nonpoint sources are those that release pollutants to the environment over a large area.

Nonrenewable resource A nonrenewable resource is one that is not replenished as fast as it is being used.

Normal fault A normal fault is an inclined break in the lithosphere where the upper block of rock has moved down relative to the lower block.

Nuclear fission Nuclear fission is the splitting of unstable atomic nuclei, accompanied by release of energy in the form of radiation and heat.

Nuclear fusion Nuclear fusion is the process of combining atomic nuclei of lighter elements to form heavier elements, accompanied by release of tremendous amounts of energy.

O

Oceanic crust Oceanic crust is the crust under the oceans. It is thin and consists mostly of the volcanic rock basalt.

Oil sand (tar sand) Oil sand is a sandstone impregnated with bitumen, a thick tar-like substance that must be heated or mixed with less viscous hydrocarbons before it will flow.

Oil shale Oil shale is an organic-rich sedimentary rock that is capable of generating oil but that hasn't experienced the temperatures and pressures needed to convert the organic material (kerogen) to oil.

Open pit An open pit is a surface excavation to extract useful rocks or minerals.

Ore Ore is the part of a mineral deposit that can be profitably mined and recovered at current and projected commodity prices.

Outcrop An outcrop is a surface exposure of the solid geosphere.

Oxbow lake An oxbow lake is a crescent-shaped body of standing water in an abandoned channel of a meandering stream.

Oxidation Oxidation is the chemical reaction that combines oxygen with other elements. It is especially common where the atmosphere, hydrosphere, and geosphere interact.

Oxisol Oxisol is a soil order characteristic of warm, wet tropical forests; these are deeply weathered soils that have been leached of much of their original mineral content.

Ozone Ozone (O_3) is a molecule consisting of three oxygen atoms.

Ozone layer The ozone layer is a part of the atmosphere that overlaps the boundary between the troposphere and stratosphere where natural ozone (O_3) is concentrated.

P

P (primary) waves P waves are seismic body waves that alternatively push and pull (that is, compress and expand) rocks along their direction of travel. They are the fastest of the seismic waves.

Pahoehoe Pahoehoe is a Hawaiian term for a type of basaltic lava flow typified by a smooth, billowy, or ropy surface.

Paleomagnetism Paleomagnetism is the magnetism preserved in magnetic minerals at the time they formed.

Pangaea Pangaea is the name of the supercontinent that formed about 250 million years ago. All of Earth's continents were amalgamated in Pangaea.

Parts per million (ppm) Parts per million is a measure of concentration, equivalent to about one drop in 50 L (13 gal) of water.

Peak oil production Peak oil production refers to the time when nothing more can be done to increase production and the amount of oil produced will begin to decline.

Period The period of a seismic wave is the interval of time required for one complete peak to peak vibration to pass by a seismometer.

Permeability Permeability is the property or capacity of a porous rock, sediment, or soil to let a fluid such as water pass through.

pH pH is the measure of a soil's acidity; a pH less than 7 is acidic, a pH of 7 is neutral, and a pH greater than 7 is alkaline.

Photosynthesis Photosynthesis is the process by which certain organisms, especially plants, use energy from sunlight to convert carbon dioxide and water into simple sugars that can be used for food. Oxygen is released by most photosynthetic reactions.

Photovoltaic cell A photovoltaic cell is a semiconductor device that converts the energy of sunlight into electric energy.

Phytoremediation Phytoremediation is the use of plants to remove contaminants from soil.

Plate (lithospheric plate, tectonic plate) A plate is a discrete piece of lithosphere that moves relative to other pieces.

Plate tectonics Plate tectonics is the unifying theory that explains the dynamic nature of the geosphere and its relation to the interaction of tectonic plates.

Plutonic igneous rock A plutonic igneous rock is a rock that crystallizes from magma within the crust.

Point sources (of pollution) Point sources are those that release pollutants from a specific site such as a pipe, ditch, tunnel, well, container, concentrated animal-feeding operation, or floating craft.

Policy Policy is a term that encompasses both specific plans of action and principles that guide future decisions.

Pollutant A pollutant is any substance that makes air unsafe or dirty.

Pore pressure Pore pressure is the pressure transmitted by the fluid that fills the voids between particles of a soil or rock mass.

Porosity Porosity is the percentage of an Earth material consisting of void spaces—small openings (pores) between particles of sand, silt, clay, or gravel, or fractures and hollows within a body of solid rock.

Potentially Responsible Party (PRP) A PRP is any individual, company, or other party potentially liable for payment of CERCLA cleanup costs, including companies that generate hazardous substances disposed of at a CERCLA site, current and former owners and operators of the site, and transporters who selected the site for disposal of hazardous substances.

Precession Precession is the change in direction of Earth's axis over long periods of time.

Precursor A precursor is a physical or seismic phenomena that may indicate a pending earthquake.

Prospect A prospect is an area that is a potential site of mineral deposits, based on geologic, geochemical, or geophysical indications.

Pumice Pumice is cavity-rich, glassy, volcanic rock that formed from magma containing abundant gas bubbles.

Pyroclastic flow A pyroclastic flow is a current of pyroclastic material, usually very hot and composed of a mixture of gases and particles of rock and lava, that flows along the ground surface.

Pyroclastics Pyroclastics are volcanic materials erupted into the atmosphere.

Q

Quartz Quartz is the mineral that contains the two most abundant elements in the crust, silicon and oxygen; it contains one atom of silicon for every two atoms of oxygen.

Quick clay Quick clay is composed of silt grains surrounded by a jumble of thin, platy clay minerals, forming a chaotic "house of cards" structure filled with water. Quick clay can lose all or nearly all internal strength upon being disturbed.

R

Radiometric dating Radiometric dating is the determination of a natural material's age by measuring the concentrations of radioactive elements and their decay products included within the material. Radiometric dating determines the absolute or numerical age of geologic materials in years before the present.

Reclamation Reclamation is the conversion of disturbed land to land suitable for habitat or other uses. It commonly involves reshaping of the land surface to resist erosion, covering it with soil, and planting new vegetation to help stabilize the land surface.

Recurrence interval A flood recurrence interval is the average time between past flood events of a similar size.

Recurrence interval Recurrence interval is the average time interval between characteristic earthquakes on a fault.

Refining Refining is the processing of petroleum to produce marketable products such as gasoline or diesel fuel.

Regional subsidence Regional subsidence is gradual lowering of the land surface over a large area.

Relative age A relative age identifies whether one material or event is older or younger than another.

Renewable resource A renewable resource is one that will continue to be available because it is naturally replenished as fast as, or faster than, it is being consumed.

Reservoir A reservoir is a place where specified matter or energy is stored in a system.

Residence time Residence time is the average amount of time that specified matter spends in a defined system reservoir.

Resisting force Resisting force is a general term for the forces that oppose gravity in surface materials.

Resource Conservation and Recovery Act (RCRA) RCRA is a federal statute that establishes policies that cover the generation, storage, transport, and disposal of solid and hazardous waste.

Reverse fault A reverse fault is an inclined break in the lithosphere where the upper block of rock has moved up relative to the lower block.

Reverse osmosis Reverse osmosis is a process for separating dissolved constituents from water by pumping it through special filters whose tiny openings allow water molecules to pass, but not salt and other minerals.

Revetment A revetment is generally a lighter-duty structure than a seawall or bulkhead, used to protect embankments or beaches from erosion by near-shore currents or light wave activity.

Richter magnitude scale The Richter magnitude scale represents the size of an earthquake based on the strongest seismic wave amplitude recorded at a standard distance of 100 kilometers (62 mi) from the epicenter with a standard torsion seismograph.

Ridge (oceanic ridge, mid-ocean ridge) An oceanic ridge is an underwater mountain range that formed at a spreading center along a divergent plate boundary.

Rift A rift is a place where crust is extending and breaking apart. Along a rift, hot material in the mantle rises and the overlying crust thins and weakens as it is warmed and stretched.

Rill A rill is a small streamlet or channel resulting from erosion of surface soil by running water.

Rip current A rip current is a plume-shaped current that conveys the water from breaking waves back through the surf zone.

Rock A rock is a solid part of the geosphere that is most commonly an aggregate of one or more minerals.

Rock cycle The rock cycle is a sequence of changes in rocks of the crust, produced by weathering, erosion, sedimentation, lithification, metamorphism, and igneous processes.

Rotational slide In a rotational slide, the unstable material slides downward and outward along a concave surface (like the bowl of a spoon) rather than a planar surface.

Runoff Runoff is a general term for surface water that flows overland to and within streams; it is the part of precipitation that accumulates in surface streams.

S

S (secondary or shear) waves S waves are seismic body waves that vibrate rock perpendicular to the direction of wave movement—that is, up and down or side to side, as in waves on a shaken rope. Typically about half as fast as P waves, they do not travel through liquids or through the outer core of the Earth.

Salination Salination is the contamination of soil by salt.

Salinity Salinity is a measure of water's concentration of dissolved salt.

Saltwater intrusion Saltwater intrusion is the movement of saltwater into a freshwater aquifer, commonly because of the withdrawal of groundwater in coastal areas and estuaries.

Scarp A scarp is a steep slope or bank on Earth's surface created by movement on a fault.

Scientific method The scientific method is a systematic approach to exploring and explaining how the natural world operates. It is an iterative process of inquiry that involves acquiring data, formulating questions, devising hypotheses, making predictions from hypotheses, and carrying out observations to support or refute hypotheses.

Seafloor spreading Seafloor spreading is the movement of newly formed oceanic crust away from mid-ocean ridges.

Seawall A seawall is a massive structure, built onshore, designed to withstand the full force of storm waves.

Sedimentary basin A sedimentary basin is a large area where sediment accumulates.

Sedimentary rock A sedimentary rock is a rock that formed from either clastic or chemical sediments.

Sedimentation Sedimentation is the process of depositing sediments.

Seismic gap A seismic gap is a segment of a fault that has not ruptured recently relative to neighboring segments.

Seismic survey A seismic survey uses seismic waves to image the subsurface character and geometry of rocks.

Seismic waves Seismic waves is a general term for elastic waves (vibrations) in Earth, typically produced by earthquakes (although they can also be generated artificially).

Seismogram A seismogram is a record of ground motion measured by a seismometer.

Seismology Seismology is the study of earthquakes, and of the structure of the Earth, by means of both natural and artificially generated seismic waves.

Seismometer A seismometer is an instrument that measures ground motions caused by passing seismic waves.

Shear strength Shear strength is a measure of how well the soil resists forces before fracturing internally.

Sheet flow Sheet flow is thin nonchannelized overland flow of water taking the form of a thin, continuous film and not concentrated into channels larger than rills.

Shield volcano A shield volcano is a volcano in the shape of a broad, low dome built by successive flows of fluid basaltic lava.

Silica Silica is the general term for the chemical compound that contains one atom of silicon for every two atoms of oxygen (SiO_2).

Silicate mineral A silicate mineral is one that includes silicon and oxygen tetrahedra in its internal structure.

Sink A sink is a system reservoir where matter has a very long residence time. Sinks isolate matter from system interactions.

Sinkhole A sinkhole is a circular depression in karst terrain.

Slag Slag is the glassy waste material left over from smelting.

Slide Slide is a general term for a mass movement that moves downslope along a sloping surface, as opposed to free falling, tumbling, or bouncing.

Slope failure Slope failure occurs when the driving force is greater than the resisting forces, resulting in the gradual or rapid downslope movement of surface material or rock.

Slump Slump is an alternative term for a rotational slide.

Smelting Smelting is a metallurgical technique that separates metal from minerals by heating them to their melting point.

Smog Smog is a dense, hazy air pollution that obscures visibility, produced by the photochemical reaction of sunlight with hydrocarbons and nitrogen oxides released into the atmosphere, especially by automotive emissions.

Soil orders Soil orders are categories of soil variations within a systematic classification scheme used to describe and compare them, based on the presence or absence of diagnostic horizons or other special properties.

Soil profile A soil profile is a vertical section of a soil through all its stratified horizons.

Spodosol Spodosol is a soil order characteristic of cool, moist coniferous forest regions; these soils are acidic and typically have a subsurface accumulation of humus that is combined with aluminum and iron oxides or hydroxides.

Stage Stage is the height of a water surface above an arbitrarily established plane, commonly the bed level of a stream.

Stakeholder A stakeholder is anyone interested in or affected by an action.

State Environmental Policy Act (SEPA) SEPA is a state statute modeled after NEPA that requires state and local agencies to consider the likely environmental consequences of a proposed action before approving or denying it.

Steady state A steady-state system is one in which the transfers of matter or energy in and out are approximately equal. As a result, the system appears not to change.

Storm surge A storm surge is a high mass of water piled up against the coast, caused primarily by strong winds offshore; it is most severe during high tide.

Stratosphere The stratosphere is the part of the atmosphere between the troposphere and mesosphere where temperatures *increase* upward.

Stratovolcano A stratovolcano is a volcano constructed of alternating layers of lava and pyroclastic deposits, along with abundant igneous intrusions. Also called a composite volcano.

Strike-slip fault A strike-slip fault is a surface between two blocks of rock along which one block slides horizontally past the other.

Subduction Subduction is the sinking of a lithospheric plate into the mantle along a convergent plate boundary.

Subduction zone A subduction zone is the inclined part of a convergent plate boundary where oceanic lithosphere is sinking into the mantle.

Subsidence Subsidence is the sudden sinking or gradual downward settling of Earth's surface with little or no horizontal motion.

Superfund Superfund is another term for CERCLA.

Surf zone The surf zone is the turbulent area along a shore created by breaking waves.

Surface waves Surface waves are seismic waves that travel along Earth's outer edges rather than through its interior.

Sustainable Sustainable means that something is capable of being continued with minimal long-term effect on the environment. In a human context, it means that the needs of the present generation are met without compromising the ability of future generations to meet their needs.

System A system is a group of interacting, interrelated, or interdependent parts that together form a whole.

T

Tailings Tailings are leftover material from mineral processing, especially the milling and flotation of sulfide minerals, that is discarded.

Talus Talus is a pile of angular rock fragments that have fallen from a cliff or steep rocky slope above.

Tephra Tephra is collective term used for all pyroclastic material, regardless of size, shape, or origin, that is ejected during an explosive volcanic eruption.

Theory In science, a theory integrates a number of extensively tested hypotheses into a well-accepted unifying framework. A theory explains a large set of observations and relationships and has been verified independently by many researchers.

Thermosphere The thermosphere is the outer atmospheric layer above the mesosphere. Temperatures decrease upward through the thermosphere and the concentrations of gases gradually decrease as the thermosphere merges with empty space.

Thrust fault A thrust fault is a low-angle reverse fault; the surface along which blocks of rock move has a low inclination.

Tide The tide is the daily rise and fall of sea level that results from the gravitational attraction of the Sun and Moon.

Topsoil Topsoil is a general term for the upper layer of a soil that typically contains high levels of organic material and the root systems of surface vegetation.

Total Maximum Daily Load (TMDL) TMDL is the amount of pollutant that a water body can receive without exceeding its water quality standard.

Transform fault A transform fault is a type of strike-slip fault that converts movements at divergent or convergent plate boundaries into lateral sliding.

Transform plate boundary A transform plate boundary is the boundary along which two tectonic plates move laterally past each other.

Translational slide A translational slide is the movement of a largely intact mass, sliding as a single coherent block or a group of blocks on a surface that is roughly parallel to the general ground surface.

Trap A trap is a place in the geosphere where impermeable barriers prevent the further movement of oil or gas, allowing them to accumulate; it consists of a reservoir rock that contains the oil or gas and an impermeable roof rock.

Trench (oceanic trench) A trench is a long, narrow depression in the seafloor. Trenches are the deepest parts of the oceans, formed where a tectonic plate sinks into the mantle along a convergent plate boundary.

Tributary A tributary is a stream feeding, joining, or flowing into a larger stream.

Troposphere The troposphere is the lowest part of the atmosphere, where most of the atmosphere's matter is concentrated and most of what people call weather occurs.

Trunk river The trunk or main river is the principal, largest, or dominating river of any given area or drainage system.

Tsunami A tsunami is a large ocean wave commonly generated by movement of the seafloor during a subduction-zone earthquake.

Tuff Tuff is lithified volcanic ash.

U

Unconfined aquifer An unconfined aquifer is an aquifer that is overlain by permeable Earth material.

Unconsolidated material Unconsolidated material is loose, nonaggregated Earth material such as soil, sand, and broken rock debris.

V

Vein A vein is a long, narrow, sheet-like concentration of minerals, commonly localized along a fault or other fracture.

Viscosity Viscosity is a substance's resistance to flow.

Volatile Organic Compound (VOC) A volatile organic compound is an easily vaporized compound of carbon and hydrogen, such as gasoline or solvents like toluene, xylene, and tetrachloroethylene.

Volcanic arc A volcanic arc is a chain of volcanoes on land that forms above a subduction zone.

Volcanic Explosivity Index (VEI) The volcanic explosivity index is a scale of numbers from 0 to 8, used to compare volcanic eruptions on the basis of the volume of material ejected by past eruptions, the height of the eruption column, the style of the eruption (lava flows versus explosive eruptions, for example), and how long the eruption lasted.

Volcanic igneous rock Volcanic igneous rocks form from lava and other material erupted at a volcano.

Volcano A volcano is a place where molten rock, or magma, rises from great depths to the uppermost levels of the crust and onto the surface.

W

Waste rock Waste rock is the rock that must be broken and disposed of in order to gain access to and excavate ore; it is valueless rock that must be removed or set aside in mining.

Watershed A watershed is the region drained by, or contributing water to, a stream, lake, or other body of water.

Water table The water table is the top of the saturated zone. From the water table to the surface, the open spaces in the ground are only partly filled with water.

Weathering Weathering is a set of physical and chemical processes that change rocks at Earth's surface.

Wetland A wetland is an area that is regularly wet or flooded and has a water table that stands at or above the land surface for at least part of the year.

PHOTO CREDITS

Front Matter Summer meadow, courtesy of Shutterstock **p xxii** AGI **p. xxiii bottom**

CHAPTER 1 NASA **xxiv opener** Steve Allen/Jupiter Images **p1** Win McNamee/Getty Images **p3 top left** Peter Harrison/Photolibrary **p3 top right** Craig Pulsifer/Getty Images **p3 bottom** Adrian Arbib/CORBIS **p4 left** Ilja Masik/Shutterstock **p4 right** James Marshall/CORBIS **p5 top** Peter Ginter **p5 bottom left** David Reed **p5 bottom right** Alan Schein/Alamy **p6 top** Jacob Silberberg/Getty Images **p6 bottom** David Cooper/Toronto Star/ZUMA/CORBIS **p7 left** parkerphotography/Alamy **p7 right** Rick Wilking/CORBIS **p8** Simon Kwong/Reuters New Media Inc./CORBIS **p9 top** Richard Bouhet/Getty Images **p9 bottom left** AP Photo **p9 bottom right** Worldsat International Inc./Photo Researchers, Inc. **p12** Jetta Productions/Getty Images **p14** Paul Kennedy/Lonely Planet Images **p16 bottom** Peter Hendrie/Lonely Planet Images **p16 top**

CHAPTER 2 Patrick Endres/Alaska Stock **p20** NASA **p23 top left** NASA **p24 left** USGS **p24 right** Bryan & Cherry Alexander Photography/Alamy **p25 left** Bryan Law/Earth Science World Image Bank - www.earthscienceworld.org **p25 right** Carsten Peter/National Geographic Image Collection **p28** Andrew Syred/Photo Researchers, Inc. **p29 top left** Christopher G. St. C. Kendall **p29 bottom right** Nick Green/PhotoLibrary **p29 right** Eye of Science/Photo Researchers, Inc. **p30 bottom** Sergio Pitamitz/Danita Delimont **p30 top** Neil Duncan/Photolibrary **p32** Joshua Strang, USAF **p33** ESA & NASA **p34 top** NASA **p34 bottom** Phil Armitage **p36** Alexander Hafemann/iStockphoto **p37 bottom right** ollirg/Shutterstock **p37 bottom left** Phil Armitage **p37 top left** Morgen Bell/iStockphoto **p37 top right** Albert Copley, University of Oklahoma/Earth Science World Image Bank - www.earthscienceworld.org **p39 top left** Andy Crawford/Dorling Kindersley Media Library **p39 top right** Sinclair Stammers/Photo Researchers, Inc. **p39 middle** Edward Kinsman/Photo Researchers, Inc. **p 39 bottom left** Fred Bavendam/Minden Pictures **p39 bottom right** Andrew Nelmerm/Dorling Kindersley Media Library **p40 bottom left** Robert Hardholt/Shutterstock **p40 bottom right** SPL/Photo Researchers, Inc. **p41 bottom middle** D. Parer & E. Parer-Cook/Auscape/Minden Pictures **p41 bottom left** George Richmond/Photo Researchers, Inc. **p41 top right** David A. Hardy/Photo Researchers, Inc. **p43 left** Wisconsin Historical Society **p43 right** Ed Reschke/Peter Arnold **p45**

CHAPTER 3 Library of Congress **p52** NASA **p53** David M. Dennis/Earth Scenes **p55 left** Art Resource/Bildarchiv Preussischer Kulturbesitz **p55 right** Woods Hole Institute **p58 bottom** Princeton University, Geosciences Department **p59** Louise A Heusinkveld/Alamy **p61** Nicholas Parfitt/Getty Images **p64** Kristin Piljay/Danita Delimont **p66** Verena Tunnicliffe **p67** Grant Dixon/Lonely Planet Images **p69** USGS **p71 left** Michael Collier **p72 left** aerialarchives.com/Alamy **p72 right** USGS **p74** Guillermo Granja/Reuters/CORBIS **p75** Bruce Molnia, USGS **p76** MODIS **p77 left** Lee Prince/Shutterstock **p77 right**

CHAPTER 4 Photograph © 2007 John C. Phillips **p80** Line de Courssou **p81** Kristin Piljay/Danita Delimont **p82 left** Michele Princigalli/iStockphoto **p82 right** Harry Taylor/Dorling Kindersley Media Library **p86 left** Mark A. Schneider/Photo Researchers, Inc. **p86 right** Amir C. Akhavan **p87 top** Mark A. Schneider/Photo Researchers, Inc. **p87 middle** Richard M. Busch **p87 bottom** Bruce Molnia, Terra Photographics **p88 left** Dr. Richard Busch **p88 right** West Virginia State Archives, Elkem Metals Collection **p89 top** Mark A. Schneider/Photo Researchers, Inc. **p89 bottom left** Mineral Information Institute **p89 bottom right** David C. Noe **p90 top** Chip Clark **p90 bottom** USGS/Cascades Volcano Observatory **p91 top** Tasa Graphic Arts, Inc. **p91 bottom left** Richard M. Busch **p91 bottom right** William E. Ferguson **p92 left** Bernd Obermann/CORBIS **p92 right** Colin Keates/Dorling Kindersley **p93 top** USGS **p93 bottom left** Maggie Craig, LibbyMT.com **p93 bottom right** Dr. Richard Busch **p94 left** Geoffrey S. Plumlee/USGS **p94 right** Stonetrust, Inc. **p95 top left** Mark A. Schneider/Photo Researchers, Inc. **p95 top right** Adam Jones/Photo Researchers, Inc. **p95 middle right 1st** G. R. 'Dick' Roberts/NSIL/Getty Images **p95 middle right 2nd** Paul H. Moser/Geological Survey of Alabama **p95 bottom right** Joel Arem/Photo Researchers, Inc. **p97 top** Scenics & Science/Alamy **p97 middle inset** Dr. Richard Busch **p97 bottom** Tiffany Chan/Shutterstock **p98 top** NOAA **p98 top right** JND Photography Canada **p98 bottom right** Larry Fellows, Arizona Geological Survey **p98 bottom** Dirk Wiersma/Photo Researchers, Inc. **p98 bottom left** Bruce Molnia, Terra Photographics **p99** Phil Morley/iStockphoto **p99 top left** Ben Kimball, NH Natural Heritage Bureau **p100 top** Duncan Heron **p100 middle** Marli Miller **p100 bottom** Marli Miller **p101 top left** Bruce Molnia, Terra Photographics **p101 middle left** Maxim Petrichuk/Shutterstock **p101 bottom left** Roger Slatt, University of Oklahoma **p101 top right** M-Sat Ltd/Photo Researchers, Inc. **p101 middle right** Marli Miller **p101 bottom right** Gary Ombler/Dorling Kindersley Media Library **p102 left** Breck P. Kent/Animals Animals - Earth Scenes **p102 top right** Breck P. Kent/Animals Animals - Earth Scenes **p102 bottom right** Prenzel Photo/Animals Animals - Earth Scenes **p103 left** USGS **p103 top right** Biophoto Associates/Photo Researchers, Inc. **p103 middle right 1st** John Shaw/Photo Researchers, Inc. **p103 middle right 2nd** Larry Fellows, Arizona Geological Survey **p103 bottom right** Sven Klaschik/iStockphoto **p105 left** Chris Dekle **p105 right** David Newham/Alamy **p107 top** Butchart Gardens **p107 bottom left** Butchart Gardens **p107 bottom right** Wendell Franks/iStockphoto **p108 top** Colin Garratt/CORBIS **p108 bottom**

CHAPTER 5 Mark Downey/Lucid Images/CORBIS **p112** David Ohlson **p113** AP Photo/William J. Smith **p117 left** Bettmann/CORBIS **p117 right** Julian J. Bommer **p118** Marli Miller **p119 left** Marli Miller **p120 right** USGS **p121** Mike Fox/ZUMA Press/Newscom **p122 left** Mike Fox/ZUMA Press/Newscom **p122 right** David Butow/CORBIS **p124** Roger Ressmeyer/CORBIS **p125** David Poole/Robert Harding **p129 top left** Ryan Pyle/CORBIS **p129 bottom left** Spencer Grant/Getty Images **p129 right** Patrick Robert/Sygma/CORBIS **p132 left** Plafker Geohazard Consultants **p132 right** Lloyd Homer, GNS Science, New Zealand **p133 left** Bettmann/CORBIS **p133 right** Getty Images **p135 top** Space Imaging/Photo Researchers, Inc. **p135 bottom left** Space Imaging/Photo Researchers, Inc. **p135 bottom right** H.D. Chadwick **p136 left** AP Photo/Chiaki Tsukumo **p136 right** Bettmann/CORBIS **p137** AP Photo/Enric Marti **p139** Tom Simondi **p140** USGS **p141 top** Ken Hadnut/USGS **p141 bottom** USGS **p143** Haruyoshi Yamaguchi/Bloomberg News/Landov **p145 top right** AP Photo/Koji Sasahara **p145 bottom right** Pornchai Kittiwongsakul/AFP/Getty Images **p145 left** SDBEnvironment/Alamy **p146**

CHAPTER 6 USGS **p150** Raj Arumugam **p151** NASA **p152** Peter French/Pacific Stock **p153** dk/Alamy **p156** Douglas Peebles/www.DanitaDelimont.com **p157 top** Kaj R. Svensson/Photo Researchers, Inc. **p157 bottom** USGS **p158 top left** USGS **p158 bottom left** First Light/Alamy **p158 right** R H Productions/Robert Harding **p159** P. Schofield/Getty Images **p160 top left** Danny Yee **p160 top right** USGS **p160 middle right** USGS **p160 bottom** Marco Fulle **p160 middle right** USGS **p161 left** Georg Gerster/Photo Researchers, Inc. **p161 right** CORBIS **p162** Frank Siteman/Photo Edit, Inc. **p164 - row 0** Photos.com/Jupiter Images Unlimited **p164 - row 1** Peter Walker/CORBIS **p164 - row 2** AP Photo **p164 - row 3** iofoto/Shutterstock **p164 - row 4** Mario Savoia/Shutterstock **p164 - row 5** National Geographic Image Collection **p164 - row 6** Stewart Behra/iStockphoto **p164 - row 7** NASA **p164 - row 8** Ted Wood/Getty Images **p165 left** Karl Kummels/SuperStock **p165 right** The Kobal Collection/20th Century Fox **p166 left** David Hardy/Photo Researchers, Inc. **p166 right** JUPITERIMAGES/Comstock Images/Alamy **p168 left** Explorer/Photo Researchers, Inc. **p169 top** Seamas Culligan/CORBIS **p169 bottom** R. Hoblitt/USGS **p170 top** USGS **p170 bottom left** USGS **p170 bottom right** USGS **p171 bottom left** Bernhard Edmaier/Photo Researchers, Inc. **p171 bottom right** The Granger Collection **p172 left** Archives de l'Academie des Sciences **p172 right** Richard Roscoe/Visuals Unlimited, Inc./Getty Images **p173 top left** USGS **p173 bottom left** Keith Ronnholm **p174** USGS **p175 top** USGS **p175 middle** USGS **p175 bottom** USGS **p178 top** Thierry Orban/CORBIS **p178 bottom left** Peter Turnley/CORBIS **p178 bottom right** Sigurgeir Jonasson/Nordic Photos **p180** USGS **p181** USGS **p182** USGS **p183 left** Les Stone/Sygma/CORBIS **p183 right** USGS **p185**

CHAPTER 7 Frank Oberle/Getty Images **p188** Brad Kellerman/Belleville News-Democrat **p189** John Zich/Time & Life Pictures/Getty Images **p190** NASA **p191** cpsnell/iStockphoto **p194 top left** Pichugin Dmitry/Shutterstock **p194 middle** Karl Johaentges/Getty Images **p194 bottom right** Bitanga87/Shutterstock **p195 left** Albo/Shutterstock **p195 right** USGS **p196 top** Bobby Haas/National Geographic/Getty Images **p196 bottom left** Visuals Unlimited/CORBIS **p196 bottom right** View Stock/Alamy **p197** David Wall/www.DanitaDelimont.com **p198 left** Mike Andrews/Animals Animals - Earth Scenes **p198 top right** NASA **p198 bottom right** Marli Miller **p199** Ben Heys/Shutterstock **p200** Perry Rahn **p201 left** David Greedy/Getty Images **p201 right** Pete Honl **p202** John Marshall Photography **p204 left** Jamie & Judy Wild/www.DanitaDelimont.com **p204 right** Garry D. McMichael/Photo Researchers, Inc. **p205 left** John Warburton-Lee/www.DanitaDelimont.com **p205 right** AP Photo/Mel Evans **p206** Marck Schmeeckle **p207 top left** Marck Schmeeckle **p207 top right** USGS **p207 bottom right** James Shaffer/Photo Edit, Inc. **p209** Ralph Harvey **p213** David McNew/Getty Images **p215 left** Du Huaju/Xinhua Press/CORBIS **p215 right** Keren Su/Getty Images **p216** NASA **p217** Ann Johansson/CORBIS **p218**

CHAPTER 8 Bruce Molnia, Terra Photographics **p222** California Coast Walk **p223** John Ballard **p225** Images of Africa Photobank/Alamy **p227 top left** David Lewis/iStockphoto **p227 top right** USGS National Landslide Information Center **p227 bottom** Stoian Markov/iStockphoto **p228** Washington State Department of Transportation/Seattle Times/MCT/Newscom **p228** Vivek Sharma/Alamy **p229** Lloyd

PHOTO CREDITS

DeForrest/USGS **p230 left** Bruce Molnia, Terra Photographics **p23 right** Kyodo via AP Images **p231 top** Jorge Uzon/Newscom **p231 bottom** USGS **p232** Greg Brooks/Geological Survey of Canada **p233 top** Marli Miller **p233 bottom** Marli Miller **p234** Plafker Geohazard Consultants **p235 top left** Earthquake Engineering Research Center - Steinbrugge Collection **p235 top right** Gerry Davis/Phototake **p235 bottom** Tony Waltham/The Anthony Blake Photo Library/Photolibrary **p237 top** Richard A. Young, Geneseo, NY **p237 bottom** Tom Holzer, USGS **p238** AP Photo/Luis Elvir **p239** AP Photo/Kevork Djansezian **p240** AP Photo/Damian Dovarganes **p241 top** AP Photo/Ric Francis **p241 bottom** USGS **p242** David McNew Reuters/Newscom **p243** USGS **p244** USGS **p245** Teresa Tamura/The Seattle Times **p246** Utah Geological Survey **p248 top** Jack Sullivan/Alamy **p248 bottom** China Photos/Getty Images **p249** Seattle Department of Transportation **p250 top left** Seattle Department of Transportation **p250 top right** qaphotos.com/Alamy **p250 bottom** Vasiliki Varvaki/iStockphoto **p251 top left** Qurren **p251 top right** Mike May **p253 top** Mike May **p253 bottom**

CHAPTER 9 Courtesy of The Rosenberg Library, Galveston, Texas **p256** Library of Congress **p257** Greg Smith/CORBIS **p258 top** Dan Bannister/Dorling Kindersley Media Library **p258 bottom** Max Earey/Shutterstock **p259** Georg Gerster/Photo Researchers, Inc. **p262 left** Karl Johaentges/Getty Images **p263** Dr. Rob Brander **p264** Alexandra Draghici/iStockphoto **p265 left** Alexandra Draghici/iStockphoto **p265 right** Mark Conlin/Alamy **p266 top** NOAA **p266 bottom** Robert E. Johnson **p269 top** Travis Hudson **p269 bottom** Frederick G. S. Clow/PictureDesk International/Newscom **p271 top** Cameron Davidson/Alamy **p271 bottom** Modern Landscapes/Alamy **p272 top** Anne Kitzman Photography/iStockphoto **p272 bottom** ZUMA Press/Newscom **p273 top right** Frontpage/Shutterstock **p273 top left** NASA **p273 bottom left** Tom Bean/Getty Images **p273 bottom right** Dennis Hallinan/Alamy **p274 top** Scott Tucker/Alamy **p274 bottom** Marli Miller **p275** Inge Johnsson/Alamy **p277** AP Photo/Bob Jordan **p279 left** Earl Robicheaux Photography/iStockphoto **p279 right** Courtesy Coastal and Hydraulics Laboratory, U.S. Army Corps of Engineers **p280 top left** Roman Krochuk/iStockphoto **p280 middle right** Arthur Gebuys - temp/Alamy **p280 middle left** Frischknecht, P./Arco Images/Peter Arnold **p283** Kyle Niemi/U.S. Coast Guard/Getty Images **p284 left** AP Photo/Ben Sklar **p284 right** UPI Photo/A.J. Sisco/Newscom **p285** Ben Maxwell/Salem Public Library **p286** Michael Dwyer/Alamy **p287 top left** Bill Brooks/Alamy **p287 top right** idp northumberland collection/Alamy **p287 bottom right** Linda Steward/iStockphoto **p288 right** Bobbé Christopherson **p288 left** U.S. Army Corps of Engineers **p289 left** U.S. Army Corps of Engineers **p289 right** Florida Department of Environmental Protection **p290 middle** Florida Department of Environmental Protection **p290 top** Koen van Weel/Reuters/CORBIS **p290 bottom** Jim Sugar/CORBIS **p291** NOAA **p292**

CHAPTER 10 Bettmann/CORBIS **p296** J.D. Fisher/Alamy **p297** Elena Elisseeva/Shutterstock **p298** Douglas Pulsipher/Alamy **p299** USBR **p300 top** AP Photo/Salt Lake Tribune, Trent Nelson **p300 middle** USBR **p300 bottom** Corbis RF/Alamy **p301** Helen & Vlad Filatov/Shutterstock **p304** Bryan Busovicki/Shutterstock **p308 left** Dave Willman/iStockphoto **p308 middle** Michael Utech/iStockphoto **p308 right** USBR **p309 left** Mark Hanauer/CORBIS **p309 right** Unknown photographer, Senate Document 973, 62-3, Plate 1, courtesy of the National Archives **p311 left** Raymond M. Turner, courtesy of the USGS Desert Laboratory Repeat Photography Collection **p311 right** Driendl Group/Jupiter Images **p312** Justin McCormack **p313** Mufty Munir/epa/CORBIS **p314** Dr. Gary Gaugler/Photo Researchers, Inc. **p315 top** Dr. Tony Brain/Photo Researchers, Inc. **p315 bottom** worldthroughthelens-DIY/Alamy **p316** Photoshot Holdings Ltd/Alamy **p318 left** Glowimages/Newscom **p318 right** Robert Brook/Photo Researchers, Inc. **p319 top** Greg Gardner/iStockphoto **p319 bottom left** Chris Knapton/Photo Researchers, Inc. **p319 bottom right** Missing35mm/iStockphoto **p321 left** Suat Irmak **p321 right** Adam Long/iStockphoto **p322** Jason Thomas of Provost & Pritchard Engineer Group on behalf of Arvin-Edison Water Storage District **p323**

CHAPTER 11 NASA **p326** Stas Volik/Shutterstock **p327** AFP/Getty Images **p328** NASA **p329 top left** Peter Skinner/Photo Researchers, Inc. **p329 middle right** moodboard/CORBIS **p329 bottom right** Motofish Images/CORBIS **p330** Gary Crabbe/Alamy **p333** USDA **p334 bottom** Hydromet/Shutterstock **p334 top** USDA **p335 bottom left** iofoto/Shutterstock **p335 top left** Soil Science Society of America **p335 bottom right** Brasilao/Shutterstock **p335 top right** USDA **p336 bottom left** Kushch Dmitry/Shutterstock **p336 top left** Grahame McConnell/Photolibrary **p336 top right** David V. Dow **p338** Eye of Science/Photo Researchers, Inc. **p340 bottom left** Mycorrhizal Applications, Inc. **p340 middle** Emanuele Biggi/Photolibrary **p340 top right** David R. Frazier/www.DanitaDelimont.com **p341** Frans Lanting **p342 top** Robert Frerck/Odyssey Productions, Inc. **p342 bottom** NOAA George E. Marsh Album **p343 top** USDA/NRCS/Natural Resources Conservation Service **p343 bottom** Lynn Betts/USDA **p344 top left** USDA **p344 top right** Oliviero Olivieri/Robert Harding **p344 bottom** USDA **p345** Richard R. Hansen/Photo Researchers, Inc. **p346** Mike Expert/Shutterstock **p347 top** vario images GmbH & Co.KG/Alamy **p347 bottom** Aerial Archives/Alamy **p348** David R. Frazier/www.DanitaDelimont.com **p349 top left** Jamie Marshall/Dorling Kindersley Media Library **p349 right** Grant Heilman Photography/Alamy **p349 bottom left** Charles Webber, California Academy of Sciences **p351**

CHAPTER 12 National Geographic Image Collection **p354** CORBIS **p355** South Tyrol Museum of Archaeology **p356 top right** South Tyrol Museum of Archaeology, Bolzano, Italy/Wolfgang Neeb/The Bridgeman Art Library **p356 bottom left** South Tyrol Museum of Archaeology, Bolzano, Italy/Wolfgang Neeb/The Bridgeman Art Library **p356 bottom right** Kaj R. Svensson/Photo Researchers, Inc. **p357 top left** Nico Tondini/Robert Harding World Imagery/CORBIS **p357 top right** Ashmolean Museum, University of Oxford, UK/The Bridgeman Art Library Nationality **p357 middle right** Steve Gorton/Dorling Kindersley Media Library **p357 bottom left** Museo Arqueologico Nacional, Madrid, Spain/Photo © AISA/The Bridgeman Art Library **p357 bottom right** Jonathan Blair/CORBIS **p360** Courtesy of Kennecott Utah Copper **p361 top left** Ken Krahulec/Utah Geological Service **p361 bottom left** Jeff Schultz/Alaska Stock **p361 right** CORBIS **p362 left** Travis Hudson **p362 right** Stephen J. Krasemann/Photo Researchers, Inc. **p363** Alaska DNR **p364 top left** Alaska DNR **p364 top right** Greenshoots Communications/Alamy **p364 bottom left** NovaGold Resources Inc. **p364 bottom right** NovaGold Resources Inc. **p365 left** Photo courtesy of Kennecott Utah Copper **p365 top right** Photo courtesy of Kennecott Utah Copper **p365 middle right** Photo courtesy of Kennecott Utah Copper **p365 bottom right** Don Despain/www.rekindlephotos.com/Alamy **p366 left** Photo courtesy of Kennecott Utah Copper **p366 right** Brian Hartshorn/Alamy **p367 left** LHB Photo/Alamy **p367 right** Anders Ryman/Alamy **p368 top** Jim West/Alamy **p368 bottom left** Greenshoots Communications/Alamy **p368 bottom right** Photo courtesy of Kennecott Utah Copper **p369 top** Travis Hudson **p369 middle** Greenshoots Communications/Alamy **p369 bottom** Travis Hudson **p370** Goran Tomasevic/Reuters **p371** Kennecott Utah Copper **p372 top** Calvin Larsen/Photo Researchers, Inc. **p372 bottom** USGS **p373 top** Travis Hudson **p373 bottom left** Travis Hudson **p373 middle right** BP Exploration Alaska **p373 bottom right** Travis Hudson **p375** USGS **p376 top** Don Despain/www.rekindlephoto.com/Alamy **p376 bottom** Xstrata **p378** Bryan & Cherry Alexander Photography/Alamy **p379** John Smierciak/Chicago Tribune/Newscom **p380**

CHAPTER 13 Ken Cedeno/CORBIS **p384** igor vorobyov/iStockphoto **p385** China Daily/Reuters **p386** John Clegg/SPL/Photo Researchers, Inc. **p389** Marli Miller/Getty Images **p391 left** Hulton Archive/Getty Images **p391 right** Ken Graham/Getty Images **p392** CGGVeritas **p393 top left** CGGVeritas **p393 bottom left** AP Photo/Arco Alaska Inc. **p393 bottom right** OERB **p394 top all** NOAA **p394 bottom** Paul Andrew Lawrence/Alamy **p396** Paul Hanna/Reuters/CORBIS **p397 top left** AP Photo/John Caps III **p397 top right** Vanessa Vick/Photo Researchers, Inc. **p397 bottom left** Natalie Fobes/CORBIS **p398** Coastwalk California **p399 left** Nick Turner/Alamy **p399 right** Craig Cozart/iStockphoto **p401 top** AP Photo/Houston Chronicle, Brett Coomer **p401 bottom** NASA **p403** Jupiter Images/Newscom **p405 top** www.mountainroadshow.com **p405 bottom** PA Dept. of Environmental Protection **p406 top left** PA Dept. of Environmental Protection **p406 top right** Central Press/Hulton Archive/Getty Images **p406 bottom right** David R. Frazier Photolibrary, Inc./Alamy **p407** Steve Allen/Jupiter Images **p410 top** Phil Degginger/Alamy **p410 bottom** Igor Kostin/CORBIS **p411** Nuclear Energy Institute **p412** Annette Soumillard/Hemis/CORBIS **p413** Dann Coffey/Getty Images **p414 left** Tim Matsui/Getty Images **p414 right** Li ming - Imaginechina/AP Photo **p415 top** Pelamis Wave Power **p415 bottom left** Owaki - Kulla/CORBIS **p415 bottom right** ARCO/Geduldig/age fotostock **p416** Kevin Foy/Alamy **p417** AP Photo/Jeff McIntosh **p418** Jonathan Blair/CORBIS **p419 left** AP Photo/Douglas C. Pizac **p419 right** Brooks Kraft/CORBIS **p421** Goss Images/Alamy **p422**

CHAPTER 14 Andre Maslennikov/age fotostock **p426** Alex Wilson/Dorling Kindersley Media Library **p427** izmostock/Alamy **p432** Bannor/Custom Medical Stock Photo/Newscom **p433** Eyecon Images/Alamy **p435 top left** Jon Arnold Images Ltd/Alamy **p435 top right** Library of Congress **p436 top** Chris Gibson/Alamy **p436 bottom** Cordelia Molloy/Photo Researchers, Inc. **p437 top** Chris Gibson/Alamy **p437 bottom** NASA **p439 left** Lamont Poole, NASA **p439 right** P. Hoffman, www.snowballearth.org **p444 left** P. Hoffman, www.snowballearth.org **p444 right** Bruce Molnia, Terra Photographics **p445 left** Dr. Claus-Dieter Hillenbrand **p445 right** NorthShoreSurfPhotos/iStockphoto **p456 top left** Dmitry Nikolaev/Shutterstock **p456 top right**

CHAPTER 15 AP Photo/David Bookstaver **p466** AP Photo/David Bookstaver **p467** Columbia River Crossing **p470** Andrew Fox/Alamy **p471** Nancy J. Pierce/Photo Researchers, Inc. **p472** Montana Historical Society Research Center Photograph Archives, Helena, MT **p474 top** USGS **p474 bottom** Joe Traver/Getty Images **p475** Montana Stock Photography/Alamy **p477** Westernmininghistory.com **p478** Bruce Molnia, Terra Photographics **p480** U.S. Fish and Wildlife Service **p481 top** John Carnemolla/Australian Picture Library/CORBIS **p483 top** AP Photo/Keystone, Walter Bieri **p483 bottom left** J & C Sohns/age fotostock **p483 bottom right** Patrick Endres/Alaska Stock **p484** Robert Nickels/Getty Images **p487 top left** Carrie Devorah/WENN.com/Newscom **p487 top right** Marli Miller **p487 bottom right** Robert Harding Picture Library Ltd/Alamy **p488**

INDEX

Note: Pages in **bold** locate figures and pages in *italic* locate tables.

D

F

Q

R

INDEX

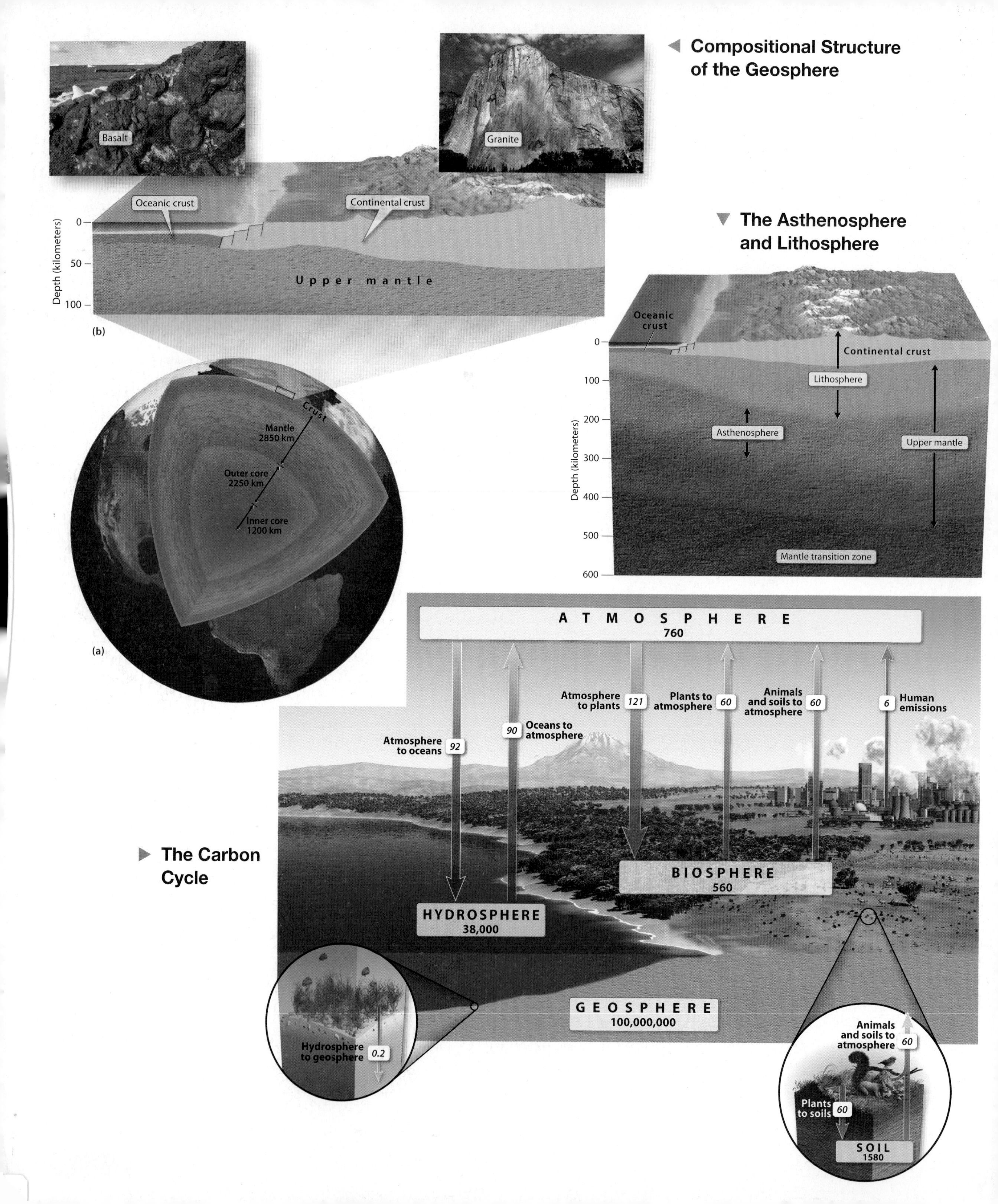
Compositional Structure of the Geosphere
Basalt
Granite
Oceanic crust
Continental crust
0
50
100
Depth (kilometers)
Upper mantle
(b)
Crust
Mantle 2850 km
Outer core 2250 km
Inner core 1200 km
(a)
The Asthenosphere and Lithosphere
Oceanic crust
Continental crust
Lithosphere
Asthenosphere
Upper mantle
Mantle transition zone
0
100
200
300
400
500
600
Depth (kilometers)
The Carbon Cycle
ATMOSPHERE
760
Atmosphere to oceans 92
Oceans to atmosphere 90
Atmosphere to plants 121
Plants to atmosphere 60
Animals and soils to atmosphere 60
Human emissions 6
BIOSPHERE
560
HYDROSPHERE
38,000
GEOSPHERE
100,000,000
Hydrosphere to geosphere 0.2
Animals and soils to atmosphere 60
Plants to soils 60
SOIL
1580